Advances in Oleogel Development, Characterization, and Nutritional Aspects

Camila Palla • Fabio Valoppi
Editors

Advances in Oleogel Development, Characterization, and Nutritional Aspects

Springer

Editors
Camila Palla
Departamento de Ingeniería Química
Universidad Nacional del Sur (UNS)
Bahía Blanca, Argentina

Planta Piloto de Ingeniería
Química – PLAPIQUI (UNS-CONICET)
Bahía Blanca, Argentina

Fabio Valoppi
University of Helsinki
Helsinki, Finland

ISBN 978-3-031-46830-8 ISBN 978-3-031-46831-5 (eBook)
https://doi.org/10.1007/978-3-031-46831-5

This Springer imprint is published by the registered company Springer Nature Switzerland AG
The registered company address is: Gewerbestrasse 11, 6330 Cham, Switzerland

Paper in this product is recyclable.

To my husband Nicolas and my loved ones, my safe place.
To all of you whose heart beats fast when discovering something new.

Camila

To my wife Tania and our children Maya and Ewan for their constant and unwavering love and inspiration, and my parents Pierino and Giacomina and grandparents Ugo and Pierina for their unconditional trust and support.
Remember to look at the sky, we spend enough time looking down at the ground.

Fabio

Preface

This book is a compendium of the work of many dedicated researchers who have focused their efforts on the design, development, understanding, and application of oleogels and oleogel-based systems over the past two decades. These novel lipid-based materials have been primarily investigated as ingredients to reduce saturated fats and eliminate *trans* fats in food products. The versatility of oleogels has expanded their potential applications beyond their initial scope, generating a growing interest in both academic and industrial communities. This interest has driven the development of the oleogel field exponentially, resulting in a growing body of scientific literature and patents. Bringing together all the scattered information available in the literature into one book was not possible without the wholehearted commitment of all authors involved in this project. This book is the result of a multidisciplinary, multicultural, and international collaboration among the top experts in the field of oleogel and related areas.

We, the editors, envisioned this book with great enthusiasm in 2021 when the world was still coping with the Covid-19 pandemic outbreak, and it finally came to life in 2024. Entitled "Advances in Oleogel Development, Characterization, and Nutritional Aspects," this comprehensive work covers a wide spectrum of topics, ranging from the basic understanding of these lipid-based materials to their practical application in real food products, highlighting the latest breakthroughs in the field. It comprises in-depth descriptions of both canonical and most advanced techniques for the physical, chemical, digestive, and physiological characterization of oleogels, as well as some regulatory and industrial considerations relevant to their production. This multifaceted approach brings new tools, standpoints, and perspectives for understanding and applying oleogels, empowering current and new generations of scientists, students, and professionals who are approaching this topic for the first time, or who already have some experience, to become experts in this field.

More in detail, the book starts with a historical overview of the evolution of the oleogel field (Chap. 1), and it continues with a thorough review of the nutritional aspects of fats and oils (Chap. 2). It then moves to different oil structuring strategies, focusing on oleogels (Chap. 3) and their classification according to preparation methods, comprising direct, indirect, and semi-direct (Chap. 4). The characteristics of oleogels obtained through direct methods using the most studied gelators, such as monoglycerides (Chap. 5), waxes (Chap. 6), ethylcellulose (Chap. 7), fatty acids, lecithin, and others, and mixtures thereof (Chaps. 8 and 9) are then discussed. A separate chapter (Chap. 10) is devoted to oleogels obtained through indirect methods. The following chapters examine methods to engineer oleogels, like for example ultrasound (Chap. 11) and 3D printing (Chap. 12), and to obtain oleogels-derived systems, such as emulsions (Chap. 13), bigels (Chap. 14), and oleofoams (Chap. 15). The book then moves toward the discussion on the physical and chemical stability of oleogels (Chap. 16), and how to use oleogels for delivery and protection of bioactive molecules (Chap. 17). To gain a deeper understanding of oleogel characterization, the main physical, physicochemical, and chemical techniques are described (Chap. 18), along with the latest and more advanced techniques based on rheology (Chap. 19), image analysis (Chap. 20), synchrotron light (Chap. 21), and computer simulations (Chaps. 22 and 23). Then, the book delves into a section devoted to gaining insight on the path of oleogels in the digestive tract, first by examining in vitro digestion model systems (Chap. 24), followed by animal experiments and human clinical trials (Chap. 25). Lastly, a broad review of the edible applications of oleogels (Chap. 26) is provided, along with an analysis of how to translate research results into products for the benefit of society by digging into the regulatory aspects, industrial feasibility, and scalability of oleogel production processes (Chap. 27). The book concludes with a final chapter (Chap. 28) presenting challenges and opportunities for future work in the field of oleogels.

We hope that this book will spark curiosity and wonder in readers, enabling them to clarify concepts, potentially solve problems, and even generate new ideas. We also hope that this book will help to pave the way to accelerate the still slow fat-to-oleogel transition and highlight the crucial role that oleogel research plays in propelling the food industry toward a healthier and more sustainable future.

Bahía Blanca, Argentina — Camila Palla
Helsinki, Finland — Fabio Valoppi

PARTICIPATING **UNIVERSITIES AND ORGANIZATIONS**

Aarhus University, Aarhus, Denmark
Australian Centre for Neutron Scattering, New South Wales, Australia
CNR – Institute of Crystallography, Trieste, Italy
Formo Bio GmbH, Berlin, Germany
Henan Agricultural University, Zhengzhou, China
Heriot-Watt University, Edinburgh, United Kingdom
INL—International Iberian Nanotechnology Laboratory, Braga, Portugal
National Institute of Technology Rourkela, Rourkela, India
Nofima AS, Ås, Norway
Planta Piloto de Ingeniería Química – PLAPIQUI (UNS-CONICET), Bahía Blanca, Argentina
Seoul National University, Seoul, Korea
South China University of Technology, Guangzhou, China
St. Francis Xavier University, Antigonish, Canada
Technion-Israel Institute of Technology, Haifa, Israel
The University of Queensland, Queensland, Australia
Toronto Metropolitan University, Toronto, Canada
U.S. Department of Agriculture, Peoria, United States
Universidad Autónoma de San Luis Potosí, San Luis Potosí, Mexico
Université Lille, Lille, France
University of California, Davis, United States
University of Campinas (UNICAMP), Campinas, Brazil
University of Edinburgh, Edinburgh, United Kingdom
University of Guelph, Guelph, Canada
University of Helsinki, Helsinki, Finland
University of Iceland, Reykjavík, Iceland
University of Illinois Urbana-Champaign, Urbana, United States
Utah State University, Utah, United States

Contents

Chapter 1
Tracing the Evolution of Oleogels: A Historical Overview

Miguel A. Cerqueira, Camila Palla, and Fabio Valoppi

Abbreviations

CAGR	Compound annual growth rate
HMWG	High molecular weight gelator
LMWG	Low molecular weight gelator
PUFA	Polyunsaturated fatty acid
SFA	Saturated fatty acid
TFA	*Trans* fatty acid

Miguel A. Cerqueira, Camila Palla and Fabio Valoppi contributed equally.

M. A. Cerqueira
INL—International Iberian Nanotechnology Laboratory, Braga, Portugal
e-mail: miguel.cerqueira@inl.int

C. Palla
Departamento de Ingeniería Química, Universidad Nacional del Sur (UNS), Bahía Blanca, Argentina

Planta Piloto de Ingeniería Química – PLAPIQUI (UNS-CONICET), Bahía Blanca, Argentina
e-mail: cpalla@plapiqui.edu.ar

F. Valoppi (✉)
Electronics Research Laboratory, Department of Physics, University of Helsinki, Helsinki, Finland

Department of Food and Nutrition, University of Helsinki, Helsinki, Finland

Helsinki Institute of Sustainability Science, Faculty of Agriculture and Forestry, University of Helsinki, Helsinki, Finland

Helsinki Institute of Life Science, University of Helsinki, Helsinki, Finland
e-mail: fabio.valoppi@helsinki.fi

C. Andrea Palla, F. Valoppi (eds.), *Advances in Oleogel Development, Characterization, and Nutritional Aspects*,
https://doi.org/10.1007/978-3-031-46831-5_1

1.1 Food Industry Needs, Challenges, Trends, and Where Oleogels Fit In

The food industry is continuously changing in response to consumer demands and regulatory guidelines, with new trends and challenges being addressed in different areas among food stakeholders.

Two of the most significant challenges are related to the growing population that needs to be fed with healthy and balanced foods, and the sustainability of the materials, processes, and products. The impact of food production on the environment is mainly related to its impact on greenhouse gas emissions, land and water use, and the loss of biodiversity. Several strategies have been proposed to optimize the agriculture and food processing processes, but in most cases, the implementation cost, related to low efficiency and high prices of ingredients, is limiting the change [1]. This implementation cost will have an impact on the other mentioned challenge, where more and more diversified production is needed.

The global population is suffering from an unbalance distribution of food, where a part of the global population suffers from undernutrition with the consumption of low amounts of food of poor quality (resulting in, for example, hunger and micronutrient deficiencies), and another part is eating too much and in an unbalanced way (resulting in, for example, obesity, diabetes, hypertension, coronary heart disease), being, therefore, a concern for both developed and developing countries [2].

Fats are one of the ingredients within the list of the so-called "Western diet" [3] that has shown an increase in consumption in the last 60 years [4]. A study of 2020, comprising 118 countries from 1960 to 2010, showed that the total calorie intake increased for all countries and the nutritional quality decreased [5]. One of the conclusions of the study was that diets are becoming less healthy and that one of the drivers for this change is the consumption of more fats (i.e., animal-based foods) in detriment of carbohydrates (i.e., plant-based foods). However, it is important to mention that people's diet can be influenced by different factors that varies across countries and regions, and households (e.g., culture, religion, climate, and traditions).

The need to improve the global diet has been highlighted by several organizations and governments, but it is consensual that it is only possible to have an effective change with the engagement of all relevant stakeholders [6]. With this in mind, the United Nations adopted 17 Sustainable Development Goals for the Global Challenges, in which the food industry plays a major role, particularly in the goals: (2) "Zero hunger," (3) "Good health and well-being," (12) "Responsible consumption and production," and (13) "Climate action" [7].

Food trends are also related to some of the presented challenges, where the consumers' awareness of the need for healthy and sustainable diets is changing the food industry. One of the main food trends is consumer engagement and empowerment. The behavior and perception of consumers, as well as food producers and retailers, impact companies' strategies and decisions. Nowadays, consumers seem to be looking for more healthy and more natural products, focused on alternative proteins, and products that can be considered clean label and sustainable. This consumers' direction is also supported by the EAT-Lancet commission, which suggests that global food consumption should shift to more plant-based foods. This is supported by the fact that plant-based foods can be healthier than the animal counterparts but also due to the environmental impact of animal-based foods [1]. They propose a universal diet based on vegetables, fruits, whole grains, legumes, nuts, and unsaturated oils, with low to moderate amounts of seafood and poultry and low quantities of red meat, processed meat, added sugar, refined grains, and starchy vegetables.

Lipids have a huge role in these healthy diets, since they are essential for the proper development of humans. While some of the essential lipids are present in food and food ingredients, they are also directly added (as animal fat or vegetable oil) conferring important technological and sensorial properties to a great variety of foods, such as spreads, bakery products, margarine, and dairy products. However, during the years, the solutions offered by the industry have been controversial, with the use of *trans* fatty acids (TFAs) that have already been banned or allowed in very low amounts in several countries [8, 9]. Another strategy is the use of saturated fatty acids (SFAs) that can be obtained by the fractionation of oils, such as palm oil. The low price, texture and rheological properties, high stability against oxidation, and shelf life are among the reasons to use palm oil. However, also in this case, studies have shown that the consumption of saturated fats (except for stearic acid) influences cholesterol levels in the blood and increases the risk of cardiovascular diseases [10]. Guidelines from the World Health Organization for a healthy diet mention that reducing the amount of total fat intake to less than 30% of total energy intake helps to prevent unhealthy weight gain in adults and that we should reduce the use of *trans* and saturated fats by replacing them by unsaturated and polyunsaturated fats [11]. The European Union has also promoted the development of foods with a healthier lipid profile by allowing companies to claim nutritional aspects on their products (e.g., low fat, fat-free, source of omega-3 fatty acids) [12].

Another problem related to the use of fats by the food industry is the environmental impact of oil production. While this is an issue that involves the entire food system, in the case of lipids, one of the great examples is palm oil. Several studies revealed the considerable impact of their production on deforestation and consequently loss of biodiversity and the release of CO_2 [13], bringing an additional problem when palm oil is used. Therefore, the removal of hydrogenated fats and the limitation of the use and consumption of saturated fats have changed the way the food industry looks at lipids.

Despite these issues, the Global Edible Oils and Fats Market is experiencing continuous expansion. The market size is projected to reach USD 831.10 billion by 2030, with a robust revenue-based compound annual growth rate (CAGR) of 7.5% from 2023 to 2030 [14]. In particular, the Margarine Market is expected to grow significantly, from USD 22.12 billion in 2023 to USD 24.93 billion in 2028, at a steady CAGR of 2.42% over the forecast period (2023–2028) [15].

Different strategies have been proposed to meet the real need for healthier, TFA-free, stable, and solid-like fats, which maintain their structure at ambient temperature, assuring a longer shelf-life. One of these strategies is fat mimetics, where the use of physically structured oils, known as oleogels, seems to be one of the most promising routes. They can guarantee the reduction of unhealthy fats or/and their replacement by mono- and polyunsaturated fats and offer technical functionalities similar to fats. This is related to the way they are produced, mostly using only food ingredients and without chemical modification of the lipids, thus presenting some benefits towards sustainability and clean label trends. Based on these advantages, it is possible to think that there is a promising potential market for oleogels. Moreover, with increasing consumer awareness regarding healthy food consumption, the Global Edible Oils and Fats Market is expected to witness an increased demand for the development of healthier products.

1.2 Oil Structuring and Oleogels—A Brief Introduction

Lipids are essential ingredients in food products, contributing to their physicochemical properties, structural integrity, stability, and overall sensory quality. However, the pressing demand for healthier food choices has given rise to a significant challenge: finding and developing healthy alternatives to solid fats while minimizing any adverse effects on product properties and consumer acceptance. When solid fat ingredients are merely replaced with oil, the resulting products generally fail to replicate the quality characteristics of the original ones, particularly in terms of textural attributes, leading to a deficiency that negatively impacts sensory perception and consumer satisfaction [16, 17]. This is one of the main reasons why the food industry and scientists have focused their attention on oil structuring and oleogelation processes, as they allow the development of soft matter structures that possess similar functionality to solid fats. The systems produced by oleogelation using edible oils are called oleogels, also referred to in the literature as organogels (see the next section for a discussion of both terms).

Oleogels are semisolid materials formed by the entrapment of oil within a three-dimensional (3D) gel network composed of structuring agents or gelators added at low concentrations [18]. Structuring agents are compounds with the ability to retain oil after specific physical processing. They are generally classified as either low molecular weight gelators (LMWGs) or high molecular weight gelators (HMWGs). Examples of the first type are fatty acids, monoglycerides, waxes, sterols, and lecithin. On the other hand, HMWGs include polymers and biopolymers such as cellulose derivatives, proteins, and starches, which typically require more processing to form stable structures that immobilize the oil phase. The choice of structuring agent depends on the specific application and the desired properties of the oleogel, allowing for versatility in the development of different oil-based products. A detailed explanation of the physical processes involved in the preparation of oleogels using LMWGs or HWMGs can be found in Chaps. 4, 5, 6, 7, 8, 9, and 10.

As oleogels are mainly formulated with oils (desirable $> 90\%$), they have a lipid profile comparable to that of the edible oil used to produce them, which is of radical importance from a nutritional point of view, as it will be highlighted in Chap. 2. This makes it possible to develop lipid materials not only with reduced SFA and free TFA content but also with a superior lipid profile by selecting oils rich in fatty acids with high biological value such as essential fatty acids. In particular, the use of oils abundant in polyunsaturated fatty acids (PUFA) allows the production of oleogels with health-promoting activity, as PUFA oils have been shown to offer numerous benefits for health, including immune system and blood pressure regulation, cholesterol metabolism, neurological and cognitive function, and insulin resistance, among others [19].

The physical properties of oleogels are the result of a combination of factors, including the chemical composition of the oil and gelator, their respective proportions, and the processing conditions employed during their production. The development of oleogels with fat-mimicking properties requires a comprehensive evaluation of thermal, rheological, textural, structural, and stability properties, as it will be discussed in Chaps. 18, 19, and 21. Additionally, oleogel characterization plays a crucial role in enhancing our understanding of the oleogel itself and determining relevant properties for its applications in various products [20].

1.3 Organogels and Oleogels in History

We are used to thinking that organogels and oleogels are relative recent terms. This is true within the food science area and to some extent in the cosmetics and pharmaceuticals areas, whereas in the case of lubricants or fundamental science these terms have been used for more than a century (organogels) and almost 75 years (oleogels).

First of all, it is important to distinguish between the two terms and define them based on the meanings that have been developed over the last decades: an organogel is simply a gel of an organic solvent where the solvent is entrapped/retained in a three-dimensional network of gelling molecules. An organogel where the organic solvent is an edible oil and gelling molecules are edible is called an oleogel.

In general, organogels have been defined by their preparation methods, where a gelator is added to an organic liquid, the mixture is heated to dissolve the gelator, and then the solution is cooled below the gelation transition temperature [21]. If the reader is already familiar with the field of oleogels, this definition can sound familiar as it is the same procedure used in the preparation of oleogels by the hot direct method; more details can be found in Chap. 4. A more detailed definition of an organogel refers to a "thermally-reversible viscoelastic liquidlike or solidlike material comprised of an organic liquid and low concentrations (typically <2 wt %) of relatively low molecular mass molecules (i.e., gelators)" [21]. However, not only LMWGs are nowadays included in the definition, but also polymer gelators [22]. Also in this case, the definition is quite similar to that of oleogels obtained by the hot direct method. Even though these two definitions could sound familiar, the development of the field and consequently the definition of organogels has changed

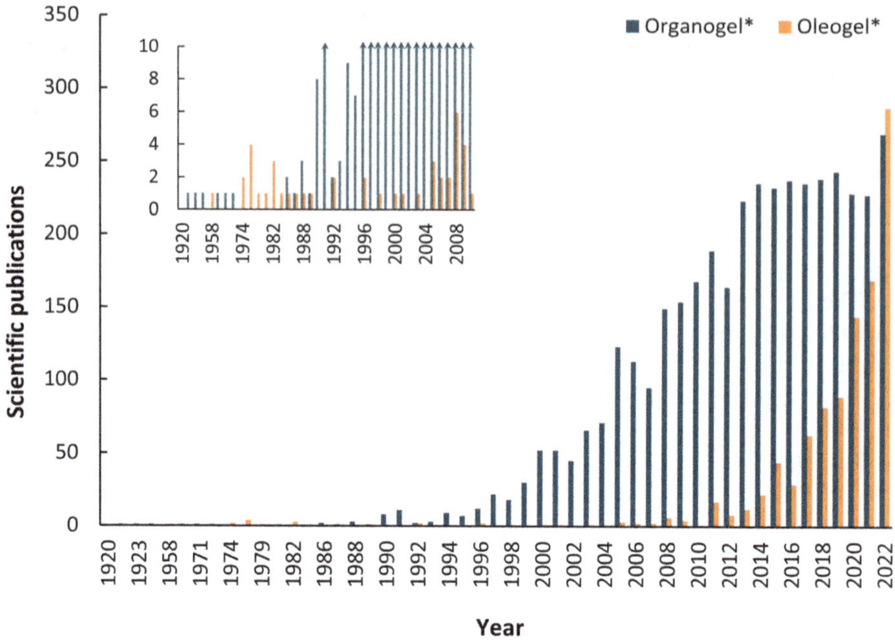

Fig. 1.1 Number of scientific publications reported by Scopus containing the term "organogel*" or "oleogel*" in the title, abstract, and keywords from 1920 to 2022. The inserted figure is the magnification of the main figure from 1920 to 2009. Note that (*) allows users to search for all terms beginning with the specified word—for example, oleogel/oleogels/oleogelation/oleogelator

over time (more information below). The field has been developing since 1920s. One of the first papers reporting the term "organogel" we could find in the literature was published by Neuhausen and Patrick [23] in 1921, although references in the same article about tests to gel organic solvents with silicic acid and a small amount of water date back to 1864. The field has continued its evolution until present days (Fig. 1.1), and during this century of research, the ability of organic gelators like metal salts of fatty acids, fatty acids derivatives, steroid derivatives, anthryl derivatives, aromatic- and steroidal-containing molecules, amino acids, and organometallic molecules to gel organic solvents like linear and cyclic hydrocarbons, alcohols, aldehydes, ketones, and aromatic solvents, to name a few, has been extensively studied [21, 24]. If the reader is interested to further deepen the topic of organogels, they can refer to the reviews by Terech and Weiss [21], Suzuki and Hanabusa [22], and Abdallah and Weiss [24], as well as to the book by Guenet [25].

From the field of organogels, edible oleogels have been developed. The first example of publications reporting on organogels made with edible oils and gelators in the food science area were those published by Ojijo et al. [26, 27] in 2004. These authors developed oleogels structured with monoglycerides by the hot direct method (mixing, heating, and cooling) using virgin olive oil. Although these structured lipid systems could be classified as oleogels, the term in the food science area was not yet

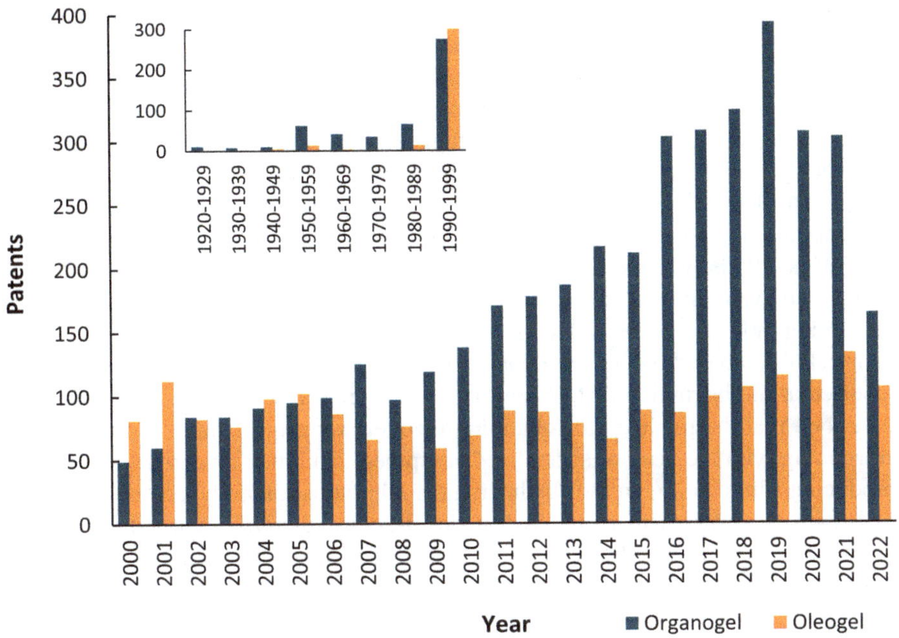

Fig. 1.2 Number of patents reported by Google Patents under the search term "organogel" or "oleogel" from 2000 to 2022 (search configuration: priority date and text in full documents). This period was selected considering that the patent lifetime is 20 years. The inserted figure shows the search results between 1920 and 1999

in use. It was not until 2007 that the word oleogel appeared as a keyword in the context of organogels made with edible oils in a paper authored by Pernetti et al. [28]. However, the term "oleogel" can be found even in an earlier paper belonging to the pharmaceutical area published in 2003, in which olive oil was gelled using ethylcellulose as a first step to develop a topical administration incorporating a drug [29]. In the pharmaceutical area, these types of gels appeared already in the 90s, but they were called lipogels [30].

As anticipated at the beginning of this section, it is interesting to know that the term oleogel is about 75 years old (Fig. 1.2). However, in the late 1930s, Steven Kistler, who is well known for developing the first aerogels [31], developed the first aerogel-templated method in which by adding aerogel particles to oils, a thickened oil/grease was obtained that could be used as a lubricant (US2260625A). This method was later brought in the edible oleogel area in 2017 by Manzocco et al. [32] and further developed by Plazzotta et al. [33]. In his patent, Kistler also suggests that "similar results [formation of an oleogel] are obtainable by replacing the swelling liquid of the gel [solvent] with a portion of the liquid to be thickened [oil]," which gives a glimpse of an oleogel preparation method, nowadays called the solvent exchange method, developed in 2015 by de Vries et al. [34]. The solvent exchange method described in Kistler's patent was further explained in the patent number US2594822A from 1949, where we could find the term oleogel making one of its first appearance: "Another means known to the prior art for the formation of greases of inorganic gels comprises formation of a hydrogel, replacement of water with a water-miscible organic solvent, replacement of the water-miscible organic solvent with an oil-soluble solvent and transference of the resulting organogel to a lubricating oil medium followed by the removal of an oil-soluble solvent to form, finally, an oleogel." From the same patent, we can also identify the definitions of oleogels and organogels that were used by the research and innovation community in the 1940s–1950s. Oleogels were referred to as gels of lubricating oils, whereas organogels were an intermediate step having the same structure as an oleogel but containing only oil-soluble solvents. It can be seen how these two terms evolved in their definition based on the evolution of the field by just comparing them with those reported at the beginning of this section. Finally, in the same patent, the inventors described that the methods for obtaining oleogels were quite time- and resources-consuming with the need of specialized equipment, and they invented a method for producing a lubricating grease "by having the lubricating oil as the continuous phase gelled by the polyvalent metal hydroxide dispersed therein," where the reported method showed features that are typical of a direct oleogelation method. From there, efforts have been made to simplify oleogelation methods, for example, by developing processes using colloidal particles and water, with the subsequent evaporation of water (DES0040177MA from 1954), or by using organogels as templates to form oleogels, where one process comprised a "very rapid evaporation of the liquid from the organogel as it is fed to the hot lubricating oil, such evaporation taking place in the presence of an ample supply of oil to replace the volatile liquid displaced by evaporation, minimizes this tendency and causes the line pore structure of the gel material to be retained and carried forward into the oleogel stage" (GB2986518XA from 1958). At this point, the first scientific publication we could find using the word "oleogel" was published by Bondi [35] in 1958.

It is interesting to note that the historical evolution of oleogel preparation methods started with very sophisticated procedures requiring specialized equipment and multiple steps and proceeded towards more simplified methods. This is the opposite that was observed in the evolution of methods for producing edible oleogels, where simple methods dominated the first decade of development (from 2004), followed by the introduction of production methods requiring multiple steps and specialized equipment (from 2013).

Through a search on Google Patents and Espacenet, we could observe that the trend to use the term "oleogels" continued in many patents. However, the field of application of oleogels in those patents shifted from lubrication (which dominated until the 1970s) to pharmaceutical applications, particularly with the development of topical creams in the 1980s. The 1990s marked further diversification and expansion of the oleogel field of application, including cosmetics, bioadhesives, fuel, veterinary, and feedstuff. During the first decade of 2000s, the use of oleogels for pharmaceutical applications intensified, and the first developments were made in the food area. In the following decade to date (2010–2023), most of the developments have been related to food applications.

This short journey taught us that the meaning of the words "oleogels" and "organogels" morphed over time. We have seen their definitions evolve from those used in the pioneering area of lubricants to the current booming in food science, pharmaceutical, and cosmetic areas. We hope that the reader finds this little excursion into the history of oleogels interesting and that the ideas and visions of the past will inspire possible new methods to obtain edible oleogels.

1.4 Oleogels: Current Areas of Research in the Food Sector

Oleogels are an active area of research with a focus on their potential to replace saturated fats in food products, as well as their use in developing functional foods and bioactive delivery systems. The research topics and knowledge groups in the field of oleogels can be effectively analyzed using Carrot[2] (search.carrot2.org), an open-source search results clustering tool. Carrot[2] utilizes a clustering process that involves analyzing the content of documents, identifying common terms and phrases, and grouping related documents together into clusters. Figure 1.3 shows the clusters resulting from the analysis of 538 documents from PubMed using the search term "oleogel." In the visualization, larger bubbles represent clusters with a

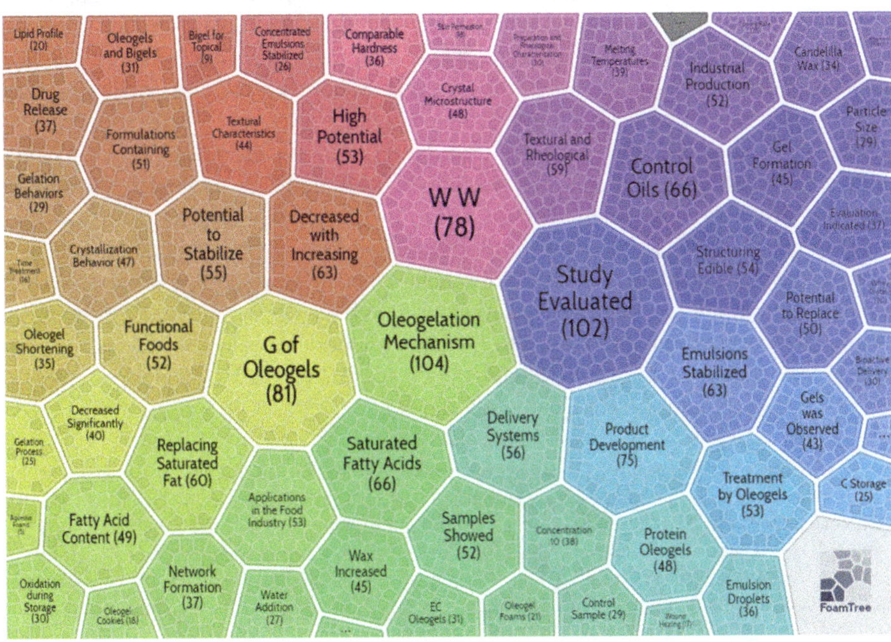

Fig. 1.3 Foam Tree Map created by Carrot[2], displaying the major topics in oleogels from an analysis of 538 bibliographic records sourced from PubMed's core database (clustering algorithm: Lingo). Generated on 20th July 2023

higher number of publications. The topics that have received more attention in the oleogel field are "Oleogelation Mechanism," "Emulsions and Stabilization," "Textural and Rheological" properties of oleogels, "Delivery Systems," "Application in the Food Industry," "Crystal Microstructure," "Protein Oleogels," and "Crystallization Behavior." Other topics that have received attention with potential to expand knowledge in the area include "Oleogel and Bigels," "Oleogel Foams," "Oxidation during Storage," "Bioactive Delivery," and "Processed Meat."

Other topics of current interest, not identified by Carrot[2], include the study of oleogels as a potential tool for modulating lipid metabolism to manage obesity, the production of oleogels using oils obtained by fermentation or extraction from microalgae and insects, the exploration of oleogels in the development of plant-based products, and their application in formulating inks for food 3D printers. While reading this book, the reader will come across both basic and detailed information on all the topics mentioned and many more.

1.5 Conclusions

The need to improve the global diet and the push for sustainable food practices have compelled the food industry and scientists to search for novel lipid materials able to confer technological and sensorial properties to products. The strategy of oleogelation, which has been widely studied in the field of lubricants for almost a century, has served as an inspiration for the development of oleogels, opening up the possibility of creating an infinite number of materials by combining edible oils with compounds capable of retaining oil after specific physical processing. Indeed, numerous studies and patents have been published on oleogels, demonstrating their remarkable potential not only to replace saturated fats in foods, but also to serve as valuable components in the creation of functional foods and efficient bioactive delivery systems. The more knowledge we acquire about oleogels, the more apparent it becomes how vast this field of study is, encompassing numerous topics and potential applications, and the need to delve deeper into each of them. A strong knowledge foundation will provide us with more tools to continue advancing in this fascinating field; thus, we will unlock the full potential of oleogels for the benefit of society and the environment.

Acknowledgments Miguel A. Cerqueira acknowledges NUTRI4ICECARE: Ice cream with incorporation of compounds of nutritional interest (NORTE-01-0247-FEDER-045380), co-financed by Norte 2020 and the European Union, through the European Regional Development Fund (ERDF). Camila Palla acknowledges the Consejo Nacional de Investigaciones Científicas y Técnicas (CONICET) of Argentina. Fabio Valoppi acknowledges the European Union's Horizon 2020 research and innovation program funding under the Marie Skłodowska-Curie grant agreement No. 836071.

References

1. Willett W, Rockström J, Loken B, Springmann M, Lang T, Vermeulen S, Garnett T, Tilman D, Declerck F (2019) The Lancet Commissions Food in the Anthropocene: the EAT – Lancet Commission on healthy diets from sustainable food systems. Lancet 393:447–492. https://doi.org/10.1016/S0140-6736(18)31788-4
2. OECD (2019) The heavy burden of obesity: The economics of prevention. OECD Health Policy Studies, OECD Publishing, Paris. https://doi.org/10.1787/67450d67-en
3. Khoury CK, Bjorkman AD, Dempewolf H, Ramirez-Villegas J, Guarino L, Jarvis A, Rieseberg LH, Struik PC (2014) Increasing homogeneity in global food supplies and the implications for food security. Proc Natl Acad Sci 111:4001 LP–4006. https://doi.org/10.1073/pnas.1313490111
4. Oberlander L, Disdier AC, Etilé F (2017) Globalisation and national trends in nutrition and health: a grouped fixed-effects approach to intercountry heterogeneity. Heal Econ (United Kingdom) 26. https://doi.org/10.1002/hec.3521
5. Le TH, Disegna M, Lloyd T (2020) National food consumption patterns: converging trends and the implications for health. EuroChoices n/a. https://doi.org/10.1111/1746-692X.12272
6. World Health Organization (2021). Obesity and overweight. https://www.who.int/news-room/fact-sheets/detail/obesity-and-overweight. Accessed 10th May 2023
7. Nations united sustainable development goals. https://www.un.org/sustainabledevelopment/
8. FDA (U.S. Food and Drug Administration) (2018) Trans fat. https://www.fda.gov/food/food-additives-petitions/trans-fat
9. European Comission (2019) Commission Regulation (EU) 2019/649. Off J Eur Union 17–20. https://doi.org/10.2903/sp.efsa.2018.en-1433
10. Zhu Y, Bo Y, Liu Y (2019) Dietary total fat, fatty acids intake, and risk of cardiovascular disease: a dose-response meta-analysis of cohort studies. Lipids Health Dis 18:91
11. WHO (World Health Organization) (2020) Healthy diet. https://www.who.int/news-room/fact-sheets/detail/healthy-diet. Accessed 10th May 2023
12. European Commission (2012) Nutrition claims. https://food.ec.europa.eu/safety/labelling-and-nutrition/nutrition-and-health-claims/nutrition-claims_en. Accessed 10th May 2023
13. Meijaard E, Brooks TM, Carlson KM, Slade EM, Garcia-Ulloa J, Gaveau DLA, Lee JSH, Santika T, Juffe-Bignoli D, Struebig MJ, Wich SA, Ancrenaz M, Koh LP, Zamira N, Abrams JF, Prins HHT, Sendashonga CN, Murdiyarso D, Furumo PR, Macfarlane N, Hoffmann R, Persio M, Descals A, Szantoi Z, Sheil D (2020) The environmental impacts of palm oil in context. Nat Plants 6:1418–1426. https://doi.org/10.1038/s41477-020-00813-w
14. GlobeNewswire (2023) Global edible oil and fats market to 2030: rising importance of vegetable oils as key functional ingredients bolsters growth. https://www.globenewswire.com/en/news-release/2023/06/02/2681143/28124/en/Global-Edible-Oil-and-Fats-Market-to-2030-Rising-Importance-of-Vegetable-Oils-as-Key-Functional-Ingredients-Bolsters-Growth.html. Accessed 27 July 2023
15. Mordor Intelligence (2023) Margarine market size & share analysis – growth trends & forecasts (2023–2028). https://www.mordorintelligence.com/industry-reports/margarine-market. Accessed 24 July 2023
16. Doan CD, Tavernier I, Okuro PK, Dewettinck K (2018) Internal and external factors affecting the crystallization, gelation and applicability of wax-based oleogels in food industry. Innov Food Sci Emerg Technol 45:42–52. https://doi.org/10.1016/j.ifset.2017.09.023
17. Gavahian M, Tiwari BK, Chu Y, Ting Y, Farahnaky A (2019) Food texture as affected by ohmic heating: mechanisms involved, recent findings, benefits, and limitations. Trends Food Sci Technol 86:328–339. https://doi.org/10.1016/j.tifs.2019.02.022
18. Patel AR, Dewettinck K (2016) Edible oil structuring: an overview and recent updates. Food Funct 7:20–29. https://doi.org/10.1039/C5FO01006C
19. Abedi E, Sahari MA (2014) Long-chain polyunsaturated fatty acid sources and evaluation of their nutritional and functional properties. Food Sci Nutr 2(5):443–463

20. Flöter E, Wettlaufer T, Conty V, Scharfe M (2021) Oleogels—their applicability and methods of characterization. Molecules 26. https://doi.org/10.3390/molecules26061673
21. Terech P, Weiss RG (1997) Low molecular mass gelators of organic liquids and the properties of their gels. Chem Rev 97. https://doi.org/10.1021/cr9700282
22. Suzuki M, Hanabusa K (2010) Polymer organogelators that make supramolecular organogels through physical cross-linking and self-assembly. Chem Soc Rev 39. https://doi.org/10.1039/b910604a
23. Neuhausen BS, Patrick WA (1921) Organogels of silicic acid. J Am Chem Soc 43. https://doi.org/10.1021/ja01441a011
24. Abdallah DJ, Weiss RG (2000) Organogels and low molecular mass organic gelators. Adv Mater 12:1237–1247. https://doi.org/10.1002/1521-4095(200009)12:17<1237::AID-ADMA1237>3.0.CO;2-B
25. Guenet J-M (2016) Organogels: thermodynamics, structure, solvent role, and properties. Springer, Cham
26. Ojijo NKO, Neeman I, Eger S, Shimoni E (2004) Effects of monoglyceride content, cooling rate and shear on the rheological properties of olive oil/monoglyceride gel networks. J Sci Food Agric 84:1585–1593. https://doi.org/10.1002/jsfa.1831
27. Ojijo NKO, Kesselman E, Shuster V, Eichler S, Eger S, Neeman I, Shimoni E (2004) Changes in microstructural, thermal, and rheological properties of olive oil/monoglyceride networks during storage. Food Res Int 37:385–393. https://doi.org/10.1016/j.foodres.2004.02.003
28. Pernetti M, van Malssen KF, Flöter E, Bot A (2007) Structuring of edible oils by alternatives to crystalline fat. Curr Opin Colloid Interface Sci 12:221–231
29. Ruíz Martínez MA, Muñoz De Benavides M, Morales Hernández ME, Gallardo Lara V (2003) Influence of the concentration of a gelling agent and the type of surfactant on the rheological characteristics of oleogels. Farmaco 58. https://doi.org/10.1016/S0014-827X(03)00180-0
30. Realdon N, Dal Zotto M, Ragazzi E, Dalla Fini G (1996) Drug release from lipogels according to gelling conditions and mechanical treatment. Drug Dev Ind Pharm 22. https://doi.org/10.3109/03639049609041982
31. Kistler S (1931) Coherent expanded aerogels and jellies. Nature 127:741
32. Manzocco L, Valoppi F, Calligaris S, Andreatta F, Spilimbergo S, Nicoli MC (2017) Exploitation of κ-carrageenan aerogels as template for edible oleogel preparation. Food Hydrocoll 71. https://doi.org/10.1016/j.foodhyd.2017.04.021
33. Plazzotta S, Calligaris S, Manzocco L (2020) Structural characterization of oleogels from whey protein aerogel particles. Food Res Int 132. https://doi.org/10.1016/j.foodres.2020.109099
34. De Vries A, Hendriks J, Van Der Linden E, Scholten E (2015) Protein oleogels from protein hydrogels via a stepwise solvent exchange route. Langmuir 31. https://doi.org/10.1021/acs.langmuir.5b03993
35. Bondi A (1958) Flow phenomena with oleogels. Trans Soc Rheol 2:303–312

Chapter 2
Nutritional Aspects of Fats and Oils

Bente Kirkhus, Gudrun V. Skuladottir, Anna-Maija Lampi, and Astrid Nilsson

Abbreviations

AA	Arachidonic acid (20:4n-6)
AI	Adequate intake
ALA	Alpha-linolenic acid (18:3n-3)
CHD	Coronary heart disease
CVD	Cardiovascular disease
DHA	Docosahexaenoic acid (22:6n-3)
E%	Energy %
EPA	Eicosapentaenoic acid (20:5n-3)
HDL	High-density lipoprotein
LA	Linoleic acid (18:2n-6)
LDL	Low-density lipoprotein
MUFA	Monounsaturated fatty acids
NCDs	Noncommunicable diseases
PUFA	Polyunsaturated fatty acids
RCT	Randomized controlled trial
SFA	Saturated fatty acids
WHO	World Health Organization

B. Kirkhus · A. Nilsson (✉)
Division Food and Health, Nofima AS, Ås, Norway
e-mail: bente.kirkhus@nofima.no; astrid.nilsson@nofma.no

G. V. Skuladottir
Department of Physiology, University of Iceland, Reykjavik, Iceland
e-mail: gudrunvs@hi.is

A.-M. Lampi
Department of Food and Nutrition, University of Helsinki, Helsinki, Finland
e-mail: anna-maija.lampi@helsinki.fi

C. Andrea Palla, F. Valoppi (eds.), *Advances in Oleogel Development, Characterization, and Nutritional Aspects*,
https://doi.org/10.1007/978-3-031-46831-5_2

2.1 Introduction

Fats and oils are important for the quality of foods (taste, smell, texture, and nutrition). Dietary fat is a major nutrient, an excellent source of energy, and provides essential fatty acids. Moreover, fats and oils provide fat-soluble vitamins and other fat-soluble compounds such as carotenoids and sterols, and many flavor-active compounds. Dietary fat may play a significant role in the prevention and treatment of noncommunicable diseases (NCDs), also known as chronic diseases. According to the Global Burden of Disease study a diet high in saturated fatty acids (SFA) and *trans* fatty acids is a leading risk factor for several NCDs, in particular metabolic diseases such as cardiovascular disease (CVD) and type 2 diabetes [1, 2]. Thus, the World Health Organization (WHO) guidelines and National Dietary Guidelines recommend limiting the intake of saturated fat and replacing solid fat high in SFA and/or *trans* fatty acids with oils high in monounsaturated fatty acids (MUFA) and polyunsaturated fatty acids (PUFA), with the challenges this presents regarding taste and texture in food products. Oleogels are promising alternatives to solid fats for food applications since they successfully replace solid fat with oils in food products without compromising on food quality. They are reported to be an innovative structured fat system used for industrial applications due to their nutritional and environmental benefits [3]. The dietary guidelines for fat, the scientific support for these advices and the rationale for altering dietary fat intake, both quantitatively and qualitatively, will be discussed in the following sections. We will describe how all dietary fats and oils are made up of saturated fatty acids (SFA), monounsaturated fatty acids (MUFA), and polyunsaturated fatty acids (PUFA) in different proportions, affecting both melting point and nutritional value.

2.2 Fatty Acid Composition of Dietary Fats, Oils, and Food Products

Edible fats and oils may be of vegetable, animal, and marine origin. In food products, fats and oils come from raw materials or as added ingredients. They consist mainly of triacylglycerols and other acylglycerols, where fatty acids are esterified with glycerol. In commercial fats and oils, the content of other lipids is, in general, less than 2% [4]. The nomenclature for fatty acids indicates number of carbon atoms (chain-length), number of double bonds (degree of saturation), and location of the first double bond at carbon counted from the omega or n end; for example, the omega-3 fatty acid α-linolenic acid (ALA; 18:3n-3) has 18 carbons, 3 double bonds, where the first double bond is located at carbon number 3 from the n end.

The fatty acid composition has a major effect on the properties of dietary fats and oils. All dietary fats and oils are made up of SFA, MUFA, and PUFA in different

Table 2.1 Typical composition of major fatty acids (% of total fatty acids) in common solid dietary fats (milk fat, beef fat/tallow, and pork fat) and semisolid dietary fats (coconut oil and palm oil) [4]

	Milk fat	Beef fat / Tallow	Pork lard	Coconut oil	Palm oil
12:0 (lauric acid)	2.9	0.2	0.1	47.5	–
14:0 (myristic acid)	10.8	4.0	1.5	18.1	1.1
16:0 (palmitic acid)	26.9	24.3	26.0	8.8	44.0
18:0 (stearic acid)	12.1	21.4	13.5	2.6	4.5
18:1n-9 (oleic acid)	28.5	33.6	43.9	6.2	39.2
18:2n-6 (linoleic acid)	3.2	1.6	9.5	1.6	10.1
18:3n-3 (α-linolenic acid)	0.4	0.6	0.4	–	0.4
20:5n-3 (eicosapentaenoic acid)	–	–	–	–	–
22:6n-3 (docosahexaenoic acid)	–	–	–	–	–
SFA	65.0	53	41.7	92.1	49.6
MUFA	31.4	37.9	47.7	6.2	39.3
PUFA (n-6)	3.2	1.6	9.5	1.6	10.1
PUFA (n-3)	0.4	–	–	–	–
Melting range	28–36 °C	45–48 °C	32–33 °C	25–28 °C	36–45 °C

SFA saturated fatty acids, *MUFA* monounsaturated fatty acids, *PUFA* polyunsaturated fatty acids

proportions (Tables 2.1 and 2.2), affecting both melting point and nutritional value. The melting point for a fatty acid depends on both chain length and number of double bounds. In general, SFA have higher melting point than MUFA and especially higher than PUFA. Most double bonds in the dietary unsaturated fatty acids are in the *cis* configuration, while *trans* double bonds can be found in ruminant fats (low levels) and in partially hydrogenated oils (higher levels). The term fat often refers specifically to triacylglycerols that are solid or semisolid at room temperature, thus excluding liquid oils. Position of fatty acids in the glycerol backbone also influences fats' physical and nutritional properties [5, 6].

Animal fats such as beef fat, pork fat and butter contain high amounts of SFA (between 25% and 65%) with relative high melting points, and they are all solid at room temperature (Table 2.1).

They also contain a relatively high level of MUFA (30–55%), but a small amount of PUFA (2–20%), mainly linoleic acid (LA; 18:2n-6). Butter fat contains approximately 4–8% of *trans* fatty acids [6]. Vegetable fats include the semisolid fats like coconut oil and palm oil, with high amounts of SFA (55–95%) (Table 2.1). Among the SFA, coconut contains mainly lauric acid (12:0) and myristic acid (14:0). In palm oil, the major SFA are palmitic acid (16:0) and stearic acid (18:0) [5, 7, 8]. Fractionation of palm oil is used to produce materials with desired melting properties with varying fatty acid compositions.

Common vegetable oils such as soybean oil, sunflower oil, olive oil, and rapeseed oil have less than 15% SFA and approximately 85% unsaturated fatty acids [7, 8]

Table 2.2 Typical composition of major fatty acids (% of total fatty acids) in common dietary oils

	Soybean oil[a]	Sunflower oil[a]	Rape seed oil[a]	Olive oil[a]	Cod liver oil[b]	Salmon oil[c]
12:0 (lauric acid)	–	–	–	–	–	–
14:0 (myristic acid)	0.1	0.1	0.1	–	3.6	3.2
16:0 (palmitic acid)	10.6	7.0	4.1	9.0	10.4	13.7
18:0 (stearic acid)	4.0	4.5	1.8	2.7	2.6	3.7
18:1n-9 (oleic acid)	23.3	18.7	60.9	80.3	16.2	29.6
18:2n-6 (linoleic acid)	53.7	67.5	21.0	6.3	1.5	16.7
18:3n-3 (α-linolenic acid)	7.6	0.8	8.8	0.7	–	3.3
20:5n-3 (eicosapentaenoic acid)	–	–	–	–	9.3	7.5
22:6n-3 (docosahexaenoic acid)	–	–	–	–	11.9	6.3
SFA	11.4	12.7	7.2	12.1	16.6	21.3
MUFA	23.4	18.9	62.9	80.9	46.4	40.5
PUFA (n-6)	53.7	67.5	21.0	6.3	1.5	17.8
PUFA (n-3)	7.6	0.8	8.8	0.7	23.6	20.5
Melting range	−20 to −23 °C	−18 to −20 °C	−9 °C	0 °C	−70 to 14 °C	−70 to 14 °C

SFA saturated fatty acids, *MUFA* monounsaturated fatty acids, *PUFA* polyunsaturated fatty acids
[a] [4]
[b] [9]
[c] [10]

(Table 2.2). In addition, olive oil and rapeseed oil are high in MUFA (60–70%) having oleic acid (18:1n-9) as the major fatty acid. The content of PUFA in rapeseed oil (ca. 30%) is higher than in olive oil (ca. 10%), and moreover rapeseed oil contains both LA (ca. 20%) and ALA (ca. 10%), which makes it a good source of both n-6 and n-3 fatty acids. Soybean oil, corn oil, and sunflower oil are high in n-6 PUFA (54–68%), mainly LA. Among them soybean oil is the only one to contain a significant amount of the n-3 PUFA ALA (8%). Different fish oils and fractions of fish oils are usually used as dietary supplements and dietary oils of marine origin are less usual. Refined cod liver oil and salmon oil are two dietary marine oils which are produced with food ingredient quality. Both are high in marine n-3 PUFA, the health-promoting eicosapentaenoic acid (EPA; 20:5n-3) and docosahexaenoic acid (DHA; 22:6n-3). The melting points of marine oils depend on refining grade since some of the refining procedure include precipitation of solid fat high in saturated fatty acid, mainly 18:0.

Typical composition of the main fatty acids in common dietary fats and oils is presented in Tables 2.1 and 2.2 and will be discussed in later sections concerning nutrition and health. Usually, fats contain more SFA than oils, which might lower their nutritional quality. However, solid fats or semisolid fats are often needed in food application to build desired textures and desired sensory properties as well as to improve storage stability. The physical state also influences the release of flavor-active compounds in the mouth. Nowadays, solid fats are obtained from natural fats, fully hydrogenated oils, and structured lipids, whereas the use of partially hydrogenated oils has diminished due to their content of *trans* fatty acids. Oleogels are a means to improve the nutritional quality of food products that need solid fats in the structure by replacing them with vegetable and marine oils.

2.3 Guidelines for Fat Intake

Dietary guidelines are statements that assist populations in choosing foods that deliver optimal nutrient intake and are associated with a reduced risk of NCDs, such as heart disease, cancer, chronic respiratory disease, and diabetes [11, 12]. The first Dietary Guidelines for Americans were released in 1980, where "avoiding too much saturated fat" was recommended [13]. In 2004, the WHO Global Strategy on Diet and Physical Activity recommended shifting consumption from saturated fat to unsaturated fats and limiting the level of saturated fat in the diet [14–16]. All Dietary Guidelines have recommended reductions in saturated fat, with the first numerical target of <10% of calories issued in 1990. Subsequent editions of the Dietary Guidelines for Americans (2010 and 2015) also introduced replacement of SFA with n-3 PUFA [17–20]. The message to decrease SFA has been supported by the American Heart Association/American College of Cardiology [21], the National Lipid Association [22], and the global recommendations issued by WHO [23].

The European Food Safety Authority (EFSA) (2010) has set dietary reference values only for a few fatty acids [24]. The content of SFA and *trans* fatty acids should be as low as possible. An adequate intake (AI) for LA was set to 4 energy% (E%) and for ALA 0.5 E%, and without reference values for upper intake. WHO recommends less than 10 E% from SFA and less than 1 E% from *trans* fatty acids [25]. There is no general official recommendation for daily intake of the marine n-3 PUFA (EPA and DHA). EFSA has suggested an AI of 250 mg of sum of EPA + DHA for adults [24], whereas WHO has recommended a daily dose of 300–500 mg EPA + DHA. However, the Global Recommendations for EPA and DHA are that healthy adults should consume a minimum of 500 mg of EPA + DHA daily to lower the risk of coronary heart disease (CHD), but higher doses of EPA + DHA (700–1000 mg/day) are often needed for individuals with metabolic risk factors, pregnant/lactating women, in infancy and during specific periods of development, and for secondary prevention of coronary heart disease (CHD) [26].

2.4 Polyunsaturated Fatty Acids and Health

2.4.1 *Health Effects of n-6 and n-3 Polyunsaturated Fatty Acids*

The n-6 PUFA LA and n-3 PUFA ALA are essential fatty acids that are vital to human health and must be provided in the diet. They have important physiological functions; for example, LA is essential for maintaining the water permeability barrier of the skin. Previous studies on intake of the LA and CHD risk have generated inconsistent results. In prospective observational studies, dietary LA intake is inversely associated with CHD risk in a dose-response manner [27], and these data provide support for current recommendations to replace SFA with LA to lower risk of CHD. On the other hand, Hoenselaar [28] raised comments regarding the review article by Farvid et al. [27], where the questions raised were: "Do different dietary sources of LA have the same influence on CHD?," and "Do different n-6 PUFA oils have various effects on CHD?." Lucas [29], also, commented on the review article by Farvid et al. [27], where Lucas hypothesized that an imbalance between n-6 and n-3 PUFA intakes may cause CVD, since LA is also the precursor of arachidonic acid (AA; 20:4n-6), from which pro-inflammatory eicosanoids and cytokines are derived. LA is present in variable quantities in many plant oils and human diets, and, therefore, specific LA deficiency does not seem to occur in the human body, and increased consumption of LA will not cause inflammation during normal metabolic conditions unless lipid peroxidation products are mixed in.

The n-3 PUFA ALA is found in certain plants, such as in dark green leafy vegetables, nuts, and oils from seeds (Table 2.2). Compared to the marine n-3 PUFA EPA and DHA, the health effects of ALA have been less studied [30]. It is well known that ALA is an essential fatty acid and a precursor of EPA and DHA in all mammals. On the other hand, the efficiency of this conversion in vivo is quite low in various species, and it is still debated whether dietary ALA can fulfil the needs of the human body or whether dietary intake of preformed DHA is necessary. The American Heart Association recommends that people without documented CHD eat a variety of fish (preferably oily) at least twice weekly (approx. 500 mg EPA + DHA) [31]. Epidemiological studies on the impact of fish consumption on CHD incidence have shown inconsistent results [32]. Previous meta-analyses showed that fish consumption reduces the risk of CHD; however, several incorporated studies show that fish consumption has no impact on CHD. In most of the published scientific papers the type of fish is not mentioned. Kris-Etherton and coworkers [33] noted that all fish are not equal in EPA and DHA content, and they raised the question "Does it matter whether the fish is fatty or lean?." The authors messages were "when studying fish intake in relation to risk of diseases it is important to have in mind that EPA and DHA content of fish varies between fish species, and that several factors such as sex, age, water temperature and season have an effect on the fatty acid composition of membrane lipids in all fish tissues." Findings strongly suggest that the data obtained from marine fatty fish-eating populations cannot be generalized to all fish-eating populations. No matter how many freshwater local catches are eaten in the communities studied, serum EPA and DHA concentrations are not affected [34].

The health benefits of marine n-3 PUFA EPA and DHA are well documented, indicating protective effects on CVD, autoimmune, and mental disorders [35]. A wide range of beneficial effects of EPA and DHA, including anti-atherothrombogenic effect, reduction in serum triglycerides, effects on arrhythmia, hypertension, and inflammation, have been suggested as possible explanations for the reduction in CVD [35, 36]. Recent research has demonstrated that EPA and DHA have distinct tissue distributions where they influence target organs in different ways [37]. EPA mainly provides the starting point for making hormones that regulate blood clotting, contraction, and relaxation of artery walls. Inflammation has been shown to improve atherosclerotic plaque stabilization in blood vessels, where it interferes with lipid oxidation and various signal transduction pathways linked to inflammation and endothelial dysfunction. These findings support a mechanistic basis for a potential benefit with dietary/supplemented EPA in reducing cardiovascular risk as is currently being tested in ongoing clinical trials [38], where health biomarkers will also be influenced by genetic variants [39] (see Sect. 2.5.3).

2.5 Saturated Fat and Cardiovascular Disease (CVD)

2.5.1 The Lipid Hypothesis in Atherogenesis

Saturated fat has been a topic of nutrition debate and dietary advice for more than a century. Consequently, the need to reduce levels of saturated fat in foods and the different ways of doing this have become one of the most important issues facing the food industry. The background is the supposed link between saturated fat intake and atherosclerosis, called the "lipid hypothesis." The lipid hypothesis postulates that (1) a high intake of saturated fat raises blood total cholesterol and (2) high total cholesterol leads to atherosclerosis and CVD, like CHD and stroke. This hypothesis was created more than 100 years ago when it was found that feeding high-fat diets and cholesterol to animals resulted in atherosclerosis [40]. Later it was observed that people with genetically high serum total cholesterol (familial hypercholesterolemia) died from myocardial infarction at young age. Since the 1950s systematic studies laid the basis for the understanding of how different fatty acids influence plasma total cholesterol [41, 42], and a strong positive association was found between intake of saturated fat and serum total cholesterol [43], as well as CHD mortality [44, 45] (Fig. 2.1). In recent years the lipid hypothesis has been debated [46], mainly due to meta-analyses of observational and randomized controlled trials (RCTs) showing no association between intake of saturated fat and CVD [47–51]. However, many of these meta-analyses (except the study of Harcombe et al. 2017 [48]), as well as other recent studies [27, 52–54] have concluded that replacing SFA with PUFA reduces risk of CVD, and hence such replacement still appears to be a useful strategy.

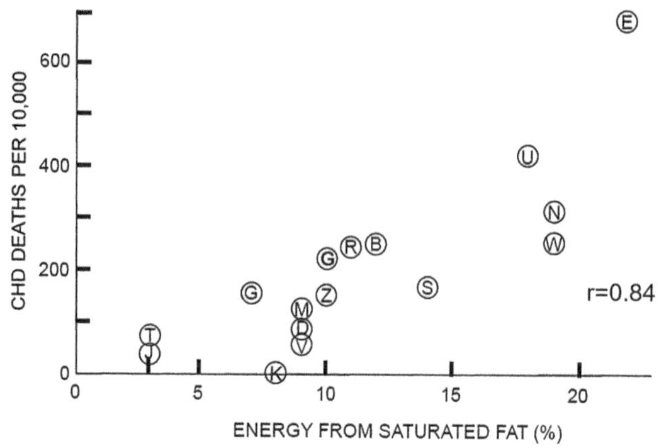

Code	Cohort
A	U.S.Railroad, USA
B	Belgrade, Serbia
C	Crevalcore, Italy
D	Dalmatia, Croatia
E	East Finland
G	Corfu, Greece
K	Crete, Greece
M	Montegiorgio, Italy
N	Zutphen, Netherlands
R	Rome, Italy
S	Slavonia, Croatia
T	Tanushimaru, Japan
U	Ushibuka, Japan
V	Velika Krsna, Serbia
W	West Finland
Z	Zrenjamin, Serbia

Fig. 2.1 Correlation between average energy intake of saturated fat (E%) and number of coronary heart disease (CHD) deaths in the Seven Countries Study after 10 year's follow-ups. (Adapted from [44] with permission from Harvard University Press)

The lipid hypothesis has been modified as the complexity of the atherosclerotic process has become evident. However, although different scientific theories about the causation of CVD have been presented over the years, the lipid hypothesis has not yet been falsified officially, and altering dietary fat intake is still seen as one of the most efficient ways of preventing CVD [25].

2.5.2 *The Role of Lipoproteins in Atherogenesis*

Dietary lipids are not soluble in water and are transported in the blood as lipoprotein particles. The lipoproteins are commonly divided into four classes according to density: chylomicrons, very low-density lipoprotein (VLDL), low-density lipoprotein (LDL), and high-density lipoprotein (HDL). The chylomicrons have the lowest density and contain about 85% triglycerides and only 2% protein, whereas HDL has the highest density and contains more than 50% protein (Fig. 2.2a). The lipoproteins have different functional and pathological significance, playing different roles in the transport of lipids (Fig. 2.2b). Chylomicrons deliver triacylglycerol (TAG) from intestinal epithelial cells to cells in the body, and VLDL deliver TAG from the liver to non-liver cells in the body. LDL and HDL transport both dietary and endogenous cholesterol in the plasma. LDL is the main transporter of cholesterol and cholesteryl esters and makes up more than half of the total lipoprotein in plasma. LDL is the most cholesterol-rich lipoprotein transporting cholesterol from the liver to peripheral cells (Fig. 2.2a, b). The LDL particle, and in particular the small dense LDL particles, is considered an important pathogenic factor in atherogenesis [55]. At high concentrations LDL can penetrate the arterial wall, possibly after some injury to the endothelial cell layer lining the inner vessel wall [56, 57], where they are

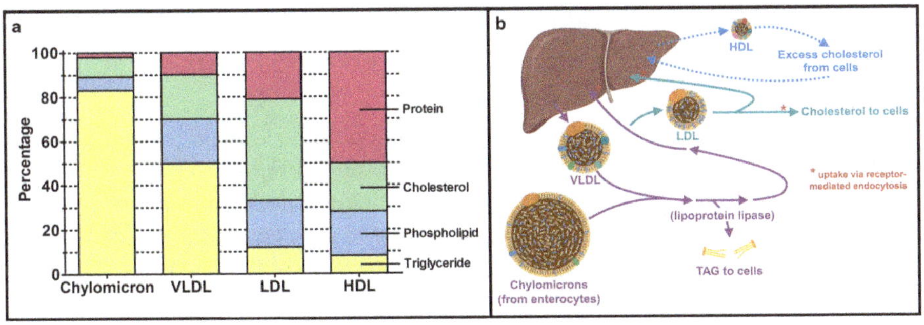

Fig. 2.2 (**a**) The cholesterol content of lipoproteins: 10% in chylomicron, 20% in very low-density lipoprotein (VLDL), 40% in low-density lipoprotein (LDL), 20% in high-density lipoprotein (HDL). Data were obtained from [151] (**b**) Transport of lipoproteins from intestinal epithelial cells to blood circulation and to cells in the body (created with BioRender.com)

modified, for example, by oxidation, and taken up in macrophages which become the so-called foam cells. The foam cells may eventually burst and leave their cholesterol to atheroma (plaque) formation, triggering the inflammatory process. The build-up of fibrous material and calcification inside the arterial walls cause the arteries to narrow and the blood flow may be inhibited. It may also end with rupture and thrombus formation. If this happens in a coronary artery, the result may be a myocardial infarction; if it happens in an artery to the brain, the result may be a stroke.

The HDL particles are supposed to transport cholesterol from peripheral cells and tissues to the liver for excretion [58] (Fig. 2.2b) and has been associated with decreased risk of CHD [59]. However, it has not been documented that altering HDL cholesterol levels by diet or drugs influences the risk of CHD [60, 61]. It appears that the association between HDL cholesterol and risk is complicated [62, 63] as there are multiple factors affecting plasma HDL cholesterol levels other than diet; for example, lack of physical activity and smoking have been associated with low HDL cholesterol. A potential role of the triglyceride-rich lipoprotein chylomicron and VLDL has been debated without reaching consensus [64, 65], although a high concentration of triglycerides is part of the metabolic risk cluster (see Sect. 2.4). Lipoprotein (a) is an LDL-like particle that is independently associated with CHD risk. It is strongly genetically determined and only to a minor degree influenced by diet. However, the primary target for risk reduction by dietary fat should still be LDL cholesterol. Total cholesterol is also considered an expedient target since it is strongly correlated to LDL cholesterol and a good marker of total amount of LDL particle mass.

2.5.3 Effects of Individual Fatty Acids on Plasma Lipoproteins

Intervention studies and meta-analyses have provided solid knowledge on the effects of individual dietary fatty acids on serum cholesterol [41, 42, 67–72] (Fig. 2.3). Short and medium-chain SFA (4:0 to 10:0) are considered to have no effect on serum cholesterol [72]. These fatty acids are partly soluble in water and not dependent on incorporation into micelles for uptake in the intestine and follow a different path of metabolism. Also, stearic acid (18:0) is considered to have no effect on serum cholesterol. In most studies lauric acid (12:0) has been found to moderately increase serum cholesterol, but most of this increase may be due to an increase in HDL cholesterol [68]. Myristic acid (14:0) is the most cholesterol-increasing fatty acid and increases both LDL and HDL cholesterol. Palmitic acid (16:0) also increases LDL cholesterol and to a minor extent HDL cholesterol. Trans fatty acids present in partially hydrogenated vegetable and hydrogenated fish oils increase total cholesterol and LDL cholesterol to about the same extent as 16:0 (Fig. 2.3), but in contrast to SFA they decrease HDL cholesterol [68, 73, 74].

A large body of evidence supports the cardioprotective effects of unsaturated fatty acids. The plant derived n-6 and n-3 PUFA, LA and ALA, have been found to reduce serum cholesterol, whereas EPA and DHA of marine origin seem to have minor effect or no effect at all [76–78]. However, EPA and DHA may exert their cardioprotective effects by reducing other risk factors than LDL cholesterol, such as serum triglycerides, blood pressure, platelet aggregation, endothelial function,

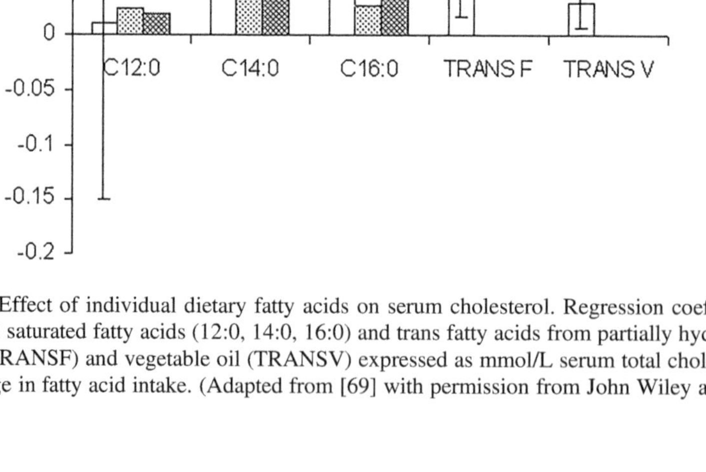

Fig. 2.3 Effect of individual dietary fatty acids on serum cholesterol. Regression coefficients of individual saturated fatty acids (12:0, 14:0, 16:0) and trans fatty acids from partially hydrogenated fish oil (TRANSF) and vegetable oil (TRANSV) expressed as mmol/L serum total cholesterol per E% change in fatty acid intake. (Adapted from [69] with permission from John Wiley and Sons)

and inflammation [35, 36] (see Sect. 2.6.1). Hence, potential detrimental effects of SFA may be counteracted by PUFA, and as observed in many studies replacing SFA with PUFA has beneficial effect on serum cholesterol as well as CVD risk.

2.5.4 Evidence Linking LDL Cholesterol to Development of Atherosclerosis and CHD

Animal Experiments
Hypercholesterolemia and atherosclerosis have been induced by feeding cholesterol and fat in virtually every species of laboratory animals [40]. Of special interest are experiments in primates where it has been possible to induce serious atherosclerosis and myocardial infarction after a relatively short time feeding with an ordinary high-fat Western diet [78].

Clinical Observations
The most well-known example of clinical observations is that of familial hypercholesterolemia, a genetic disorder affecting the LDL receptor [79, 80]. The reduced uptake of cholesterol in the liver results in high serum cholesterol levels and development of CHD. Those that are homozygote for the defect may die from myocardial infarction in early childhood [79].

Intervention Studies
Since the 1960s several secondary and primary prevention studies have demonstrated that reducing SFA and increasing PUFA in the diet may decrease serum cholesterol as well as reduce atherosclerotic events [81, 82]. In the Oslo Diet and Smoking Study, a 5-year randomized intervention study including men with elevated serum cholesterol levels, a diet low in total and saturated fat resulted in a 47% reduction in CHD events [83]. It was estimated that most of the effect could be accounted for by reduction in serum cholesterol. A recent follow-up study showed that the difference in mortality between the intervention group and controls was still significant after 40 years [84]. Interventions at national level, often promoted by national nutrition policy programs, provide further evidence of a potential link between LDL cholesterol and development of atherosclerosis. In most Western countries CHD mortality has declined during the last 50 years, and studies suggest that reduction in serum cholesterol may have contributed to about 35% of this decline [85, 86]. In Finland where CHD mortality shifted from being the highest in the world to being among the lowest, reduction in serum cholesterol due to reduced intake of SFA (mainly dairy) and increased intake of PUFA can explain most of the decline [87]. Similarly, in Norway reduction in cholesterol level due to decrease in dietary SFA and trans fatty acids and increase in PUFA can explain a major part of the decline in CHD mortality [88]. Also, interventions with cholesterol-reducing drugs (e.g., statins, bile acid sequestrants) support the hypothesis of serum cholesterol being a causative factor in CVD [89, 90]. A reduction in CHD mortality has been observed for different statins in long-term clinical intervention studies both in individuals with previous CHD and in individuals free of disease at entry [92–94].

Observational Studies (Cohort Studies and Case Control Studies)
Many prospective cohort studies in different parts of the world have identified CVD risk factors such as serum cholesterol, smoking, and high blood pressure. Meta-analyses of such studies have provided evidence that serum LDL cholesterol concentration is strongly and log-linearly associated with a dose-dependent increase in CVD risk [89]. However, although the relative risk associated with serum cholesterol (after correcting for smoking and blood pressure) is the same in different populations, the absolute risk may be very different. As an example, in the 1990s the risk of dying from a myocardial infarction was five times higher in northern Europe than in the Mediterranean region or in Japan at the same serum cholesterol level [45, 94]. This shows that there must be other factors in addition to serum cholesterol that influence CVD risk. Variations in habitual lifestyle could explain some of the discrepancies observed between populations, for example, variations in dietary intake of fruit and vegetables (antioxidants), fish (marine n-3 PUFA EPA and DHA), and physical activity. Dietary antioxidants are believed to have a significant impact, as oxidation of LDL has been related to increased CVD risk in many studies [95, 96]. Other factors that may influence the pathogenesis of CVD are arrhythmia and thrombotic processes [97, 98], stress [99], and inflammation [56, 57, 100]. The role of dietary fat in these processes is, however, not well documented and awaits further studies.

Considering the well-documented effect of SFA on serum cholesterol level in short-term randomized controlled trials (RCTs), it is reasonable to assume that there would be a strong association between intake of SFA and risk of CHD. In the Seven Countries Study [43, 45] as well as in international comparisons [101], this correlation has been very high. However, some recent meta-analyses of prospective cohort studies have failed to find significant associations between intake of SFA and CHD risk [48–51]. A challenge with these studies is the adjustment for confounders [102], as well as difficulties in obtaining reliable data for food intake [103, 104]. The importance of small dense LDL particles that are particularly pathogenic [55] has been put forward as a possible explanation for the lack of association between SFA intake and CHD risk, as many studies claim that saturated fat increases the level of large LDL particles that are much less strongly related to CVD risk. Moreover, the food matrix in which the fat is embedded may have significant impact but is usually not taken into consideration [18]. The lack of association between dietary SFA and serum cholesterol in many of these studies also needs an explanation. Clearly, it can be hard to demonstrate an association between fat intake and serum cholesterol since the biological variability in serum cholesterol is much higher than the changes that can possibly be induced by changes in dietary fat [105]. Furthermore, many of the studies that show no association between total intake of SFA and CHD risk very often show reduced risk when replacing SFA with PUFA, which is in line with published predictive equations and studies showing that the dietary SFA/PUFA ratio is a better predictor of serum cholesterol levels than total intake of SFA [106].

2.6 Saturated Fat and Other Noncommunicable Diseases

2.6.1 *Metabolic Syndrome and Type 2 Diabetes*

The metabolic syndrome is defined as a cluster of risk factors that predisposes for type 2 diabetes and CVD. These include insulin resistance, impaired blood lipids (e.g., increased triglycerides, decreased HDL), abdominal obesity, inflammation, and hypertension [107]. Some studies suggest that high intakes of total fat and SFA increase the risk of metabolic syndrome and that high intakes of MUFA and PUFA may reduce the risk [108, 109]. Others conclude that there is not sufficient evidence to establish an association between metabolic syndrome and intake of any fat, and that prevention should primarily focus on correcting overweight by reducing the total intake of fat.

When it comes to type 2 diabetes it is well documented that both the quantity and quality of dietary fat influence insulin resistance. Intervention studies have shown that reducing dietary SFA and replacing them with MUFA and PUFA improve insulin sensitivity [110, 111], and it has been discussed whether specific fatty acids, in particular palmitic acid (16:0), may promote insulin resistance more efficiently than other fatty acids [112]. Epidemiological evidence also suggests that replacing SFA with PUFA and MUFA has beneficial effects on insulin sensitivity and is likely to reduce risk of type 2 diabetes [109, 110, 113].

2.6.2 *Cancer*

The high incidence of certain cancers like colorectal cancer, prostate cancer, and breast cancer in Western countries suggests that lifestyle factors may play an important role. For most cancer types, existing data are not sufficient to conclude on any significant effect of dietary fat [114, 115], although there is some evidence that SFA moderately increase breast cancer risk and that progression of prostate cancer is more rapid on high SFA intake. It was also recently shown that 16:0, but not 18:1 or 18:2, promotes metastasis in oral carcinomas and melanoma in mice [116]. A recent review and meta-analysis on olive oil (high in 18:1) and cancer risk showed that increasing intake of olive oil lowered the risk of overall cancer, including breast cancer and colorectal cancer [117]. It is, however, not known to what extent this was due to changes in dietary fatty acid composition, for example, intake of SFA, or other beneficial components in olive oil.

The World Cancer Research Fund [118, 119] has stated that there is limited, but suggestive evidence that foods containing animal fat high in SFA, such as processed meat, increase colorectal cancer risk. It has been suggested that the association between high dietary fat and colorectal cancer may be due to increased bile acid secretion into the gastrointestinal tract with increased microbial formation of secondary bile acids that are carcinogenic [120]. Human data are scarce, but animal studies have shown that a diet high in saturated fat increases the gut microbial conversion of primary to secondary bile acids that causes inflammation and may play a role in colorectal cancer. Furthermore, high dietary fat, in particular saturated fat, has been shown to increase the synthesis of taurine conjugated bile acids that promote the growth of the gut bacteria Bilophila, generating H_2S gas which is genotoxic [121].

2.7 Benefits of Replacing Saturated with Unsaturated Fat in Foods

2.7.1 Beneficial Effects on Serum Cholesterol Levels

Healthier food products can be obtained by removing saturated fat and/or adding polyunsaturated liquid oils (vegetable oils, fish oils) to processed products using emulsion technology and oleogels that mimic the function of solid fats in food products. Replacing saturated with unsaturated fat in food products has the potential to lower serum cholesterol. In a recent study, replacing dietary SFA with PUFA showed a significant reduction in serum cholesterol of 8% already after 3 days [122]. It has been shown that if a group of individuals are given the same diet over a certain time, cholesterol levels will be almost normally distributed. The serum cholesterol level, as well as the response to changes in diet, is dependent on our genes. Hence, when we are talking about the effects of fatty acids on serum cholesterol, we are talking about average changes in the cholesterol observed in a group of individuals.

Predictive equations based on how individual fatty acids affect serum cholesterol can be a valuable tool in product development, making it possible to optimize the fatty acid composition of the products [69, 123]. Several predictive equations have been published during the years. They are all based on regression analyses of intervention studies and show similar trends; SFA (12:0, 14:0, 16:0) and trans fatty acids increase serum cholesterol, whereas PUFA (LA, ALA) and MUFA (18: 1n-9) are cholesterol-reducing (see Table 2.3).

In the Western diet, myristic acid (14:0) makes up about 10% of the SFA, while palmitic acid (16:0) contributes to more than 50%. Even if 14:0 is more cholesterol increasing (Fig. 2.3), the higher amount of 16:0 makes it the most important cholesterol-increasing fatty acids in the Western diet. Animal fat accounts for the majority of 16:0 today, but palm oil is also an important source. Palm oil is semi-solid at room temperature and fractionated and/or interesterified palm oil is commonly used in industrially processed foods, for example, as a replacement of trans fat, in order to achieve desired sensory quality and product consistency. However, fully hydrogenated and interesterified vegetable oils to increase the content of the "cholesterol neutral" stearic acid (18:0) at the expense of 16:0 may be a better alternative [124].

Table 2.3 Published predictive equations for estimating changes (Δ) in serum cholesterol (Chol) (mmol/L) in response to changes in dietary fatty acids (in E%)

Predictive equations	References
$\Delta Chol = 0.062 \times \Delta SFA - 0.03 \times \Delta PUFA$	Keys 1965 [42]
$\Delta Chol = 0.054 \times \Delta SFA - 0.032 \times \Delta PUFA - 0.003 \times \Delta MUFA$	Hegsted 1965 [41]
$\Delta Chol = 0.056 \times \Delta SFA - 0.016 \times \Delta PUFA - 0.003 \times \Delta MUFA$	Mensink 1992 [67]
$\Delta Chol = 0.052 \times \Delta SFA - 0.001 \times 18:0 - 0.025 \times \Delta PUFA - 0.012 \times \Delta MUFA$	Yu 1995 [71]
$\Delta Chol = 0.052 \times \Delta SFA - 0.026 \times \Delta PUFA - 0.005 \times \Delta MUFA$	Clarke 1997 [66]
$\Delta Chol = 0.01 \times \Delta 12:0 + 0.12 \times \Delta 14:0 + 0.06 \times \Delta 16:0 + 0.03 \times \Delta TFA - 0.017 \times \Delta PUFA - 0.004 \times \Delta MUFA$	Müller 2001 [69]
$\Delta Chol = 0.07 \times \Delta 12:0 + 0.06 \times \Delta 14:0 + 0.04 \times \Delta 16:0 + 0.03 \times \Delta TFA - 0.021 \times \Delta PUFA - 0.006 \times \Delta MUFA$	Sanders 2009 [70]

SFA saturated fatty acids (12:0+14:0+16:0), MUFA monounsaturated fatty acids (18:1), PUFA polyunsaturated fatty acids (18:2+18:3), TFA trans fatty acids ($18:1_{trans}$)

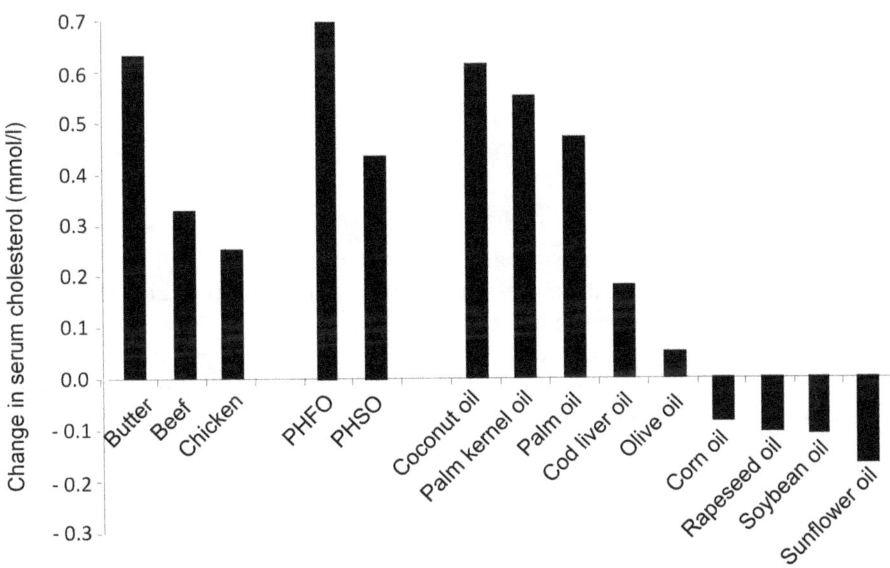

Fig. 2.4 Predicted changes in serum cholesterol (mmol/L) when 20E% carbohydrates (or "neutral" fat) is replaced by the fat in various products, using the predictive equation of [69]. PHFO = partially hydrogenated fish oil, PHSO = partially hydrogenated soybean oil

Figure 2.4 gives an indication of how different fat-containing products and oils affect the level of serum cholesterol using the predictive equation of Müller et al. [69]. Butter, partially hydrogenated oils, refined coconut oil, and palm oil are the most hypercholesterolemic, whereas oils from soybean, rapeseed, and sunflower are the most cholesterol-reducing. The estimated cholesterol effects are in good agreement with recent meta-analyses that compare effects of butter [125, 126], palm oil [127], coconut oil [128, 129], olive oil [130], and rapeseed oil [131], which taken together show that oils rich in unsaturated fatty acids (e.g., olive oil, rapeseed oil) are preferable to oils and fats rich in saturated fatty acids (e.g., butter, palm oil, coconut oil).

2.7.2 *Benefits of Bioactive Compounds Naturally Occurring in Liquid Oils*

Although the degree of saturation is assumed to be the primary mediator in terms of health effects of dietary fats and oils, it is not the only factor. Liquid oils may also contain health beneficial n-3 PUFA (see Sect. 2.4.1), polar lipids, and micronutrients such as vitamins, phytosterols, polyphenols, and carotenoids that may contribute to an even more favorable nutritional profile.

Vitamins

Vitamin E (α-tocopherol) is commonly found in all vegetable oils. Tocopherols are known to be efficient antioxidants. Sunflower oil has significant higher level of α-tocopherol compared to other commonly used vegetable oils [132]. Soybean oil is particularly high in vitamin K1, whereas fish oils are high in the fat-soluble vitamins A and D.

Phytosterols

It is well documented that phytosterols lower serum cholesterol [133, 134]. The underlying mechanisms for the cholesterol reduction are sterols replacing cholesterol in the micelles, thereby inhibiting cholesterol absorption in the small intestine. Phytosterols are naturally occurring in the cell membranes of plants and may be found in varying amounts (approximately 1%) in vegetable oils depending on the processing conditions used during refining and frying [135, 136]. In recent years advances in food technology have made it possible to add sterols to a variety of food products including margarines, yogurts, fruit juices, and cereal bars. Phytosterols may also be introduced in edible oleogels as building blocks (oleogelators) [137].

Polar Lipids (Phospholipids and Galactolipids)

Polar lipids (phospholipids and galactolipids) are major lipid constituents of plant cell membranes and may be present in small amounts in vegetable oils. Beneficial effects of dietary phospholipids have been proposed since the 1990s in relation to, for example, coronary heart disease, inflammation, or cancer; however, more research is needed to understand the impact of phospholipids supplementation in humans [138]. Oat oil is particularly high in polar lipids (about 15%) and commercial oat oil fractions containing even higher amounts (40%) can be obtained [139]. The main component of polar lipids in oats is the galactolipid digalactosyldiacylglycerol (DGDG) containing one or two fatty acids linked to a glycerol moiety. Galactolipids have been shown to possess anti-inflammatory and anti-tumor promoting activities [140], and influence fat digestion [141], and when used in food emulsions galactolipids prolong fat digestion and increase satiety [142, 143].

Polyphenols

Polyphenols are natural antioxidants found in fruits, vegetables, and cereals. The biological activity of polyphenols is strongly related to their antioxidant properties. Polyphenols found in virgin olive oil (hydroxytyrosol and oleuropein complex) are well documented to protect against harmful oxidation of LDL [144] and have been associated with lower risk of CVD in some studies [145].

Carotenoids

Edible oils may contain carotenoids with antioxidant capacity. Red palm oil contains high concentrations of beta- and alpha-carotene (approximately 500–800 μg/g) [135] that also have provitamin A activity. Astaxanthin is a powerful antioxidant found in krill oil, but not in most fish oils.

Processing, like heating and refining, have no or minor effect on fatty acid composition, but it may impact the preservation of bioactive compounds in vegetable oils [132, 146]. Cold-pressed oils (virgin oils) are believed to ensure better preservation of bioactive compounds. Hence, depending on the source, raw material quality and processing conditions, edible liquid oils may contain varying levels of compounds that are beneficial to health, and replacing SFA in foods with oleogels containing liquid oils may improve the nutritional profile beyond the expected cholesterol-lowering potential.

2.8 The Impact of the Food Matrix on Nutritional Value

Dietary patterns are composed of foods and nutrients, and this inter-relationship must be acknowledged in dietary research [147]. Based on several recent studies, mass media reports are suggesting that full fat dairy is better for consumers [148]. Rice [149] reviewed 18 epidemiological studies that showed that total dairy intake did not contribute to higher CVD risk, and that consuming milk or fermented dairy products such as yogurt and cheese may reduce CVD risk. The impact of full fat dairy products is difficult to determine, while dairy products contain other compounds that may reduce CVD risk [148]. Dietary studies [18, 150] have shown that despite high content of SFA, dairy fat does not promote atherogenesis in the order of magnitude that could be expected. However, while milk and cheese are associated with a slightly lower CVD risk compared to meat, dairy fat results in a significantly greater CVD risk relative to vegetable sources of fats and unsaturated fatty acids. Whether the food matrix may modify the effect of dairy fat on health outcomes warrants further investigation.

Nutrition research must consider dietary patterns and the complex inter-relationship between dietary components to ensure translation into meaningful and health-promoting dietary recommendations. Each type of fat has its own unique effects on the body. While most nutrition studies look at effects of individual nutrients, even the same specific type of fat may have different effects and varying nutritional value depending on its origin and the food matrix in which it is embedded.

2.9 Conclusions

Dietary fat is an excellent source of energy and provides essential fatty acids. Moreover, fats and oils provide fat-soluble vitamins and compounds such as carotenoids and sterols. The lipid hypothesis postulates that a high intake of saturated fat raises blood total cholesterol, and that high total cholesterol leads to atherosclerosis, CHD, and stroke. Several predictive equations based on regression analyses of intervention studies have shown that SFA (12:0, 14:0, 16:0) and *trans* fatty acids increase serum cholesterol, whereas LA, ALA, and MUFA (18:1n-9) are cholesterol-reducing. These data provide support for current recommendations to replace SFA with MUFA and PUFA to lower risk of CHD. LA and ALA are essential fatty acids and must be provided in diet. ALA is found in different amounts in plant oils and is a precursor of n-3 PUFA EPA and DHA in mammals. The health benefits of EPA and DHA, provided by marine oils, are well documented, indicating protective effects on CVD, autoimmune, reduction in serum triglycerides, arrhythmia, hypertension, and inflammation. The efficiency of the conversion of ALA to EPA and DHA in humans is low, and it is still debated whether dietary ALA can fulfil the needs of the human body or whether dietary intake of preformed EPA and DHA is necessary.

According to dietary guidelines, healthier food products can be obtained by removing saturated fat and/or adding polyunsaturated liquid oils (vegetable oils, fish oils) to processed foods using emulsion technology and oleogels that mimic the function of solid fats in food products. Fat contributes highly to the food quality (taste, smell, texture, and nutrition) and the choice of oils in oleogels is of high importance for the nutrition quality and health.

References

1. Afshin AEA (2019) Health effects of dietary risks in 195 countries, 1990–2017: a systematic analysis for the Global Burden of Disease Study 2017. Lancet 393:1958–1972
2. Schulze MB et al (2018) Food based dietary patterns and chronic disease prevention. BMJ 361:k2396. https://doi.org/10.1136/bmj.k2396
3. Manzoor S, Masoodi FA, Rashid R, Nasqashi F, Ahmad M (2022) Oleogels for the development of healthy meat products: a review. Applied. Food Res 2:100212
4. O'Brien RD (2009) Fats and oils. Formulating and processing for applications, 3rd edn. CRC Press, Boca Raton

5. Haas MJ (2005) Animal fats. In: S. F (ed) Bailey's industrial oil and fat products, 6th edn. John Wiley & Sons Inc, USA, pp 161–212
6. Padley FB, Gunstone FD, Harwood JL (1994) Occurrence and characteristics of oils and fats. In: Harwood HJ, Gunstone FD, Padley FB (eds) The lipid handbook, 2nd edn. Chapman & Hall, London
7. Dubois V, Breton S, Linder M (2007) Fatty acid profiles of 80 vegetable oils with regard to their nutritional potential. Eur J Lipid Sci Technol 109:710–732
8. Gunstone FD (2005) Vegetable oils. In: S. F (ed) Bailey's industrial oil and fat products. John Wiley & Sons Inc, USA
9. Ilievska B et al (2016) Topical formulation comprising fatty acid extract from cod liver oil: development, evaluation and stability studies. Mar Drugs 14. https://doi.org/10.3390/md14060105
10. Dovale-Rosabal G et al (2019) Concentration of EPA and DHA from refined salmon oil by optimizing the urea(−)fatty acid adduction reaction conditions using response surface methodology. Molecules 24. https://doi.org/10.3390/molecules24091642
11. Reedy J et al (2014) Higher diet quality is associated with decreased risk of all-cause, cardiovascular disease, and cancer mortality among older adults. J Nutr 144:881–889. https://doi.org/10.3945/jn.113.189407
12. Russell J et al (2013) Adherence to dietary guidelines and 15-year risk of all-cause mortality. Br J Nutr 109:547–555. https://doi.org/10.1017/S0007114512001377
13. US Department of Health and Human Services and US Department of Agriculture (1980) Dietary guidelines for Americans. Available from: https://www.dietaryguidelines.gov/about-dietary-guidelines/previous-editions/1980-dietary-guidelines-americans. Access date 13 Mar 2023
14. Jahns L et al (2018) The history and future of dietary guidance in America. Adv Nutr 9:136–147. https://doi.org/10.1093/advances/nmx025
15. Kris-Etherton PM, Krauss RM (2020) Public health guidelines should recommend reducing saturated fat consumption as much as possible: YES. Am J Clin Nutr 112:13–18. https://doi.org/10.1093/ajcn/nqaa110
16. WHO (2004) Global strategy on diet, physical activity and health (WHA57.17). World Health Organization, Geneva
17. US Department of Health and Human Services and US Department of Agriculture (2015) Dietary guidelines for Americans, 8th edn. US Government Printing Office, Washington, DC
18. Astrup A et al (2020) Saturated fats and health: a reassessment and proposal for food-based recommendations: JACC state-of-the-art review. J Am Coll Cardiol 76:844–857. https://doi.org/10.1016/j.jacc.2020.05.077
19. Lenighan YM, McNulty BA, Roche HM (2019) Dietary fat composition: replacement of saturated fatty acids with PUFA as a public health strategy, with an emphasis on alpha-linolenic acid. Proc Nutr Soc 78:234–245. https://doi.org/10.1017/S0029665118002793
20. US Department of Health and Human Services and US Department of Agriculture (2010) Dietary guidelines for Americans, 7th edn. US Government Printing Office, Washington, DC
21. Arnett DK et al (2019) 2019 ACC/AHA guideline on the primary prevention of cardiovascular disease: a report of the American College of Cardiology/American Heart Association task force on clinical practice guidelines. Circulation 140:e596–e646. https://doi.org/10.1161/CIR.0000000000000678
22. Jacobson TA et al (2015) National lipid association recommendations for patient-centered management of dyslipidemia: part 2. J Clin Lipidol 9:S1–122. https://doi.org/10.1016/j.jacl.2015.09.002
23. WHO (2020) Healthy diet. Available from: https://www.who.int/news-room/fact-sheets/detail/healthy-diet. Access date 13 Mar 2023
24. EFSA (2010) EFSA panel on dietetic products, nutrition, and allergies (NDA); scientific opinion on dietary reference values for fats, including saturated fatty acids, polyunsaturated fatty acids, monounsaturated fatty acids, trans fatty acids, and cholesterol. EFSA J 8:1461

25. WHO (2018) Guidelines: saturated fatty acids and trans fatty acids intake for adults and children. Fact sheet no 394. Available from: https://cdn.who.int/media/docs/default-source/healthy-diet/healthy-diet-fact-sheet-394.pdf. Access date 13 Mar 2023
26. GOED (2016) Recommendations for daily intake of EPA and DHA omega-3 fatty acids. Available from: https://goedomega3.com/storage/app/media/press-releases/press_release_Intake_Recommendations.pdf. Accessed 13 Mar 2023
27. Farvid MS et al (2014) Dietary linoleic acid and risk of coronary heart disease: a systematic review and meta-analysis of prospective cohort studies. Circulation 130:1568–1578. https://doi.org/10.1161/CIRCULATIONAHA.114.010236
28. Hoenselaar R (2015) Letter by Hoenselaar regarding article, "Dietary linoleic acid and risk of coronary heart disease: a systematic review and meta-analysis of prospective cohort studies". Circulation 132:e20. https://doi.org/10.1161/CIRCULATIONAHA.114.014510
29. Lucas M (2015) Letter by Lucas regarding articles, "Dietary linoleic acid and risk of coronary heart disease: a systematic review and meta-analysis of prospective cohort studies" and "Circulating omega-6 polyunsaturated fatty acids and total and cause-specific mortality: the cardiovascular health study". Circulation 132:e21. https://doi.org/10.1161/CIRCULATIONAHA.114.013446
30. Schwab U et al (2014) Effect of the amount and type of dietary fat on cardiometabolic risk factors and risk of developing type 2 diabetes, cardiovascular diseases, and cancer: a systematic review. Food Nutr Res 58. https://doi.org/10.3402/fnr.v58.25145
31. American Heart Association (2014) Frequently asked questions about fish. Available at: https://www.heart.org/en/healthy-living/healthy-eating/eat-smart/fats/fish-and-omega-3-fatty-acids. Access date 12 Dec 2022
32. He K (2009) Fish, long-chain omega-3 polyunsaturated fatty acids and prevention of cardiovascular disease--eat fish or take fish oil supplement? Prog Cardiovasc Dis 52:95–114. https://doi.org/10.1016/j.pcad.2009.06.003
33. Kris-Etherton PM et al (2000) Polyunsaturated fatty acids in the food chain in the United States. Am J Clin Nutr 71:179S–188S. https://doi.org/10.1093/ajcn/71.1.179S
34. Philibert A et al (2006) Fish intake and serum fatty acid profiles from freshwater fish. Am J Clin Nutr 84:1299–1307. https://doi.org/10.1093/ajcn/84.6.1299
35. Calder PC (2014) Very long chain omega-3 (n-3) fatty acids and human health. Eur J Lipid Sci Technol 116:1280–1300
36. Innes JK, Calder PC (2020) Marine Omega-3 (N-3) fatty acids for cardiovascular health: an update for 2020. Int J Mol Sci 21. https://doi.org/10.3390/ijms21041362
37. Mason RP (2019) New insights into mechanisms of action for omega-3 fatty acids in atherothrombotic cardiovascular disease. Curr Atheroscler Rep 21:2
38. Borow KM, Nelson JR, Mason RP (2015) Biologic plausibility, cellular effects, and molecular mechanisms of eicosapentaenoic acid (EPA) in atherosclerosis. Atherosclerosis 242:357–366. https://doi.org/10.1016/j.atherosclerosis.2015.07.035
39. Minihane AM (2016) Impact of genotype on EPA and DHA status and responsiveness to increased intakes. Nutrients 8:123. https://doi.org/10.3390/nu8030123
40. Stamler J (1992) Established major coronary risk factors. In: Elliott P, Marmot M (eds) Coronary heart disease epidemiology: from aetiology to public health. Oxford University Press, London, pp 32–70
41. Hegsted DM et al (1965) Quantitative effects of dietary fat on serum cholesterol in man. Am J Clin Nutr 17:281–295. https://doi.org/10.1093/ajcn/17.5.281
42. Keys A, Anderson JT, Grande F (1965) Serum cholesterol response to changes in the diet: IV. Particular saturated fatty acids in the diet. Metabolism 14:776–787. https://doi.org/10.1016/0026-0495(65)90004-1
43. Keys AE (1970) Coronary heart disease in seven countries. Circulation 41:1–211
44. Keys A (1980) Seven countries. A multivariate analysis of death and coronary heart disease. Harvard University Press, Cambridge, Massachusetts and London

45. Kromhout D et al (1995) Dietary saturated and trans fatty acids and cholesterol and 25-year mortality from coronary heart disease: the seven countries study. Prev Med 24:308–315. https://doi.org/10.1006/pmed.1995.1049
46. Bier DM (2016) Saturated fats and cardiovascular disease: interpretations not as simple as they once were. Crit Rev Food Sci Nutr 56:1943–1946. https://doi.org/10.1080/10408398.2014.998332
47. de Souza RJ et al (2015) Intake of saturated and trans unsaturated fatty acids and risk of all cause mortality, cardiovascular disease, and type 2 diabetes: systematic review and meta-analysis of observational studies. BMJ 351:h3978. https://doi.org/10.1136/bmj.h3978
48. Harcombe Z, Baker JS, Davies B (2017) Evidence from prospective cohort studies does not support current dietary fat guidelines: a systematic review and meta-analysis. Br J Sports Med 51:1743–1749. https://doi.org/10.1136/bjsports-2016-096550
49. Ramsden CE et al (2016) Re-evaluation of the traditional diet-heart hypothesis: analysis of recovered data from Minnesota coronary experiment (1968–73). BMJ 353:i1246. https://doi.org/10.1136/bmj.i1246
50. Siri-Tarino PW et al (2010) Meta-analysis of prospective cohort studies evaluating the association of saturated fat with cardiovascular disease. Am J Clin Nutr 91:535–546. https://doi.org/10.3945/ajcn.2009.27725
51. Skeaff CM, Miller J (2009) Dietary fat and coronary heart disease: summary of evidence from prospective cohort and randomised controlled trials. Ann Nutr Metab 55:173–201. https://doi.org/10.1159/000229002
52. Astrup A et al (2011) The role of reducing intakes of saturated fat in the prevention of cardiovascular disease: where does the evidence stand in 2010? Am J Clin Nutr 93:684–688. https://doi.org/10.3945/ajcn.110.004622
53. Hooper L et al (2020) Reduction in saturated fat intake for cardiovascular disease. Cochrane Database Syst Rev 8. https://doi.org/10.1002/14651858.CD011737.pub3
54. Mozaffarian D, Micha R, Wallace S (2010) Effects on coronary heart disease of increasing polyunsaturated fat in place of saturated fat: a systematic review and meta-analysis of randomized controlled trials. PLoS Med 7:e1000252. https://doi.org/10.1371/journal.pmed.1000252
55. Whitney E, Rolfes SR (2005) Understanding nutrition (Tenth edition). Thomson Learning Inc
56. Pichler G et al (2018) LDL particle size and composition and incident cardiovascular disease in a south-European population: the Hortega-Liposcale follow-up study. Int J Cardiol 264:172–178. https://doi.org/10.1016/j.ijcard.2018.03.128
57. Libby P et al (2009) Inflammation in atherosclerosis: from pathophysiology to practice. J Am Coll Cardiol 54:2129–2138. https://doi.org/10.1016/j.jacc.2009.09.009
58. Ross R (1999) Atherosclerosis--an inflammatory disease. N Engl J Med 340:115–126. https://doi.org/10.1056/NEJM199901143400207
59. Cuchel M, Rader DJ (2006) Macrophage reverse cholesterol transport: key to the regression of atherosclerosis? Circulation 113:2548–2555. https://doi.org/10.1161/CIRCULATIONAHA.104.475715
60. Castelli WP (1984) Epidemiology of coronary heart disease: the Framingham study. Am J Med 76:4–12. https://doi.org/10.1016/0002-9343(84)90952-5
61. Briel M et al (2009) Association between change in high density lipoprotein cholesterol and cardiovascular disease morbidity and mortality: systematic review and meta-regression analysis. BMJ 338:b92. https://doi.org/10.1136/bmj.b92
62. Voight BF et al (2012) Plasma HDL cholesterol and risk of myocardial infarction: a mendelian randomisation study. Lancet 380:572–580. https://doi.org/10.1016/S0140-6736(12)60312-2
63. Rohatgi A et al (2021) HDL in the 21st century: a multifunctional roadmap for future HDL research. Circulation 143:2293–2309. https://doi.org/10.1161/CIRCULATIONAHA.120.044221
64. van der Steeg WA et al (2008) High-density lipoprotein cholesterol, high-density lipoprotein particle size, and apolipoprotein A-I: significance for cardiovascular risk: the IDEAL and

EPIC-Norfolk studies. J Am Coll Cardiol 51:634–642. https://doi.org/10.1016/j.jacc.2007.09.060
65. Emerging Risk Factors, C et al (2009) Major lipids, apolipoproteins, and risk of vascular disease. JAMA 302:1993–2000. https://doi.org/10.1001/jama.2009.1619
66. Heidemann BE et al (2021) The relation between VLDL-cholesterol and risk of cardiovascular events in patients with manifest cardiovascular disease. Int J Cardiol 322:251–257. https://doi.org/10.1016/j.ijcard.2020.08.030
67. Clarke R et al (1997) Dietary lipids and blood cholesterol: quantitative meta-analysis of metabolic ward studies. BMJ 314:112–117. https://doi.org/10.1136/bmj.314.7074.112
68. Mensink RP, Katan MB (1992) Effect of dietary fatty acids on serum lipids and lipoproteins. A meta-analysis of 27 trials. Arterioscler Thromb 12:911–919. https://doi.org/10.1161/01.atv.12.8.911
69. Mensink RP et al (2003) Effects of dietary fatty acids and carbohydrates on the ratio of serum total to HDL cholesterol and on serum lipids and apolipoproteins: a meta-analysis of 60 controlled trials. Am J Clin Nutr 77:1146–1155. https://doi.org/10.1093/ajcn/77.5.1146
70. Muller H, Kirkhus B, Pedersen JI (2001) Serum cholesterol predictive equations with special emphasis on trans and saturated fatty acids. An analysis from designed controlled studies. Lipids 36:783–791. https://doi.org/10.1007/s11745-001-0785-6
71. Sanders TA (2009) Fat and fatty acid intake and metabolic effects in the human body. Ann Nutr Metab 55:162–172. https://doi.org/10.1159/000229001
72. Yu S et al (1995) Plasma cholesterol-predictive equations demonstrate that stearic acid is neutral and monounsaturated fatty acids are hypocholesterolemic. Am J Clin Nutr 61:1129–1139. https://doi.org/10.1093/ajcn/61.4.1129
73. St-Onge MP et al (2008) Medium chain triglyceride oil consumption as part of a weight loss diet does not lead to an adverse metabolic profile when compared to olive oil. J Am Coll Nutr 27:547–552. https://doi.org/10.1080/07315724.2008.10719737
74. Almendingen K et al (1995) Effects of partially hydrogenated fish oil, partially hydrogenated soybean oil, and butter on serum lipoproteins and Lp[a] in men. J Lipid Res 36:1370–1384
75. Mensink RP, Katan MB (1990) Effect of dietary trans fatty acids on high-density and low-density lipoprotein cholesterol levels in healthy subjects. N Engl J Med 323:439–445. https://doi.org/10.1056/NEJM199008163230703
76. Allaire J et al (2018) High-dose DHA has more profound effects on LDL-related features than high-dose EPA: the ComparED study. J Clin Endocrinol Metab 103:2909–2917. https://doi.org/10.1210/jc.2017-02745
77. Innes JK, Calder PC (2018) The differential effects of eicosapentaenoic acid and docosahexaenoic acid on cardiometabolic risk factors: a systematic review. Int J Mol Sci 19. https://doi.org/10.3390/ijms19020532
78. Jacobson TA et al (2012) Effects of eicosapentaenoic acid and docosahexaenoic acid on low-density lipoprotein cholesterol and other lipids: a review. J Clin Lipidol 6:5–18. https://doi.org/10.1016/j.jacl.2011.10.018
79. Taylor CB et al (1962) Atherosclerosis in rhesus monkeys. II. Arterial lesions associated with hypercholesteremia induced by dietary fat and cholesterol. Arch Pathol 74:16–34
80. Brown MS, Goldstein JL (1986) A receptor-mediated pathway for cholesterol homeostasis. Science 232:34–47. https://doi.org/10.1126/science.3513311
81. Yuan G, Wang J, Hegele RA (2006) Heterozygous familial hypercholesterolemia: an underrecognized cause of early cardiovascular disease. CMAJ 174:1124–1129. https://doi.org/10.1503/cmaj.051313
82. Dayton S, Pearce ML (1969) Diet high in unsaturated fat. A controlled clinical trial. Minn Med 52:1237–1242. https://doi.org/10.1161/01.cir.40.1s2.ii-1
83. Miettinen M et al (1972) Effect of cholesterol-lowering diet on mortality from coronary heart-disease and other causes. A twelve-year clinical trial in men and women. Lancet 2:835–838. https://doi.org/10.1016/s0140-6736(72)92208-8

84. Hjermann I et al (1981) Effect of diet and smoking intervention on the incidence of coronary heart disease. Report from the Oslo study group of a randomised trial in healthy men. Lancet 2: 1303–1310. https://doi.org/10.1016/s0140-6736(81)91338-6
85. Holme I et al (2016) Lifelong benefits on myocardial infarction mortality: 40-year follow-up of the randomized Oslo diet and antismoking study. J Intern Med 280:221–227. https://doi.org/10.1111/joim.12485
86. Ford ES et al (2007) Explaining the decrease in U.S. deaths from coronary disease, 1980–2000. N Engl J Med 356:2388–2398. https://doi.org/10.1056/NEJMsa053935
87. Kuulasmaa K et al (2000) Estimation of contribution of changes in classic risk factors to trends in coronary-event rates across the WHO MONICA project populations. Lancet 355:675–687. https://doi.org/10.1016/s0140-6736(99)11180-2
88. Puska P (2009) Fat and heart disease: yes we can make a change--the case of North Karelia (Finland). Ann Nutr Metab 54(Suppl 1):33–38. https://doi.org/10.1159/000220825
89. Pedersen JI, Tverdal A, Kirkhus B (2004) Diet changes and the rise and fall of cardiovascular disease mortality in Norway. Tidsskr Nor Laegeforen 124:1532–1536
90. Ference BA et al (2017) Low-density lipoproteins cause atherosclerotic cardiovascular disease. 1. Evidence from genetic, epidemiologic, and clinical studies. A consensus statement from the European Atherosclerosis Society Consensus Panel. Eur Heart J 38:2459–2472. https://doi.org/10.1093/eurheartj/ehx144
91. Thompson GR (2009) History of the cholesterol controversy in Britain. QJM 102:81–86. https://doi.org/10.1093/qjmed/hcn158
92. Baigent C et al (2005) Efficacy and safety of cholesterol-lowering treatment: prospective meta-analysis of data from 90,056 participants in 14 randomised trials of statins. Lancet 366:1267–1278. https://doi.org/10.1016/S0140-6736(05)67394-1
93. Shepherd J (1995) Statin therapy in clinical practice: new developments. Curr Opin Lipidol 6: 254–255
94. Scandinavian Sivastatin Survival Study Group (1994) Randomised trial of cholesterol lowering in 4444 patients with coronary heart disease: the Scandinavian simvastatin survival study (4S). Lancet 344:1383–1389
95. Kromhout D (1999) Serum cholesterol in cross-cultural perspective. The seven countries study. Acta Cardiol 54:155–158
96. Gao S, Liu J (2017) Association between circulating oxidized low-density lipoprotein and atherosclerotic cardiovascular disease. Chronic Dis Transl Med 3:89–94. https://doi.org/10.1016/j.cdtm.2017.02.008
97. Holvoet P et al (2003) Association of high coronary heart disease risk status with circulating oxidized LDL in the well-functioning elderly: findings from the health, aging, and body composition study. Arterioscler Thromb Vasc Biol 23:1444–1448. https://doi.org/10.1161/01.ATV.0000080379.05071.22
98. Lefevre M et al (2004) Dietary fatty acids, hemostasis, and cardiovascular disease risk. J Am Diet Assoc 104:410–419; quiz 492. https://doi.org/10.1016/j.jada.2003.12.022
99. Renaud S et al (1986) Nutrients, platelet function and composition in nine groups of French and British farmers. Atherosclerosis 60:37–48. https://doi.org/10.1016/0021-9150(86)90085-7
100. Rosch PJ (2008) Cholesterol does not cause coronary heart disease in contrast to stress. Scand Cardiovasc J 42:244–249. https://doi.org/10.1080/14017430801993701
101. Sorriento D, Iaccarino G (2019) Inflammation and cardiovascular diseases: the most recent findings. Int J Mol Sci 20. https://doi.org/10.3390/ijms20163879
102. Artaud-Wild SM et al (1993) Differences in coronary mortality can be explained by differences in cholesterol and saturated fat intakes in 40 countries but not in France and Finland. A paradox. Circulation 88:2771–2779. https://doi.org/10.1161/01.cir.88.6.2771
103. Stamler J (2010) Diet-heart: a problematic revisit. Am J Clin Nutr 91:497–499. https://doi.org/10.3945/ajcn.2010.29216

104. Bingham SA et al (2003) Are imprecise methods obscuring a relation between fat and breast cancer? Lancet 362:212–214. https://doi.org/10.1016/S0140-6736(03)13913-X
105. Prentice RL (2003) Dietary assessment and the reliability of nutritional epidemiology reports. Lancet 362:182–183. https://doi.org/10.1016/S0140-6736(03)13950-5
106. Keys A (1988) Diet and blood cholesterol in population surveys--lessons from analysis of the data from a major survey in Israel. Am J Clin Nutr 48:1161–1165. https://doi.org/10.1093/ajcn/48.5.1161
107. Muller H et al (2003) The serum LDL/HDL cholesterol ratio is influenced more favorably by exchanging saturated with unsaturated fat than by reducing saturated fat in the diet of women. J Nutr 133:78–83. https://doi.org/10.1093/jn/133.1.78
108. Alberti KG et al (2009) Harmonizing the metabolic syndrome: a joint interim statement of the International Diabetes Federation Task Force on Epidemiology and Prevention; National Heart, Lung, and Blood Institute; American Heart Association; World Heart Federation; International Atherosclerosis Society; and International Association for the Study of Obesity. Circulation 120:1640–1645. https://doi.org/10.1161/CIRCULATIONAHA.109.192644
109. Melanson EL, Astrup A, Donahoo WT (2009) The relationship between dietary fat and fatty acid intake and body weight, diabetes, and the metabolic syndrome. Ann Nutr Metab 55:229–243. https://doi.org/10.1159/000229004
110. Riccardi G, Giacco R, Rivellese AA (2004) Dietary fat, insulin sensitivity and the metabolic syndrome. Clin Nutr 23:447–456. https://doi.org/10.1016/j.clnu.2004.02.006
111. Riserus U (2008) Fatty acids and insulin sensitivity. Curr Opin Clin Nutr Metab Care 11:100–105. https://doi.org/10.1097/MCO.0b013e3282f52708
112. Vessby B et al (2001) Substituting dietary saturated for monounsaturated fat impairs insulin sensitivity in healthy men and women: the KANWU study. Diabetologia 44:312–319. https://doi.org/10.1007/s001250051620
113. Vessby B et al (2002) Desaturation and elongation of fatty acids and insulin action. Ann N Y Acad Sci 967:183–195. https://doi.org/10.1111/j.1749-6632.2002.tb04275.x
114. Luukkonen PK et al (2018) Saturated fat is more metabolically harmful for the human liver than unsaturated fat or simple sugars. Diabetes Care 41:1732–1739. https://doi.org/10.2337/dc18-0071
115. Bojkova B, Winklewski PJ, Wszedybyl-Winklewska M (2020) Dietary fat and cancer-which is good, which is bad, and the body of evidence. Int J Mol Sci 21. https://doi.org/10.3390/ijms21114114
116. Gerber M (2009) Background review paper on total fat, fatty acid intake and cancers. Ann Nutr Metab 55:140–161. https://doi.org/10.1159/000229000
117. Pascual G et al (2021) Dietary palmitic acid promotes a prometastatic memory via Schwann cells. Nature 599:485–490. https://doi.org/10.1038/s41586-021-04075-0
118. Markellos C et al (2022) Olive oil intake and cancer risk: a systematic review and meta-analysis. PLoS One 17. https://doi.org/10.1371/journal.pone.0261649
119. WCRF/AICR (2007) Diet, nutrition, physical activity and cancer: a global perspective. American Institute for Cancer Research, Washington, DC
120. WCRF/AICR (2018) Diet, nutrition and physical activity and colorectal cancer. Continuous Update Project Expert Report
121. Bernstein C et al (2011) Carcinogenicity of deoxycholate, a secondary bile acid. Arch Toxicol 85:863–871. https://doi.org/10.1007/s00204-011-0648-7
122. Devkota S et al (2012) Dietary-fat-induced taurocholic acid promotes pathobiont expansion and colitis in Il10−/− mice. Nature 487:104–108. https://doi.org/10.1038/nature11225
123. Gaundal L et al (2021) Beneficial effect on serum cholesterol levels, but not glycaemic regulation, after replacing SFA with PUFA for 3 d: a randomised crossover trial. Br J Nutr 125:915–925. https://doi.org/10.1017/S0007114520003402
124. Pedersen JI, Kirkhus B, Muller H (2003) Serum cholesterol predictive equations in product development. Eur J Med Res 8:325–331

125. Pedersen JI, Kirkhus B (2008) Fatty acid composition of post trans margarines and their health implications. Lipid Technol 20:132–135
126. Duarte C et al (2021) Dairy versus other saturated fats source and cardiometabolic risk markers: systematic review of randomized controlled trials. Crit Rev Food Sci Nutr 61:450–461. https://doi.org/10.1080/10408398.2020.1736509
127. Schwingshackl L et al (2018) Effects of oils and solid fats on blood lipids: a systematic review and network meta-analysis. J Lipid Res 59:1771–1782. https://doi.org/10.1194/jlr.P085522
128. Hisham MDB et al (2020) The effects of palm oil on serum lipid profiles: a systematic review and meta-analysis. Asia Pac J Clin Nutr 29:523–536. https://doi.org/10.6133/apjcn.202009_29(3).0011
129. Neelakantan N, Seah JYH, van Dam RM (2020) The effect of coconut oil consumption on cardiovascular risk factors: a systematic review and meta-analysis of clinical trials. Circulation 141:803–814. https://doi.org/10.1161/CIRCULATIONAHA.119.043052
130. Teng M et al (2020) Impact of coconut oil consumption on cardiovascular health: a systematic review and meta-analysis. Nutr Rev 78:249–259. https://doi.org/10.1093/nutrit/nuz074
131. Ghobadi S et al (2019) Comparison of blood lipid-lowering effects of olive oil and other plant oils: a systematic review and meta-analysis of 27 randomized placebo-controlled clinical trials. Crit Rev Food Sci Nutr 59:2110–2124. https://doi.org/10.1080/10408398.2018.1438349
132. Amiri M et al (2020) The effects of canola oil on cardiovascular risk factors: a systematic review and meta-analysis with dose-response analysis of controlled clinical trials. Nutr Metab Cardiovasc Dis 30:2133–2145. https://doi.org/10.1016/j.numecd.2020.06.007
133. Fine F, Brochet C, Gaud M, Carre P, Simon N, Ramli F, Joffre F (2016) Micronutrients in vegetable oils: the impact of crushing and refining processes on vitamins and antioxidants in sunflower, rapeseed, and soybean oils. Eur J Lipid Sci Technol 118:680–697
134. Gupta AK et al (2011) Role of phytosterols in lipid-lowering: current perspectives. QJM 104: 301–308. https://doi.org/10.1093/qjmed/hcr007
135. Katan MB et al (2003) Efficacy and safety of plant stanols and sterols in the management of blood cholesterol levels. Mayo Clin Proc 78:965–978. https://doi.org/10.4065/78.8.965
136. Kamal-Eldin A (2005) Minor components of fats and oils. In: Shahidi (ed) Bailey's industrial oil and fat products, 6th edn. John Wiley & Sons, Inc., New York
137. Phillips KM (2002) Free and esterified sterol composition of edible oils and fats. J Food Compos Anal 15:123–142
138. Matheson A et al (2018) Phytosterol-based edible oleogels: a novel way of replacing saturated fat in food. Nutr Bull 43:189–194. https://doi.org/10.1111/nbu.12325
139. Kullenberg D et al (2012) Health effects of dietary phospholipids. Lipids Health Dis 11:3. https://doi.org/10.1186/1476-511X-11-3
140. Banás K, Harasym J (2021) Current knowledge of content and composition of oat oil—future perspectives of oat as oil source. Food Bioprocess Technol 14:232–247
141. Christensen LP (2009) Galactolipids as potential health promoting compounds in vegetable foods. Recent Pat Food Nutr Agric 1:50–58. https://doi.org/10.2174/2212798410901010050
142. Chu BS et al (2009) Modulating pancreatic lipase activity with galactolipids: effects of emulsion interfacial composition. Langmuir 25:9352–9360. https://doi.org/10.1021/la9008174
143. Burns AA et al (2002) Dose-response effects of a novel fat emulsion (Olibra) on energy and macronutrient intakes up to 36 h post-consumption. Eur J Clin Nutr 56:368–377. https://doi.org/10.1038/sj.ejcn.1601326
144. Ohlsson L et al (2014) Postprandial effects on plasma lipids and satiety hormones from intake of liposomes made from fractionated oat oil: two randomized crossover studies. Food Nutr Res 58. https://doi.org/10.3402/fnr.v58.24465
145. EFSA (2011) Scientific opinion on the substantiation of health claims related to polyphenols in olive and protection of LDL particles from oxidative damage, Parma, Italy

146. Estruch R et al (2018) Primary prevention of cardiovascular disease with a mediterranean diet supplemented with extra-virgin olive oil or nuts. N Engl J Med 378:e34. https://doi.org/10.1056/NEJMoa1800389
147. Gorzynik-Debicka M et al (2018) Potential health benefits of olive oil and plant polyphenols. Int J Mol Sci 19. https://doi.org/10.3390/ijms19030686
148. Tapsell LC et al (2016) Foods, nutrients, and dietary patterns: interconnections and implications for dietary guidelines. Adv Nutr 7:445–454. https://doi.org/10.3945/an.115.011718
149. Briggs MA, Petersen KS, Kris-Etherton PM (2017) Saturated fatty acids and cardiovascular disease: replacements for saturated fat to reduce cardiovascular risk. Healthcare (Basel) 5. https://doi.org/10.3390/healthcare5020029
150. Rice BH (2014) Dairy and cardiovascular disease: a review of recent observational research. Curr Nutr Rep 3:130–138. https://doi.org/10.1007/s13668-014-0076-4
151. Chen M et al (2016) Dairy fat and risk of cardiovascular disease in 3 cohorts of US adults. Am J Clin Nutr 104:1209–1217. https://doi.org/10.3945/ajcn.116.134460

Chapter 3
Novel Strategies for Structuring Liquid Oils, Their Applications, and Health Implications

Yasamin Soleimanian, Rachel Tanti, Nicole Shaw, and Alejandro G. Marangoni

Abbreviations

12-HAS	12-hydoxystearic acid
A_w	Water activity
BW	Beeswax
CLW	Candelilla wax
CRW	Carnauba wax
CVD	Cardiovascular disease
DAGs	Diacylglycerols
EC	Ethyl-cellulose
FDA	Food and Drug Administration
GL	Glycerolysis
GP	Glycerolysis product
HDL	High-density lipoprotein
HIPE	High-internal phase emulsion
HLB	Hydrophilic-lipophilic balance
HOAO	High oleic algal
HOCO	High oleic canola
HOSO	High oleic sunflower oil
HPMC	Hydroxypropyl methylcellulose
LDL	Low-density lipoprotein
LMOG	Low molecular weight gelators
MAGs	Monoacylglycerols

Y. Soleimanian · R. Tanti · N. Shaw · A. G. Marangoni (✉)
Department of Food Science, University of Guelph, Guelph, ON, Canada
e-mail: ysoleima@uoguelph.ca; amarango@uoguelph.ca

C. Andrea Palla, F. Valoppi (eds.), *Advances in Oleogel Development, Characterization, and Nutritional Aspects*,
https://doi.org/10.1007/978-3-031-46831-5_3

MC	Methylcellulose
PHOs	Partially hydrogenated oils
RBW	Rice bran wax
SFC	Solid fat content
SFW	Sunflower wax
SLC	Starch-lipid composites
SOSA	Stearyl alcohol/stearic acid
TAGs	Triacylglycerol
T_g	Glass transition temperature
WHO	World Health Organization
XG	Xanthan gum

3.1 Introduction

Edible fats are composed of triacylglycerol molecules (TAGs). These TAGs consist of a glycerol backbone with three fatty acid chains attached by ester linkages. The chemical nature of the fatty acids gives a solid-like or liquid-like character to a particular fat at room temperature, and the interaction between polycrystalline TAG colloids is what traditionally structures fat [1–5]. Typically, a bulk fat containing a large proportion of long-chain fatty acids, with a high degree of saturation or high content of *trans* unsaturated fatty acids, results in fat with a higher melting point that is solid at room temperature. In contrast, fats containing high amounts of short-chain fatty acids, or fatty acids with a high degree of *cis* unsaturation, have lower melting points and are liquid at room temperature. This is the main difference between fats and oils, and what dictates the suitability of fat in a particular food application. Semi-solid fats provide a variety of characteristics to different baked goods that oils cannot. For example, solid fats can trap small air bubbles that stay in cake batters longer, producing more reliably fluffy cakes. Solid fats can also keep dough layers apart in pastries resulting in increased volume or lift, whereas, using oil for these applications would result in lower-volume cakes and dough layers that adhere together and leak oil [6]. As the reader might have noticed, fat functionality in many industrial applications requires high contents of saturated or *trans* fat. This functional requirement, however, goes against most health recommendations.

The health effects of fat consumption and the associated risk of cardiovascular disease (CVD) have been a controversial area of research over the past several decades [7–10]. The health controversy surrounding saturated fat can be traced back to an epidemiological study conducted by Keys [11], commonly known as the Seven Countries Study. Keys [11] showed that the percentage of caloric intake from saturated fat was positively associated with coronary-related deaths, whereas a negative association was found from monounsaturated fat consumption [11]. Since this time, many concerns regarding the validity of the Seven Countries Study have been raised [12], as well as multiple observational and experimental trials have since shown contradictory results regarding the deleterious health effects of saturated fat consumption [7, 8, 13, 14]. Some studies have shown positive associations between the consumption of saturated fat and the risk of CVD [8, 14], whereas others have not [7, 13] a single study has shown an inverse association between the consumption of polyunsaturated fat and CVD [13], whereas again, several others have not [7, 14]. The aforementioned collection of studies on this matter indicates that there is no consensus in the literature regarding the consumption of saturated fat and an increased risk of CVD.

The major sources of dietary fat have changed within the past century [15], with the use of butter and lard being slowly replaced by inexpensive and more versatile plant-based oils and shortenings. With the introduction of mechanization, vegetable oils were becoming less expensive than animal fat sources. As a result, researchers had been putting forth tremendous effort toward changing the characteristics of fats, specifically changing the melting point of inexpensive oils to manufacture margarines and shortenings to act as lard and butter substitutes [6]. In certain cases, animal-derived shortenings, such as lard and compound lard (vegetable oil mixed with lard), became perceived as being unhealthy, unhygienic, and adulterated due to their association with the unsavory conditions in the meat-packing industry and popular marketing of all-vegetable shortenings [6].

A major technological breakthrough in the fats and oils industry was hydrogenation. Vegetable oils could be stabilized by hydrogenation and made more functional for inclusion into food products, by saturating double bonds and turning liquid oil into solid fat. Fully hydrogenated oils yielded brittle fats; however, partial hydrogenation created semi-solid, plastic fats that could be used for many food applications. Partial hydrogenation transforms *cis* unsaturated double bonds to *trans* double bonds to create triglycerides of higher melting points, which are solid, but plastic, at room temperature. This was the method of choice for transforming oils into fats, while still claiming high unsaturated fatty acid levels. This is achieved by varying the temperature, pressure, agitation, and catalyst concentration of the hydrogenation reaction, thereby creating a number of different hydrogenated fats with tailored characteristics in structure, texture, lubrication, tenderness, and aeration for the intended food product [6]. It was thought that since these hydrogenated fats came from healthful vegetable oils, that originally contained low saturated fat and cholesterol contents, they were a healthy alternative to animal-derived fats. Hydrogenated oils became ubiquitous in the food industry, with the replacement of butter and lard with margarines and shortenings in cakes, cookies, chips, breads, icings, fillings, etc. However, partially hydrogenated oils (PHOs) contain *trans* fats, which have now been identified to cause negative health effects [14, 16, 17]. It is important to note that *trans* fats also naturally occur in certain foods, such as meats and dairy products. These ruminant *trans* fats are present in very low amounts and have not been conclusively associated with the negative health effects of industrial-produced *trans* fats [18, 19].

In 2003, the US Food and Drug Administration (FDA) mandated that all packaged food manufacturers must label *trans* fat content on their products [20]. In 2007, Heath Canada stated that *trans* fats should be limited to 5% of the total fat content in the food product and that spreadable margarines should be limited to 2% of total fat content [21]. The Minister of Health called on the food industry to comply with these recommendations within 2 years and began a monitoring program to track the *trans* fat content in specific food products. In 2009, the World Health Organization (WHO) deemed *trans* fat produced by partial hydrogenation of fats and oils to be considered an industrial food additive that demonstrated no health benefits and a clear risk to human health, and recommended *trans* fat should consist of less than 1% of an individuals' daily caloric intake [22]. Due to the continuing mounting evidence confirming the negative health effects of *trans* fatty acid consumption, the generally recognized as safe (GRAS) status of PHOs was removed by the FDA with a 3-year compliance period [17]. Scientific evidence has shown that the consumption of *trans* fat from PHOs is associated with increased levels of low-density lipoprotein (LDL) cholesterol, decreased levels of high-density lipoprotein (HDL) cholesterol, and increased levels of plasma triglyceride. All three factors are well-defined risk markers for CVD and promote insulin resistance in humans, which leads to the onset of pre- and type 2 diabetes [14, 16, 17]. In 2017 Health Canada banned PHOs in foods completely (https://www.canada.ca/en/health-canada/services/food-nutrition/public-involvement-partnerships/modification-prohibiting-use-partially-hydrogenated-oils-in-foods.html), followed by the U.S.A. in 2018 (https://www.fda.gov/food/food-additives-petitions/final-determination-regarding-partially-hydrogenated-oils-removing-trans-fat). It is rather remarkable this took so long. Since the early 1990s [23], it has been known that *trans* fats increase LDL, triacylglycerols (TAGs), and insulin levels, and reduce beneficial HDL, all factors which increase the risk of developing CVD [24]. Only in 2006, 16 years after the deleterious effects of *trans* fatty acids on human health were conclusively shown, it became mandatory to claim *trans* fatty acids on food labels. The processed food industry remained heavily reliant on the partial hydrogenation of vegetable oils to produce solid fats until 2017–2018, and many continue under grandfather clauses. Now the focus has changed to saturated fat. The reduction of saturated fat in the diet has been a long-standing recommendation. The first dietary guidelines in the United States were published by the American Heart Association in 1957, and later by the US Senate in 1977. In regard to fat consumption, they recommended overall fat consumption should account for a total of 30–40% of total caloric intake with a reduction of saturated fat consumption to upwards of 10% of total caloric intake and balance the remainder with poly and monounsaturated fat sources [25]. Since Keys' early work in the 1950s, it has been known that increased SFA consumption is positively correlated with increased plasma total and LDL cholesterol levels, a marker of risk in CVD [26]. Replacing SFA with *cis*-PUFA or *cis*-MUFA is more favorable in terms of serum lipid levels than replacing carbohydrates with SFA [27]. For total and LDL cholesterol, the most favorable effects were seen in *cis*-PUFA, and overall health benefits have been correlated to increased consumption of mono and poly-unsaturated fatty acids which can be seen in Fig. 3.1.

Mensink [27] also found that increasing intake of steric acid did not appear to have a significant effect on these or other serum lipid values, concluding it does not negatively affect cardiovascular health, shown in Fig. 3.2.

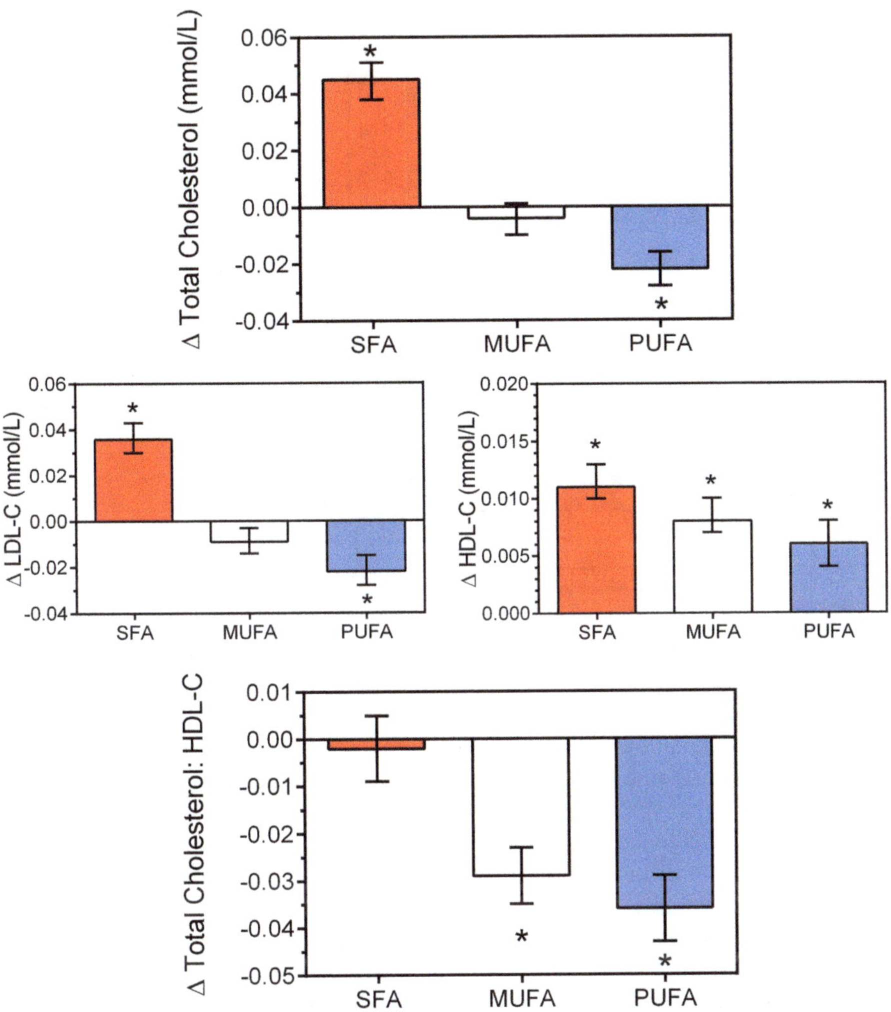

Fig. 3.1 Effect of saturated- (SFA), mono- (MUFA), and poly-unsaturated (PUFA) fat on serum lipoprotein levels, data from [27]. HDL: high density lipoprotein, LDL, low density lipoprotein

Based on the American Heart Association general guidance and recommendation, only 5–6% of total daily calories should be achieved by saturated fat. This means for a diet of 2000 calories a day only about 13 g of saturated fat per day is recommended (https://www.heart.org/en/healthy-living/healthy-eating/eat-smart/fats/saturated-fats).

Removing *trans* fat and PHOs from processed foods has already been undertaken by many companies; however, *trans* fat can still be abundantly found in products such as frozen packaged baked desserts, baked goods, coffee creamers/whiteners, spreads, fried foods, refrigerated doughs, vegetable shortening, etc. [21]. Removal of

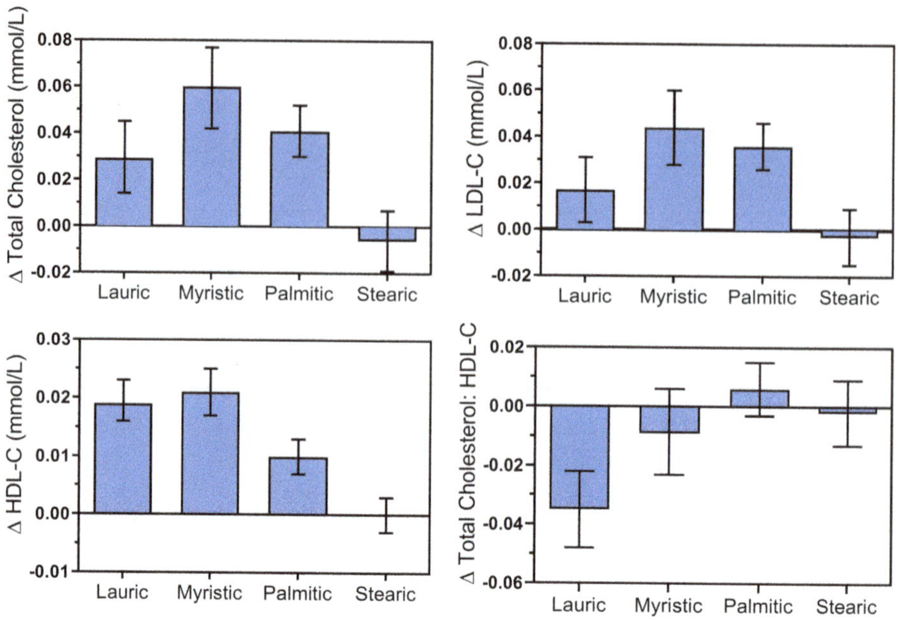

Fig. 3.2 Effect of consumption of saturated fatty acids on serum lipoprotein levels, data from [27]. HDL: high-density lipoprotein, LDL, low-density lipoprotein

saturated fats from animal sources has also been undertaken with the introduction of low-fat milk and meat products; however, much of the time, these saturated fats are replaced with carbohydrates. This strategy of saturated fat replacement, however, has been a cause of concern because low-fat, high-carbohydrate diets are associated with a different set of health issues including dyslipidemia, a component of metabolic syndrome [28–30].

Many food products are formulated with high amounts of saturated fat and/or *trans* fat because they contribute to the product's flavor, texture, mouthfeel, and functionality [31]. Research on how to replace these fats in food products without negatively affecting the organoleptic properties has been a topic of concern for the food industry and researchers alike. From an industrial perspective, a "drop-in replacement" or an ingredient that is completely interchangeable in a process is of the highest value. Ideally, changes to a product recipe should have minimal effects on product processing and the final properties of the product. Any changes to how a product must be processed upon the addition of the replacement ingredient can affect the quality of the final product and may incur losses, financial or other. Researchers must then be able to develop a suitable replacement that is both functionally and economically feasible. It is also important to consider other factors when seeking replacements for fats and oils, such as environmental sustainability, which is why the use of tropical oils, such as palm oil, is of concern [32]. In the last couple of years, a lot of attention has been drawn to the biodiversity loss in tropical rainforests due to the cultivation of both palm and coconut oil, predominantly in Malaysia and

Indonesia. These rainforests are biodiversity havens and act as massive carbon sinks. When these tropical forests are cut down and turned into palm oil plantations, a large increase in carbon emissions results. Conversely, palm is also the crop that has the highest yield per hectare and could be a more sustainable oil if produced responsibly [33]. Notable here are the recent Roundtable on Sustainable Palm Oil initiatives (https://rspo.org). Coconut oil is another prominent oil, whose production has expanded due to the demise of PHOs and the rise in the plant-based food market in which coconut oil is a main ingredient. Coconut oil is cultivated in more than 90 countries, but more than 80% of virgin or refined coconut oil is produced in Indonesia and the Philippines. Similar to the palm tree, the coconut tree grows in the same tropical forests with a high degree of biodiversity.

Besides environmental concerns about coconut oil and palm oil production, during these years, a big shift in food sources for the human diet has been observed, and one of the main routes to change this situation is the transition from animal-based foods to plant-based food products. Many scientists have been focused on mimicking the physicochemical, rheological, and textural properties of the food's target components (proteins, fats, and carbohydrates) and overcoming challenges introduced through the native organoleptic and palatability properties commonly associated with plant-based foods. However, as mentioned earlier both palm oil and coconut oil are not sustainable solutions for this change at the moment.

Recently, the use of edible organogels, or oleogels, has become a popular strategy to replace saturated fats in a variety of food products [31, 34–38]. Traditionally, fats are structured with TAGs, which provide a hierarchical crystalline structure that binds and traps oil that in turn provides structure and desirable functionality to numerous food products such as cheese, butter, and lard [1, 2, 39–41]. It is within these TAG structuring agents that *trans* and saturated fatty acids are present. Organogelation is a novel strategy that involves adding a gelator or a structuring agent to liquid oils, which subsequently forms a network that gels the oil and imparts a solid-like character that is analogous to fat [35, 42]. These networks differ from the structures arising from TAG crystallization, but have similar functionalities in numerous food applications, making them promising alternatives [43]. Oleogels can mimic animal fats and if necessary, minimize saturated fat content while obtaining the same functionality that saturated fats provide [35].

Many structuring agents have been identified that can be added or prepared with oil in order to mimic the functionality of hardstock fats that are both free of *trans* and have reduced saturated fat content. Direct approaches for obtaining organogels include structuring agents such as monoacylglycerols (MAGs) and diacylglycerols (DAGs), ricinelaidic acid, 12-hydoxystearic acid (12-HSA), ethyl-cellulose (EC), ceramides, and wax esters [34, 35]. Another recent method found is the use of indirect structuring agents, which include foam and emulsion templates made from modified polysaccharides and proteins to form functional structured oils [43].

One of the benefits of organogels is their high degree of versatility. For example, different oils, structuring agents, and mixtures of the two can be used to replace *trans* and saturated fats. These combinations result in structured oils with different functionalities, which then can be applied to different food products that have specific fat

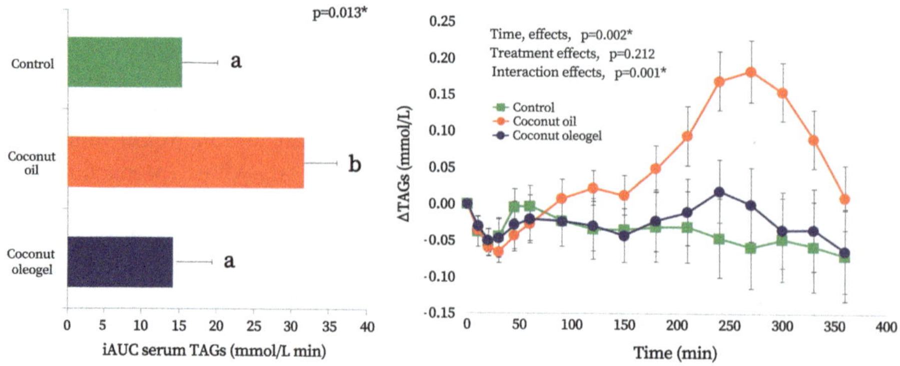

Fig. 3.3 Serum triacylglycerol changes following consumption of a carbohydrate-rich breakfast alone (control), with coconut oil, or with coconut oleogel. Bar graph corresponds to the area under the curve for each of the treatments. (Reproduced from [52] under the terms of the Creative Commons Attribution-NonCommercial 3.0 Unported Licence (CC BY-NC 3.0))

requirements. Multiple studies have been conducted in order to understand how these structured oils behave and affect the properties of different food applications including frankfurters [44], ice cream [45], chocolate [46, 47], cookies [48], and cream filling [49]. The understanding of food systems has become increasingly important in order to design its functional components, which in this case, is its fat content. As a result, further research needs to examine individual food systems in order to understand the effectiveness of replacing saturated and *trans* fat with structured oils. Furthermore, certain oleogels alter the physiological response of lipid metabolism (by decreasing rate of lipolysis) and can be used as an improved delivery system for lipid-soluble bioactive molecules and micronutrients [50, 51]. Oleogels open an exciting area of research whereby fats and oils that have demonstrated negative health effects can be structured in such a way to alter the metabolism of lipids, and minimize these effects [52]. For example, Tan et al. [52] demonstrated that ingesting oleogelated coconut oil (with EC) almost entirely eliminated the serum triglyceride peak after consuming a high-fat meal. These results were not seen in unstructured coconut oil (Fig. 3.3). Similar results have also been seen in structured emulsions [53, 54]. Rice bran wax (RBW) oleogels have also shown hypolipidemic and hypocholesterolemia affects post-consumption [55].

Figure 3.4 outlines the different mechanisms used to structure fat mimetics (oleogels). This section will review the most promising methods used to structure edible oils, their applications in food systems, and their limitations.

Many strategies of oil structuring have been reported using multiple different structuring agents that are food-grade, economically viable, and are able to impart desirable physical properties at low concentrations [31, 34, 35]. These structuring agents can be loosely defined under the following categories: crystalline particles, crystalline fibers, structured emulsions, polymeric strands [34, 35], and more recently inorganic particles [43]. The following section will provide a brief summary of the aforementioned novel oil structuring methods, while the section after will

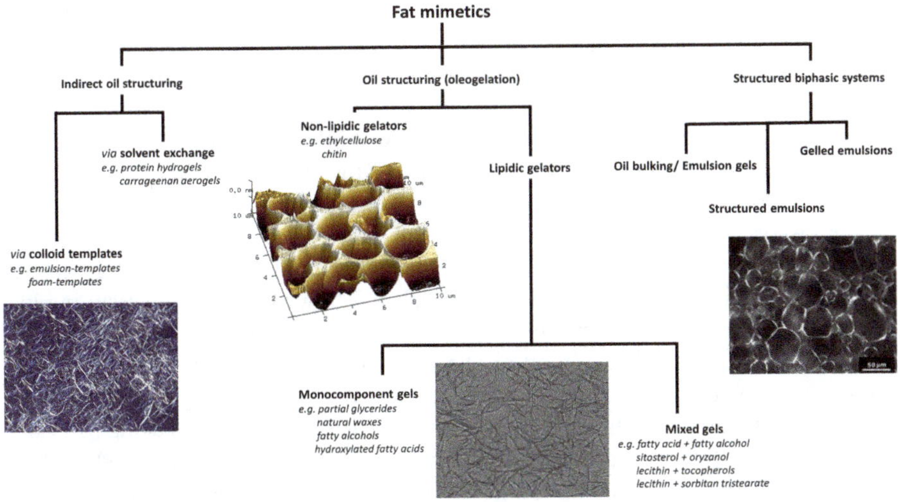

Fig. 3.4 Different fat mimetics strategies. (Reproduced from [42] with permission from Elsevier)

focus on more specific systems. In addition, novel direct structuring of oil into solid fat through enzymatic conversion of native TAGs to MAGs and DAGs and the potential of enzymatically modified systems as the oil phase of oleogels to produce adipose tissue mimetics will be discussed.

3.2 Oleogels and Oleogelation Strategies

3.2.1 Crystalline Particles

3.2.1.1 Partial Glycerides

Monoglycerides and diglycerides are a very promising class of organogelators as these molecules are natural components of edible oils and are already in common use in foods. MAGs consist of a single fatty acid esterified with a hydroxyl group attached to the end or middle carbon of the glycerol. Thus, MAGs are generally described as amphipathic molecules, having a polar "head" with no charged functional groups and a hydrophobic "tail" formed by the alkyl chain of the fatty acid. This gives them the ability to efficiently structure oil at concentrations of 10 wt% [56–58]. There are no restrictions for the use of MAGs in food products, making them an attractive structuring agent. Further, only a low concentration of MAG is needed to structure oils [56, 57]. MAGs are metabolized in a more favorable manner compared to TAGs leading to decreased body fat accumulation [59, 60].

MAGs are optically active molecules that exist as two chiral isomers, denoted by *D* and *L*. The application of MAGs in food systems is mostly associated with their emulsifying capacity, favoring the formation of emulsions and foams as well as preventing starch retrogradation in bakery products through the development of MAG–amylose complexes [35, 61]. However, an important property of MAGs in pure and commercial forms is their capacity of structuring vegetable oils and oil in water emulsions, through the formation of several liquid crystalline structures [61]. MAGs have higher crystallization and melting temperatures (both saturated and unsaturated) compared to triglycerides (TAGs) [62]. Therefore, the use of MAGs as organogelators has merited the attention of many research groups to produce fat mimetics. MAG-based oleogels are prepared by direct dispersion and melting of a relatively low amount of MAG in oil followed by cooling, resulting in a complex solid network of crystalline particles that can entrap the oil. Although a 2–3 wt% of saturated MAGs in oil has been reported as the critical gelling concentration, greater amounts (5–10%) are required to obtain rheological and textural properties comparable to those of commercial plastic fats [63]. When using MAGs, it is also possible to incorporate large amounts of water, which in some food applications comes as a great advantage, particularly in biphasic systems containing both oil and water phases [35, 64]. The crystallization and polymorphic properties of MAG crystals are very important in determining their functional properties in food products. Although at least five polymorphic forms can be found for triglycerides, MAGs crystalize in α, sub-α, and thermodynamically stable β forms. During cooling, the MAGs form an inverse lamellar phase (at a temperature above the Krafft or chain melting temperature) with a hexagonal packing similar to the α-gel phase in the aqueous systems. Further cooling of the system (below the crystallization point of MAGs) results in the formation of the sub-α phase with an orthorhombic crystalline packing. Both phases are metastable and change into a triclinic subcell packing with a higher melting point (β polymorph) with time [35]. Processing parameters such as MAG composition (saturated/unsaturated and chain length) and concentration, heating and cooling temperatures, agitation, shear rate during cooling, and storage time affect the kinetics of crystallization and the size of MAG crystals [63]. These changes in microstructure translate into the macroscopic mechanical features as evident in the rheological behavior of MAG oleogels [65, 66].

Most of the research about MAGs as gelators is based on MAGs of palmitic and stearic acids. Monostearin and monopalmitin have been successfully applied to gel a variety of oils including high oleic sunflower oil (HOSO) [66–73], rapeseed oil [74], coconut oil [72], flaxseed oil [75], canola oil [76, 77], safflower oil [78], hazelnut oil [79, 80], olive oil [56, 57], corn oil [81], and cod liver oil [82]. However, very recent studies have reported the use of MAGs of shorter-chain fatty acids such as caprylic, capric, and laureate acids [71, 74, 83] and longer-chain fatty acids such as behenic acid [71, 74]. Ferro et al. [72] and Valoppi et al. [84] studied the effect of oil type on the structure and physical properties of MAG-based oleogels. Stronger gel networks were formed with the increase in oil viscosity and reduction of oil polarity [84]. Furthermore, the degree of saturation of oils affected oleogel properties. Long-chain monounsaturated fatty acids favored the packing of MAG crystals in a cohesive gel [72].

The main concern related to using MAGs as structuring agents is their lack of structural stability. MAG structural destabilization is due to a progressive rearrangement and separation of chiral MAGs (*D*- and *L*-isomers) [63] and a tendency to transform from the α to β-polymorph form over storage time whereby hardness and brittleness increase, while oil binding capacity is lost [85], causing a gritty texture and oiling off [43]. However, their stability can be enhanced by the addition of other co-emulsifiers and changes in processing conditions. Polyglycerol fatty acid esters (PGE), sodium stearyl lactylate (SSL), and propylene glycol monostearate (PGMS) are all examples of surfactants that help stabilize the α-gel state and can be used to enhance stability in food products including low-fat shortenings, and aeration properties of cake batter [86]. Anionic, mono-valent, and small head structures are all characteristics of coemulsifiers that enhance the stability of MAG-structured oils. Recent investigations have also indicated improved rheological properties and physical stability of oleogels based on a mixture of MAGs and phytosterols [87–89]. For example, a 3:1 monoglyceride:phytosterol in sunflower oil can replace 50% of pork fat in Frankfurt sausages ending in a product with strong gel strength, and similar elasticity, cohesiveness, and oxidation levels [88]. Phytosterols can prevent the aggregation of MAGs into larger structures. This results in a system with plenty of small crystals, which typically produce a harder gel [87]. Moreover, oleogels formulated with MAGs and waxes from different sources [75, 90, 91], hydrogenated fat [92], lecithin [93], a mixture of wax and EC [94] and EC [78] have also been studied to evaluate their synergistic properties and ability to limit the sub-α to β polymorphic transition of MAGs. Palla et al. [95] attempted to match the texture profile and oil-binding capacity of a MAG-oleogel with that of a commercial margarine by studying the effect of preparation variables (MAG concentration, heating temperature, stirring speed, and cooling temperature) on oleogel characteristics. MAG concentration and cooling temperature were found to be the most contributing factors to oil binding capacity, and textural and rheological parameters [95]. In addition, new technologies such as high-intensity ultrasonic have been found to improve the physical properties of MAG-based oleogels [68, 69]. Shear and cooling rates can be used to alter the structure of the crystalline network and ultimately influence their physiochemical properties in food [57, 96]. When using MAGs, it is possible to incorporate large amounts of water, which in some food applications comes as a great advantage, particularly in biphasic systems containing both an oil and a water phase [58, 97]. Their strong emulsifying ability makes them excellent candidates for cream-like products [65]. Another partial glyceride-stabilized oleogel strategy exploits the use of DAGs. Similar to MAGs, DAGs have high melting points and desirable health benefits, making them an interesting candidate for oil structuring. DAGs are diesters of fatty acids with glycerin and can exist in two isomers: 1,2- (or 2,3) DAGs and 1,3-DAGs. While 1,2-DAGs have a melting temperature similar to that of TAGs, 1,3 isomers show a melting temperature that is approximately 10 °C higher than TAGs of the same fatty acid composition. In addition, due to different metabolic pathways, 1,3-DAGs inhibit body fat accumulation and reduce postprandial TAG levels in human serum [62]. The crystallization behavior of 1,3-DAGs and 1,2-DAGs differs significantly. 1,3-DAGs crystallize in β configuration either a high-melting polymorphic form β_1, or a lower-melting polymorphic form β_2 [98]. Although β crystals are related to oil loss and sandiness in fat products, they are not necessarily an issue if small β crystals are formed. In fact, small β crystals are also capable of forming a stable crystal network structure with desired rheological properties [98]. Xu et al. [99] reported that fat blends containing DAGs had a stronger network structure with a more solid-like nature compared to the fats structured purely with TAGs. These merits suggest the promising use of DAGs in the fabrication of oleogels. Tavernier et al. [98] investigated the effect of MAGs (1, 2, 4 wt%) on the crystallization behavior of 20 wt% DAG sunflower oil oleogel. MAGs delayed the onset of nucleation and retarded the crystal network development by 1,3-DAGs. The effect of MAG-DAG interactions on the microstructure and physicochemical properties of the oleogels was further explored by Wang et al. [100]. In this study, the minimum concentration for DAG to form a solid-like gel was found to be 8 wt%. Different MAG/DAG weight ratios (0:10, 1:9, 3:7,5:5, 7:3, 9:1, and 10:0) were then applied to prepare mixed oleogels with total gelator concentrations of 8 and 12 wt%. MAG-based oleogels produced significantly higher hardness, possibly due to the higher crystallization rate, higher melting point, and higher oil binding capacity of MAG molecules. The mixture showed reduced crystallization enthalpies compared to mono-component oleogels, indicating eutectic behavior and co-crystallization of MAG and DAG molecules. However, the oleogels with higher MAG ratios showed lower solid fat content at low temperatures. Oleogels with good plasticity and crystal morphology were obtained using MAG: DAG ratio greater than 5:5 [100].

The interfacial activity and strong network structure formed by DAG have also led to research on DAGs as a structuring agent in emulsions or foams by forming Pickering particles. Oleofoams can be fabricated by whipping the suitable oil phase and have shown great potential as low-calorie aeration systems. However, saturated lipids or a large amount of surfactants are commonly required [101]. In work conducted by Lei et al. [102], oleogel prepared with 10 wt% long-chain DAG was found to be a feasible substitute for hydrogenated palm oil (HPO), showing comparable rheological and thermos-sensitive properties. The incorporation of 2 wt% HPO into the system contributed to ultra-stable oil foams with higher viscoelasticity and smaller bubble size. To provide desirable oil structuring and whipping ability, Qiu et al. [101] suggested the combination of DAG and β sitosterol as gelators. The gelators interacted through hydrogen bonding, which had a synergistic effect on the formation of a stable physical barrier and the prevention of air bubble destabilization. In subsequent work, the authors compared the performance of pure diglycerol laurate in an oleogel and oleofoam with lauric acid and its other glycerides (glycerol monolaurate and triglyceride laurate). Lipid crystal size and polymorphism largely affected the whipping ability and foam stability. DAG crystals exhibited smaller size and β and β' polymorphs with uniform distribution at the bubble surface, which contributed to the formation of ultra-stable oleofoams. The lauric acid stabilized system had a comparable over-run to that of DAG oleofoams while the MAG and TAG mixtures showed lower over-run. According to the results, a variety of plant oil foams with high stability (up to 8 months) can be stabilized by DAG crystals [103].

In conclusion, considering the natural origin of MAGs and DAGs and their unique physicochemical and physiological functions, partial glyceride-based structured fats can be designed not only for solid fat replacement purposes but also to modulate lipid digestion. These clean-label multifunctional systems have been used in a wide range of food products with promising results. Future work in the partial glyceride-oleogel field requires a focus on technological and health-promoting applications as well as scaling up laboratory experiments to pilot and industrial scale.

3.2.1.2 Waxes

Plant-based waxes are also capable of structuring oil at low concentrations (1–4 wt%) by forming crystalline particles that aggregate and form a network that entraps oil [34, 35]. Some major sources of wax that have been studied include RBW [104], sunflower wax (SFW) [105], candelilla wax (CLW) [106], carnauba wax (CRW) [107], shellac wax [108], beeswax (BW) [109], and sugarcane wax [107]. Natural waxes are promising gelling agents for vegetable oils because of their strong oil-binding capacity, economic feasibility, and wide availability to be used in the food industry [42, 106, 110]. Chemically, waxes are composed of mainly long-chain esters from fatty acids and fatty alcohols along with minor components such as n-alkanes and sterol esters [104, 111, 112]. Low concentration of wax (1–4%) is needed in an oleogel formulation to obtain gelling properties [104, 113]. This makes them economically viable along with the fact that waxes are often by-products of oil refining [113]. Table 3.1 shows the most promising waxes, their chemical composition, and their melting temperature. All of the following waxes included in Table 3.1 have gained GRAS status from the FDA, excluding SFW.

There is a strong relationship between wax composition and crystal morphology. Smaller crystals form upon faster cooling and increased shear rates. Wax esters crystalize in mono-molecular chain packing with alkyl chains either in a tilted or orthogonal orientation [115]. Melting points and molar heats of fusion are related to the position of the ester bonds, the more asymmetrical the bonds the lower the melting points and molar heats. Crystals that are uniform in size and range between 1 and 5 μm produce a smoother texture and soft mouth feel [116, 117]. Waxes tend

Table 3.1 Chemical composition and melting temperature of vegetable waxes and beeswax [109, 113, 114]

	Esters (%)	Free fatty acids (%)	Free fatty alcohol (%)	Hydrocarbons (%)	Resins (%)	Melting temperature (T_m)
RBW	92–97	0–2	–	–	3–8	78–82
SFW	97–100	0–1	–	–	0–3	74–77
CLW	27–35	7–10	10–15	50–65	–	60–73
CRW	84–85	3–3.5	2–3	2.5–3	6.5–10	80–85
BW	71	12	–	14	6	63.15

BW beeswax, *CLW* candelilla wax, *CRW* carnauba wax, *RBW* rice bran wax, *SFW* sunflower wax to leave a waxy taste in the mouth and can be gritty, therefore, tailoring the processing parameters to make the wax crystal structure optimal is essential to limit undesirable sensory properties. Synergistic mixtures of waxes have also been reported. Ghazani et al. [118]. discovered synergism between SFW and RBW, as well as CLW and BW, resulting in increased hardness, elastic constant, plasticity, and oil binding capacity of oleogels compared to each counterpart on their own. These mixtures with olive oil demonstrated similar physical characteristics to commercial margarines. Wax oleogels have also had many successful applications in the bakery sector, replacing margarine and spread products with minimal sensory differences [48, 119, 120]. Wax oleogels have also been used to lower saturated fats in deep fat frying mediums [121].

Although wax oleogels are great structuring agents, they do come with some limitations, namely, sustainability issues with the extraction of certain waxes, such as CLW [122], changes in rheological properties in systems over time (stability), cloudy appearance, and more importantly, waxy mouthfeel in food products. Further questions also exist regarding digestibility and health risks/benefits associated with long-term consumption of these waxes [123].

3.2.1.3 Fatty Acids and Fatty Alcohols

Fatty acids and fatty alcohols were some of the first molecules studied to structure oils. These low-molecular-weight compounds are able to form supramolecular structures through non-covalent interactions that structure oil very effectively at low concentrations (~1% w/w) [124–126]. An attractive attribute of these systems is their similarity of crystallization behavior to TAGs [127]. To make fatty acid /fatty alcohol gels, the mixture is simply heated, and gelation occurs upon cooling. Stearyl alcohol /stearic acid (SOSA) has the highest oil structuring capacity in edible oils compared to any other fatty acid/fatty alcohol mixture [125]. Gandolfo et al. [125] demonstrated that a specific mixture of fatty acids and fatty alcohols of similar chain lengths displays a synergistic effect in terms of oleogel strength. The 7:3 (w/w) SOSA mixture formed smaller crystals with needle-like morphology than the other proportions. These small crystals had a greater surface area, higher oil binding, and formed a stronger network [125, 128]. Blach et al. [124] revisited this system and demonstrated that the desirable synergistic effect was due to an increase in the spatial distribution of network crystalline mass. Both studies used a 5% structurant concentration, and the same effect was observed using sunflower oil [125, 128]. Rapeseed oil's optimal ratio was slightly different than for soybean and sunflower oil, with an optimal ratio of 8:2 w/w [125]. These optimal ratios displayed no oil loss upon centrifugation and had great stability, making them excellent candidates for food applications [124]. The lowest concentration of these molecules that can form a crystal network was 3.5% [128].

Food applications for SOSA include margarine, sausages, and Dutch croquettes [128]. SOSA can also be used in combination with other structuring agents (such as EC), which results in enhanced plastic flow behavior to resemble commercial margarine [126]. Lecithin has also been shown to display synergism when mixed with various fatty acids and/or fatty alcohols [127]. Lecithin and tocopherol [129], β-sitosterol and oryzanol [130], and lecithin-stearic acid-water [131] provide flexibility to alter desirable properties for specific commercial applications at low gel concentrations [132].

3.2.2 Crystalline Fibers

Oil structuring can be achieved by self-assembled fibrillar networks (SAFIN) of low molecular weight gelators (LMOG), for example, 12-HSA, ricinelaidic acid, and phytosterols with oryzanol [35]. These gelators are capable of forming helical and twisted crystalline ribbons, hundreds of micrometers long, which immobilize oil.

3.2.3 Structured Emulsions

One of the first oil gels that demonstrated viability for use in food applications were high-internal phase emulsion (HIPE) gels, consisting of oil encapsulated in multilamellar crystalline MAG vesicles [58, 133, 134]. The properties of MAG gels were exploited to create a simple and highly stable oil-in-water emulsion gel with 27–70% (v/v) water, 66–27% (v/v) oil, zero *trans*, and low saturated fat content [58]. The resultant structured emulsion exhibits a fat-like appearance with highly versatile shortening functionality and is commercially available today as AAK's AkoBiscGO (https://www.aak.com/applications/bakery/biscuits-and-cookies/AkoBiscGO/), trademarked under the name Coasun™ (https://www.coasun.com/). The addition of a large percentage of water to make CoaSun™ can significantly cut production costs [44].

The gel preparation involves dispersing of hot oil-monoglyceride-charged co-surfactant mixture in hot water by vigorous mixing followed by cooling, resulting in effective microencapsulation of oil within crystalline shells of monoglyceride and a solid-like structure.

Once emulsified, a monoglyceride monolayer surrounds the oil droplets and self-assembles to form a highly hydrated, multilamellar La state at temperatures slightly above the Krafft temperature of the monoglycerides. Upon cooling, the fatty acid chains of the monoglycerides crystallize in a hexagonal conformation (a subcell polymorph), referred to as an alpha gel, with high amounts of water entrapped between stacked bilayers. This, so called, a-gel state dehydrates and eventually transforms into an anhydrous crystalline phase (b' or b polymorph), referred to as the coagel or β-gel [58]. Although MAGs are incorporated at low levels (a few percent), yet they are able to stabilize the emulsion. A co-surfactant enhances the formation and stabilization of the a-gel state and adjusts the vesicle surface charge, affecting vesicle–vesicle interactions and rheological behavior of the emulsion

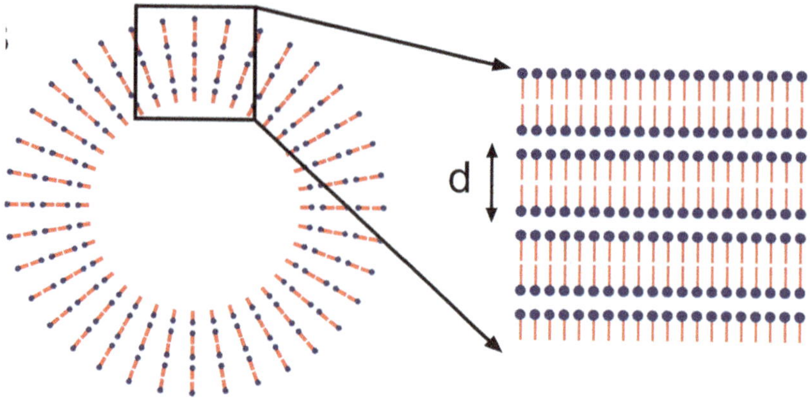

Fig. 3.5 Schematic of monoglyceride bilayers surrounding the oil droplets. Each of multiple bilayers could hold a layer of water between the hydrophilic head groups

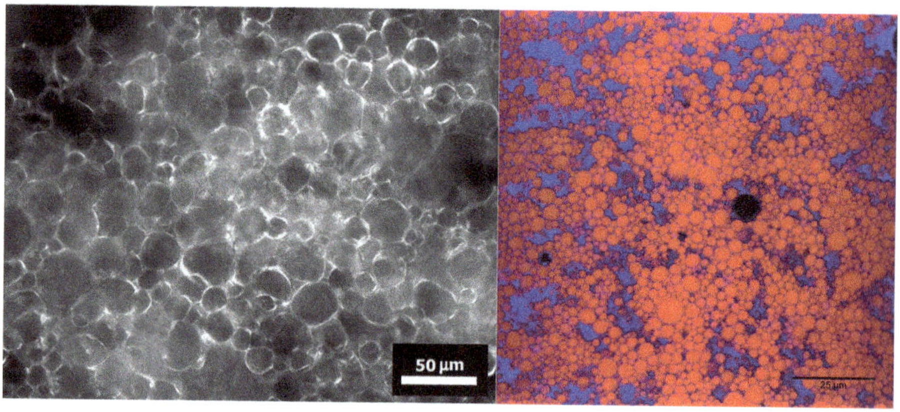

Fig. 3.6 Polarized light (left) and confocal microscopy (right) images of a structured oil-in-water emulsion (MAG gel)

[58]. Thus, the surface charge can be adjusted to convert a simple oil-water mixture into a fat-like material with desirable functional properties [58]. Figure 3.5 depicts a schematic of the outer walls of the droplet composed of stacked crystalline monoglyceride and water bilayers surrounding the oil. These droplets can be easily observed when using polarized light or cryo-scanning electron microscopy, as shown in Fig. 3.6.

Having very low SFA and zero *trans* fat content is one of the greatest advantages of these materials, making it possible to manufacture healthier food products, when prepared with canola oil (55.25%) and 40% water, CoaSun™ will contain 7–8% total saturates and can usually replace traditional shortenings on a 1:1 basis [135]. Furthermore, the unique structure of Coasun™ has shown the ability to regulate blood lipid and insulin response in humans. There is also a potential to incorporate nutraceuticals such as phytosterols, omega-3 fatty acids, and fat-soluble antioxidants into the water or oil phase of these structured emulsions, which increases their nutritional benefits [44].

Another method of creating structured emulsions was achieved by forming an organogel using shellac [136] or BW [109] in oil, and incorporating water into the molten organogel continuous phase in either the presence or absence of varying amounts of emulsifiers (e.g., Tween20, Tween80) and hydrocolloids (e.g., xanthan gum, XG). The molten organogel is subsequently cooled to entrap water droplets in the resultant crystalline organogel network. The shellac emulsions, which were prepared in the absence of any emulsifiers, consisted of 20 wt% water with either 1.6, 3.2, or 4.8 wt% shellac. This emulsion is stabilized by a combination of physical entrapment and steric stabilization as shellac both imparts structure to the continuous oil phase and accumulates at the water–oil interface, which prevents coalescence and phase separation [136]. The BW structured emulsions contained 3.75–4.50 wt% BW in the final emulsion and required the presence of emulsifiers and XG to remain stable [109]. It was speculated that Tween 20 or Tween 80 incorporation increased the hydrophilic-lipophilic balance (HLB) value, making the mixture more suitable for a water-in-oil emulsion. The addition of XG increased the viscosity of the water phase and decreased the sample melting point [109].

Alternatively, various food-grade hydrocolloids have been used to create oil continuous HIPE gels [137]. To develop such systems, the hydrocolloid solution is finely dispersed in an oil continuous phase at >70 °C and subsequently cooled to produce a high phase volume of closely packed gelled droplets that physically entrap oil and exhibit gel-like behavior [137]. The characteristics of used hydrocolloids affect the gelling efficiency of the internal water droplet phase and usually a combination of hydrocolloids is preferred. For example, galactomannans are known to show synergistic interactions with carrageenan, XG, and locust bean gum, leading to the formation of gel at low polymer concentrations [137].

3.2.4 *Inorganic Particles*

Hydrophilic fumed silica was able to structure sunflower oil to produce viscoelastic gels at concentrations of 10 and 15 wt% when silica particles were dispersed uniformly in oil under high shear [138]. At lower concentrations (2.5 and 5 wt%), the samples were viscous sols with negligible yield stresses. This was attributed to the highly concentrated samples having a higher degree of particle–particle interactions leading to a stronger fractal-like network formation which entrapped oil. In addition, a weak hydrogel containing locust bean gum and carrageenan could be mixed with the 15 wt% silica organogel to produce a bigel, which showed a higher gel strength than either component in isolation [138].

3.2.5 Hydrophobic Polymers/Oleocolloids

The only known direct polymeric organogelator and one that has shown great potential for use in food applications is EC. EC, a derivative of cellulose, is a non-water soluble, non-toxic, linear polysaccharide that is used as a thickener in foods. EC is made through a chemical modification process whereby cellulose is introduced to an alkaline environment and then exposed to chloroethane gas [139]. This induces the ethylation of the hydroxyl groups, which are exposed on the backbone of cellulose. The degree of ethylation is dictated by % ethoxy content which for most commercial food grades of EC lies between 48% and 49.5% (the range in which EC is soluble in organic solvents) [140]. Substituting the hydroxyl groups with ethoxy groups renders EC soluble in oil above its glass transition temperature ($T_g \sim 135$ °C) and upon cooling forms a three-dimensional entangled polymer network and a resultant gel. The functionality, microstructure, and processing conditions of these gels have been extensively studied [141–143]. Other food-grade polymers have been used to structure oils via an indirect methodology involving creating polymeric templated scaffoldings which are also capable of physically entrapping oil. EC oleogels are good substitutes to replace solid fats because of their similar fat globule size and microstructure [44, 144]. The strong mechanical strength of EC oleogels allows the gels to decrease the rate of lipolysis and serum TG postprandially compared to regular oils [51].

As mentioned, the process for preparing organogels using EC involves heating the oil-polymer mixture above its T_g, approximately 135 °C, and mixing it until the EC is dissolved (Gravelle and Marangoni, 2018). It is important to consider that the temperature of the mixture does not exceed 185 °C as there is a melting phase transition around this point and decomposition of the polymer starts. Upon cooling, EC will return to its rigid state creating gaps within the polymer network that physically entrap the liquid oil phase [145]. Various factors that affect the strength of the oleogel include temperature, cooling rate, the presence of surfactants, the type of liquid oil, the molecular weight of EC being used, and holding time during preparation. An increased holding time at T_g results in a stronger gel [146]. Because a high temperature is needed to create these gels, heat-induced oxidative degradation can occur, decreasing the shelf-life of the oil [147]. The molecular weight of EC also influences final gel strength, cross-over behavior, and gel point temperature [148]. EC is sold based on its molecular weight such that different molecular weights are tailored to different applications [149].

Many food applications have been studied including replacing some animal fat with EC oleogels in breakfast sausage [150], liver pate [151], salami [152], and frankfurters [143, 147]. It should be emphasized that in most of these applications only part of the animal fats was replaced in order to maintain consumer acceptability. EC oleogels have also been used to make heat-resistant chocolate [46]. EC can also be mixed with different molecules to manipulate the desired flow behavior and demonstrate synergism with respect to enhanced crystallization structure and increased plasticity. Gravelle et al. [126] demonstrated how the brittle behavior of

SOSA oleogels can be plasticized by the addition of EC allowing for plasticized behavior under deformation, mimicking a commercial margarine. Barbut et al. [153] also demonstrated that adding 1.5% or 3% sorbitan monostearate (SMS) to an 8% EC oleogel showed enhanced similarities in the hardness and color of beef fat in frankfurter sausages. In the USA, EC has obtained a "generally regarded as safe" (GRAS) status from the US Food and Drug Administration (FDA) for indirect food usage (21 Code of Federal Regulations section: 182.90) for use up to 5% (w/w) EC in oil. Depending on the type of oil used, the required amount of EC needed to form a gel differs. For canola and soybean oil a minimum of 6 and 4% w/w respectively are needed [143]. In the future, further research is needed on the consumer acceptability and sensory analysis of food products made with EC [154]. Another limitation of EC is it is not a natural ingredient. This stops many food companies from using it as many consumers steer away from long chemical names on food labels, seen frequently in plant-based products. The trend toward clean-label products has drastically increased in the last decade [155]. Ideally, a natural polymer could be found in the future with similar properties to EC.

3.2.6 Hydrocolloid-Stabilized Oil-Filled Gels, Foams, and Powders

Hydrocolloids are attractive agents to structure oils due to the lack of regulatory restrictions and because of their high molecular weight allowing them to form gel networks at low concentrations [42]. The majority of these hydrocolloids are proteins and modified polysaccharides and are thus hydrophilic and unable to be dispersed in oil like EC. Therefore, these polymers are added through indirect methods including emulsion and foam templates, which will be discussed. Finding proteins to use as structurants in oils has become an expanding area of research in the last couple of years. Proteins possess some key advantages including their price compared to other structurants, nutritional composition, clean label, and eligibility to be used in vegetarian and vegan products which are on the rise [156].

3.2.6.1 Emulsion-Templated Oleogels

Emulsion-templated oleogels are obtained by first creating an emulsion in which the aqueous phase (containing the hydrophilic polymer, a polysaccharide, or protein) is the continuous phase. The hydrocolloid stabilizes the oil-water interface and can get crosslinked to provide elasticity to the interface [157]. The continuous phase is then removed through oven drying, spray drying, or freeze drying, followed by shearing, resulting in a semi-solid material with a high concentration of lipid globules coated in the hydrocolloid creating the emulsion template [158, 159]. Gels structured through freeze drying resulted in harder gels when compared to oven-dried gels

[160, 161]. The structure and characteristics of the emulsion are closely related to the final properties of the oleogel [42]. Therefore, the structure of the oleogel is highly influenced by the functionality of the protein or polysaccharide. To use proteins in this structuring method they must be surface active and have the ability to coagulate [156]. Thankfully protein functionality can be modified through various processing parameters, more research needs to be done in this area to understand what is the structure of an optimal protein or polysaccharide for this application. Soy protein isolate thus far has been the most promising plant-based protein to form these gels in terms of its rheological properties [162]. Many protein oleogels tend to be much softer than the counterparts they are aiming to replace. As more research is done in this area and more proteins are explored in the upcoming years, they could make an excellent candidate for fat replacement with economic, environmental, and nutritional advantages.

In terms of polysaccharides, XG was used as a thickening agent and either hydroxypropyl methylcellulose (HPMC) or methylcellulose (MC) was used as emulsifying agents [158, 159]. Emulsions prepared with HPMC or MC alone resulted in coalescence and oil leaking upon drying. XG used in isolation could not emulsify, as it is nonsurface active, whereas, when incorporated in combination with HPMC/MC a more uniform oil droplet size distribution and stable dried emulsion was achieved [158]. In another study, it was found that the XG could also be used in combination with gelatin (a protein) to produce similarly structured oils where an increase of polymer concentration at the oil–water interface resulted in increased gel strength [159]. The microstructure of these structured oils shows distinct tightly packed oil droplets resembling that of a HIPE [158].

3.2.6.2 Foam-Templated Oleogels

The basic steps for obtaining a foam-templated oleogel are similar to those for emulsion-templated oleogels. To make a foam-templated gel, protein or polysaccharide is first mixed into an aqueous solution. Air is then introduced into the system (i.e., through homogenization) to create an aqueous foam. The aqueous foam is freeze-dried, becoming a "cryogel." Oil is introduced into the cryogel transforming it into an oleogel. Foam-templated gels tend to be softer compared to HIPE gels [108]. Advantage of this approach of forming oleogels is the environmental aspects of the processing steps, no high temperatures, harsh chemicals, or additives are used [163].

Patel and coworkers, the pioneers in this area, originally incorporated oil into a HPMC scaffold, [163]. In this work, an aqueous solution of HPMC was foamed mechanically to incorporate air and freeze-dried to create a porous structure that is able to adsorb a high amount of oil; however, since freeze-drying creates an open-celled structure, oil will be released under compression. To prevent this, the oil-saturated foams were sheared to disperse the polymer sheets and trap the oil [163]. This templating concept has also been used previously to partially (60%) and fully substitute pork backfat content in fermented sausages by adsorbing extra-virgin olive oil to a whey protein-based crumb and white pan bread [164]. Alessandro et al. [164] found no significant difference in weight loss of the oil-soaked whey protein-based crumb, white pan bread, and full-fat control products. In addition, taste, color, and odor characteristics of the 60% replacement with oil-soaked whey protein crumb were comparable to the commercial product; however, long-term stability was not examined. Another application for this templating concept is the use of cellulose-based freeze-dried foams as oil absorbents for cleaning oil spills in water [165]. The freeze-dried foams used for oil spill applications, typically are not food grade, which makes HPMC foams unique and useful for food applications. HPMC and MC have been used in the food, cosmetic, and pharmaceutical industries for decades and both have been granted a GRAS status by the US FDA. HPMC and MC are well known for their ability to create gels in aqueous solutions, but the literature is limited on their use in oil-based systems.

The functional properties of the final oleogel are directly related to the cryogel structure [166]. These gels can be used to decrease the density and caloric effects of food products [167]. Further, they can be tailored to change the texture, mouthfeel, appearance, and rheological properties of food products. Scaling up and stability of these foams need to be further researched [168]; however, they provide great potential for the replacement of *trans* and if desirable saturated fatty acids in edible applications.

3.3 Enzymatic Glycerolysis

Enzymatic glycerolysis (GL) is another technological alternative for obtaining zero trans, low saturated plastic fats, and improving the sustainability of the fat sources. The process converts oil-native TAGs into partial glycerides in the presence of glycerol and lipase, exploiting their higher crystallization and melting temperatures to directly structure the liquid oil [169, 170]. This is beneficial to consumers as partial glycerides are natural components of fats and oils and there is no change in the fatty acid composition (i.e., no increase in *trans* or SFA) of the lipid system. Therefore, readily available plant oils rich in unsaturated fatty acids can be modified to replace palm oil and other high-saturated-fat [169]. In addition to lower SFA intake, glycerolysis products (GPs) are of great interest due to the health benefits of DAGs on lipid metabolism such as inhibiting body fat accumulation and reducing serum postprandial triglyceride and LDL level, as stated earlier [62, 171]. Therefore, an edible unsaturated oil structured using GL would provide the nutritional benefits of MUFAs, PUFAs, and DAGs, while exhibiting the required solidity [169]. Another interesting property of GPs is their self-dispersion and emulsifying properties, showing their promise for use in emulsified products [169, 172] and as bioavailable lipid carriers in the formulation of functional foods and nutritional supplements [173].

Traditionally, GL is applied for simply producing partial glycerides for use as emulsifiers and other food ingredients. However, Marangoni group investigated the effectiveness of lipase-catalyzed GL for the novel purpose of fat structuring, where partial glycerides are not extracted after production [169]. This study was performed using cottonseed oil and non-specific *Candida antarctica* lipase B (2% w/w relative to the oil) at 65 °C for different time lengths and with varying glycerol to oil molar ratios (0.10:1, 0.25:1, 0.50:1, 1:1, 2:1, 4:1). A rapid increase of MAGs and DAGs content was observed during the first 18 h of reaction, which reached almost a plateau after 48 h. DAG content was maximized at 0.5:1 glycerol:oil ratio and decreased slightly at higher glycerol content while the 1:1 ratio resulted in the highest MAG production. Reaction conditions also had an impact on MAG species. The longer reaction time and the higher glycerol content increased monoolein and monopalmitin while having a limited effect on monolinolein content. Overall, results showed dramatic changes in solid fat content (SFC), crystallization behavior, microstructure, and oil binding properties of the oil followed by GL. At 5 °C, SFC increased from almost 8% in native oil to 29% in optimally structured cottonseed oil GP, with 21.7% solids remaining at 20 °C. Microstructure and network properties obtained at 1:1 glycerol:oil ratio resulted in high oil binding capabilities and reduced the oil loss [169].

The type of lipase (1,3-specific lipase from *Rhizomucor miehei* vs non-specific *Candida antarctica* lipase B) also affected the structuring ability of the process, which could be attributed to either differing stereo specificities or the overall power of the enzyme to produce partial glycerides. *Rhizomucor miehei* enzyme showed lower enzyme activity (longer reaction time) and substantially reduced capacity for producing MAGs, resulting in reductions of attainable SFC and structuring capability. Rather than enzyme characteristics, enzyme quantity can also affect process time. Higher enzyme concentrations can substantially reduce the reaction time, making it much more feasible for industrial usage. Additionally, lipase could be separated upon completion of an industrial-scale GL reaction and reused for additional reactions [62].

The ability to alter the physical properties of the oil simply by changing the reaction conditions highlighted the high potential of this process for producing alternatives for tropical oils and/or interesterified blends of hydrogenated oils, with improved sustainability and health properties [169].

In follow-up research, GL was performed on a variety of liquid oils such as tigernut, peanut, cottonseed, rice bran, olive, soybean, sesame, canola, high oleic algal (HOAO), and high oleic canola (HOCO) oils using optimized condition (*Candida antarctica* lipase B; 65 1C; 48 h; 1:1 glycerol:TAG molar ratio) [172]. The fatty acid composition (SFA and oleic acid content) of the native oil and thus the composition of partial glycerides formed after GL substantially contributed to the crystalline material and physical properties. No crystalline material at 5 °C was observed in HOAO due to having low SFA contents (2.8%) and more than 90% oleic acid. However, canola oil and HOCO with 7% SFAs and much less oleic acid (63.1% and 72.5%, respectively) achieved a SFC of 8.3% and 18.7% at 5 °C, respectively. Overall, structuring ability was highest when native oil with high oleic acid content (>60%) contained more than 10% SFA (ideally 14–25%). In most systems, oleic acid, as monoolein and diolein was responsible for the structuring potential as observed in tigernut, peanut, olive, and HOCO. While monopalmitin was the major MAG species in cottonseed, soybean, and rice bran systems [172].

In a recent unpublished work from the Marangoni group, the GL process of different plant oils was modified and scaled up. Product characteristics were comparable at laboratory and pilot plant scales, supporting the commercial viability of the process. In addition, the potential of structuring oil systems containing more saturated fat such as shea olein and palm olein was investigated for the first time for the purpose of shortening and animal fat mimetic production.

Shea olein and palm olein GPs displayed high plasticity and broad melting peaks with high melting temperatures, similar to animal fat. GPs were then converted to oleogels, using EC with different molecular weights (20 cP, and 45 cP) at 5% w/w to emulate desirable rheological and textural attributes of adipose tissue. A combination of GL and oleogelation produced a more deformable and cohesive product. Comparison of the melting, rheological and mechanical properties of the EC oleogels with those of whole pork, beef, and lamb adipose tissue showed the potential for EC oleogels of oil GPs to be used as adipose tissue mimetics in the new generation plant-based meat analogs.

3.4 Food Applications

Previous works on organogels have been predominantly focused on understanding fundamental properties including microstructure, polymorphism, and rheological behavior [31, 34, 35]. This understanding is critical to applying organogels to food systems as the behavior of traditional fats needs to be mimicked in order to impart similar if not equal functionality in the final food product. In addition, "drop-in ingredient" replacement is highly preferred in the food industry to avoid incurred costs. The desired functional characteristics of the fat content are variable depending on the specifications of the food application. Replacing highly saturated hard stock fats with liquid oil often causes increased oil migration resulting in oil leakage and other quality defects. In response to these problems, an increasing amount of research has focused on exploring the potential use of various types of organogels in a wide range of food applications including spreads, bakery and pastries, cream fillings, ice cream, meat, and dairy products which is overviewed in the following.

Plant wax in soybean oil organogels was used for margarine production, including CLW, RBW, and SFW. It was found that although CLW and RBW were able to form stable organogels, in a margarine formulation CLW showed phase separation and RBW had low firmness; however, SFW was able to form the most desirable margarine characteristics of the three waxes studied [174].

Margarines made using CRW and monoglycerides in virgin olive oil at 3, 7, and 10 wt% concentration levels were studied, all samples, in particular the 7% monoglyceride sample, resembled the textural and thermal properties of a commercial margarine [175].

Shellac organogels were able to fully replace a commercial hydrogenated vegetable oil stabilizer and partially replace (~27%) palm oil in a chocolate paste, and create an emulsifier-free margarine with up to 60 wt% water incorporation [176].

Margarines made using virgin olive oil organogels containing 5 wt% of BW or SFW demonstrated consumer hedonic liking scores of appearance, odor, flavor, and spreadability above neutrality, demonstrating consumer acceptability of these margarine replacements [120].

Structured MAG gels, or Coasun™, are used as an all-purpose shortening replacement in cookies. Traditional all-purpose shortening had superior shortening functionality compared to MAG gels; however, incorporation of the unstructured components of the emulsion resulted in lower cookie dough firmness and increased spread value when compared to its structured emulsion counterpart. Therefore, the structured emulsion maintained a similar dough firmness to the all-purpose shortening dough, indicating that the structured emulsion had a greater ability to act as a structural component in the cookie dough [177].

The ability of MAG gels to act similar to a shortening such as development time and prevention of gluten aggregation was demonstrated [135]. These characteristics are beneficial for baked products that do not rely on gluten network formation for structure, for example, cookies, pie crusts, and pastries.

Traditional shortening was replaced with MAG gel, modified with added 3 to 5 wt % EC [178]. Oil loss was monitored at 37 °C using filter papers to promote oil loss; it was found that cookies made with added EC resulted in lower oil loss overtime than the structured emulsion.

A standard cookie mixture was used to compare cookies containing approximately 24% fat content from organogels prepared with 5 wt% SFW and BW in hazelnut oil and a commercial bakery shortening [48]. Cookies prepared with organogels had a lower aeration (indicated by higher diameter: thickness ratio), lower hardness, higher factorability, and equal or higher positive sensory attribute scores than the cookies prepared with commercial bakery shortening. Therefore, cookies prepared with wax organogels can be considered to exhibit good sensory quality. This was supported by consumer hedonic liking data, which indicated that the organogel cookies were preferred and better accepted than the control cookies [48].

Candelilla wax (3 and 6 wt%)- canola oil organogels resulted in cookies that had more spread upon baking, and lower snapping force contributing to softer eating characteristics than shortening control cookies [179].

MAG organogels were prepared with sunflower oil and were used in sweet bread [180]. Incorporation of MAG organogel resulted in an inhomogeneous lipid distribution in the dough, contributing to low leavening and firmer structure than a palm oil control.

An emulsion-templated structured oil was made with MC and XG as a replacement for traditional shortening in cake batters. It was found that the batter properties and cake attributes of the structured oil were comparable to commercial shortenings and showed significant differences from liquid oil controls [181].

Shellac wax-based emulsions were used as a cake margarine replacement in a basic sponge cake recipe. It was found that though the shellac emulsion batter properties were not comparable to the control in terms of density and flow behavior, the resultant baked cakes showed mostly comparable textural and sensorial properties [176].

A model cream was prepared using a mix of 60% organogel or oil and 40% intersterified hydrogenated palm oil [178]. The organogel was prepared with 6% EC and 2% sorbitan monostearate with either canola or HOSO. An oil binding test conducted at 20 °C using filter papers to promote oil loss showed that creams made with canola oil and HOSO organogel had very low oil loss compared to creams made with oil, demonstrating organogelation is an effective way to slow if not prevent oil migration in cream fillings.

A standard ice cream mix was prepared using 10% RBW in HOSO organogel to replace the solid fat content traditionally used in ice creams such as milk fat, or non-dairy sources such as palm kernel oil, palm oil, coconut oil, or hydrogenated oil. The RBW organogel was successfully emulsified in the ice cream mix and resulted in characteristics similar to that of crystallized fat droplets than of liquid oil. However, the sufficient structure was not developed to cause a delay in a structural collapse during meltdown [45].

Replacement of saturated fat with vegetable oil in comminuted meat products, such as frankfurters, resulted in products that were nearly 3x chewier and had an unacceptable rubbery texture [143, 182]. Frankfurters made with organogels of 10% EC in oil were not significantly different in hardness or chewiness to the beef fat controls [143, 178]. Pork fat in breakfast sausages was also replaced with EC organogels (~20% of the product); it was found the textural qualities of the pork fat control could be matched [183].

Organogels composed of a mixture of RBW, EC, and vegetable oil were used in a cream cheese product giving comparable spreadability, hardness, and had an approximate 25% reduction in total fat content than the full-fat control product [184]. Similarly, a mixture of organogelators in soybean oil was used to make a processed cheese product [185].

Successful application of GL-structured oils in the formulation of margarine and peanut butter has been reported [169, 172]. Emulsifier-free margarine produced with tigernut and cottonseed oils GPs displayed a smooth force-deformation curve like that of commercial margarine, indicative of plastic flow behavior. This was attributed to the thermal softening behavior of these systems, showing high SFC at 5 °C and gradual reduction of solids with an increase of temperature [169, 172]. The authors also suggested rice bran, olive, and HOCO glycerolized systems as other viable options for trans-free spread applications [172]. Similarly, peanut butter made with peanut oil GP presented high stability and comparable firmness to the commercial one [169].

3.5 Formulation Considerations

Whey protein isolate-xanthan gum complexes (WPXC) were used as a fat replacement to partially replace shortening (50 or 75 wt%) in cake frosting and sandwich cookie filling [186]. Response surface methodology was used to optimize the low-fat formulations based on WPXC viscosity and textural properties. WPXC cake frosting and sandwich cookie filling textural, rheological, and melting profiles similar to those of full-fat controls were achieved; however, there were difficulties with water activity for the latter application. The low-fat WPXC had a water activity (A_w) of 0.82 ± 0.02, this A_w is suitable for cake frostings; however, commercial sandwich cookie fillings had an A_w ranging 0.4–0.53 [186]. Since cookie biscuits are hygroscopic, expand with moisture uptake, and have an A_w between 0.2 and 0.6, the moisture would migrate to equilibrium, resulting in the separation of the cookie-cream interface, and a moist cookie without the desirable crisp texture [186, 187]. This is also a problem with cream fillings formulated with highly saturated hard stock fat and unsaturated oils, where migration of liquid oil leads to softening of the cookies and textural defects [178]. In the case of the cream with significant water content, humectants and bulking agents may be used to maintain a low A_w, and for a high-fat-content cream, a suitable structuring agent capable of slowing or preventing oil migration is required. In addition, there are textural differences among cake frostings and sandwich cookie fillings that need to be considered. A highly adhesive product is undesirable because it would have a sticky mouthfeel, a highly viscous product is difficult to pump and extrude during processing, and a fluid-like product would not be suitable to maintain its shape on the product [186]. Soft textures are optimal in the case of cake frostings in order to be spread thinly without breaking the cake crumb, whereas, firmer textures are more desirable for sandwich cookie fillings in order to prevent cookie misalignment, smearing, and shape maintenance, such that cream does not squeeze out when handled or bitten [186, 188]. It is also advantageous for cookie creams to rapidly solidify after spreading to prevent these defects during packaging, storage, and transportation [189].

Starch-lipid composites (SLC) have been a common replacement for the fat content in high-fat foods as the starch component imparts texture and viscosity and the lipid component provides the taste, melting properties, and mouthfeel characteristic of full-fat products [190]. SLC was used as a shortening replacement in cake icing; the SLC samples had hydrogenated fat contents of 1.2–12.8% fat and a constant 0.85 A_w, compared to the 21.5% fat content and 0.84–0.85 A_w of the shortening control. It was found that 16% and 24% SLC content can be used to prepare cake icings with as low as 6% fat content with similar characteristics as the shortening control. Melting behavior of the SLC icings was observed at higher temperatures than the shortening control icing, indicating that these low-fat icings would hold their shape in the mouth when consumed instead of melting like the shortening control [190]. This high-temperature melting behavior, although advantageous for shelf life, may result in a waxy mouthfeel.

3.6 Conclusions

Oleogels provide both environmental and health advantages to animal fats. Most successful food applications thus far involve only partial replacement of animal or highly saturated fats; however, a significant reduction in these fats can still be seen without significantly altering sensory properties. There is a gap in the laboratory structuring of these gels and the commercialization of them. Table 3.2 Summarizes

Table 3.2 Summary of promising direct and indirect methods to structure edible oils

Method	Key points	Limitations	Promising food applications
Wax oleogels	Low concentration required	Waxy taste Time-dependent change of rheological properties Sensitive to water in system Low firmness	Cookies [48, 119] Confectionary fillings [191] Ice cream [45] Cream cheese [184] Spreads and margarine [174, 176] Frying medium [121] Olive oil spread [118]
Fatty acids Fatty alcohols	Similar solid-state structure to TAGs High stability	High amount needed	Margarine [126] Peanut butter [192]
Monoglycerides	No restrictions as food additives Economic Forms crystalline network at low concentrations	Unstable Strength of network is limited compared to TAGs	Emulsifiers in many food products Cake batter [193] Low fat shortening [177] Frankfurter sausage [88]
Diglycerides	No restrictions as food additives Economic	Low crystallization rate Strength of network is limited compared to TAGs	Oleofoams [101–103]
Ethylcellulose	Easy manipulation of gel strength Serum TG lowering effect on oil	Synthetic ingredient Off flavor Limited concentrations permitted	Replacing animal fats in meat products [143, 147, 194] Ice cream [178] Cream cheese [144] Margarine [126] Heat-resistant chocolate [46, 47]
Emulsion template	Can use with biphasic food products Clean label Low gelation concentration	Protein oleogels tend to increase hardens of the final product Limited sensory studies reported	Butter [195]
Foam template			Cake [196] Muffins [197] Sandwich cookie cream [198] Stabilized Peanut Butter [199]

the promising oleogel strategies and their food applications. Further research needs to go into consumer acceptability, economic feasibility, and digestion of oleogels; however, they provide great promise for improving the health and environmental impacts of the food system for the future. On the other hand, enzymatic oil structuring through GL carries the added benefits associated with partial glyceride consumption and commercial viability. So far, GL structured oils are successfully utilized for margarine and peanut butter production. There is also great potential for these products to be used as shortening and as the lipid source in meat analogs.

References

1. Marangoni AG, Acevedo N, Maleky F et al (2012) Structure and functionality of edible fats. Soft Matter 8:1275–1300. https://doi.org/10.1039/C1SM06234D
2. Marangoni AG, van Duynhoven JPM, Acevedo NC et al (2020) Advances in our understanding of the structure and functionality of edible fats and fat mimetics. Soft Matter 16:289–306. https://doi.org/10.1039/C9SM01704F
3. Marangoni AG, Wesdorp L (2013) Fat crystal networks, 2nd edn. CRC Press, Boca Raton
4. Marangoni AG (2012) Structure-function analysis of edible fats. AOCS Press, Urbana
5. Marangoni AG (2018) Structure-function analysis of edible fats, 2nd edn. AOCS Press, Urbana
6. Schleifer D (2012) The perfect solution: how trans fat became the replacement for saturated fats. Technol Cult 53:94–119
7. Ascherio A, Rimm EB, Giovannucci EL et al (1996) Dietary fat and risk of coronary heart disease in men: cohort follow up study in the United States. Br Med J 313:84–90
8. McGee DL, Reed DM, Yano K et al (1984) Ten-year incidence of coronary heart disease in the Honolulu Heart Program: relationship to nutrient intake. Am J Epidemiol 119:667–676
9. Siri-Tarino PW, Sun Q, Hu FB, Krauss RM (2010) Meta-analysis of prospective cohort studies evaluating the association of saturated fat with cardiovascular disease. Am J Clin Nutr 91:535–546. https://doi.org/10.3945/ajcn.2009.27725
10. Harika RK, Eilander A, Alssema M et al (2013) Intake of fatty acids in general populations worldwide does not meet dietary recommendations to prevent coronary heart disease: a systematic review of data from 40 countries. Ann Nutr Metab 63:229–238. https://doi.org/10.1159/000355437
11. Keys A (1970) Coronary heart disease in seven countries. Circulation 41:186–195
12. Reiser R (1973) Saturated fat in the diet and serum cholesterol concentration: a critical examination of the literature. Am J Clin Nutr 26:524–555
13. Shekelle RB, Shryock AM, Paul O et al (1981) Diet, serum cholesterol, and death from coronary heart disease. N Engl J Med 304:65–70. https://doi.org/10.1056/NEJM198101083040201
14. Hu FB, Manson JAE, Willett WC (2001) Types of dietary fat and risk of coronary heart disease: a critical review. J Am Coll Nutr 20:5–19. https://doi.org/10.1080/07315724.2001.10719008
15. Blasbalg TL, Hibbeln JR, Ramsden CE et al (2011) Changes in consumption of omega-3 and omega-6 fatty acids in the United States during the 20th century. Am J Clin Nutr 93:950–962. https://doi.org/10.3945/ajcn.110.006643
16. AHA (2014) What your cholesterol levels mean. American Heart Association. http://www.heart.org/HEARTORG/Conditions/Cholesterol/AboutCholesterol/What-Your-Cholesterol-Levels-Mean_UCM_305562_Article.jsp. Accessed 4 Apr 2015

17. FDA (2015) Final determination regarding partially hydrogenated oils (Removing trans fat). http://www.fda.gov/Food/IngredientsPackagingLabeling/FoodAdditivesIngredients/ucm449162.htm. Accessed 16 June 2015
18. Brouwer IA, Wanders AJ, Katan MB (2010) Effect of animal and industrial trans fatty acids on HDL and LDL cholesterol levels in humans: a quantitative review. PLoS One 5:1–10. https://doi.org/10.1371/journal.pone.0009434
19. Jakobsen MU, Overvad K, Dyerberg J, Heitmann BL (2008) Intake of ruminant trans fatty acids and risk of coronary heart disease. Int J Epidemiol 37:173–182. https://doi.org/10.1093/ije/dym243
20. FDA (2003) Guidance for industry: trans fatty acids in nutrition labeling, nutrient content claims, health claimcs; small entity compliance guide. http://www.fda.gov/Food/GuidanceRegulation/GuidanceDocumentsRegulatoryInformation/LabelingNutrition/ucm053479.htm. Accessed 16 June 2015
21. Health Canada (2009) Archived Trans Fat Monitoring Program. http://www.hc-sc.gc.ca/fn-an/nutrition/gras-trans-fats/tfa-age_four-data_quatr-donn-eng.php. Accessed 8 Aug 2015
22. Uauy R, Aro A, Clarke R et al (2009) WHO Scientific Update on trans fatty acids: summary and conclusions. Eur J Clin Nutr 63:S68–S75. https://doi.org/10.1038/ejcn.2009.15
23. Mensink P, Katan M (1990) Effect of dietary trans fatty acids on high-density and low-density lipoprotein cholesterol levels in healthy subjects. N Engl J Med 323:638–639. https://doi.org/10.1056/NEJM199008163230703
24. Dhaka V, Gulia N, Singh K et al (2011) Trans fats-sources, health risks and alternative approach-a review. J Food Sci Technol 48:534–541. https://doi.org/10.1007/s13197-010-0225-8
25. Kritchevsky D (1998) History of recommendations to the public about dietary fat. J Nutr 128: 449S–452S
26. Keys A (1950) Physiology in diseases of the heart and lungs. Am J Public Health 40:346–347. https://doi.org/10.2105/AJPH.40.3.346-C
27. Mensink RP (2016) Effects of saturated fatty acids on serum lipids and lipoproteins: a systematic review and regression analysis. World Health Organization. https://apps.who.int/iris/handle/10665/246104
28. Forsythe CE, Phinney SD, Fernandez ML et al (2008) Comparison of low fat and low carbohydrate diets on circulating fatty acid composition and markers of inflammation. Lipids 43:65–77. https://doi.org/10.1007/s11745-007-3132-7
29. USDA (2015) Scientific Report of the 2015 Dietary Guidelines Advisory Committee. Part D. Chapter 6: cross-cutting topics of public health importance
30. Volek JS, Fernandez ML, Feinman RD, Phinney SD (2008) Dietary carbohydrate restriction induces a unique metabolic state positively affecting atherogenic dyslipidemia, fatty acid partitioning, and metabolic syndrome. Prog Lipid Res 47:307–318. https://doi.org/10.1016/j.plipres.2008.02.003
31. Co E, Marangoni AG (2012) Organogels: an alternative edible oil- structuring method. J Am Oil Chem Soc 89:749–780. https://doi.org/10.1007/s11746-012-2049-3
32. Parsons S, Raikova S, Chuck CJ (2020) The viability and desirability of replacing palm oil. Nat Sustain 3:412–418. https://doi.org/10.1038/s41893-020-0487-8
33. Beyer R, Rademacher T (2021) Species richness and carbon footprints of vegetable oils: can high yields outweigh palm oil's environmental impact? Sustainability 13:1813. https://doi.org/10.3390/SU13041813
34. Marangoni AG, Garti N (2011) Edible oleogels structure and health implications, 1st edn. AOCS Press, Urbana
35. Marangoni AG, Garti N (2018) Edible oleogels: structure and health implications, 2nd edn. Elsevier, Amsterdam
36. Patel AR (2017) Edible oil structuring: concepts, methods and applications. Royal Society of Chemistry, Cambridge

37. Rogers MA (2009) Novel structuring strategies for unsaturated fats – meeting the zero-trans, zero-saturated fat challenge: a review. Food Res Int 42:747–753. https://doi.org/10.1016/J.FOODRES.2009.02.024
38. Rogers MA, Strober T, Bot A et al (2014) Edible oleogels in molecular gastronomy. Int J Gastron Food Sci 2:22–31. https://doi.org/10.1016/j.ijgfs.2014.05.001
39. Acevedo NC, Marangoni AG (2010) Characterization of the nanoscale in triacylglycerol crystal networks. Cryst Growth Des 10:3327–3333. https://doi.org/10.1021/cg100468e
40. Acevedo NC, Marangoni AG (2010) Toward nanoscale engineering of triacylglycerol crystal networks. Cryst Growth Des 10:3334–3339. https://doi.org/10.1021/cg100469x
41. Acevedo NC, Marangoni AG (2015) Nanostructured fat crystal systems. Annu Rev Food Sci Technol 6:71–96. https://doi.org/10.1146/annurev-food-030713-092400
42. Patel AR, Nicholson RA, Marangoni AG (2020) Applications of fat mimetics for the replacement of saturated and hydrogenated fat in food products. Curr Opin Food Sci 33:61–68. https://doi.org/10.1016/J.COFS.2019.12.008
43. Patel AR (2015) Alternative routes to oil structuring, 1st edn. Springer International Publishing, New York
44. Zetzl AK, Marangoni AG (2014) Structured emulsions and edible oleogels as solutions to trans fat. In: Kodali DR (ed) Trans fats replacement solutions. AOCS Press, Urbana, pp 215–243
45. Zulim Botega DC, Marangoni AG, Smith AK, Goff HD (2013) The potential application of rice bran wax oleogel to replace solid fat and enhance unsaturated fat content in ice cream. J Food Sci 78:1334–1339. https://doi.org/10.1111/1750-3841.12175
46. Stortz TA, Marangoni AG (2013) Ethylcellulose solvent substitution method of preparing heat resistant chocolate. Food Res Int 51:797–803. https://doi.org/10.1016/J.FOODRES.2013.01.059
47. Stortz TA, de Moura DC, Laredo T, Marangoni AG (2014) Molecular interactions of ethylcellulose with sucrose particles. RSC Adv 4:55048–55061. https://doi.org/10.1039/C4RA12010H
48. Yılmaz E, Öğütcü M (2015b) The texture, sensory properties and stability of cookies prepared with wax oleogels. Food Funct 6:1194–1204. https://doi.org/10.1039/C5FO00019J
49. Hughes NE, Marangoni AG, Wright AJ et al (2009) Potential food applications of edible oil organogels. Trends Food Sci Technol 20:470–480. https://doi.org/10.1016/J.TIFS.2009.06.002
50. O'Sullivan CM, Barbut S, Marangoni AG (2016) Edible oleogels for the oral delivery of lipid soluble molecules: composition and structural design considerations. Trends Food Sci Technol 57:59–73. https://doi.org/10.1016/J.TIFS.2016.08.018
51. O'sullivan CM, Davidovich-Pinhas M, Wright AJ et al (2017) Ethylcellulose oleogels for lipophilic bioactive delivery – effect of oleogelation on in vitro bioaccessibility and stability of beta-carotene. Food Funct 8:1438–1451. https://doi.org/10.1039/C6FO01805J
52. Tan SY, Wan-Yi Peh E, Marangoni AG, Henry CJ (2017) Effects of liquid oil vs. oleogel co-ingested with a carbohydrate-rich meal on human blood triglycerides, glucose, insulin and appetite. Food Funct 8:241–249. https://doi.org/10.1039/C6FO01274D
53. Rush JWE, Jantzi PS, Dupak K et al (2008) Effect of food preparation on the structure and metabolic responses to a monostearin–oil–water gel-based spread. Food Res Int 41:1065–1071. https://doi.org/10.1016/J.FOODRES.2008.07.017
54. Rush JWE, Jantzi PS, Dupak K et al (2009) Acute metabolic responses to butter, margarine, and a monoglyceride gel-structured spread. Food Res Int 42:1034–1039. https://doi.org/10.1016/J.FOODRES.2009.04.013
55. Limpimwong W, Kumrungsee T, Kato N et al (2017) Rice bran wax oleogel: a potential margarine replacement and its digestibility effect in rats fed a high-fat diet. J Funct Foods 39:250–256. https://doi.org/10.1016/J.JFF.2017.10.035
56. Ojijo NKO, Kesselman E, Shuster V et al (2004) Changes in microstructural, thermal, and rheological properties of olive oil/monoglyceride networks during storage. Food Res Int 37:385–393. https://doi.org/10.1016/J.FOODRES.2004.02.003

57. Ojijo NKO, Neeman I, Eger S, Shimoni E (2004) Effects of monoglyceride content, cooling rate and shear on the rheological properties of olive oil/monoglyceride gel networks. J Sci Food Agric 84:1585–1593. https://doi.org/10.1002/JSFA.1831
58. Marangoni AG, Idziak SHJ, Vega C et al (2007) Encapsulation-structuring of edible oil attenuates acute elevation of blood lipids and insulin in humans. Soft Matter 3:183. https://doi.org/10.1039/b611985a
59. Kondo H, Hase T, Murase T, Tokimitsu I (2003) Digestion and assimilation features of dietary dag in the rat small intestine. Lipids 38:25–30. https://doi.org/10.1007/S11745-003-1027-7/METRICS
60. Osaki N, Meguro S, Yajima N et al (2005) Metabolites of dietary triacylglycerol and diacylglycerol during the digestion process in rats. Lipids 40:281–286. https://doi.org/10.1007/S11745-005-1383-3
61. Toro-Vazquez JF, Aguilar-Zárate M, López-Martinez A, Charó-Alonso M (2022) Structuring vegetable oils with monoglycerides and monoglyceride-lecithin or monoglyceride-ethylcellulose mixtures. In: Food chemistry, function and analysis. Royal Society of Chemistry, Cambridge, pp 201–234
62. Nicholson RA, Marangoni AG (2022) Glycerolysis-structured lipid systems. In: Toro-Vazquez JF (ed) Development of trans-free lipid systems and their use in food products, 1st edn. Royal Society of Chemistry, Cambridge, pp 1–362
63. Palla CA, Dominguez M, Carrín ME (2022) An overview of structure engineering to tailor the functionality of monoglyceride oleogels. Compr Rev Food Sci Food Saf 21:2587–2614. https://doi.org/10.1111/1541-4337.12930
64. Mao L, Calligaris S, Barba L, Miao S (2014) Monoglyceride self-assembled structure in O/W emulsion: formation, characterization and its effect on emulsion properties. Food Res Int 58: 81–88. https://doi.org/10.1016/J.FOODRES.2014.01.042
65. Chen C-H, Terentjev EM (2018) Monoglycerides in oils. In: Edible oleogels. Elsevier, Cambridge, pp 103–131
66. Palla C, de Vicente J, Carrín ME, Gálvez Ruiz MJ (2019) Effects of cooling temperature profiles on the monoglycerides oleogel properties: a rheo-microscopy study. Food Res Int 125: 108613. https://doi.org/10.1016/J.FOODRES.2019.108613
67. Giacomozzi AS, Carrín ME, Palla CA (2018) Muffins elaborated with optimized monoglycerides oleogels: from solid fat replacer obtention to product quality evaluation. J Food Sci 83: 1505–1515. https://doi.org/10.1111/1750-3841.14174
68. Giacomozzi AS, Palla CA, Carrín ME, Martini S (2019) Physical properties of monoglycerides oleogels modified by concentration, cooling rate, and high-intensity ultrasound. J Food Sci 84:2549–2561. https://doi.org/10.1111/1750-3841.14762
69. Giacomozzi A, Palla C, Carrín ME, Martini S (2020) Tailoring physical properties of monoglycerides oleogels using high-intensity ultrasound. Food Res Int 134:109231. https://doi.org/10.1016/J.FOODRES.2020.109231
70. Giacomozzi AS, Carrín ME, Palla CA (2021) Storage stability of oleogels made from monoglycerides and high oleic sunflower oil. Food Biophys 16:306–316. https://doi.org/10.1007/S11483-020-09661-9/FIGURES/4
71. Li J, Guo R, Bi Y et al (2021) Comprehensive evaluation of saturated monoglycerides for the forming of oleogels. LWT 151:112061. https://doi.org/10.1016/J.LWT.2021.112061
72. Ferro AC, Okuro PK, Badan AP, Cunha RL (2019) Role of the oil on glyceryl monostearate based oleogels. Food Res Int 120:610–619. https://doi.org/10.1016/J.FOODRES.2018.11.013
73. Cerqueira MA, Fasolin LH, Picone CSF et al (2017) Structural and mechanical properties of organogels: role of oil and gelator molecular structure. Food Res Int 96:161–170. https://doi.org/10.1016/J.FOODRES.2017.03.021
74. Monié A, Franceschi S, Balayssac S et al (2022) Study of rapeseed oil gelation induced by commercial monoglycerides using a chemometric approach. Food Chem 369:130870. https://doi.org/10.1016/J.FOODCHEM.2021.130870

75. Barroso NG, Okuro PK, Ribeiro APB, Cunha RL (2020) Tailoring properties of mixed-component oleogels: wax and monoglyceride interactions towards flaxseed oil structuring. Gels 6:5. https://doi.org/10.3390/GELS6010005
76. Rosen-Kligvasser J, Davidovich-Pinhas M (2021) The role of hydrogen bonds in TAG derivative-based oleogel structure and properties. Food Chem 334:127585. https://doi.org/10.1016/J.FOODCHEM.2020.127585
77. Golodnizky D, Rosen-Kligvasser J, Davidovich-Pinhas M (2021) The role of the polar head group and aliphatic tail in the self-assembly of low molecular weight molecules in oil. Food Struct 30:100240. https://doi.org/10.1016/J.FOOSTR.2021.100240
78. Lopez-Martínez A, Charó-Alonso MA, Marangoni AG, Toro-Vazquez JF (2015) Monoglyceride organogels developed in vegetable oil with and without ethylcellulose. Food Res Int 72: 37–46. https://doi.org/10.1016/J.FOODRES.2015.03.019
79. Chen CH, Terentjev EM (2009) Aging and metastability of monoglycerides in hydrophobic solutions. Langmuir 25:6717–6724. https://doi.org/10.1021/LA9002065/ASSET/IMAGES/MEDIUM/LA-2009-002065_0001.GIF
80. Chen CH, Terentjev EM (2011) Colloid–monoglyceride composites in hydrophobic solutions. Colloids Surf A Physicochem Eng Asp 384:536–542. https://doi.org/10.1016/J.COLSURFA.2011.05.020
81. Kesselman E, Shimoni E (2007) Imaging of oil/monoglyceride networks by polarizing near-field scanning optical microscopy. Food Biophys 2:117–123. https://doi.org/10.1007/S11483-007-9038-3/TABLES/1
82. da Pieve S, Calligaris S, Panozzo A et al (2011) Effect of monoglyceride organogel structure on cod liver oil stability. Food Res Int 44:2978–2983. https://doi.org/10.1016/J.FOODRES.2011.07.011
83. Pan J, Tang L, Dong Q et al (2021) Effect of oleogelation on physical properties and oxidative stability of camellia oil-based oleogels and oleogel emulsions. Food Res Int 140:110057. https://doi.org/10.1016/J.FOODRES.2020.110057
84. Valoppi F, Calligaris S, Barba L et al (2017) Influence of oil type on formation, structure, thermal, and physical properties of monoglyceride-based organogel. Eur J Lipid Sci Technol 119. https://doi.org/10.1002/ejlt.201500549
85. Krog N (1997) Association of emulsifiers in aqueous systems. In: Dickinson E, Bergenstahl D (eds) Food colloids: proteins, lipids and polysaccharides. Google Books
86. Wang FC, Marangoni AG (2016) Advances in the application of food emulsifier α-gel phases: saturated monoglycerides, polyglycerol fatty acid esters, and their derivatives. J Colloid Interface Sci 483:394–403. https://doi.org/10.1016/J.JCIS.2016.08.012
87. Bin Sintang MD, Rimaux T, van de Walle D et al (2017) Oil structuring properties of monoglycerides and phytosterols mixtures. Eur J Lipid Sci Technol 119:1500517. https://doi.org/10.1002/EJLT.201500517
88. Kouzounis D, Lazaridou A, Katsanidis E (2017) Partial replacement of animal fat by oleogels structured with monoglycerides and phytosterols in frankfurter sausages. Meat Sci 130:38–46. https://doi.org/10.1016/J.MEATSCI.2017.04.004
89. Truong T, Prakash S, Bhandari B (2019) Effects of crystallisation of native phytosterols and monoacylglycerols on foaming properties of whipped oleogels. Food Chem 285:86–93. https://doi.org/10.1016/J.FOODCHEM.2019.01.134
90. Choi KO, Hwang HS, Jeong S et al (2020) The thermal, rheological, and structural characterization of grapeseed oil oleogels structured with binary blends of oleogelator. J Food Sci 85: 3432–3441. https://doi.org/10.1111/1750-3841.15442
91. Chen C, Zhang C, Zhang Q et al (2021) Study of monoglycerides enriched with unsaturated fatty acids at sn-2 position as oleogelators for oleogel preparation. Food Chem 354:129534. https://doi.org/10.1016/J.FOODCHEM.2021.129534
92. da Silva TLT, Danthine S (2021) Effect of high-intensity ultrasound on the oleogelation and physical properties of high melting point monoglycerides and triglycerides oleogels. J Food Sci 86:343–356. https://doi.org/10.1111/1750-3841.15589

93. Aguilar-Zárate M, de la Peña-Gil A, Álvarez-Mitre FM et al (2019) Vegetable and mineral oil organogels based on monoglyceride and lecithin mixtures. Food Biophys 14:326–345. https://doi.org/10.1007/S11483-019-09583-1/FIGURES/13
94. Rodríguez-Hernández AK, Pérez-Martínez JD, Gallegos-Infante JA et al (2021) Rheological properties of ethyl cellulose-monoglyceride-candelilla wax oleogel vis-a-vis edible shortenings. Carbohydr Polym 252:117171. https://doi.org/10.1016/J.CARBPOL.2020.117171
95. Palla C, Giacomozzi A, Genovese DB, Carrín ME (2017) Multi–objective optimization of high oleic sunflower oil and monoglycerides oleogels: searching for rheological and textural properties similar to margarine. Food Struct 12:1–14. https://doi.org/10.1016/j.foostr.2017.02.005
96. da Pieve S, Calligaris S, Co E et al (2010) Shear nanostructuring of monoglyceride organogels. Food Biophys 5:211–217. https://doi.org/10.1007/S11483-010-9162-3/FIGURES/4
97. Calligaris S, da Pieve S, Arrighetti G, Barba L (2010) Effect of the structure of monoglyceride–oil–water gels on aroma partition. Food Res Int 43:671–677. https://doi.org/10.1016/J.FOODRES.2009.10.011
98. Tavernier I, Moens K, Heyman B et al (2019) Relating crystallization behavior of monoacylglycerols-diacylglycerol mixtures to the strength of their crystalline network in oil. Food Res Int 120:504–513. https://doi.org/10.1016/j.foodres.2018.10.092
99. Xu Y, Wei C, Zhao X et al (2016) A comparative study on microstructure, texture, rheology, and crystallization kinetics of palm-based diacylglycerol oils and corresponding palm-based oils. Eur J Lipid Sci Technol 118:1179–1192. https://doi.org/10.1002/EJLT.201500369
100. Wang X, Ma D, Liu Y et al (2022) Physical properties of oleogels fabricated by the combination of diacylglycerols and monoacylglycerols. J Am Oil Chem Soc. https://doi.org/10.1002/aocs.12622
101. Qiu C, Lei M, Lee WJ et al (2021) Fabrication and characterization of stable oleofoam based on medium-long chain diacylglycerol and β-sitosterol. Food Chem 350. https://doi.org/10.1016/j.foodchem.2021.129275
102. Lei M, Zhang N, Lee WJ et al (2020) Non-aqueous foams formed by whipping diacylglycerol stabilized oleogel. Food Chem 312. https://doi.org/10.1016/j.foodchem.2019.126047
103. Qiu C, Wang S, Wang Y et al (2022) Stabilisation of oleofoams by lauric acid and its glycerol esters. Food Chem 386. https://doi.org/10.1016/j.foodchem.2022.132776
104. Dassanayake LSK, Kodali DR, Ueno S, Sato K (2009) Physical properties of rice bran wax in bulk and organogels. J Am Oil Chem Soc 86:1163–1173. https://doi.org/10.1007/s11746-009-1464-6
105. Hwang HS, Kim S, Singh M et al (2012) Organogel formation of soybean oil with waxes. J Am Oil Chem Soc 89:639–647. https://doi.org/10.1007/s11746-011-1953-2
106. Toro-Vazquez JF, Morales-Rueda JA, Dibildox-Alvarado E et al (2007) Thermal and textural properties of organogels developed by candelilla wax in safflower oil. J Am Oil Chem Soc 84: 989–1000. https://doi.org/10.1007/s11746-007-1139-0
107. Rocha JCB, Lopes JD, Mascarenhas MCN et al (2013) Thermal and rheological properties of organogels formed by sugarcane or candelilla wax in soybean oil. Food Res Int 50:318–323. https://doi.org/10.1016/j.foodres.2012.10.043
108. Patel AR, Dewettinck K (2015) Comparative evaluation of structured oil systems: shellac oleogel, HPMC oleogel, and HIPE gel. Eur J Lipid Sci Technol 117:1772–1781. https://doi.org/10.1002/EJLT.201400553
109. Öğütcü M, Arifoğlu N, Yılmaz E (2015) Preparation and characterization of virgin olive oil-beeswax oleogel emulsion products. J Am Oil Chem Soc 92:459–471. https://doi.org/10.1007/s11746-015-2615-6
110. Blake AI, Toro-Vazquez JF, Hwang H-S (2018) Wax oleogels structure and health implications. In: Marangoni AG, Garti N (eds) Edible oleogels, 2nd edn. AOCS Press, Urbana, pp 133–171

111. Martini AS, Carelli AA, Lee J (2008) Effect of the addition of waxes on the crystallization behavior of anhydrous milk fat. J Am Oil Chem Soc 85:1097–1104. https://doi.org/10.1007/s11746-008-1310-2
112. Doan CD, Tavernier I, Okuro PK, Dewettinck K (2018) Internal and external factors affecting the crystallization, gelation and applicability of wax-based oleogels in food industry. Innovative Food Sci Emerg Technol 45:42–52. https://doi.org/10.1016/J.IFSET.2017.09.023
113. Blake AI, Co ED, Marangoni AG (2014) Structure and physical properties of plant wax crystal networks and their relationship to oil binding capacity. J Am Oil Chem Soc 91:885–903. https://doi.org/10.1007/s11746-014-2435-0
114. Jana S, Martini S (2016) Phase behavior of binary blends of four different waxes DSC differential scanning calorimetry PLM polarized light microscopy. J Am Oil Chem Soc 93: 543–554. https://doi.org/10.1007/s11746-016-2789-6
115. Brykczynski H, Wettlaufer T, Flöter E (2022) Revisiting pure component wax esters as basis of wax-based oleogels. J Am Oil Chem Soc. https://doi.org/10.1002/AOCS.12589
116. Kerr RM, Tombokan X, Ghosh S, Martini S (2011) Crystallization behavior of anhydrous milk fat-sunflower oil wax blends. J Agric Food Chem 59:2689–2695. https://doi.org/10.1021/jf1046046
117. Ramírez-Gómez NO, Acevedo NC, Toro-Vázquez JF et al (2016) Phase behavior, structure and rheology of candelilla wax/fully hydrogenated soybean oil mixtures with and without vegetable oil. Food Res Int 89:828–837. https://doi.org/10.1016/J.FOODRES.2016.10.025
118. Ghazani SM, Dobson S, Marangoni AG (2022) Hardness, plasticity, and oil binding capacity of binary mixtures of natural waxes in olive oil. Curr Res Food Sci 5:998–1008. https://doi.org/10.1016/J.CRFS.2022.06.002
119. Hwang HS, Singh M, Winkler-Moser JK et al (2014) Preparation of margarines from organogels of sunflower wax and vegetable oils. J Food Sci 79:C1926–C1932. https://doi.org/10.1111/1750-3841.12596
120. Yılmaz E, Öğütcü M (2015) Oleogels as spreadable fat and butter alternatives: sensory description and consumer perception. RSC Adv 5:50259–50267. https://doi.org/10.1039/C5RA06689A
121. Lim J, Jeong S, Oh IK, Lee S (2017) Evaluation of soybean oil-carnauba wax oleogels as an alternative to high saturated fat frying media for instant fried noodles. LWT 84:788–794. https://doi.org/10.1016/J.LWT.2017.06.054
122. Núñez-García IC, Rodríguez-Flores LG, Guadiana-De-Dios MH et al (2022) Candelilla wax extracted by traditional method and an ecofriendly process: assessment of its chemical, structural and thermal properties. Molecules 27:3735. https://doi.org/10.3390/MOLECULES27123735
123. Soleimanian Y, Goli SAH, Shirvani A et al (2020) Wax-based delivery systems: preparation, characterization, and food applications. Compr Rev Food Sci Food Saf 19:2994–3030. https://doi.org/10.1111/1541-4337.12614
124. Blach C, Gravelle AJ, Peyronel F et al (2016) Revisiting the crystallization behavior of stearyl alcohol: stearic acid (SO: SA) mixtures in edible oil. RSC Adv 6:81151–81163. https://doi.org/10.1039/c6ra15142f
125. Gandolfo FG, Bot A, Flöter E (2004) Structuring of edible oils by long-chain FA, fatty alcohols, and their mixtures. J Am Oil Chem Soc 81:1–6
126. Gravelle AJ, Davidovich-Pinhas M, Barbut S, Marangoni AG (2017) Influencing the crystallization behavior of binary mixtures of stearyl alcohol and stearic acid (SOSA) using ethylcellulose. Food Res Int 91:1–10. https://doi.org/10.1016/J.FOODRES.2016.11.024
127. Bot A, Flöter E (2018) Structuring edible oil phases with fatty acids and alcohols. In: Patel AR (ed) Edible oil structuring: concepts, methods and applications. Royal Society of Chemistry, Cambridge, pp 95–105
128. Schaink HM, van Malssen KF, Morgado-Alves S et al (2007) Crystal network for edible oil organogels: possibilities and limitations of the fatty acid and fatty alcohol systems. Food Res Int 40:1185–1193. https://doi.org/10.1016/J.FOODRES.2007.06.013

129. Nikiforidis CV, Scholten E (2013) Self-assemblies of lecithin and α-tocopherol as gelators of lipid material. RSC Adv 4:2466–2473. https://doi.org/10.1039/C3RA46584E
130. Bot A, Ruud AE, Ae A, Roijers EC (2008) Fibrils of γ-oryzanol + ß-sitosterol in edible oil organogels. J Am Oil Chem Soc 85:1127–1134. https://doi.org/10.1007/s11746-008-1298-7
131. Gaudino N, Ghazani SM, Clark S et al (2019) Development of lecithin and stearic acid based oleogels and oleogel emulsions for edible semisolid applications. Food Res Int 116:79–89. https://doi.org/10.1016/J.FOODRES.2018.12.021
132. Patel AR (2019) Oleogelation for food structuring based on synergistic interactions among food components. In: Encyclopedia of food chemistry, pp 715–718. https://doi.org/10.1016/B978-0-08-100596-5.21520-5
133. Batte HD, Wright AJ, Rush JW et al (2007) Effect of processing conditions on the structure of monostearin-oil-water gels. Food Res Int 40:982–988. https://doi.org/10.1016/J.FOODRES.2007.05.001
134. Batte HD, Wright AJ, Rush JW et al (2007) Phase behavior, stability, and mesomorphism of monostearin-oil-water gels. https://doi.org/10.1007/s11483-007-9026-7
135. Huschka B, Challacombe C, Marangoni AG, Seetharaman K (2011) Comparison of oil, shortening, and a structured shortening on wheat dough rheology and starch pasting properties. Cereal Chem 88:253–259. https://doi.org/10.1094/CCHEM-03-10-0041
136. Patel AR, Schatteman D, De Vos WH et al (2013) Preparation and rheological characterization of shellac oleogels and oleogel-based emulsions. J Colloid Interface Sci 411:114–121. https://doi.org/10.1016/j.jcis.2013.08.039
137. Patel AR, Rodriguez Y, Lesaffer A, Dewettinck K (2014) High internal phase emulsion gels (HIPE-gels) prepared using food-grade components. RSC Adv 4:18136. https://doi.org/10.1039/c4ra02119c
138. Patel AR, Mankoč B, Bin Sintang MD et al (2015) Fumed silica-based organogels and 'aqueous-organic' bigels. RSC Adv 5:9703–9708. https://doi.org/10.1039/C4RA15437A
139. Koch W (1937) Properties and uses of ethylcellulose. Ind Eng Chem 29:687–690
140. Gravelle AJ, Marangoni AG (2018) Ethylcellulose oleogels: structure, functionality, and food applications. In: Toldrá F (ed) Advances in food and nutrition research. Academic Press, Cambridge, pp 1–56
141. Davidovich-Pinhas M, Barbut S, Marangoni AG (2015) The gelation of oil using ethyl cellulose. Carbohydr Polym 117:869–878. https://doi.org/10.1016/J.CARBPOL.2014.10.035
142. Gravelle AJ, Barbut S, Quinton M, Marangoni AG (2014) Towards the development of a predictive model of the formulation-dependent mechanical behaviour of edible oil-based ethylcellulose oleogels. J Food Eng 143:114–122. https://doi.org/10.1016/j.jfoodeng.2014.06.036
143. Zetzl AK, Marangoni AG, Barbut S (2012) Mechanical properties of ethylcellulose oleogels and their potential for saturated fat reduction in frankfurters. Food Funct 3:327–337. https://doi.org/10.1039/c2fo10202a
144. Bemer HL, Limbaugh M, Cramer ED et al (2016) Vegetable organogels incorporation in cream cheese products. Food Res Int 85:67–75. https://doi.org/10.1016/J.FOODRES.2016.04.016
145. Laredo T, Barbut S, Marangoni AG (2011) Molecular interactions of polymer oleogelation. Soft Matter 7:2734. https://doi.org/10.1039/c0sm00885k
146. Gravelle AJ, Barbut S, Marangoni AG (2012) Ethylcellulose oleogels: manufacturing considerations and effects of oil oxidation. Food Res Int 48:578–583. https://doi.org/10.1016/J.FOODRES.2012.05.020
147. Barbut S, Marangoni A (2019) Organogels use in meat processing – effects of fat/oil type and heating rate. Meat Sci 149:9–13. https://doi.org/10.1016/J.MEATSCI.2018.11.003
148. Davidovich-Pinhas M, Gravelle AJ, Barbut S, Marangoni AG (2015) Temperature effects on the gelation of ethylcellulose oleogels. Food Hydrocoll 46:76–83. https://doi.org/10.1016/j.foodhyd.2014.12.030

149. Davidovich-Pinhas M, Barbut S, Marangoni AG (2016) Development, characterization, and utilization of food-grade polymer oleogels. Annu Rev Food Sci Technol 7:65–91. https://doi.org/10.1146/annurev-food-041715-033225
150. Barbut S, Wood J, Marangoni A (2016) Quality effects of using organogels in breakfast sausage. Meat Sci 122:84–89. https://doi.org/10.1016/J.MEATSCI.2016.07.022
151. Barbut S, Tiensa BE, Marangoni AG (2021) Partial fat replacement in liver pate using canola oil organogel. LWT 139:110428. https://doi.org/10.1016/J.LWT.2020.110428
152. Woern C, Marangoni AG, Weiss J, Barbut S (2021) Effects of partially replacing animal fat by ethylcellulose based organogels in ground cooked salami and frankfurters. Food Res Int 143–147. https://doi.org/10.1016/J.FOODRES.2021.110431
153. Barbut S, Wood J, Marangoni AG (2016) Effects of organogel hardness and formulation on acceptance of frankfurters. J Food Sci 81:183–188. https://doi.org/10.1111/1750-3841.13409
154. Mattice KD, Marangoni AG (2018) Edible applications of ethylcellulose oleogels. In: Patel AR (ed) Edible oil structuring: concepts, methods and applications. Royal Society of Chemistry, Cambridge, pp 250–274
155. Asioli D, Aschemann-Witzel J, Caputo V et al (2017) Making sense of the "clean label" trends: a review of consumer food choice behavior and discussion of industry implications. Food Res Int 99:58–71. https://doi.org/10.1016/J.FOODRES.2017.07.022
156. Scharfe M, Flöter E (2020) Oleogelation: from scientific feasibility to applicability in food products. Eur J Lipid Sci Technol 122:2000213. https://doi.org/10.1002/EJLT.202000213
157. Mezzenga R, Ulrich S (2010) Spray-dried oil powder with ultrahigh oil content. Langmuir 26: 16658–16661. https://doi.org/10.1021/LA103447N/SUPPL_FILE/LA103447N_SI_001.PDF
158. Patel AR, Cludts N, Bin Sintang MD et al (2014) Polysaccharide-based oleogels prepared with an emulsion-templated approach. ChemPhysChem 15:3435–3439. https://doi.org/10.1002/cphc.201402473
159. Patel AR, Rajarethinem PS, Cludts N et al (2015) Biopolymer-based structuring of liquid oil into soft solids and oleogels using water-continuous emulsions as templates. Langmuir 31: 2065–2073
160. Chen X-W, Yang X-Q (2019) Characterization of orange oil powders and oleogels fabricated from emulsion templates stabilized solely by a natural triterpene saponin. J Agric Food Chem 67:2637–2646. https://doi.org/10.1021/acs.jafc.8b04588
161. Qiu C, Huang Y, Li A et al (2018) Fabrication and characterization of oleogel stabilized by gelatin-polyphenol-polysaccharides nanocomplexes. J Agric Food Chem 66:13243–13252. https://doi.org/10.1021/ACS.JAFC.8B02039/ASSET/IMAGES/LARGE/JF-2018-02039S_0010.JPEG
162. Feichtinger A, Scholten E (2020) Preparation of protein oleogels: effect on structure and functionality. Foods 9:1745. https://doi.org/10.3390/foods9121745
163. Patel AR, Schatteman D, Lesaffer A, Dewettinck K (2013) A foam-templated approach for fabricating organogels using a water-soluble polymer. RSC Adv 3:22900. https://doi.org/10.1039/c3ra44763d
164. Alessandro M, Nobile D, Conte A et al (2009) New strategies for reducing the pork back-fat content in typical Italian salami. Meat Sci 81:263–269. https://doi.org/10.1016/j.meatsci.2008.07.026
165. Korhonen JT, Kettunen M, Ras RHA, Ikkala O (2011) Hydrophobic nanocellulose aerogels as floating, sustainable, reusable, and recyclable oil absorbents. ACS Appl Mater Interfaces 3: 1813–1816. https://doi.org/10.1021/am200475b
166. Jiang Q, Du L, Li S et al (2021) Polysaccharide-stabilized aqueous foams to fabricate highly oil-absorbing cryogels: application and formation process for preparation of edible oleogels. Food Hydrocoll 120:106901. https://doi.org/10.1016/J.FOODHYD.2021.106901
167. Heymans R, Tavernier I, Dewettinck K, van der Meeren P (2017) Crystal stabilization of edible oil foams. Trends Food Sci Technol 69:13–24. https://doi.org/10.1016/J.TIFS.2017.08.015

168. Zhi Z, Dewettinck K, van Bockstaele F (2022) Alternative oil structuring techniques: oil powders, double emulsions and oil foams. In: Toro-Vazquez JF (ed) Development of trans-free lipid systems and their use in food products. Royal Society of Chemistry, Cambridge, pp 21–52
169. Nicholson RA, Marangoni AG (2020) Enzymatic glycerolysis converts vegetable oils into structural fats with the potential to replace palm oil in food products. Nat Food 1(11):684–692. https://doi.org/10.1038/s43016-020-00160-1
170. Nicholson RA, Marangoni AG (2022) Glycerolysis structured oils as natural fat replacements. Curr Opin Food Sci 43:1–6. https://doi.org/10.1016/J.COFS.2021.09.002
171. Lo SK, Tan CP, Long K et al (2008) Diacylglycerol oil-properties, processes and products: a review. Food Bioprocess Technol 1:223–233. https://doi.org/10.1007/s11947-007-0049-3
172. Nicholson RA, Marangoni AG (2021) Lipase-catalyzed glycerolysis extended to the conversion of a variety of edible oils into structural fats. Curr Res Food Sci 4:163–174. https://doi.org/10.1016/J.CRFS.2021.03.005
173. Corzo-Martínez M, Vázquez L, Arranz-Martínez P et al (2016) Production of a bioactive lipid-based delivery system from ratfish liver oil by enzymatic glycerolysis. Food Bioprod Process 100:311–322. https://doi.org/10.1016/J.FBP.2016.08.003
174. Hwang HS, Singh M, Bakota EL et al (2013) Margarine from organogels of plant wax and soybean oil. J Am Oil Chem Soc 90:1705–1712. https://doi.org/10.1007/s11746-013-2315-z
175. Öğütcü M, Yilmaz E (2014) Oleogels of virgin olive oil with carnauba wax and monoglyceride as spreadable products. Grasas Aceites 65:e040
176. Patel AR, Rajarethinem P, Grędowska A et al (2014) Edible applications of shellac oleogels: spreads, chocolate paste and cakes. Food Funct 5:645–652. https://doi.org/10.1039/c4fo00034j
177. Goldstein A, Seetharaman K (2011) Effect of a novel monoglyceride stabilized oil in water emulsion shortening on cookie properties. Food Res Int 44:1476–1481. https://doi.org/10.1016/j.foodres.2011.03.029
178. Stortz TA, Zetzl AK, Barbut S et al (2012) Edible oleogels in food products to help maximize health benefits and improve nutritional profiles. Lipid Technol 24:151–154. https://doi.org/10.1002/lite.201200205
179. Jang A, Bae W, Hwang H-S et al (2015) Evaluation of canola oil oleogels with candelilla wax as an alternative to shortening in baked goods. Food Chem 187:525–529. https://doi.org/10.1016/j.foodchem.2015.04.110
180. Calligaris S, Manzocco L, Valoppi F, Nicoli MC (2013) Effect of palm oil replacement with monoglyceride organogel and hydrogel on sweet bread properties. Food Res Int 51:596–602. https://doi.org/10.1016/j.foodres.2013.01.007
181. Patel AR, Cludts N, Bin SMD et al (2014) Edible oleogels based on water soluble food polymers: preparation, characterization and potential application. Food Funct 5:2833–2841. https://doi.org/10.1039/c4fo00624k
182. Youssef MK, Barbut S (2010) Physicochemical effects of the lipid phase and protein level on meat emulsion stability, texture, and microstructure. J Food Sci 75:S108–S114. https://doi.org/10.1111/j.1750-3841.2009.01475.x
183. Wood JM (2013) Reduction of saturated fat in finely comminuted and ground meat products by use of canola oil organogels and the effect on organoleptic qualities, texture and microstructure. University of Guelph
184. Limbaugh MD (2015) Organogellators as fat replacement in cream cheese products. In: Oral presentation presented at the 106th American Oil Chemist's Society annual meeting and industry showcases, Orlando, FL
185. Huang H, Hallinan R, Maleky F (2018) Comparison of different oleogels in processed cheese products formulation. Int J Food Sci Technol 53:2525–2534. https://doi.org/10.1111/IJFS.13846

186. Laneuville SI, Paquin P, Turgeon SL (2005) Formula optimization of a low-fat food system containing whey protein isolate – xanthan gum complexes as fat replacer. J Food Sci 70:513–519
187. Battaiotto LL, Lupano CE, Bevilacqua AE (2013) Optimization of basic ingredient combination for sandwich cookie filling using response surface methodology. Food Bioprocess Technol 6:1847–1855. https://doi.org/10.1007/s11947-012-0853-2
188. Kanagaratnam S, Mansor I, Sahri MM et al (2009) Palm-based trans fatty acid-free biscuit cream fat. Malaysian Palm Oil Board
189. Shamsudin SY (2009) Non-lauric fats for cream filling. Malaysian Palm Oil Board
190. Singh M, Byars JA (2011) Jet-cooked high amylose corn starch and shortening composites for use in cake icings. J Food Sci 76:530–535. https://doi.org/10.1111/j.1750-3841.2011.02364.x
191. Doan CD, Patel AR, Tavernier I et al (2016) The feasibility of wax-based oleogel as a potential co-structurant with palm oil in low-saturated fat confectionery fillings. Eur J Lipid Sci Technol 118:1903–1914. https://doi.org/10.1002/ejlt.201500172
192. Valoppi F, Calligaris S, Marangoni AG (2016) Structure and physical properties of oleogels containing peanut oil and saturated fatty alcohols. Eur J Lipid Sci Technol 119:1600252. https://doi.org/10.1002/EJLT.201600252
193. Lee LY, Chin NL, Lim CH et al (2014) Saturated distilled monoglycerides variants in gel-form cake emulsifiers. Agric Agric Sci Proc 2:191–198. https://doi.org/10.1016/J.AASPRO.2014.11.027
194. Barbut S, Wood J, Marangoni A (2016) Potential use of organogels to replace animal fat in comminuted meat products. Meat Sci 122:155–162. https://doi.org/10.1016/J.MEATSCI.2016.08.003
195. Gutiérrez-luna K, Ansorena D, Astiasarán I (2022) Use of hydrocolloids and vegetable oils for the formulation of a butter replacer: optimization and oxidative stability. LWT 153:112538. https://doi.org/10.1016/J.LWT.2021.112538
196. Mohanan A, Tang YR, Nickerson MT, Ghosh S (2020) Oleogelation using pulse protein-stabilized foams and their potential as a baking ingredient. RSC Adv 10:14892–14905. https://doi.org/10.1039/C9RA07614J
197. Oh IK, Lee S (2018) Utilization of foam structured hydroxypropyl methylcellulose for oleogels and their application as a solid fat replacer in muffins. Food Hydrocoll 77:796–802. https://doi.org/10.1016/J.FOODHYD.2017.11.022
198. Tanti R, Barbut S, Marangoni AG (2016a) Hydroxypropyl methylcellulose and methylcellulose structured oil as a replacement for shortening in sandwich cookie creams. Food Hydrocoll 61:329–337. https://doi.org/10.1016/J.FOODHYD.2016.05.032
199. Tanti R, Barbut S, Marangoni AG (2016b) Oil stabilization of natural peanut butter using food grade polymers. Food Hydrocoll 61:399–408. https://doi.org/10.1016/J.FOODHYD.2016.05.034

Chapter 4
Oleogel Preparation Methods and Classification

Tiago C. Pinto, Saman Sabet, Afsane Kazerani García, Satu Kirjoranta, and Fabio Valoppi

Abbreviations

12-HSA	12-hydroxystearic acid
CMCS	Carboxymethyl chitosan
DE	Degree of esterification
DG	Diglyceride
DSC	Differential scanning calorimetry
EC	Ethylcellulose
FD	Freeze drying
GRAS	Generally recognized as safe
HMWG	High molecular weight gelator

T. C. Pinto · A. Kazerani García · S. Kirjoranta
Department of Food and Nutrition, University of Helsinki, Helsinki, Finland
e-mail: tiago.pinto@helsinki.fi; afsane.a.kazerani@helsinki.fi; satu.kirjoranta@helsinki.fi

S. Sabet
Department of Food and Nutrition, University of Helsinki, Helsinki, Finland

Helsinki Institute of Sustainability Science, Faculty of Agriculture and Forestry, University of Helsinki, Helsinki, Finland
e-mail: saman.sabet@helsinki.fi

F. Valoppi (✉)
Department of Food and Nutrition, University of Helsinki, Helsinki, Finland

Helsinki Institute of Sustainability Science, Faculty of Agriculture and Forestry, University of Helsinki, Helsinki, Finland

Electronics Research Laboratory, Department of Physics, University of Helsinki, Helsinki, Finland

Helsinki Institute of Life Science, University of Helsinki, Helsinki, Finland
e-mail: fabio.valoppi@helsinki.fi

C. Andrea Palla, F. Valoppi (eds.), *Advances in Oleogel Development, Characterization, and Nutritional Aspects*,
https://doi.org/10.1007/978-3-031-46831-5_4

HPMC	Hydroxypropyl methyl cellulose
HSP	Hansen solubility parameter
LMWG	Low molecular weight gelator
MC	Methylcellulose
MCT	Medium chain triacylglycerol
MG	Monoglyceride
SC-CO_2	Supercritical carbon dioxide
TAG	Triacylglyceride
USES	Ultrasound-enhanced electrospinning
WPI	Whey protein isolate

4.1 Introduction

Oleogels are edible, semi-solid, self-supporting, often anhydrous lipid-based materials. They belong to the class of fat mimetics and have been developed to substitute solid fats (at room temperature) such as palm oil, coconut oil, lard, tallow, butter, and margarine in food products, possibly without jeopardizing their structure, appearance, and sensory attributes [1, 2]. As described in Chaps. 2 and 3, excessive consumption of solid fats which contain saturated fatty acids has been associated with deleterious effects on human health, like cardiovascular diseases, diabetes type 2, cancer, and onset of obesity [3, 4]. To reduce the risk factors associated with saturated fats, their substitution with liquid oils containing unsaturated fatty acids is recommended. However, direct substitution in food products is not possible, as it can result in the loss of the structure provided by solid fats, which plays a vital role in food's sensorial attributes and physical appearance [1]. Oleogels can overcome this problem because their texture and appearance are tailorable and vary from an opaque spreadable lipid gel like margarine to an elastic or brittle translucent/transparent material like silicone or rubber, and usually, they contain more than 70% of edible liquid oils. Oil is converted to oleogels using structuring/gelling molecules called oleogelators (or organogelators). The term oleogel is typically used as a synonym for organogel, even though the latter term refers to any gel containing an organic solvent that can be either edible like vegetable, marine, and animal oils, or non-edible like acetone and hexadecane; oleogels contain only edible oils, being more appropriately considered a subclass of organogels.

Oleogels can be obtained using small molecules or (bio)polymers. In the first case, the oleogelators are called *low molecular weight gelators* (LMWGs), whereas in the second case, the term used is *high molecular weight gelators* (HMWGs). Examples of LMWGs are monoglycerides, fatty alcohols, fatty acids, waxes from plant and animal origin, plant sterols and their esters, and mixtures thereof [1]. On the other hand, proteins, polysaccharides, and their chemically derived counterparts, like cellulose, methylcellulose, ethylcellulose, xanthan gum, carrageenans, chitosan, chitin, and milk proteins are examples of HMWGs used to fabricate oleogels [1, 5]. A list of the most studied gelators can be found in Table 4.1 (Sect. 4.2).

To successfully obtain an oleogel, one needs to understand the physical form and physico-chemical properties of the gelators. For example, an important factor to consider is the balance between solubility and insolubility of the gelator in the solvent (oil, in this case). When there is balance, oleogels can be formed, whereas imbalances lead the gelator to precipitate or to fully solubilize, generating in both cases unstructured oils [60]. The majority of LMWGs and some HMWGs can be dissolved in oils at temperatures above their melting or glass transition temperature, and during cooling, they self-assemble forming a network that entraps the oil and gels the system [1]. We refer to this case as a *hot direct method*. However, even if the gelator is insoluble in oil, it can still be used for oleogel preparation. For example, the majority of HMWGs are insoluble in oil but soluble in water, ethanol, or aqueous solutions. Therefore, these gelators can be solubilized in molecular form or dispersed as aggregates. Usually, they are added in a solvent other than oil to fabricate structures used as templates, where oil is retained or absorbed to form an oleogel. In this pre-oleogel stage, we can find hydrogels, emulsions, foams, fibers, or encapsulated lipids. Hydrogels are usually converted into solvent-gels (by exchanging the water phase with an organic solvent, *via* a stepwise solvent substitution) or aerogels (using for example supercritical carbon dioxide drying) before being further converted into an oleogel [61, 62]. On the other hand, emulsions (with or without droplet crosslinking) and foams are dried before being converted into an oleogel [63, 64]. Fibers obtained through electrospinning are cut through wet milling/shearing before being converted into an aerogel and then to an oleogel, or directly used in oil to form oleogels [66–68]. Finally, encapsulated lipids are obtained by coating solid lipid droplets with biopolymers and then melted in oil to obtain during cooling an oleogel structured through a hybrid network composed of biopolymers and solid lipids [68]. Therefore, any methods where a second solvent (like water) is necessary to solubilize or disperse gelators, but it is removed before obtaining the final oleogel, are called *indirect methods*. Finally, there are some cases where gelators are insoluble both in oil and water and can be used as particles. If the particles can structure the oil by direct dispersion at, or below room temperature, we refer to this method as a *cold direct method* [69]. Whereas, if the particles do not possess any ability to structure the oil after dispersion, but a secondary liquid like water at low concentration needs to be added to the dispersion to obtain an oleogel but cannot be removed from the system, we refer to *semi-direct methods* [70].

This chapter aims to provide a comprehensive review and discussion of the different classes of gelators, their properties, and give some insights on their production methods, followed by a description of the oleogelation strategies and their current classification, along with some observations of their advantages and disadvantages. The chapter concludes with the proposal of new classification developed in our research group based on thermal treatment, electrical energy, and time required to form oleogels.

4.2 Production of Oleogelators

A non-exhaustive list of the most studied gelators is presented below, along with a brief description of their extraction/production methods and some of their properties (Table 4.1). The properties reported in the table may be of interest to the reader for understanding the methods described in Sect. 4.3 or when considering new research paths or industrial applications. For example, the degree of esterification (DE) is a facet specifically relevant for polysaccharides, as their gelling properties depend on the percentage or degree of esterified carboxyl groups [40, 71, 72]. Other properties such as the molecular weight of a gelator will affect the physical characteristics of a gel, as well as their interaction with different solvents and their time-dependent reactions with other materials in the system. Understanding the temperature dependence of a gelator's structure and its phase change (melting, crystallization, glass transition temperatures) is of great value, especially when considering the resulting oleogel's application, the limitations of temperature-labile compounds, or the sensory performance of the oleogel in food products.

During the early stages of experimentation and product development, it is recommended to familiarize oneself with the technical data sheet of the selected ingredients. Data sheets are commonly provided by the manufacturer and will aid the researchers, providing helpful information.

4.3 Oleogelation Methods

Among the available structuring agents for edible oils, a distinction can be made according to the type of strategy that is necessary for their dispersion in oil. In this section, we discuss the different oleogelation methods, grouping them into three categories: (1) addition of one or more gelators to oil—direct methods, (2) addition of gelators to a secondary solvent, typically water, that needs to be removed from the system to obtain an oleogel—indirect methods, (3) addition of gelators to oil that, not being able to structure oil themselves, require the presence of a secondary liquid at a low concentration to obtain an oleogel—semi-direct methods.

4.3.1 *Direct Methods*

The first developed and most studied oleogelation method involves the direct addition of the gelators to the oil, drawing inspiration from the traditional oil structuring method of having a colloidal network of triacylglyceride crystals, as present in most fat products that are available to the consumer [60]. Certain gelator molecules have the ability of crystallizing/self-assembling in the oil, allowing it to become semi-solid at room temperature. These gelators are typically LMWGs and can be added to the oil either as a single component or in mixtures, being addressed

Table 4.1 A list of the most studied oleogelators, their most common production methods, and their properties

Oleogelator type	Name	Extraction	Properties	References
Fatty acids	Mainly saturated fatty acids with carbon chain length between 16 and 22	Chemical hydrolysis using KOH followed by acid treatment. Enzymatic reaction of oils and fats with lipase	Amphiphilic. Melting temperature is dependent on saturation level, carbon chain length, and polymorphic form. The melting ranges for the neat components mainly used in oleogels are between 60 °C and 80 °C	[6–8]
Fatty alcohols (FAs)	Cetyl alcohol, stearyl alcohol, and other FAs	Obtained from natural sources rich in triacylglycerides (TAGs) or rich in wax esters through hydrogenation of fractioned and hydrolyzed TAGs	Amphiphilic but water insoluble when the molecular chain length is above 10 atoms. Melting temperature in FAs with 12 or lower atom chain length is around 16 °C, if above 12 atom chain length the melting temperature rises up to 72 °C (as in the case of 1-Docosanol)	[9–12]
Mono, di and triacylglycerides	Mainly saturated with carbon chain length of the esterified FA between 14 and 20	Glycerolysis through enzymatic reactions or through glycerol reaction with vegetable oils and fats, as well as hydrolysis with NaOH at high temperatures, followed by distillation	Amphiphilic. Its melting temperature is dependent on saturation level, carbon chain length, and polymorphic form. The melting ranges for the neat components mainly used in oleogels are MGs from 36 to70 °C, DGs from 60 to 67 °C, and TAGs from 54 to 68 °C	[13–15]
Phospholipids	Soy and egg lecithin	Extracted from crude oil through degumming, drying, and cooling. Egg yolk lecithin is obtained through several methods, including solvent, sub- and supercritical extraction.	Amphiphilic. Many phase transitions can be registered when applying heat from 40 to 200 °C	[16–19]

(continued)

Table 4.1 (continued)

Oleogelator type	Name	Extraction	Properties	References
Phytosterols and their esters	β-sitosterol	Phytosterol obtained from a wide array of plants (sunflower, pine bark, cocoa husks). Traditionally obtained through inefficient solvent extraction, apart from generating toxic waste. Other novel, more efficient, and economical methods exist, including microwave, supercritical fluid, ultrasonication, and Soxhlet extractions	Of special health appeal, due to its cholesterol-lowering and anti-cancer properties	[20–23]
	γ -oryzanol	Obtained from rice (*Oryza sativa L.*) bran. Extraction methods are comparable to those of β-sitosterol	Of special health interest, as it lowers intestinal cholesterol absorption, allowing for its elimination through feces	[23, 24]
Polysaccharides	Alginate	Extracted after a treatment with an alkaline solution from milled brown algae (*Phaeophyceae* genus)	Formed by a linear backbone of mannuronic and guluronic acids. Forms gels in the presence of cations	[25, 26]
	Carrageenan	Obtained from red algae by alkaline extraction and precipitation	Formed by sulfated and nonsulfated galactose and anhydrogalactose units. All carrageenans can be dissolved in water to form highly viscous solutions with pseudoplastic behavior. Forms gels with cations	[27, 28]
	Chitin	Chemically extracted from fungi, bacteria and shells of arthropods such as crabs and shrimps	Formed by linear backbone of N-acetylglucosamine units. Insoluble in water at acidic and neutral pH	[30–32]

	Ethylcellulose (EC)	Produced from alkali cellulose with ethyl chloride	Water insoluble. Undergoes a glass transition temperature (onset) around 130 °C	[33–35]
	Hydroxypropyl-methylcellulose (HPMC)	Product of the reaction of alkali cellulose with both propylene oxide and methyl chloride.	Soluble in cold water (below 40 °C) and surface-active. HPMC exhibits reversible thermal gelation from 55 to77 °C. Being non-digestible, acts as dietary fiber and improves the health of intestinal microbiota	[36–39]
	Methylcellulose (MC)	Produced from alkali cellulose treated with methyl chloride	Soluble in cold water (below 40 °C) and surface-active. MC has reversible thermal gelation at about 65 °C. Their solutions form a gel when heated up; the gels reliquefy when cooled down. Being non-digestible, acts as dietary fiber	[35, 39]
	Pectin	Obtained by acid hot extraction followed by precipitation of fruit by-products such as citrus peels and apple pomace	Water soluble. Formed by a linear backbone of galacturonic acid units and having a variable (high, above 50% and low, below 50%) degree of methylation	[26, 40, 41]
	Xanthan gum	Extracted from the precipitate of aerobic bacteria (*Xanthomonas campestris* and other *Xanthomonas* spp.) fermentation	Soluble both in cold and hot water. Formed by a linear backbone of glucose units with some branching	[26, 42, 43]
Proteins	Pea isolate	Obtained through alkaline solubilization of pea flour followed by isoelectric precipitation and recovery of the protein fractions	Mainly composed of albumins (water-soluble and of high nutritional value) and globulins (good emulsifying and gelling properties). Overall, pea proteins have lower gelling capacity and form a weaker gel structure than soy proteins	[45–49]

(continued)

Table 4.1 (continued)

Oleogelator type	Name	Extraction	Properties	References
	Whey protein isolate	Obtained from bovine's milk through ultrafiltration, ion exchange chromatography, and spray drying processes	Complex mixture mainly comprising β-lactoglobulin and α-lactalbumin, and immunoglobulins. Amphiphilic, with an isoelectric point of ~4.7	[49, 50]
Waxes	Beeswax	Obtained from bee (Apis mellifera) honeycombs through melting or chemical extraction with solvents)	Water insoluble. Heterogeneous composition comprising mainly wax esters (~60%, containing C16 fatty acids esterified with C24-C32 fatty alcohols), C27, C29, or C31 hydrocarbons (27%), C16 and C24 fatty acids (~8%), and C30 and C32 fatty alcohols (~6%). Broad melting event (from 40 to 70 °C) with melting peaks at 52 and 65 °C	[51, 52]
	Candelilla wax	Obtained from *Euphorbia antisyphilitica. Traditional method: 0.3% (v/v) sulfuric acid heat treatment. This method produces toxic gases and yields samples of lower quality that require further processing. Ecofriendly method: 2.4% (v/v) citric acid heat treatment*	Water insoluble. Heterogeneous composition comprising mainly C31 hydrocarbons (70–75%), wax esters (15%, containing C16, C18, or C22 fatty acids esterified with C18, C28, or C30 fatty alcohols), and C16 or C30 free fatty acids (~10%). Broad melting event (from 40 to 75 °C) with several melting peaks at 65 to 72.5 °C	[51, 53, 54]

as

	Carnauba wax	Obtained from the leaves of *Copernicia prunifera*. Depending on the quality of the leaves: through physical extraction (scrapping off or boiling and centrifuging the leaves) or solvent extraction	Low solubility in water. Heterogeneous composition comprising mainly wax esters (60%, containing C16, C18, C20, or C24 fatty acids esterified with C18, C30, or C32 fatty alcohols), C30, C32, and C34 free fatty alcohols (30–35%), and C16, C24, or C28 free fatty acids (around 6%). Broad melting event (from 79 to 87.5 °C) with melting peaks around 80 to 85 °C	[51, 56–58]
	Rice bran wax	Obtained from rice bran oil, through solvent dewaxing, defatting and bleaching	Water insoluble. Mainly composed of wax esters (>93%) containing C20–C24 fatty acids esterified with C24–C28 fatty alcohols, small percentage (<6%) of free fatty acids with carbon chain length between C16 and C24. A melting point can appear from 78 to 82 °C	[51, 57, 58]
	Sunflower wax	Obtained by the winterization sunflower (*Helianthus annuus*) oil refining.	Water insoluble. Mainly composed of wax esters (>95%) containing C20–C22 fatty acids esterified with C24–C28 fatty alcohols, small percentage (<5%) of free fatty acids with carbon chain length between C16 and C20. A melting point can appear from 70 to 77 °C	[51, 57, 59]

multi-component gelators. Up to date, the only HMWG which can structure oil through direct dispersion is ethylcellulose, due to the polarity of most HMWGs not being compatible with unsaturated fatty acid-rich oils. There are two types of direct oleogelation strategies: the hot direct method and the cold direct method.

4.3.1.1 Hot Direct Method

The hot direct method is a process in which the mixture of the oil with the gelator is heated to a temperature above the melting point of the gelator (or in the case of ethylcellulose, above its glass transition temperature), under constant agitation, to ensure full dissolution (usually 10–30 min). This is followed by a cooling phase, below the gelation transition temperature, in which the thermoreversible three-dimensional continuous network is formed, trapping the oil within the gel structure. The mechanical properties of the oleogel can be affected not only by the type of oil [73] and the type of gelator [51] and the ratio between them, but also by factors such as the cooling rate [74], and the mixture ratio when using multi-component gelators [75], among other factors. If an insufficient amount of gelators is used, or when the combination of gelators is incompatible and fails to work together, a gel is not obtained, but rather a viscous liquid or a dispersion of crystals in oil. The temperature at which the oleogel should be prepared depends highly on the type of gelator. For instance, monoglycerides usually require 70–80 °C heating. Waxes are a range of gelators with very different properties when retrieved from different sources and whose composition is highly variable, requiring heating in the 50–90+ °C range. Multi-component systems of β-sitosterol and γ-oryzanol require a heating phase of up to 80–90 °C [76]. Wang et al. [76] published a comprehensive review of the textural and rheological properties of LWMG-based oleogels, providing a useful knowledge pool about the preparation and properties of this type of oleogels. Ethylcellulose, as extensively explored in Chap. 7, requires a heating stage of above 140 °C to be dissolved in oil.

In order to structure oil, the gelator molecules that can be used via the hot direct method must first self-assemble through highly specific noncovalent interactions into primary particles through precipitation and/or crystallization ("bottom-up" nanofabrication) [77]. This phenomenon is followed by the assembly of individual molecules into supramolecular structures like crystal lattices, liquid crystals, micelles, bilayers, fibrils, and agglomerates, forming 3D networks capable of entraining the oil, resulting in a semi-solid material [78]. Typically, the intermolecular forces that drive aggregation are non-covalent, such as hydrogen-bonding, π–π stacking, dipole-dipole, and London dispersion forces [79]. These interactions stabilize the clusters and lead to the formation of a continuous network, which is influenced by anisotropic growth of supramolecular structures and anisotropic diffusion of molecules or clusters of molecules [78]. The balance between solubility and insolubility is the most important requirement and the key to obtaining a gel. To clarify, for a material to be used as a gelator, there must be a suitable balance between its affinity to the solvent (edible oil) and enough insolubility in the solvent so that self-organization and assembly between gelator molecules are triggered (solubility limit of the gelator) [78]. The hot direct method relies on the interaction between solvent–gelator and gelator–gelator, as they play a central role in the formation of oleogels. Achieving optimal gelation seems to be the result of an optimal solvent-to-gelator ratio, although the direct effects of the solvent on the physical properties of the final oleogel are not well understood [79]. In general, a weak solvent–gelator interaction results in the prevalence of gelator–gelator interactions, which may lead to the formation of a continuous network, which is essential for the formation of an oleogel. Imbalances in this interaction can lead to the formation of other types of systems, *i.e.,* solutions or precipitates. The ability of a LMWG to gel a solvent has mostly been empirically explored, generally with a trial-and-error approach being used, which results in several failed trials.

Therefore, it is important to understand the interactions between the gelator and the solvent and reach an equilibrium of solubility and insolubility. The term "solubility parameter" was first coined by Hildebrand and Scott, in an attempt to predict solubility relations [80, 81]. The Hildebrand solubility parameter is a measure of the cohesive energy density of a substance, which is the energy required to separate the molecules in a material from each other. As such, the difference in solubility parameters for solvent–solute combinations is important in determining the solubility of the system. If the Hildebrand parameters of a solute and the solvent are similar, then the substance is likely to be soluble in that solvent. Adversely, if the Hildebrand parameters of the solute and the solvent are dissimilar, then the solute is likely to be insoluble in that solvent [82]. A limitation of the Hildebrand parameter is that it is unable to quantify the specific intermolecular interactions, which is part of the reason why the Hansen solubility parameters (HSP) were developed. These measure the total cohesive energy of a species as the sum of three individual energetic components (i.e., dispersion interactions, dipole–dipole interactions, and hydrogen-bonding interactions). In the same way as the Hildebrand parameter, the cohesive energy densities describe the ability of the solvent to solubilize the solute, or in this case, they describe the ability of a certain oil to solubilize a gelator, because they quantify the intermolecular forces that are required to overcome the gelator–gelator and oil–oil interactions. This equilibrium can be studied through the HSP and conveniently shown using the Hansen space, which allows for the visualization of the three fundamental intermolecular forces: dispersion, polar, and hydrogen bonding [83, 84]. If the HSP values of two compounds are close together in the HSP space, then the gelator is likely to be soluble in the oil. Some studies have been conducted using 12-hydroxystearic acid (12-HSA) and its derivatives as gelators, with an assessment of the impact of the hydroxyl group position in the formation of hydrogen bonds, as well as some other gelators [77, 79, 86–88]. These studies have helped to understand the gelation behavior of these compounds, by predicting their solubility in certain solvents. In addition, researchers have used these parameters to optimize known gelation systems, identify new ones, and develop tailored materials for the target applications [88, 89].

The diversity of forces acting on both gelators and solvents makes it difficult to accurately predict the outcome of a gelator–solvent combination. Though the use of

HSP can come up as a suitable way of predicting gelation behavior, other efforts are being made toward this goal. An interesting study by Cuello et al. [90] has proposed a Big Data solution to uniformize the existing data about LMWG and enhance the way knowledge is stored in this field. The authors collected data from heterogeneous sources and combined them into a unique, homogeneous platform, to allow the unification of current and newly obtained sources, as well as providing computing solutions to try and unravel the relationships between solvents and gelators. They demonstrated that the platform can be used to identify and characterize the key factors that influence the behavior of these gelators in food systems, and can provide insights on their future use. Essentially, this would be an application of data science and engineering to analyze and predict the most valuable cases to try in a laboratory setting, drifting away from the empirical approach that can overly consume time and resources.

4.3.1.2 Cold Direct Method

The cold direct method is carried out at ambient temperature or below, requiring only agitation for the dispersion of the gelator. The fact that it takes place at room temperature constitutes an advantage in comparison to the hot direct method, since temperature-induced oil oxidation is avoided. The method was first introduced by Patel et al. [91], who prepared oil dispersions of fumed silica by mixing hydrophilic colloidal silica particles with sunflower oil, followed by shearing the dispersion using a high-speed homogenizer, at room temperature. Oleogels were successfully formed at a concentration of 10 and 15% fumed silica, resulting in a 3D network based on the fractal aggregation of silica particles and consequent entrapment of oil. In contrast, hydrophobic fumed silica in oil exhibited low gelling behavior in oil, possibly due to the decrease in hydrogen bonding sites [92]. Since hydrophilic fumed silica has low solubility in oil, the attractive forces between silica particles are the drivers for the formation of the network and oil entrapment with minimal leakage [93].

Another approach explored the use of mercerized cellulose in the gelation of rapeseed oil [69]. Dispersions of cellulose powders from different botanical sources were prepared, with mass concentrations ranging from 5 to 40% in rapeseed oil. The dispersions were manually stirred for 2 min at room temperature. The intent of the authors when using such a wide concentration range was to conduct an empirical attempt at unveiling the critical value at which the dispersion exhibits solid-like behavior. It was concluded that though oleogelation critically depends on the cellulose content of the vegetable powders, the minimum gelling concentration and the textural properties are mainly governed by the size of the cellulose fibers, regardless of the botanical source.

4.3.2 *Indirect Methods*

Most food-approved polymers are inherently hydrophilic and are therefore unable to directly structure oil. Materials such as proteins and polysaccharides are regularly used in food applications and generally do not constitute an issue regarding legislation, depending on the specific type of protein or polysaccharide being used and its source. Indirect oleogelation methods are beneficial because they allow hydrophilic and amphiphilic polymers to be applied in oil structuring applications, usually starting out with water-rich systems and involving a series of procedures to remove water from the system.

4.3.2.1 Emulsion-Templated Methods

The indirect oleogelation strategies that start with the preparation of an emulsion are often grouped into the category of emulsion-templated methods.

Original Emulsion-Templated Method

The first indirect oleogelation approach was developed by Romoscanu and Mezzenga [63]. The authors suggested the application of a percolating 3D network of proteins for transforming oil into an elastic solid, without chemical modification. This method kicks off with the preparation of a monodisperse oil-in-water emulsion, where the oil droplets are stabilized by a cross-linked protein monolayer adsorbed at their interface. The procedure involves pumping the oil phase in the form of droplets from a pressurized tank into a β-lactoglobulin solution of 1% (w/w) that is coflowing via a glass capillary, creating a controlled environment. Due to the constant characteristics of the flow and break-off of the droplets, this method allows the fabrication of emulsions with a very high degree of monodispersity; however, there is the drawback of having a low output rate due to the sequential creation of the droplets. The emulsion is then left for one hour to complete protein adsorption onto the interface, and a washing procedure follows, with the aim of removing unadsorbed protein from the system. The cross-linking of the adsorbed β-lactoglobulin is then performed: this can be achieved either thermally, through the acceleration of cross-linking kinetics by keeping the emulsion at 80 °C for 10 min, or chemically, using glutaraldehyde. When applying chemical cross-linking with glutaraldehyde, the emulsion is poured into the same volume of 1% (w/w) glutaraldehyde, to avoid interparticle cross-linking. Another washing procedure follows to remove nonreacted glutaraldehyde, similar to the removal of unadsorbed protein. At this stage, there is the addition of a small amount of glycerol to increase its chemical potential and safeguard any internal stresses that can happen during the drying phase. This takes place at room temperature, where the emulsion is allowed to dry until the totality of the water has evaporated from the system, resulting in a fully

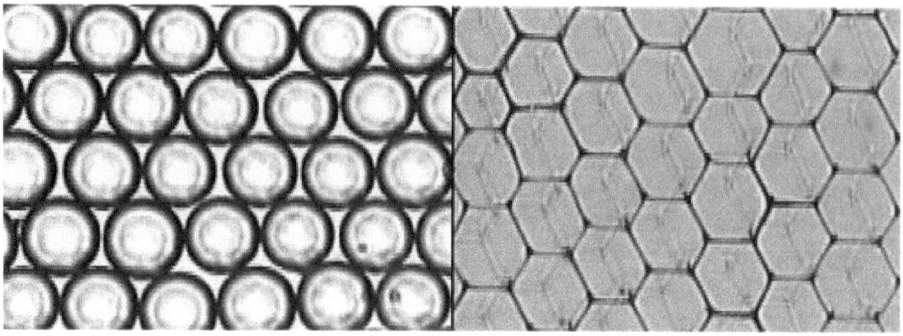

Fig. 4.1 Optical micrographs of an emulsion template (24 μm droplet radius) and a resulting two-layered thin film of gel exhibiting a polyhedral structure. (Figure reproduced from [63] with permission from the American Chemical Society)

transparent oleogel. The final structure resembles that of a dried foam, with a protein bilayer acting as the walls and the air being replaced by the chemically unmodified oil. The analysis of the oleogel's microstructure reveals a polyhedral arrangement with sizes comparable to the droplet size in the emulsion, as depicted in (Fig. 4.1) [63].

While the original method for preparing oleogels using emulsion templating was a breakthrough in the field, it has some disadvantages due to the use of chemically cross-linked proteins. Such proteins face many hurdles in being accepted for use in food products, including concerns about their safety, potential adverse effects on human health, and their impact on food texture and flavor [94]. Thermal cross-linking has comparable issues and also exhibits very slow cross-link kinetics, resulting in longer processing times and increased energy costs. Additionally, the cross-linking reaction can be difficult to control, and the complexity of the method made other authors shift to different approaches. Therefore, the method has undergone further developments with a focus on eliminating cross-linkers, leading to newly developed emulsion-templated methods, which are now more commonly cited as the reference, compared to the original emulsion-templated method here described.

Modified Emulsion-Templated Method

Unveiling the potential of cross-linked proteins for oil structuring by Romoscanu and Mezzenga [63] was the necessary foundation to discover what hydrophilic and amphiphilic polymers really have to offer in the structuring of hydrophobic oils. The procedure of using an emulsion template for the preparation of oleogels was later adopted and modified by Patel et al. [64], using the combination of a surface-active and a non-surface-active polysaccharides to generate oleogels. First, the authors attempted to establish a stable oil-in-water emulsion using only one surface-active polysaccharide (hydroxypropyl methyl cellulose—HPMC—or methylcellulose—

MC), with the results exhibiting a large droplet size and oil separation phenomena upon drying. On the other hand, a non-surface-active polysaccharide like xanthan gum was unable to stabilize the emulsion on its own, owing to its non-surface-active nature. However, the combination of one of the cellulose derivatives (HPMC or MC) with xanthan gum enhanced emulsion stability, exhibiting a more uniform droplet size with smaller droplets. These emulsions were prepared by dispersing oil in an HPMC or MC solution using a high-speed homogenizer followed by the addition of xanthan gum solution under continuous shearing. The emulsions were then oven-dried at temperatures between 50 °C and 80 °C until complete removal of water. The dried samples were then briefly sheared for 30 s to obtain an oleogel sample, which consisted of clusters of tightly packed oil droplets in oil continuous medium, with over 97% (w/w) of oil. The study investigated both the cellulose derivative type and its grade, observing their effect on the rheology of the emulsions, where lower viscosity polymer solutions resulted in stronger gels. This type of functionality was associated with the ability that the polysaccharides have to increase the stiffness of the interface, and hence contribute to the overall consistency of the emulsion (and, later, of the oleogel). This approach resulted in oleogels with a unique microstructure, featuring tightly packed oil droplets and no signs of oil leakage, as well as interesting rheological properties, such as high storage modulus, shear sensitivity, good thixotropic recovery, and thermostability.

While the use of polysaccharides provides advantages in terms of using GRAS ingredients, incorporating other biopolymers such as proteins has been acknowledged as advantageous. This is because proteins are label-friendly and have well-known additional health benefits that consumers value. Patel et al. [95] continued the work by modifying the original method by Romoscanu and Mezzenga [63], with the aim of using proteins and developing an alternative that does not require crosslinking. Instead, this alternative methodology by Patel et al. [95] benefits from the strong molecular complexes that result from protein–polysaccharide interactions at the interface. This was achieved by combining gelatin and xanthan gum, two edible natural materials that are commonly used in the food industry and are known for establishing hydrophobic interactions and non-Coloumbic interactions with the involvement of –NH and –OH groups. The oleogels were obtained by homogenizing sunflower oil in a gelatin solution, followed by the immediate addition of xanthan gum solution, under continuous shearing. The drying of the emulsion was carried out through both oven-drying and freeze-drying (FD), followed by a short step of shearing to create an oleogel (Fig. 4.2).

Once again, the stiffened interfacial membranes provide the oil droplets with better stability against stresses that the emulsion faces while drying. This kind of system was obtained by other authors using protein-polysaccharide complexes or multi-polysaccharide complexes, and in some cases, such a stable interface was already obtained by using proteins only. Tavernier et al. [96] first achieved an efficient structuring of oil using only unmodified proteins (soy protein isolate), and further attempts using 2% sodium caseinate also proved to be successful [97, 98]. However, the interactions between proteins and polysaccharides and the

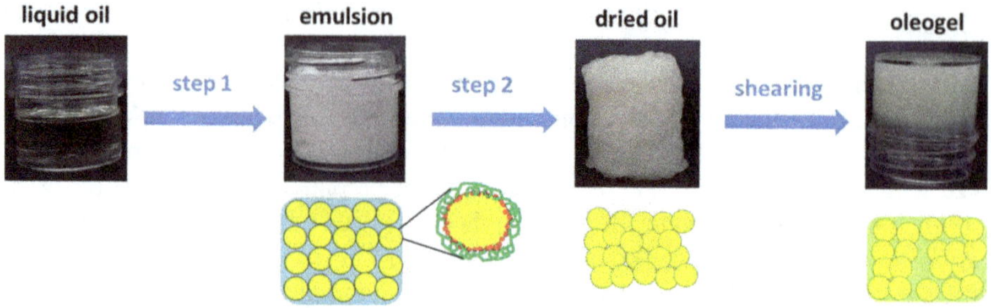

Fig. 4.2 Visual representation of the modified emulsion-templated method using protein-polysaccharide stabilization, through a combination of photographs depicting the result of each step of the process and sketches depicting the microstructure of the samples. Liquid oil is used to prepare a 60% (w/w) oil-in-water emulsion (step 1), followed by the removal of water through oven-drying or freeze-drying (step 2), and further shearing of the dried product (step 3), resulting in the formation of an oleogel. (Figure reproduced from [95] under the open access article as ACS AuthorChoice)

parameters that need to be fulfilled to achieve a sufficiently stable interface are still not fully elucidated [99].

Microcapsule-Templated Approach

The type of oleogels that are originated by LMWGs versus the type of oleogels that are originated by polymers is significantly different in terms of properties that they confer to food products. For example, LMWGs components that form crystalline conformations tend to result in oleogels with mouthfeel and processing properties that are more resemblant to fats. Adversely, polymeric gelators can provide other interesting textural properties such as firmness and consistency [68]. Coming up with a way that can combine these types of properties into one single oleogel has been attempted via direct strategies using multi-component gelator systems such as ethylcellulose-glycerol monoleate, [100], stearyl alcohol-stearic acid-ethylcellulose [101], and via indirect strategies using Pickering emulsions stabilized by zein–stearate complexes [102]. However, most of these attempts encompass harsh or energy-intensive processing (namely, heating at high temperatures or FD), which is not beneficial nor sustainably feasible in terms of process scaling-up and further commercialization. Thus, an effort was made by Patel [68] to execute a new approach using coated crystalline microcapsules to fabricate oleogels where both fat crystals and polymer sheets cooperate in providing structure to the system (Fig. 4.3). Microcapsules were prepared using palm stearin, a common hard stock fat; the procedure starts with the development of an oil-in-water emulsion stabilized by MC and an immediate dilution of the emulsion in ice-cold water to trigger immediate solidification and creaming of the oil droplets, coated with methylcellulose. These microcapsules were then recovered and dried at mild temperatures (30–32 °C). The microcapsules were then dispersed in oil for the preparation of

Fig. 4.3 Visual depiction of the oleogel preparation process, represented through a combination of photographs, sketches, and polarized light microscopy images (**a**–**c**). The comparison showcased in the figure is between two types of oleogels: one prepared from unprocessed palm stearin, on the left (**a** left, **b** left, and **c**), and the other from palm stearin capsules made using an emulsion that contained 10% (w/w) oil phase, on the right (**a** right, **b** right, and **d**). The crystal structures in (**b**) left are representative of palm stearin crystals, whereas (**b**) right shows the same crystals and MC strands. (Figure reproduced from [68] with permission from Elsevier)

oleogels; the dispersion was heated up to 70 °C, leading to the melting of the capsules and the dispersion of the capsule coats as polymer strands. The system was then cooled down to room temperature, forming a network of crystallized palm stearin and MC strands. When compared to an oleogel structured using only palm stearin, the microcapsule approach presented advantages such as a controlled crystallization of palm stearin in discrete spherical units, as opposed to uncontrolled growth and aggregation characteristic of pure palm stearin oleogels.

Oleosome-Templated Approach

Oleosomes are microbodies that can be found abundantly in oleaginous seeds and fruits, essentially being naturally pre-emulsified oil. Functionally, they are important for lipidic storage, serving as an energy source and protection against environmental stresses. Structurally, they consist of oil droplets that are encapsulated and stabilized by a unique protein/phospholipid membrane [103]. These can provide an excellent medium for forming gel-like structures; despite that, there are very few applications of oleosomes in the oleogel field.

Mert and Vilgis [104] first explored this possibility by extracting natural oleosome structures from oil-bearing plant materials (particularly, hazelnut) and stabilizing them with xanthan gum or pectin. These hydrocolloids were used first for the stabilization of natural oleosome suspensions using an electrostatic deposition technique, that has been shown to improve aggregation stability of emulsion droplets when exposed to environmental stresses such as dehydration. Then the water was driven off the system via FD, obtaining a dry product that was further sheared, until a gel structure was obtained [104]. The driving force behind the establishment of this type of oleogel is the complexation between the oleosin proteins that are present in the oleosome membrane and the polysaccharides, following the trend of other similar oleogelation systems that have been mentioned earlier in this section (4.3.2.1). This could be considered a more sustainable and economic option, considering the use of natural oil bodies that do not have to be artificially fabricated but instead extracted from oleaginous plants. The electrostatic interactions between oleosomes and polysaccharides have been studied in systems containing soybean oleosomes, sodium alginate, and xanthan gum, to further understand said interactions and amplify their application in the oleogel field [105].

4.3.2.2 Foam-Templated Method

During the same period that the modified emulsion-templated method was developed, the same research group explored the development of an alternative method. Patel et al. [106] reported for the first time a foam-templated approach using a water-soluble polymer and low-temperature processing. The foam template is turned into a porous cryogel via FD to remove water, resulting in a material with excellent oil sorption properties. Cellulose derivatives such as HPMC or MC can be very intuitively suitable candidates for this kind of approach: these derivatives are synthesized by substituting the hydroxyl group of cellulose with hydroxyl propyl or methyl groups. This imbues the molecules with a certain degree of hydrophobicity, conferring them an amphiphilic character and a certain degree of surface activity, which makes it easy to incorporate air into the HPMC solution [107].

The method developed by Patel required the preparation of an HPMC solution first, by dissolving it in water in a concentration of 1–2% and overnight mixing. The solution was further processed using a high-speed homogenizer operating at 11,000 rpm; this process aerates the system and results in an aqueous foam, presenting an average bubble size of less than 150 μm. The obtained foam was then subjected to FD, originating a porous cryogel. The oleogel formation ensues, by absorbing high, weighted quantities of sunflower oil into the dried cryogels and allowing the system to rest overnight. At this stage, as the cryogel was formed only by HPMC and without the use of any crosslinking, the oil was expected to flow through the entire structure and hence be quickly absorbed, but not tightly bind to the

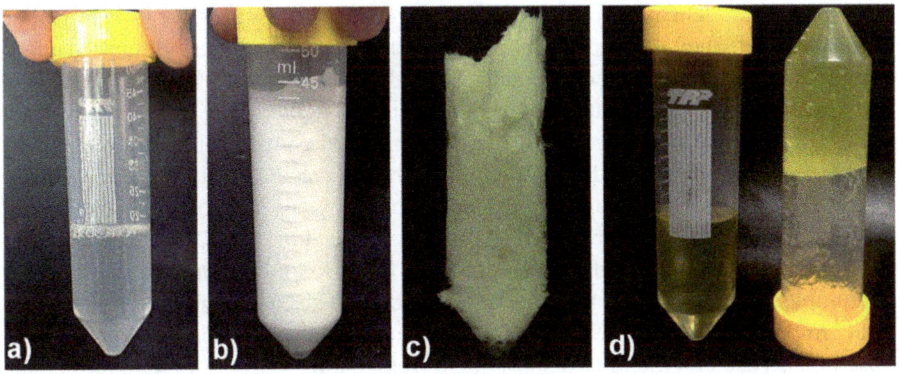

Fig. 4.4 Sequential photographs of the foam-templated method stages: (**a**) 1% (w/w) hydroxypropyl methylcellulose (HPMC) (4000 cps) solution; (**b**) aqueous foam formed by aerating HPMC solution; (**c**) porous cryogel obtained by removal of water by freeze-drying; and (**d**) comparative pictures of sunflower oil on the left and organogel (with 98% (w/w) oil) formed using the cryogel on the right. (Figure reproduced from [106] with permission from the Royal Society of Chemistry)

structure, having a high risk of oozing out with minimal pressure. As such, the material was then sheared at 11,000 rpm using a high-speed homogenizer to obtain oleogels and prevent the release of oil (Fig. 4.4). This way, the polymer sheets were uniformly dispersed in the oil-continuous phase, physically trapping the oil, and preventing leakage. The weight of the added oil was calculated at about 98–99 times the weight of the cryogel. Later approaches used this foam-templated method as a guideline and focused on broadening the range of gelators suitable for this kind of methodology to include proteins. So far, a combination of gelatin and xanthan gum [108] and a combination of pea/faba protein and xanthan gum [109] have been successfully used to prepare oleogels through a foam template. These two studies have in common the fact that a polysaccharide was necessary to improve foam stability. Another study with rice bran protein successfully resulted in oleogels [110]; this approach required a pH adjustment step to adjust surface activity and allow for a faster adsorption on the interface.

The advantage of the foam-templated method is that high temperature is not needed, which could lower the risk of lipid oxidation and non-desired flavors. However, it is a method that consumes high amounts of energy and time, due to the FD process.

4.3.2.3 Hydrogel-Templated Method

Solvent Exchange Method

After the potential of proteins was revealed through the emulsion-templated method with crosslinking, as described above, de Vries et al. [62] further developed the applicability of proteins as sole gelators for oil structuring *via* a solvent exchange procedure. First, a heat-set protein hydrogel was created using whey protein isolate (WPI) powder; they were prepared by heat denaturation of the proteins, using a temperature-controlled water bath at 85 °C for 30 min. The gels were then allowed to cool down to room temperature and stored overnight at 4 °C. The exchange of water retained in the protein matrices for sunflower oil was made following a stepwise approach, using an intermediate solvent. For the intermediate solvent, tetrahydrofuran (THF) or acetone was used, considering that they are miscible both with water and sunflower oil. The hydrogel was cut into cylindrical pieces that were placed onto mesh metal buckets and then immersed for 8–12 h into the next solvent under continuous stirring of the solvent, and so forth. The succession of immersions goes from a solution of 30% (v/v) intermediate solvent in water, proceeding with 50% (v/v), 70% (v/v), and two subsequent immersions in 100% (v/v) intermediate solvent, following the above-mentioned stepwise approach. After the full replacement of the water with the intermediate solvent, a similar stepwise approach was applied using solutions of 30% (v/v), 50% (v/v), and 70% (v/v) sunflower oil in intermediate solvent and 100% (v/v) pure sunflower oil, to reach full substitution of the intermediate solvent with oil. The result oleogels were removed from the mesh metal buckets and blotted dry with tissue paper (Fig. 4.5). Up to 91% (w/w) oil could

Fig. 4.5 On top, the appearance of 15% WPI hydrogels (left) and respective oleogels (right) after the solvent exchange. (Reprinted from [62] with permission from the American Chemical Society). On the bottom, (**a**) the appearance of heat-set WPI aggregates after centrifugation; (**b**) dispersion of freeze-dried protein aggregates in sunflower oil; and (**c**) resultant protein aggregate-based oleogels obtained through a solvent exchange method. (Reprinted from [111] with permission from Elsevier)

be incorporated into the system with a residual amount of water (under 1% (w/w)), which exhibited a Young's modulus above the respective hydrogels by 2 orders of magnitude.

However, the research group acknowledged that the preparation method, though effective, had limited flexibility to alter the rheological properties of the final protein oleogels. The oleogels were much stiffer and more brittle than the original gelling system (hydrogel). In an attempt to create a system that allowed for better tuning of its rheological properties, they decided to extend their previous work and explore the possibility of using protein aggregates of colloidal size to structure oil, while comparing them to their hydrogel counterparts [111] (Fig. 4.5).

This alternative process developed by de Vries et al. [111] starts with the preparation of a protein stock solution of 4% (w/w) WPI. The mixture was then refrigerated overnight to ensure that the protein was fully hydrated and the pH was adjusted to 5.7 the next day using a 1 M HCl solution. The solution was heated at 85 °C for 15 min to denature the protein, resulting in a weak gel, which was easily broken into smaller pieces. These pieces were homogenized using a high-speed homogenizer for 3 min at 13,000 rpm. The protein aggregates were collected by centrifugation and washed twice with demineralized water to remove any remaining soluble protein. For preparation of the oleogels, the WPI aggregates were dispersed in acetone and then centrifuged to collect the pellet containing the protein. This process was repeated once more using acetone to ensure water removal and twice in sunflower oil. The resulting pellet was diluted with sunflower oil, and excess acetone was allowed to evaporate overnight. The mixture was then centrifuged to increase the concentration of protein aggregates and form a gel. An alternative method of dispersing the protein in oil was also tested using FD. The resulting powder was dispersed in oil and centrifuged to increase protein concentration. As opposed to the first approach, instead of using a protein backbone to structure oil, the proteins are viewed as building blocks to the system rather than a fixed network. This study found that efficient network formation was achieved through hydrophilic interactions between the aggregates, even when in a hydrophobic oil medium. Typically, the network formation of colloidal particles depends on both particle–particle and particle–solvent interactions, as explored in Sect. 4.3.1. Therefore, it must be considered that changes at the level of the solvent can manipulate the network formation and its resulting rheological properties. Further works focused on using this protein aggregate-solvent exchange while varying the polarity of the oil, establishing a relationship between the polarity of the oil and the gel strength [112].

Aerogel-Templated Method

The aerogel-templated method was first described by Manzocco et al. [61]. In this method, a hydrogel made of κ-carrageenan was converted to an alcohol gel via a stepwise solvent substitution (water to ethanol), using the approach described by de Vries et al. [62] and in the previous section. The obtained alcohol gel was then dried under a continuous flow of supercritical carbon dioxide (SC-CO_2) at 11 ± 1 MPa and 45 °C. After 8 h SC-CO_2 drying, a monolith aerogel was obtained. Subsequently, the aerogel was soaked in sunflower oil which diffused inside the aerogel porous structure and an oleogel was formed. Carbon dioxide transits from liquid or gas to a supercritical fluid above 31.1 °C and 73.8 bar (or 7.38 MPa), giving it low viscosity and high diffusivity like gasses and high density like liquids. These properties make supercritical fluids excellent and versatile solvents. Upon returning to ambient pressure, CO_2 transits to a gas state, leaving the material that has been in contact with, without any traces. The κ-carrageenan aerogel obtained by Manzocco et al. [61] was able to absorb a mass of oil between two and four times its original mass, leading to an oleogel composed of around 20–25% biopolymer and the remaining fraction of sunflower oil. These oleogels were characterized by high firmness (between 100 and 300 N upon compression) due to the dense and compact aerogel structure, and their oil holding capacity was between 60% and 80% (expressed oil upon centrifugation compared to the original oil mass in the oleogel).

From this original work, the same research group has published a series of articles demonstrating that the properties of oleogels obtained with SC-CO_2 dried aerogels can be greatly tailored. Indeed, by adding lettuce as a filler into κ-carrageenan gels, one can reduce the firmness of the resulting oleogel and increase the mass of absorbed oil up to 15 times the mass of the initial aerogel [113]. The same authors showed that fresh-cut salad waste-based aerogels, cellulose aerogels, as well as WPI aerogel particles can be used to form an oleogel [115–117]. In the latter case, the resulting oleogel contained 15% particles and 85% sunflower oil and resulted in a moldable material with rheological properties that are typical of a gel [115] (Fig. 4.6). Finally, these authors also demonstrated that oleogels structured through protein aerogel particles one can steer oil and protein digestibility [117]. Particles of aerogels made of starch were recently employed by Alavi and Ciftci [118] to obtain a moldable oleogel. The addition of chitosan to starch prior to forming an aerogel improved considerably the oil structuring ability and the mechanical properties of the resulting system.

Although SC-CO_2 drying is a great tool to obtain aerogels for oleogel preparation, the need for specialized and tailor-made equipment, as well as long processing times for the aerogel preparation (48 h to one week to replace water with ethanol to obtain an alcohol gel and 8 h of drying to obtain an aerogel), led researchers to look for alternative aerogel production methods like FD. Even if there is still a debate on the

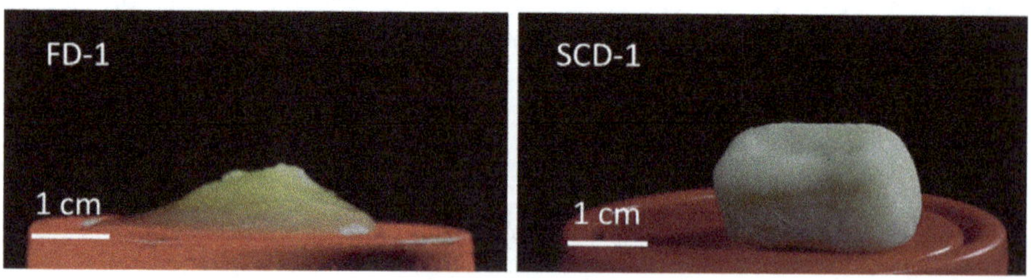

Fig. 4.6 Oleogels obtained with freeze-dried (FD-1) and supercritical CO_2 dried (SCD-1) whey protein aerogel particles dispersed in sunflower oil. The composition of FD-1 is 31% particles and 69% oil, and that of SCD-1 is 15% particles and 85% oil. (Figure reproduced from [115] with permission from Elsevier)

definition of aerogels based on the preparation technique [119], in this paragraph, we consider porous materials as defined by Smirnova and Gurikov [120] as aerogels, regardless of the preparation method. In FD, water-based gels/materials are directly converted into porous materials after a drying step under vacuum at low temperature (usually −20 – −60 °C for 24–72 h), preceded by the freezing of the original material (usually at −80 °C to promote the formation of small ice crystals, which lead to smaller pores during drying). Some studies highlighted that if the same hydrogel is used for FD and SC-CO_2 drying, the structure of the resulting aerogels is more collapsed in FD aerogels, leading to mechanically weaker oleogels or showing a lower ability to absorb oil [114, 115] (Fig. 4.6).

However, these limits can be solved by improving the structure of the starting hydrogel. On this topic, Zhao et al. [121] added carboxymethyl chitosan (CMCS) to WPI to form heat-set modified hydrogels, which were subsequently FD to form an aerogel. The presence of up to 0.75% CMCS in the aerogel led to the formation of oleogels (after immersion of aerogel for 6 h in soybean oil) which were able to absorb oil up to five times the mass of the initial aerogel and retain more than 96% of the initial oil upon centrifugation. In addition, the oleogel obtained had better oxidative stability and high astaxanthin bioaccessibility compared to bulk oil. In another study, Chen and Zhang [122] developed aerogels by applying FD to hydrogels composed of alginate/soy protein conjugates obtained through the Maillard reaction. Oleogels were obtained by immersing the aerogel for 6 h in corn oil, which was able to absorb oil at 10.9 times its initial mass and retain 40% of it upon centrifugation. Finally, Li and Zhang [65] developed oleogels by dispersing gelatin-based aerogels in camellia oil. Aerogels were obtained through FD of either gelatin hydrogels or short gelatin electrospun nanofibers in water. However, we report here data related to the first case, whereas data related to oleogels obtained using electrospun nanofibers will be discussed in the following section. The authors showed that all the studied characteristics and properties of the oleogels were dependent on the concentration of gelatin in the starting hydrogel. In particular, aerogels were able to absorb a mass of oil comprised between 20 and 100 times their initial mass and retain between 30% and 60% of the initial oil upon centrifugation. Moreover, the obtained oleogels showed thixotropic recovery between 76% and 94%, and a free fatty acid release during in vitro digestion between 40% and 60%.

4.3.2.4 Electrospun Nanofiber-Templated Method

The last indirect method developed for oleogel production we present in this chapter is based on electrospun nanofibers. Electrospinning is an electrohydrodynamic process where micro- and nanofibers are drawn out from an electrified polymer solution and dried during their travel to a grounded collector. A typical electrospinning setup consists of a high-voltage power source, a syringe pump, a spinneret (typically a needle with a blunt tip), and a conductive collector [123]. However, needleless electrospinning devices are also emerging [124]. Different food grade high molar mass biopolymers like zein, gelatin, whey protein, starch, cellulose derivatives, carrageenans, alginate, pullulan, dextran, chitin, and chitosan, to name a few, have been electrospun forming nanofibers with diameters in the range of few hundreds of nm (generally below 1 μm) [123, 126–128]. Biopolymers used in electrospinning do not need to possess any emulsifying, foaming, and thickening properties like those used in the other indirect methods, making the production of electrospun-based oleogels a more versatile indirect method for oil structuring.

Most of the research on developing oleogels from electrospun nanofibers is on the structuration of castor oil for lubrication purposes [66, 129–131]. In these works, the electrospun nanofibers were obtained by applying an electric field of 0.4–1.5 kV/cm and using solutions containing mixtures at different concentrations and ratios of Kraft lignin and cellulose acetate, and of polyvinylpyrrolidone and Kraft lignin [128, 130]. Electrospun nanofibers were dispersed at a concentration of 5–30% (w/w) in castor oil using gentle mixing. The resulting oleogels showed rheological and tribological properties dependent on the concentration of the nanofibers in the oleogel and the composition of the nanofiber. However, in general, the obtained oleogels showed rheological and tribological properties like commercially available lubricating greases made from metallic soaps and mineral oils [66, 129–131].

On the other hand, to the best of our knowledge, the only published example of edible oleogels obtained using electrospun nanofibers was recently published by Li and Zhang [65]. The authors developed gelatin-based electrospun nanofibers using an electric field of around 1.3 kV/cm applied to gelatin in an acetic acid solution. Following, nanofibers were added to liquid tert-butanol at different concentrations and homogenized using a high-speed homogenizer to reduce their length. The solvent was then removed using FD. The final system was an aerogel formed by short electrospun nanofibers, which were immersed in camellia oil and formed an oleogel. Although this method is the result of the application of two indirect methods and could be classified as a hybrid method, here we decided to group it in this category since the main structure of the oleogel is given by the nanofibers. The authors proved that oil absorption and retention, as well as rheological properties and digestion profiles of oleogels, were correlated with the concentration of nanofibers dispersed in tert-butanol before forming the aerogels. More specifically, aerogels were able to absorb a mass of oil between 60 and 125 times their initial weight and retain between 60% and 80% oil upon centrifugation, which exhibited better oil absorption and retention compared to aerogels obtained using gelatin hydrogel as a starting system for aerogel production (more details in the previous paragraph). On the other hand, the oleogel obtained using nanofibers showed a thixotropic recovery between 36 and 79% and a release of free fatty acids during in vitro digestion between 40 and 50%.

Although electrospinning is a promising technology to obtain nanofibers for oil structuration, it is still affected by some limitations such as possible blockages in the spinneret, residual solvents in the nanofibers, and long processing times [131]. To overcome these problems a novel open-surface needle-free electrospinning device featuring one or multiple focused ultrasonic transducers, namely ultrasound-enhanced electrospinning (USES) has been developed [132–135]. We recently developed a new, cold method to obtain oleogels using a mat of USES nanofibers. The nanofibers were formed using a standard polymer, polyethylene oxide. After dispersing the nanofibers in oil and subjecting the mixture to cryo-milling, oleogels could be formed in rapeseed, walnut, and flaxseed oils at nanofiber concentrations above 10%. The oleogels were formed by a jammed dispersion of nanofiber mat fragments, exhibited excellent thixotropic recovery, and the stiffness of the oleogel was proportional to the nanofiber concentration and the unsaturation level of the fatty acids composing the oil [67].

4.3.3 Semi-Direct Method

The semi-direct oleogelation method is a recently developed technique that uses gelators or particles that can be dispersed in oil but require a secondary liquid to form an oleogel. The term "semi-direct" was chosen because this method features characteristics of both direct and indirect methods: direct addition of gelators or particles to oil, and the use of a secondary liquid like water. However, unlike in the indirect methods, the water cannot be removed from the system, or the self-supporting structure of the oleogel is lost. Therefore, in the semi-direct method, oleogels are formed by mixing the particles and the two liquids. Currently, there is only one type of method that can be classified as "semi-direct," i.e., capillary suspensions. In this method, insoluble particles are first added to the oil at concentrations ranging from 10% to 50–60%, forming an oil-particle dispersion. Water or aqueous solution is then added, usually, in the concentration range of 2–20% of the final system and an oleogel is formed upon mixing. Capillary suspensions are a particular case of ternary particle–liquid–liquid systems. Indeed, depending on the relative ratio among particles and liquids, other systems like Pickering emulsions, bigels (bicontinuous gels), spherical agglomerates, and granular materials, can be obtained (Fig. 4.7) [135].

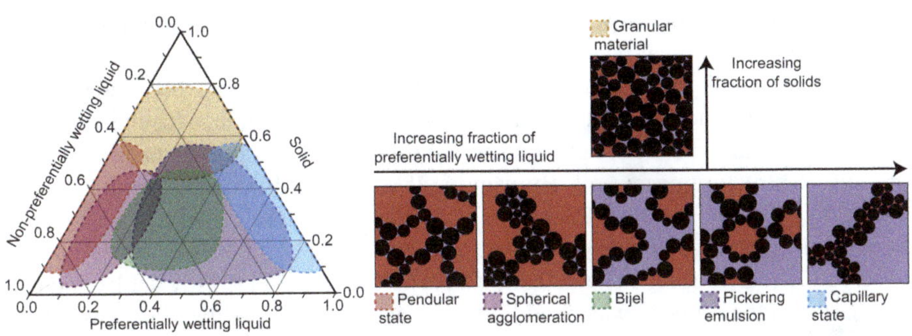

Fig. 4.7 On the left, a ternary diagram of particle–liquid–liquid systems depicting estimated areas of stability for different states based on the relative volume fractions. On the right, a schematic representation of each state depicted in the ternary diagram. Capillary suspensions occupy the perimeter of the ternary diagram, where one of the secondary liquids appears as a minor phase across various particle volume fractions. (Figure reproduced from [135] with permission from Elsevier)

The formation of liquid bridges among particles in capillary suspensions results in the structuring of the system through capillary forces. The strength of these forces is influenced by the dimension of the particles, their distance, the interfacial tension between the two fluids in contact with the particles, and the three-phase wetting angle that the secondary liquid forms against the solid surface in presence of the bulk liquid [135, 136]. Capillary suspensions comprise two different states: pendular and capillary, which depend on the wettability of the particles with respect to the two immiscible liquids used in the ternary system. In particular, the pendular state refers to ternary systems where the secondary liquid preferentially wets the particles that are dispersed in a non-preferentially wetting liquid. The secondary liquid binds individual particles together forming pendular bridges (Fig. 4.7). In the examples reported below, most of the oleogels are formed through capillary suspensions in the pendular state, since the particles are mainly hydrophilic and are dispersed in oil, where water is used as a secondary liquid. On the other hand, capillary state refers to systems where particles are dispersed in a bulk liquid that preferentially wets them, and a secondary liquid that does not preferentially wet the particles is added to the system. The secondary liquid fills the gaps among particles forming clusters that are kept together by the capillary forces from the bulk liquid (Fig. 4.7) [135, 137]. The capillary suspension states can be differentiated based on the saturation (S) level of the wetting liquid, expressed as the ratio of the volume of wetting liquid to the total liquid volume of the system. In the capillary state, the S value approaches unity, while in the pendular state, it is close to zero. Regardless of the capillary suspension state, a transition from a liquid-like unstructured suspension to a gel-like material is always observed [135].

One of the earliest studies on edible oil structuring through capillary suspensions was described by Hoffmann et al. [136]. In this study, the authors used starch granules and cocoa particles at 30–35% volume fraction and 10–30% water to structure sunflower oil. The rheological properties of the system were dependent on particle and water volume fractions, and the addition of glycerol to water increased gel strength. However, the order of secondary liquid addition did not influence the rheological behavior of the system, leading to similar results whether the water was added to the starch-oil suspension, or it was absorbed onto the dry starch granule surface. In another early study on oil structuration, Mustafa et al. [70] used particles derived from agri-food waste such as tomato peels and spent coffee grounds to structure peanut oil. By using a 25% volume fraction of particles in oil and adding 17–57% volume fraction of water relative to the oil using a high-speed homogenizer, a transition from a dispersion to a semi-solid gel-like material was observed. The hydrophilic character of the particle surface led to the establishment of capillary bridges upon water addition, arranging the particles into a three-dimensional network that entrapped the oil phase. The authors demonstrated that the resulting material stiffened with increasing water content (up to a certain point) and with decreasing particle size through high-pressure homogenization [70].

Following these first studies on the use of capillary bridges in oil structuring, other recent works showed the effectiveness of proteins, cellulose particles, and fibers in semi-direct oleogelation methods, although proteins and fibers have been typically used in indirect methods, whereas cellulose particles have been used in hot and cold direct methods [69, 138]. In particular, hydrophilic modified zein particles (size of ~200 nm), heat-set whey protein isolate particles, cellulose particles with an average size of 25 μm, particles from fiber-rich fractions of yellow pea such as epidermal pea cell wall (average size of 20 μm) and pea hull (average size of 300 μm) were used to structure soybean, sunflower, algal, castor, and medium chain triacylglycerol (MCT) oils [140–142]. In general, 10% to 40% particles were dispersed in oils, followed by the addition of water at concentrations between 2% and 30% and the system was mixed using ball milling, high shear mixing, or magnetic stirring. Eventually, all systems formed oleogels at an oil-water-particle ratio that allowed a capillary suspension in the pendular state to be formed. When full pendular bridging (also called funicular state) among particles was formed, *i.e.,* each particle was interconnected with neighbor particles through water bridges and participated in network formation, stiff oleogels were produced [139]. Oleogel rheological properties could be further modulated by modifying ionic strength, solid content, and pH of the secondary liquid [139, 140].

Finally, as seen from the above examples, the selection of particles and liquids is fundamental to obtain an oleogel through capillary suspensions and to be able to tailor its rheological properties. On this topic, Jarray et al. [142] introduced a new approach that can predict the formation of capillary suspensions in the pendular state and their rheological properties using the HSP computed from molecular dynamics simulations. The authors elucidated through simulations and experimental work that the gel strength of capillary suspensions obtained with hydrophilic silica particles arises from the intermolecular interactions of its components, where the interfacial tension between the bulk and secondary liquid drives the gel strength up to a certain limit, after which the secondary liquid–particle polar interactions and hydrogen bond formation play a major role. The approach proposed by Jarray et al. [142] could potentially be extended to any particles. Furthermore, by using the HSP theory, one can select the proper secondary liquid, which can lead to the formation of capillary suspensions and calculate the resulting gel strength, reducing the need for extensive experimental work.

4.4 New Classification of Oleogels

As discussed in Sect. 4.3, oleogel preparation methods can involve water or other solvents, heating and cooling cycles, drying procedures, and gelators with different molecular weights. However, the difference among oleogelation methods can also be explained by taking into consideration (a) heat energy that the oil is subjected to, (b) the overall electrical energy consumed by all devices during oleogelation, and (c) overall oleogelation time. These three factors are imperative in oleogel production since they are important criteria affecting oxidative and/or storage stability of the oleogels, as well as their sustainability, upscaling ability, and overall production cost. To make use of the benefits of oleogels in society, their production must be scaled up from the laboratory to an industrial level. Achieving this requires contextualizing the classification of oleogel preparation methods within an industrial framework. However, the conventional classification of oleogels based on either the molecular weight of the gelators (LMWGs *vs.* HMWGs) or the oleogelation methods (direct, indirect, and semi-direct), does not provide pertinent information for industrial applications.

In a recently published article [143], we discussed the importance of these three industrially relevant factors—overall heat, electrical energy, and time—in obtaining a new classification system for oleogel preparation methods. To this aim, we calculated these three parameters for 216 laboratory-conducted oleogelation cases retrieved from the literature. The overall heat that the oil is subjected to during oleogelation was calculated by integrating the time-temperature (TT) profiles. The overall electrical energy consumption was calculated by summing the electrical consumption of all devices used during oleogel preparation. The overall time was calculated by summing the time necessary for every single action during oleogel preparation. Each oleogel preparation procedure was then plotted in a 3D space where the three axes corresponded to heat, electrical energy, and time. Each oleogel preparation case was distributed in a different position within the scatter plot and different groups were visible. By applying the K-means clustering algorithm followed by the scree plot analysis, we were able to determine an optimal number of clusters. From this clustering, we developed a new oleogel classification where each oleogel preparation case (type of method, gelator concentration, etc.) was assigned to a new class based on its level of input: low, medium, or high (Fig. 4.8). The low-input approaches require low inputs (heat, electrical energy, and time), and are the optimal cases in terms of oxidative stability, sustainability, and industrial relevance. The medium-input approaches need a medium amount of at least one input, thereby making them potentially unattractive; however, they can still be considered when other options from the low-input approaches are not applicable. The high-input approaches require a high amount of at least one input. These methods are currently the least attractive ones.

Our new classification challenges the commonly held belief that oil subjected to hot direct methods undergoes more severe thermal cycles and higher heat exposure than those processed *via* indirect methods. For example, the emulsion-templated approach using oven-drying (a common indirect method) subjects the oil to higher heat treatment than some hot direct methods. Instead, we propose that oleogelation methods should be evaluated on a case-by-case basis.

The results of our novel classification also highlighted that scaling up oleogelation requires considering additional aspects that usually were not considered in the laboratory-conducted oleogelation cases, such as proper control of the cooling rate, as it significantly affects the heat energy that the oil is subjected to. The cooling rate is also a crucial factor in achieving consistency in physical properties (e.g., minimum gelation concentration, melting temperature, melting enthalpy, yield stress, solid phase content, and oil binding capacity) of oleogel during scale-up [144]. A constant surface area-to-volume ratio was recently suggested to be a key factor in scaling up oleogel production. By keeping the ratio constant, authors

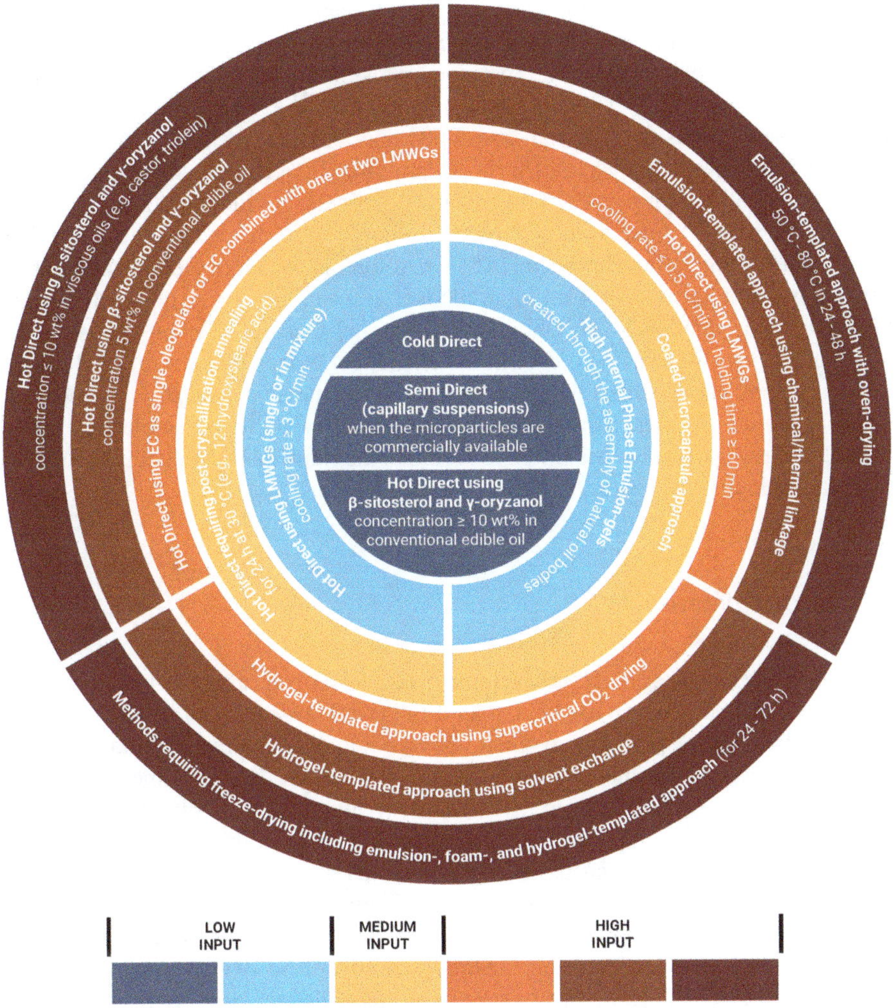

Fig. 4.8 Schematic visualization of a novel oleogel classification to high-, medium-, and low-input methods based on the overall heat, electrical energy consumption, and time input for each oleogel preparation method. EC and LMWGs stand for ethylcellulose and low molecular weight gelators, respectively

proved that a uniform heat dissipation could be achieved, leading to homogeneous gels with consistent physical properties (comparing small and large oleogel batches) [144].

It should be noted that the sustainability and feasibility of oleogelation approaches depend on the commercial availability of the gelators or ingredients that are required in the methods. For example, a semi-direct oleogelation approach may be challenging to be scaled up if the microparticles used in the method are not commercially available or if preparing them requires a high energy input. However, when microparticles are commercially available, this method requires very low inputs, and therefore can be considered a gentle and sustainable way to produce oleogels at large scale in the food industry. Moreover, various factors beyond the technical aspects of oleogelation must also be taken into account when considering an industrial-scale aim. These factors may include the cost and availability of gelators, as well as the capital and operating costs associated with different oleogelation approaches. The health implications of the gelators, the physicochemical and viscoelastic properties of the resulting oleogels, possible regulatory barriers, and logistics and transportation expenses should also be accounted for when selecting an oleogelation approach [145]. Therefore, a low-input oleogel preparation method will not necessarily lead to an oleogel that can be used in all food applications at any production level. Obtaining an oleogel that is applicable across many food categories still remains a significant challenge. Nonetheless, our novel classification system can potentially help in understanding the effects of different methods on the oxidative stability, sustainability, and industrial viability of the oleogel. It also provides a fresh perspective and tool to facilitate the transition of oleogel preparation from lab to industrial scale.

4.5 Conclusions

In this chapter, oleogel preparation methods developed until early 2023 were reviewed and described giving the reader a full overview of the topic. The development of the oleogel preparation methods was initially inspired by traditional oil structuring techniques and solid fats present in nature. The first methods to be developed were straightforward and intuitive, involving the direct addition of gelators, oil-soluble compounds that can directly structure oil based on solvent–solvent, solvent–particle, and particle–particle interactions. These early works successfully discovered most of the regularly used gelators in the field, establishing the foundation of oleogel research. As the potential of oleogels became apparent, researchers began to explore new methods based on the development of one or more intermediate stages before obtaining an oleogel, including hydrogels, emulsions, foams, aerogels, fibers, and particle dispersions. These efforts have led to breakthroughs aimed at (i) broadening the range of gelators to increase consumer acceptability, (ii) addressing the disadvantages of previous oleogelation methods, and (iii) tailoring oleogel properties envisioning specific applications. In this chapter, a new classification of oleogels based on industrially relevant parameters has also been discussed, in an attempt to bring the benefits of oleogels a step closer to society. The research on oleogels is still ongoing and new methods are expected to emerge in the future.

Acknowledgments The authors acknowledge Jane and Aatos Erkko Foundation (grant number 200075), the University of Helsinki (decision letter number HY/217/05.01.07/2020), and Business Finland (project number 1871/31/2021), for their funding. Fabio Valoppi also acknowledges the European Union's Horizon 2020 research and innovation program funding under the Marie Skłodowska-Curie grant agreement No. 836071.

References

1. Marangoni AG, van Duynhoven JPM, Acevedo NC, Nicholson RA, Patel AR (2020) Advances in our understanding of the structure and functionality of edible fats and fat mimetics. Soft Matter 16:289–306. https://doi.org/10.1039/c9sm01704f
2. Singh A, Auzanneau FI, Rogers MA (2017) Advances in edible oleogel technologies – a decade in review. Food Res Int 97:307–317. https://doi.org/10.1016/j.foodres.2017.04.022
3. Meldrum DR, Morris MA, Gambone JC (2017) Obesity pandemic: causes, consequences, and solutions-but do we have the will? Fertil Steril 107:833–839. https://doi.org/10.1016/j.fertnstert.2017.02.104
4. Azaïs-Braesco V, Brighenti F, Paoletti R, Peracino A, Scarborough P, Visioli F, Vögele C (2009) Healthy food and healthy choices: a new european profile approach. Atheroscler Suppl 10:1–11. https://doi.org/10.1016/j.atherosclerosissup.2009.09.001
5. Kavya M, Udayarajan C, Fabra MJ, Lopez-Rubio A, Nisha P (2022) Edible oleogels based on high molecular weight oleogelators and its prospects in food applications. Crit Rev Food Sci Nutr:1–24. https://doi.org/10.1080/10408398.2022.2142195
6. Gunstone FD (2004) The chemistry of oils and fats: sources, composition, properties, and uses. CRC Press, Boca Raton
7. Tambun R, Ferani DG, Afrina A, Tambun JAA, Tarigan IAA (2019) Fatty acid direct production from palm kernel oil. IOP Conf Ser Mater Sci Eng 505:012115. https://doi.org/10.1088/1757-899X/505/1/012115
8. Gandolfo FG, Bot A, Flöter E (2004) Structuring of edible oils by long-chain fa, fatty alcohols, and their mixtures. J Am Oil Chem Soc 81:1–6. https://doi.org/10.1007/s11746-004-0851-5
9. Kreutzer UR (1984) Manufacture of fatty alcohols based on natural fats and oils. J Am Oil Chem Soc 61:343–348. https://doi.org/10.1007/bf02678792
10. Monick JA (1979) Fatty alcohols. J Am Oil Chem Soc 56:853A–860A. https://doi.org/10.1007/BF02667462
11. Marangoni AG, Garti N (n.d.) 9.2 fatty alcohols. Edible oleogels – structure and health implications, 2nd edn. AOCS Press
12. Valoppi F, Calligaris S, Marangoni AG (2016) Phase transition and polymorphic behavior of binary systems containing fatty alcohols and peanut oil. Cryst Growth Des 16:4209–4215. https://doi.org/10.1021/acs.cgd.6b00145
13. Clogston J, Rathman J, Tomasko D, Walker H, Caffrey M (2000) Phase behavior of a monoacylglycerol: (myverol 18–99k)/water system. Chem Phys Lipids 107:191–220. https://doi.org/10.1016/S0009-3084(00)00182-1
14. Nicholson RA, Marangoni AG (2020) Enzymatic glycerolysis converts vegetable oils into structural fats with the potential to replace palm oil in food products. Nat Food 1:684–692. https://doi.org/10.1038/s43016-020-00160-1
15. Wagner K, Davidovich-Pinhas M (2023) Di-acylglycerides as oil structuring agents. Food Struct 36. https://doi.org/10.1016/j.foostr.2023.100320
16. Bot F, Cossuta D, O'Mahony JA (2021) Inter-relationships between composition, physicochemical properties and functionality of lecithin ingredients. Trends Food Sci Technol 111: 261–270. https://doi.org/10.1016/j.tifs.2021.02.028
17. Small DM (1967) Phase equilibria and structure of dry and hydrated egg lecithin. J Lipid Res 8:551–557. https://doi.org/10.1016/S0022-2275(20)38874-X
18. Montalvo G, Pons R, Zhang G, Díaz M, Valiente M (2013) Structure and phase equilibria of the soybean lecithin/peg 40 monostearate/water system. Langmuir 29:14369–14379. https://doi.org/10.1021/la402764w
19. Zhao, F., Li, R., Liu, Y. & Chen, H. (2022) Perspectives on lecithin from egg yolk: Extraction, physicochemical properties, modification, and applications. Front. Nutr. 9: 1082671. https://doi.org/10.3389/fnut.2022.1082671

20. Karaoglu O, Alpdogan G, Ozdemir IS, Ertas E (2020) Solid phase extraction of β-sitosterol and α-tocopherol from sunflower oil deodorizer distillate using desilicated zeolite. Grasas Aceites 71:370–e370. https://doi.org/10.3989/gya.0570191
21. Ionin VA, Kazachenko AS, Skripnikov AM, Veprikova EV, Belash MY, Taran OP (2021) Experimental and mathematical optimization of the β-sitosterol extraction from mechanically activated pine bark. J Sib Fed Univ: Chem 14:302–314. https://doi.org/10.17516/1998-2836-0248
22. Ibrahim NH, Mahmud MS, Nurdin S (n.d.) Microwave-assisted extraction of β-sitosterol from cocoa shell waste. IOP Conf Ser Mater Sci Eng 2020
23. Bin Sayeed MS, Ameen SS (2015) Beta-sitosterol: a promising but orphan nutraceutical to fight against cancer. Nutr Cancer 67:1216–1222. https://doi.org/10.1080/01635581.2015.1087042
24. Kumar P, Yadav D, Kumar P, Panesar PS, Bunkar DS, Mishra D, Chopra HK (2016) Comparative study on conventional, ultrasonication and microwave assisted extraction of γ-oryzanol from rice bran. J Food Sci Technol 53:2047–2053. https://doi.org/10.1007/s13197-016-2175-2
25. Wang L, Lin Q, Yang T, Liang Y, Nie Y, Luo Y, Shen J, Fu X, Tang Y, Luo F (2017) Oryzanol modifies high fat diet-induced obesity, liver gene expression profile, and inflammation response in mice. J Agric Food Chem 65:8374–8385. https://doi.org/10.1021/acs.jafc.7b03230
26. Sabet S, Rashidinejad A, Melton LD, Zujovic Z, Akbarinejad A, Nieuwoudt M, Seal CK, McGillivray DJ (2021) The interactions between the two negatively charged polysaccharides: gum arabic and alginate. Food Hydrocoll 112:106343. https://doi.org/10.1016/j.foodhyd.2020.106343
27. BeMiller JN (2018) Carbohydrate chemistry for food scientists. Wiley-Blackwell, Hoboken
28. Azevedo G, Torres MD, Sousa-Pinto I, Hilliou L (2015) Effect of pre-extraction alkali treatment on the chemical structure and gelling properties of extracted hybrid carrageenan from chondrus crispus and ahnfeltiopsis devoniensis. Food Hydrocoll 50:150–158. https://doi.org/10.1016/j.foodhyd.2015.03.029
29. Jiang J-L, Zhang W-Z, Ni W-X, Shao J-W (2021) Insight on structure-property relationships of carrageenan from marine red algal: a review. Carbohydr Polym 257:117642. https://doi.org/10.1016/j.carbpol.2021.117642
30. Song EH, Shang J, Ratner DM (2012) Polysaccharides. In: Matyjaszewski K, Möller M (eds) Polymer science: a comprehensive reference, 1st edn. Elsevier, Amsterdam, pp 137–155
31. Pighinelli L (2019) Methods of chitin production a short review. Am J Biomed Sci Res 3:307–314. https://doi.org/10.34297/ajbsr.2019.03.000682
32. National Center for Biotechnology Information (2023) Pubchem compound summary for cid 6857375 [Online]. Available: https://pubchem.ncbi.nlm.nih.gov/compound/6857375. Accessed 20 Mar 2023
33. Winholz M (1983) The merck index: an encyclopedia of chemicals, drugs, and biologicals. Merck & Co, Rahway
34. National Center for Biotechnology Information (2023) Pubchem compound summary for cid 24832091, ethyl cellulose [Online]. Available: https://pubchem.ncbi.nlm.nih.gov/compound/Ethyl-cellulose. Accessed 20 Mar 2023
35. Gravelle AJ, Barbut S, Marangoni AG (2012) Ethylcellulose oleogels: manufacturing considerations and effects of oil oxidation. Food Res Int 48:578–583. https://doi.org/10.1016/j.foodres.2012.05.020
36. Sarkar N (1979) Thermal gelation properties of methyl and hydroxypropyl methylcellulose. J Appl Polym Sci 24:1073–1087. https://doi.org/10.1002/app.1979.070240420
37. Dürig T, Karan K (2019) Binders in wet granulation. In: Narang AS, Badawy SIF (eds) Handbook of pharmaceutical wet granulation. Academic, Cambridge, MA, pp 317–349
38. Lundqvist R (1999) Molecular weight studies on hydroxypropyl methylcellulose ii. Intrinsic viscosity. Int J Polym 5:61–84. https://doi.org/10.1080/10236669908014174

39. Cox LM, Cho I, Young SA, Anderson WHK, Waters BJ, Hung SC, Gao Z, Mahana D, Bihan M, Alekseyenko AV, Methé BA, Blaser MJ (2013) The nonfermentable dietary fiber hydroxypropyl methylcellulose modulates intestinal microbiota. FASEB J 27:692–702. https://doi.org/10.1096/fj.12-219477
40. National Center for Biotechnology Information (2023 Pubchem compound summary for cid 525128, methyl cellulose [Online]. Available: https://pubchem.ncbi.nlm.nih.gov/compound/Methyl-cellulose. Accessed 20 Mar 2023
41. Singthong J, Cui SW, Ningsanond S, Douglas Goff H (2004) Structural characterization, degree of esterification and some gelling properties of krueo ma noy (cissampelos pareira) pectin. Carbohydr Polym 58:391–400. https://doi.org/10.1016/j.carbpol.2004.07.018
42. Chalapud MC, Baümler ER, Carelli AA, Salgado-Cruz MD, Morales-Sánchez E, Rentería-Ortega M, Calderón-Domínguez G (2022) Pectin films with recovered sunflower waxes produced by electrospraying. Membranes 12. https://doi.org/10.3390/membranes12060560
43. McArdle R, Hamill R (2011) Utilisation of hydrocolloids in processed meat systems. In: Kerry JP, Kerry JF (eds) Processed meats: improving safety, nutrition and quality. Woodhead Publishing, Chichester, pp 243–269
44. Miranda AL, Costa SS, Assis DDJ, Jesus CS, Guimarães AG, Druzian JI (2020) Influence of strain and fermentation time on the production, composition, and properties of xanthan gum. J Appl Polym Sci 137:48557. https://doi.org/10.1002/app.48557
45. Yang J, Zamani S, Liang L, Chen L (2021) Extraction methods significantly impact pea protein composition, structure and gelling properties. Food Hydrocoll 117:106678. https://doi.org/10.1016/j.foodhyd.2021.106678
46. Emkani M, Oliete B, Saurel R (2021) Pea protein extraction assisted by lactic fermentation: impact on protein profile and thermal properties. Foods 10:549. https://doi.org/10.3390/foods10030549
47. Djemaoune Y, Cases E, Saurel R (2019) The effect of high-pressure microfluidization treatment on the foaming properties of pea albumin aggregates. J Food Sci 84:2242–2249. https://doi.org/10.1111/1750-3841.14734
48. Shand PJ, Ya H, Pietrasik Z, Wanasundara PKJPD (2007) Physicochemical and textural properties of heat-induced pea protein isolate gels. Food Chem 102:1119–1130. https://doi.org/10.1016/j.foodchem.2006.06.060
49. Berghout JAM, Boom RM, van der Goot AJ (2015) Understanding the differences in gelling properties between lupin protein isolate and soy protein isolate. Food Hydrocoll 43:465–472. https://doi.org/10.1016/j.foodhyd.2014.07.003
50. Walstra P, Jenness R (1984) Dairy chemistry & physics. Wiley, Chichester
51. Sabet S, Seal CK, Swedlund PJ, McGillivray DJ (2020) Depositing alginate on the surface of bilayer emulsions. Food Hydrocoll 100:105385. https://doi.org/10.1016/j.foodhyd.2019.105385
52. Doan CD, To CM, De Vrieze M, Lynen F, Danthine S, Brown A, Dewettinck K, Patel AR (2017) Chemical profiling of the major components in natural waxes to elucidate their role in liquid oil structuring. Food Chem 214:717–725. https://doi.org/10.1016/j.foodchem.2016.07.123
53. Winkler-Moser JK, Anderson J, Felker FC, Hwang H-S (2019) Physical properties of beeswax, sunflower wax, and candelilla wax mixtures and oleogels. J Am Oil Chem Soc 96:1125–1142. https://doi.org/10.1002/aocs.12280
54. Rojas Molina R, de León Zapata MA, Jasso Cantú D, Aguilar N, C. b. (2019) Pasado, presente y futuro de la candelilla. Rev Mex Cienc Forestales 2:7–18. https://doi.org/10.29298/rmcf.v2i6.571
55. Núñez-García IC, Rodríguez-Flores LG, Guadiana-De-Dios MH, González-Hernández MD, Martínez-Ávila GCG, Gallegos-Infante JA, González-Laredo R, Rosas-Flores W, Martínez-Gómez VJ, Rojas R, Villanueva-Fierro I, Rutiaga-Quiñones M (2022) Candelilla wax extracted by traditional method and an ecofriendly process: assessment of its chemical,

structural and thermal properties. Molecules 27:3735. https://doi.org/10.3390/molecules27123735
56. Steinle JV (1936) Carnauba wax an expedition to its source. Ind Eng Chem 28:1004–1008. https://doi.org/10.1021/ie50321a003
57. de Freitas CAS, de Sousa PHM, Soares DJ, da Silva JYG, Benjamin SR, Guedes MIF (2019) Carnauba wax uses in food – a review. Food Chem 291:38–48. https://doi.org/10.1016/j.foodchem.2019.03.133
58. Blake AI, Co ED, Marangoni AG (2014) Structure and physical properties of plant wax crystal networks and their relationship to oil binding capacity. J Am Oil Chem Soc 91:885–903. https://doi.org/10.1007/s11746-014-2435-0
59. Vali SR, Ju Y-H, Kaimal TNB, Chern Y-T (2005) A process for the preparation of food-grade rice bran wax and the determination of its composition. J Am Oil Chem Soc 82:57–64. https://doi.org/10.1007/s11746-005-1043-z
60. Leibovitz Z, Ruckenstein C (1984) Winterization of sunflower oil. J Am Oil Chem Soc 61: 870–872. https://doi.org/10.1007/BF02542153
61. Marangoni AG (2012) Organogels: an alternative edible oil-structuring method. J Am Oil Chem Soc 89:749–780. https://doi.org/10.1007/s11746-012-2049-3
62. Manzocco L, Valoppi F, Calligaris S, Andreatta F, Spilimbergo S, Nicoli MC (2017) Exploitation of κ-carrageenan aerogels as template for edible oleogel preparation. Food Hydrocoll 71:68–75. https://doi.org/10.1016/j.foodhyd.2017.04.021
63. de Vries A, Hendriks J, van der Linden E, Scholten E (2015) Protein oleogels from protein hydrogels via a stepwise solvent exchange route. Langmuir 31:13850–13859. https://doi.org/10.1021/acs.langmuir.5b03993
64. Romoscanu AI, Mezzenga R (2006) Emulsion-templated fully reversible protein-in-oil gels. Langmuir 22:7812–7818. https://doi.org/10.1021/la060878p
65. Patel AR, Cludts N, Bin Sintang MD, Lewille B, Lesaffer A, Dewettinck K (2014) Polysaccharide-based oleogels prepared with an emulsion-templated approach. ChemPhysChem 15:3435–3439. https://doi.org/10.1002/cphc.201402473
66. Li J, Zhang H (2023) Efficient fabrication, characterization, and in vitro digestion of aerogel-templated oleogels from a facile method: electrospun short fibers. Food Hydrocoll 135: 108185. https://doi.org/10.1016/j.foodhyd.2022.108185
67. Rubio-Valle JF, Valencia C, Sánchez M, Martín-Alfonso JE, Franco JM (2023) Oil structuring properties of electrospun kraft lignin/cellulose acetate nanofibers for lubricating applications: influence of lignin source and lignin/cellulose acetate ratio. Cellulose (Lond):1553–1566. https://doi.org/10.1007/s10570-022-04963-2
68. Valoppi F, Schavikin J, Lassila P, Laidmäe I, Heinämäki J, Hietala S, Haeggström E, Salmi A (2023) Formation and characterization of oleogels obtained via direct dispersion of ultrasound-enhanced electrospun nanofibers and cold milling. Food Struct 37:100338. https://doi.org/10.1016/j.foostr.2023.100338
69. Patel AR (2017) Methylcellulose-coated microcapsules of palm stearine as structuring templates for creating hybrid oleogels. Mater Chem Phys 195:268–274. https://doi.org/10.1016/j.matchemphys.2017.03.059
70. David A, David M, Lesniarek P, Corfias E, Pululu Y, Delample M, Snabre P (2021) Oleogelation of rapeseed oil with cellulose fibers as an innovative strategy for palm oil substitution in chocolate spreads. J Food Eng 292. https://doi.org/10.1016/j.jfoodeng.2020.110315
71. Mustafa W, Pataro G, Ferrari G, Donsì F (2018) Novel approaches to oil structuring via the addition of high-pressure homogenized agri-food residues and water forming capillary bridges 236:9–18
72. Barros AS, Mafra I, Ferreira D, Cardoso S, Reis A, Lopes da Silva JA, Delgadillo I, Rutledge DN, Coimbra MA (2002) Determination of the degree of methylesterification of pectic polysaccharides by ft-ir using an outer product pls1 regression. Carbohydr Polym 50:85–94. https://doi.org/10.1016/S0144-8617(02)00017-6

73. Manrique GD, Lajolo FM (2002) Ft-ir spectroscopy as a tool for measuring degree of methyl esterification in pectins isolated from ripening papaya fruit. Postharvest Biol Technol 25:99–107. https://doi.org/10.1016/S0925-5214(01)00160-0
74. Yang S, Zhu M, Wang N, Cui X, Xu Q, Saleh ASM, Duan Y, Xiao Z (2018) Influence of oil type on characteristics of β-sitosterol and stearic acid based oleogel. Food Biophys 13:362–373. https://doi.org/10.1007/s11483-018-9542-7
75. Yao Y, Zhou H, Liu W, Li C, Wang S (2021) The effect of cooling rate on the microstructure and macroscopic properties of rice bran wax oleogels. J Oleo Sci 70:135–143. https://doi.org/10.5650/jos.ess20112
76. Sawalha H, Venema P, Bot A, Flöter E, Adel RD, van der Linden E (2015) The phase behavior of gamma-oryzanol and beta-sitosterol in edible oil. J Am Oil Chem Soc 92:1651–1659. https://doi.org/10.1007/s11746-015-2731-3
77. Wang Z, Chandrapala J, Truong T, Farahnaky A (2022) Oleogels prepared with low molecular weight gelators: texture, rheology and sensory properties, a review. Crit Rev Food Sci Nutr:1–45. https://doi.org/10.1080/10408398.2022.2027339
78. Rogers MA (2018) Hansen solubility parameters as a tool in the quest for new edible oleogels. J Am Oil Chem Soc 95:393–405. https://doi.org/10.1002/aocs.12050
79. Patel AR (2017) A colloidal gel perspective for understanding oleogelation. Curr Opin Food Sci 15:1–7. https://doi.org/10.1016/j.cofs.2017.02.013
80. Gao J, Wu S, Rogers MA (2012) Harnessing hansen solubility parameters to predict organogel formation. J Mater Chem 22. https://doi.org/10.1039/c2jm32056h
81. Hildebrand J, Scott RL (1950) The solubility of nonelectrolytes. Reinhold, New York
82. Hildebrand J, Scott RL (1962) Regular solutions. Prentice-Hall, Englewood Cliffs
83. Hansen CM (2007) Hansen solubility parameters: a user's handbook. CRC Press, Boca Raton
84. Raynal M, Bouteiller L (2011) Organogel formation rationalized by hansen solubility parameters. Chem Commun 47:8271–8273. https://doi.org/10.1039/c1cc13244j
85. Diehn KK, Oh H, Hashemipour R, Weiss RG, Raghavan SR (2014) Insights into organogelation and its kinetics from hansen solubility parameters. Toward a priori predictions of molecular gelation 10:2632–2640. https://doi.org/10.1039/c3sm52297k
86. Abraham S, Lan Y, Lam RS, Grahame DA, Kim JJ, Weiss RG, Rogers MA (2012) Influence of positional isomers on the macroscale and nanoscale architectures of aggregates of racemic hydroxyoctadecanoic acids in their molecular gel, dispersion, and solid states. Langmuir 28:4955–4964. https://doi.org/10.1021/la204412t
87. Lan Y, Rogers MA (2015) 12-hydroxystearic acid safins in aliphatic diols – a molecular oddity. CrystEngComm 17:8031–8038. https://doi.org/10.1039/c5ce00652j
88. Rogers MA, Marangoni AG (2016) Kinetics of 12-hydroxyoctadecanoic acid safin crystallization rationalized using hansen solubility parameters. Langmuir 32:12833–12841. https://doi.org/10.1021/acs.langmuir.6b03476
89. Lan Y, Corradini MG, Liu X, May TE, Borondics F, Weiss RG, Rogers MA (2014) Comparing and correlating solubility parameters governing the self-assembly of molecular gels using 1,3:2,4-dibenzylidene sorbitol as the gelator. Langmuir 30:14128–14142. https://doi.org/10.1021/la5008389
90. Bonnet J, Suissa G, Raynal M, Bouteiller L (2015) Organogel formation rationalized by hansen solubility parameters: influence of gelator structure 11:2308–2312. https://doi.org/10.1039/c5sm00017c
91. Cuello V, Corradini MG, Rogers M, Zarza G (2020) Data science & engineering into food science: a novel big data platform for low molecular weight gelators' behavioral analysis. J Comput Sci Technol 20:72–79. https://doi.org/10.24215/16666038.20.e08
92. Patel AR, Mankoč B, Bin Sintang MD, Lesaffer A, Dewettinck K (2015) Fumed silica-based organogels and 'aqueous-organic' bigels. RSC Adv 5:9703–9708. https://doi.org/10.1039/c4ra15437a

93. Nordström J, Matic A, Sun J, Forsyth M, MacFarlane DR (2010) Aggregation, ageing and transport properties of surface modified fumed silica dispersions:6. https://doi.org/10.1039/b921488g
94. Whitby CP (2020) Structuring edible oils with fumed silica particles. Front Sustain Food Syst 4. https://doi.org/10.3389/fsufs.2020.585160
95. Sulaiman NS, Sintang MD, Mohd Zaini H, Munsu E, Matajun P, Pindi W (2022) Applications of protein crosslinking in food products. Int Food Res J 29:723–739. https://doi.org/10.47836/ifrj.29.4.01
96. Patel AR, Rajarethinem PS, Cludts N, Lewille B, De Vos WH, Lesaffer A, Dewettinck K (2015) Biopolymer-based structuring of liquid oil into soft solids and oleogels using water-continuous emulsions as templates. Langmuir 31:2065–2073. https://doi.org/10.1021/la502829u
97. Tavernier I, Patel AR, Van der Meeren P, Dewettinck K (2017) Emulsion-templated liquid oil structuring with soy protein and soy protein: K-carrageenan complexes. Food Hydrocoll 65: 107–120. https://doi.org/10.1016/j.foodhyd.2016.11.008
98. Abdolmaleki K, Alizadeh L, Nayebzadeh K, Hosseini SM, Shahin R (2020) Oleogel production based on binary and ternary mixtures of sodium caseinate, xanthan gum, and guar gum: optimization of hydrocolloids concentration and drying method. J Texture Stud 51:290–299. https://doi.org/10.1111/jtxs.12469
99. Alizadeh L, Abdolmaleki K, Nayebzadeh K, Hosseini SM (2020) Oleogel fabrication based on sodium caseinate, hydroxypropyl methylcellulose, and beeswax: effect of concentration, oleogelation method, and their optimization. J Am Oil Chem Soc 97:485–496. https://doi.org/10.1002/aocs.12341
100. Feichtinger A, Scholten E (2020) Preparation of protein oleogels: effect on structure and functionality. Foods 9:1745. https://doi.org/10.3390/foods9121745
101. Stortz TA, Marangoni AG (2014) The replacement for petrolatum: thixotropic ethylcellulose oleogels in triglyceride oils. Green Chem 16:3064–3070. https://doi.org/10.1039/c4gc00052h
102. Gravelle AJ, Davidovich-Pinhas M, Barbut S, Marangoni AG (2017) Influencing the crystallization behavior of binary mixtures of stearyl alcohol and stearic acid (sosa) using ethylcellulose. Food Res Int 91:1–10. https://doi.org/10.1016/j.foodres.2016.11.024
103. Gao ZM, Yang XQ, Wu NN, Wang LJ, Wang JM, Guo J, Yin SW (2014) Protein-based pickering emulsion and oil gel prepared by complexes of zein colloidal particles and stearate. J Agric Food Chem 62:2672–2678. https://doi.org/10.1021/jf500005y
104. Nikiforidis CV (2019) Structure and functions of oleosomes (oil bodies). Adv Colloid Interf Sci 274:102039. https://doi.org/10.1016/j.cis.2019.102039
105. Mert B, Vilgis TA (2021) Hydrocolloid coated oleosomes for development of oleogels. Food Hydrocoll 119:106832. https://doi.org/10.1016/j.foodhyd.2021.106832
106. Zambrano JC, Vilgis TA (2023) Tunable oleosome-based oleogels: influence of polysaccharide type for polymer bridging-based structuring. Food Hydrocoll 137. https://doi.org/10.1016/j.foodhyd.2022.108399
107. Patel AR, Schatteman D, Lesaffer A, Dewettinck K (2013) A foam-templated approach for fabricating organogels using a water-soluble polymer. RSC Adv 3:22900–22903. https://doi.org/10.1039/c3ra44763d
108. Shibata T, Chemistry RS, o. (2011) Cellulose and its derivatives in medical use. In: Williams P (ed) Renewable resources for functional polymers and biomaterials. The Royal Society of Chemistry
109. Abdollahi M, Goli SAH, Soltanizadeh N (2020) Physicochemical properties of foam-templated oleogel based on gelatin and xanthan gum. Eur J Lipid Sci Technol 122:1900196. https://doi.org/10.1002/ejlt.201900196
110. Mohanan A, Tang YR, Nickerson MT, Ghosh S (2020) Oleogelation using pulse protein-stabilized foams and their potential as a baking ingredient. RSC Adv 10:14892–14905. https://doi.org/10.1039/c9ra07614j

111. Wei F, Lu M, Li J, Xiao J, Rogers MA, Cao Y, Lan Y (2022) Construction of foam-templated oleogels based on rice bran protein. Food Hydrocoll 124:107245. https://doi.org/10.1016/j.foodhyd.2021.107245
112. de Vries A, Wesseling A, van der Linden E, Scholten E (2017) Protein oleogels from heat-set whey protein aggregates 486:75–83. https://doi.org/10.1016/j.jcis.2016.09.043
113. de Vries A, Gomez YL, van der Linden E, Scholten E (2017) The effect of oil type on network formation by protein aggregates into oleogels. RSC Adv 7:11803–11812. https://doi.org/10.1039/c7ra00396j
114. Plazzotta S, Calligaris S, Manzocco L (2019) Structure of oleogels from κ-carrageenan templates as affected by supercritical-co2-drying, freeze-drying and lettuce-filler addition. Food Hydrocoll 96:1–10. https://doi.org/10.1016/j.foodhyd.2019.05.008
115. Plazzotta S, Calligaris S, Manzocco L (2018) Application of different drying techniques to fresh-cut salad waste to obtain food ingredients rich in antioxidants and with high solvent loading capacity. LWT 89:276–283. https://doi.org/10.1016/j.lwt.2017.10.056
116. Plazzotta S, Calligaris S, Manzocco L (2020) Structural characterization of oleogels from whey protein aerogel particles. Food Res Int 132:109099. https://doi.org/10.1016/j.foodres.2020.109099
117. Ciuffarin F, Négrier M, Plazzotta S, Libralato M, Calligaris S, Budtova T, Manzocco L (2023) Interactions of cellulose cryogels and aerogels with water and oil: structure-function relationships. Food Hydrocoll 140:108631. https://doi.org/10.1016/j.foodhyd.2023.108631
118. Plazzotta S, Alongi M, De Berardinis L, Melchior S, Calligaris S, Manzocco L (2022) Steering protein and lipid digestibility by oleogelation with protein aerogels. Food Funct 13:10601–10609. https://doi.org/10.1039/d2fo01257j
119. Alavi F, Ciftci ON (2023) Effect of starch type and chitosan supplementation on physicochemical properties, morphology, and oil structuring capacity of composite starch bioaerogels. Food Hydrocoll 141:108637. https://doi.org/10.1016/j.foodhyd.2023.108637
120. Vareda JP, Lamy-Mendes A, Duraes L (2018) A reconsideration on the definition of the term aerogel based on current drying trends. Microporous Mesoporous Mater 258:211–216. https://doi.org/10.1016/j.micromeso.2017.09.016
121. Smirnova I, Gurikov P (2018) Aerogel production: current status, research directions, and future opportunities. J Supercrit Fluids 134:228–233. https://doi.org/10.1016/j.supflu.2017.12.037
122. Zhao WJ, Wei ZH, Xue CH, Meng Y (2023) Development of food-grade oleogel via the aerogel-templated method: oxidation stability, astaxanthin delivery and emulsifying application. Food Hydrocoll 134:108058. https://doi.org/10.1016/j.foodhyd.2022.108058
123. Chen K, Zhang H (2020) Fabrication of oleogels via a facile method by oil absorption in the aerogel templates of protein-polysaccharide conjugates. ACS Appl Mater Interfaces 12:7795–7804. https://doi.org/10.1021/acsami.9b21435
124. Xue J, Wu T, Dai Y, Xia Y (2019) Electrospinning and electrospun nanofibers: methods, materials, and applications. Chem Rev 119:5298–5415. https://doi.org/10.1021/acs.chemrev.8b00593
125. Partheniadis I, Nikolakakis I, Laidmäe I, Heinämäki J (2020) A mini-review: needleless electrospinning of nanofibers for pharmaceutical and biomedical applications. Processes 8:673
126. Nieuwland M, Geerdink P, Brier P, van den Eijnden P, Henket JTMM, Langelaan MLP, Stroeks N, van Deventer HC, Martin AH (2013) Food-grade electrospinning of proteins. Innov Food Sci Emerg Technol 20:269–275. https://doi.org/10.1016/j.ifset.2013.09.004
127. Zhong J, Mohan SD, Bell A, Terry A, Mitchell GR, Davis FJ (2018) Electrospinning of food-grade nanofibres from whey protein. Int J Biol Macromol 113:764–773. https://doi.org/10.1016/j.ijbiomac.2018.02.113
128. Stijnman AC, Bodnar I, Tromp RH (2011) Electrospinning of food-grade polysaccharides. Food Hydrocoll 25:1393–1398. https://doi.org/10.1016/j.foodhyd.2011.01.005
129. Rubio-Valle JF, Sanchez MC, Valencia C, Martin-Alfonso JE, Franco JM (2021) Electrohydrodynamic processing of pvp-doped kraft lignin micro- and nano-structures and

application of electrospun nanofiber templates to produce oleogels. Polymers 13:2206. https://doi.org/10.3390/polym13132206
130. Rubio-Valle JF, Sanchez MC, Valencia C, Martin-Alfonso JE, Franco JM (2022) Production of lignin/cellulose acetate fiber-bead structures by electrospinning and exploration of their potential as green structuring agents for vegetable lubricating oils. Ind Crop Prod 188:115579. https://doi.org/10.1016/j.indcrop.2022.115579
131. Borrego M, Martin-Alfonso JE, Sanchez MC, Valencia C, Franco JM (2021) Electrospun lignin-pvp nanofibers and their ability for structuring oil. Int J Biol Macromol 180:212–221. https://doi.org/10.1016/j.ijbiomac.2021.03.069
132. Hakkarainen E, Korkjas A, Laidmae I, Lust A, Semjonov K, Kogermann K, Nieminen HJ, Salmi A, Korhonen O, Haeggstrom E, Heinamaki J (2019) Comparison of traditional and ultrasound-enhanced electrospinning in fabricating nanofibrous drug delivery systems. Pharmaceutics 11:495. https://doi.org/10.3390/pharmaceutics11100495
133. Nieminen HJ, Laidmäe I, Salmi A, Rauhala T, Paulin T, Heinämäki J, Hæggström E (2018) Ultrasound-enhanced electrospinning. Sci Rep 8:4437. https://doi.org/10.1038/s41598-018-22124-z
134. Österberg H, Mäkinen J, Schavikin J, Valoppit F, Nikolaev D, Laidmäe I, Heinamaki J, Salmi A, Hægström E (n.d.) Scaling-up the ultrasound-enhanced electrospinning device. In: 2022 IEEE International Ultrasonics Symposium (IUS), Venice, Italy, 10–13 October 2022
135. Hunnakko J, Laidmäe I, Puranen T, Mäkinen J, Helander P, Nieminen HJ, Nolvi A, Kogermann K, Heinämäki J, Salmi A, Hæggström E (n.d.) Towards chronic wound pads: gradient nanofiber structure generated by ultrasound enhanced electrospinning (uses). In: 2019 IEEE International Ultrasonics Symposium (IUS)
136. Koos E (2014) Capillary suspensions: particle networks formed through the capillary force. Curr Opin Colloid Interface Sci 19:575–584. https://doi.org/10.1016/j.cocis.2014.10.004
137. Hoffmann S, Koos E, Willenbacher N (2014) Using capillary bridges to tune stability and flow behavior of food suspensions. Food Hydrocoll 40:44–52. https://doi.org/10.1016/j.foodhyd.2014.01.027
138. Koos E, Willenbacher N (2011) Capillary forces in suspension rheology. Science 331:897–900. https://doi.org/10.1126/science.1199243
139. Bhattarai M, Penttilä P, Barba L, Macias-Rodriguez B, Hietala S, Mikkonen KS, Valoppi F (2022) Size-dependent filling effect of crystalline celluloses in structural engineering of composite oleogels. LWT 160:113331. https://doi.org/10.1016/j.lwt.2022.113331
140. Wang GS, Chen HY, Wang LJ, Zou Y, Wan ZL, Yang XQ (2022) Formation of protein oleogels via capillary attraction of engineered protein particles. Food Hydrocoll 133:107912. https://doi.org/10.1016/j.foodhyd.2022.107912
141. Gao ZM, Zhang C, Li YL, Wu YH, Deng QC, Ni XW (2023) Edible oleogels fabricated by dispersing cellulose particles in oil phase: effects from the water addition. Food Hydrocoll 134: 108040. https://doi.org/10.1016/j.foodhyd.2022.108040
142. Calabrese V, Gunes DZ, Farrés IF (2021) Rheological control of pea fibre dispersions in oil: the role of particle and water volume fractions. Food Hydrocoll 121:106988. https://doi.org/10.1016/j.foodhyd.2021.106988
143. Jarray A, Feichtinger A, Scholten E (2022) Linking intermolecular interactions and rheological behaviour in capillary suspensions. J Colloid Interface Sci 627:415–426. https://doi.org/10.1016/j.jcis.2022.07.067
144. Sabet S, Pinto TC, Kirjoranta S, Kazerani Garcia A, Valoppi F (2023) Clustering of oleogel production methods reveals pitfalls and advantages for sustainable, upscalable, and oxidative stable oleogels. J Food Eng 357:111659. https://doi.org/10.1016/j.jfoodeng.2023.111659
145. Sagiri SS, Samateh M, Pan S, Maldarelli C, John G (2023) A heat transfer model and supporting experiments to guide the uniform gelation of molecular oleogels during scale-up. J Am Oil Chem Soc. https://doi.org/10.1002/aocs.12689
146. Scharfe M, Flöter E (2020) Oleogelation: from scientific feasibility to applicability in food products. Eur J Lipid Sci Technol 122:2000213. https://doi.org/10.1002/ejlt.202000213

Chapter 5
Monoglyceride Oleogels

Camila Palla and Maria Elena Carrín

Abbreviations

CA	Canola oil
CL	Cod liver oil
CO	Corn oil
CS	Castor oil
EVO	Extra-virgin olive oil
FA	Fatty acid
FX	Flaxseed
G′	Elastic modulus
G″	Viscous modulus
HA	Hazelnut oil
HIU	High-intensity ultrasound
HOSA	High oleic safflower oil
HOSO	High oleic sunflower oil
LMWG	Low-molecular-weight gelator
MG	Monoglyceride
MUFA	Monounsaturated fatty acid
OBC	Oil binding capacity
PE	Peanut oil
RA	Rapeseed oil
SF	Sunflower oil

C. Palla (✉) · M. E. Carrín
Departamento de Ingeniería Química, Universidad Nacional del Sur (UNS), Bahía Blanca, Argentina

Planta Piloto de Ingeniería Química - PLAPIQUI (UNS-CONICET), Bahía Blanca, Argentina
e-mail: cpalla@plapiqui.edu.ar; mcarrin@plapiqui.edu.ar

C. Andrea Palla, F. Valoppi (eds.), *Advances in Oleogel Development, Characterization, and Nutritional Aspects*,
https://doi.org/10.1007/978-3-031-46831-5_5

SMG	Saturated monoglyceride
SY	Soybean oil
TPA	Texture profile analysis
UFA	Unsaturated fatty acid
UMG	Unsaturated monoglyceride
VO	Virgin olive oil
W1	First deformation work
W2	Second deformation work

5.1 Introduction

Monoglycerides (MGs) are important emulsifiers extensively used in the food industry due to their great ability to stabilize oil and water phases. The utilization of MGs in oil structuring is relatively recent, with the earliest studies emerging around 2000 [1, 2]. This line of research has been mainly driven by the necessity of having solid and semisolid fats without *trans* fatty acids and with reduced levels of saturated fatty acids, while exhibiting properties comparable to conventional fats [3]. MGs fall within the classification of low-molecular-weight gelators (LMWGs), which usually self-assemble into crystalline structures that form supramolecular systems able to entrap oil. The properties and functionality of LMWG oleogels are directly related to their structural building blocks [4]. Numerous studies have demonstrated that the MG crystalline structure can provide oleogels with optimal levels of hardness, stability, spreadability, plasticity, and elasticity [5, 6].

By acquiring a thorough understanding of the factors that impact MG crystallization and gelation, it becomes possible to strategically manipulate physical process parameters. This level of control allows for the engineering of structures with tailored functionality to meet the desired requirements for specific applications. This chapter provides an overview of the fundamental characteristics of MG oleogels. First, the pivotal attributes of MGs that have been studied as gelators are presented, followed by a description of the process parameters that influence MG oleogel formation. Afterward, an examination of the physicochemical properties of these gels is conducted, emphasizing the variables that can be changed to optimize these properties. Finally, the challenges and future research directions related to MG oleogels are identified.

5.2 Production of Monoglyceride Oleogels

5.2.1 *Sources and Types of Monoglycerides*

MGs are lipid molecules formed by a glycerol backbone and a fatty acid (FA) attached to one of its hydroxyl groups. The esterification position determines their classification as either 1-MG or 2-MG, with the former referring to an attachment at an extreme position and the latter at an internal position. In addition, MGs are further classified based on the length and saturation level of the carbon chain. The presence of two hydroxyl and one acyl groups confers amphiphilic properties upon MGs, making them nonionic surfactants with exceptional emulsifying capabilities. Therefore, they find widespread application across diverse industries that require the stabilization of mixtures comprising oily and aqueous phases [7].

MGs offer several advantages when incorporated into food formulations, specifically in oil structuring. First, MGs are considered safe food additives that are already approved for use [8, 9]. Second, they are predominantly derived from vegetable oils, such as palm, soybean, sunflower, or rapeseed oil. Third, they are usually obtained in highly pure form, with purities ranging from 90% to 95% [10]. Lastly, but not less importantly, MGs are widely available in the market at a relatively affordable cost [11]. In fact, the Global Distilled Monoglyceride market was valued at USD 869.9 million in 2021–2022 and is expected to grow at a compound annual growth rate of 4.4% during 2021–2026 [12] and 7.92% for 2023–2030 [13].

Table 5.1 presents data on representative commercial MG mixtures used for oil structuring purposes. These mixtures are usually characterized by their purity, FA composition, and MG isomer identification. Although the specific isomer ratio is not usually reported, it is known that commercial MGs predominantly consist of over 90% saturated 1-MG [14]. Saturated monoglycerides (SMGs) have been the primary focus of research as gelators, with only a limited number of contributions in the literature regarding unsaturated monoglycerides (UMGs) [3]. In general, oil structuring experiments have involved testing SMG blends comprising approximately equal proportions of palmitic (C16:0) and stearic (C18:0) acid (~40–60% each) as well as blends with a high stearic acid content (>90%) [3]. It has been reported that concentrations between 2% and 5% of these gelators in oil are sufficient to obtain a nonflowing soft solid, while concentrations between 5% and 10% are required for oleogels with a high level of structuration (oil binding capacity >90%) [3]. On the other hand, it has been observed that SMGs with medium-chain FA (≤C10:0) or UMGs with high 2-UMG content are ineffective in structuring oils, even when used at concentrations of up to 20% SMG [15] or 10% UMG [14], respectively. Table 5.1 also provides the melting point of the different MGs, which depends on their purity degree and FA composition. This information is essential to establish a suitable heating temperature for oleogel production. Another factor to consider in oleogel production is the source of the MGs, particularly when MGs are derived from hydrogenated palm oil and meeting sustainability criteria is a requirement.

Table 5.1 Composition, purity grade, and melting point of some commercial monoglycerides (MGs) tested for oil structuring

Commercial name/brand	Composition	Purity grade	Melting point (°C)	References
C8-10 MG intermediate Type I	C8:0 (81%) C10:0 (18.6%)	89.8%	n.s	[15]
Hangzhou Kangyuan Food Science and Technology	C12:0 (n.s)	>90%	58.35	[16]
Myverol™	C16:0 (59.8%) C18:0 (38.85%) C14:0 (1.4%)	n.s	68.05	[17]
Natural word	C16:0 (45%) C18:0 (8%) C18:1 *trans* (21%) C18:1 *cis* (20%)	n.s	72.1	[18]
Myverol™ 18-08 NP	C18:0 (90.65%) C16:0 (6.43%) C20:0 (1.53%)	>90%	72.0	[19]
Vandermoortele Lipids N.V.	C18:1 (77.44%) C18:2 (12.01%)	n.s	n.s	[20]
No commercial MGs[a]	C18:2 (66.4%) C18:1 (22.2%) C18:3 (5.27%)	87.6%	n.s	[14]
Danisco Co.	C22:0 (90.0%) C18:0 (3.4%)	>90%	n.s	[21]

C8:0 = caprylic acid; C10:0 = capric acid; C12:0 = lauric acid; C14:0 = myristic acid; C16:0 = palmitic acid; C18:0 = stearic acid; C18:1 *cis* = oleic acid; C18:1 *trans* = elaidic acid; C18:2 = linoleic acid; C18:3 = linolenic acid; C20:0 = arachidic acid; C22:0 = behenic acid
[a]Obtained from a two-step enzymatic ethanolysis reaction (ethanol:soybean oil weight ratio = 2:1, 10% Lipozyme 435, 4 h, ambient temperature)

5.2.2 *Factors Affecting Monoglyceride Oleogel Formation*

MG oleogels are physical gels prepared by a simple and direct method, which involves dispersing a relatively low amount of MG in oil, followed by heating and stirring to allow solubilization, and subsequent cooling to form a complex solid network of crystalline particles that entraps the oil [3]. As MGs are cooled in the presence of a hydrophobic solvent, they undergo various molecular reorganizations that lead to the formation of inverse bilayers. These bilayers then grow and organize into lamellar platelet microstructures, which serve as the fundamental building blocks of the network [22]. The microplatelets intersect and create junction zones that contribute to the formation of the oleogel's structure as well as its mechanical and rheological properties. These microstructures are held together by noncovalent interactions, such as van der Waals forces and hydrogen bonds [23]. The oil is retained in the three-dimensional (3D) crystalline structure because of a complex interplay of factors not fully identified, such as surface forces (adsorption) [24–26].

The ultimate structure of the gel is significantly influenced by size, shape, clustering, and interaction of the crystals. Therefore, all factors that can affect the formation and characteristics of the crystalline network, as well as the interaction between MGs and oil, influence the physical properties of MG oleogels. Several factors have been identified as having an impact on the properties of MG oleogels, among them: MG type and concentration [27, 28], oil type [17, 29], heating temperature [30], stirring speed [31], cooling rate/temperature [19], shear rate during cooling [22], and storage time and temperature [32, 33]. It has also been reported that MG oleogel formation can be modified by the presence of additional components [34–36] as well as by using ultrasound technologies [37, 38], which have the potential to alter MG crystallization kinetics. For a comprehensive overview of how that technology has been applied for this purpose, please see Chap. 11.

5.3 Physicochemical Properties of Monoglyceride Oleogels

The following subsections focus on the analysis of SMG oleogels prepared by the direct method and cooled under static conditions, where their properties were examined after a short aging period (12–72 h after preparation), unless explicitly stated otherwise.

5.3.1 Structural Characteristics

The nanoscale structure of SMG oleogels can be characterized by the elucidation of the polymorphic forms of their crystals, using small- and wide-angle X-ray scattering techniques. Multiple studies have reported the presence of β and, to a lesser extent, sub-α phases [17, 39, 40]. The β form is the most thermodynamically stable, while the sub-α form is indicative of an unset gel [40, 41]. According to previous studies, the crystalline molecular organization of SMG (mixtures of palmitic and stearic acid) in oils (from medium- and long-chain triglycerides) remained unchanged when the concentration varied from 5% to 10% or even from 10% to 70% [17, 37, 42, 43]. In this concentration range, SMG molecules crystallized in the triclinic β form and exhibited a lamellar thickness ranging from 46 to 49 Å. However, when the amount of a monostearin-rich SMG was increased from 10% or 15% to 20%, the interplane spacing expanded to 66 Å [44]. These results demonstrate that the molecular organization is highly dependent on the MG composition and concentration.

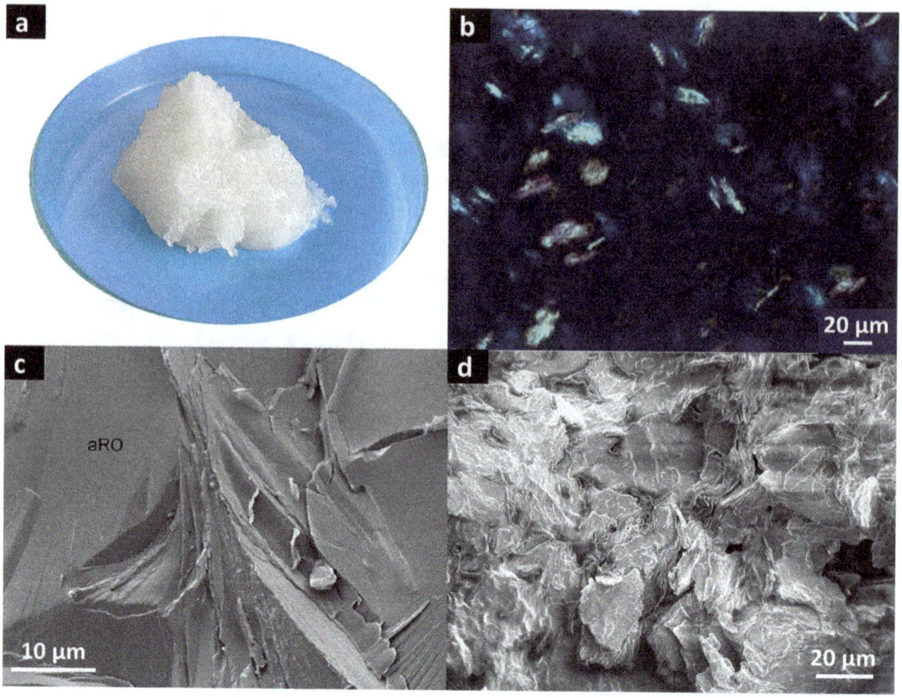

Fig. 5.1 Images of saturated monoglyceride oleogels showing (**a**) the macroscopic structure, (**b**) the microstructure as observed by polarized light microscopy (reproduced from [48], with permission from John Wiley & Sons), (**c**) crystal platelets embedded in amorphous rapeseed oil (aRO) visualized by cryo-scanning electron microscopy (reproduced from [40], with permission from Elsevier), and (**d**) crystal platelets from a chia oleogel sample de-oiled with isobutanol—following the procedure described by [49] with slight modifications—visualized by scanning electron microscopy (micrograph from our laboratory)

The image in Fig. 5.1a shows the visual characteristics of an MG oleogel, revealing its homogeneous appearance and semi-solid consistency. The crystalline morphology of SMG oleogels is characterized by platelets that exhibit a needle-like appearance under polarized light microscopy due to the orientation of the platelets respect to the glass cover slip and projection of the crystals in the focal plane (Fig. 5.1b–d). These crystals tend to aggregate and form clusters. Thus, the resulting architecture of the crystal network may exhibit sandwich-like or spherulitic/rosette-like structures, or irregular patterns, depending on factors such as the type of MG, and the presence of impurities, traces of water, or particles [19, 28, 45–47]. Additionally, the processing conditions play a crucial role in crystal nucleation and size. By implementing conditions of either high supercooling or high supersaturation, a fast nucleation rate can be achieved that facilitates the formation of numerous smaller and uniform crystals, thereby enhancing the spatial distribution of mass.

5.3.2 Rheological Properties

Rheological properties analysis in SMG oleogels provided valuable insights into their viscoelastic behavior and stability. The elastic and viscous moduli, G′ and G″, are usually determined by frequency sweep tests in the linear viscoelastic region. Rheological measurements have revealed that SMG oleogels had a solid-like behavior (G′ > G″), with G′ remaining constant throughout the frequency sweep when concentrations of 5% SMG or higher were used [2, 31, 44]. This indicated that SMG oleogels had a stable gel structure with a predominantly elastic nature. Figure 5.2 shows the maximum G′ values of SMG oleogels reported in various studies, including unpublished data from our laboratory. These values were recorded from oscillatory tests conducted at or around 20 °C on oleogels formulated with various oils—arranged in decreasing order of unsaturated fatty acid (UFA) content—and SMG concentrations within the typical range. With increasing SMG concentration, a

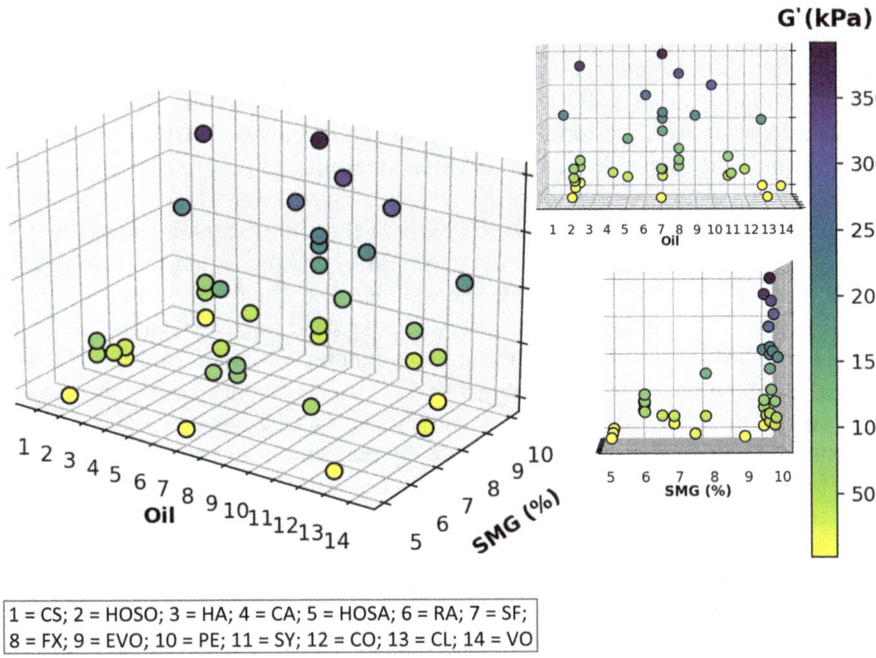

Fig. 5.2 Elastic moduli (G′) values of oleogels from saturated monoglyceride (SMG) (C16:0 and C18:0 mixtures) and different oils: castor oil (CS), high oleic sunflower oil (HOSO), hazelnut oil (HA), canola oil (CA), high oleic safflower oil (HOSA), rapeseed oil (RA), sunflower oil (SF), flaxseed oil (FX), extra-virgin olive oil (EVO), peanut oil (PE), soybean oil (SY), corn oil (CO), cod liver oil (CL), and virgin olive oil (VO). The oils were labeled from 1 to 14, with 1 being the oil with the highest content of unsaturated fatty acids and 14 being the lowest. The small figures on the right side are the projections of G′ in the oil plane (top) and the SMG plane (bottom). (Data from [2, 4, 6, 15, 17–20, 22, 27–32, 38, 42, 44, 45, 50, 51] and unpublished data from our laboratory)

global trend can be observed in Fig. 5.2, indicating a corresponding increase in G′. This rise in gel elasticity can be attributed to the presence of a higher solid mass that is distributed more evenly throughout the material, resulting in enhanced crystal-crystal interactions, as suggested by previous studies [1, 42]. Furthermore, the region exhibiting the highest G′ values seems to correspond to oils with higher UFA content, suggesting that SMG may have improved structuring capabilities when combined with these oils. In particular, it has been reported that long-chain monounsaturated FAs (MUFAs) favored the packing of SMG crystals in a cohesive gel [29]. Alternatively, it is possible that the cocrystallization of saturated triglycerides from oils could be hindering the formation of the SMG crystalline network. In addition to FA composition, other oil-related factors such as polarity and viscosity have been demonstrated to exert a significant influence on the ability of SMG to form strong networks [17]. While the type of oil does impact the characteristics of the gel, it would be conceivable to attain materials with similar values of G′, given appropriate adjustments in the SMG concentration and other variables not considered in this particular analysis (e.g., cooling temperature, storage time, etc., and disregarding disparities in test conditions). This finding instills optimism, as it suggests the feasibility of producing oleogels with desirable attributes using oils readily obtainable in different geographical regions worldwide.

The gelation point of SMG oleogels determined by temperature sweep tests was found to coincide with the formation of the crystalline network [19]. Furthermore, the gel point increased with increasing SMG concentration [2, 28, 47] and was also influenced by the composition of SMG [28]. For example, an oleogel made from a high-purity monostearin (94%) showed a higher gel point temperature than that of an oleogel made from a mixture of monostearin (38%) and monopalmitin (54%) [28].

The thixotropic behavior of SMG oleogels has been analyzed to evaluate their capacity for recovering their original structure after shear force application. Remarkably, when SMG was used in concentrations ranging from 8% to 15%, the gels showed a thixotropic recovery of up to 94% [52]. According to authors, this finding suggests that the crystalline network within the gels has the capacity to dissociate in response to high shear forces and subsequently reassemble and reorganize once the shear force is removed. Conversely, recoveries between 49% and 61% were recently reported for 6% SMG oleogels dynamically produced on a margarine pilot plant [48]. This disparity between results underscores the necessity for conducting pilot plant studies under diverse conditions. Specifically, it is crucial to investigate gelation under static conditions, which has shown promising outcomes in terms of improved recovery properties in laboratory-scale experiments. Such findings will have significant implications for industrial applications.

5.3.3 Textural Properties

Although textural characteristics in SMG oleogels have been studied to a lesser extent, they are of paramount importance. These characteristics are closely linked to the consistency, stability, and sensory attributes of the final product, which in turn have significant implications for food applications and consumer satisfaction. Hardness, to a greater extent, and adhesiveness, cohesiveness, spreadability, and fracturability have been the properties generally investigated in SMG oleogels [3]. Hardness values of SMG oleogels from various studies, along with unpublished data from our laboratory, are presented in Fig. 5.3. These values were obtained through texture profile analysis (TPA)—two cycles of penetration—or penetrometer tests carried out at sample temperatures between 5 and 20 °C in oleogels formulated with different oils, arranged in a descending order based on their UFA content, and SMG concentrations within the typical range. Compared to commercial products, SMG oleogels were softer than stick margarines (21.5 N and 15.5 N [53]), but had a consistency similar to spreadable margarines (1.6 N [31]). Increasing the SMG content allowed for gels to exhibit enhanced resistance to deformation, which can be ascribed to the greater amount of crystallized material and greater number of gelator-gelator interactions. The highest hardness values (~5–5.5 N) have been

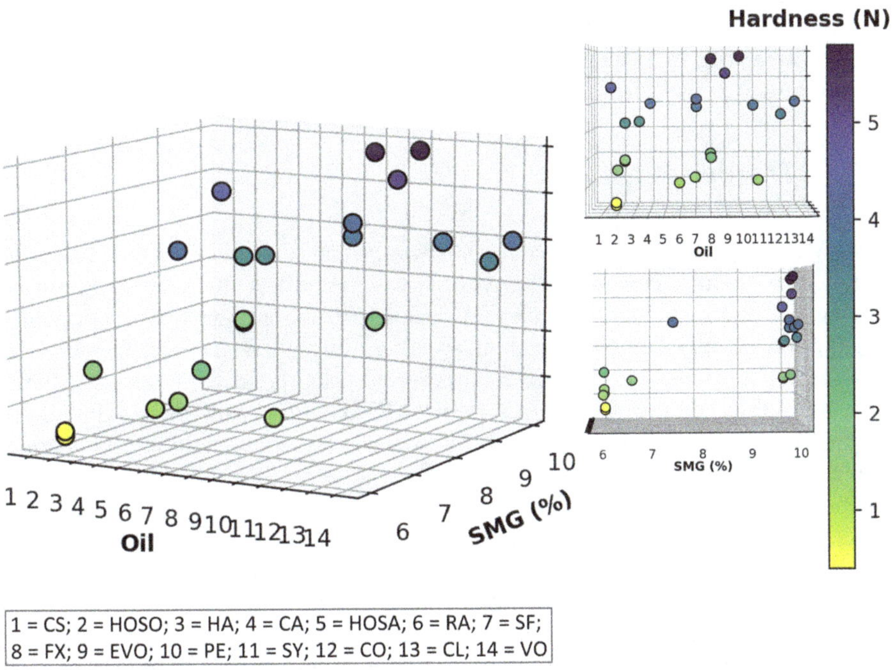

Fig. 5.3 Hardness values for oleogels from saturated monoglyceride (SMG) (C16:0 and C18:0 mixtures) and different oils: castor oil (CS), high oleic sunflower oil (HOSO), hazelnut oil (HA), canola oil (CA), high oleic safflower oil (HOSA), rapeseed oil (RA), sunflower oil (SF), flaxseed oil (FX), extra-virgin olive oil (EVO), peanut oil (PE), soybean oil (SY), corn oil (CO), cod liver oil (CL), and virgin olive oil (VO). The oils were labeled from 1 to 14, with 1 being the oil with the highest content of unsaturated fatty acids and 14 being the lowest. The small figures on the right side are the projections of hardness in the oil plane (top) and the SMG plane (bottom). (Data from [6, 17, 19, 23, 27, 31, 32, 38, 45, 51, 54] and unpublished data from our laboratory)

reported for 10% SMG in flaxseed, peanut, and extra-virgin olive oils (9–11 in Fig. 5.3) with UFA contents ranging from 83% to 89%. On the other hand, it has been observed that variations in the cooling temperature during the gelation process have a significant impact on hardness of SMG oleogels. Cooling the samples at 30 °C, as opposed to 5 °C, resulted in a substantial increase in hardness of up to 39% [19]. The authors found that higher cooling temperatures led to the formation of longer crystal aggregates that conferred greater resistance to the first penetration cycle in the TPA test, whereas lower cooling temperatures produced gels composed of smaller crystals that exhibited enhanced resistance to deformation in the second penetration cycle.

In addition, the adhesiveness of oleogels formulated with hazelnut oil, olive oil, and high oleic sunflower oil (HOSO) has shown a positive correlation with SMG concentrations, indicating the development of stronger attractive forces between the gel and probe surfaces due to increased crystalline material [6, 27]. Conversely, as the SMG content increases, the cohesiveness of oleogels tends to decrease [27, 31, 45]. This behavior can be explained by considering that the cohesiveness parameter determines the material's resistance to a second deformation work (W2) relative to the first deformation work (W1). If the increase in W2 is smaller than the increase in W1 due to the rise in SMG, the cohesiveness of the oleogel is reduced [31].

5.3.4 *Thermal Properties*

Melting and crystallization temperatures play a critical role in defining the functionality of a fat product. This makes it necessary to determine the thermal events during heating and cooling cycles of oleogels using differential scanning calorimetry. Heating of oleogel samples—without prior erasing their thermal memory—is required to determine intrinsic material properties. Different thermogram patterns have been observed during the melting of SMG oleogels, with some exhibiting a single endothermic peak [4, 16, 33, 45, 54], while others show multiple endothermic events [29, 55, 56]. These variations in thermogram profiles may be due to SMG composition and/or differences in temperature-time analysis conditions. On the other hand, increases in SMG content have been observed to correlate with significant increases in both onset and peak melting temperatures, as well as melting enthalpies [4, 38, 56]. It is well-established that higher levels of supersaturation can lead to greater amounts, interactions, and organization of crystalline material. The enthalpy value reflects the energy needed to induce thermal transformations in the material, including crystal melting/formation and polymorphic changes. Therefore, it is positively affected by the amount and organization level of crystallized material [57]. For example, when the SMG concentration in HOSO oleogels was increased from 3% to 6%, a twofold increase in melting enthalpy was observed [45]. In contrast, variations in the cooling rate during oleogel formation have not shown significant effects on the aforementioned variables. For example, the onset, peak, and enthalpy of melting of 6% SMG oleogels were observed to change from 57.4 °C,

61.6 °C, and 11.8 J/g to 51.7 °C, 62.8 °C, and 11.5 J/g, respectively, when the cooling rate was changed from 0.1 to 10 °C/min. It is interesting to note that some commercial products, which are targeted for replacement by oleogels, exhibit melting points in the range of 30–32 °C (interesterified products similar to butter), 38–42 °C (pastry and puff pastry margarines), 44–47 °C (refined bovine fats and regular shortenings), 40–50 °C (roll-in shortenings), and 44–54 °C (roll-in margarines) [58, 59]. Given the comparatively lower melting points of these materials in relation to SMG oleogels, it is mandatory to corroborate the suitability of SMG oleogels in the desired application to establish whether they meet the intended requirements.

On the other hand, the introduction of perturbations during the stage of oleogel formation has been proposed as a strategy to modify its final properties. However, minor or no differences in melting behavior have been observed when oleogels were dynamically formed or when high-intensity ultrasound (HIU) was applied during gelation [38, 48, 54].

5.3.5 Stability and Shelf Life

The structural stability of SMG oleogels is directly related to the ability of the crystalline network to retain oil. This property, usually referred to as oil binding capacity (OBC), quantifies the amount of oil that remains bound within the oleogel after it has been subjected to centrifugation or other oil release techniques. Several studies have shown that SMG concentration notably affects OBC. For example, when SMG concentration was increased from 3–5% to 10%, OBC increased from 69.3–80.9% to 93.8–99.8% [18, 27, 31, 51]. Therefore, using 10% of SMG in the oleogel ensures a network strong enough to hold and retain the whole volume of oil [3]. Moreover, OBC has been found to improve in the order monolaurin < monostearin < glycerol monobehenate [21]. Notably, the highest OBC values were consistently reported when using HOSO compared to more saturated oils such as coconut oil or polyunsaturated oils like sunflower oil [29–31]. Oleogel formation conditions also affect its final stability. The presence of a homogenous structure comprising small crystals, achieved through the use of high cooling rates or HIU, has been related to high OBC [19, 27, 31, 38]. In addition, oleogelation under static conditions, rather than using shear, resulted in the formation of stronger structures with higher OBC [22].

When discussing the stability of SMG oleogels, time is one of the most important factors. It is known that, after the cooling stage, a stabilization or annealing period of typically 12–72 h at temperatures between 4 and 25 °C is required to achieve semi-stable properties. Nevertheless, it is important to note that the network built by MG crystals has been shown to be a dynamic structure [3, 48]. As a result, some structural and chemical instabilities may occur in oleogels during prolonged storage [6, 32, 41]. SMG oleogels of hazelnut oil stored at 26 °C were shown to undergo a gradual rearrangement from sub-α phase toward more stable β-crystal structures over a period of about 7 days. This transformation involves the segregation of chiral MGs, resulting in denser structures and the release of initially trapped oil [39, 41]. In general, storage of SMG oleogels at low temperatures can delay changes in physicochemical properties [6, 32, 41]. For instance, no significant variations in the main physicochemical parameters were observed in SMG oleogels of HOSO during 21 days of storage at 5 °C [32]. After 2 months, crystal size, hardness, and elasticity were the properties most negatively affected, with variations over 30%. Nevertheless, polymorphism and melting behavior were not changed throughout the whole storage time, while OBC and fractal dimension slightly decreased. Similarly, hardness and adhesiveness of SMG oleogels from virgin olive oil gradually decreased as the storage time increased, with higher incidence at higher temperature (20 °C vs. 4 °C) for a period of up to 90 days [6]. These findings suggest that it is imperative to evaluate oleogel properties according to the storage conditions and intended end use.

In terms of chemical stability, one of the major concerns when using unsaturated oils is their potential oxidative deterioration. However, it has been shown that SMG oleogels exhibit slower primary and secondary oxidation processes than the corresponding oils [16, 32, 42, 55]. This suggests that SMG structural arrangements could act as physical barriers capable of hindering the movement of oxidation reagents, thereby extending the useful life of unsaturated oils.

5.4 Potential Applications of Monoglyceride Oleogels in the Food Industry

SMG oleogels have been extensively studied as a full or partial solid fat replacement in a wide range of food products, such as spreadable, bakeries, confectioneries, and meat products, in addition to being evaluated for specific technological functionalities like oil migration inhibition, emulsion stabilization, and lipid oxidation stability enhancement [5]. In addition, they have been investigated as carrier matrices for liposoluble bioactive compounds in the development of health-promoting food systems. For more detailed information on food applications and uses as delivery and protection of bioactive molecules, see Chaps. 26 and 17, respectively.

5.5 Nutritional Aspects of Monoglyceride Oleogels

MG oleogels are part of the wide range of lipid materials developed to address the elimination of *trans* FAs and the reduction of saturated FAs in industrially processed foods. This implies that a significant aspect of reformulating food products with oleogels involves evaluating the improvement in the nutritional profile of these products. The degree of improvement depends on the oleogel composition (including oil type and type and amount of gelator), fat composition of the replaced material, and level of substitution (partial or complete). Specifically, when SMG oleogels are used, there is a contribution to the saturated FA content due to the chemical nature of SMG that must be considered. However, based on the available evidence regarding the biological effects of FAs, opting for an SMG mixture rich in stearic acid may offer an additional advantage when choosing a gelator. This is supported by findings indicating that stearic acid, in contrast to palmitic acid, does not elevate total cholesterol and low-density lipoprotein levels [60, 61].

Additionally, SMG oleogels can be designed to provide health benefits beyond replacing SFAs with UFAs. Some studies have explored the possibility of incorporating lipid-soluble bioactive compounds, such as phytosterols, curcumin, β-carotene, lupeol, and quercetin, into the oleogel [5, 43]. This allows for the development of controlled release systems with potential anticancer, anticholesterol, anti-inflammatory, and other beneficial activities.

5.6 Conclusions

This chapter provides an overview of MG oleogels, specifically examining the factors influencing crystallization and gel formation, along with their key physicochemical properties. Given the remarkable advantages offered by MGs as gelators and the favorable mechanical properties of oleogels formulated with them, it is imperative to devote more attention to improving the weak properties of these materials to resemble more closely those of traditional solid fats while improving long-term stability. Exploring innovative technologies and incorporating additional components capable of modifying the structural characteristics of the crystalline network are currently viable strategies to enhance oleogel properties that need to be addressed in more depth. In this regard, it would be interesting to advance in the use of bioactive compounds, which can provide added health value to fat substitutes. Finally, the transition from laboratory to pilot plant experiments is essential to gain valuable insight into the technical and economic feasibility of scaling up the MG oleogel production process for industrial applications.

Acknowledgments The authors acknowledge the financial support by the Consejo Nacional de Investigaciones Científicas y Técnicas (CONICET) (PIP 2021–2023 No. 101968), the Agencia Nacional de Promoción Científica y Tecnológica (ANPCyT) (PICT-2021-GRF-TI-00273), and the Universidad Nacional del Sur (UNS) (PGI 24/M172) in Argentina. The authors would also like to thank Ing. Martina Domínguez for her help in preparing the samples for SEM analysis.

References

1. Ojijo NKO, Kesselman E, Shuster V, Eichler S, Eger S, Neeman I, Shimoni E (2004) Changes in microstructural, thermal, and rheological properties of olive oil/monoglyceride networks during storage. Food Res Int 37:385–393. https://doi.org/10.1016/j.foodres.2004.02.003

2. Ojijo NKO, Neeman I, Eger S, Shimoni E (2004) Effects of monoglyceride content, cooling rate and shear on the rheological properties of olive oil/monoglyceride gel networks. J Sci Food Agric 84:1585–1593. https://doi.org/10.1002/jsfa.1831
3. Palla CA, Dominguez M, Carrín ME (2022) An overview of structure engineering to tailor the functionality of monoglyceride oleogels. Compr Rev Food Sci Food Saf 21:2587–2614. https://doi.org/10.1111/1541-4337.12930
4. Fasolin LH, Cerqueira MA, Pastrana LM, Vicente AA, Cunha RL (2018) Thermodynamic, rheological and structural properties of edible oils structured with LMOGs: influence of gelator and oil phase. Food Struct 16:50–58. https://doi.org/10.1016/j.foostr.2018.03.003
5. Palla CA, Dominguez M, Carrín ME (2022) Recent advances on food-based applications of monoglyceride oleogels. JAOCS, J Am Oil Chem Soc 99:985–1006. https://doi.org/10.1002/aocs.12617
6. Öğütcü M, Yılmaz E (2014) Oleogels of virgin olive oil with carnauba wax and monoglyceride as spreadable products. Grasas Aceites 65:e040. https://doi.org/10.3989/gya.0349141
7. Nitbani FO, Tjitda PJP, Nurohmah BA, Wogo HE (2020) Preparation of fatty acid and monoglyceride from vegetable oil. J Oleo Sci 69:277–295. https://doi.org/10.5650/jos.ess19168
8. U.S. Food & Drug Administration (2021) CFR - Code of Federal Regulations title 21 - part 172 -Food additives permitted for direct addition to food for human consumption. In: U.S. Food Drug. https://www.accessdata.fda.gov/scripts/cdrh/cfdocs/cfcfr/CFRSearch.cfm?fr=172.840. Accessed 14 Dec 2021
9. EFSA Panel on Food Additives and Nutrient Sources added to Food (ANS) (2017) Re-evaluation of mono- and di-glycerides of fatty acids (E 471) as food additives. EFSA J 15. https://doi.org/10.2903/j.efsa.2017.5045
10. Barfod NM, SparsØ FV (2007) Structure and function of emulsifiers and their role in microstructure formation in complex foods. In: Understanding and controlling the microstructure of complex foods
11. Rocha-Guzmán NE, Cháirez-Ramírez MH, Pérez-Martínez JD, Rosas-Flores W, de Ornelas-Paz JJ, Moreno-Jiménez MR, González-Laredo RF, Gallegos-Infante JA (2021) Use of organogel-based emulsions (o/w) as a tool to increase the bioaccessibility of lupeol, curcumin, and quercetin. JAOCS, J Am Oil Chem Soc 98:1177–1188. https://doi.org/10.1002/aocs.12528
12. 360ResearchReports (2020) Global distilled monoglyceride market research report 2020. https://www.360researchreports.com/global-distilled-monoglyceride-market-15046641. Accessed 5 Jun 2021
13. Research reports world (2023) Global food grade distilled monoglyceride market research report 2023
14. Chen C, Zhang C, Zhang Q, Ju X, Wang Z, He R (2021) Study of monoglycerides enriched with unsaturated fatty acids at sn-2 position as oleogelators for oleogel preparation. Food Chem 354. https://doi.org/10.1016/j.foodchem.2021.129534
15. Monié A, Franceschi S, Balayssac S, Malet-Martino M, Delample M, Perez E, Garrigues JC (2022) Study of rapeseed oil gelation induced by commercial monoglycerides using a chemometric approach. Food Chem 369. https://doi.org/10.1016/j.foodchem.2021.130870
16. Pan J, Tang L, Dong Q, Li Y, Zhang H (2021) Effect of oleogelation on physical properties and oxidative stability of camellia oil-based oleogels and oleogel emulsions. Food Res Int 140. https://doi.org/10.1016/j.foodres.2020.110057
17. Valoppi F, Calligaris S, Barba L, Šegatin N, Poklar Ulrih N, Nicoli MC (2017) Influence of oil type on formation, structure, thermal, and physical properties of monoglyceride-based organogel. Eur J Lipid Sci Technol 119. https://doi.org/10.1002/ejlt.201500549
18. Fayaz G, Polenghi O, Giardina A, Cerne V, Calligaris S (2021) Structural and rheological properties of medium-chain triacylglyceride oleogels. Int J Food Sci Technol 56:1040–1047. https://doi.org/10.1111/ijfs.14757

19. Palla C, de Vicente J, Carrín ME, Gálvez Ruiz MJ (2019) Effects of cooling temperature profiles on the monoglycerides oleogel properties: a rheo-microscopy study. Food Res Int J 125. https://doi.org/10.1016/j.foodres.2019.108613
20. Bin Sintang MD, Rimaux T, Van De WD, Dewettinck K, Patel AR (2016) Studying the oil structuring properties of monoglycerides and phytosterols mixtures. Eur J Lipid Sci Technol A 119:1500517
21. Li J, Guo R, Bi Y, Zhang H, Xu X (2021) Comprehensive evaluation of saturated monoglycerides for the forming of oleogels. LWT 151. https://doi.org/10.1016/j.lwt.2021.112061
22. da Pieve S, Calligaris S, Co E, Nicoli MC, Marangoni AG (2010) Shear Nanostructuring of monoglyceride organogels. Food Biophys 5:211–217. https://doi.org/10.1007/s11483-010-9162-3
23. Rosen-Kligvasser J, Davidovich-Pinhas M (2021) The role of hydrogen bonds in TAG derivative-based oleogel structure and properties. Food Chem 334. https://doi.org/10.1016/j.foodchem.2020.127585
24. Blake AI, Marangoni AG (2015) The use of cooling rate to engineer the microstructure and oil binding capacity of wax crystal networks. Food Biophys 10:456–465. https://doi.org/10.1007/s11483-015-9409-0
25. Abdallah DJ, Weiss RG (2000) n-Alkanes gel n-alkanes (and many other organic liquids). Langmuir 16:352–355. https://doi.org/10.1021/la990795r
26. Patel AR (2017) A colloidal gel perspective for understanding oleogelation. Curr Opin, Food Sci, p 15
27. Giacomozzi AS, Carrín ME, Palla CA (2018) Muffins elaborated with optimized Monoglycerides Oleogels: from solid fat replacer obtention to product quality evaluation. J Food Sci 83: 1505–1515. https://doi.org/10.1111/1750-3841.14174
28. López-Martínez A, Morales-Rueda JA, Dibildox-Alvarado E, Charó-Alonso MA, Marangoni AG, Toro-Vazquez JF (2014) Comparing the crystallization and rheological behavior of organogels developed by pure and commercial monoglycerides in vegetable oil. Food Res Int 64:946–957. https://doi.org/10.1016/j.foodres.2014.08.029
29. Ferro AC, Okuro PK, Badan AP, Cunha RL (2019) Role of the oil on glyceryl monostearate based oleogels. Food Res Int 120:610–619. https://doi.org/10.1016/j.foodres.2018.11.013
30. Rocha-Amador OG, Gallegos-Infante JA, Huang Q, Rocha-Guzman NE, Rociomoreno-Jimenez M, Gonzalez-Laredo RF (2014) Influence of commercial saturated monoglyceride, mono-/diglycerides mixtures, vegetable oil, stirring speed, and temperature on the physical properties of organogels. Int J Food Sci 2014. https://doi.org/10.1155/2014/513641
31. Palla C, Giacomozzi A, Genovese DB, Carrín ME (2017) Multi–objective optimization of high oleic sunflower oil and monoglycerides oleogels: searching for rheological and textural properties similar to margarine. Food Struct 12:1–14. https://doi.org/10.1016/j.foostr.2017.02.005
32. Giacomozzi AS, Carrín ME, Palla CA (2021) Storage stability of Oleogels made from Monoglycerides and high Oleic sunflower oil. Food Biophys 16:306–316. https://doi.org/10.1007/s11483-020-09661-9
33. Zampouni K, Soniadis A, Moschakis T, Biliaderis CG, Lazaridou A, Katsanidis E (2022) Crystalline microstructure and physicochemical properties of olive oil oleogels formulated with monoglycerides and phytosterols. LWT 154. https://doi.org/10.1016/j.lwt.2021.112815
34. Cotabarren IM, Cruces S, Palla CA (2019) Extrusion 3D printing of nutraceutical oral dosage forms formulated with monoglycerides oleogels and phytosterols mixtures. Food Res Int 126. https://doi.org/10.1016/j.foodres.2019.108676
35. Rodríguez-Hernández AK, Pérez-Martínez JD, Gallegos-Infante JA, Toro-Vazquez JF, Ornelas-Paz JJ (2021) Rheological properties of ethyl cellulose-monoglyceride-candelilla wax oleogel vis-a-vis edible shortenings. Carbohydr Polym 252. https://doi.org/10.1016/j.carbpol.2020.117171
36. Bhattarai M, Penttilä P, Barba L, Macias-Rodriguez B, Hietala S, Mikkonen KS, Valoppi F (2022) Size-dependent filling effect of crystalline celluloses in structural engineering of composite oleogels. LWT 160. https://doi.org/10.1016/j.lwt.2022.113331

37. Valoppi F, Salmi A, Ratilainen M, Barba L, Puranen T, Tommiska O, Helander P, Heikkilä J, Haeggström E (2020) Controlling oleogel crystallization using ultrasonic standing waves. Sci Rep 10. https://doi.org/10.1038/s41598-020-71177-6
38. Giacomozzi A, Palla C, Carrín ME, Martini S (2020) Tailoring physical properties of monoglycerides oleogels using high-intensity ultrasound. Food Res Int 134. https://doi.org/10.1016/j.foodres.2020.109231
39. Chen CH, Terentjev EM (2011) Colloid-monoglyceride composites in hydrophobic solutions. Colloids Surfaces A Physicochem Eng Asp 384:536–542. https://doi.org/10.1016/j.colsurfa.2011.05.020
40. Verstringe S, Moens K, De Clercq N, Dewettinck K (2015) Crystallization behavior of monoacylglycerols in a hydrophobic and a hydrophilic solvent. Food Res Int 67. https://doi.org/10.1016/j.foodres.2014.10.027
41. Chen CH, Terentjev EM (2009) Aging and metastability of monoglycerides in hydrophobic solutions. Langmuir 25:6717–6724. https://doi.org/10.1021/la9002065
42. da Pieve S, Calligaris S, Panozzo A, Arrighetti G, Nicoli MC (2011) Effect of monoglyceride organogel structure on cod liver oil stability. Food Res Int 44:2978–2983. https://doi.org/10.1016/j.foodres.2011.07.011
43. Calligaris S, Valoppi F, Barba L, Anese M, Nicoli MC (2018) β-Carotene degradation kinetics as affected by fat crystal network and solid/liquid ratio. Food Res Int 105. https://doi.org/10.1016/j.foodres.2017.11.062
44. Cerqueira MA, Fasolin LH, Picone CSF, Pastrana LM, Cunha RL, Vicente AA (2017) Structural and mechanical properties of organogels: role of oil and gelator molecular structure. Food Res Int 96:161–170. https://doi.org/10.1016/j.foodres.2017.03.021
45. Giacomozzi AS, Palla CA, Carrín ME, Martini S (2019) Physical properties of Monoglycerides Oleogels modified by concentration, cooling rate, and high-intensity ultrasound. J Food Sci 84: 2549–2561. https://doi.org/10.1111/1750-3841.14762
46. Kesselman E, Shimoni E (2007) Imaging of oil/monoglyceride networks by polarizing near-field scanning optical microscopy. Food Biophys 2:117–123. https://doi.org/10.1007/s11483-007-9038-3
47. Lupi FR, Greco V, Baldino N, de Cindio B, Fischer P, Gabriele D (2016) The effects of intermolecular interactions on the physical properties of organogels in edible oils. J Colloid Interface Sci 483:154–164. https://doi.org/10.1016/j.jcis.2016.08.009
48. Rondou K, De Witte F, Rimaux T, Dewinter W, Dewettinck K, Verwaeren J, Van Bockstaele F (2022) Multiscale analysis of monoglyceride oleogels during storage. JAOCS, J Am Oil Chem Soc 99:1019–1031. https://doi.org/10.1002/aocs.12645
49. Sánchez-Becerril M, Marangoni AG, Perea-Flores MJ, Cayetano-Castro N, Martínez-Gutiérrez H, Andraca-Adame JA, Pérez-Martínez JD (2018) Characterization of the micro and nanostructure of the candelilla wax organogels crystal networks. Food Struct 16:1–7. https://doi.org/10.1016/j.foostr.2018.02.001
50. Fayaz G, Calligaris S, Nicoli MC (2020) Comparative study on the ability of different Oleogelators to structure sunflower oil. Food Biophys 15:42–49. https://doi.org/10.1007/s11483-019-09597-9
51. Yılmaz E, Öğütcü M (2014) Properties and stability of hazelnut oil organogels with beeswax and monoglyceride. JAOCS, J Am Oil Chem Soc 91:1007–1017. https://doi.org/10.1007/s11746-014-2434-1
52. Ashkar A, Laufer S, Rosen-Kligvasser J, Lesmes U, Davidovich-Pinhas M (2019) Impact of different oil gelators and oleogelation mechanisms on digestive lipolysis of canola oil oleogels. Food Hydrocoll 97. https://doi.org/10.1016/j.foodhyd.2019.105218
53. Merchán Sandoval J, Carelli A, Palla CA, Baümler E (2020) Preparation and characterization of oleogel emulsions: a comparative study between the use of recovered and commercial sunflower waxes as structuring agent. J Food Sci 85:2866–2878. https://doi.org/10.1111/1750-3841.15361

54. da Silva TLT, Danthine S (2021) Effect of high-intensity ultrasound on the oleogelation and physical properties of high melting point monoglycerides and triglycerides oleogels. J Food Sci 86:343–356. https://doi.org/10.1111/1750-3841.15589
55. Kamali E, Sahari MA, Barzegar M, Ahmadi Gavlighi H (2019) Novel oleogel formulation based on amaranth oil: physicochemical characterization. Food Sci Nutr 7:1986–1996. https://doi.org/10.1002/fsn3.1018
56. Si H, Cheong LZ, Huang J, Wang X, Zhang H (2016) Physical properties of soybean Oleogels and oil migration evaluation in model praline system. JAOCS, J Am Oil Chem Soc 93:1075–1084. https://doi.org/10.1007/s11746-016-2846-1
57. Toro-Vazquez JF, Morales-Rueda JA, Dibildox-Alvarado E, Charó-Alonso M, Alonzo-Macias-M, González-Chávez MM (2007) Thermal and textural properties of organogels developed by candelilla wax in safflower oil. JAOCS, J Am Oil Chem Soc 84:989–1000. https://doi.org/10.1007/s11746-007-1139-0
58. RDC (2023) http://www.refineriadelcentro.com.ar/es/catalogo. Accessed 18 Jun 2023
59. Rodriguez-Velazquez S Fat. In: Rodriguez-Velazquez S (ed) Chemistry of cooking. https://chem.libretexts.org/Bookshelves/Biological_Chemistry/Chemistry_of_Cooking_(Rodriguez-Velazquez). Accessed 18 June 2023
60. Kris-Etherton PM, Griel AE, Psota TL, Gebauer SK, Zhang J, Etherton TD (2005) Dietary stearic acid and risk of cardiovascular disease: intake, sources, digestion, and absorption. Lipids 40:1193
61. FAO (2010) Fats and fatty acids in human nutrition: report of an expert consultation

Chapter 6
Wax-Based Oleogels

Hong-Sik Hwang and Jill K. Winkler-Moser

Abbreviations

AFM	Atomic force microscopy
cryo-SEM	Cryogenic-scanning electron microscopy
DDGs	Distiller's dried grains with solubles
FDA	Food and Drug Administration
GRAS	Generally recognized as safe
PUFA	Polyunsaturated fatty acids
SEM	Scanning electron microscopy

6.1 Introduction

Wax-based oleogels (wax oleogel) are one of the most common types of oleogel that have been studied. In a search of the Scopus® database for the keyword "oleogel" within article titles, abstracts, and keywords in December 2022, the keyword "wax" co-occurred with the highest number of research articles (217 articles) followed by "monoglyceride" (71 articles) and "ethylcellulose" (49 articles). Waxes are hydrophobic mixtures composed of long carbon-chain containing components with melting points above room temperature, and they can be derived from crude petroleum

H.-S. Hwang (✉) · J. K. Winkler-Moser
U.S. Department of Agriculture, Agricultural Research Service, National Center for Agricultural Utilization Research, Functional Food Research, Peoria, IL, USA
e-mail: hongsik.hwang@usda.gov; jill.moser@usda.gov

C. Andrea Palla, F. Valoppi (eds.), *Advances in Oleogel Development, Characterization, and Nutritional Aspects*,
https://doi.org/10.1007/978-3-031-46831-5_6

oil, or are produced naturally from plants and animals. Waxes are often made up of mixtures of components, the most common of which are long-chain hydrocarbons, wax esters (ester of a fatty alcohol and a fatty acid), free fatty alcohols, free fatty acids, resins, and various other components [1–3]. In addition, within each wax component, there may be a diversity of carbon chain lengths, and the wax esters may include di-esters, tri-esters, as well as hydroxy fatty acid moieties [1, 4]. Most oleogel studies have focused on natural waxes derived from plants and animals, and the results of these studies are summarized in this chapter. So far, the most commonly studied waxes have been beeswax, candelilla wax, carnauba wax, rice bran wax, and sunflower wax. Beeswax, candelilla wax, and carnauba wax are diverse mixtures of hydrocarbons, wax esters, fatty acids, and alcohols and other wax components, while the commercially available pure forms of rice bran wax and sunflower wax are mainly (>90% w/w) composed of wax esters with small percentages of other components such as free fatty acids [3, 5, 6].

Wax oleogels are usually prepared using the direct method [7, 8]. In this case, waxes are melted completely in an oil, and upon cooling, the waxes crystallize, forming a porous network that traps liquid oil, seemingly through a combination of adhesive and capillary forces, depending on the nature of the oil and the wax types [5, 9]. While most studies have focused on the directly prepared oleogels, a recent study also evaluated rice bran wax oleogels as part of bigel systems, prepared by emulsifying the molten wax oleogel mixtures with a hydrogel to form hydrogel-in-oleogel, oleogel-in-hydrogel, or bicontinuous emulsions [10]. In addition, wax oleogels have been studied in a variety of food products where they are directly substituted for a conventional fat, or are used as the base hard stock to form margarines [11, 12]. Some of the promising results as well as the pitfalls or further work needed for the application of wax oleogels into food products are also summarized in this chapter.

6.2 Advantages of Waxes as Oleogelators

Based on many studies, it can be concluded that natural waxes have many advantages over other oleogelators studied. Natural waxes are inexpensive and nontoxic, and many of them are already listed as generally recognized as safe (GRAS) by the Food and Drug Administration (FDA) for some applications. They can form strong gels at relatively low concentrations, have high physical and oxidative stability, and have high potential for practical application in foods [13–15]. Total cholesterol level in rats fed on 0.5–1.0% sunflower wax in their diets was lower than that in rats fed control diets [16], and antilipogenic and antiadipogenic effects were observed with beeswax and carnauba wax in an in vitro study [17] indicating that wax oleogels may have additional health benefits beyond the general health benefits of oleogels by replacing saturated fats with unsaturated fatty acids [18] and controlled release of oil [19–21].

Gelation ability of waxes and physical properties of wax oleogels have been reported to be excellent. For example, sunflower wax formed a gel in soybean, almond, canola, corn, grapeseed, safflower, and sunflower oils at concentrations as low as 0.3% (w/w) [22]. Oleogels of virgin olive oil with 7 and 10% carnauba wax had greater firmness than those with monoglyceride [13]. Oleogels formed with rice bran wax in extra virgin olive oil had higher firmness than those with monoglycerides and ethylcellulose, but lower firmness than an oleogel prepared with a mixture of γ-oryzanol and β-sitosterol [23]. Similarly, oleogels prepared with beeswax, sunflower wax, or rice bran wax in sunflower oil had higher G' and G'' moduli than those prepared with monoglyceride, octadecanol, or stearic acid, but they were not as high as those of oleogels made with a γ-oryzanol and β-sitosterol mixture [24].

Wax oleogels may also have less of an effect on sensory properties and higher physical and oxidative stability than other oleogels. A sensory acceptability test showed that burgers made with beeswax oleogels had no significant effect on any of the sensory parameters evaluated while those made with ethyl cellulose oleogels had negative effects [25]. Öğütcü et al. [14] reported that sunflower wax-fish oil oleogels had higher oxidation stability than monoglyceride oleogels during storage at 4 and 20 °C. Another study found that carnauba wax oleogels had higher physical and oxidative stability than monoglyceride oleogels during storage at 4 and 20 °C [13].

6.3 Limitations of Wax Oleogels

In general, the melting point of wax oleogels is higher than conventional solid fats, which may result in an undesirable waxy mouthcoating when consumed [2, 11, 26]. The melting point of wax oleogels decreases with decreasing amount of wax [11], but the low level of wax will result in decreased firmness. Efforts have been made to lower the melting point of wax oleogels. Among the possible solutions to the problem, binary mixture systems having eutectic properties seemed to be very promising, which will be discussed in Sect. 6.4.2.

Since the strength of wax oleogels is basically attributed to the dense network of thin platelet wax crystals in oil [27, 28], and this network can be easily destroyed by shearing and other mechanical forces, wax oleogels have strong shear sensitivity and weak thixotropic recovery [29]. This property may limit the application of wax oleogels for some food products, and efforts are being made to overcome this limitation. For example, improvements in the elasticity and mechanical recovery after shear of candelilla wax oleogels were observed by adding tripalmitin [30], saturated fat [31], monoglycerides [32], and mixtures of ethylcellulose and monoglycerides [33].

6.4 Properties of Wax Oleogels

6.4.1 Needle Versus Platelet Crystals

Earlier studies typically used polarized light microscopy to investigate the morphology of wax crystals in wax oleogels and reported that waxes formed needle-shaped crystals. However, when needle-shaped crystals were scrutinized with phase-contrast microscopy, scanning electron microscope (SEM), and atomic force microscopy (AFM), it was found that needle-shaped objects under polarized light microscope were actually the edges of platelet-shaped crystals [28]. Figure 6.1a shows a microphotograph of sunflower wax oleogel taken by polarized light microscope, which is very similar to photographs in the previous studies that claimed needle-shaped crystals. However, when the same spot of the sample was observed under a phase-contrast microscope, these needle-shaped objects were found to be the edges of platelet crystals (Fig. 6.1b). SEM and AFM further confirmed platelet crystals in sunflower wax oleogels. Hwang et al. [28] also reported that these platelet crystals formed a dense network, with increased network density observed with higher concentrations of sunflower wax or with faster cooling rates, resulting in higher firmness.

Blake and Marangoni [34] also demonstrated that rice bran wax, sunflower wax, and candelilla wax formed platelet-shaped crystals in peanut oil by cryogenic-SEM (cryo-SEM). Later studies with cryo-SEM further proved that sunflower, rice bran, berry, and candelilla waxes formed platelet-shaped crystals in oleogels [5, 27, 35]. Fig. 6.2 shows cryo-SEM pictures of de-oiled oleogels formed by sunflower, candelilla, and berry waxes. The major components in waxes, wax esters, also formed platelet-shaped crystals [36]. Therefore, polarized light microscopy is not as useful for studying the morphology of wax crystals in wax oleogels, and phase-contrast microscopy and cryo-SEM are recommended.

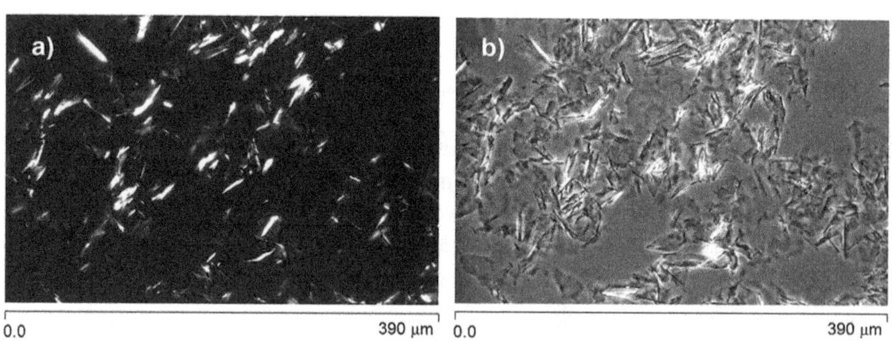

Fig. 6.1 Microphotographs taken by (**a**) polarized light microscope and (**b**) phase-contrast microscope at the same spot of the oleogel sample made with 5% sunflower wax in soybean oil. (Reproduced from [28] with permission from Elsevier)

Fig. 6.2 Cryo-SEM of rice bran oil oleogels with (**a**) sunflower wax, (**b**) candelilla wax, and (**c**) berry wax. (Reproduced from [5] with permission from Elsevier)

6.4.2 *Effect of Wax Type*

As expected, since different waxes have different compositions [5], their gelation abilities are significantly different. For example, Hwang et al. [2] reported that sunflower wax, candelilla wax, rice bran wax, beeswax, Japan wax, and shellac wax had significantly different minimum gelation concentrations (or critical concentration) ranging from 0.5 to 5% (w/w) in soybean oil. Shi et al. [37] found that beeswax, candelilla wax, and carnauba wax had minimum gelation concentrations of 3, 2, and 4%, respectively, in camellia seed oil.

The minimum gelation concentration of wax is also largely dependent on the origin of wax as well as the detailed composition of individual wax components. For example, rice bran wax samples obtained from three different suppliers had 0.5, 1, and 5% minimum gelation concentrations in soybean oil [2]. This was likely due to differences in the composition and/or purity of waxes. Doan et al. [5] summarized the literature values from composition analyses of seven different waxes, which were shown to differ widely across different studies. For example, four separate studies reported the content of wax esters in carnauba wax to be 75–85 [38], 34.3 [39], 84–85 [40], and 59–62% [3]. The wax ester content in sunflower wax was reported to be 66% by Kanya et al. [41] while another study reported 96–97% wax esters

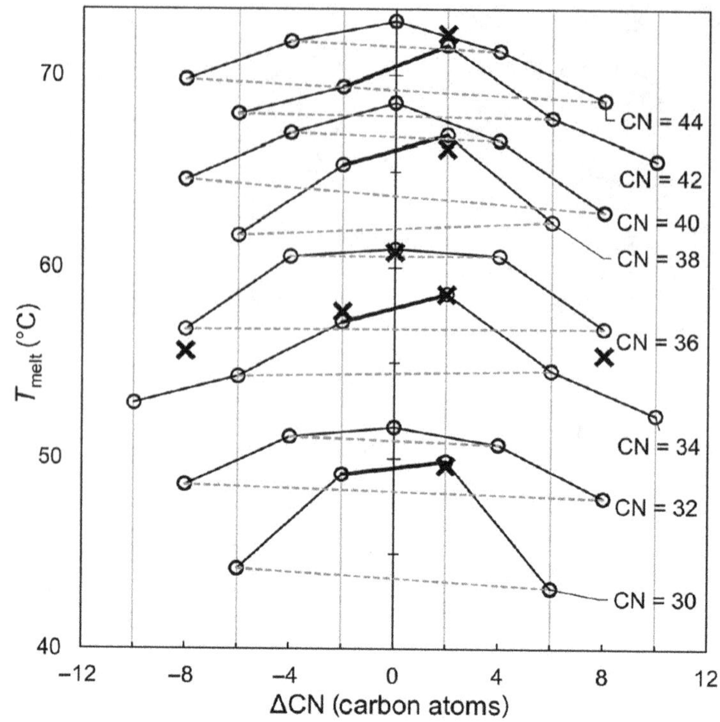

Fig. 6.3 Melting temperature of wax esters with different symmetries and total carbon number (CN). $\Delta CN = CN_{FaOH} - CN_{FA}$, where CN_{FaOH} is the carbon number of the fatty alcohol (FaOH) and CN_{FA} is that of the fatty acid (FA). (x): determined by [42]. (o): determined by other studies. (Reproduced from [42] under open access agreement)

[3]. The detailed composition of natural waxes may depend on growing environments, separation processes, and purification methods.

Wax esters are the major components in many natural waxes. Waxes with a high content of wax esters can produce strong but brittle oleogels [3]. The chain length and structure of wax esters also significantly affected the gelation ability and the properties of oleogels. In a study with wax esters with narrow chain length distributions, it was found that the minimum gelation concentration was lower with longer-chain wax esters [2]. However, in other studies with pure wax esters, lower G′ and G″ moduli were observed with oleogels with longer wax esters [42, 43]. Longer wax esters had higher melting points than shorter wax esters, and their oleogels also had the same trend [2, 42]. Symmetric wax esters (wax esters with the same chain lengths of fatty acid and fatty alcohol moieties) had higher melting points than asymmetric wax esters [42]. Fig. 6.3 shows the melting point of wax esters with different chain lengths and symmetries. Therefore, properties of wax oleogels are affected not only by the amounts of wax esters, but also by their length and structure.

Comparisons between waxes: Although as mentioned above, the gelation ability of waxes and properties of wax oleogels can vary with the origin of wax samples, several comparative studies give information on general characteristics of a few waxes on oleogelation. Hwang et al. [11] compared gelation properties of sunflower, rice bran, and candelilla waxes and found that oleogels and margarines made with sunflower wax had higher firmness than those made with rice bran and candelilla waxes. Oil-binding capacities of sunflower wax-hazelnut oil oleogels were higher than carnauba wax oleogels [44]. Oleogel with 1% sunflower wax in peanut oil had the highest oil-binding capacity, followed by candelilla wax and rice bran wax oleogels [9]. Sunflower wax and beeswax at 5% (w/w) were able to form gels in medium-chain triacylglycerols while rapeseed wax did not make a gel at the concentration [45]. Canola oil oleogel with candelilla wax had the highest hardness, followed by carnauba and beeswax oleogels at 10% wax concentration while beeswax oleogel had the highest adhesiveness and cohesiveness [46]. Oleogels formed with 2.5–10% (w/w) candelilla wax in microbial oil or sunflower oil had higher oil-binding capacity than the corresponding carnauba wax oleogels [47]. In another comparative study, oleogel with 10% beeswax in sunflower oil had the highest storage (G') and loss (G'') moduli followed by sunflower wax oleogel and then rice bran wax oleogel [24]. Oleogel with 10% monoglyceride was not as firm as wax oleogels in this study. In general, sunflower wax had greater gelation ability and provided stronger oleogels than other waxes.

New waxes studied: Some new waxes have recently been studied in efforts to develop oleogels with better properties and to utilize waxes obtained as by-products. Gelation abilities of tea, rapeseed, orange peel, rose, and berry waxes were evaluated in sunflower oil, and their minimum gelation concentrations were determined to be 7.0, 25.0, 15.0, 8.0, and 5.0%, respectively [48]. Minimum gelation concentrations of sorghum waxes from sorghum bran, sorghum distiller's dried grains with solubles (DDGs), and sorghum kernels were measured to be 6, 6, and 4% in fish oil, respectively [49]. The highest oil-binding capacity was observed with sorghum DDG wax oleogels followed by those with sorghum bran wax and sorghum kernel wax. Propolis wax was also evaluated as oleogelators in canola, sesame, sunflower, and flaxseed oils, and it was found that this wax could form a gel in these oils at 2% concentration [50]. Another study [51] compared propolis wax-pomegranate seed oil oleogels with the corresponding beeswax oleogels and found that propolis wax oleogels had lower firmness and G' and G'' moduli compared to beeswax oleogels, which was attributed to the larger propolis wax crystals with a less organized network. Cetyl wax esters produced by the enzymatic reaction of cetyl alcohol with carotenoid-rich microbial oil or soybean fatty acid distillate, a by-product during soybean oil production, were evaluated as oleogelators in soybean oil or extra virgin olive oil [52, 53]. These wax esters at 20% could form an oleogel while 7 and 10% wax esters could not form an oleogel. Gelation ability of animal waxes including whale spermaceti wax, lanolin wax, shellac wax, and beeswax was compared in sunflower oil, and their minimum gelation concentrations were

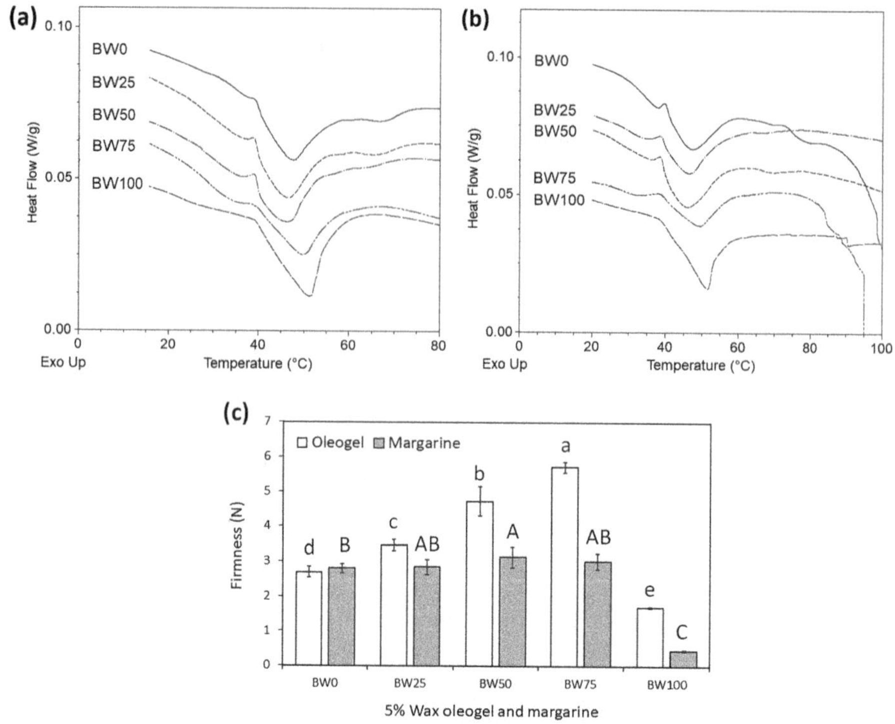

Fig. 6.4 Melting profiles of (**a**) oleogels and (**b**) margarines with 5% mixture of candelilla wax (CLW) and beeswax (BW) in soybean oil and (**c**) their firmness. BW0, BW25, BW50, BW75, and BW100 stand for the ratios of 0:100, 25:75, 50:50, 75:25, and 100:0 BW:CLW. (Reproduced from [56] with permission from Wiley)

determined to be 2, 20, 2, and 1%, respectively [54]. The order of the firmness of oleogels with 5% wax was whale spermaceti wax ≈ beeswax > shellac wax.

Binary systems: Due to the potential problem of wax oleogels' melting points that are higher than those of conventional solid fats, efforts have been made to lower their melting points by use of binary systems of waxes. Jana and Martini [55] reported the eutectic behavior of the mixture of beeswax with sunflower wax or rice bran wax. Winkler-Moser et al. [26] also conducted studies on binary wax mixtures and observed that oleogels made with the mixture of beeswax and candelilla wax had not only lower melting points, but also higher firmnesses than those made with either wax alone. Margarines prepared from these oleogels also showed eutectic melting properties and increased firmness than those with each wax (Fig. 6.4) [56]. In addition, oleogels made with mixtures of candelilla wax and glyceryl monostearate in grapeseed oil or canola oil had lower melting points and higher firmness than those with candelilla wax or glyceryl monostearate alone [57, 58].

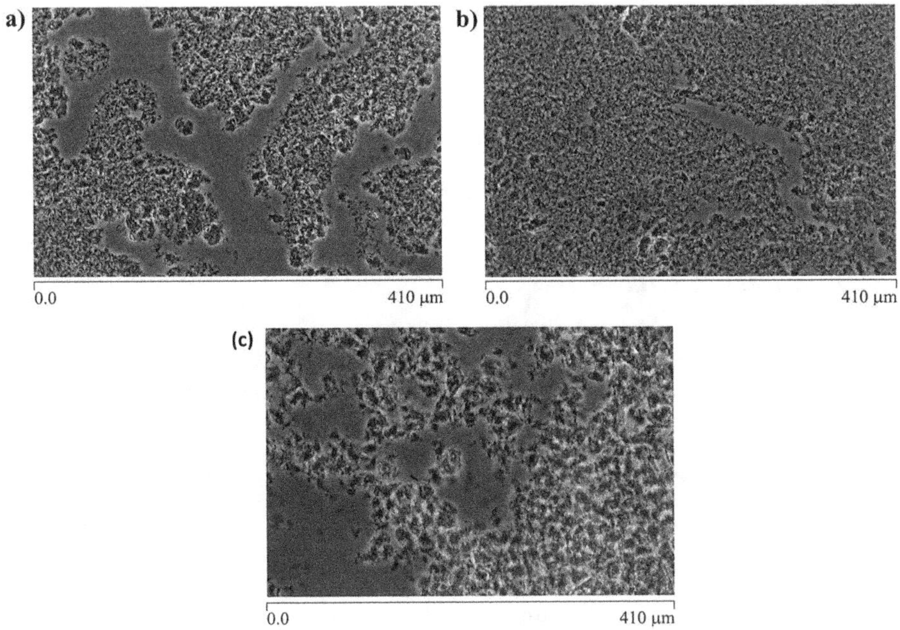

Fig. 6.5 Phase-contrast micrographs of grapeseed oil oleogels with (**a**) candelilla wax, (**b**) 75:25 mixture of candelilla wax and glyceryl monostearate, and (**c**) glyceryl monostearate. The dimension of each picture is 410 μm × 275 μm. (Reproduced from [57] with permission from Wiley)

The major components in candelilla wax are hydrocarbons (72.9%) while those in beeswax are wax esters (58.0%) [3]. The eutectic properties of oleogels with the binary wax system of candelilla wax and beeswax were thought to be attributed to the interaction between hydrocarbons and wax esters [26]. This explanation was supported by another study conducted with the fractions separated from beeswax [4]. Eutectic behaviors were observed when the fraction containing 94.8% hydrocarbons was mixed with the other fraction rich in wax esters, free fatty acids, or fatty alcohols. The increased gel strength by mixtures of beeswax and candelilla wax was also thought to due to a synergistic interaction between hydrocarbons and wax esters at an optimum ratio, which resulted in smaller crystals and a denser network [26]. Firmness of oleogels made with these beeswax and candelilla wax combinations were recently shown to be enhanced greatly by the addition of 0.5% sunflower wax, without impacting the solid fat content at temperatures near body temperature [6]. Choi et al. [57] explained the higher firmness of grapeseed oil oleogel with the mixture (75:25) of candelilla wax and glyceryl monostearate than that with each component by the formation of smaller crystals and a denser crystal network (Fig. 6.5).

Although the melting point was not lowered, mixtures of rice bran wax and berry wax at the ratios of 3:7 and 2:8 and those of sunflower wax and berry wax at the ratios of 9:1 to 5:5 provided the firmer oleogels in rice bran oil than each wax alone

[35]. While the major components in rice bran and sunflower waxes are wax esters, berry wax is primarily composed of fatty acids [3]. It was hypothesized that rice bran wax or sunflower wax crystallized first, and then berry wax formed solid bridges between the already present wax crystals to increase the hardness. Several studies have also demonstrated synergistic firmness of rice bran wax and sunflower wax mixtures in oleogels [59, 60].

6.4.3 *Effect of Oils*

Beeswax had minimum gelation concentrations of 3, 3, 3, and 4% in camellia oil, soybean oil, sunflower oil, and flaxseed oil, respectively [61]. In another study, the minimum gelation concentration of rice bran wax in eight different vegetable oils ranged from 8% to 10 % (w/v) [62]. Hwang et al. [22] reported that the minimum gelation concentration of sunflower wax in 13 different vegetable oils varied from 0.3% to 1% (w/w). Firmness of 5% sunflower wax oleogels with these oils was also significantly different ranging from 3.81 (flaxseed oil oleogel) to 7.72 N (soybean oil oleogel). Margarines prepared with these oleogels also had significantly different firmness ranging from 2.62 N with flaxseed oil oleogel to 6.67 N with sesame oil oleogel. Pang et al. [61] reported that oleogels with 6% beeswax in four different oils had the firmness order of soybean oil > flaxseed oil > sunflower oil > camellia oil. These studies did not find any specific correlations between the firmness of oleogels or margarines and fatty acids composition.

Some recent studies reported that waxes had greater gelation ability in oils with a higher degree of unsaturation, and their oleogels had higher gel strength. In a study conducted with oleogels of beeswax in different oils including camellia, sunflower, corn, and linseed oils, it was found that beeswax could form a gel at 1% (w/w) in linseed oil while it formed a gel at 1.2, 1.3 and 1.4% in corn, sunflower, and camellia oils, respectively [63]. It was postulated that an oil with a higher degree of unsaturation accelerated the crystallization rate, and formed denser and more stable networks of wax crystals, and produced oleogels with stronger van der Waals forces. In another study with oleogels made with sunflower wax or rice bran wax in sunflower, mustard, soybean, sesame, groundnut, rice bran, palm, and coconut oils, it was found that, in general, oils with higher polyunsaturated fatty acids (PUFAs) or higher unsaturation had lower minimum gelation concentrations, and their oleogels had higher oil-binding capacity [62]. Studies with other oleogels than wax oleogels have reported similar results. For example, 10% (w/w) ethylcellulose oleogels were firmer with oils rich in highly unsaturated fatty acids, which was explained by a greater degree of conformation freedom with oils with higher unsaturation that resulted in tight packing of molecules [64]. In contrast, the opposite trend was observed in a study with 6% carnauba wax oleogels prepared with hempseed, almond, rice, sesame, grapeseed, and pumpkin seed oils [65]. In this study, oils with higher saturated fatty acid contents provided oleogels with higher strength and yield stress. Therefore, it is obvious that oil affects the oleogel properties, but more studies are needed for better understanding of the effects of oil composition.

The fatty acid chain length of oil also significantly affects oleogel properties. In a study with beeswax oleogels prepared with medium-chain triglycerides and long-chain triglycerides, it was found that long-chain triglyceride oleogels had higher G′ and G″ values, which was thought to be due to lower mobility of the gelator and the formation of a strengthened network of gels by longer chains [66]. Another study with beeswax oleogels made with hempseed, almond, rice, sesame, grapeseed, and pumpkin seed oils found that, in general, oils with more long-chain saturated fatty acids had a more interconnected structure providing higher complex viscosity [65].

6.4.4 Effect of Polar Compounds in Oil

Very weak negative correlations ($R^2 = 0.34$, 0.20, and 0.16 for 3, 5, and 7% sunflower wax, respectively) between the firmness of margarine and the amount of polar compounds in oil were observed in a study with margarine samples prepared with sunflower wax oleogels of 13 different vegetable oils [22]. However, a later study reported that the firmness of sunflower wax oleogels increased when polar compounds were removed from high stearic soybean oils indicating that polar compounds in these oils negatively affected the firmness of oleogels [67]. Sunflower wax, rice bran wax, candelilla wax, and beeswax formed highly ordered wax crystal structures and had lower wax solubility in oils when polar compounds were removed from sunflower oil and canola oil, but they also had higher minimum gelation concentration and their oleogels had lower resistance to deformation [68].

6.4.5 Effect of Cooling Rate

Sunflower wax-soybean oil oleogels had about three- to sixfold higher firmness when the cooling rate increased from 1 to 4 °C/min, which was thought to be due to the larger number of smaller crystals that formed a denser crystal network during faster cooling [2]. Blake and Marangoni [9] also reported that rapid cooling increased oil-binding capacity and decreased crystal length in wax-peanut oil oleogels. Many other studies reported that the faster cooling rate during the formation of oleogels increased the gel strength. Therefore, firmness and other physical properties of oleogels can be tailored by the cooling rate during the production of oleogels and oleogel-containing food products.

6.4.6 Sensory Properties

Many studies on sensory properties of wax oleogels or food products containing wax oleogel reported that they were acceptable or had high potential to be accepted with some modifications. For example, a study with oleogels made with 5% beeswax or sunflower wax in hazelnut oil or olive oil found that their sensory attributes including appearance, odor, and flavor were scored about 3.0 where the score of 2.5 was the neutral point, but they were lower than those of commercial breakfast margarine and butter. Hazelnut oleogels had higher spreadability scores than the commercial breakfast margarine [69]. When cooked burgers prepared with beeswax oleogel were compared to control burger made with animal fat, no significant effects on any of the sensory parameters including texture, color, flavor, and overall acceptability were observed by replacing animal fat with oleogel [25].

Virgin olive oil oleogels with 1% (w/w) sunflower wax and 1% (w/w) thyme or cumin spice had high acceptability scores by consumers [70]. Consumer preference scores of the odor and flavor of fish oil oleogels were enhanced by adding strawberry or lemon flavor [71]. Spermaceti wax-virgin olive oil oleogels flavored with turmeric or red pepper had very high appearance scores, and their aroma and flavor were also liked by consumers [72]. Therefore, adding a flavor in oleogels helps increase their acceptability by consumers. More sensory properties of food products containing wax oleogels will be discussed in the next section.

6.5 Properties of Food Products Containing Wax Oleogels

While earlier studies focused on the basic understanding on the properties of oleogels, many recent studies have been conducted on the application of wax oleogels in actual food products. The success in achieving the desired physical and sensory properties seems to have encouraged scientists to pursue the practical applications of wax oleogels.

Table 6.1 summarizes the most recent studies of food products prepared with wax oleogels and their properties compared to those with conventional fats.

In some studies, cookie samples with wax oleogels had similar spread factor, hardness, fracturability, appearance, flavor, smell, and acceptability to those with commercial margarine or shortening [73–75]. Frolova et al. [76] also reported that cookies made with oleogels of beeswax, fractions of beeswax, or candelilla wax in sunflower oil had high scores on taste, texture, surface condition, shape, and fracture view although no direct comparisons with conventional cookies were made. However, other studies found that cookie samples had somewhat different textual and sensory properties compared to control cookies. Cookies with wax oleogels had higher breaking force than those with commercial shortening [77, 78], and some properties, including snapping force, surface cracks, color, smell, and apparency, were somewhat different from the control [79–82]. Cake samples prepared with wax

Table 6.1 Recent studies of the application of wax oleogels in food products

Food	Wax (%, w/w)	Oil	Properties compared to the control with commercial fat	Reference
Cookie	Sunflower wax, rice bran wax, candelilla wax, beeswax (8%)	Olive, soybean, and flaxseed oils	Similar spread factor, hardness, and fracturability	[73]
	Sunflower wax, beeswax (5%)	Hazelnut oil	Equal (or slightly higher) scores on appearance, texture, flavor, smell, and acceptability	[74]
	Rice bran wax, candelilla wax, beeswax, carnauba wax (3, 5, 7, 9%)	Corn oil	Cookies with oleogels of 5% rice bran wax, 7 and 9% beeswax, 5% candelilla wax, and 5, 7 and 9% carnauba wax had similar hardness	[75]
	Beeswax and its fractions (6%)	Sunflower oil	No comparisons were made: high scores on taste, texture, surface condition, shape, and fracture view	[76]
	Candelilla wax (3, 6%)	Canola oil	Cookies with partial replacement (30 and 60%) had higher breaking force and lower spread ratio	[77]
	Carnauba wax, candelilla wax (2.5, 5%)	Sunflower oil	Slightly higher breaking force and higher spread ratio	[78]
	Candelilla wax, carnauba wax, beeswax (10%)	Insect oil	Candelilla wax and beeswax: higher spread factor and lower snapping force. Carnauba wax: similar spread factor and snapping force	[79]
	Candelilla wax (3, 6%)	Canola oil	G′ and G″ of dough were lower; Cookies had similar shape, but lower snapping force	[80]
	Rice bran wax, beeswax (6%)	Corn oil	Rice bran wax: lower scores in surface cracks and color; similar scores in appearance, flavor, crispiness, mouthfeel, and overall acceptability. Beeswax: lower scores in surface cracks, color crispiness, mouthfeel, and overall acceptability; similar in appearance and flavor	[81]
	Sunflower wax (10%)	Rapeseed oil	Lower height, higher compactness, similar hardness and browning, and slightly different in smell and appearance	[82]

(continued)

Table 6.1 (continued)

Food	Wax (%, w/w)	Oil	Properties compared to the control with commercial fat	Reference
Cake	Sunflower wax, rice bran wax, beeswax (5%)	Canola oil	Higher cohesiveness, similar resilience, and lower hardness, adhesiveness, springiness, gumminess, and chewiness	[84]
	Candelilla wax (5%)	Canola oil	Higher chewiness, resilience, adhesiveness, and cohesiveness and lower hardness as 25, 50, 75, and 100% butter was replaced with oleogel	[83]
	Carnauba wax (5%)	Cotton seed oil, high oleic sunflower oil, and their mixtures	Four out of five samples had higher or similar overall acceptability in sensory evaluation	[86]
	Rice bran wax, beeswax, candelilla wax (10%)	Sunflower oil	Higher hardness and chewiness and lower specific volume and cohesiveness	[85]
Muffin	Candelilla wax (10%)	Sunflower oil	Lower specific volume and cohesiveness, higher hardness and chewiness, and similar porosity and springiness	[87]
	Candelilla wax (10%)	Grapeseed oil	25% replacement: similar fragmentation index, number of closed pored, total porosity, and relaxation time and slightly lower firmness. 50–100% replacement: lower fragmentation index and total porosity and higher number of closed pores, firmness, and relaxation time	[88]
Bread	Rice bran wax (10%)	Expeller-pressed high oleic soybean oil	Similar fermentation height, loaf volume, and crumb firmness and lower specific volume	[89]
	Candelilla wax (10%)	Rice bran oil	75% replacement: similar specific volume, hardness, adhesiveness, springiness, cohesiveness, and chewiness	[90]
Meat products	Rice bran wax (2%)	Rice bran oil	(Compared to Thai sweet sausage with 20% pork back fat) 50% and 75% replacement: similar hardness and higher oiliness, overall acceptance, and appearance scores	[92]

(continued)

Table 6.1 (continued)

Food	Wax (%, w/w)	Oil	Properties compared to the control with commercial fat	Reference
	Beeswax (10%)	Mixture of olive, linseed, and fish oils	(Compared to pork burgers with 6% pork backfat) Similar flavor, texture, and overall acceptability scores	[93]
	Beeswax (10%)	Mixture of olive, linseed, and fish oils	(Compared to pates 15% pork backfat) Similar color, taste, texture, and overall acceptability scores	[25]
	Rice bran wax (2.5%)	Soybean oil	(Compared to Frankfurter-type sausages with 22% pork backfat) Similar aroma and off-flavor, lower flavor, and higher firmness and chewiness scores	[94]
	Beeswax (10%)	Sesame oil	(Compared to beef burger with 16% fat) Higher flavor and overall acceptance and similar texture and color scores	[95]
	Sunflower wax, beeswax (10%)	Flaxseed oil	(Compared to sucuk with tallow fat) Sunflower wax: lower appearance, aroma, flavor, and hardness, higher chewiness and juiciness, and similar fattiness scores. Beeswax: lower appearance, hardness, aroma, and flavor and higher chewiness, fattiness, and juiciness scores	[96]
Cheese	Rice bran wax, sunflower wax (5, 10%)	Soybean oil	(Compared to commercial cheese with 40% solid milk fat) Similar hardness, storage modulus and meltability	[97]
	Rice bran wax (10%)	Soybean oil	(Compared to commercial cream cheese) Similar hardness, spreadability, mouthfeel, and sweetness and higher flavor and bitterness scores	[98]
	Carnauba wax (3, 6, 9%)	Canola oil	(Compared to imitation cheese with palm oil) Higher firmness and elasticity and lower mobility of oil and meltability	[99]

(continued)

Table 6.1 (continued)

Food	Wax (%, w/w)	Oil	Properties compared to the control with commercial fat	Reference
Ice cream	Carnauba wax (6%)	Soybean oil	50% replacement: similar flavor, texture, appearance, and overall consumer scores and slower melting rate. 100% replacement: lower flavor and overall scores and slower melting rate	[100]

oleogels were somewhat different from the control samples prepared with commercial shortening in cohesiveness, resilience, adhesiveness, hardness, chewiness, and specific volume [83–85] while some samples had higher overall acceptability than the control samples [86]. The feasibility of wax oleogels in muffin and bread samples was also evaluated [87–90]. In general, physical properties of muffin and bread samples made with wax oleogels were slightly different or similar to those with commercial shortening, and the slight differences could be overcome by partial replacement.

Meat products are one of the major contributors to dietary intake of saturated fats, and many studies have been conducted to incorporate oleogels in meat products such as meat batters, patties, burgers, frankfurters, sausages, and pates. Although it is straightforward that replacing saturated fats in meat products with healthy oil-based oleogels will promote health, it seems it is still very challenging to achieve acceptable texture, spreadability, flavor, and oxidative stability with products with oleogels [91]. As shown in Table 6.1, it is encouraging that recent studies with meat products containing wax oleogels reported that their appearance, taste, texture, smell, and overall acceptance scores were similar to those of the control samples containing animal fat [25, 92–94]. Even higher scores on flavor and overall acceptance than the control samples with animal fat were observed with beef burger samples where beef fat was replaced with beeswax-sesame oil oleogel [95]. It was thought to be due to preferred taste and smell of sesame oil, indicating that sensory scores can be improved by using tasty oils or by adding a flavor. In contrast, a study with sucuk (a Turkish style of beef sausage) made with 10% sunflower wax or beeswax-flaxseed oil oleogels reported that further improvements were needed [96].

Some cheese products with wax oleogels were reported to have similar hardness, storage modulus, meltability, spreadability, mouthfeel, and sweetness and higher flavor and bitterness scores than those made with milk fat [97, 98]. Another study with cheese products made with 3–9% carnauba wax in canola oil oleogels reported that they had higher firmness and elasticity and lower oil mobility and meltability than those with palm oil [99]. Ice cream prepared by replacing 50% milk fat with 6% carnauba wax-soybean oil oleogel had similar flavor, texture, appearance, and overall consumer scores and slower melting rate than the original ice cream [100]. When 100% milk fat was replaced with this oleogel, lower flavor and overall scores were observed indicating that only partial replacement was recommended for this application.

6.6 Other Applications of Wax Oleogels

Peanut butter stabilizer: Stabilizers are used to prevent oil separation and to improve texture and spreadability of peanut butter. Conventional peanut butter stabilizers are typically composed of hydrogenated oils, and studies were conducted to replace these stabilizers with wax oleogels. Waxes can be directly added in peanut butter and heated to form oleogels and also wax oleogels can be prepared by dissolving wax in peanut oil with heat before adding in peanut butter. Desired firmness and spreadability as well as prevention of oil separation could be achieved with wax oleogels, depending on the wax type and the amount added [12, 101, 102].

Frying medium: Chicken samples fried in 5 or 10% (w/v) carnauba wax-canola oil oleogel had lower oil uptake (8.53 and 9.15%, respectively) compared to samples fried in canola oil (15.10%) [103]. Similarly, *mathri*, an Indian traditional fried snack cracker made with wheat flour, was prepared by frying in soybean oil oleogels with 5, 10, or 15% carnauba wax and compared with that fried in soybean oil [104]. Oil uptake was 27.1, 19,6, 21.0, and 21.2% for the samples fried in oil and oleogels with 5, 10, and 15% carnauba wax, respectively. The sensory analysis of these samples revealed that overall acceptability and taste scores were similar between the mathri samples fried in oil and 10% wax oleogel while the sample fried with 15% wax oleogel had a waxy feel and taste. These studies showed that wax oleogels can be used in place of oil to reduce oil uptake of fried foods.

6.7 Conclusions

The number of studies on wax oleogels has increased, the fastest among other oleogels due to their advantages of high gel strength, low critical gelation concentrations, high physical and oxidative stability during storage, and relatively low effects on sensory properties of reformulated foods. Waxes form a dense network of many small platelet-shaped crystals that provide high gel strength. The type and origin of waxes significantly affect the properties of wax oleogels. Sunflower wax, rice bran wax, candelilla wax, carnauba wax, and beeswax have been the most studied among other waxes due to their strong gelation ability. Some binary or ternary systems of waxes and mixtures of wax and monoglyceride could provide oleogels with a lower melting point and a higher firmness than those with either component alone. Although more studies are needed, based on studies thus far, it appears that polar compounds in oil negatively affect the physical properties of wax oleogels and that oils, while higher levels of oil triacylglycerol unsaturation produce stronger gels. Many recent studies have reported that food products where conventional fats were replaced with wax oleogels had similar physical and sensory properties to commercial products while others have reported that further improvements were needed. Acceptable physical and sensory properties of food products containing wax oleogels can be achieved by finding right types of wax and oil and using the optimized wax amount. From the review of recent studies, it can be concluded that wax oleogels have high potential for the development of healthy food products.

References

1. Tulloch AP (1973) Comparison of some commercial waxes by gas liquid chromatography. J Am Oil Chem Soc 50(9):367–371. https://doi.org/10.1007/BF02640842
2. Hwang HS, Kim S, Singh M et al (2012) Organogel formation of soybean oil with waxes. J Am Oil Chem Soc 89(4):639–647. https://doi.org/10.1007/s11746-011-1953-2
3. Doan CD, To CM, De Vrieze M et al (2017) Chemical profiling of the major components in natural waxes to elucidate their role in liquid oil structuring. Food Chem 214:717–725. https://doi.org/10.1016/j.foodchem.2016.07.123
4. Sarkisyan V, Sobolev R, Frolova Y et al (2021) Beeswax fractions used as potential oil gelling agents. J Am Oil Chem Soc 98(3):281–296. https://doi.org/10.1002/aocs.12451
5. Doan CD, Tavernier I, Okuro PK et al (2018) Internal and external factors affecting the crystallization, gelation and applicability of wax-based oleogels in food industry. Innov Food Sci Emerg Technol 45:42–52. https://doi.org/10.1016/j.ifset.2017.09.023
6. Winkler-Moser JK, Hwang HS, Felker FC et al (2023) Increasing the firmness of wax-based oleogels using ternary mixtures of sunflower wax with beeswax:candelilla wax combinations. J Am Oil Chem Soc. (in press). https://doi.org/10.1002/aocs.12679
7. Barroso NG, Santos MAS, Okuro PK et al (2022) Composition and process approaches that underpin the mechanical properties of oleogels. J Am Oil Chem Soc 99(11):971–984. https://doi.org/10.1002/aocs.12635
8. Silva PM, Cerqueira MA, Martins AJ et al (2022) Oleogels and bigels as alternatives to saturated fats: a review on their application by the food industry. J Am Oil Chem Soc 99(11): 911–923. https://doi.org/10.1002/aocs.12637
9. Blake AI, Marangoni AG (2015) The use of cooling rate to engineer the microstructure and oil binding capacity of wax crystal networks. Food Biophys 10(4):456–465. https://doi.org/10.1007/s11483-015-9409-0
10. Saffold AC, Acevedo NC (2022) The effect of mono-diglycerides on the mechanical properties, microstructure, and physical stability of an edible rice bran wax–gelatin biphasic gel system. J Am Oil Chem Soc 99(11):1033–1043. https://doi.org/10.1002/aocs.12640
11. Hwang HS, Singh M, Bakota EL et al (2013) Margarine from organogels of plant wax and soybean oil. J Am Oil Chem Soc 90(11):1705–1712. https://doi.org/10.1007/s11746-013-2315-z
12. Winkler-Moser JK, Anderson J, Byars JA et al (2019) Evaluation of beeswax, candelilla wax, rice bran wax, and sunflower wax as alternative stabilizers for peanut butter. J Am Oil Chem Soc 96(11):1235–1248. https://doi.org/10.1002/aocs.12276
13. Öğütcü M, Yilmaz E (2014) Oleogels of virgin olive oil with carnauba wax and monoglyceride as spreadable products. Grasas Aceites 65(3):e040. https://doi.org/10.3989/gya.0349141

14. Öğütcü M, Temizkan R, Arifoğlu N et al (2015) Structure and stability of fish oil organogels prepared with sunflower wax and monoglyceride. J Oleo Sci 64(7):713–720. https://doi.org/10.5650/jos.ess15053
15. Demirkesen I, Mert B (2020) Recent developments of oleogel utilizations in bakery products. Crit Rev Food Sci Nutr 60(14):2460–2479. https://doi.org/10.1080/10408398.2019.1649243
16. Kanya TCS, Lokesh BR, Sastry MCS (2008) Effect of sunflower wax supplemented diet on lipid profiles in albino rats. J Food Biochem 32(1):60–71. https://doi.org/10.1111/j.1745-4514.2007.00146.x
17. Issara U, Park S, Park S (2019) Determination of fat accumulation reduction by edible fatty acids and natural waxes in vitro. Food Sci Anim Resour 39(3):430–445. https://doi.org/10.5851/kosfa.2019.e38
18. Limpimwong W, Kumrungsee T, Kato N et al (2017) Rice bran wax oleogel: A potential margarine replacement and its digestibility effect in rats fed a high-fat diet. J Func Foods 39: 250–256. https://doi.org/10.1016/j.jff.2017.10.035
19. Marangoni AG, Idziak SHJ, Vega C et al (2007) Encapsulation-stucturing of edible oil attenuates acute elevation of blood lipids and insulin in humans. Soft Matter 3(2):183–187. https://doi.org/10.1039/b611985a
20. Rush JWE, Jantzi PS, Dupak K et al (2009) Acute metabolic responses to butter, margarine, and a monoglyceride gel-structured spread. Food Res Int 42(8):1034–1039. https://doi.org/10.1016/j.foodres.2009.04.013
21. Tan SY, Wan-Yi Peh E, Marangoni AG et al (2017) Effects of liquid oil vs. oleogel co-ingested with a carbohydrate-rich meal on human blood triglycerides, glucose, insulin and appetite. Food Funct 8(1):241–249. https://doi.org/10.1039/c6fo01274d
22. Hwang HS, Singh M, Winkler-Moser JK et al (2014) Preparation of margarines from organogels of sunflower wax and vegetable oils. J Food Sci 79(10):C1926–C1932. https://doi.org/10.1111/1750-3841.12596
23. Alongi M, Lucci P, Clodoveo ML et al (2022) Oleogelation of extra virgin olive oil by different oleogelators affects the physical properties and the stability of bioactive compounds. Food Chem 368:130779. https://doi.org/10.1016/j.foodchem.2021.130779
24. Fayaz G, Calligaris S, Nicoli MC (2020) Comparative study on the ability of different oleogelators to structure sunflower oil. Food Biophys 15(1):42–49. https://doi.org/10.1007/s11483-019-09597-9
25. Gómez-Estaca J, Herrero AM, Herranz B et al (2019) Characterization of ethyl cellulose and beeswax oleogels and their suitability as fat replacers in healthier lipid pâtés development. Food Hydrocoll 87:960–969. https://doi.org/10.1016/j.foodhyd.2018.09.029
26. Winkler-Moser JK, Anderson J, Felker FC et al (2019) Physical properties of beeswax, sunflower wax, and candelilla wax mixtures and oleogels. J Am Oil Chem Soc 96(10): 1125–1142. https://doi.org/10.1002/aocs.12280
27. Doan CD, Tavernier I, Sintang MDB et al (2017) Crystallization and gelation behavior of low- and high melting waxes in rice bran oil: a case-study on berry wax and sunflower wax. Food Biophys 12(1):97–108. https://doi.org/10.1007/s11483-016-9467-y
28. Hwang HS, Kim S, Evans KO et al (2015) Morphology and networks of sunflower wax crystals in soybean oil organogel. Food Struc 5:10–20. https://doi.org/10.1016/j.foostr.2015.04.002
29. Patel AR, Babaahmadi M, Lesaffer A et al (2015) Rheological profiling of organogels prepared at critical gelling concentrations of natural waxes in a triacylglycerol solvent. J Agric Food Chem 63(19):4862–4869. https://doi.org/10.1021/acs.jafc.5b01548
30. Chopin-Doroteo M, Morales-Rueda JA, Dibildox-Alvarado E et al (2011) The effect of shearing in the thermo-mechanical properties of candelilla wax and candelilla wax-tripalmitin organogels. Food Biophys 6(3):359–376. https://doi.org/10.1007/s11483-011-9212-5

31. Ramírez-Gómez NO, Acevedo NC, Toro-Vázquez JF et al (2016) Phase behavior, structure and rheology of candelilla wax/fully hydrogenated soybean oil mixtures with and without vegetable oil. Food Res Int 89:828–837. https://doi.org/10.1016/j.foodres.2016.10.025
32. Toro-Vazquez JF, Mauricio-Pérez R, González-Chávez MM et al (2013) Physical properties of organogels and water in oil emulsions structured by mixtures of candelilla wax and monoglycerides. Food Res Int 54(2):1360–1368. https://doi.org/10.1016/j.foodres.2013.09.046
33. Rodríguez-Hernández AK, Pérez-Martínez JD, Gallegos-Infante JA et al (2021) Rheological properties of ethyl cellulose-monoglyceride-candelilla wax oleogel vis-a-vis edible shortenings. Carbohydr Polym 252:117171. https://doi.org/10.1016/j.carbpol.2020.117171
34. Blake AI, Marangoni AG (2015) Plant wax crystals display platelet-like morphology. Food Struc 3:30–34. https://doi.org/10.1016/j.foostr.2015.01.001
35. Tavernier I, Doan CD, Van De Walle D et al (2017) Sequential crystallization of high and low melting waxes to improve oil structuring in wax-based oleogels. RSC Adv 7(20): 12113–12125. https://doi.org/10.1039/c6ra27650d
36. Brykczynski H, Hetzer B, Flöter E (2022) An attempt to relate oleogel properties to wax ester chemical structures. Gels 8(9):579. https://doi.org/10.3390/gels8090579
37. Shi Z, Cao L, Kang S et al (2022) Influence of wax type on characteristics of oleogels from camellia oil and medium chain triglycerides. Int J Food Sci Technol 57(4):2003–2014. https://doi.org/10.1111/ijfs.15344
38. Endlein E, Peleikis KH (2011) Natural waxes–properties, compositions and applications. SOFW J 137(4):1–8
39. Hwang KT, Cuppett SL, Weller CL et al (2002) HPLC of grain sorghum wax classes highlighting separation of aldehydes from wax esters and steryl esters. J Sep Sci 25(9): 619–623. https://doi.org/10.1002/1615-9314(20020601)25:9<619::AID-JSSC619>3.0.CO;2-J
40. Akoh CC, Min DB (2008) Food lipids: chemistry, nutrition, and biotechnology. CRC Press, Boca Raton. https://doi.org/10.1201/9781420046649
41. Kanya TCS, Rao LJ, Sastry MCS (2007) Characterization of wax esters, free fatty alcohols and free fatty acids of crude wax from sunflower seed oil refineries. Food Chem 101(4): 1552–1557. https://doi.org/10.1016/j.foodchem.2006.04.008
42. Brykczynski H, Wettlaufer T, Flöter E (2022) Revisiting pure component wax esters as basis of wax-based oleogels. J Am Oil Chem Soc 99:925–941. https://doi.org/10.1002/aocs.12589
43. Avendaño-Vásquez G, De la Peña-Gil A, Charó-Alvarado ME et al (2020) Self-assembly of symmetrical and asymmetrical alkyl esters in the neat state and in oleogels. Front Sustain Food Syst 4:132. https://doi.org/10.3389/fsufs.2020.00132
44. Öğütcü M, Yilmaz E (2015) Characterization of hazelnut oil oleogels prepared with sunflower and carnauba waxes. Int J Food Prop 18(8):1741–1755. https://doi.org/10.1080/10942912.2014.933352
45. Fayaz G, Polenghi O, Giardina A et al (2021) Structural and rheological properties of medium-chain triacylglyceride oleogels. Int J Food Sci Technol 56(2):1040–1047. https://doi.org/10.1111/ijfs.14757
46. Lim J, Hwang HS, Lee S (2017) Oil-structuring characterization of natural waxes in canola oil oleogels: rheological, thermal, and oxidative properties. Appl Biol Chem 60(1):17–22. https://doi.org/10.1007/s13765-016-0243-y
47. Mavria A, Tsouko E, Protonotariou S et al (2022) Sustainable production of novel oleogels valorizing microbial oil rich in carotenoids derived from spent coffee grounds. J Agric Food Chem 70(35):10807–10817. https://doi.org/10.1021/acs.jafc.2c03478
48. Yilmaz E, Keskin Uslu E, Öz C (2021) Oleogels of some plant waxes: Characterization and comparison with sunflower wax oleogel. J Am Oil Chem Soc 98(6):643–655. https://doi.org/10.1002/aocs.12490
49. Liu L, Ramirez ISA, Yang J et al (2020) Evaluation of oil-gelling properties and crystallization behavior of sorghum wax in fish oil. Food Chem 309:125567. https://doi.org/10.1016/j.foodchem.2019.125567

50. Fayaz G, Goli SAH, Kadivar M (2017) A novel propolis wax-Based organogel: effect of oil type on its formation, crystal structure and thermal properties. J Am Oil Chem Soc 94(1): 47–55. https://doi.org/10.1007/s11746-016-2915-5
51. Fayaz G, Goli SAH, Kadivar M et al (2017) Pomegranate seed oil organogels structured by propolis wax, beeswax, and their mixture. Eur J Lipid Sci Technol 119(10):1700032. https://doi.org/10.1002/ejlt.201700032
52. Papadaki A, Kopsahelis N, Mallouchos A et al (2019) Bioprocess development for the production of novel oleogels from soybean and microbial oils. Food Res Int 126:108684. https://doi.org/10.1016/j.foodres.2019.108684
53. Papadaki A, Kopsahelis N, Freire DMG et al (2020) Olive oil oleogel formulation using wax esters derived from soybean fatty acid distillate. Biomolecules 10(1):106. https://doi.org/10.3390/biom10010106
54. Yılmaz E, Uslu EK, Toksöz B (2020) Structure, rheological and sensory properties of some animal wax based oleogels. J Oleo Sci 69(10):1317–1329. https://doi.org/10.5650/jos.ess20081
55. Jana S, Martini S (2016) Phase behavior of binary blends of four different waxes. J Am Oil Chem Soc 93(4):543–554. https://doi.org/10.1007/s11746-016-2789-6
56. Hwang HS, Winkler-Moser JK (2020) Properties of margarines prepared from soybean oil oleogels with mixtures of candelilla wax and beeswax. J Food Sci 85(10):3293–3302. https://doi.org/10.1111/1750-3841.15444
57. Choi KO, Hwang HS, Jeong S et al (2020) The thermal, rheological, and structural characterization of grapeseed oil oleogels structured with binary blends of oleogelator. J Food Sci 85(10):3432–3441. https://doi.org/10.1111/1750-3841.15442
58. Kim M, Hwang HS, Jeong S et al (2022) Utilization of oleogels with binary oleogelator blends for filling creams low in saturated fat. LWT 155:112972. https://doi.org/10.1016/j.lwt.2021.112972
59. Jana S, Martini S (2016) Physical characterization of crystalline networks formed by binary blends of waxes in soybean oil. Food Res Int 89:245–253. https://doi.org/10.1016/j.foodres.2016.08.003
60. Ghazani SM, Dobson S, Marangoni AG (2022) Hardness, plasticity, and oil binding capacity of binary mixtures of natural waxes in olive oil. Curr Res Food Sci 5:998–1008. https://doi.org/10.1016/j.crfs.2022.06.002
61. Pang M, Shi Z, Lei Z et al (2020) Structure and thermal properties of beeswax-based oleogels with different types of vegetable oil. Grasas Aceites 71(4):e380. https://doi.org/10.3989/GYA.0806192
62. Holey SA, Sekhar KPC, Mishra SS et al (2021) Effect of oil unsaturation and wax composition on stability, properties and food applicability of oleogels. J Am Oil Chem Soc 98(12): 1189–1203. https://doi.org/10.1002/aocs.12536
63. Han W, Chai X, Liu Y et al (2022) Crystal network structure and stability of beeswax-based oleogels with different polyunsaturated fatty acid oils. Food Chem 381:131745. https://doi.org/10.1016/j.foodchem.2021.131745
64. Laredo T, Barbut S, Marangoni AG (2011) Molecular interactions of polymer oleogelation. Soft Matter 7(6):2734–2743. https://doi.org/10.1039/c0sm00885k
65. Borriello A, Antonella Miele N, Masi P et al (2022) Effect of fatty acid composition of vegetable oils on crystallization and gelation kinetics of oleogels based on natural wax. Food Chem 375:131805. https://doi.org/10.1016/j.foodchem.2021.131805
66. Martins AJ, Cerqueira MA, Fasolin LH et al (2016) Beeswax organogels: Influence of gelator concentration and oil type in the gelation process. Food Res Int 84:170–179. https://doi.org/10.1016/j.foodres.2016.03.035
67. Hwang HS, Gillman JD, Winkler-Moser JK et al (2018) Properties of oleogels formed with high-stearic soybean oils and sunflower wax. J Am Oil Chem Soc 95(5):557–569. https://doi.org/10.1002/aocs.12060

68. Scharfe M, Niksch J, Flöter E (2022) Influence of minor oil components on sunflower, rice Bran, candelilla, and beeswax oleogels. Eur J Lipid Sci Technol 124(7):2100068. https://doi.org/10.1002/ejlt.202100068
69. Yilmaz E, Ögütcü M (2015) Oleogels as spreadable fat and butter alternatives: Sensory description and consumer perception. RSC Adv 5(62):50259–50267. https://doi.org/10.1039/c5ra06689a
70. Yilmaz E, Demirci Ş (2021) Preparation and evaluation of virgin olive oil oleogels including thyme and cumin spices with sunflower wax. Gels 7(3):95. https://doi.org/10.3390/gels7030095
71. Yılmaz E, Öğütcü M, Arifoglu N (2015) Assessment of thermal and textural characteristics and consumer preferences of lemon and strawberry flavored fish oil organogels. J Oleo Sci 64(10):1049–1056. https://doi.org/10.5650/jos.ess15113
72. Yilmaz E, Demirci Ş, Uslu EK (2022) Red pepper and turmeric-flavored virgin olive oil oleogels prepared with whale Spermaceti wax. J Oleo Sci 71(2):187–199. https://doi.org/10.5650/jos.ess21167
73. Hwang HS, Singh M, Lee S (2016) Properties of cookies made with natural wax-vegetable oil organogels. J Food Sci 81(5):C1045–C1054. https://doi.org/10.1111/1750-3841.13279
74. Yilmaz E, Ögütcü M (2015) The texture, sensory properties and stability of cookies prepared with wax oleogels. Food Funct 6(4):1194–1204. https://doi.org/10.1039/c5fo00019j
75. Li S, Zhu L, Li X et al (2022) Determination of characteristic evaluation indexes for novel cookies prepared with wax oleogels. J Sci Food Agric 102(12):5544–5553. https://doi.org/10.1002/jsfa.11909
76. Frolova YV, Sobolev RV, Kochetkova AA (2021) Comparative analysis of the properties of cookies containing oleogel based on beeswax and its fractions. IOP Conf Ser Earth Environ Sci 941(1):012033. https://doi.org/10.1088/1755-1315/941/1/012033
77. Mert B, Demirkesen I (2016) Reducing saturated fat with oleogel/shortening blends in a baked product. Food Chem 199:809–816. https://doi.org/10.1016/j.foodchem.2015.12.087
78. Mert B, Demirkesen I (2016) Evaluation of highly unsaturated oleogels as shortening replacer in a short dough product. LWT 68:477–484. https://doi.org/10.1016/j.lwt.2015.12.063
79. Kim D, Oh I (2022) The characteristic of insect oil for a potential component of oleogel and its application as a solid fat replacer in cookies. Gels 8(6):355. https://doi.org/10.3390/gels8060355
80. Jang A, Bae W, Hwang HS et al (2015) Evaluation of canola oil oleogels with candelilla wax as an alternative to shortening in baked goods. Food Chem 187:525–529. https://doi.org/10.1016/j.foodchem.2015.04.110
81. Li S, Wu G, Li X et al (2021) Roles of gelator type and gelation technology on texture and sensory properties of cookies prepared with oleogels. Food Chem 356:129667. https://doi.org/10.1016/j.foodchem.2021.129667
82. Schubert M, Erlenbusch N, Wittland S et al (2022) Rapeseed oil based oleogels for the improvement of the fatty acid profile using cookies as an example. Eur J Lipid Sci Technol 124(11):2200033. https://doi.org/10.1002/ejlt.202200033
83. Alvarez-Ramirez J, Vernon-Carter EJ, Carrera-Tarela Y et al (2020) Effects of candelilla wax/canola oil oleogel on the rheology, texture, thermal properties and in vitro starch digestibility of wheat sponge cake bread. LWT 130:109701. https://doi.org/10.1016/j.lwt.2020.109701
84. Wettlaufer T, Flöter E (2022) Wax based oleogels and their application in sponge cakes. Food Funct 13(18):9419–9433. https://doi.org/10.1039/d2fo00563h
85. Oh IK, Amoah C, Lim J et al (2017) Assessing the effectiveness of wax-based sunflower oil oleogels in cakes as a shortening replacer. LWT 86:430–437. https://doi.org/10.1016/j.lwt.2017.08.021
86. Pehlivanoglu H, Ozulku G, Yildirim RM et al (2018) Investigating the usage of unsaturated fatty acid-rich and low-calorie oleogels as a shortening mimetics in cake. J Food Process Preserv 42(6):e13621. https://doi.org/10.1111/jfpp.13621

87. Jeong S, Lee S, Oh I (2021) Development of antioxidant-fortified oleogel and its application as a solid fat replacer to muffin. Foods 10(12):3059. https://doi.org/10.3390/foods10123059
88. Lim J, Jeong S, Lee J et al (2017) Effect of shortening replacement with oleogels on the rheological and tomographic characteristics of aerated baked goods. J Sci Food Agric 97(11): 3727–3732. https://doi.org/10.1002/jsfa.8235
89. Zhao M, Rao J, Chen B (2022) Effect of high oleic soybean oil oleogels on the properties of doughs and corresponding bakery products. J Am Oil Chem Soc 99(11):1071–1083. https://doi.org/10.1002/aocs.12594
90. Jung D, Oh I, Lee J et al (2020) Utilization of butter and oleogel blends in sweet pan bread for saturated fat reduction: dough rheology and baking performance. LWT 125:109194. https://doi.org/10.1016/j.lwt.2020.109194
91. López-Pedrouso M, Lorenzo JM, Gullón B et al (2021) Novel strategy for developing healthy meat products replacing saturated fat with oleogels. Curr Opin Food 40:40–45. https://doi.org/10.1016/j.cofs.2020.06.003
92. Issara U (2022) Improvement of Thai sweet sausage (Goon Chiang) properties by oleogel made of rice bran wax and rice bran oil: a textural, sensorial, and nutritional aspect. IOP Conf Ser Earth Environ Sci 995(1):012045. https://doi.org/10.1088/1755-1315/995/1/012045
93. Gómez-Estaca J, Pintado T, Jiménez-Colmenero F et al (2020) The effect of household storage and cooking practices on quality attributes of pork burgers formulated with PUFA- and curcumin-loaded oleogels as healthy fat substitutes. LWT 119:108909. https://doi.org/10.1016/j.lwt.2019.108909
94. Wolfer TL, Acevedo NC, Prusa KJ et al (2018) Replacement of pork fat in frankfurter-type sausages by soybean oil oleogels structured with rice bran wax. Meat Sci 145:352–362. https://doi.org/10.1016/j.meatsci.2018.07.012
95. Moghtadaei M, Soltanizadeh N, Goli SAH (2018) Production of sesame oil oleogels based on beeswax and application as partial substitutes of animal fat in beef burger. Food Res Int 108: 368–377. https://doi.org/10.1016/j.foodres.2018.03.051
96. Yılmaz E, Toksöz B (2022) Flaxseed oil-wax oleogels replacement for tallowfat in sucuk samples provided higher concentrations of polyunsaturated fatty acids and aromatic volatiles. Meat Sci 192:108875. https://doi.org/10.1016/j.meatsci.2022.108875
97. Huang H, Hallinan R, Maleky F (2018) Comparison of different oleogels in processed cheese products formulation. Int J Food Sci Technol 53(11):2525–2534. https://doi.org/10.1111/ijfs.13846
98. Bemer HL, Limbaugh M, Cramer ED et al (2016) Vegetable organogels incorporation in cream cheese products. Food Res Int 85:67–75. https://doi.org/10.1016/j.foodres.2016.04.016
99. Moon K, Choi KO, Jeong S et al (2021) Solid fat replacement with canola oil-carnauba wax oleogels for dairy-free imitation cheese low in saturated fat. Foods 10(6):1351. https://doi.org/10.3390/foods10061351
100. Airoldi R, da Silva TLT, Ract JNR et al (2022) Potential use of carnauba wax oleogel to replace saturated fat in ice cream. J Am Oil Chem Soc 99(11):1085–1099. https://doi.org/10.1002/aocs.12652
101. Ferdaus MJ, Blount RJS, Silva RCD (2022) Assessment of natural waxes as stabilizers in peanut butter. Foods 11(19):3127. https://doi.org/10.3390/foods11193127
102. Huang Z, Guo B, Deng C et al (2020) Stabilization of peanut butter by rice bran wax. J Food Sci 85(6):1793–1798. https://doi.org/10.1111/1750-3841.15176
103. Adrah K, Adegoke SC, Nowlin K et al (2022) Study of oleogel as a frying medium for deep-fried chicken. J Food Meas Charact 16(2):1114–1123. https://doi.org/10.1007/s11694-021-01237-6
104. Chauhan DS, Khare A, Lal AB et al (2022) Utilising oleogel as a frying medium for deep fried Indian traditional product (Mathri) to reduce oil uptake. J Indian Chem Soc 99(3):100378. https://doi.org/10.1016/j.jics.2022.100378

Chapter 7
Direct Oil Structuring Using Ethylcellulose

Andrew J. Gravelle

Abbreviations

CGC	Critical gelation concentration
DCS	Differential scanning calorimetry
EC	Ethylcellulose
G^*	Complex modulus
G'	Elastic modulus
G''	Viscous modulus
LMOG	Low-molecular-weight oleogelator
MG	Monoglyceride
MW	Molecular weight
T_g	Glass transition temperature
T_{gel}	Gel point temperature
T_m	Melting temperature

7.1 General Properties of EC

Cellulose is an unbranched linear homopolymeric carbohydrate made up of repeating β(1→4)-linked D-glucose residues. The highly linear conformation allows cellulose to form extensive inter- and intramolecular hydrogen bonds and assemble into hierarchical structures known as microfibrils, which provide structural support in the cell walls of plants. Ethylcellulose (EC) is a chemically modified derivative of

A. J. Gravelle (✉)
Department of Food Science and Technology, University of California, Davis, Davis, CA, USA
e-mail: agravelle@ucdavis.edu

C. Andrea Palla, F. Valoppi (eds.), *Advances in Oleogel Development, Characterization, and Nutritional Aspects*,
https://doi.org/10.1007/978-3-031-46831-5_7

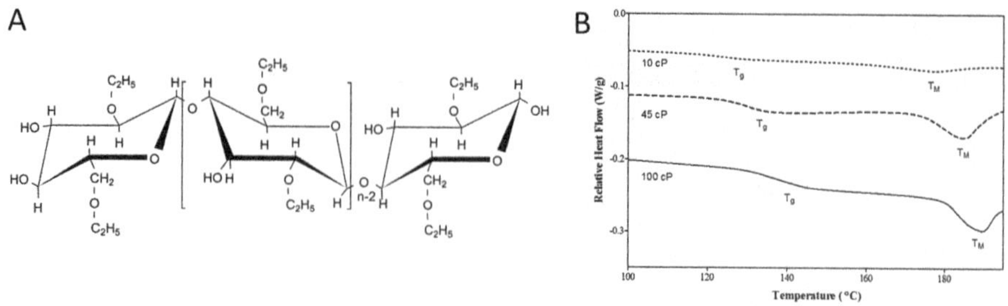

Fig. 7.1 Ethylcellulose (EC) representative chemical structure (**a**) and thermal behavior with varying molecular weight (**b**). T_g —glass transition temperature; T_m —melting temperature

cellulose used in a wide range of applications across numerous industries, including manufacturing, pharmaceuticals, cosmetics, and agriculture [1, 2]. It possesses excellent flexibility, extensibility, film forming and water-barrier properties and is considered physiologically inert. As such, EC finds uses in tablet coatings and binders, drug delivery and controlled release applications, food-printable inks, and as a component of barrier coatings [3–6]. It has also been explored for several biomedical [7] and drug delivery applications [8].

Industrially, EC is produced by reacting alkalized cellulosic biomass with ethyl chloride gas in the presence of a catalyst, inducing the formation of ethyl ether bonds at the hydroxyl moieties of carbons C2, C3, and C6 (Fig. 7.1a). The alkaline treatment opens the cellulose crystal structure, increasing accessibility to the hydroxyl groups, and causes the formation of alkoxide ions [9]. The extent of ethoxylation will impact the polymer's physical properties and technical performance. Common commercial grades are prepared with an ethoxy content of ~48–49.5% [10], corresponding to a degree of substitution of ~2.4–2.5 [11]. This level of substitution renders the polymer insoluble in water, but soluble in a variety of organic solvents, and produces a minimum in the glass transition temperature (T_g), occurring in the range of 125–140 °C [11].

EC is commercially classified according to its viscosity, measured as a 5 wt% solution in an 80/20 mixture of toluene and ethanol at 25 °C. These values are reported in units of centipoise, and will be abbreviated to denote this classification (e.g., 10 cP EC; EC^{10}). It has been shown that EC viscosity is positively correlated to the average molecular weight (MW) of the polymer [12, 13], which has a direct impact on the performance of EC as an oleogelator. Figure 7.1b demonstrates that viscosity is also positively correlated to T_g, which progressively increases from ~125 to ~140 °C for 10, 45, and 100 cP varieties [12]. It has also been shown that microcrystalline domains remain in the EC polymer, which are expected to be unsubstituted. Both differential scanning calorimetry (DSC) [14] and X-ray scattering [12] approaches have confirmed there is a reduction in the extent of crystallinity after heating the polymer above its melting temperature (T_m; ~180–190 °C). This suggests a greater number of unsubstituted hydroxyl groups would be available for gel formation upon melting, and firmer oleogels have been reported when heating to

T_m, as compared to those heated slightly above T_g (190 and 140 °C, respectively) [15]. However, standard protocols for producing EC oleogels generally restrict heating to ≤160 °C to minimize thermally induced lipid oxidation and EC depolymerization [16–18]. It is thus expected that the crystalline regions would remain in EC-based oleogels, and the breakdown in crystallinity is not expected to influence gel properties.

7.2 Structuring Edible Oils

7.2.1 EC as a Single-Component Gelator

Commercial-grade ethylcellulose is known to be compatible with various lipid-based compounds, including natural waxes, fatty acids, fatty alcohols, and triglyceride oils [11]. The concept of using EC to structure liquid vegetable oils was first explored by Aiache and coworkers for cosmetic applications [19]. In this study, a triglyceride oil was reacted with polyoxyethylene glycol to obtain a mixture of both ethoxylated mono- and diglycerides, and mono-, di-, and triglycerides. It was reported that heating mixtures of EC powder (2–9 wt%) in the ethoxylated oil to the polymer's T_g formed clarified dispersions. Upon cooling, these mixtures produced viscous solutions to paste-like suspensions, depending on the polymer concentration.

The use of EC in lipid-based systems was expanded to edible vegetable oils by Laredo et al. [20], who demonstrated that self-supporting gels could be produced using an equivalent heating/cooling procedure. It was proposed that heating above the T_g facilitated dissolution of the polymer due to an increase in molecular flexibility. Upon cooling, the individual chains return to a more rigid state, promoting the formation of inter- and intramolecular hydrogen bonds between the unsubstituted hydroxyl groups. At sufficiently high concentrations, a system-spanning, self-supporting network will form, entrapping the liquid oil. The physical nature of the polymer-polymer junction zones was demonstrated through a shift in the OH-stretching band of EC by Fourier transform infrared (FTIR) spectroscopy [20]. The mechanical strength of these gels has also been shown to decrease with increasing temperature [15], providing further indirect evidence for the physical nature of the polymer network. In general, the minimum polymer concentration to form a self-supporting gel is ~4–6 wt% EC, depending on the polymer MW and composition of the oil phase [21, 22]. The mechanical strength of vegetable oil–based EC oleogels has been extensively characterized and depends on a variety of factors, including EC viscosity and concentration, oil type, processing conditions, and the presence of small MW surfactants. These are discussed in greater detail below.

7.2.1.1 Polymer MW and Processing Conditions

Commercial grades of EC are available in viscosities ranging from 4 to 300 cP, but the practical range for structuring triglyceride oils is generally accepted to be those ranging from 10 to 45 cP [21, 23]. Lower viscosity varieties form a less entangled network with fewer intermolecular junction zones and thus require higher concentrations to form self-supporting gels. Higher MW versions form stronger gels, but the corresponding increase in T_g (Fig. 7.1b) means they must be heated to a higher temperature to sufficiently disperse the polymer for gel formation [16, 24]. This process can contribute to both thermal oxidation of the oil phase and also result in depolymerization and browning of the polysaccharide when heated above 160 °C due to caramelization reactions.

As a potential strategy to reduce the T_g and expand the utility of EC oleogels, Mashhadi and coworkers [25] subjected EC^{100} to an acid hydrolysis process. Partial hydrolysis was achieved using ethanol and acetic acid in varying ratios and under different heating conditions. They reported a reduction in T_g of ~10–30 °C, with an additional thermal event at lower temperatures, which was attributed to smaller polymer fragments. Firmer oleogels resulted for all hydrolysis conditions, but varied in the extent of this increase. It is noteworthy that the observed reinforcement may have been in part due to the improved dispersibility of lower MW chains when following the same heating protocol required to fully disperse EC^{100} [21]. Therefore, further investigation would be required to fully characterize the MW distribution of the hydrolysates and explore the potential benefits of this approach relative to using existing commercially available varieties of EC with lower MW.

The mechanical strength of EC oleogels is also positively correlated to polymer concentration and has been shown to follow a power-law scaling behavior [21]. However, the scaling relation depends on numerous factors, including oil type, polymer MW, and the presence of surfactant molecules in the oil phase [21–23]. In general, gel strength is positively correlated with the viscosity grade (i.e., MW) [21, 23, 24]. The effect of oil type has been shown to be more complex, as development of thermally induced oxidative byproducts and the presence of polar compounds can dramatically alter solvent/polymer interactions [16, 18, 22]. These factors are discussed in greater detail in the following section.

Processing conditions and gel-setting temperature have also been shown to impact gel strength. Davidovich-Pinhas and coworkers [15] demonstrated that canola oil oleogels prepared with 15 wt% EC^{20} had a gel strength of ~50 N when allowed to set at temperatures ranging from −20 to 50 °C, while this increased to ~120 N when gelation was allowed to occur at 80–100 °C. A similar trend was seen when incorporating the surfactants sorbitan monostearate and glycerol monooleate, but produced a less dramatic increase in gel strength, with a maximum occurring when gels were allowed to equilibrate at 50–80 °C. An increase in gel strength has also been noted when gels were subjected to a thermal annealing procedure; i.e., initially setting a gel at 20 °C and subsequently reheating to 80 °C for 1 h produced an equivalent enhancement in gel strength to that achieved by initially gelling at 80 °

C (unpublished data). These effects were attributed to a temperature-induced weakening and rearrangement of the hydrogen-bonding network supporting the gel. Maintaining the gels at a temperature slightly below their gel point temperature (T_{gel}) may have provided more opportunity for the polymer chains to take on more extended conformations and establish new interchain interactions, ultimately increasing crosslink density [15]. Therefore, controlling the gelation conditions may provide an effective approach to manipulate the mechanical performance of EC oleogels; however, the impact of thermal history should also be considered if applications require heating/cooling cycles, such as the melting and recrystallization of fats.

7.2.1.2 Solvent Composition

Oil composition has been widely recognized as having a major impact on the mechanical and textural properties of EC oleogels. Laredo and coworkers [20] first reported an increase in gel strength for oleogels produced using canola, soybean, and flaxseed oils, respectively. Due to the lack of polymer-solvent interactions observed by (FTIR) spectroscopy, these differences were attributed to a moderate increase in density of the lipid phase due to the higher degree of acyl chain unsaturation (canola > soy > flaxseed) [20, 21]. While this conclusion was supported by microstructural analysis (i.e., decreased pore size) [26], subsequent work suggested this phenomenon resulted from the build-up of thermally induced polar oxidation by-products [16, 27]. The observed trend in gel strength was therefore associated with a greater susceptibility of the oil phase to undergo oxidative rancidity during the polymer dissolution process. To minimize the impact of the heating procedure, antioxidants and highly controlled and reproducible heating protocols are generally employed.

The polarity of the oil phase and the presence of additional polar compounds have also been shown to have a dramatic impact on both the mechanical and textural properties of EC oleogels. Gravelle and coworkers [22] evaluated the effect of bulk solvent polarity by supplementing soybean oil-based oleogels (11 wt% EC^{45}) with either mineral oil or castor oil. The former is a mixture of alkane hydrocarbons with no dipole moment. In contrast, castor oil is a triglyceride oil consisting primarily of ricinoleic acid (>85%), a derivative of oleic acid with a hydroxyl functional group at carbon 12. While both solvents are miscible in triglyceride oils, only the latter would contribute to hydrogen bonding with the polymer network. Mineral oil reduced the bulk solvent polarity and weakened the resulting gels, which could not be formed above 10 wt% supplementation (Fig. 7.2a).

The addition of castor oil caused a linear increase in gel strength up to ~10 wt% supplementation, above which a plateau was observed (Fig. 7.2a). Increasing solvent polarity also improved the functional performance of these oleogels. When prepared exclusively with refined, bleached, and deodorized vegetable oils, the resulting oleogels exhibit highly brittle fracture behavior and exude liquid oil at the site of fracture due to the weak solvent-polymer interactions (Fig. 7.2c). The presence of

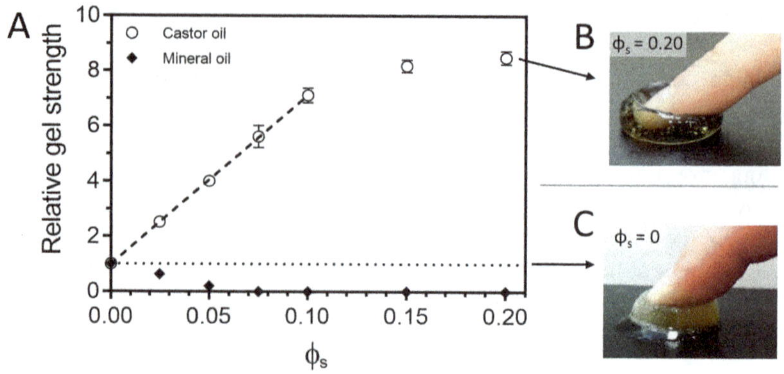

Fig. 7.2 Mechanical properties of ethylcellulose oleogels supplemented with either castor oil (open circles) or mineral oil (filled diamonds) at varying wt% (φ_s). Relative gel strength determined by texture profile analysis (**a**) and macroscopic deformation for oleogels supplemented with 20 wt% castor oil (**b**) or no supplementation (**c**). (Panel (**a**) reproduced from [22], with permission from Elsevier)

castor oil dramatically increased the gel elasticity and eliminated the occurrence of syneresis (Fig. 7.2b). This suggested the hydroxyl groups in the lipid phase directly interacted with EC, improving solvent-polymer interactions, producing an increase in gel strength and enhanced oil binding. This interpretation was supported by applying the formalism of Hansen solubility parameters to the dose-response effect on molten oleogel viscosity. Optimal solvent-polymer interactions (i.e., maximum viscosity) occurred at ~80 wt% castor oil, corresponding to a maximum in hydrogen bonding interactions, while intermolecular and dispersive forces had little impact on EC solubility and gelation behavior. Similar conclusions have been drawn for EC in other solvents, such as mixtures of vegetable oils and glycerol monooleate [28].

The microstructure of EC oleogels has been visualized using both cryo-scanning electron microscopy (Fig. 7.3a) [26] and atomic force microscopy (Fig. 7.3b–d) [18, 29]. The internal polymeric network appears as a 2-dimensional porous mat or 3-dimensional closed-cell foam with porous walls, depending on the extent of oil removal. Wall thickness has been reported to change based on extraction conditions, but may also be influenced by the composition of the oil phase. Giacintucci and coworkers observed thicker walls in oleogels prepared with extra virgin olive oil (Fig. 7.3b), while these became thinner after deodorization (Fig. 7.3c) and neutralization and bleaching (Fig. 7.3d), which removed naturally present, polar minor compounds. Thicker walls and variations in pore size were proposed to be correlated to higher gel strength; however, variability in the partial solvent extraction procedure should not be dismissed. In spite of this, the thicker pore walls in oleogels prepared with less refined oils (which contain polar minor compounds) may again be indicative of stronger solvent-polymer interactions, which would contribute to enhanced gel strength, as discussed above.

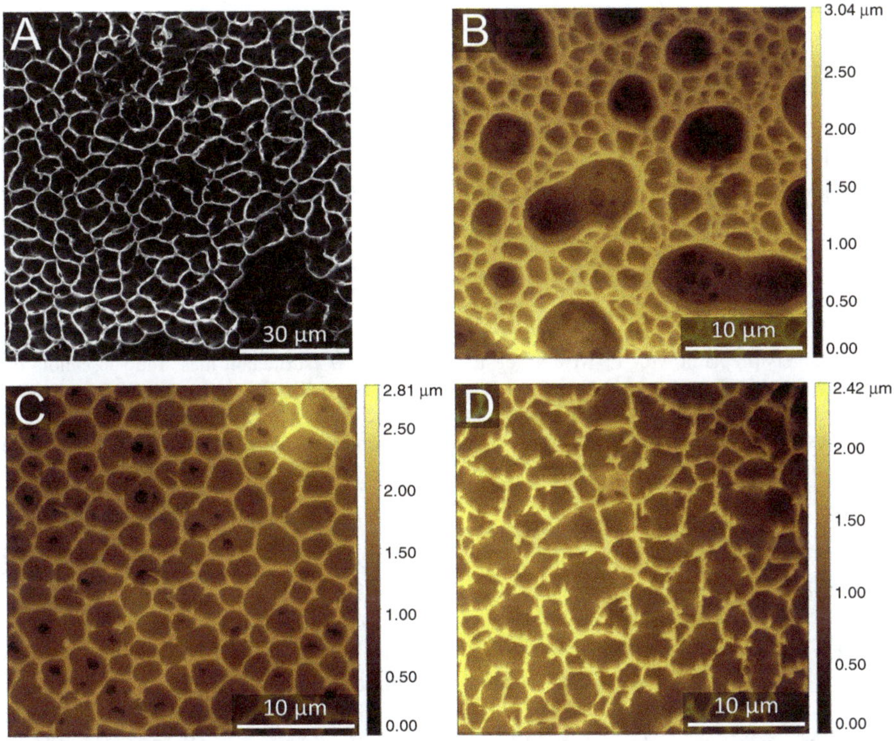

Fig. 7.3 Microstructure of partially de-oiled ethylcellulose oleogels: (**a**) Scanning electron micrograph of soybean oil-based oleogel (reproduced from [26], with permission from Elsevier); (**b–d**) atomic force micrographs of an oleogel prepared with extra virgin olive oil (adapted from [18], with permission from Elsevier), which were untreated (**b**), degummed (**c**), or degummed, neutralized, and bleached (**d**)

7.2.1.3 Low-Molecular-Weight Additives

In addition to direct interactions with the bulk solvent, various amphiphilic small molecules with polar functional groups can also have a positive effect on gel strength. These additives generally produced a more dramatic change in gel strength at low concentrations, due to their inherent surface activity. However, their impact is generally dependent on the particular chemical structure [23, 30]. The common food-grade surfactants glycerol monooleate, sorbitan monooleate, and sorbitan monostearate were shown to have a similar effect on the gelation behavior and rheological and mechanical properties of EC oleogels [15]. When incorporated at a 3:1 EC-surfactant molar ratio, both molecules produced an equivalent increase in gel strength and corresponding depression in the gel point temperature (Fig. 7.4a).

The addition of oleic acid and oleyl alcohol has also been shown to elicit a dramatic increase in gel strength and elasticity at low concentrations (≤1 wt%; Fig. 7.4b) in soybean oil-based oleogels prepared with 11 wt% EC[45]. This

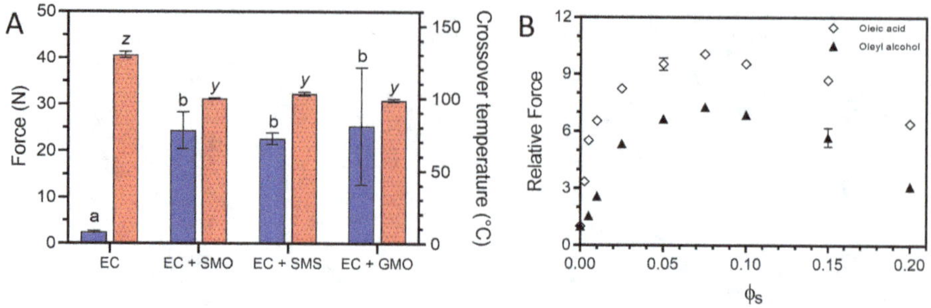

Fig. 7.4 (**a**) Gel strength (filled bars) and crossover temperature (patterned bars) of ethylcellulose (EC) oleogels prepared with 11 wt% 45 cP EC, and a 3:1 EC/surfactant mass ratio in canola oil (adapted from [30], with permission from Elsevier). (**b**) Normalized gel strength of EC oleogels prepared with canola oil and supplemented with oleic acid (open diamonds) or oleyl alcohol (filled diamonds). X-axis indicates the mass fraction additive supplemented in the oil phase (adapted from [22], with permission from Elsevier)

pronounced effect was attributed to the migration of the polar headgroups to the hydrophilic regions of the EC backbone, enhancing oil-polymer interactions at much lower concentrations than that seen with increasing bulk solvent polarity. The plateau and subsequent decrease in gel strength at high surfactant concentrations (≥10 wt%) can be attributed to further plasticization of the polymer [11]. These formulations also displayed a high level of elasticity and complete lack of syneresis upon deformation, analogous to the gels prepared with castor oil [22]. Fatty acid derivatives such as glycol esters are also commonly employed as plasticizing agents in products formulated with EC to reduce brittleness and improve suppleness and thermoplasticity (e.g., in film and coating applications) [9]. It has been further shown that removing free fatty acids from cold-pressed flaxseed oil caused a significant decrease in the strength of the resulting gel [22]. Taken together, these findings demonstrate that in addition to elevated solvent polarity (e.g., due to thermal oxidation), polar minor compounds can enhance oil-polymer interactions, and serve as effective plasticizers, which can have a dramatic impact on the mechanical and textural properties of EC oleogels.

Incorporating surfactants into EC oleogels may also expand their utility in applications beyond food systems, such as lipid-based printable inks. Kavimughill and coworkers [31] evaluated the gelation properties and hot-melt extrusion performance of EC oleogels (11 wt% EC^{100}) prepared with medium-chain triglycerides and the nonionic surfactant polyethylene glycol monostearate (PGMS40; 40 polyethylene glycol units per stearic acid molecule). The addition of surfactant (0–5 wt %) was positively correlated to gel strength and viscosity, and imparted greater plasticity, which improved printability. Optimum printing and multilayer stacking were achieved with 5 wt% surfactant extruded at 45 °C. The formation of a surfactant crystal network also contributed to the improved printing characteristics.

Taken together, these results demonstrate both the bulk polarity of the lipid phase and presence of surface-active minor components play a strong role in determining the mechanical, rheological, and functional behavior of EC oleogels. Building on this approach, the following section discusses relevant studies which have combined EC with surface-active small molecules that structure oil via the formation of lipid-based crystalline networks.

7.2.2 *Multicomponent Oleogels Based on EC*

Traditional fats provide a broad range of desirable technofunctional properties in foods. These traits arise through a complex interplay between their chemical composition and processing conditions, which ultimately determines the structural characteristics of the underlying hierarchical fat crystal network and its associated macroscopic performance [32]. Single-component oleogelator systems may be able to replicate certain attributes (e.g., firmness, melting behavior, and fracture stress), but often lack the ability to mimic the full range of behaviors provided by even a single fat source [33]. To expand the functional properties of oleogels and develop improved fat mimetics, the use of binary or ternary oil structuring systems has become an increasingly popular approach. Those systems which produce cooperative networks that synergistically enhance desirable attributes are of particular interest [34, 35].

The ability of EC to directly disperse in edible oil and form an extended polymer network makes it uniquely compatible with other lipid-soluble oleogelators, the majority of which are high-melting point lipid-based compounds [36, 37]. These low-molecular-weight oleogelators (LMOGs) are most commonly incorporated in the oil phase by heating above their melting point. Upon cooling, they assemble into a 3-dimensional crystalline network which entraps the liquid oil phase, analogous to traditional triglyceride fats [32]. Direct oil structuring by crystalline networks has been extensively investigated using gelators such as natural waxes, monoglycerides, and fatty acids, fatty alcohols, and their mixtures [37]. Furthermore, interactions between the polar regions of these crystalline gelators and EC may produce synergistic binary networks. Therefore, introducing a secondary crystalline oil-structuring network could provide functional benefits to EC oleogels, such as promoting favorable melting behavior and plastic flow characteristics. Additionally, EC has been incorporated in some LMOG-based oleogels below its critical gelation concentration (CGC), providing an alternative route to modulate crystallization behavior and performance of the resulting oleogel.

7.2.2.1 Monoglycerides (MGs)

MG-based oleogels have been extensively studied and can impart many of the desirable properties associated with traditional fats [38]. The performance of these LMOGs is intrinsically linked to the polymorphic form and microstructural arrangement of the crystal network, which are largely dictated by the chemical composition of the MGs and processing conditions used during gelation. Incorporating EC has been explored as an alternate route to modulate the rheological and mechanical performance of MG-based oleogels. Lopez-Martínez and coworkers [39] were the first to investigate the impact of EC on crystalline gelator interactions. This group evaluated oleogels using two varieties of MG (~80% palmitic or 50/50 palmitic-stearic) prepared with 2 wt% or 8 wt% MG, either as the sole oleogelator or with 6 wt % EC[4]. While the authors noted the latter was below the CGC of EC based on previous reports, control gels prepared with EC alone had a complex modulus (G*) of >5000 Pa, suggesting a system-spanning polymer network was able to form. In all cases, the addition of EC caused a dramatic increase in G*. The effect was particularly pronounced for the 8 wt% MG gels, which saw a two-decade increase in G* with EC addition. This was attributed to the formation of a synergistic binary network resulting from a combination of polymer-crystalline gelator interactions and the formation of a crystalline network in the bulk liquid oil entrapped within the EC polymer matrix. During a 2-week storage period, the authors noted the combined system delayed the transition of MG from the sub-α to the β-polymorph [39]. This transition has been shown to cause deleterious effects on MG oleogel performance, including syneresis and the formation of larger crystals, which promote brittleness [40, 41]. Introducing an EC polymer network could therefore potentially be used to modulate the structural properties of crystalline oleogelators at varying length scales to improve the stability and performance of the multicomponent system.

7.2.2.2 Fatty Acid/Fatty Alcohol Mixtures

Both fatty acids [42, 43] and fatty alcohols [44, 45] have been independently explored as single-component crystalline oleogelators. When used in appropriate ratios, these two classes of molecules are also able to co-crystallize, allowing such combinations to serve as a distinct oleogelator system [46]. The acid/alcohol ratio plays a major role in determining the crystal morphology and associated oil binding, mechanical, and rheological properties [47, 48]. It has generally been reported that a 1:3 acid/alcohol ratio minimizes interfacial tension in the crystal unit cell, and the resulting crystal network has been shown to provide improved oil binding and enhanced gel strength [47, 49].

Considering the strong interaction between EC and these low molecular weight amphiphilic molecules (e.g., Fig. 7.4b), multicomponent oleogels combining mixtures of stearic acid and stearyl alcohol with EC have also been explored. Using a 3:7 acid/alcohol ratio, Gravelle and coworkers [50] reported evidence of a direct interaction between the crystalline and polymer networks for oleogels structured with 6 wt% EC[45] and 5 wt% small molecules. The addition of EC had a dramatic impact on the microstructural arrangement of these mixed crystals, which appeared smaller and assembled into fan-like structures (Fig. 7.5a). It was proposed this morphology was caused by crystal nucleation along the polymer backbone. However, these interactions did not impact the polymorphic form of the mixed acid/alcohol crystals.

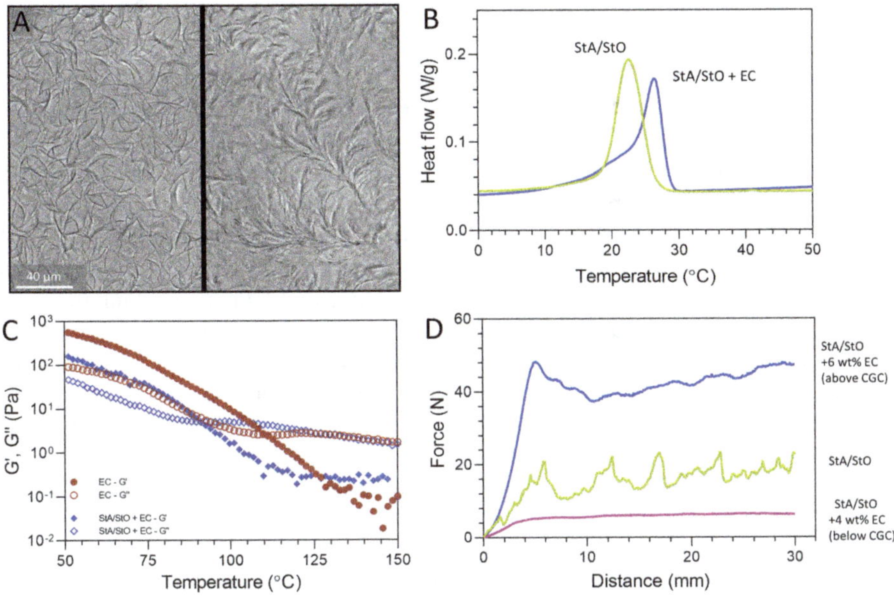

Fig. 7.5 Physical properties of oleogels prepared from a 3:7 mixture of stearic acid/stearyl alcohol (StA/StO). (**a**) Microstructure of StA/StO oleogels without (Panel **a**, *left*) or with (Panel **a**, *right*) 6 wt% 45 cP ethylcellulose (EC); (**b**) DSC thermogram of StA/StO crystallization; (**c**) rheological characterization of EC crossover temperature; (**d**) large deformation flow behavior evaluated by back extrusion. *CGC—critical gelation concentration. (Adapted from [50], with permission from Elsevier)

The presence of EC also produced an increase in the onset of crystallization for these mixed crystals (Fig. 7.5b). The LMOG mixture similarly caused a ~20 °C reduction in the polymer's crossover temperature during cooling (i.e., the temperature where the shear elastic modulus and shear storage modulus are equal; G′ = G″), which is commonly taken as a measure of the gel point temperature (Fig. 7.5c). Thermal analysis also demonstrated a dose-response reduction in the T_g of EC with increasing content of the small molecules. This lends further support to the hypothesis that direct polymer-small molecule interactions contributed to the observed mechanical response. Finally, the combined system produced a synergistic enhancement in gel strength, while the change in crystal morphology also reduced brittle flow behavior associated with larger crystals that undergo catastrophic failure during mechanical deformation (Fig. 7.5d). In contrast, when EC was incorporated below its CGC (~4 wt%), the gels were softer, and displayed no brittle fracture during penetration. Subsequent work by the same group further explored the effect of acid/alcohol ratio [51]. Both the crystal type and composition were reported to have a dramatic impact on the resulting crystal morphology and associated performance of these multicomponent oleogelator systems.

7.2.2.3 Free Fatty Acids

Lauric acid is a 12-carbon medium-chain saturated fatty acid found in high quantities in coconut oil (~50%), and contributes to its desirable melting properties. Lauric acid has also been explored as a secondary structuring agent used in combination with EC^{20} to produce canola oil-based oleogels [52]. When combined, these gelators produced a synergistic enhancement in gel strength. While the polymorphic form of lauric acid was not impacted by the addition of EC, the peak crystallization and melting temperature was depressed as EC concentration increased. This trend was hypothesized to result from steric hinderance induced by EC/lauric acid interactions. Further, the presence of lauric acid dramatically reduced the crossover temperature of EC, consistent with that observed for mixtures of fatty acids and fatty alcohols (noted above). The latter was attributed to EC plasticization and enhanced solvent-polymer interactions promoted by lauric acid. The combined polymer-fatty acid system also showed improved thixotropy relative to oleogels independently structured with EC or lauric acid. Oleogels prepared with a combination of stearic acid and EC have also been shown to produce a synergistic increase in gel strength and high oil-binding capacity [51]. The use of medium- or long-chain fatty acids may thus be a promising way to both modulate the rheological properties of EC oleogels and impart desirable textural characteristics and melting behavior.

EC has also been used to produce oleogels in combination with behenic acid, a 22-carbon saturated fatty acid commonly used in cosmetic and lubrication applications. Ahmadi and coworkers [53] evaluated soybean oil-based oleogels prepared with 6 wt% total structurant at varying EC^{20}/behenic acid ratios. While the addition of EC below its CGC enhanced the rheological response of oleogels predominantly structured by behenic acid, no synergistic enhancement in gel strength was observed for those gels supported primarily by the EC polymer network. Further studies would be required to evaluate the direct impact of behenic acid on EC, such as by fixing EC concentration and varying the content of fatty acid. This work further demonstrated that a 1:5 EC/behenic acid ratio was able to produce stable water-in-oleogel emulsions (5–45 wt% water) without the need for an additional surfactant. In contrast, stable emulsions could not be formed using oleogels structured with behenic acid alone. This suggests EC may modify the crystallization behavior of behenic acid to enhance emulsification activity, and/or directly contribute to the emulsification properties. Further studies will also be needed to explore the emulsification properties of these multicomponent oleogels.

7.2.2.4 Lecithin

Food-grade soy lecithin is a complex mixture of phospholipids and has been shown to form oleogels via spontaneous assembly of tubular reverse micelles in the presence of small amounts of water (~0.2–2.0 wt%) [54]. Multicomponent oleogels have also been produced by combining 10 wt% EC^{20} with soy lecithin predominantly composed of either unsaturated (18:2) or saturated (18:0) fatty acid acyl chains [55]. The combined systems were reported to produce up to a tenfold increase in G′. Interestingly, this synergistic enhancement was most pronounced at the lowest lecithin concentrations evaluated (1 wt%), while higher concentrations diminished the viscoelastic response. The stress-strain response and yield stress of both EC and EC/lecithin hybrid gels (10 wt% EC, 1 wt% lecithin) were comparable to a commercial lard shortening, but exhibited a higher stress overshoot. This effect was attributed to the polymer network's ability to withstand higher loading or undergo greater deformation prior to yielding compared to a fat crystal network. Large amplitude oscillatory shear analysis further indicated the solid- to fluid-like response of the binary oleogels was similar to those structured with EC alone, indicating the polymer network dominated the nonlinear viscoelastic response. Both oleogel systems underwent fluidization at higher strain than the commercial shortening. This further indicated the polymer network dominated the rheological behavior and may require a high degree of plasticization or lower crosslink density to accurately mimic the rheological performance of a fat crystal network. Additional studies will be required to confirm the molecular interactions responsible for the dynamic viscoelastic behavior of the EC/lecithin gelator system.

7.3 EC-Based Oleogels as Shortening Mimetics

The development of multicomponent oleogels capable of mimicking the performance of commercial fats has been a particular area of interest across the field of edible oleogels [56, 57]. Several studies have undertaken screening of multicomponent oleogelator systems incorporating EC as an approach to develop edible shortenings with reduced saturated fat and satisfactory rheological and technofunctional behavior. Naeli and coworkers [58] used a response surface methodology to evaluate the rheological performance of oleogels having EC and MG as a base structuring system. The linear viscoelastic behavior was optimized to mimic commercial shortenings using various combinations of palm stearin (as a fat hardstock), hydroxypropyl methylcellulose, and gum arabic (as a thickener). All oleogels were prepared in a laboratory-scale scraped-surface heat exchanger to more accurately mimic the production of commercial shortenings. The optimized oleogels displayed comparable rheological behavior to the commercial standards with only ~16% saturated fatty acids, as compared to ~50–60% in the controls.

Rodríguez-Hernández and coworkers [59] also explored the rheological behavior of EC used in combination with MG and candelilla wax as a potential shortening replacement. Both the binary EC/MG and ternary systems exhibited a similar sharp increase in the complex modulus (G*) during cooling due to the formation of the MG crystal network. Although no distinct rheological response could be attributable to the wax crystallization, the ternary system produced stiffer oleogels (i.e., higher G*). The presence of EC in both the binary and ternary gelator systems dramatically improved the shear recovery, with select formulations displaying a comparable response to commercial shortenings.

The direct formation of water-in-oleogel emulsions structured with EC and MG has also been reported using a hot emulsification approach [60]. Emulsions containing 20 wt% water were prepared by separately forming concentrated glycerol monostearate emulsions and a hot EC[4]/oil dispersion. The two systems were then combined under shear at 70 °C to produce the final emulsion. Stable emulsions with solid-like rheological properties were achieved with 1 wt% glycerol monostearate and 7 wt% EC[4] in high oleic safflower oil. These gels also remained stable after freeze-thaw cycling, which can encourage surfactant desorption and droplet destabilization [61]. It was concluded that EC promoted emulsion stability by enhancing the oil viscosity and/or through direct interaction with the interfacial glycerol monostearate crystals. Although these emulsions had a lower G′ than commercial low-fat spreads, this could potentially be addressed by optimizing the oleogel formulation and processing conditions (e.g., to reduce water droplet size). Thus, various approaches using EC in combination with additional structuring agents have shown considerable potential for developing accurate fat mimetics, and these could be further expanded and optimized for specific applications.

7.4 Conclusion

There has been growing interest in EC as an oleogelator due to its unique ability to be directly dispersed in edible oils and form a physically crosslinked polymer network. The free hydroxyl groups along the polymer backbone also provide a means of forming synergistic interactions with a variety of directly dispersible oleogelators, including monoglycerides, mixtures of fatty acids and fatty alcohols, and inverse lecithin micelles. Due to their unique structuring properties, both single- and multicomponent EC-based oleogels have been incorporated into numerous food systems. EC has been used in chocolate to confer heat resistance [62, 63], modify viscosity and flow behavior [64], and serve as a cocoa butter substitute for saturated fat reduction [65, 66]. Single- and multicomponent EC oleogels have been incorporated into a wide variety of meat products, including frankfurters [21, 67, 68], salami [69], breakfast sausages [70], burgers [71, 72], and liver pâté [73–75]. They have also been explored as a milk fat substitute in cream cheese [76], and to develop shortening mimetics for use in baked goods such as cakes [72], breads [77], and biscuits [78, 79].

Beyond direct fat substitution, EC oleogels may serve as a means of modulating lipid digestion and serve as delivery vehicles for lipid-soluble nutraceuticals. O'Sullivan and coworkers [80] demonstrated EC oleogel strength can directly impact both the rate and extent of lipolysis, as well as the bioavailability of a model lipid-soluble nutraceutical (β-carotene) using a static in vitro digestion model. Ashkar and coworkers [81, 82] similarly reported that lipid bioaccessibility is restricted by the EC polymer network and the rate of lipolysis is negatively correlated to gel firmness, which could serve as a means of regulating digestion behavior. Human in vivo trials have also shown that EC-based oleogels may provide an effective means of moderating the elevation of plasma triglycerides in response to a carbohydrate-rich meal [83, 84]. The ability of EC to regulate accessibility to the entrapped lipid phase could also be harnessed for drug delivery applications [8]. The wide adaptability of EC to be combined with complementary gelators thus provides considerable opportunity to further develop EC-based multicomponent oleogels for use as fat mimetics, delivery vehicles for nutraceuticals, and controlled-release applications in both the food and pharmaceutical industries.

Acknowledgments Andrew Gravelle acknowledges financial support from the USDA National Institute of Food and Agriculture Hatch/Multi-State project.

References

1. Dow Cellulosics (2005) Ethocel - Ethylcellulose polymers technical handbook, pp 1–27
2. Ahmadi P, Jahanban-Esfahlan A, Ahmadi A et al (2020) Development of ethyl cellulose-based formulations: a perspective on the novel technical methods. Food Rev Int. https://doi.org/10.1080/87559129.2020.1741007
3. Yang D, Peng X, Zhong L et al (2014) "Green" films from renewable resources: properties of epoxidized soybean oil plasticized ethyl cellulose films. Carbohydr Polym 103:198–206. https://doi.org/10.1016/j.carbpol.2013.12.043
4. Homaee Borujeni S, Mirdamadian SZ, Varshosaz J, Taheri A (2020) Three-dimensional (3D) printed tablets using ethyl cellulose and hydroxypropyl cellulose to achieve zero order sustained release profile. Cellulose 27:1573–1589. https://doi.org/10.1007/s10570-019-02881-4
5. Wasilewska K, Winnicka K (2019) Ethylcellulose – a pharmaceutical excipient with multidirectional application in drug dosage forms development. Materials (Basel) 12:3386. https://doi.org/10.3390/ma12203386
6. Adeleke OA (2019) Premium ethylcellulose polymer based architectures at work in drug delivery. Int J Pharm X 1:100023. https://doi.org/10.1016/j.ijpx.2019.100023
7. Seddiqi H, Oliaei E, Honarkar H et al (2021) Cellulose and its derivatives: towards biomedical applications. Cellulose 28:1893–1931. https://doi.org/10.1007/s10570-020-03674-w
8. Ashkar A, Sosnik A, Davidovich-Pinhas M (2022) Structured edible lipid-based particle systems for oral drug-delivery. Biotechnol Adv 54:107789. https://doi.org/10.1016/j.biotechadv.2021.107789
9. Rekhi GSGS, Jambhekar SSSS (1995) Ethylcellulose - a polymer review. Drug Dev Ind Pharm 21:61–77
10. Dow Pharma & Food Solutions (2016) ETHOCEL ™ Ethylcellulose a technical review a portfolio of versatile solutions to help address a variety of formulation and processing needs
11. Koch W (1937) Properties and uses of ethylcellulose. Ind Eng Chem 29:687–690
12. Davidovich-Pinhas M, Barbut S, Marangoni AG (2014) Physical structure and thermal behavior of ethylcellulose. Cellulose 21:3243–3255. https://doi.org/10.1007/s10570-014-0377-1
13. Sánchez R, Franco JM, Delgado MA et al (2011) Thermal and mechanical characterization of cellulosic derivatives-based oleogels potentially applicable as bio-lubricating greases: influence of ethyl cellulose molecular weight. Carbohydr Polym 83:151–158. https://doi.org/10.1016/j.carbpol.2010.07.033

14. Lai HL, Pitt K, Craig DQM (2010) Characterisation of the thermal properties of ethylcellulose using differential scanning and quasi-isothermal calorimetric approaches. Int J Pharm 386:178–184. https://doi.org/10.1016/j.ijpharm.2009.11.013
15. Davidovich-Pinhas M, Gravelle AJ, Barbut S, Marangoni AG (2015) Temperature effects on the gelation of ethylcellulose oleogels. Food Hydrocoll 46:76–83. https://doi.org/10.1016/j.foodhyd.2014.12.030
16. Gravelle AJ, Barbut S, Marangoni AG (2012) Ethylcellulose oleogels: manufacturing considerations and effects of oil oxidation. Food Res Int 48:578–583. https://doi.org/10.1016/j.foodres.2012.05.020
17. Fu H, Lo YM, Yan M et al (2020) Characterization of thermo-oxidative behavior of ethylcellulose oleogels. Food Chem 305:125470. https://doi.org/10.1016/j.foodchem.2019.125470
18. Giacintucci V, Di Mattia CD, Sacchetti G et al (2018) Ethylcellulose oleogels with extra virgin olive oil: the role of oil minor components on microstructure and mechanical strength. Food Hydrocoll 84:508–514. https://doi.org/10.1016/j.foodhyd.2018.05.030
19. Aiache J-M, Gauthier P, Aiache S (1992) New gelification method for vegetable oils I: cosmetic application. Int J Cosmet Sci 14:228–234. https://doi.org/10.1111/j.1467-2494.1992.tb00056.x
20. Laredo T, Barbut S, Marangoni AG (2011) Molecular interactions of polymer oleogelation. Soft Matter 7:2734–2743. https://doi.org/10.1039/c0sm00885k
21. Zetzl AK, Marangoni AG, Barbut S (2012) Mechanical properties of ethylcellulose oleogels and their potential for saturated fat reduction in frankfurters. Food Funct 3:327–337. https://doi.org/10.1039/c2fo10202a
22. Gravelle AJ, Davidovich-Pinhas M, Zetzl AK et al (2016) Influence of solvent quality on the mechanical strength of ethylcellulose oleogels. Carbohydr Polym 135:169–179. https://doi.org/10.1016/j.carbpol.2015.08.050
23. Gravelle AJ, Barbut S, Quinton M, Marangoni AG (2014) Towards the development of a predictive model of the formulation-dependent mechanical behaviour of edible oil-based ethylcellulose oleogels. J Food Eng 143:114–122. https://doi.org/10.1016/j.jfoodeng.2014.06.036
24. Davidovich-Pinhas M, Barbut S, Marangoni AG (2015) The gelation of oil using ethyl cellulose. Carbohydr Polym 117:869–878. https://doi.org/10.1016/j.carbpol.2014.10.035
25. Mashhadi H, Tabibiazar M, Nourabi A, Roufegarinejad L (2023) Evaluation of the effect of partial hydrolysis ethyl cellulose on physicochemical properties of soybean oil oleogel. Int J Food Sci Technol 58:1195–1203. https://doi.org/10.1111/ijfs.16267
26. Zetzl AK, Gravelle AJ, Kurylowicz M et al (2014) Microstructure of ethylcellulose oleogels and its relationship to mechanical properties. Food Struct 2:27–40. https://doi.org/10.1016/j.foostr.2014.07.002
27. Gravelle AJ, Barbut S, Marangoni AG (2013) Fractionation of ethylcellulose oleogels during setting. Food Funct 4:153–161. https://doi.org/10.1039/c2fo30227f
28. Stortz TA, Marangoni AG (2014) The replacement for petrolatum: thixotropic ethylcellulose oleogels in triglyceride oils. Green Chem 16:3064–3070. https://doi.org/10.1039/c4gc00052h
29. O'Sullivan CM (2016) In-vitro bioaccessibility and stability of beta-carotene in ethylcellulose oleogels. 1–95
30. Davidovich-Pinhas M, Barbut S, Marangoni AG (2015) The role of surfactants on ethylcellulose oleogel structure and mechanical properties. Carbohydr Polym 127:355–362. https://doi.org/10.1016/j.carbpol.2015.03.085
31. Kavimughil M, Leena MM, Moses JA, Anandharamakrishnan C (2022) Effect of material composition and 3D printing temperature on hot-melt extrusion of ethyl cellulose based medium chain triglyceride oil oleogel. J Food Eng 111055. https://doi.org/10.1016/j.jfoodeng.2022.111055
32. Marangoni AG, Van Duynhoven JPM, Acevedo NC et al (2020) Advances in our understanding of the structure and functionality of edible fats and fat mimetics. Soft Matter 16:289–306. https://doi.org/10.1039/c9sm01704f

33. Li L, Liu G, Bogojevic O et al (2022) Edible oleogels as solid fat alternatives: composition and oleogelation mechanism implications. Compr Rev Food Sci Food Saf:1–28. https://doi.org/10.1111/1541-4337.12928
34. Sivakanthan S, Fawzia S, Madhujith T, Karim A (2022) Synergistic effects of oleogelators in tailoring the properties of oleogels: a review. Compr Rev Food Sci Food Saf 1–32. https://doi.org/10.1111/1541-4337.12966
35. Shakeel A, Farooq U, Gabriele D et al (2021) Bigels and multi-component organogels: an overview from rheological perspective. Food Hydrocoll 111:106190. https://doi.org/10.1016/j.foodhyd.2020.106190
36. Martins AJ, Vicente AA, Cunha RL, Cerqueira MA (2018) Edible oleogels: an opportunity for fat replacement in foods. Food Funct 9:758–773. https://doi.org/10.1039/c7fo01641g
37. Wang Z, Chandrapala J, Truong T, Farahnaky A (2022) Oleogels prepared with low molecular weight gelators: texture, rheology and sensory properties, a review. Crit Rev Food Sci Nutr 0:1–45. https://doi.org/10.1080/10408398.2022.2027339
38. Palla CA, Dominguez M, Carrín ME (2022) An overview of structure engineering to tailor the functionality of monoglyceride oleogels. Compr Rev Food Sci Food Saf 21:2587–2614. https://doi.org/10.1111/1541-4337.12930
39. Lopez-Martínez A, Charó-Alonso MA, Marangoni AG, Toro-Vazquez JF (2015) Monoglyceride organogels developed in vegetable oil with and without ethylcellulose. Food Res Int 72:37–46. https://doi.org/10.1016/j.foodres.2015.03.019
40. Chen CH, Terentjev EM (2009) Aging and metastability of monoglycerides in hydrophobic solutions. Langmuir 25:6717–6724. https://doi.org/10.1021/la9002065
41. Ojijo NKO, Neeman I, Eger S, Shimoni E (2004) Effects of monoglyceride content, cooling rate and shear on the rheological properties of olive oil/monoglyceride gel networks. J Sci Food Agric 84:1585–1593. https://doi.org/10.1002/jsfa.1831
42. Sagiri SS, Singh VK, Pal K et al (2015) Stearic acid based oleogels: a study on the molecular, thermal and mechanical properties. Mater Sci Eng C 48:688–699. https://doi.org/10.1016/j.msec.2014.12.018
43. Uvanesh K, Sagiri SS, Senthilguru K et al (2016) Effect of span 60 on the microstructure, crystallization kinetics, and mechanical properties of stearic acid oleogels: an in-depth analysis. J Food Sci 81:E380–E387. https://doi.org/10.1111/1750-3841.13170
44. Valoppi F, Calligaris S, Marangoni AG (2017) Structure and physical properties of oleogels containing peanut oil and saturated fatty alcohols. Eur J Lipid Sci Technol 119:1–11. https://doi.org/10.1002/ejlt.201600252
45. Lupi FR, Gabriele D, Greco V et al (2013) A rheological characterisation of an olive oil/fatty alcohols organogel. Food Res Int 51:510–517. https://doi.org/10.1016/j.foodres.2013.01.013
46. Gandolfo FG, Bot A, Flöter E (2004) Structuring of edible oils by long-chain FA, fatty alcohols, and their mixtures. JAOCS, J Am Oil Chem Soc 81:1–6. https://doi.org/10.1007/s11746-004-0851-5
47. Blach C, Gravelle AJ, Peyronel F et al (2016) Revisiting the crystallization behavior of stearyl alcohol: stearic acid (SO : SA) mixtures in edible oil. RSC Adv 6. https://doi.org/10.1039/c6ra15142f
48. Schaink HM, van Malssen KF, Morgado-Alves S et al (2007) Crystal network for edible oil organogels: possibilities and limitations of the fatty acid and fatty alcohol systems. Food Res Int 40:1185–1193. https://doi.org/10.1016/j.foodres.2007.06.013
49. Callau M, Sow-Kébé K, Nicolas-Morgantini L, Fameau AL (2020) Effect of the ratio between behenyl alcohol and behenic acid on the oleogel properties. J Colloid Interface Sci 560:874–884. https://doi.org/10.1016/j.jcis.2019.10.111
50. Gravelle AJ, Davidovich-Pinhas M, Barbut S, Marangoni AG (2017) Influencing the crystallization behavior of binary mixtures of stearyl alcohol and stearic acid (SOSA) using ethylcellulose. Food Res Int 91:1–10. https://doi.org/10.1016/j.foodres.2016.11.024

51. Gravelle AJ, Blach C, Weiss J et al (2017) Structure and properties of an ethylcellulose and stearyl alcohol/stearic acid (EC/SO:SA) hybrid oleogelator system. Eur J Lipid Sci Technol 119:1700069. https://doi.org/10.1002/ejlt.201700069
52. Haj Eisa A, Laufer S, Rosen-Kligvasser J, Davidovich-Pinhas M (2020) Stabilization of ethylcellulose oleogel network using lauric acid. Eur J Lipid Sci Technol 122:1–10. https://doi.org/10.1002/ejlt.201900044
53. Ahmadi P, Tabibiazar M, Roufegarinejad L, Babazadeh A (2020) Development of behenic acid-ethyl cellulose oleogel stabilized Pickering emulsions as low calorie fat replacer. Int J Biol Macromol 150:974–981. https://doi.org/10.1016/j.ijbiomac.2019.10.205
54. Bodennec M, Guo Q, Rousseau D (2016) Molecular and microstructural characterization of lecithin-based oleogels made with vegetable oil. RSC Adv 6:47373–47381. https://doi.org/10.1039/c6ra04324k
55. Aguilar-Zárate M, Macias-Rodriguez BA, Toro-Vazquez JF, Marangoni AG (2019) Engineering rheological properties of edible oleogels with ethylcellulose and lecithin. Carbohydr Polym 205:98–105. https://doi.org/10.1016/j.carbpol.2018.10.032
56. Puşcaş A, Mureşan V, Socaciu C, Muste S (2020) Oleogels in food: a review of current and potential applications. Foods 9:1–27. https://doi.org/10.3390/foods9010070
57. Scharfe M, Flöter E (2020) Oleogelation: from scientific feasibility to applicability in food products. Eur J Lipid Sci Technol 122:1–24. https://doi.org/10.1002/ejlt.202000213
58. Naeli MH, Milani JM, Farmani J, Zargaraan A (2022) Developing and optimizing low-saturated oleogel shortening based on ethyl cellulose and hydroxypropyl methyl cellulose biopolymers. Food Chem 369:130963. https://doi.org/10.1016/j.foodchem.2021.130963
59. Rodríguez-Hernández AK, Pérez-Martínez JD, Gallegos-Infante JA et al (2021) Rheological properties of ethyl cellulose-monoglyceride-candelilla wax oleogel vis-a-vis edible shortenings. Carbohydr Polym 252:117171. https://doi.org/10.1016/j.carbpol.2020.117171
60. García-Ortega ML, Toro-Vazquez JF, Ghosh S (2021) Development and characterization of structured water-in-oil emulsions with ethyl cellulose oleogels. Food Res Int 150:110763. https://doi.org/10.1016/j.foodres.2021.110763
61. Ghosh S, Rousseau D (2011) Fat crystals and water-in-oil emulsion stability. Curr Opin Colloid Interface Sci 16:421–431. https://doi.org/10.1016/j.cocis.2011.06.006
62. Stortz TA, Marangoni AG (2013) Ethylcellulose solvent substitution method of preparing heat resistant chocolate. Food Res Int 51:797–803. https://doi.org/10.1016/j.foodres.2013.01.059
63. Stortz TA, Laredo T, Marangoni AG (2015) The role of lecithin and solvent addition in ethylcellulose-stabilized heat resistant chocolate. Food Biophys 10:253–263. https://doi.org/10.1007/s11483-014-9379-7
64. Ceballos MR, Bierbrauer KL, Faudone SN et al (2016) Influence of EC-medium chain triglycerides blend on the flow behaviour and beta-V polymoprh retention of dark chocolate. Food Struct 10:1–9. https://doi.org/10.1016/j.foostr.2016.10.004
65. Li L, Liu G (2019) Corn oil-based oleogels with different gelation mechanisms as novel cocoa butter alternatives in dark chocolate. J Food Eng 263:114–122. https://doi.org/10.1016/j.jfoodeng.2019.06.001
66. Li L, Liu G, Lin Y (2021) Physical and bloom stability of low-saturation chocolates with oleogels based on different gelation mechanisms. LWT - Food Sci Technol 140:110807. https://doi.org/10.1016/j.lwt.2020.110807
67. Barbut S, Wood J, Marangoni AG (2016) Effects of organogel hardness and formulation on acceptance of frankfurters. J Food Sci 81:C2183–C2188. https://doi.org/10.1111/1750-3841.13409
68. Barbut S, Wood J, Marangoni AG (2016) Potential use of organogels to replace animal fat in comminuted meat products. Meat Sci 122:155–162. https://doi.org/10.1016/j.meatsci.2016.08.003
69. Woern C, Marangoni AG, Weiss J, Barbut S (2021) Effects of partially replacing animal fat by ethylcellulose based organogels in ground cooked salami. Food Res Int 147:110431. https://doi.org/10.1016/j.foodres.2021.110431

70. Barbut S, Wood J, Marangoni AG (2016) Quality effects of using organogels in breakfast sausage. Meat Sci 122:84–89. https://doi.org/10.1016/j.meatsci.2016.07.022
71. Gómez-Estaca J, Herrero AM, Herranz B et al (2019) Characterization of ethyl cellulose and beeswax oleogels and their suitability as fat replacers in healthier lipid pâtés development. Food Hydrocoll 87:960–969. https://doi.org/10.1016/j.foodhyd.2018.09.029
72. Adili L, Roufegarinejad L, Tabibiazar M et al (2020) Development and characterization of reinforced ethyl cellulose based oleogel with adipic acid: its application in cake and beef burger. LWT - Food Sci Technol 126:109277. https://doi.org/10.1016/j.lwt.2020.109277
73. Barbut S, Marangoni AG, Thode U, Tiensa BE (2019) Using canola oil organogels as fat replacement in liver Pâté. J Food Sci 84:2646–2651. https://doi.org/10.1111/1750-3841.14753
74. Barbut S, Tiensa BE, Marangoni AG (2021) Partial fat replacement in liver pâté using canola oil organogel. LWT - Food Sci Technol 139:110428. https://doi.org/10.1016/j.lwt.2020.110428
75. Gómez-Estaca J, Pintado T, Jiménez-Colmenero F, Cofrades S (2019) Assessment of a healthy oil combination structured in ethyl cellulose and beeswax oleogels as animal fat replacers in low-fat, PUFA-enriched pork burgers. Food Bioprocess Technol 12:1068–1081. https://doi.org/10.1007/s11947-019-02281-3
76. Bemer HL, Limbaugh M, Cramer ED et al (2016) Vegetable organogels incorporation in cream cheese products. Food Res Int 85:67–75. https://doi.org/10.1016/j.foodres.2016.04.016
77. Ye X, Li P, Lo YM et al (2019) Development of novel shortenings structured by ethylcellulose oleogels. J Food Sci 84:1456–1464. https://doi.org/10.1111/1750-3841.14615
78. Onacik-Gür S, Zbikowska A (2020) Effect of high-oleic rapeseed oil oleogels on the quality of short-dough biscuits and fat migration. J Food Sci Technol 57:1609–1618. https://doi.org/10.1007/s13197-019-04193-8
79. Onacik-Gür S, Zbikowska A (2022) The effect of green tea extract and oleogels on the physico-chemical properties and oxidative stability of short-dough biscuits during storage. LWT 172: 114197. https://doi.org/10.1016/j.lwt.2022.114197
80. O'Sullivan CM, Davidovich-Pinhas M, Wright AJ et al (2017) Ethylcellulose oleogels for lipophilic bioactive delivery-effect of oleogelation on: in vitro bioaccessibility and stability of beta-carotene. Food Funct 8:1438–1451. https://doi.org/10.1039/c6fo01805j
81. Ashkar A, Laufer S, Rosen-Kligvasser J et al (2019) Impact of different oil gelators and oleogelation mechanisms on digestive lipolysis of canola oil oleogels. Food Hydrocoll 97: 105218. https://doi.org/10.1016/j.foodhyd.2019.105218
82. Ashkar A, Rosen-Kligvasser J, Lesmes U, Davidovich-Pinhas M (2020) Controlling lipid intestinal digestibility using various oil structuring mechanisms. Food Funct 11:7495–7508. https://doi.org/10.1039/d0fo00223b
83. Tan SY, Peh E, Siow PC et al (2017) Effects of the physical-form and the degree-of-saturation of oil on postprandial plasma triglycerides, glycemia and appetite of healthy Chinese adults. Food Funct 8:4433–4440. https://doi.org/10.1039/C7FO01194F
84. Tan SY, Peh E, Lau E et al (2017) Physical form of dietary fat alters postprandial substrate utilization and glycemic response in healthy Chinese men. J Nutr 147:1138–1144. https://doi.org/10.3945/jn.116.246728

Chapter 8
Oleogels Produced by Direct Methods Using as Gelator: Fatty Acids (Including 12-HSA), Fatty Alcohols, Ceramides, Lecithins, Sterols, Cellulose Fibers, and Fumed Silica

Linlin Li, Guoqin Liu, and Zheng Guo

Abbreviations

12-HSA	12-hydroxystearic acid
FDA	Food and Drug Administration
GRAS	Generally Recognized as Safe
LDL	Low-density lipoprotein
LMWO	Low molecular weight oleogelators
PC	Phosphatidylcholine
PE	Phosphatidylethanolamine
PO	Polymeric oleogelators
PS	Phosphatidylserine
WHO	World Health Organization

L. Li
College of Food Science and Technology, Henan Agricultural University, Zhengzhou, China

G. Liu
School of Food Science and Engineering, South China University of Technology, Guangzhou, China

Z. Guo (✉)
Department of Biological and Chemical Engineering, Faculty of Technical Science, Aarhus University, Aarhus, Denmark
e-mail: guoqin@scut.edu.cn; guo@bce.au.dk

C. Andrea Palla, F. Valoppi (eds.), *Advances in Oleogel Development, Characterization, and Nutritional Aspects*,
https://doi.org/10.1007/978-3-031-46831-5_8

8.1 Introduction

Vegetable oil is the main source of fats consumed by the human body, which occupies an important part of daily food consumption. In recent years, as consumers' demand for the nutritional and safety of food has increased, the demand for food has undergone a transition from sensory to nutrition and safety. In the food industry, semi-solid and solid fats carry vital roles, as they are responsible for desired structural, functional, and sensory attributes of many food products [1]. It is therefore of great interest to explore various oil structuring strategies, including hydrogenation, interesterification, and fractionation, to obtain the desired textural and sensory properties. The solid fat formed by the conventional structuring method has been widely applied in baking, dairy, spreads, margarine, chocolate, and confectionery products [2]. However, several epidemiological studies have reported the link between the consumption of traditional solid fat and a higher risk of coronary heart disease, obesity, and type II diabetes due to *trans*- and saturated fats [3, 4]. The reason is, the consumption of saturated and *trans*-fats has been proven to increase both total and low-density lipoprotein (LDL) cholesterol (a typical biomarker for cardiovascular diseases) [5]. Within this context, the World Health Organization (WHO) has released a "REPLACE" plan, aiming to remove *trans*-fat in the global food supply chain. One strategy of this plan is to "promote the development of healthier alternatives for traditional solid fats" [6]. Therefore, interest in seeking novel alternatives to structure bulk oil without saturated fats is rising.

As an interesting family of soft materials, oleogels can structure liquid vegetable oils without the use of *trans* and/or saturated fats, providing useful and novel functions in food systems [7]. The main function of oleogels in the food industry can be summarized into two aspects: one being alternatives to traditional solid fat; one being improvers for functional properties. In terms of the first aspect, even though the textural and microstructural properties of oleogels are different from conventional fat, the end products prepared with oleogels do exhibit comparable qualities to the standard formulation [2]. In terms of the second aspect, the addition of specific oleogels in chocolate can increase its thermal stability and delay the development of fat bloom [8–10].

Oleogels consist of oleogelators and liquid oil encapsulated in them. The oleogelator can form a three-dimensional network by self-assembling to form a fibrous, laminar, or polymer chain structure, which entraps the liquid oil in it, and such a semi-solid system is defined as an oleogel [11, 12]. The promising prospects have kick-started the investigation of oleogels, including the exploration of oleogelator, oleogelation mechanism, and practical aspects concerning food and pharmaceutical applications [13]. In order to conduct a comprehensive analysis of the current status of oleogel research, the keyword clustering analysis was performed using CiteSpace software on the literature with "oleogel" or "organogel" as the topic in the Web of Science data for the past 20 years, and the results are shown in Fig. 8.1.

From Fig. 8.1, it can be seen that the research hotspots of oleogels in the food field focus on the physicochemical properties (crystallization and rheology), oleogelator

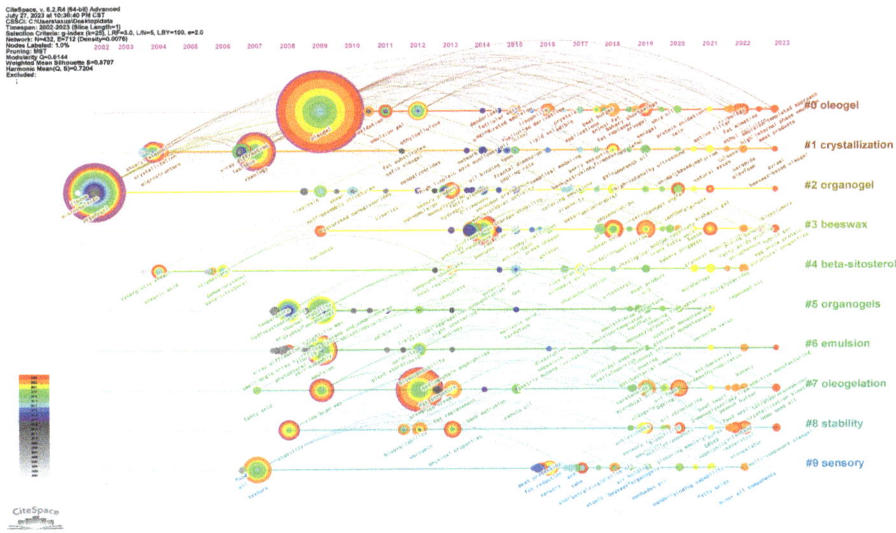

Fig. 8.1 Research advances of oleogels by citespace (data from Web of Science by searching oleogels or organogels in Food Science Technology)

(beeswax and beta-sitosterol), and the application of oleogels (stability and sensory). In terms of development trend, the oleogels research was mainly focused on the development of oleogelator (rice bran wax, fatty acid, hydroxystearic acid, oryzanol, etc.) and characterization of oleogelation behavior (XRD, SAXS, gelation, kinetics, etc.) between 2000 and 2010 year. In the following decade, the research focused on oleogels shifted to the application of oleogels (bakery product, beef burger, meat products, fat replacer, etc.) and the study of oleogel-derived systems (aerogel, foam, emulsion, etc.). Well-known and widely studied oleogelator such as monoglycerides, glycerides, fatty acids, fatty alcohols, plant waxes, ethyl cellulose, and phytosterols are both edible oleogelators [14–17]. On the other hand, 12-hydroxystearic acid (12-HSA) and 12-hydroxy ricinoleic acid are representatives of non-edible oleogelator and cannot be directly applied in food because of their low toxicity [18, 19]. According to the molecular weight, oleogelator can be classified into low molecular weight oleogelators (LMWO) and Polymeric Oleogelators (PO). All of the above-mentioned oleogelator are LMWO except ethyl cellulose, which can be dissolved directly in vegetable oil by shear and heat treatment to prepare oleogels after cooling. Ethyl cellulose is the only PO that can be added directly to prepare oleogels [20]. In recent years, some hydrophilic macromolecular polymers such as chitosan, soy protein, gelatin, β-lactoglobulin, and cellulose derivatives have similarly achieved the construction of oleogels by indirect methods (emulsion template method, hydrogel template method, solvent substitution method, etc.) [21–24].

There are various materials capable of structuring vegetable oil; however, materials need to be legally approved for their use in the food industry. Every country or area has its regulations for the approved food additives with a limited range of

Table 8.1 Summary of oleogelators used to produce oleogels discussed in this chapter

Category	Oleogelators	Approved for food applications	References
Fatty acid	Stearic acid	FDA	[26]
	12-HSA	N.A.	[19]
Fatty alcohol	Stearyl alcohol	FDA	[27]
Sterol	β-Sitosterol	N.A.	[28]
Others	Lecithin	FAO, EU (additive E322), and FDA	[29]
	Fumed silica	EU (additive E551) and FDA	[30] [31]

Adapted from [25] with permission from John Wiley and Sons
N.A. not applicable

concentrations. Table 8.1 summarizes the relevant materials (discussed in this chapter) used for the production of oleogels regarding their approval by FAO, EFSA, and FDA. Though these regulations might restrict the application of the oleogel system in the food field, careful consideration should be given in the investigation of these oleogelators.

8.2 Fatty Acids and Fatty Alcohols

One typical class of components with great potential as structuring agents are long-chain fatty acids and fatty alcohols. They are usually used as raw materials in the cosmetic and food industry. Fatty acids are usually obtained by the hydrolysis of triglycerides. Fatty alcohols can be derived from triglycerides by transesterification with methanol followed by high-pressure hydrogenation [32]. They can form a crystalline network fabricated by platelets at low concentrations, resulting in a colloidal gel system. In addition, fatty acid and fatty alcohol with varying chain lengths are perfect candidates to investigate the influence of oleogelator structure on the gelling properties of oleogels due to their similar structure [13]. The most studied fatty acid and fatty alcohols in oleogelation are summarized in Table 8.2.

8.2.1 Fatty Acids

The gelling capacity of saturated fatty acid was first described in the previous research of Daniel and Rajasekharan [34], where saturated FA with chain lengths of C-10 to C-31 gelled the vegetable oil successfully without changing the chemical properties of vegetable oil. Additionally, the structuring feasibility and melting point of saturated fatty acids with longer chain lengths were higher than those with shorter ones. This phenomenon was also observed by Valoppi et al. [35]. The stronger gel strength was attributed to the aggregation of smaller crystals. The oleogelation

Table 8.2 Summary of physical parameters of common fatty acid and fatty alcohols

Category	Chain length	Common name	Systematic name	Melting point (°C)	Melting enthalpy (kJ/mol)
Fatty acid	16	Palmitic acid	Hexadecanoic acid	61.8	54.9
	18	Stearic acid	Octadecanoic acid	68.8	61.3
	20	Arachidic acid	Eicosanoic acid	75.4	69.2
	22	Behenic acid	Docosanoic acid	81.0	N.A.
Fatty alcohols	16	Palmityl alcohol	1-Hexadecanol	49.3	34.7
	18	Stearyl alcohol	1-Octadecanol	59.5	66.6
	20	Arachyl alcohol	1-Eicosanol	66.1	73.7
	22	Behenyl alcohol	1-Docosanol	72.5	86.0

Reproduced from [33] with permission from John Wiley and Sons
N.A. not applicable

mechanism of fatty acid concluded that the linear structure (head to tail) of fatty acid enables them to create a lattice structure to entrap triglycerides, resulting in the solidification of liquid oil [34]. In a followed systematic study conducted by Gandolfo et al. [33], a series of fatty alcohols and fatty acids with chain lengths ranging from C-16 to C-22 were evaluated for their potential use in oil structuring. The minimum concentration for palmitic acid, stearic acid, eicosanoic acid, and behenic acid to produce solid gel was 2%. The microstructural characterization confirmed that well-defined lozenge-shaped crystals (200 μm) formed by fatty acid were the building blocks for the construction of oleogels.

Among the saturated fatty acids, stearic acid is the second most naturally abundant saturated fatty acid found in both plants and animals [34]. Stearic acid has been shown to have nutraceutical properties (lowering of LDL cholesterol levels), making stearic acid hold great promise as a healthy oleogelator for the development of oleogels [36]. Its derivatives have been utilized in the preparation of food, pharmaceuticals, and other formulations as stabilizers, surface activators, plasticizers, waterproofing agents, polishing agents, and softening agents [37]. The stearic acid-based oleogels can be achieved by heating the mixture of stearic acid and vegetable oil to a temperature over the melting point of the stearic acid (69–72 °C) followed by a cooling process for crystallization [38]. In recent research by Jiang et al. [26], the gelling capacity of stearic acid and its derivatives (12-hydroxy stearic acid, methyl stearate, 2-hydroxyethyl stearate, sorbitan stearate) was systematically investigated by physical–chemical characterization. For the first time, the relationship between "gelator structure"–"assembly behavior"–"oleogel properties" was studied by molecular dynamics and an independent gradient model. From the simulating self-assembly processes of stearic acid and its derivatives (presented in

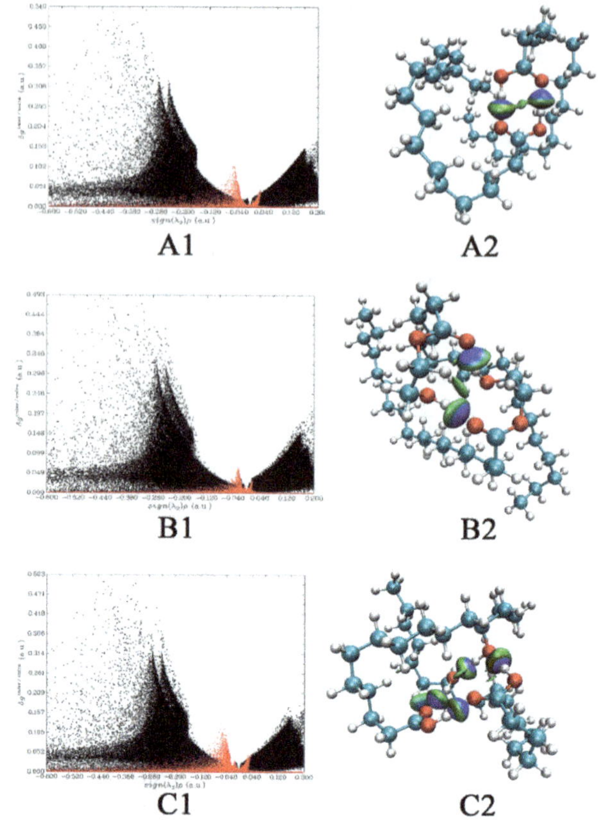

Fig. 8.2 Interaction between monomers in a dimer of stearic acid (a1: scatter graphs of δ_{ginter} and δ_{gintra} vs. sign(λ2)p; a2: visual weak interactions), hydroxyethyl stearate (b1: scatter graphs of δ_{ginter} and δ_{gintra} vs. sign(λ2)p; b2: visual weak interactions), and hydroxy stearic acid (c1: scatter graphs of δ_{ginter} and δ_{gintra} vs. sign(λ2)p; c2: visual weak interactions). (Reproduced from [26] with permission from Elsevier)

Fig. 8.2), it was found that stearic acid and 2-hydroxyethyl stearate dimers exhibited similar interaction but much lower than those of the 12-hydroxy stearic acid dimer. This phenomenon explained the stronger mechanical strength of 12-hydroxy stearic acid oleogels.

8.2.2 *Fatty Acid Derivatives (12-Hydroxystearic Acid)*

8.2.2.1 Structure and Production of 12-Hydroxystearic Acid

12-Hydroxystearic acid (12-HSA) is among the most widely studied low-molecular-weight oleogelators (LMWGs) because it can structure organic solvents at remarkably low concentrations (less than 1 wt.%) [39]. As shown in Fig. 8.3a, 12-HSA behaves like a white flake or needle-like crystal at room temperature with a melting point of 74–78 °C [40]. The peculiar self-assembly performance of 12-HSA contributes to its unique molecular structure [41]. 12-HSA is a saturated fatty acid (18-carbon) with a hydroxyl substitution at C12 in its molecular structure. Within

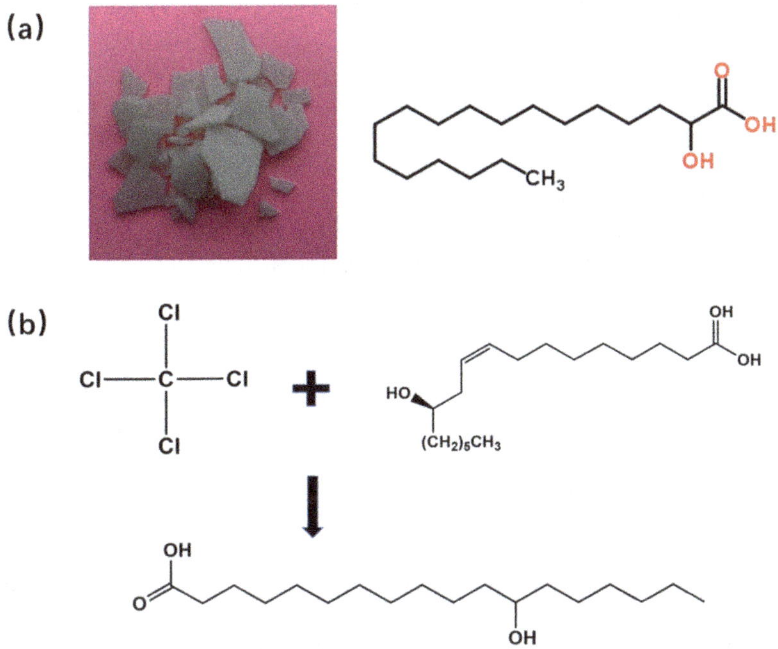

Fig. 8.3 (**a**) Molecular formula and (**b**) reaction involved in the production of 12-HSA. (Reproduced from [40] with permission from John Wiley & Sons)

the internal molecules of 12-HSA, the hydrogen-bonding interactions formed by the interaction between hydroxyl and carboxyl groups help to self-assemble into molecular aggregates and form crystal network structures to entrap liquid oils and fats [42]. Compared to stearic acid, the presence of a hydroxyl group at C12 in 12-HSA offers an additional hydrogen bonding center and introduces molecular chirality which is reflected in the supramolecular morphology [41]. In addition, the London dispersion forces between the polymethylene chains further improved the affinity for self-assembly [43, 44]. Owing to these advantages, 12-HSA is among the most effective oleogelator available in various commercial products in industrial production (lubricants, coatings, and rubber), pharmaceuticals, and cosmetics due to its great efficient gelation ability.

12-HSA can be obtained from the hydrogenation of ricinoleic acid found in renewable, biodegradable material (castor oil). The synthesis route is presented in Fig. 8.3b: ricinoleic acid was fully hydrogenated at 150 °C in the presence of a nickel catalyst, followed by the saponification to cleave free fatty acid by NaOH; after that, the products were separated by mineral acid [19, 45]. At present, 12-HSA is not approved as GRAS in the food industry. It is the main shortcoming to broaden the potential applications of 12-HSA in foods and nutraceutical applications [41].

8.2.2.2 Application of 12-HSA in Oleogelation

The development of 12-HSA organogels can be tracked back to 1946 when the lithium soap of 12-HSA was employed in lubricating greases [46]. In 1975, 12-HSA was the first ingredient reported to be incorporated into peanut butter to prevent oil separation [47]. Since then, increasing attention has been paid to the self-assembly mechanism and application of 12-HSA organogels, making 12-HSA organogels become well-documented supramolecular systems [48]. The mechanism involved in the formation of self-assembled fiber networks in supramolecular systems was well clarified in previous research [49, 50]. During the crystallization process of 12-HSA, supersaturation allows 12-HSA molecules to self-assemble into primary crystalline structures that grow into rods, tubes, or sheets at the one-dimensional scale; and then further assembly to form three-dimensional networks through non-covalent interactions [41]. The elastic properties of 12-HSA organogels were provided by the transient junctions (the entanglement of fibers) or permanent junctions (branching of fibers) within the fibrous network [50].

Multiple studies of the 12-HSA oleogels have been focused so far on their crystallization behavior and macro–micro properties. For example, Rogers and Marangoni [51] reported that the nucleation rate of a self-assembling fibrillar network of 12-HSA in canola oil was found to be inversely proportional to the undercooling-time exposure of the system during non-isothermal crystallization. The physical performance of 12-HSA oleogel was highly dependent on the organogelator concentration, storage temperature, and solvent type [52, 53]. The reason can be concluded that the self-assembled fibrillar network structure of 12-HSA can be manipulated or tailored by its thermal history (e.g., cooling temperature, cooling rate, and storage temperature), giving rise to different macroscopic properties (i.e., rheological behavior, oil binding capacity, storage stability, and so on) [54]. Recently, 12-HSA organogels have been developed as injectable drug delivery systems with virgin coconut oil [55] and mineral oil [56]. Agogué et al. [57] developed 12-HSA organogel to replace mineral waxes can thus be substituted in the formulation of candles to provide useful reference and guidance for other studies on the application of 12-HSA oragnogels. Yang et al. [48] investigated the effect of the alkyl chain lengths in HSA-modified molecular structures on thermal stabilities and supramolecular structures of the organogels. 12-HSA oleogels are attracting interest as oil structuring agents, nutrients, synthesis templates, and to develop functional materials (responsive gels, and aqueous foams) [41]. The macroscopic properties of 12-HSA oleogels can be tailored by modifying the chemical structure of 12-HSA, allowing for the rational design of oleogel-based products with improved functional profiles.

8.2.3 Fatty Alcohols

With regard to fatty alcohols, some authors have explored their potential application as structuring agents. Compared to fatty acids with similar chain lengths, saturated fatty alcohols showed lower efficiencies in oleogelation [34]. Similarly, the chain lengths of fatty alcohols also significantly influence the physical properties of the oleogels. Valoppi et al. [35] proved that a longer chain length of fatty alcohol (from $C_{16}OH$ to $C_{22}OH$) led to the formation of oleogels with smaller crystals and a higher ability to retain oil (as shown in Fig. 8.4). $C_{14}OH$ and $C_{16}OH$ were not able to be efficiently used to develop stable oleogels, while $C_{20}OH$ and $C_{22}OH$ fatty alcohols could constitute efficient gelators of oleogelation. It can be concluded that the solubility of fatty alcohol decreased with the chain length of fatty alcohol increased, and therefore, more crystals were formed by the fatty alcohol.

Most studies about fatty alcohols in the food industry focused on stearyl alcohol, as stearyl alcohol is the only fatty alcohol that is registered as self-affirmed and generally recognized as a safe (GRAS) oleogelator for edible oil structuring [58]. It is thus considered an edible compound that can be directly applied to food formulation [59]. Another attractive example of fatty alcohol as an oleogelator is policosanol, which is a mixture of fatty alcohols (mainly octacosanol). Policosanol attracted increasing attention owing to its gelation efficiency (oleogelation threshold:

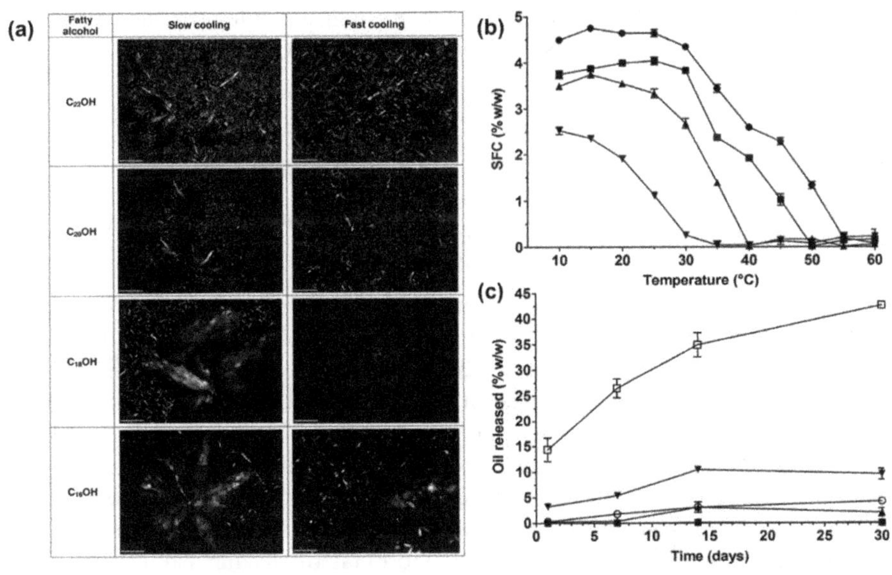

Fig. 8.4 Characterization of oleogels containing 5% $C_{22}OH$, $C_{20}OH$, $C_{18}OH$, and $C_{16}OH$: (**a**) polarized light micrographs; (**b**) solid fat content as a function of temperature; (**c**) Oil released content as a function of storage time. (Reproduced from [35] with permission from John Wiley & Sons)

0.5 wt%) and nutrition value (lowering low-density lipoprotein and enhancing athletic performances) [60, 61].

8.2.4 Fatty Acid +Fatty Alcohols

Long-chain fatty acids and fatty alcohols can structure edible oils not only by themselves but also in combination [62]. A mixed system of fatty acids + fatty alcohols has been applied in oleogels, emulsions, and foam systems [33, 63, 64]. Gandolfo et al. [33] set a precedent for research on the mixed system of fatty acids and fatty alcohols as oleogelator. They confirmed the synergistic effect on the mechanical strength when fatty acids and fatty alcohols are mixed at specific ratios [33]. The synergistic enhancement resulted from the smaller crystals (needle-like) formed by the combined crystallization of fatty acid + fatty alcohol, which were different from those formed by either fatty acid or fatty alcohol alone (platelet-like) [65]. After that, a series of research followed the research on the investigation of the synergistic effect of stearyl alcohol + stearic acid [27, 66] and behenyl alcohol + behenic acid [67]. As shown in Fig. 8.5a, b, the ratio of fatty acids and fatty alcohols had a significant influence on the microstructure and the oleogel properties. In another interesting research (Fig. 8.5c), ethylcellulose was added into stearyl alcohol + stearic acid oleogels. Results indicated that the original higher-ordered crystal network had undergone the change from needle-like, oriented platelets to branched, feather-like structures. Also, the onset of crystallization temperature of stearyl alcohol + stearic acid was increased from 22.7 to 26.5 °C [68].

8.3 Ceramides

8.3.1 Structure and Health Benefits

As a functional lipid, ceramide plays an important role in regulating cell growth (differentiation, proliferation, and apoptosis), inhibiting colon cancer, and reducing total serum cholesterol, and improving the serum lipoprotein [69, 70]. Besides this, ceramides have a significant influence on the permeability of the epidermis [71]. Ceramide belongs to a subclass of sphingolipid compounds and is composed of long-chain sphingoid bases (dihydrosphingosine, sphingosine, phytosphingosine, or 6-hydroxysphingosine), which are linked to long-chain fatty acid through amide bonding [72]. The structure of ceramide is presented in Fig. 8.6. The hydroxyl groups in the head group enable ceramide to form intermolecular and intramolecular hydrogen bonding [73]. Ceramides are varied in different acyl chain lengths (C16–C33), unsaturation degree of fatty acid, and hydroxylation patterns [72, 74].

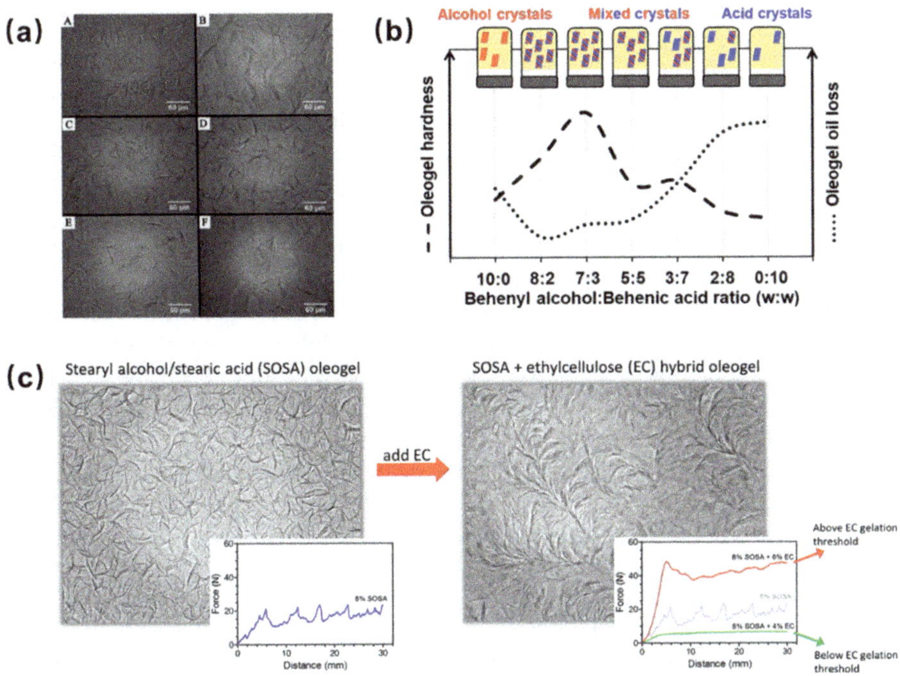

Fig. 8.5 Oleogels structured by fatty acid + fatty alcohols: (**a**) Microscopy images of oleogels with 5% stearyl alcohol: stearic acid (SO: SA) ratios of 10: 0 (A), 9: 1 (B), 8: 2 (C), 7: 3 (D), 4: 6 (E), 0: 10 (F), (reproduced from [27], with permissions from The Royal Society of Chemistry); (**b**) schematic of oleogel properties influenced by the ratio between behenyl alcohol and behenic acid (reproduced from [67] with permission from Elsevier); and (**c**) oleogels structured by stearyl alcohol and stearic acid with the addition of ethylcellulose (reproduced from [68] with permission from Elsevier)

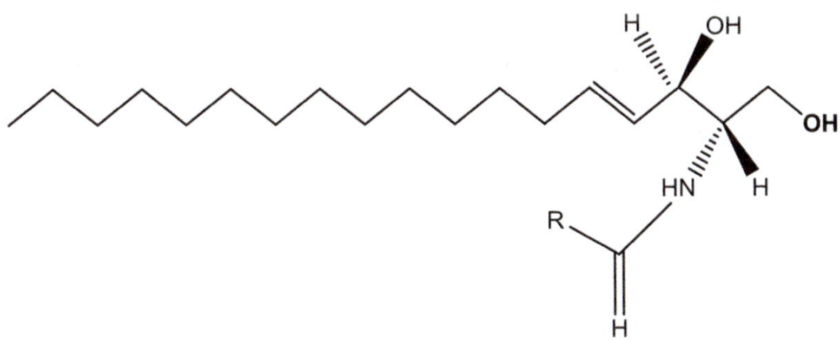

Fig. 8.6 The schematic diagram for the molecular structure of ceramide

8.3.2 Application in Oleogelation

Ceramides are mostly used in practical applications to prepare liposomes, which have been widely used in the cosmetic industry. In the pharmaceutical and food fields, Chemin et al. used sphingomyelin liposomes as carriers to encapsulate an anticancer drug (S12363) with an encapsulation rate of 90% [75]; Yilmaz et al. prepared microemulsions containing ceramide with high stability during the storage of 6 months [76]. In addition to being a raw material for liposomes, ceramides sphingolipids used as gelling agents to structure liquid vegetable oils have been reported by Rogers et al. [77]. It was reported that ceramide is the first oleogelator that can form oleogels under 2% of vegetable oil. Unlike the translucent oleogels formed by ceramide and canola oil, egg sphingomyelin developed by enzymatic conversion formed opaque oleogels with canola oil at 7% levels. These results indicated that ceramides from synthetic ways are not as efficient as pure ceramides. As mentioned above, the great structural diversity (chain length and fatty acid groups) of ceramides makes it possible to regulate network functionality. In the previous study, a series of ceramides with different chain lengths, including C2 ceramide, C6 ceramide, C10 ceramide, C14 ceramide, and C18 ceramide were investigated to structure canola oil [78]. Results suggested that the gelling efficiency of ceramide was decreased with the increasing chain length as C2 ceramides produced "true oleogel" and C18 ceramide only produced an opaque viscous solution (Fig. 8.7a). The microstructure of the oleogels network can explain the reason because ceramides with shorter chain lengths assembled into fibrillar crystals, while ceramides with longer chains exhibited spherulitic crystals (Fig. 8.7b).

Recently, the multi-component system containing ceramide attracted increasing attention in oleogels as well as oleogel-based emulsions due to their higher efficiency for oleogelation [79]. In the ternary system of ceramides +β-sitosterol +stearic acid, the ratio of oleogelator significantly influenced the morphologies and physicochemical properties of oleogels [80]. The co-assembling of lecithin and ceramides was confirmed by Guo et al. [81]. In another work conducted by Liao et al. [82], the addition of NaOH and H_3PO_4 can tailor the properties of water-induced ceramides + lecithin oleogels, including microstructure and rheological properties. To be specific, the addition of NaOH reduced the crystal size and H_3PO_4 reduced the strength of oleogels. Either ceramide or lecithin would interfere with the interactions between lecithin and water molecules [82]. Subsequently, the crystallization kinetics study of ceramides + lecithin oleogels under non-isothermal was systematically investigated. The macroscopic properties and crystallization kinetics progress were significantly influenced by cooling rates [83]. The lower cooling rate allowed for the formation of stronger oleogels with larger crystal sizes.

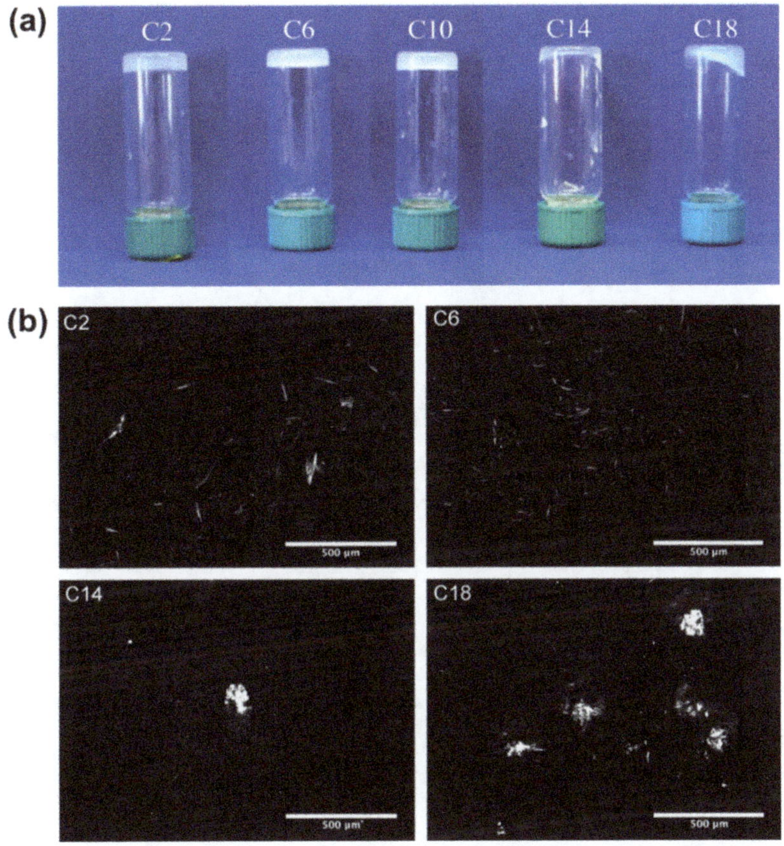

Fig. 8.7 Visual appearance (**a**) and polarized light micrographs (**b**) of different chain length ceramides used as oleogelators (2 wt%) for canola oil. (Reproduced from [78], with permission from Wiley under the terms of the Creative Commons Attribution 3.0 Unported license (CC BY 3.0))

8.4 Lecithin

8.4.1 Source and Composition

The phospholipids found in nature can be divided into two categories: animal phospholipids and plant phospholipids. Plant-based phospholipids are mainly found in the seeds of oilseeds (soy, sunflower, and rapeseed), with soybeans being the most important source of plant-based phospholipids. As presented in Table 8.3, phospholipids in biological vegetable cells are mainly classified as phosphatidylcholine (PC), phosphatidylinositol (PI), phosphatidylethanolamine (PE), and phosphatidylserine (PS) [84, 85]. Among them, phosphatidylcholine was the first phospholipid that was isolated and first identified from a mixture of phospholipids

Table 8.3 Structures and characteristics of phospholipids

Phospholipid	Position Sn_1/Sn_2	Chemical structure
Phosphatidylcholine	18:2/18:2	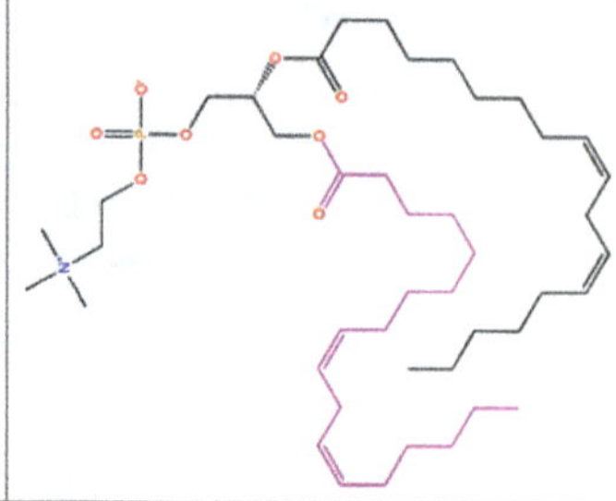
Phosphatidylethanolamine	16:0/18:2	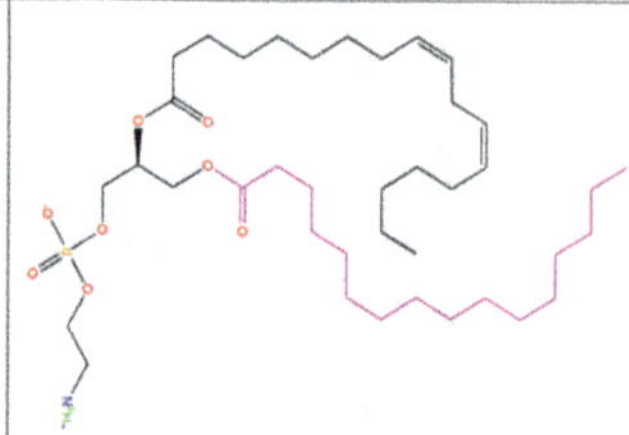
Phosphatidylinositol (PI)	18:1/18:1	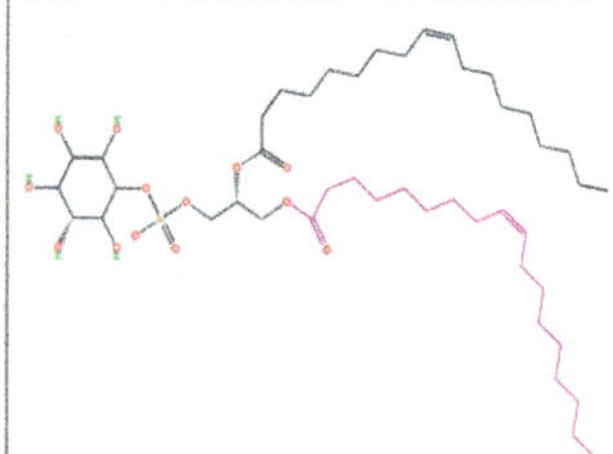
Phosphatidylserine (PS)	18:1/18:1	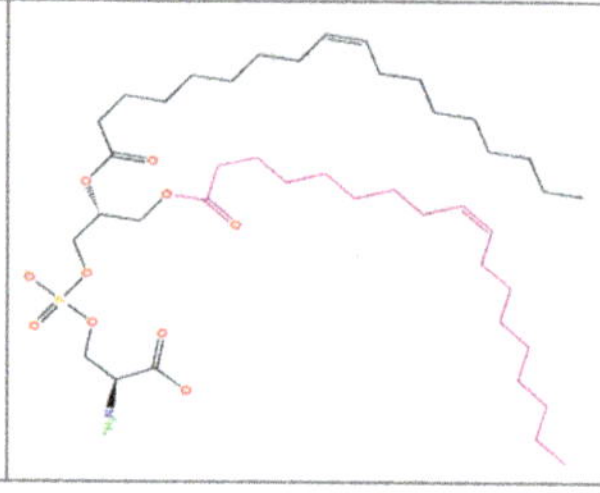

Reproduced from [86], with permission from John Wiley & Sons

and it is also known as lecithin. In many studies in the field of medicine, the terms "lecithin" and "phosphatidylcholine" are used interchangeably [85]. In the food industry, the term "lecithin" usually refers to the mixture of lipids containing more than 50% phospholipids derived from animal or vegetable origin.

The most common commercial source of lecithin is the refining process of crude vegetable oils, including physical or enzymatic degumming. Soybean lecithin accounts for 40–70% of soybean oil and oil feet processed from soybeans, which are abundant sources and relatively inexpensive. The lecithin molecule is composed of choline, phosphate, glycerol, and fatty acids, with a molecular weight of

647–863 g/mol [86]. The structure of the lecithin molecule is characterized by amphiphilicity: a hydrophilic head composed of choline linked to phosphate and a hydrophobic tail composed of fatty acid chains. Due to its structural amphiphilicity, lecithin can absorb on interfacial membranes, decreasing the interfacial tension between the oil and water phase [87]. Therefore, lecithin is a natural surfactant, which has a wide range of applications and a promising future in the field of functional food research. In the current food industry, lecithin is often used as a healthy food additive as an emulsifier, surfactant, wetting agent, dispersant, and crystallization modifier [88, 89]. Interestingly, lecithin can also be added to food systems as an antioxidant [90]. According to the report of the World Health Organization (WHO), the daily consumption of 22–83 g of lecithin per person can increase the total nutritional potency and lower blood cholesterol without any side effects [91].

8.4.2 Oil-Gelling Properties of Lecithin

Depending on the food matrix, and physical and colloidal state, lecithin can serve as a stabilizer or crystal habit modifier (changing crystal morphology) [92]. Due to its amphiphilic structures, lecithin can be used as an oleogelator in a non-polar organic solvent (hexane and isooctane), self-assembling into reverse spherical micelles (with uniaxial growth) in the presence of a small amount of polar solvent (water) [93]. The molecular and microstructural organization of lecithin oleogels was further studied by Bodennec et al. [94]. The schematic representation of the structural organization of lecithin is shown in Fig. 8.8. Small-angle X-ray diffraction revealed the reverse hexagonal (H_{II}) array with a *d*-spacing of 52 Å (Fig. 8.8a) and strands with a diameter of 1 μm (Fig. 8.8b). Figure 8.8c, d shows the fibrous network of lecithin oleogels (Fig. 8.8e). The proposed structuring mechanism of lecithin oleogels was that the fibrous network formed by branching and overlapping of bundles at junction zones along the reverse micellar chains realized the entrapment of liquid oil [94]. However, the three-dimensional network formed by these "worm-like" reverse micelles is sensitive to shear and water concentration, which is not beneficial for its universal application.

To address the mentioned concerns, lecithin has been applied as a multi-component oleogel system leading to an improvement in the strength of oleogels. Unlike mono-component oleogels, multi-component oleogel is fabricated by two or more oleogelators through synergistic interactions. As shown in Fig. 8.9, multi-component oleogels are divided into three groups [49]: (i) both gelators can self-assemble independently (self-sorting) or co-assemble through synergistic interactions (co-assembly) to form oleogels, such as stearic acid + stearyl alcohol and β-sitosterol + γ-oryzanol systems [95, 96]; (ii) both gelators are not capable of gelation but can co-assemble to form oleogels through synergistic interactions, such as the sucrose ester + lecithin and lecithin + α-tocopherol systems [29, 97]; (iii) one gelator with gelation capability and a modifying component that can

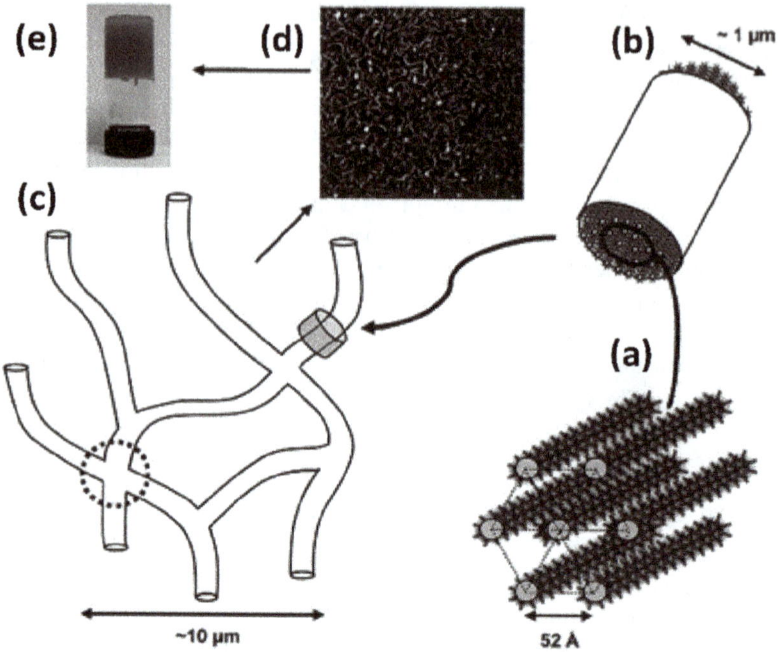

Fig. 8.8 Schematic diagram of the structural organization for a lecithin oleogel at ambient temperature (**a**: cylinders organized into reverse hexagonal (H_{II}) array; **b**: fiber consists of a bundle of micellar nanochannels across; **c**: the junction zones in the cross-crossing fibers; **d**: fibrous gel network; **e**: opaque gel). (Reproduced from [94] with permission from the Royal Society of Chemistry)

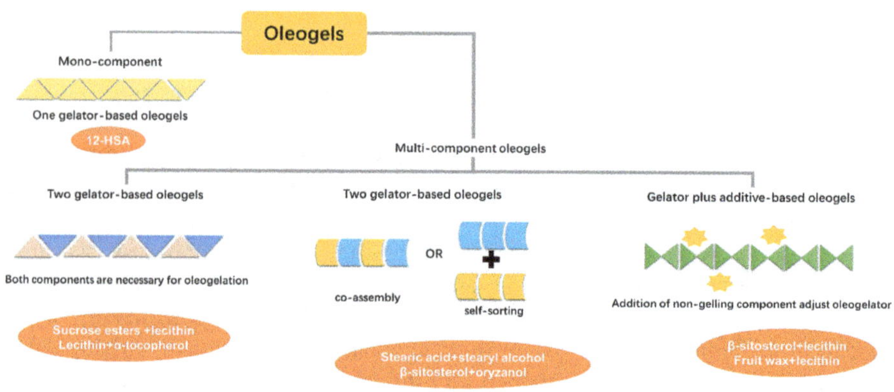

Fig. 8.9 Schematic representations of mono-component and multi-component oleogels. (Reproduced from [13] with permission from John Wiley & Sons)

Table 8.4 Overview of literature evaluating lecithin-based multi-component oleogels

Multi-component oleogel system	Role of lecithin	Solvent phase	Findings	References
Lecithin + stearic acid	Co-gelator	Canola oil	Oleogels were primarily formed through the entanglement of bundles of reverse worm-like micelles of lecithin with stearic acid crystals	[102]
Lecithin + β-sitosterol	Co-gelator	Sunflower oil	β-sitosterol had an enhanced structuring capacity in the presence of lecithin	[16]
Lecithin + monoglyceride	Non-gelling agent	Vegetable oil/mineral oil	The lecithin + monoglyceride oleogel had higher highest elasticity and longer storage stability without phase separation than monoglyceride oleogel	[100]
Lecithin + sucrose esters	Co-gelator	Sunflower oil	The 7:3 (sucrose esters: lecithin) combination showed enhanced rheological properties	[29]
Lecithin + ethyl cellulose	Non-gelling agent	High oleic canola oil	1% Lecithin was enough to improve the mechanical strength of ethyl cellulose polymer oleogels	[103]
Lecithin +12-HSA	Non-gelling agent	Cyclohexane	Intermolecular between lecithin and 12-HSA (1:1) caused a structural change in the fibrous network in the 12-HSA oleogel and consequently induced gel strength	[101]
Lecithin + α-tocopherol	Co-gelator	Sunflower oil	When the gelators were combined at a 1:1 ratio in sunflower oil, oleogel was obtained	[97]
Lecithin + fruit wax	Non-gelling agent	Sunflower oil	The synergistic effect on the oleogel strength was observed at ratios of 75:25 and 50:50 (fruit wax: lecithin)	[1]
Lecithin + sorbitan tri-stearate	Co-gelator	Sunflower oil	A synergetic effect is observed with their mixture, at specific ratios of lecithin: STS, between 40:60 and 60:40, when firm gels are obtained	[89]
Lecithin + ceramide	Co-gelator	Sunflower oil	Strong oleogels with superior oil binding capacity were formed at specific ratios (6:4 and 5:5)	[81]

modulate the self-assembly behavior of the host gelator, such as β-sitosterol + lecithin and fruit wax + lecithin systems [1, 98].

It is clear that lecithin plays an important role in multi-component oleogel systems, acting as a co-gelator or non-gelling agent through non-covalent interactions such as hydrogen bonds and van der Waals [99]. Many researchers have reported the benefits from lecithin incorporation in multi-component such as improvement in gelation, either by promoting the gel formation or indirectly by acting as a crystal modifier, modulating the packing geometry and intermolecular interactions of both gelling agents and non-gelling agents (Okuro et al. 2020).

Table 8.4 summarizes the examples of research in literature that investigated the lecithin-based multi-component. For instance, the shear moduli of ethyl cellulose oleogel increased tenfold after the addition of % lecithin due to the promotion of a more gradual thickening response (Aguilar-Zárate et al. 2019). A similar effect was also observed in sucrose esters + lecithin and monoglyceride + lecithin. The elasticity of monoglyceride was improved with the addition of lecithin due to the synergistic effect that existed between the monoglyceride and the lecithin [100]. Sucrose esters cannot structure liquid oil individually because its hydrophilic monomers would aggregate in a non-polar medium. The incorporation of lecithin can modify the self-assembly mode of sucrose esters by interrupting the interaction (H-bonding) between their hydrophilic monomers [29]. However, there is an exception in that lecithin exerted a negative influence on the oleogelation process of 12-hydroxystearic acid [101]. A structural change from fibrous aggregates to a spherical form was observed after the addition of lecithin, resulting in the reduction of interaction and mechanical strength of 12-HSA oleogels. In our previous work, β-sitosterol + lecithin oleogels exhibited a positive effect on delaying the fat bloom development in the chocolate system due to the existence of lecithin hindered the crystallization of cocoa butter.

8.5 Sterols

8.5.1 *Source and Health Benefits*

Sterols, also called phytosterols, are molecules with a similar chemical structure to cholesterol, differing in the side chain. They can be divided into three categories considering their different sources: animal sterols, plant sterols, and fungal sterols [104]. There are over 250 kinds of sterols and related compounds found in various plant and marine materials, including free sterols, sterol esters, sterol glucosides, and acylated sterol glucosides [105]. Among them, sitosterol, stigmasterol, and campesterol (4-desmethyl sterols) are the most common ones. Generally, the natural source of sterols is vegetable oils and products derived from oils, followed by cereal grains, cereal products, and nuts. Most vegetable oils contain 1–5 g phytosterols per kg of oil. However, more sterols were detected in corn oil (8.09–15.57 g/kg), rapeseed oil (5.13–9.79 g/kg), rice bran oil (2.25 g/kg), and wheat germ oils (19.70 g/kg) [106]. It should be noted that the refining process (e.g., alkali refining, decolorization, deodorization) would greatly reduce the sterol content in vegetable oils. In addition to their anti-inflammatory functions, phytosterols also show positive effects on cholesterol-lowering, anti-cancer, and aging retardation. The US Food and Drug Administration (FDA) has given phytosterols a "Generally Recognized as Safe (GRAS)" designation. In 2008, the European Union Food Safety Law Commission issued an announcement further confirming the safety of phytosterols [107]. Phytosterols were first used in food products in margarine and then widely used in dairy products, bakery products, sausages, vegetable oils, and fruit juices.

8.5.2 Application in Oleogelation

Mixtures of plant sterols with specific molecular structures can be potentially applied as a solid fat alternative like triacylglycerols for structuring the liquid oil in some food formulations [96]. Bot and Agterof [28] evaluated the gel-forming capacity of different plant sterols (5α-cholestane, dihydrocholesterol, cholesterol, β-sitosterol, stigmasterol, and ergosterol) with γ-Oryzanol. They discovered that the essential structure determining the formation of oleogels is the hydroxyl group and ring structure instead of the alkyl residue [28]. Multi-component oleogels formed by β-sitosterol and γ-oryzanol are one of the most investigated examples. In recent years, extensive studies investigated the utilization of γ-oryzanol + β-sitosterol for the development of oleogels and emulsions [108–110].

Neither β-sitosterol nor γ-oryzanol can be considered efficient oleogelators when used individually. However, when β-sitosterol + γ-oryzanol are blended at a 1:1 molar ratio, they can self-assemble into fibrils or double-walled hollow tubules (10 nm in diameter) due to the intermolecular hydrogen bonding between β-sitosterol and γ-oryzanol molecules. From the analysis of the chemical bond formation mechanism, sitosterol can form intermolecular hydrogen bonds (R-O-H...O-R′) through the hydroxyl group (R-O-H) with the oxygen atom (-C=O) in the strongly electronegative carbonyl group of the γ-oryzanol molecule [28, 96]. Both the macroscopic properties and microstructure of γ-oryzanol+-β-sitosterol have been investigated in detail.

Bot et al. [111] investigated the tubule structure fabricated by plant sterols and γ-oryzanol in sunflower oil by X-ray scattering experiments. As presented in Fig. 8.10a, the tubules have a diameter of 6.7–8.0 nm and a wall thickness of 0.8–1.2 nm. The sunflower oil was entrapped both inside and outside the tubule.

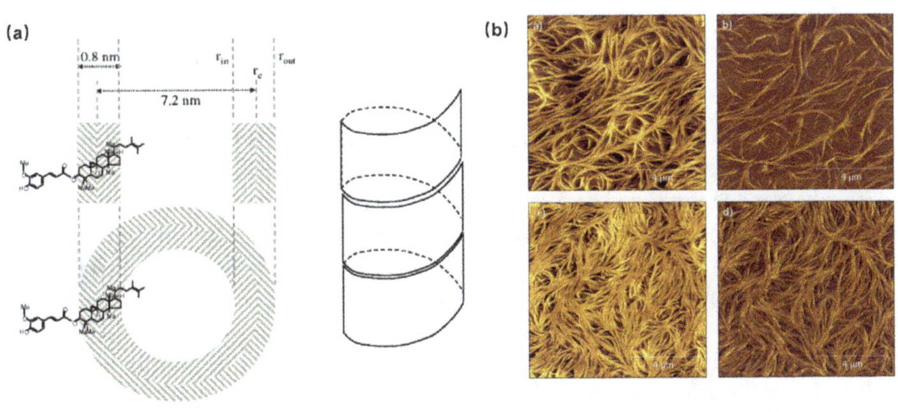

Fig. 8.10 Microstructure of sterol+γ-oryzanol oleogel: (**a**) schematic representation of the tubular structure (reproduced from [111] with permission from Springer-Nature); (**b**) atomic Force Microscope images, height (left column) and phase (right column) (reprinted with permission from [110], Copyright (2017) American Chemical Society)

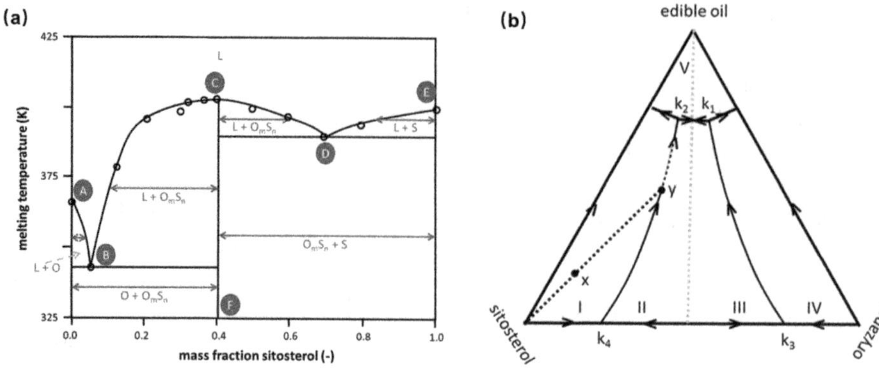

Fig. 8.11 The phase diagram for (**a**) the binary γ-oryzanol + β-sitosterol system and (**b**) the ternary γ-oryzanol + β-sitosterol + sunflower oil system. (Reproduced from [115] with permission from Springer, under the terms of the Creative Commons Attribution 4.0 International License)

In the following study conducted by [110], the microstructure of γ-oryzanol+-β-sitosterol was directly observed by Atomic Force Microscope (Fig. 8.10b). Interestingly, they found that the self-assembled structure in the oleogels systems behaves like ribbons instead of twisted strands with a diameter of ~9.8 nm.

With regard to the oleogelation mechanism of γ-oryzanol+β-sitosterol, the Avrami model (Eq. (8.1) or Eq. (8.2)) was applied to investigate the nucleation and growth of crystals [112].

$$X_{\mathrm{crit}} = \frac{G'(t)}{G'(\infty)} = 1 - \exp\left[-k\left(t - t_{\mathrm{g}}\right)^{n}\right] \tag{8.1}$$

$$\ln\left(-\ln\left(1 - \frac{G'(t)}{G'(\infty)}\right)\right) = \ln(k) + n\ln\left(t - t_{\mathrm{g}}\right) \tag{8.2}$$

where X_{crit} is the degree of oleogelation in the system, G'(t) is the storage modulus at time t, $G'(\infty)$ is the plateau value when the oleogelation is complete, n is the Avrami exponent, t is the time, and t_{g} k, and k is the rate constant of growth.

The Avrami index (n) is an important parameter for characterizing the kinetics of fat crystallization and is a dimensional function of crystalline growth, reflecting the intrinsic mechanism of crystal nucleation and growth [113]. The values of n correspond to the different types of nucleation and growth [114]. Upon fitting the rheological data of γ-oryzanol+β-sitosterol systems into the Avrami model, a value of $n = 0.9$ was obtained at the initial stage of growth. The result indicated that the crystallization of γ-oryzanol+β-sitosterol followed a one-dimensional growth pattern with purely instantaneous nucleation [110]. The phase behavior of β-sitosterol and γ-oryzanol in sunflower oil/canola oil was investigated [115, 116]. As shown in Fig. 8.11a, the "W" shape curve indicated that eutectic behavior existed in the binary system of β-sitosterol and γ-oryzanol, with point B and point D being the eutectic points. When the temperature was lower than point B, the solid in the binary system depended on the composition of β-sitosterol and γ-oryzanol [115]. Compared to the binary system, the addition of sunflower oil decreased the melting point of γ-oryzanol + β-sitosterol due to the dilution effect. Though the phase behavior of ternary systems did not exhibit the transformation from binary mixture to tubular structure, a gradual transition seems to occur when the volume of edible oil increased (as shown in Fig. 8.11b).

Another similar system of sterols-based multi-component system worthy of attention is β-sitosterol +lecithin. The gelation capacity of β-sitosterol + lecithin was first reported by Han et al. [98]. They found that the addition of lecithin generally promoted the formation of needle-like crystals due to the changes in the self-assembly mode of β-sitosterol [16]. After the development in the last decades, a novel application of the β-sitosterol + lecithin oleogels has been extensively explored, including cocoa butter alternatives [117], delivery system for resveratrol [118], and curcumin [119].

8.6 Cellulose Fibers

Increasing attention has been aroused to search for novel oleogelator from a natural source. Plant biomass (cellulose, hemicelluloses, and lignin) are ideal candidates because they are abundant and highly distributed from a renewable source. Cellulose-based fibers are considered as a novel stabilizer with sustainability, biodegradability, and nontoxicity to replace traditional synthetic stabilizers in Pickering emulsions, which are widely used in the delivery system and functional food [120]. However, little information is available on the application of cellulose fiber into the oleogel system. To date, ethyl cellulose is known as the only polymer oleogelator to structure oil through the direct addition method. In addition, the preparation of ethyl cellulose needs to be conducted at a high temperature (>140 °C), which might cause the oxidation of vegetable oil.

In the latest study, David et al. [121] reported exciting progress in the utilization of cellulose fibers to structure rapeseed oil at ambient temperature without any thermal treatment. Cellulose fibers from bamboo, wheat, pea, acacia, and chicory were obtained from the mercerization process. As shown in Fig. 8.12a, b, clear evidence of a structuring effect of the rapeseed oil was obtained for bamboo and wheat cellulose powders at the concentration of 30%. However, phase separation was observed in the 30 wt% dispersions of pea, acacia, and chicory, indicating that they lacked texturing ability. The rheology result (Fig. 8.12c) confirmed the solid-like behavior of 30 wt% bamboo and 30 wt% wheat dispersions. In addition, oleogels prepared by cellulose fibers can be used in chocolate spreads production without any additives to improve both the rheological properties and the nutritional qualities of food products [121]. It should be noted that the application of cellulose-based fibers into oleogelation is still in its infancy. It is worthy to explore the further

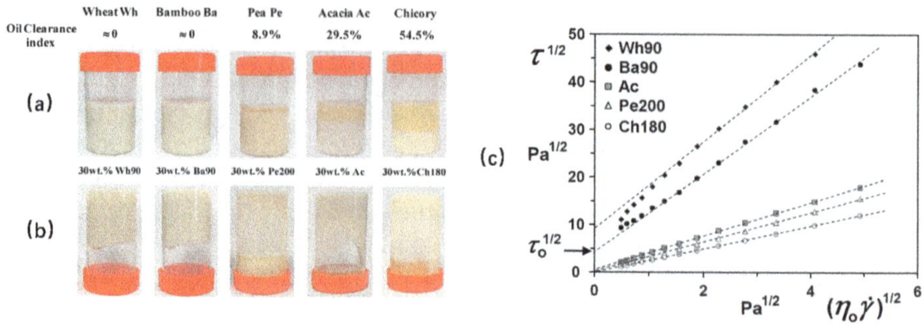

Fig. 8.12 Photographs of 30 wt% dispersions of cellulose fibers from wheat, bamboo, pea, acacia, and chicory from left to right in rapeseed oil (**a** and **b**); square root of the shear stress vs. the square root of viscosity η_o and shear rate for several 30 wt% dispersions of cellulose fibers from wheat, bamboo, pea, acacia, and chicory (**c**). (Reproduced from [121] with permission from Elsevier)

application of cellulose in the oleogelation field, and more in-depth research is needed to elucidate the structural properties of cellulose fibers.

8.7 Fumed Silica

8.7.1 Preparation and Multi-scale Structure of Fumed Silica

Fumed silica (silicon dioxide) is among the most researched inorganic particles that have been widely utilized in foods, pharmaceuticals, cosmetics, ceramics, and electronics [122]. It is an amorphous SiO_2 obtained by flame hydrolysis of $SiCl_4$ at extremely high temperatures (>1000 °C) [123]. $SiCl_4$ reacts with H_2 and O_2 to obtain spherical SiO_2 with a diameter between 5 and 30 nm, which is also called primary particles [31]. Then the primary particles collide together in the flame and fuse, forming irregularly shaped aggregates (Fig. 8.13a). When the aggregates leave the flame environment and cool down, the joining process ends and the primary aggregates with a length from 100 to 1000 nm are formed [124]. As the fusing process is irreversible, the structure of aggregates cannot be destroyed by physical treatment. The primary aggregates are the "smallest structural unit" of fumed silica. When the fumed silica was further cooled, the primed aggregates are stacked with each other by chlorine bonds and other physical interactions to form larger size agglomerates (Fig. 8.13b). The final agglomerates are white powers with a large specific surface area of 100–400 m^2/g, bulk density of 0.02–0.05 g/cm^3, and aggregate density of 0.7 g/cm^3 [31, 124].

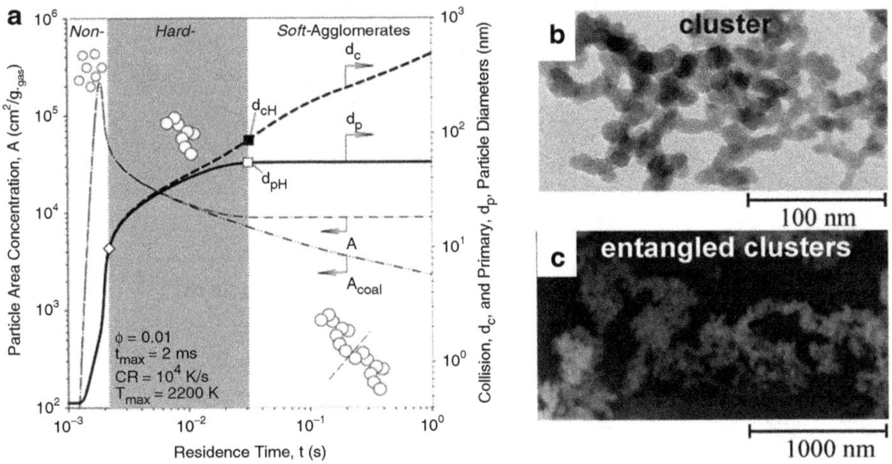

Fig. 8.13 Evolution and multi-scale structure of SiO_2: (**a**) total particle area concentration for partial sintering (dash-dotted line) and full coalescence (dash-double-dotted line) and primary particle (solid line) and agglomerate collision (dashed line) diameters (reproduced with permission from [125], Copyright (2004) American Chemical Society); (**b**) electron microscopy images of the primary nanoparticles are fused together into clusters with irregular, chain-like geometry (reproduced from [31] with permission from Elsevier); and (**c**) electron microscopy images of the dry powder consist of fused clusters that physically entangle together into porous, non-spherical agglomerates (reproduced from [31] with permission from Elsevier)

8.7.2 Fumed Silica-Based Oleogels

The surface chemistry of fumed silica powders significantly affects the interactions of particle–particle and particle–substrate, which is an important factor to determine the performance of fumed silica [126]. The smooth surface with abundant Si-O-Si groups of fumed silica can provide sites for particle-particle and particle-substrate van der Waals forces and hydrogen bonding [127, 128]. The behavior of fumed silica in both aqueous and organic systems has been investigated for several years. In earlier studies, it was reported that hydrophilic fumed silica, as well as partially methylated fumed silica is able to develop stable dispersions in organic media (methanol, ethanol, acetone, and dioxane) [129]. In organic media, fumed silica can enhance the viscosity and further generate a colloidal gel (i.e., a three-dimensional network of particles) when the interparticle linkage is formed extensively and spans the sample volume [128]. Similar behavior of fumed silica was also observed in edible oils.

For the first time, Patel et al. [122] reported the utilization of fumed silica as a structurant to gel sunflower oil. The microstructure and rheological behavior of fumed silica oleogels are shown in Fig. 8.14. The transparent appearance of fumed silica oleogels indicated that the building blocks (silica particles) in the oleogels are considerably smaller than the wavelength of visible light with a wavelength range between 400 and 750 nm. It can be seen from cryo-SEM results of fumed silica

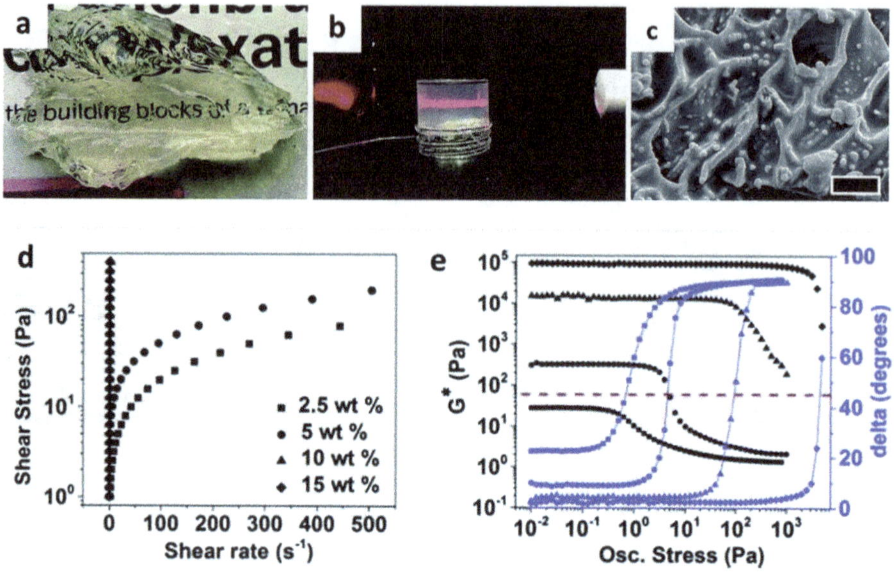

Fig. 8.14 (**a**) Photograph of transparent organogel prepared at 15 wt% of fumed silica; (**b**) Faraday–Tyndall effect observed when a laser beam is passed through the organogel sample; and (**c**) cryo-SEM image of 15 wt% fumed silica organogel where surface oil was removed using butanol (scale bar = 1 μm). (Reproduced from [122] with permission from the Royal Society of Chemistry under the terms of the Creative Commons Attribution 3.0 Unported license (CC BY 3.0))

oleogel (de-oiled) that the vegetable oil was entrapped by the colloidal network fabricated by silicon particles. The rheological measurement showed that the oleogels structured by 15 wt% fumed silica exhibited the highest mechanical strength [122]. In the serious work of Chauhan et al., fumed silica oleogels were incorporated into a model fat-structured food system and showed similar rheological and thermal behavior with the full-fat systems [130, 131]. Whitby et al. [31] investigated the relationship between the rheological behavior and the morphology of the networks of fumed silica oleogels. They proposed that the gel strength was reduced with the silanol groups on the particle surfaces due to the formation of a tenuous network structure with larger pore sizes. These results demonstrated the potential of fumed silica oleogels to partially replace solid fat in the food industry.

8.8 Conclusions

Research on oleogels have been advanced greatly in the last 20 years, particularly in the exploration of emerging oleogelators and novel applications as fat-mimicking agents. So far, many kinds of structuring agents have been used for creating viscoelastic "gel-like" structures. This chapter enumerates and exemplifies the common low molecular weight oleogelators, such as fatty acid (includes 12-HSA), fatty alcohols, ceramides, lecithin, sterols, and fumed silica. In addition, an emerging group of cellulose fibers from vegetal source was also included. Most low molecular weight oleogelators showed high nutrition value and adjustable self-assembly behavior, enabling oleogels with tunable functionality in varied application scenarios. Generally, the relationship between the structural characteristics of oleogelator and gelling properties of oleogels was discussed in detail. The discussion in this chapter helps to understand self-assembling mechanisms involved in oleogelation process; and the relationship between oleogelator molecules and physical properties of oleogels (i.e., rheological behavior, thermal profiles, microstructural characteristics, temperature). Due to self-assembly behavior of oleogelator being complex and prone to be influenced by the gelator-solvent balance, prediction of the structuring capacity of oleogelators in advance is still challenging. Given the current trend in the exploitation of oleogelators, several potential research prospects are rooted in multicomponent oleogels systems and exploration of more oleogelation systems through biomimicry investigation.

Acknowledgments This study was financed by Danmarks Frie Forskningsfond | Teknologi og Produktion (0136-00206B), the Novo Nodisk Fonden (NNF16OC0021740), National Natural Science Fund of China [grant numbers: U21A20270, 31972002 and 31771895], Ministry of Education Engineering Research Center of Starch & Protein Processing, Guangdong Province Laboratory for Green Processing of Natural Products and Product Safety.

References

1. Okuro PK, Tavernier I, Bin Sintang MD et al (2018) Synergistic interactions between lecithin and fruit wax in oleogel formation [J]. Food Funct 9(3):1755–1767
2. Patel AR, Nicholson RA, Marangoni AG (2020) Applications of fat mimetics for the replacement of saturated and hydrogenated fat in food products [J]. Curr Opin Food Sci 33:61–68
3. Teegala SM, Willett WC, Mozaffarian D (2009) Consumption and health effects of trans fatty acids: a review [J]. J AOAC Int 92(5):1250–1257
4. Restrepo BJ, Rieger M (2016) Trans fat and cardiovascular disease mortality: evidence from bans in restaurants in New York [J]. J Health Econ 45:176–196
5. Siri-Tarino PW, Sun Q, Hu FB et al (2010) Saturated fat, carbohydrate, and cardiovascular disease [J]. Am J Clin Nutr 91(3):502–509
6. Ghebreyesus TA, Frieden TR (2018) Replace: a roadmap to make the world trans fat free by 2023 [J]. Lancet 391(10134):1978–1980
7. De la Peña-Gil A, Álvarez-Mitre FM, González-Chávez MM et al (2017) Combined effect of shearing and cooling rate on the rheology of organogels developed by selected gelators [J]. Food Res Int 93:52–65
8. Patel AR, Rajarethinem PS, Gredowska A et al (2014) Edible applications of shellac oleogels: spreads, chocolate paste and cakes [J]. Food Funct 5(4):645–652
9. Li L, Liu G, Lin Y (2021) Physical and bloom stability of low-saturation chocolates with oleogels based on different gelation mechanisms [J]. LWT 140:110807
10. Stortz TA, Marangoni AG (2013) Ethylcellulose solvent substitution method of preparing heat resistant chocolate [J]. Food Res Int 51(2):797–803

11. Toro-Vazquez JF, Morales-Rueda J, Torres-Martínez A et al (2013) Cooling rate effects on the microstructure, solid content, and rheological properties of organogels of amides derived from stearic and (r)-12-hydroxystearic acid in vegetable oil [J]. Langmuir 29(25):7642–7654
12. Terech P, Weiss RG (1997) Low molecular mass gelators of organic liquids and the properties of their gels [J]. Chem Rev 97(8):3133–3160
13. Li L, Liu G, Bogojevic O et al (2022) Edible oleogels as solid fat alternatives: composition and oleogelation mechanism implications [J]. Compr Rev Food Sci Food Saf 21(3):2077–2104
14. Patel AR, Dewettinck K (2016) Edible oil structuring: an overview and recent updates [J]. Food Funct 7(1):20–29
15. Gómez-Estaca J, Herrero AM, Herranz B et al (2019) Characterization of ethyl cellulose and beeswax oleogels and their suitability as fat replacers in healthier lipid pâtés development [J]. Food Hydrocoll 87:960–969
16. Han L, Li L, Li B et al (2014) Structure and physical properties of organogels developed by sitosterol and lecithin with sunflower oil [J]. J Am Oil Chem Soc 91(10):1783–1792
17. Shah DK, Sagiri SS, Behera B et al (2013) Development of olive oil based organogels using sorbitan monopalmitate and sorbitan monostearate: a comparative study [J]. J Appl Polym Sci 129(2):793–805
18. Wright AJ, Marangoni AG (2006) Formation, structure, and rheological properties of ricinelaidic acid-vegetable oil organogels [J]. J Am Oil Chem Soc 83(6):497–503
19. Rogers MA (2018) 12-hydroxystearic acid oleogels [M]. In: Edible oleogels. AOCS Press, London 85–102
20. Gravelle AJ, Davidovich-Pinhas M, Zetzl AK et al (2016) Influence of solvent quality on the mechanical strength of ethylcellulose oleogels [J]. Carbohydr Polym 135:169–179
21. Meng Z, Qi K, Guo Y et al (2018) Effects of thickening agents on the formation and properties of edible oleogels based on hydroxypropyl methyl cellulose [J]. Food Chem 246:137–149
22. Patel AR, Rajarethinem PS, Cludts N et al (2015) Biopolymer-based structuring of liquid oil into soft solids and oleogels using water-continuous emulsions as templates [J]. Langmuir 31(7):2065–2073
23. Tavernier I, Patel AR, Van der Meeren P et al (2017) Emulsion-templated liquid oil structuring with soy protein and soy protein: K-carrageenan complexes [J]. Food Hydrocoll 65:107–120
24. Chen X-W, Yang X-Q (2019) Characterization of orange oil powders and oleogels fabricated from emulsion templates stabilized solely by a natural triterpene saponin [J]. J Agric Food Chem 67(9):2637–2646
25. Silva PM, Cerqueira MA, Martins AJ et al (2022) Oleogels and bigels as alternatives to saturated fats: a review on their application by the food industry [J]. J Am Oil Chem Soc 99(11):911–923
26. Jiang Z, Lu X, Geng S et al (2020) Structuring of sunflower oil by stearic acid derivatives: experimental and molecular modelling studies [J]. Food Chem 324:126801
27. Blach C, Gravelle AJ, Peyronel F et al (2016) Revisiting the crystallization behavior of stearyl alcohol : stearic acid (so : Sa) mixtures in edible oil [J]. RSC Adv 6(84):81151–81163
28. Bot A, Agterof WGM (2006) Structuring of edible oils by mixtures of γ-oryzanol with β-sitosterol or related phytosterols [J]. J Am Oil Chem Soc 83(6):513–521
29. Bin Sintang MD, Danthine S, Patel AR et al (2017) Mixed surfactant systems of sucrose esters and lecithin as a synergistic approach for oil structuring [J]. J Colloid Interface Sci 504:387–396
30. Whitby CP (2020) Structuring edible oils with fumed silica particles [J]. Front Sustainable Food Syst 4:585160
31. Whitby CP, Krebsz M, Booty SJ (2018) Understanding the role of hydrogen bonding in the aggregation of fumed silica particles in triglyceride solvents [J]. J Colloid Interface Sci 527:1–9
32. Bot A, Flöter E (2018) Structuring edible oil phases with fatty acids and alcohols [J]. In: Edible oil structuring: concepts methods and applications, AOCS press, Urbana 95–105

33. Gandolfo FG, Bot A, Flöter E (2004) Structuring of edible oils by long-chain fa, fatty alcohols, and their mixtures [J]. J Am Oil Chem Soc 81(1):1–6
34. Daniel J, Rajasekharan R (2003) Organogelation of plant oils and hydrocarbons by long-chain saturated fa, fatty alcohols, wax esters, and dicarboxylic acids [J]. J Am Oil Chem Soc 80(5): 417–421
35. Valoppi F, Calligaris S, Marangoni AG (2017) Structure and physical properties of oleogels containing peanut oil and saturated fatty alcohols [J]. Eur J Lipid Sci Technol 119(5):1600252
36. Hunter JE, Zhang J, Kris-Etherton PM (2010) Cardiovascular disease risk of dietary stearic acid compared with trans, other saturated, and unsaturated fatty acids: a systematic review [J]. Am J Clin Nutr 91(1):46–63
37. Hu F-Q, Zhao M-D, Yuan H et al (2006) A novel chitosan oligosaccharide–stearic acid micelles for gene delivery: properties and in vitro transfection studies [J]. Int J Pharm 315(1–2):158–166
38. Sagiri SS, Singh VK, Pal K et al (2015) Stearic acid based oleogels: a study on the molecular, thermal and mechanical properties [J]. Mater Sci Eng C Mater Biol Appl 48:688–699
39. Weiss R.G., Terech P. Molecular gels [J]. Materials with self-assembled fibrillar networks, 2006:
40. Şahan N, Paksoy H (2018) Developing microencapsulated 12-hydroxystearic acid (hsa) for phase change material use [J]. Int J Energy Res 42(10):3351–3360
41. Fameau A-L, Rogers MA (2020) The curious case of 12-hydroxystearic acid — the dr. Jekyll & mr. Hyde of molecular gelators [J]. Curr Opin Colloid Interface Sci 45:68–82
42. Alvarez-Mitre FM, Mallia VA, Weiss RG et al (2017) Self-assembly in vegetable oils of ionic gelators derived from (r)-12-hydroxystearic acid [J]. Food Struct 13:56–69
43. Espinosa-Dzib A, Vyazovkin S (2021) Nanoconfined gelation in systems based on stearic and 12-hydroxystearic acids: a calorimetric study [J]. J Mol Liq 335:116191
44. Mallia VA, George M, Blair DL et al (2009) Robust organogels from nitrogen-containing derivatives of (r)-12-hydroxystearic acid as gelators: comparisons with gels from stearic acid derivatives† [J]. Langmuir 25(15):8615–8625
45. Maskaev AK, Man'kovskaya NK, Lend'el IV et al (1971) Preparation of 12-hydroxystearic acid, the raw material for plastic greases [J]. Chem Technol Fuels Oils 7(2):109–112
46. Fraser HM (1946) Production of lubricants, Google Patents
47. Elliger CA, Guadagni DG, Dunlap CE (1972) Thickening action of hydroxystearates in peanut butter [J]. J Am Oil Chem Soc 49(9):536–537
48. Yang H-K, Zhang C, He X-N et al (2021) Effects of alkyl chain lengths on 12-hydroxystearic acid derivatives based supramolecular organogels [J]. Colloids Surf A Physicochem Eng Asp 616:126319
49. Patel AR (2017) A colloidal gel perspective for understanding oleogelation [J]. Curr Opin Food Sci 15:1–7
50. Li J-L, Liu X-Y (2010) Architecture of supramolecular soft functional materials: from understanding to micro-/nanoscale engineering [J]. Adv Funct Mater 20(19):3196–3216
51. Rogers MA, Marangoni AG (2008) Non-isothermal nucleation and crystallization of 12-hydroxystearic acid in vegetable oils [J]. Cryst Growth Des 8(12):4596–4601
52. Rogers MA, Wright AJ, Marangoni AG (2008) Crystalline stability of self-assembled fibrillar networks of 12-hydroxystearic acid in edible oils [J]. Food Res Int 41(10):1026–1034
53. Rogers MA, Wright AJ, Marangoni AG (2009) Nanostructuring fiber morphology and solvent inclusions in 12-hydroxystearic acid / canola oil organogels [J]. Curr Opin Colloid Interface Sci 14(1):33–42
54. Gao J, Wu S, Emge TJ et al (2013) Nanoscale and microscale structural changes alter the critical gelator concentration of self-assembled fibrillar networks [J]. CrystEngComm 15(22): 4507–4515
55. Tantishaiyakul V, Ouiyangkul P, Wajasat M et al (2018) A supramolecular gel based on 12-hydroxystearic acid/virgin coconut oil for injectable drug delivery [J]. Eur J Lipid Sci Technol 120(10):1800178

56. Esposito CL, Tardif V, Sarrazin M et al (2020) Preparation and characterization of 12-hsa-based organogels as injectable implants for the controlled delivery of hydrophilic and lipophilic therapeutic agents [J]. Mater Sci Eng C 114:110999
57. Agogué M.C., Loisel C., Gonçalves O., et al. (2022) Multi-scale study of the structuration of candle blends with a high content of vegetable fats to replace paraffins: effect of 12-hydroxystearic acid content [J]. J Am Oil Chem Soc n/a(n/a) 99(12):1137–1150
58. Marangoni AG, Garti N (2018) Edible oleogels: structure and health implications [M]. Elsevier, AOCS, Urbana
59. Wang FC, Gravelle AJ, Blake AI et al (2016) Novel trans fat replacement strategies [J]. Curr Opin Food Sci 7:27–34
60. Lupi FR, Gabriele D, Baldino N et al (2013) Olive oil/policosanol organogels for nutraceutical and drug delivery purposes [J]. Food Funct 4(10):1512–1520
61. Xu Z, Fitz E, Riediger N et al (2007) Dietary octacosanol reduces plasma triacylglycerol levels but not atherogenesis in apolipoprotein e–knockout mice [J]. Nutr Res 27(4):212–217
62. Pernetti M, van Malssen KF, Flöter E et al (2007) Structuring of edible oils by alternatives to crystalline fat [J]. Curr Opin Colloid Interface Sci 12(4–5):221–231
63. Callau M, Sow-Kébé K, Jenkins N et al (2020) Effect of the ratio between fatty alcohol and fatty acid on foaming properties of whipped oleogels [J]. Food Chem 333:127403
64. Zhoh C-K, Lee K-Y, Kim D-N (2009) The influences of fatty alcohol and fatty acid on rheological properties of o/w emulsion [J]. J Soc Cosmet Sci Korea 35(2):103–110
65. Schaink HM, van Malssen KF, Morgado-Alves S et al (2007) Crystal network for edible oil organogels: possibilities and limitations of the fatty acid and fatty alcohol systems [J]. Food Res Int 40(9):1185–1193
66. Gravelle AJ, Marangoni AG (2018) Chapter 8: Vegetable oil oleogels structured using mixtures of stearyl alcohol and stearic acid (so:Sa) [M]. In: Marangoni AG, Garti N (eds) Edible oleogels, second edn. AOCS Press, Urbana, 193–217
67. Callau M, Sow-Kébé K, Nicolas-Morgantini L et al (2020) Effect of the ratio between behenyl alcohol and behenic acid on the oleogel properties [J]. J Colloid Interface Sci 560:874–884
68. Gravelle AJ, Davidovich-Pinhas M, Barbut S et al (2017) Influencing the crystallization behavior of binary mixtures of stearyl alcohol and stearic acid (sosa) using ethylcellulose [J]. Food Res Int 91:1–10
69. Ogretmen B, Hannun YA (2004) Biologically active sphingolipids in cancer pathogenesis and treatment [J]. Nat Rev Cancer 4(8):604–616
70. Singh A, Auzanneau FI, Rogers MA (2017) Advances in edible oleogel technologies - a decade in review [J]. Food Res Int 97:307–317
71. Imokawa G, Abe A, Jin K et al (1991) Decreased level of ceramides in stratum corneum of atopic dermatitis: an etiologic factor in atopic dry skin? [J]. J Investig Dermatol 96(4):523–526
72. Tessema EN, Gebre-Mariam T, Neubert RHH et al (2017) Potential applications of phyto-derived ceramides in improving epidermal barrier function [J]. Skin Pharmacol Physiol 30(3): 115–138
73. Bouwstra JA, Honeywell-Nguyen PL (2002) Skin structure and mode of action of vesicles [J]. Adv Drug Deliv Rev 54:S41–S55
74. Masukawa Y, Narita H, Shimizu E et al (2008) Characterization of overall ceramide species in human stratum corneum [J]. J Lipid Res 49(7):1466–1476
75. Chemin C, Péan J-M, Bourgaux C et al (2009) Supramolecular organization of s12363-liposomes prepared with two different remote loading processes [J]. Biochim Biophys Acta Biomembr 1788(5):926–935
76. Yilmaz E, Borchert H-H (2005) Design of a phytosphingosine-containing, positively-charged nanoemulsion as a colloidal carrier system for dermal application of ceramides [J]. Eur J Pharm Biopharm 60(1):91–98
77. Rogers MA, Wright AJ, Marangoni AG (2009) Oil organogels: the fat of the future? [J]. Soft Matter 5(8):1594–1596

78. Rogers MA, Spagnuolo PA, Wang T-M et al (2017) A potential bioactive hard-stock fat replacer comprised of a molecular gel [J]. Food Sci Nutr 5(3):579–587
79. Zhang J., Dong L., Zheng Q., et al. Surfactant-free oleogel-based emulsion stabilized by co-assembled ceramide/lecithin crystals with controlled digestibility [J]. J Sci Food Agric, 2022, n/a(n/a):
80. Wang TM, Rogers MA (2015) Biomimicry–an approach to engineering oils into solid fat s [J]. Lipid Technol 27(8):175–178
81. Guo S, Song M, Gao X et al (2020) Assembly pattern of multicomponent supramolecular oleogel composed of ceramide and lecithin in sunflower oil: self-assembly or self-sorting? [J]. Food Funct 11(9):7651–7660
82. Liao Z, Guo S, Lu M et al (2022) Tailoring water-induced multi-component (ceramide and lecithin) oleogels: influence of solute added in water [J]. Food Biophys 17(1):84–92
83. Hu B, Zheng Q, Weng Z et al (2022) Non-isothermal crystallization kinetics study of multi-component oleogels [J]. Food Chem 389:133123
84. Dowhan W, Bogdanov M, Mileykovskaya E (2008) Chapter 1: Functional roles of lipids in membranes [M]. In: Vance DE, Vance JE (eds) Biochemistry of lipids, lipoproteins and membranes, Fifth edn. Elsevier, San Diego, pp 1–37
85. Robert C, Couëdelo L, Vaysse C et al (2020) Vegetable lecithins: a review of their compositional diversity, impact on lipid metabolism and potential in cardiometabolic disease prevention [J]. Biochimie 169:121–132
86. Gutiérrez-Méndez N, Chavez-Garay DR, Leal-Ramos MY (2022) Lecithins: a comprehensive review of their properties and their use in formulating microemulsions [J]. J Food Biochem 46(7):e14157
87. Spernath A, Aserin A, Garti N (2006) Fully dilutable microemulsions embedded with phospholipids and stabilized by short-chain organic acids and polyols [J]. J Colloid Interface Sci 299(2):900–909
88. List GR (2015) Chapter 1: Soybean lecithin: food, industrial uses, and other applications [M]. In: Ahmad MU, Xu X (eds) Polar lipids. Elsevier, Urbana, 1–33
89. Pernetti M, van Malssen K, Kalnin D et al (2007) Structuring edible oil with lecithin and sorbitan tri-stearate [J]. Food Hydrocoll 21(5–6):855–861
90. Pan Y, Tikekar RV, Nitin N (2013) Effect of antioxidant properties of lecithin emulsifier on oxidative stability of encapsulated bioactive compounds [J]. Int J Pharm 450(1–2):129–137
91. Cobb M, Turkki P, Linscheer W et al (1980) Lecithin supplementation in healthy volunteers effect on cholesterol esterification and plasma, and bile lipids [J]. Ann Nutr Metab 24(4): 228–237
92. Okuro PK, Malfatti-Gasperini AA, Vicente AA et al (2018) Lecithin and phytosterols-based mixtures as hybrid structuring agents in different organic phases [J]. Food Res Int 111:168–177
93. Kumar R, Katare OP (2005) Lecithin organogels as a potential phospholipid-structured system for topical drug delivery: a review [J]. AAPS PharmSciTech 6(2):E298–E310
94. Bodennec M, Guo Q, Rousseau D (2016) Molecular and microstructural characterization of lecithin-based oleogels made with vegetable oil [J]. RSC Adv 6(53):47373–47381
95. Yadav I, Kasiviswanathan U, Soni C et al (2017) Stearic acid modified stearyl alcohol oleogel: analysis of the thermal, mechanical and drug release properties [J]. J Surfactant Deterg 20(4): 851–861
96. Bot A, den Adel R, Roijers EC (2008) Fibrils of γ-oryzanol + β-sitosterol in edible oil organogels [J]. J Am Oil Chem Soc 85(12):1127–1134
97. Nikiforidis CV, Scholten E (2014) Self-assemblies of lecithin and α-tocopherol as gelators of lipid material [J]. RSC Adv 4(5):2466–2473
98. Han LJ, Li L, Zhao L et al (2013) Rheological properties of organogels developed by sitosterol and lecithin [J]. Food Res Int 53(1):42–48
99. Okuro PK, Martins AJ, Vicente AA et al (2020) Perspective on oleogelator mixtures, structure design and behaviour towards digestibility of oleogels [J]. Curr Opin Food Sci 35:27–35

100. Aguilar-Zárate M, De la Peña-Gil A, Álvarez-Mitre FM et al (2019) Vegetable and mineral oil organogels based on monoglyceride and lecithin mixtures [J]. Food Biophys 14(3):326–345
101. Tamura T, Ichikawa M (1997) Effect of lecithin on organogel formation of 12-hydroxystearic acid [J]. J Am Oil Chem Soc 74(5):491–495
102. Gaudino N, Ghazani SM, Clark S et al (2019) Development of lecithin and stearic acid based oleogels and oleogel emulsions for edible semisolid applications [J]. Food Res Int 116:79–89
103. Aguilar-Zárate M, Macias-Rodriguez BA, Toro-Vazquez JF et al (2019) Engineering rheological properties of edible oleogels with ethylcellulose and lecithin [J]. Carbohydr Polym 205: 98–105
104. Hu Z, He B, Ma L et al (2017) Recent advances in ergosterol biosynthesis and regulation mechanisms in saccharomyces cerevisiae [J]. Indian J Microbiol 57(3):270–277
105. Du Y, Fu X, Chu Y et al (2022) Biosynthesis and the roles of plant sterols in development and stress responses [J]. Int J Mol Sci 23(4):2332
106. Weihrauch JL (1978) Sterol content of foods of plant origin [J]. J Am Diet Assoc 73(1):39–47
107. Gupta E (2020) Β-sitosterol: predominant phytosterol of therapeutic potential [M]. In: Mishra P, Mishra RR, Adetunji CO (eds) Innovations in food technology. Springer, Singapore, pp 465–477
108. Pinto TC, Martins AJ, Pastrana L et al (2022) Water-in-oleogel emulsion based on γ-oryzanol and phytosterol mixtures: challenges and its potential use for the delivery of bioactives [J]. J Am Oil Chem Soc 99(11):1045–1053
109. Martins AJ, Cerqueira MA, Pastrana LM et al (2019) Sterol-based oleogels' characterization envisioning food applications [J]. J Sci Food Agric 99(7):3318–3325
110. Matheson AB, Koutsos V, Dalkas G et al (2017) Microstructure of β-sitosterol:Γ-oryzanol edible organogels [J]. Langmuir 33(18):4537–4542
111. Bot A, Adel RD, Roijers EC et al (2009) Effect of sterol type on structure of tubules in sterol + γ-oryzanol-based organogels [J]. Food Biophys 4(4):266–272
112. Huang X, Raghavan SR, Terech P et al (2006) Distinct kinetic pathways generate organogel networks with contrasting fractality and thixotropic properties [J]. J Am Chem Soc 128(47): 15341–15352
113. Metin S, Hartel RW (1998) Thermal analysis of isothermal crystallization kinetics in blends of cocoa butter with milk fat or milk fat fractions [J]. J Am Oil Chem Soc 75(11):1617–1624
114. Wright AJ, Hartel RW, Narine SS et al (2000) The effect of minor components on milk fat crystallization [J]. J Am Oil Chem Soc 77(5):463–475
115. Sawalha H, Venema P, Bot A et al (2015) The phase behavior of γ-oryzanol and β-sitosterol in edible oil [J]. J Am Oil Chem Soc 92(11–12):1651–1659
116. Alhasawi FM, Rogers M (2013) Ternary phase diagram of β-sitosterol–γ-oryzanol–canola oil [J]. J Am Oil Chem Soc 90(10):1533–1540
117. Li L, Liu G (2019) Corn oil-based oleogels with different gelation mechanisms as novel cocoa butter alternatives in dark chocolate [J]. J Food Eng 263:114–122
118. Qu K, Qiu H, Zhang H et al (2022) Characterization of physically stable oleogels transporting active substances rich in resveratrol [J]. Food Biosci 49:101830
119. Li L, Wan W, Cheng W et al (2019) Oxidatively stable curcumin-loaded oleogels structured by β-sitosterol and lecithin: physical characteristics and release behaviour in vitro [J]. Int J Food Sci Technol 54(7):2502–2510
120. Yuan T, Zeng J, Wang B et al (2021) Pickering emulsion stabilized by cellulosic fibers: morphological properties-interfacial stabilization-rheological behavior relationships [J]. Carbohydr Polym 269:118339
121. David A, David M, Lesniarek P et al (2021) Oleogelation of rapeseed oil with cellulose fibers as an innovative strategy for palm oil substitution in chocolate spreads [J]. J Food Eng 292: 110315
122. Patel A, Mankoč B, Sintang MB et al (2015) Fumed silica-based organogels and 'aqueous-organic' bigels [J]. RSC Adv 5(13):9703–9708

123. Hurd AJ, Flower WL (1988) In situ growth and structure of fractal silica aggregates in a flame [J]. J Colloid Interface Sci 122(1):178–192
124. Barthel H, Rösch L, Weis J (2005) Fumed silica - production, properties, and applications [M]. In: Organosilicon chemistry set, pp 761–778
125. Tsantilis S, Pratsinis SE (2004) Soft- and hard-agglomerate aerosols made at high temperatures [J]. Langmuir 20(14):5933–5939
126. Ranka M, Varkey N, Ramakrishnan S et al (2015) Impact of small changes in particle surface chemistry for unentangled polymer nanocomposites [J]. Soft Matter 11(8):1634–1645
127. Hooper JB, Schweizer KS, Desai TG et al (2004) Structure, surface excess and effective interactions in polymer nanocomposite melts and concentrated solutions [J]. J Chem Phys 121(14):6986–6997
128. Raghavan SR, Walls HJ, Khan SA (2000) Rheology of silica dispersions in organic liquids: new evidence for solvation forces dictated by hydrogen bonding [J]. Langmuir 16(21): 7920–7930
129. Benitez R, Contreras S, Goldfarb J (1971) Dispersions of methylated silica in mixed solvents [J]. J Colloid Interface Sci 36(1):146–150
130. Chauhan RR, Dullens RP, Velikov KP et al (2017) Exploring concentration, surface area and surface chemistry effects of colloidal aggregates on fat crystal networks [J]. RSC Adv 7(46): 28780–28787
131. Chauhan RR, Dullens RP, Velikov KP et al (2017) The effect of colloidal aggregates on fat crystal networks [J]. Food Funct 8(1):352–359

Chapter 9
Vegetable Waxes as Multicomponent Gelator Systems

Jorge F. Toro-Vazquez, Mayra Aguilar-Zárate, and Miriam A. Charó-Alonso

Abbreviations

AE	Asymmetrical ester
BW	Berry wax
C16	Hexadecanoic acid or palmitic acid
C18	Octadecanoic acid or stearic acid
C20	Eicosanoic acid or arachidic acid
C22	Docosanoic acid or behenic acid
C24	Tetracosanoic acid or lignoceric acid
C26	Hexacosanoic acid
C28	Octacosanoic acid
CO	Coconut oil
CRW	Carnauba wax
CW	Candelilla wax
Da	Daltons, molecular mass unit
DSC	Differential scanning calorimetry
FHS	Fully hydrogenated soybean oil
G^*	Complex modulus
G'	Elastic modulus
G''	Viscous modulus
MGC	Minimum gelator concentration
RBW	Rice bran wax
SE	Symmetrical ester
SFO	Safflower oil high in triolein

J. F. Toro-Vazquez (✉) · M. Aguilar-Zárate · M. A. Charó-Alonso
Facultad de Ciencias Químicas-CIEP, Laboratorio de Fisicoquímica de Alimentos, Universidad Autónoma de San Luis Potosí, San Luis, Mexico
e-mail: toro@uaslp.mx; mayra.aguilar@uaslp.mx; miriamto@uaslp.mx

C. Andrea Palla, F. Valoppi (eds.), *Advances in Oleogel Development, Characterization, and Nutritional Aspects*,
https://doi.org/10.1007/978-3-031-46831-5_9

SFW	Safflower wax
TAGs	Triacylglycerides or triglycerides
$T_{Cr\ onset}$	Temperature for the onset of crystallization determined by DSC
W/O	Water in oil
w/w	Weight per weight percentage

9.1 Introduction

An organogel, also known as molecular gel, is a particular class of gel composed of an organic solvent physically trapped within a supramolecular structure developed through the spontaneous self-assembly of low molecular weight (<3000 Da) molecules (i.e., gelator molecules). The molecular self-assembly usually occurs at temperatures below the gelator's solubility limit in the solvent. Different studies have shown that the molecular self-assembly of edible and non-edible gelator molecules also might occur when the organic solvent is vegetable or mineral oil. In those cases, the gels developed are usually referred to as oleogels. The potential use of edible oleogels resides in their use as replacement of the solid phase provided by saturated and *trans* fats to different food systems (i.e., margarines, confectionery and table spreads, shortenings). Nowadays, the food industry faces the need to develop thermodynamically stable systems with functional properties accepted by consumers without the use of *trans* fatty acids and limited concentration of saturated fatty acids. This mainly because there is a well-documented effect of *trans* fats consumption with an increased risk in developing cardiovascular heart disease, even above the risk associated with the consumption of saturated fat [1]. The oleogelation is a useful and novel alternative to structure vegetable oils, providing them with elastic properties without the use of *trans* and/or saturated fats [2–4]. One of the most promising alternatives for the development of edible oleogels is the use of natural vegetable waxes [5].

Advances in oleogel research have identified several vegetable waxes that can structure large volumes of vegetable oils, producing edible oleogels with similar functionality to *trans* and saturated fats. Vegetable waxes have excellent gelation properties providing structure to vegetable oils with as little as 1–4% (w/w), show high oil binding capacity, and have reversible thermomechanical properties. Consequently, vegetal waxes have gained considerable attention in the development of oleogels. Additionally, the major cosmetic companies are moving away from animal-based (e.g., tallow) and petroleum-derived materials (e.g., petroleum jelly), looking now for renewable and functional vegetable-based materials. We can also develop vegetable wax-based oleogels with useful and novel functional properties for the cosmetics industry [6–8]. In this chapter, we will discuss the gelling capacity of some commercial vegetable waxes considering them naturally multicomponent gelling systems. Within this framework, we will discuss examples associated with the molecular interactions occurring between wax components that determine the gelling capacity of some vegetable waxes. Finally, we will discuss information regarding the particular interactions of vegetable waxes with triacylglycerides in the development of oleogels. Particular remarks will be made to recent results obtained by our group in the use of candelilla wax (CW) for the development of oleogels alone and in interaction with saturated triglycerides.

9.2 Vegetable Waxes as Naturally Multicomponent and Polyfunctional Systems

For a given temperature of a gelator in an organic solvent solution, the balance between the gelator–gelator and the gelator–organic solvent (i.e., the vegetable oil) interactions determine whether the gelator would be soluble (i.e., no gelation), not soluble in the oil (i.e., the gelator precipitates), or able of developing a supramolecular structure that physically traps the liquid phase forming a self-supporting structure. The development of the self-supporting structure responsible for the oleogel properties (i.e., oil-binding, thermorevesibility, and viscoelasticity) requires a minimum gelator concentration. From this concentration and above, the supramolecular structure developed by the gelator has a large solid-to-liquid interfacial area that results in a phase change of the system, i.e., a gel structure is formed. Within this context, it is important to note that when using pure molecules (i.e., 12-hydroxystearic acid, glyceryl monostearate), the minimum gelator concentration (MGC) required to develop a self-supporting structure depends, essentially, on the gelators' molecular structure and its functional groups [9]. However, in gelators naturally formed by a mixture of molecules (i.e., vegetable waxes) or gelator mixtures intentionally designed, the MGC resulting in the formation of a self-supporting structure depends on the capability of each molecule to develop independent self-assembly structures, or co-assembled organizations among them or with native components present in the oil. This is affected by the heat (i.e., temperature and cooling rate) and mass transfer conditions (i.e., static or extent of shear rate) applied to the system. The relevance of this resides in that the thermomechanical properties and thermodynamic stability of the oleogels developed by mixtures of gelator molecules, usually exceeds the ones achieved by the oleogels developed independently by each gelator [10–13]. However, most studies done using formulated mixtures of molecules just describe the final physical or functional properties of the hybrid oleogel (i.e., higher elasticity, improved stability for phase separation), but do not study the mechanisms and/or conditions that determine the synergistic interaction between the molecules that structure the oleogel [14–16]. Other studies fail to recognize the native multicomponent and polyfunctional character of the gelator system used. This is generally the case when vegetable waxes are used for the development of oleogels and other soft-matter systems like bi-gels (i.e., structures formed by a mixture of an oleogel and a hydrogel) and water-in-oil (W/O) emulsions [14, 17]. The microstructure of the bi-gels and W/O emulsions mimic the functionality of *trans* and saturated fats [18, 19], but the polyfunctional character of the vegetable waxes is usually overlooked. The functionality of the vegetable waxes is usually associated just with the components in major concentration (i.e., *n*-alkanes in candelilla wax; long chain esters in carnauba and rice waxes) disregarding the physical and/or chemical properties of other native components (i.e., free fatty acids and fatty alcohols, alcohols and esters of penta-cyclic triterpenoids, sterol esters). Some of these minor components also have self-assembly properties in organic solvents or vegetable oils [20, 21], others have surface-active properties and/or have functional groups that might hinder (i.e., antagonistic effect) or enhance (i.e., synergistic effect) the development of self-supporting structures [22].

9.3 Gelling Properties of Some Commercial Vegetable Waxes and Their Relationship with Wax Composition

The number of studies characterizing the thermomechanical and microstructural properties of edible (and non-edible) oleogels and related soft-matter systems developed with vegetable waxes had increased considerably during the last 10–15 years [17, 23–25]. The vegetable waxes have a natural appeal to the consumer and have commercial availability and, in several cases, "safe status" provided by regulatory agencies (i.e., the CW and the carnauba wax). Remarkably, in contrast with the saturated and *trans* fats that usually require the use of a concentration exceeding 30% to trap unsaturated vegetable oils, the formation of oleogels by vegetable wax requires concentrations lower than 5%. This behavior is associated with the large solid-to-liquid interfacial area resulting from the supramolecular structure developed through the molecular self-assembly of the gelator or mixture of gelators [26] (i.e., vegetable waxes). Within this context, Dassanayake et al. [27] established that at 20 °C just 0.5% of rice bran wax (RBW) was required to gel olive oil in 10.7 min, 2% of CW in 13.8 min, and 4% of carnauba wax (CRW) in 13.08 min. Lower vegetable wax concentrations did not gel the olive oil even after 2 days at 20 °C. The authors established the gelation time as the time until the system stopped flowing when manual handshaking was applied to test tubes containing the oil–gelator solution [27]. In agreement with these results, we recently determined that the MGC required to gel safflower oil high in triolein was just 1.0% for RBW, 1.5% for CW, and 5% for CRW (Toro-Vazquez, Aguilar-Zárate, and Charó-Alonso, unpublished results). These MGC were established after heating (90 °C for 30 min) 3 g of gelator oil solutions in glass vials (28 mm i.d., 57 mm height), followed by storage for 3 h at 15 °C, and then inverting the vials during 1 h at 15 °C to then visually determining the flow (no gel formation) or no-flow (gel formation) of the system. Under these conditions, the solid content (i.e., crystal mass) in the oleogels was 0.89% (± 0.10%), 1.27% (± 0.15%), and 2.62% (± 0.17%) for the RBW, the CW, and the CRW oleogels, respectively. These results pointed out that, independent of the methodology used to determine the oil gelling, the RBW shows the higher oil gelling capacity followed by CW and then CRW. The high gelling capacity of the RBW is just comparable with the gelling properties of sunflower wax (SFW), with MGC as low as 0.5–1.0% in soybean oil, salad oil, olive oil, tea seed oil, and sunflower oil [2, 28]. However, it is important to note that in the case of the RBW, there are reports showing through rheological measurements that in 5% RBW solutions in rice bran oil, no gelation occurred even after 48 h storage at 5 °C [29]. From our own studies using the inverted vial methodology, we had found several RBW batches with MGC as high as 3% in soybean oil at 25 °C (unpublished results). Within this framework, we had to consider, as Scharfe, Niksch, and Flöter [30] pointed out, the effect that TAGs composition (i.e., type of vegetable oil) and minor polar components of the vegetable oil have on the crystal habit and gelling properties of vegetable waxes. Additionally, the agronomic and environmental conditions prevalent during the growing of the vegetal source of the wax (i.e., carnauba palms, rice, candelilla shrubs, sunflower seed), and variations in the processing conditions used to extract and refine the wax, ought to affect the vegetable wax's composition and, subsequently, the vegetable waxes' gelling performance. Using SFW, RBW, CW, CRW, beeswax (BW), and berry wax (BEW) to gel rice bran oil, Doan et al. [31] showed that independent of the type of vegetable wax, the oleogel strength evaluated through rheological measurements (i.e., oscillation yield stress and the storage modulus) had a negative correlation with the wax's concentration of free fatty alcohols and a positive relationship with the *n*-alkanes, free fatty acids, and long chain esters concentrations. The results of this study also showed that although the presence of higher concentrations of long chain esters increased the oleogels' strength and the initial flow yield stress, the gels were more brittle. On the other hand, the wax's concentration of *n*-alkanes had a stronger correlation with the oleogels yield stress and storage modulus than the free fatty acids and long chain esters concentration [31]. Evidently, addressing the role of wax components or wax fractions in determining the gelling capacity of vegetable waxes, might support further investigations that eventually establish the relationship between the effect of growing conditions with the vegetable wax composition and their corresponding gelling capacity. Vegetable waxes are complex mixtures of components and establishing the composition that optimizes their gelling capacity is a difficult and long-term task.

9.3.1 The Multicomponent Gelator Character of Vegetable Waxes

Given the high content of long-chain esters of the RBW, and SFW, the gelling mechanism of these chemically homogeneous vegetable waxes and corresponding gelling capacity has been trying to establish by studying the self-assembly of pure long chain saturated esters and their mixtures. Within this context,

Avendaño-Vásquez et al. [32] studied the relationship between the thermomechanical properties and the crystal microstructure developed by saturated symmetrical esters (SE: 14:14, 16:16, 18:18, 20:20, and 22:22) and asymmetrical esters (AE: 18:14, 18:16, 18:20, 18:22) in both the neat state and in 3% alkyl ester in high oleic safflower oil. The X-ray and microscopy analysis done in this investigation showed that in the neat state and in the oleogels developed, the SE self-assembled developing plate crystals, while the AE developed acicular crystals. There were three relevant conclusions from this study: (1) The melting temperature of the AE and the SE in the neat state and in the 3% oleogels increased linearly as the carbon number of the esters increased; in contrast, the molar heat of melting of the esters in the neat state and in the 3% oleogels increased quadratically as the carbon number increased, (2) the AE developed oleogels with higher elasticity (G') than the SE oleogels, and (3) as the ester carbon number increased in the AE or in the SE, the G' of the corresponding oleogels decreased while crystal size increased [32]. In a recent study that involved the use of pure alkyl esters with carbon number from 30 to 46, the authors concluded that despite the microstructural changes observed in 10% oleogels developed in medium chain TAGs using different cooling rates (i.e., 0.8, 5, 10 °C/min), alkyl ester with smaller total carbon numbers yielded oleogels with higher rigidity (i.e., G^*) than the oleogels developed with alkyl esters with higher total carbon. These results agree with the conclusions obtained previously by Avendaño-Vásquez et al. [32]. Within this framework, other studies done by Brykczynski, Wettlaufer, and Flöter [33] involving the development of oleogels by 10% binary mixtures of alkyl esters (i.e., 1:2, 1:1, 2:1 wt:wt ratios) in medium-chained triacylglycerides showed that chain length differences of 12 and 16 carbons between the alkyl esters resulted in separate crystallization of the esters (i.e., no mixed crystals). Independent of the alkyl esters ratio used this crystallization behavior was observed in the oleogels formulated with binary mixtures of palmityl myristate (16:14) and behenyl arachidate (22:20) and with 16:14 and behenyl lignocerate (22:24). Evidently, the 16:14–22:20 mixture provided the 12 total carbon difference, and the 16:14–22:24 mixture provided the 16 total carbon difference. In contrast, developing oleogels with alkyl esters differing in just two carbons (i.e., 18:18–20:18) resulted in the formation of mixed crystals. Nevertheless, the authors pointed out that the rheological measurements indicated that the strongest oleogels (i.e., highest complex modulus, G*) were obtained by mixtures of alkyls esters that crystallize independently (i.e., 16:14–22:20 and 16:14–22:24). These results were explained considering the difference in crystallization temperatures between the alkyl esters present in the binary mixture that resulted in a sequential crystallization (see Sect. 9.4) and, subsequently, in a sintering effect that reinforced the three-dimensional crystal network of the oleogel [33]. Although RBW and SFW are considered chemically homogeneous vegetable waxes, the studies discussed above show that their gelling capacity and oleogel properties are not just determined by their alkyl ester content. When studying the RBW and SFW gelling properties, besides the ester content we must consider the alkyl ester structure (i.e., symmetry or asymmetry) and carbon length difference between the major alkyl ester components of the wax. Therefore, despite the high extent of compositional homogeneity to

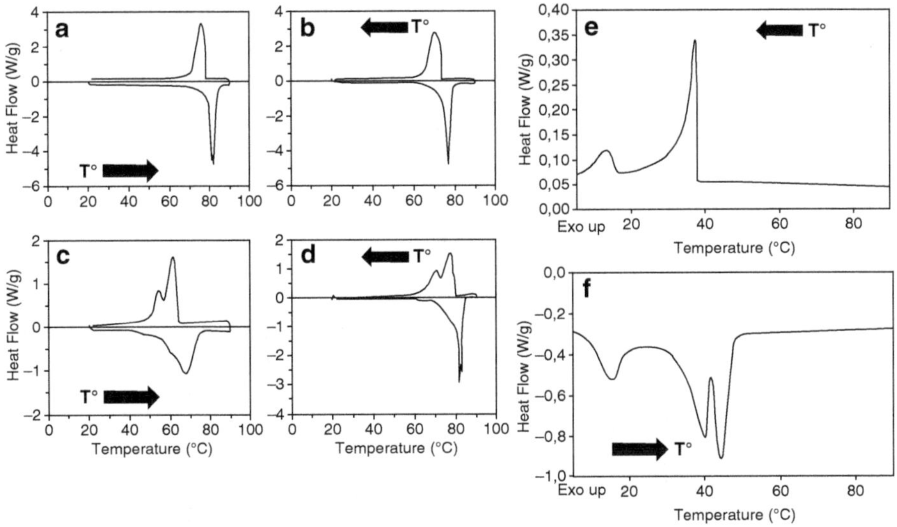

Fig. 9.1 Crystallization and melting thermograms of neat vegetable waxes obtained by DSC. (**a**) rice bran wax; (**b**) sunflower wax; (**c**) candelilla wax; (**d**) carnauba wax; (**e** and **f**) correspond to cooling and heating thermograms, respectively, of berry wax. The thermograms shown in (**a**) to (**d**) were obtained in a DSC Q-2000 (TA Instruments, New Castle, DL, USA) by initially storing the samples for 24 h at room temperature and then heating them from 20 to 90 °C (5 °C/min) followed by cooling from 90 to 20 °C (5 °C/min) (thermograms slightly modified from [34]; with permission from John Wiley & Sons, Inc.). The thermograms shown in (**e**) and (**f**) were obtained in a DSC Q-1000 (TA Instruments, New Castle, DL, USA) by initially cooling the vegetable wax from 90 to 0 °C (5 °C/min) maintaining the sample for 10 min at 0 °C and then heating to 90 °C (5 °C/min) (thermograms slightly modified from [35]; with permission from the Royal Society of Chemistry)

understand better, and eventually control the crystallization/melting behavior and rheological properties of the RBW and SFW oleogels, these vegetable waxes must be considered multicomponent gelator systems.

In contrast with the RBW and SFW, the CW is constituted mainly by *n*-alkanes (72.92% ± 2.23%) and, in lower concentration by long chain esters (15.76% ± 0.35%). In the same way, the CRW is constituted by 62.05% (± 3.03%) of long chain esters and 30.74% (± 2.48%) of long chain alcohols. The cooling thermograms of the neat CW and CRW obtained by DSC showed just one major exotherm with peak temperatures at ≈67 °C (Fig. 9.1c) and ≈83 °C (Fig. 9.1d), respectively. However, the exotherms showed a larger spread between the onset and end of crystallization than the corresponding exotherms for RBW (Fig. 9.1a) and SFW (Fig. 9.1b) [34]. This crystallization behavior was associated with the higher compositional heterogeneity of CW and CRW, in contrast with the lower compositional heterogeneity of RBW and SFW (i.e., less than 6% of free alkyl acids and free alkyl alcohols) [28, 31]. The compositional effect on the thermal behavior of these vegetal waxes was more evident when we compare the corresponding melting thermograms (Fig. 9.1) [34]. Within this context, based on the results above discussed, the vegetable waxes with lower compositional heterogeneity have higher gelling capacity than the CW and CRW. Consequently, their corresponding MGC (see discussion of Sect. 9.3) indicated that the vegetable waxes with high long chain ester content and higher homogeneous composition are more efficient gelators than CW and CRW. However, it is important to note that although CRW has higher concentration of long chain esters than the CW, the CRW requires higher wax concentration (i.e., higher MGC) than the CW to develop oleogels. We explained this behavior considering that the self-assembly of the long chain esters is mainly associated with the structural and self-assembly properties of the alkyl chains of the esterified fatty alcohols and fatty acids [32, 33]. Therefore, it is easy to envision that the alkyl chains of the long chain esters and the hydrocarbon backbone of the *n*-alkanes, both main components of the CW, could easily interact through Van der Waal forces resulting in efficient mixed molecular packing. In contrast, the interactions between the major components of CRW, i.e., long chain esters and long chain alcohols, might be limited. Given the polar character of the primary -OH group of the alkyl alcohols, in an organic solvent the alkyl alcohols self-assembly occurs through the development of linear chains of the alcohols held together through a zig-zag pattern of strong hydrogen bonding between neighboring -OH groups and by dispersive interactions between the alkyl tails [36]. This alkyl chain packing occurs particularly in those alkyl alcohols with carbon number higher than 30 [36], as those present in CRW. Within this context, we consider that in vegetable waxes as CRW, the alkyl alcohol-alkyl alcohol interaction is favored compromising the alkyl alcohol-alkyl ester and alkyl alcohol-vegetable oil interactions, and subsequently limiting the CRW gelling capacity. The overall result is that the CRW has lower gelling capacity (i.e., higher MGC) than the CW. This analysis again points out the multicomponent gelator character of the vegetable waxes and the relevant effect of the molecular interactions occurring between their native components in determining the vegetable waxes' gelling capacity.

9.4 Fundament of Sequential Crystallization in the Development of Vegetable Wax-Based Oleogels

Some of the vegetable waxes already in the commercial market with high gelling capacity, particularly the SFW and the RBW, are characterized by high content of long chain saturated alkyl esters (i.e., greater than 93%) [37]. Thus, the SFW and RBW are considered mono-component waxes with low compositional heterogeneity, since they are essentially constituted by long chain esters (i.e., 96.23% ± 0.19% and 93.49% ± 2.63%, respectively) and less than 6% of free alkyl acids and free alkyl alcohols. In these vegetable waxes the long chain esters are mainly constituted by arachidic acid (C20), behenic acid (C22), lignoceric acid (C24) esterified with 1-tetracosanol (C24), 1-hexacosanol (C26), and 1-octacosanol (C28) [31]. When neat RBW and SFW are analyzed by DSC given their large concentration of long-chain esters the cooling thermogram shows just one single narrow crystallization exotherm (Fig. 9.1a and Fig.9.1b, respectively), associated with the development of large needle-like crystals that crystallize at ≈82 and ≈77 °C, respectively [34]. Another vegetable wax considered a mono-component wax is the berry wax extracted from the fruits of the *Rhus verniciflua* tree. This vegetable wax is constituted by more than 95% of fatty acids, essentially palmitic acid (≈79%; C16) and stearic acid (≈13%; C18), and less than 5% of fatty alcohols esterified in diglyceride and TAGs molecules [31]. The overall result is that berry wax show three crystallization exotherm (Fig. 9.1f) associated with its major components: 36.40 °C (± 0.16 °C) for the free fatty acids, 35.64 °C (± 0.19 °C) for the diacylglycerides, and 14.25 °C (± 0.07 °C) for the TAGs [24]. As already indicated, the RBW, the SFW, and the berry wax are essentially mono-component waxes. Nevertheless, their crystallization behavior and the oleogels thermomechanical properties and microstructure are different. For instance, Doan et al. [24] showed that the kinetics of crystallization and subsequent oil gelling took place in a shorter time and at lower temperature in 5% berry wax vegetable oil solutions than in 5% SFW solutions. Additionally, these authors noted that SFW crystallized at high temperature developing platelet crystals resulting in stronger oleogels than the berry wax oleogels. The berry wax crystallized at low temperature developing oleogels structured by small crystals [24]. Since the RBW has wax ester composition similar to SFW, the RBW also develops large platelet crystals that in the neat state and in oil solution crystallize and melt at lower temperatures than the SFW, but still at higher temperature than the berry wax [31, 33]. The formulation of oleogels with binary mixtures of vegetable waxes, one having high crystallization temperature (that results in oleogels with high melting temperatures like SFW and RBW oleogels) and other with low crystallization temperature (that forms oleogels with low melting temperatures like the berry wax oleogels), could be useful when developing structured oil systems. Within this framework, Tavenier et al. [35] developed oleogels using binary mixtures of SFW: berry wax and RBW:berry wax in rice bran oil, and observed that a certain RBW: berry wax (i.e., 30:70 and 20:80) and SFW:berry wax (90:10, 80:20, 70:30, 60:40) ratios, the resulting 5.0% (wt/wt) oleogels were more cohesive and harder than the oleogels obtained just with the use of SFW or RBW [35]. The authors concluded that the small crystals developed at low temperatures by the berry wax's components reinforced the crystal network microstructure, tentatively by forming solid crystal bridges (i.e., sintering effect) with the crystals developed at higher crystallization temperature by the SFW or the RBW. This is because of the sequential crystallization occurring during cooling of the binary mixture of vegetable waxes. Thus, the crystal structure initially developed by the wax with higher crystallization temperature could be reinforced through the formation of solid crystal bridges developed by the low crystallization temperature wax. The use of sequential crystallization using mixtures of waxes with high and low crystallization temperatures is a useful tool to tailor the thermomechanical properties of vegetable wax-based oleogels.

9.5 Interactions of Vegetable Waxes with TAGs and Other Edible Gelators

Butter, margarine, and table spreads are the most preferred spread products worldwide. However, to have the functional properties required by the consumers these products must use high content of saturated fatty acids and/or *trans* fatty acids. As already indicated, there is a well-documented increased risk of developing cardiovascular heart diseases associated with the consumption of saturated and, particularly, *trans* fats. Hwang et al. [38] and Chopin-Doroteo et al. [10] have pointed out that although the oleogels are a useful alternative to structure vegetable oils, these systems have limited rheological properties to mimic the properties required by the consumer in margarines and table spreads. Within this context, the interactions between the vegetable wax and the TAGs during the development of oleogels are of utmost importance to evaluate the potential use of vegetable wax-based oleogels in the formulation of low-saturated and *trans*-free margarines and spreads. Additionally, the vegetable wax-TAGs interaction may result in the development of oleogels using lower vegetable wax concentrations, thus diminishing the characteristic waxy flavor of vegetable wax oleogels. The relevance of the edible oleogels using vegetable waxes resides in their use in food systems as functional and organoleptically acceptable replacement of saturated and *trans* fats. Within this context, our group showed that the mixed oleogels developed with 3% CW and 1% tripalmitin had higher elasticity and yield stress than CW oleogels [10]. Given the structural similarities between the aliphatic chain of hentriacontane (i.e., the major *n*-alkane in CW) and the alkyl chains of the palmitic acid in the tripalmitin, and based on the X-ray and DSC analysis Chopin-Doroteo et al. [10] concluded that during cooling the tripalmitin co-crystallized in a *2L* packing through the orthorhombic organization of the hentriacontane (Fig. 9.2). This study also showed that after applying shear under particular time and temperature conditions during cooling, the 3% CW and the 3% CW-1% tripalmitin systems developed larger crystals (i.e., fiber-like and plate-like) forming a well-structured crystal network in comparison with the smaller crystals obtained under static conditions. In consequence, the 3% CW oleogels and particularly the 3% CW-1% tripalmitin oleogels had higher resistance to deformation and higher thixotropy behavior than the oleogels developed under static conditions [10]. Other studies using mixtures of 0.5–1.0% of SFW with anhydrous milk fat, a fat constituted mainly of saturated fatty acids ($\approx$63–70%), showed that the mixed oleogel had harder texture than the SFW oleogels [39, 40]. The results of these studies showed that the crystal network developed by the CW-tripalmitin and the SFW-anhydrous milk fat provides higher rheological and textural properties than the one developed just by the vegetable wax or the saturated TAGs.

Within the previous framework, we recently studied the gelling behavior of CW and RBW in safflower oil high in triolein (SFO; 63.3% $\pm$ 0.06% of triolein), in contrast with the CW and RBW gelling behavior in SFO added with 15% of coconut oil (SFO + 15% CO) (Toro-Vazquez, Aguilar-Zarate and Charó-Alonso,

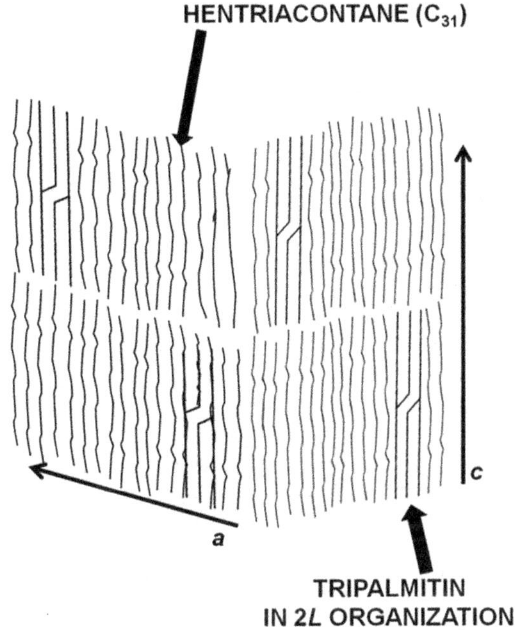

Fig. 9.2 Tentative organization of hentriacontane and tripalmitin co-crystals in the 3% candelilla wax -1% tripalmitin oleogels developed in sunflower high in triolein. The *a* and *c* axes of the co-crystal subcell are shown for orientation purposes (Figure slightly modified from [10], with permission from John Wiley & Sons)

unpublished results). The coconut oil (CO) has high concentration of TAGs esterified with medium-chain and long-chain saturated fatty acids (44.0–52.0% of lauric acid, 13.0–19.0% of myristic acid, 8.0–11.0% of palmitic acid, and 1.0–3.0% of stearic acid) [41]. Therefore, the addition of 15% of coconut oil to the SFO increased the concentration of saturated TAGs in the oil. Thus, the SFO and the SFO + 15% CO were considered model systems (i.e., edible organic solvents) useful to evaluate the CW and RBW gelling behavior in an oil predominantly with unsaturated TAGs composition (i.e., SFO) and in a vegetable oil with limited concentration of saturated TAGs (i.e., SFO + 15% CO). As discussed above, RBW is considered a chemically homogeneous wax with long alkyl esters as the major component, while CW has a more heterogeneous composition with *n*-alkanes (≈ 67%) and triterpenoid alcohols (≈ 23%) as major components and long alkyl esters (≈ 7.6%) as the minor component. Table 9.1 shows the gelling behavior of CW and RBX, and the textural parameters of the corresponding oleogels developed in SFO and SFO + 15% CO. The results showed that the CW and the RBW had different gelling behavior. Thus, the MGC and the temperature for the onset of crystallization ($T_{Cr\ onset}$) of the CW were not affected by the presence of the saturated TAGs in the SFO. Additionally, the crystallization thermal profile for the CW was essentially the same in the SFO and the SFO + 15% CO (Table 9.1 and Fig. 9.3a). Nevertheless, the crystal mass developed by the CW in the SFO + 15% CO, measured as the solid content, was significantly higher than the one developed by the CW in the SFO (Table 9.1). The solid content of the CW oleogel in the SFO + 15% CO was even above the amount of solid that could be developed should full crystallization of the CW added

Table 9.1 Percentage of solid content and textural parameters of oleogels developed by vegetable waxes at the corresponding minimal gelling concentrations in safflower oil high in triolein (SFO) and SFO added with 15% of coconut oil (SFO + 15% CO) (Toro-Vazquez, Aguilar-Zárate, and Charó-Alonso, unpublished results)

	Candelilla wax		Rice bran wax	
	SFO	SFO + 15% CO	SFO	SFO + 15% CO
Minimal gelling concentration[a] (%)	1.50	1.50	1.00	3.00
$T_{Cr\ onset}$ (°C)	32.85[d, e] (0.21)	31.20[d, e] (0.00)	44.35[d, e] (1.20)	55.42[d, e] (0.18)
Solid content[b] (%)	1.27[d, e] (0.15)	2.20[d, e] (0.12)	0.89[d, e] (0.10)	5.63[d, e] (0.18)
Hardness[c] (g-force)	36.00[d, e] (3.00)	149.20[d, e] (6.00)	14.00[d, e] (0.00)	129.00[d, e] (7.00)
Adhesivity[c] (g-force)	35.60[d, e] (2.10)	74.00[d, e] (11.00)	8.00[d, e] (0.70)	64.20[d, e] (3.50)

[a]Minimal gelling concentration determined by the inverted vial test after 3 h at 15 °C ($n = 2$)
[b]Solid content determined in the oleogels by nuclear magnetic resonance (mq20 Minispec Bruker; Bruker Analytic; Rheinstetten, Alemania) after 24 h at 5 °C ($n = 2$). The values are reported as the mean and, between brackets, the corresponding standard deviation
[c]Determined in the oleogels with a texture analyzer TA-XT plus (Stable Microsystems Ltd; Surrey, England), after 24 h at 5 °C ($n = 5$). The values are reported as the mean and, between brackets, the corresponding standard deviation
[d]For the same parameter and gelling agent, a different first super index indicates that the effect of the addition of 15% of coconut oil to the safflower oil had a significant effect ($P < 0.05$)
[e]For the same parameter and vegetable oil, a different second super index indicates that the gelators had a different effect ($P < 0.05$) in the parameter measured in the oleogel

(i.e., 1.5%) occur during oleogelation. These results showed that in the SFO + 15% CO, at a temperature below the $T_{Cr\ onset}$ the saturated TAGs crystallized along with the CW. This could occur through the co-crystallization of the alkyl chains of trilaurin, the major TAG present in coconut oil (20–22%), with the hentriacontane of CW through a packing organization like the one occurring for the CW-TP oleogels (Fig. 9.2). The oleogels developed by CW in the SFO and SFO + 15% CO were structured in both cases by small crystals (i.e., microplatelets) of similar shape and size (Fig. 9.4). Nevertheless, the CW oleogels in the SFO + 15% CO also showed higher textural parameters than the CW oleogels in SFO, surely associated with the higher solid content present in the CW oleogels in the SFO + 15% CO (Table 9.1). In contrast with the CW, the presence of saturated TAGs of the coconut oil limited the development of the oleogel by the RBW, this is evaluated by the microstructure of the RBW oleogels in SFO and in SFO + 15% CO (Fig. 9.5). Thus, in the SFO + 15% CO the RBW and the saturated TAGs crystallized independently, developing a heterogeneous crystal network formed mainly by small crystals (microplatelets) that surrounded larger fibrillar crystals (Fig. 9.5b, d). The smaller crystals were associated to the RBW and the larger fibrillar crystals with the saturated TAGs. We concluded that in the SFO + 15% CO the saturated TAGs were not structurally compatible with the main components of the RBW (i.e., the alkyl esters). Subsequently, the amount of RBW required to develop the oleogel (i.e.,

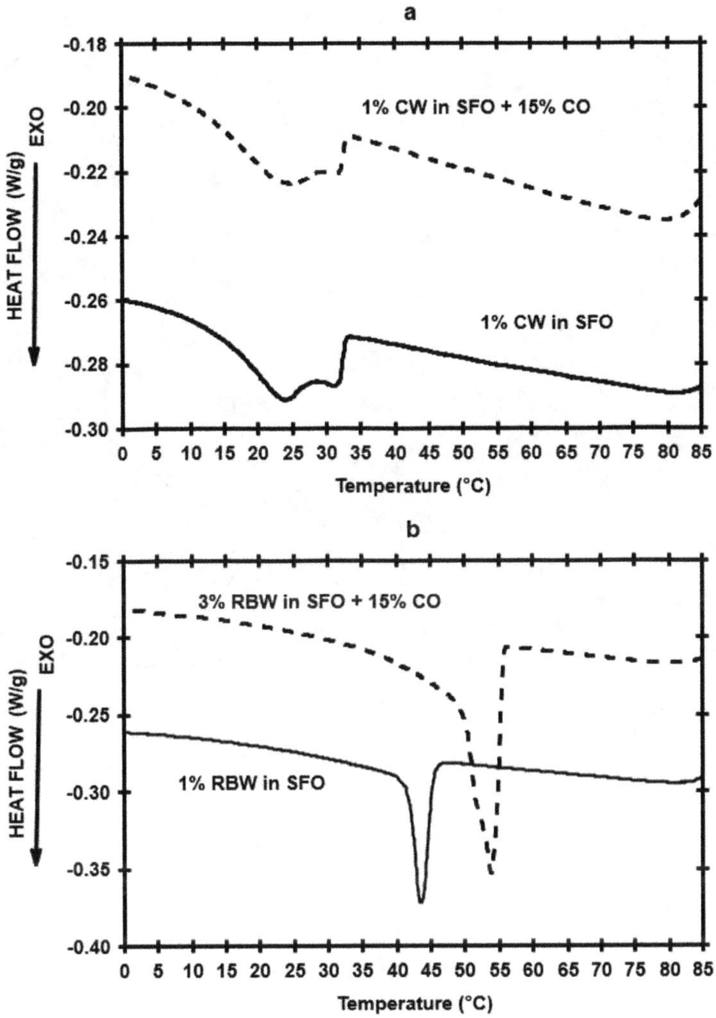

Fig. 9.3 Crystallization thermograms of candelilla wax (CW) and rice bran wax (RBW) solutions in safflower oil high in triolein (SFO) and SFO + 15% coconut oil (CO) at the corresponding minimal gelling concentration. The vegetable wax solution in the corresponding oil was initially heated at 90 °C for 20 min in the DSC, then cooled at 10 °C/min until achieving 0 °C. The crystallization thermograms were obtained in a DSC Discovery Series (TA Instruments, New Castle, DL, USA) (Toro-Vazquez, Aguilar-Zárate, and Charó-Alonso, unpublished results)

the MGC) in the SFO + 15% CO was almost three times higher than in the SFO (Table 9.1). Because of this, the RBW in the SFO + 15% CO crystallized at higher $T_{Cr\ onset}$ and the oleogels had higher solid content and higher textural parameters than the RBW oleogels in the SFO (Table 9.1, Fig. 9.3b). The melting thermograms corresponding to CW and RBW oleogels developed in each type of oil at the MGC are shown in Fig. 9.6. All these results show the effect of the compositional

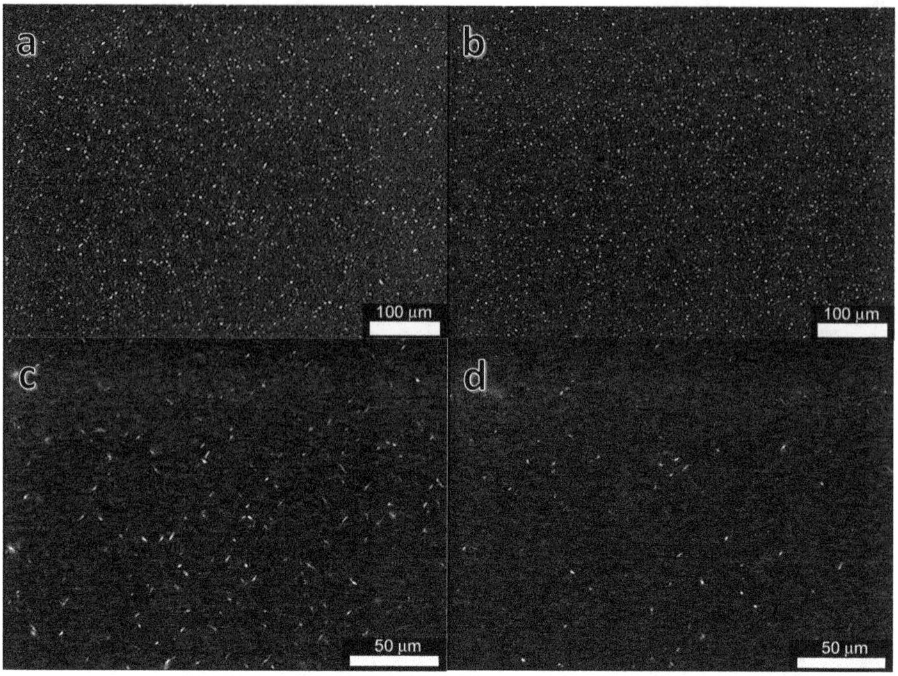

Fig. 9.4 Polarized light photographs of candelilla wax oleogels developed in safflower oil high in triolein (**a**, **c**; obtained at 20× and 50×, respectively) and in safflower oil high in triolein +15% coconut oil (**b**, **d** obtained at 20× and 50×, respectively). The oleogels were developed at the minimal gelling concentration indicated in Table 9.1. The photographs were obtained after 30 min at 5 °C using an Olympus BX5 polarized light microscope (Olympus Optical Co; Ltd., Tokyo, Japan) (Toro-Vazquez, Aguilar-Zárate, and Charó-Alonso, unpublished results)

differences between the CW and the RBW on their interaction with saturated TAGs and the subsequent gelling behavior of these waxes. It is relevant to note that the CW oleogels in SFO developed at the MGC, had a microstructure that resulted in oleogels with higher mechanical properties than the corresponding RBW oleogels. This behavior was equally evident when we compared the CW and RBW oleogels developed in the presence of low concentrations of saturated TAGs (i.e., in the SFO + 15% CO) (Table 9.1). Therefore, despite the higher solid content present in the RBW oleogels developed in the SFO and the SFO + 15% CO, the corresponding CW oleogels had higher textural parameters (Table 9.1).

In another study involving mixtures of CW (0–3%) and fully hydrogenated soybean oil (FHS; 0–15%) in sunflower oil high in triolein, the results showed that CW decreased the size of the spherulitic crystals develop by the FHS, a behavior that was associated with the nucleation of FHS on the surface of CW crystals [42]. The overall result was that in the CW-FHS systems, the elasticity and the structural recovery of the oleogels increased above the rheological properties obtained with the crystallized systems developed just with CW or FHS. The authors explained this considering that the structural recovery was hindered by the characteristic

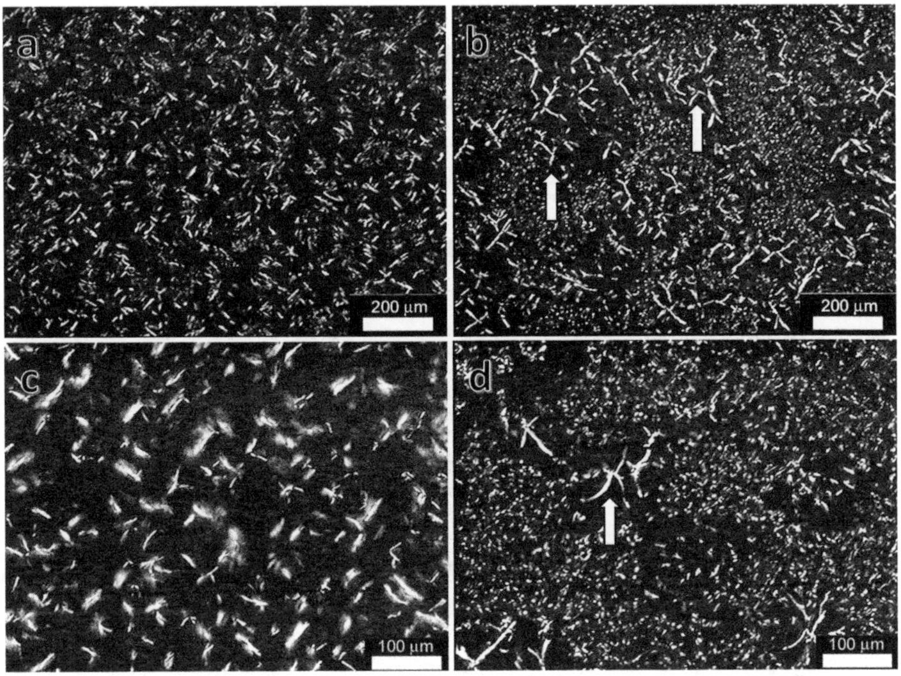

Fig. 9.5 Polarized ligh photographs of rice bran wax (RBW) oleogels developed in safflower oil high in triolein (**a**, **c**; obtained at 10× and 20×, respectively) and in safflower oil high in triolein +15% coconut oil (**b**, **d** obtained at 10× and 20×, respectively). The oleogels were developed at the minimal gelling concentration indicated in Table 9.1. The white arrow shown in **b** and **d** indicates the saturated TAGs crystals; in these panes, the RBW crystals appear as small birrefrigent platelets. The photographs were obtained after 30 min at 5 °C using an Olympus BX5 polarized light microscope (Olympus Optical Co; Ltd., Tokyo, Japan) (Toro-Vazquez, Aguilar-Zárate, and Charó-Alonso, unpublished results)

microplatelets intertwining present in the CW oleogels and in CW-FHS oleogels with the higher CW proportion. In contrast, the elasticity and structural recovery increased as the crystals in the CW-FHS oleogel were smaller (i.e., at lower CW proportion) [42]. On the other hand, da Silva, Arellano, and Martini [43] studied the interaction between CW, monoacylglycerols, and a hard fat (i.e., saturated TAGs) to develop oleogels in soybean and high-oleic sunflower oils using a total structuring agents concentration of 5%. The 5% CW oleogels were the hardest, but when CW was partially replaced by monoglycerides (1%) and FHS (1%) while maintaining at 5% concentration of all structuring agents (i.e., CW 3%), the oleogels obtained (25 °C for 24 h) were softer but with improved elasticity and thixotropy than with 5% CW oleogels. The authors observed similar results even though the total content of the structuring agent increased to 6 or 7% [43]. All these results show that CW develops oleogels with useful thermomechanical properties alone or in interaction with saturated TAGs. When CW is in mixture with saturated TAGs during cooling some of the CW components (i.e., *n*-alkanes) interact with the alkyl chains of the

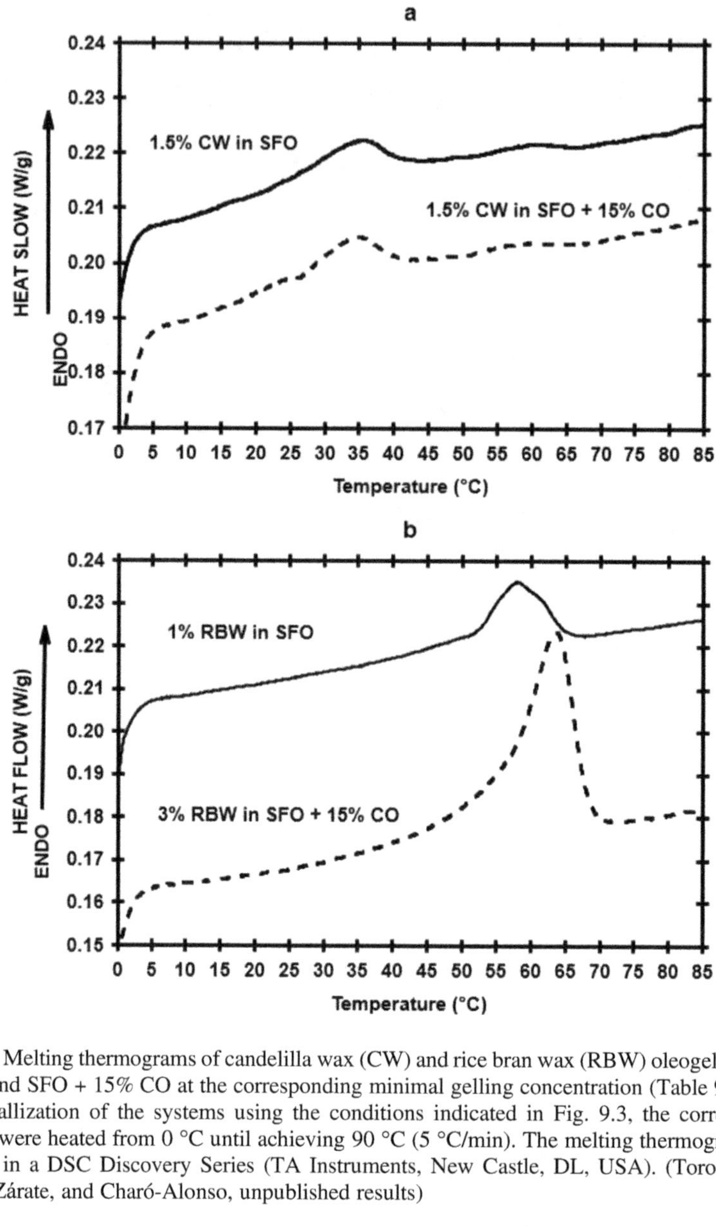

Fig. 9.6 Melting thermograms of candelilla wax (CW) and rice bran wax (RBW) oleogels obtained in SFO and SFO + 15% CO at the corresponding minimal gelling concentration (Table 9.1). After the crystallization of the systems using the conditions indicated in Fig. 9.3, the corresponding oleogels were heated from 0 °C until achieving 90 °C (5 °C/min). The melting thermograms were obtained in a DSC Discovery Series (TA Instruments, New Castle, DL, USA). (Toro-Vazquez, Aguilar-Zárate, and Charó-Alonso, unpublished results)

TAGs developing oleogels with thermal and structural recovery properties relevant for their potential use in products like whipped cream and laminated dough. Nevertheless, when CW oleogels in soybean oil (1–6% of CW) were incorporated at 80% (wt/wt) into a margarine formulation containing skim milk (19.46%), mono- and di-acylglycerides (0.2%), lecithin (0.1%), salt (0.10%), and other minor ingredients, the mixture showed phase separation after making the emulsion by shearing at

3000 rpm for 7 min at 75 °C [38]. In this study, from the plant waxes evaluated (i.e., CW, RBW, and SFW), the SFW developed the oleogels and margarines with the highest firmness. Remarkably, even the margarine formulations using the 2–6% SFW oleogels had firmness similar to that of the margarines containing 18–30% of hydrogenated soybean oil [38]. This study contrasts with the results recently obtained in our laboratory with 0.75–3% CW oleogels in safflower oil high in triolein. These CW oleogels were used in the development of water-in-oil (W/O) emulsion at 25 °C (water:oil of 40:60, 50:50, 60:40) [44]. The microscopic and X-ray analysis showed that, at all CW concentrations and water:oil ratios used, the CW developed structured W/O emulsions stable for phase separation at least during 20 days at 25 °C. In particular, the W/O emulsions developed with 1.5% CW using 40:60 and 50:50 water:oil ratios showed similar rheological behavior that commercial mayonnaise and spreads, showing a percentage of recovery >80% after shearing. The difference between the CW behavior observed in the study of Hwang et al. [38] and our study might reside mainly in the temperature used to develop the emulsion. At 75 °C the CW components would be mainly soluble in the oil, a too high temperature to develop a structured oil phase that might stabilize the W/O emulsion. On the other hand, at 25 °C the CW components would be close to achieve full crystallization, thus providing a structured oil phase that stabilizes the W/O emulsion. Based on this study we concluded that CW is a poly-functional material able to develop structured W/O emulsions, where the stabilization of the water droplets in the oil phase occurs via the adsorption of surface-active CW's components (i.e., terpenoid alcohols) while the oil phase is gelled by another CW's component (i.e., *n*-alkanes). The overall result is the development of structured low-calories, *trans*-free highly stable W/O emulsion with elastic properties useful in the formulation of food systems [44].

9.6 Conclusions

The discussion of results obtained by different groups with oleogels developed by SFW, RBW, CRW, and CW show the multicomponent gelator character of the vegetable waxes, and the relevant effect of the molecular interactions occurring between their native components in determining the waxes' gelling capacity. Although some vegetable waxes are constituted essentially by just one type of compound (i.e., long chain alkyl esters in SFW and RBW), these waxes are gelator naturally formed by a complex mixture of components. Thus, in vegetable waxes like SFW and RBW, their gelling behavior and oleogel properties are not just determined by the wax ester content. Within this context, when determining the gelling capacity of SFW and RBW, the alkyl ester structure (i.e., symmetry or asymmetry) and the carbon length difference between the major ester components of the wax must be also considered. In other cases, minor native components present in the wax might also have self-assembly properties or surface-active properties. The molecular interaction of these components with the major components of the wax might hinder or enhance the development of the self-supporting structures. Additional factors to consider are the interactions between mixtures of vegetable waxes and with TAGs, particularly saturated TAGs. For instance, the use of mixtures of waxes, one with high and the other with low crystallization temperatures, can result in sequential crystallization during oleogelation. During this process, the crystal structure initially developed by the wax with higher crystallization temperature would be reinforced through the formation of solid crystal bridges (i.e., sintering effect) developed by the low crystallization temperature wax. In other cases, the molecular interaction between the vegetable wax (i.e., CW) and minor proportions of saturated TAGs can result in the development of co-crystals that develop three-dimensional crystal structures with improved mechanical properties. In both cases, the overall result is that the mixed oleogels have improved elasticity and structural recovery above the rheological properties obtained with the oleogels developed independently by each component. The results discussed show that the use of sequential crystallization and the interaction of some vegetable waxes with saturated TAGs are useful tools to tailor the thermomechanical properties of vegetable wax-based oleogels, favoring their use in the formulation of low-saturated fats and *trans*-free food systems (i.e., margarines, vegetable spreads, confectionary products, shortenings).

Acknowledgments The researches reported in this chapter as "unpublished results" were done with funds provided by the Consejo Nacional de Ciencia y Tecnología (CONACYT) through the grant CB-280981-2018.

References

1. Ascherio A, Katan MB, Zock PL, Stampfer MJ, Willett WC (1999) Trans fatty acids and coronary heart disease. N Engl J Med 340:1994–1998
2. Dassanayake LSK, Kodali DR, Ueno S (2011) Formation of oleogels based on edible lipid materials. Curr Opin Colloid Interface Sci 16:432–439. https://doi.org/10.1016/j.cocis.2011.05.005
3. Marangoni AG, Garti N (2018) Edible oleogels: structure and health implications, 2nd edn. Elsevier, AOCS Press and Academic, Urbana, IL
4. Rogers MA, Strober T, Bot A, Toro-Vazquez JF, Stortz T, Marangoni AG (2014) Edible oleogels in molecular gastronomy. Int J Gastron Food Sci 2:22–31. https://doi.org/10.1016/j.ijgfs.2014.05.001
5. Blake AI, Toro-Vazquez JF, Hwang H-S (2018) Wax oleogels. In: Edible oleogels, pp 133–171. https://doi.org/10.1016/B978-0-12-814270-7.00006-X
6. Ferrari V, Mondet J (2003) Care and/or make-up cosmetic composition structured with silicone polymers and organogelling agents, in rigid form. Patent number WO 2003/105788 A3, pp 8
7. Morales ME, Gallardo V, Clarés B, García MB, Ruiz MA (2010) Study and description of hydrogels and organogels as vehicles for cosmetic active ingredients. Int J Cosmet Sci 32:314–314. https://doi.org/10.1111/j.1468-2494.2010.00580_4.x
8. Perez-Nowak V (2011) Fluid cosmetic composition, useful for making up and/or caring the skin and/or lips, comprises polyester, non-volatile silicone oil, organogelling agent comprising

N-acylglutamides e.g. N-lauroylglutamic acid dibutylamide, and wax. Patent number 1154327, pp 1–74
9. Toro-Vazquez JF, Morales-Rueda J, Torres-Martínez A, Charó-Alonso MA, Mallia VA, Weiss RG (2013) Cooling rate effects on the microstructure, solid content, and rheological properties of organogels of amides derived from stearic and (R)-12- hydroxystearic acid in vegetable oil. Langmuir 29:7642–7654. https://doi.org/10.1021/la400809a
10. Chopin-Doroteo M, Morales-Rueda JA, Dibildox-Alvarado E, Charó-Alonso MA, de la Peña-Gil A, Toro-Vazquez JF (2011) The effect of shearing in the thermo-mechanical properties of Candelilla wax and Candelilla wax-tripalmitin organogels. Food Biophys 6:359–376. https://doi.org/10.1007/s11483-011-9212-5
11. Aguilar-Zárate M, De la Peña-Gil A, Álvarez-Mitre FM, Charó-Alonso MA, Toro-Vazquez JF (2019) Vegetable and mineral oil organogels based on monoglyceride and lecithin mixtures. Food Biophys 14(3):326–345. https://doi.org/10.1007/S11483-019-09583-1
12. Pernetti M, van Malssen K, Kalnin D, Flöter E (2007) Structuring edible oil with lecithin and sorbitan tri-stearate. Food Hydrocoll 21:855–861. https://doi.org/10.1016/j.foodhyd.2006.10.023
13. Gandolfo FG, Bot A, Flöter E (2004) Structuring of edible oils by long-chain FA, fatty alcohols, and their mixtures. J Am Oil Chem Soc 81:1–6. https://doi.org/10.1007/s11746-004-0851-5
14. Hwang HS, Winkler-Moser JK (2020) Properties of margarines prepared from soybean oil oleogels with mixtures of candelilla wax and beeswax. J Food Sci 85:3293–3302. https://doi.org/10.1111/1750-3841.15444
15. Kavya M, Udayarajan C, Fabra MJ, López-Rubio A, Nisha P (2022) Edible oleogels based on high molecular weight oleogelators and its prospects in food applications. pp 1–24. https://doi.org/10.1080/10408398.2022.2142195
16. Sivakanthan S, Fawzia S, Madhujith T, Karim A (2022) Synergistic effects of oleogelators in tailoring the properties of oleogels: a review. Compr Rev Food Sci Food Saf 21:3507–3539. https://doi.org/10.1111/1541-4337.12966
17. da Silva TLT, Barrera Arellano D, Martini S (2019) Effect of water addition on physical properties of emulsion gels. Food Biophys 14:30–40. https://doi.org/10.1007/s11483-018-9554-3
18. Shakeel A, Farooq U, Gabriele D, Marangoni AG, Lupi FR (2021) Bigels and multi-component organogels: an overview from rheological perspective. Food Hydrocoll 111:1–25. https://doi.org/10.1016/J.FOODHYD.2020.106190
19. Wijarnprecha K, de Vries A, Sonwai S, Rousseau D (2021) Water-in-oleogel emulsions—from structure design to functionality. Front Sustain Food Syst 4:1–6. https://doi.org/10.3389/fsufs.2020.566445
20. Turina A d V, Nolan MV, Zygadlo JA, Perillo MA (2006) Natural terpenes: self-assembly and membrane partitioning. Biophys Chem 122:101–113. https://doi.org/10.1016/j.bpc.2006.02.007
21. Gopal Bag B, Chandan Barai A (2020) Self-assembly of naturally occurring stigmasterol in liquids yielding a fibrillar network and gel. RCS Adv 10:4755–4762. https://doi.org/10.1039/c9ra10376g
22. Liu Y, Xia H, Guo S, Lu X, Zeng C (2022) Development and characterization of a novel naturally occurring pentacyclic triterpene self-stabilized pickering emulsion. Colloids Surf A Physicochem Eng Asp 634:127908. https://doi.org/10.1016/j.colsurfa.2021.127908
23. Martini S, Tan CY, Jana S (2015) Physical characterization of wax/oil crystalline networks. J Food Sci 80:C989–C997. https://doi.org/10.1111/1750-3841.12853
24. Doan CD, Tavernier I, Dona M, Sintang B, Danthine S, Van De Walle D, Rimaux T, Dewettinck K (2017) Crystallization and gelation behavior of low- and high melting waxes in rice bran oil: a case-study on berry wax and sunflower wax. Food Biophys 12:97–108. https://doi.org/10.1007/s11483-016-9467-y

25. Wettlaufer T, Brykczynski H, Flöter E (2022) Wax-based oleogels—properties in medium chain triglycerides and canola oil. Eur J Lipid Sci Technol 124:1–12. https://doi.org/10.1002/EJLT.202100114
26. Toro-Vazquez JF, Aguilar-Zárate M, López-Martinez A, Charó-Alonso M (2022) Chapter 8: Structuring vegetable oils with monoglycerides and monoglyceride–lecithin or monoglyceride–ethylcellulose mixtures. In: Food Chem Funct Anal 2022-January, pp 201–234. https://doi.org/10.1039/9781839166532-00201
27. Dassanayake LSK, Kodali DR, Ueno S, Sato K (2009) Physical properties of rice bran wax in bulk and organogels. J Am Oil Chem Soc 86:1163–1173. https://doi.org/10.1007/s11746-009-1464-6
28. Hwang H-S, Kim S, Singh M, Winkler-Moser JK, Liu SX (2012) Organogel formation of soybean oil with waxes. J Am Oil Chem Soc 89:639–647. https://doi.org/10.1007/s11746-011-1953-2
29. Doan CD, Van De Walle D, Dewettinck K, Ashok PR (2015) Evaluating the oil-gelling properties of natural waxes in rice bran oil: rheological, thermal, and microstructural study. J Am Oil Chem Soc 92:801–811. https://doi.org/10.1007/s11746-015-2645-0
30. Scharfe M, Niksch J, Flöter E (2022) Influence of minor oil components on sunflower, rice bran, candelilla, and beeswax oleogels. Eur J Lipid Sci Technol 124(7):1–19. https://doi.org/10.1002/EJLT.202100068
31. Diem Doan C, Ming TC, De Vrieze M, Lynen F, Danthine S, Brown A, Dewettinck K, Patel AR (2017) Chemical profiling of the major components in natural waxes to elucidate their role in liquid oil structuring. Food Chem 214:717–725. https://doi.org/10.1016/j.foodchem.2016.07.123
32. Avendaño-Vásquez G, De la Peña-Gil A, Charó-Alvarado ME, Charó-Alonso MA, Toro-Vazquez JF (2020) Self-assembly of symmetrical and asymmetrical alkyl esters in the neat state and in oleogels. Front Sustain Food Syst 4:1–21. https://doi.org/10.3389/fsufs.2020.00132
33. Brykczynski H, Wettlaufer T, Flöter E (2022) Revisiting pure component wax esters as basis of wax-based oleogels. J Am Oil Chem Soc 99:925–941. https://doi.org/10.1002/AOCS.12589
34. Blake AI, Co ED, Marangoni AG (2014) Structure and physical properties of plant wax crystal networks and their relationship to oil binding capacity. J Am Oil Chem Soc 91:885–903. https://doi.org/10.1007/s11746-014-2435-0
35. Tavernier I, Doan CD, Van De Walle D, Danthine S, Rimaux T, Dewettinck K (2017) Sequential crystallization of high and low melting waxes to improve oil structuring in wax-based oleogels. RSC Adv 7:12113–12125. https://doi.org/10.1039/C6RA27650D
36. Zangi R (2019) Self-assembly of alcohols adsorbed on graphene. J Phys Chem C 123:16902–16910. https://doi.org/10.1021/ACS.JPCC.9B04839/SUPPL_FILE/JP9B04839_SI_001.PDF
37. Blake AI, Toro-Vazquez JF, Hwang H-S (2018) Wax oleogels. In: Marangoni GA, Garti N (eds) Edible oleogels, Second edn. AOCS Press, pp 133–171. https://doi.org/10.1016/B978-0-12-814270-7.00006-X
38. Hwang H-S, Singh M, Bakota EL, Winkler-Moser JK, Kim S, Liu SX (2013) Margarine from organogels of plant wax and soybean oil. J Am Oil Chem Soc 90:1705–1712. https://doi.org/10.1007/s11746-013-2315-z
39. Kerr RM, Tombokan X, Ghosh S, Martini S (2011) Crystallization behavior of anhydrous milk fat−sunflower oil wax blends. J Agric Food Chem 59:2689–2695. https://doi.org/10.1021/JF1046046
40. Martini S, Carelli AA, Lee J (2008) Effect of the addition of waxes on the crystallization behavior of anhydrous milk fat. J Am Chem Soc 85:1097–1104. https://doi.org/10.1007/s11746-008-1310-2
41. Canapi EC, Agustin YTV, Moro EA, Pedrosa E, Bendaño MLJ (2005) Edible oil and fat products: edible oils. Coconut. In: Shahidi F (ed) Bailey's industrial oil and fat products, vol 2, 6th edn. Wiley, p 121
42. Ramírez-Gómez NO, Acevedo NC, Toro-Vázquez JF, Ornelas-Paz JJ, Dibildox-Alvarado E, Pérez-Martínez JD (2016) Phase behavior, structure and rheology of candelilla wax/fully

hydrogenated soybean oil mixtures with and without vegetable oil. Food Res Int 89:828–837. https://doi.org/10.1016/J.FOODRES.2016.10.025

43. da Silva TLT, Arellano DB, Martini S (2018) Physical properties of candelilla wax, monoacylglycerols, and fully hydrogenated oil oleogels. J Am Oil Chem Soc 95:797–811. https://doi.org/10.1002/AOCS.12096
44. de la Peña-Gil A, Charo-Alonso M, Toro-Vazquez JF (2023). Development of structured W/O emulsions with the use of only candelilla wax. J Am Oil Chem Soc. https://doi.org/10.1002/aocs.12753

Chapter 10
Oleogels Produced by Indirect Methods

Andrew J. Gravelle, Graziele Grossi Bovi Karatay, and Miriam Dupas Hubinger

Abbreviations

G′	Shear elastic modulus
HIPE	High internal phase emulsion
HPMC	Hydroxypropyl methylcellulose
LMOG	Low molecular weight oleogelator
MAG	Monoacylglycerol
MC	Methylcellulose
OBC	Oil binding capacity
SCD	Supercritical CO_2 drying
SPI	Soy protein isolate
WPI	Whey protein isolate
XG	Xanthan gum

Andrew J. Gravelle and Graziele Grossi Bovi Karatay contributed equally to this work.

A. J. Gravelle (✉)
Department of Food Science & Technology, University of California, Davis, CA, USA
e-mail: agravelle@ucdavis.edu

G. G. Bovi Karatay · M. D. Hubinger
Department of Food Engineering, School of Food Engineering, University of Campinas (UNICAMP), Campinas, SP, Brazil
e-mail: ggrossi@unicamp.br; mhub@unicamp.br

C. Andrea Palla, F. Valoppi (eds.), *Advances in Oleogel Development, Characterization, and Nutritional Aspects*,
https://doi.org/10.1007/978-3-031-46831-5_10

10.1 Introduction

Oleogels are soft matter, solid-like structures which exhibit similar physicochemical properties to solid fats, but have the nutritional profile of liquid oil [1, 2]. These characteristics are of high importance to the food industry from a product development perspective, as they would enable the replacement of solid fats containing a significant proportion of *trans*- or saturated fatty acids with mono- and polyunsaturated fatty acids [3–6]. In addition, the use of domestically grown liquid oils can have a positive impact on environmental sustainability, as the harvesting of exotic fats such as palm oil and shea butter has been linked to the deforestation of tropical forests, causing irreparable losses in pristine wilderness, native vegetation, and ecological biodiversity [7, 8]. Oleogels also provide an opportunity to incorporate lipid-soluble nutraceuticals [9] and are being actively investigated as novel drug delivery vehicles [10]. In this regard, oleogels have the potential to be used as ingredients in a wide variety of applications, thereby contributing to the development of novel, more environmentally friendly, and healthier food products [6, 11].

While oleogels represent a promising alternative to traditional solid fats, developing true mimetics that not only provide analogous techno-functional performance (e.g., plasticity, melting behavior) but are also economically viable and nutritious are all critical aspects to their wider adaptation in foods. The appropriate selection of an oleogelator depends on many factors, such as cost and availability, oil structuring efficiency, and capacity to meet the desired physical properties of the final intended product [12]. Ideally, oleogelators will be food-grade materials that are inexpensive, widely available, and contribute to a healthy diet [13]. Oleogelators can broadly be grouped into two categories: low molecular weight oleogelators (LMOGs), which include waxes, sterols, fatty acids and fatty alcohols, oligopeptides, and lecithin, among others [14], and biopolymeric gelators, such as proteins and polysaccharides [1, 10, 15]. Even though a wide variety of oleogelator systems have been developed using LMOGs, their use is still limited as they often have high melting points, show brittle yielding behavior, and many would not be considered 'clean label' ingredients [1]. On the other hand, proteins and polysaccharides are abundant, widely consumed, and generally inexpensive, making them ideal candidates for oleogelators. Furthermore, proteins (e.g., chickpea, pea, lentil, gelatin, and whey proteins) are generally considered nutritive, and their use as functional ingredients aligns with the clean labeling trend [1]. Polysaccharides (e.g., xanthan, carrageen, locust gum, alginate, and cellulose derivatives) have also shown their value as potential ingredients for producing oleogels due to their functional properties, such as thickening, stabilizing, emulsifying, and dispersing agents [16]. However, proteins and polysaccharides do not readily mix with oil, so indirect production methods are needed. Oleogels can also be produced using individual biopolymers, LMOGs, or combinations of these to produce multicomponent gelator systems, which may provide a synergistic enhancement in some desirable attributes [17]. In short, various combinations of oleogelators can produce oleogels, allowing them to be tailor-made according to the desired nutritional, techno-functional, and rheological properties.

The conversion of liquid oil into a solid-like structure is generally achieved by one of two routes, depending on the type of structuring agents employed. The so-called direct method disperses the oleogelator directly into the liquid oil by heating above the gelator melting temperature, followed by cooling to ambient temperature [13, 18]. This approach is generally restricted to LMOGs which form space-filling crystal networks or self-assembling fibrillar structures [19] that bind or entrap the liquid oil [20]. To date, the only food-grade polymer capable of structuring oil by the direct method is the cellulose derivative ethylcellulose [21–23]. A variety of proteins, polysaccharides, and combinations thereof have also been used to produce oleogels; however, these have all been achieved through indirect approaches, due to the inherent hydrophilic nature of these compounds. While some proteins and polysaccharides possess amphiphilic properties, even these generally cannot be directly suspended in triglyceride oils [1, 24]. However, several alternative indirect methods have been recently identified to cope with this deadlock of producing oleogels using proteins and polysaccharides. Briefly, indirect methods can be divided into four main approaches: (i) emulsification, (ii) oil absorption, (iii) solvent exchange, and (iv) protein aggregates. In each of these categories, there is a need to produce an initial structure that can serve as a so-called template (e.g., emulsion, hydrogel, foam, or protein aggregates) which can subsequently be converted into an oleogel by various means, depending on the approach employed.

This chapter aims to describe and exemplify indirect methods to produce protein- and polysaccharide-based oleogels (Sect. 10.2), highlight the potential of using specific proteins and polysaccharides as structuring agents for the development of oleogels using indirect methods (Sect. 10.3), and summarize advances made in the application of these oleogels in food products (Sect. 10.4).

10.2 Indirect Oil Structuring Methods

The main approaches used to produce oleogels include emulsification (Fig. 10.1a), oil absorption (Fig. 10.1b, c), solvent exchange (Fig. 10.1d, path 1a), and protein aggregation (Fig. 10.1d, paths 1b and 2). In some of these approaches, apart from the initial system, it is necessary to produce an intermediate network (i.e., the formation of an aerogel or cryogel), which serves as a template capable of absorbing oil to generate the final oleogel system. It is worth mentioning that in this context, a cryogel refers to a specific case of an aerogel produced by freeze-drying, but is used to differentiate these two distinct templating strategies.

10.2.1 Emulsification

In the emulsification approach, oleogels are produced using emulsions as the initial "templating" system. This strategy can be divided into two main steps: (i) the

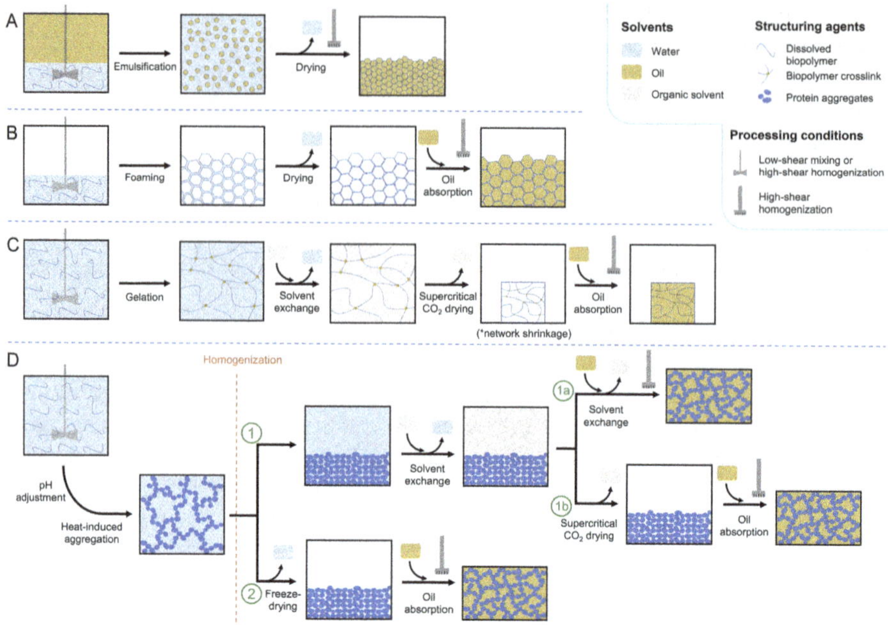

Fig. 10.1 Schematic representation of select indirect templating strategies: (**a**) Emulsion-templating; (**b**) oil absorption via foam-templating; (**c**) oil absorption via hydrogel-templating; and (**d**) methods using protein aggregates. The latter are generally formed using pH adjustment and heat denaturation, and further used to produce oleogels by (1a) direct solvent exchange; (1b) solvent exchange via supercritical CO_2 drying and passive oil absorption; or (2) freeze-drying protein aggregates for direct suspension in oil. For all strategies, the final stage involves high-shear homogenization to fully disperse the templated structure, producing an oil-continuous system (indicated by the homogenizer shaft). The mixing paddle (**a**–**c**) may represent either high-shear homogenization or low-shear mixing

production of the initial template, mainly a conventional or concentrated emulsion, and (ii) the dehydration of the initial system, which is often followed by a final shearing step that leads to the formation of the oil-continuous gel structure. This process is shown schematically in Fig. 10.1a.

10.2.1.1 Oleogels from Emulsions

The first report of oleogels produced using an indirect templating approach with potential food applications was carried out by Romoscanu and Mezzenga [25]. In this study, β-lactoglobulin (the major protein in milk whey) was used as a stabilizer to prepare water-continuous oil-in-water emulsions containing either paraffin oil or olive oil. The emulsions were then used as templates to obtain oleogels by freeze-drying. This procedure led to excess unadsorbed protein at the surface of the oil droplets. To remove the excess protein, several washing steps had to be carried out.

Further, to withstand the aggressive drying procedure, the protein films were strengthened using either thermal or chemical crosslinking. The need for these additional processing steps is generally considered a drawback to this methodology [26]. To address these shortcomings, subsequent investigations have favored using polysaccharide thickeners to provide added stability and explored more passive thermal dehydration techniques. This approach was first demonstrated by Patel and coworkers, who prepared emulsions using a surface-active hydrocolloid (either gelatin or methylcellulose (MC)) in combination with the non-surface-active hydrocolloid xanthan gum (XG) [2, 27, 28]. These combinations resulted in higher interfacial film elasticity, leading to higher resistance to rupture during the drying of the emulsion templates.

Tavernier et al. later employed the emulsion-templated approach to produce oleogels using soy protein isolate (SPI) alone, or in combination with the charged polysaccharide κ-carrageenan [29]. The authors reported that the protein-polysaccharide ratio, pH, and ionic strength strongly influenced the microscopic appearance and stability of the resulting electrostatic complexes. In a follow-up study, Tavernier et al. [30] investigated how to enhance the structure recovery of SPI oleogels by combining the emulsion-templated approach with a direct method using low concentrations of candelilla wax in the oil phase (i.e., forming structured emulsions). The authors reported that the wax did not significantly influence the gel strength, but reduced the shear sensitivity of the oleogels. Another example of an animal-derived protein used to produce oleogel by this approach is gelatin. Patel et al. prepared oleogels using gelatin and XG, which were reported to have high gel strength and thixotropic recovery [2].

Oleogels using solely polysaccharides by the emulsion-templated approach have also been reported. In the study of Patel et al., oleogels were prepared using a concentrated oil-in-water emulsion stabilized with a combination of either MC or HPMC with XG, followed by evaporation of the continuous water phase to drive network formation [28]. The resulting oleogels had high gel strength ($G' > 4000$ Pa), shear sensitivity, good thixotropic recovery, and good thermostability. Solely polysaccharide-based oleogels have been more extensively studied recently [31–35].

The above examples indicate that a wide variety of proteins and polysaccharides, as well as their combination, could be used for the emulsion-templated approach. Still, their effectiveness will mainly depend on factors such as surface activity, solubility, and robustness of the interfacial film that is formed. As for the type of oil, there is no limitation; therefore, a wide range of plant sources can be used. A drawback of this methodology is that it requires long drying times to dehydrate the emulsion, which is undesirable as it requires batch processing and can cause thermally-induced lipid oxidation. It is also worth mentioning that although it is not as commonly reported, oxidative degradation of protein is also possible, which may occur through reactions with secondary lipid oxidation products or via Maillard reactions with reducing carbohydrates and could impact functionality, sensory properties, and nutritional quality of the final product [23, 36, 37].

10.2.1.2 Oleogels from High Internal Phase Emulsions (HIPEs)

One approach to addressing the long drying times inherent to the emulsion-templated strategy has been to directly treat highly concentrated emulsions as structured lipid systems. With this approach, the initial emulsion is concentrated without fully removing the water phase, thus providing shorter drying times. HIPEs can be defined as concentrated emulsions where the volume fraction of the oil phase is >74%, which is the upper limit packing density of monodispersed hard spheres [38, 39]. These materials often exhibit interesting semi-solid mechanical and rheological properties, but due to the residual water content, they are not classified as oleogels themselves [40]. Partial water removal leads to a system with a dense interfacial polymeric network where the oil droplets are very tightly packed and are commonly forced to adopt a non-spherical morphology due to the depleted continuous phase. The resulting product should be further sheared to break the interfacial polymer network into smaller fragments [1] and produce what is referred to as an emulgel, or a concentrated emulsion that exhibits some solid-like properties [35].

The physicochemical properties of HIPEs are defined by the properties of the interface. When proteins are used as stabilizers, the emulsion properties can be manipulated to form a viscoelastic network by strengthening the interfacial layer through the formation of salt bridges, heat-induced aggregation, or by incorporating polysaccharides [1]. Such a combined protein-polysaccharide system was explored by Wijaya et al. [40] using whey protein isolate/low methoxyl pectin electrostatic colloidal complexes. These HIPEs were formed with 80 wt% oil and subsequently thermally dehydrated to form oleogels. With these HIPEs, the drying process could be completed at lower temperatures and required a shorter time (24 h at 35 °C) than oleogels obtained using conventional emulsions as templates. However, these biopolymer complexes resulted in poor oleogel stabilization, as indicated by evidence of oiling off after drying and shearing. Using an analogous approach, this same group characterized the changes in structure and physical properties of HIPE-templated oleogels stabilized with a combination of sodium caseinate and alginate produced under the same processing conditions [41]. It was shown that both the pH (5.5, 6.0, and 7.0) and protein/polysaccharide ratio (6:1, 8:1, 10:1, and 12:1) of the mixtures strongly affected the observed structural changes. These effects were particularly apparent in the microstructure and viscoelastic behavior of the network formed after drying and shearing. It was noted that pH was important in obtaining ultra-stable HIPEs, which was paramount to producing stable oleogels. An increase in pH produced smaller oil droplets, improving the physical stability of the HIPEs and thus improved oil retention in the corresponding oleogels. The ratio of protein to polysaccharide was reported to play a significant role in thickening the interface, producing firmer oleogels with improved oil retention and stronger rheological gel properties at a higher protein/polysaccharide ratio (at a fixed alginate concentration). Specifically for oil loss, oil binding capacity (OBC) improved with increasing pH for all casein/alginate ratios, indicating a stronger electrostatic interaction between the structuring agents.

Vélez-Erazo et al. [42] produced an initial emulsion containing 60% oil and 40% aqueous phase (containing 2 wt% hydrocolloids) containing pea protein as an emulsifier and a variety of polysaccharides at a fixed 4:1 protein-polysaccharide ratio. The final HIPE gels were produced by concentrating the initial emulsion using thermal dehydration at 65 °C for 48 h. The final HIPE gels had a water content ranging from 2% to 25%. Results showed that after 4 weeks of storage at 5 °C, only those systems stabilized with XG or tara gum had a creamy and uniform appearance, with minimal oil loss (0.89 and 1.72%, respectively). All other pea protein-polysaccharide combinations were less stable, with oil loss between ~35 and 75%. In a follow-up study, Vélez-Erazo et al. [4] explored the influence of pH (3, 5, and 7) and pea protein/XG ratio (i.e., 4:1, 8:1, and 12:1) on their emulsion-templated oleogels. It was reported that complete water removal was not possible using oven dehydration and homogenization under the same processing conditions as the previously mentioned study. The water content in the final system remained between 2% and 26%. These authors reported that the water loss was directly associated with the higher water holding capacity of XG compared to pea protein, as less water was retained as the ratio of protein increased relative to XG. In both these studies, the initial goal was to produce oleogels from emulsions, but as the system could not be completely dried, HIPEs were formed via incomplete drying of the initial network.

Despite the apparent advantages of the HIPE-templated approach, particularly with respect to milder processing conditions for water removal, some drawbacks persist. Water retention can be highly system-specific, and complete water removal cannot always be achieved. Further, oil retention after shearing is strongly dependent on the viscoelastic properties of the emulsion interface, as shearing can lead to excessive oil loss if the oil/emulsifier interaction is weak [42].

10.2.2 Oil Absorption

An alternative approach to structuring edible oils with proteins and polysaccharides is by passively absorbing liquid oil into a suitably conditioned polymer network. This generally takes the form of a porous scaffold [6], which may be formed by initially preparing a stable polymer network through either foaming (Fig. 10.1b) or inducing gelation and subsequently removing the water phase (Fig. 10.1c). The water may be removed by either directly freeze-drying or introducing an intermediate solvent that is purged using supercritical CO_2 drying. While both approaches technically produce a solid foam aerogel structure, the former is referred to as a cryogel approach to distinguish the methodology [18, 43].

In the oil absorption approach, there is direct physical sorption of oil in the porous structures of cryogel or aerogel scaffold [44]. This approach can be divided into three main steps: (i) the production of an initial system, most commonly a foam (Fig. 10.1b) or a hydrogel (Fig. 10.1c); (ii) dehydration of the initial system, which leads to the formation of an intermediate solid foam template, and (iii) passive absorption of oil into the intermediate scaffold, generally followed by shearing to produce the final oil-continuous oleogel.

10.2.2.1 Foam-Templated Oil Absorption (or Cryogel-Templated)

The foam-templated approach is known for its simplicity in preparation and for being an eco-friendly method, as it avoids high temperatures or harsh chemical crosslinkers. The foaming procedure exposes hydrophobic regions to the air/water interface, while hydrophilic regions stabilize the aqueous phase before drying (Fig. 10.1b). As structuring agents, the amphiphilic cellulose derivative hydroxypropyl methylcellulose (HPMC) has been explored most extensively, due to its ability to stabilize air cells (high foamability). XG is often used to enhance the lifetime of foams (i.e., foam stability) and avoid collapse during the water removal process [43, 45]. Based on this principle, surface-active proteins could also be promising candidates for producing foam templates, provided the interfacial film of the foams is sufficiently robust to withstand the drying procedure.

The feasibility of this approach was first demonstrated by Patel et al. [43] using HPMC. Foams were first generated by applying high-shear mixing to HPMC aqueous solutions (1–5 wt%) to incorporate air, followed by freeze-drying. Oleogels were then produced by allowing liquid oil to passively absorb into the dried templates and shearing the oil-absorbed cryogels. These cryogel templates could absorb up to 100 times their weight of liquid oil in a few seconds. A modification to the foaming procedure was investigated by Tanti et al. using spray-dried powders produced from 1 wt% HPMC solutions [46, 47]; however, the resulting powders lacked the porous cryogel template structure and thus had comparatively poor oil binding capacity. This methodology has thus not been further investigated.

Additional secondary thickening agents have also been explored as an approach to expand the functionality of HPMC-based cryogel templates. Jiang et al. demonstrated that adding XG modulated the microstructure, oil absorption properties, and rheological behavior of the resulting polymer scaffolds [48]. A subsequent investigation compared the effect of various thickening polysaccharides, including κ-carrageenan, carboxy methylcellulose, flaxseed gum, guar gum, and gum Arabic [49]. In this work, κ-carrageenan and flaxseed gum altered the microstructure, producing smaller pores and improved oil absorption capacity. As an alternative approach, the same group introduced glyceryl monostearate into HPMC-based foam templates as a secondary crystalline oleogel network, with the aim of providing the thermally responsive and plastic deformation characteristics of traditional fats [50]. This approach showed promise to mimic the use of butter in baking applications. Overall, these works demonstrate that suitable combinations of polysaccharides may provide synergistic enhancement in HPMC-based oleogels, which could potentially serve as fat mimetics; however, further work will be needed to optimize the most critical functional properties for the intended application.

While polysaccharide-based systems have proven to be effective foaming agents for producing oleogels, foam templating has also been explored using proteins as the interfacially active polymer. Abdollahi et al. explored combinations of gelatin (5 or 6% w/v) with varying amounts of XG (0, 0.1, or 0.2% w/v) as a stabilizing agent [18]. They reported that gelatin alone could not produce stable foams during aeration, whereas adding XG increased the solution viscosity and provided sufficient foam stability to withstand freeze-drying. Crosslink density and firmness of the cryogels increased with biopolymer concentration, but had no significant effect on oil sorption. The gelatin/XG templates displayed considerable oil binding capacity (>92%), high gel strength, and thixotropic behavior, thus showing strong potential as fat mimetics.

As an alternative to animal-derived proteins, Mohanan et al. explored the potential of using pulse proteins as foaming agents to produce oleogels [51]. This work compared the performance of cryogel templates made from commercially sourced pea or faba bean protein concentrates and isolates prepared at varying pH values (3, 5, 7, and 9) and stabilized with XG. Foams consisting of 5% protein and 0.25% XG were characterized as a function of pH to evaluate their foaming behavior and oil structuring properties. While the porous network of the freeze-dried foams could bind liquid oil, the OBC was dependent on the size, number, and connectivity of the cryogel pores. Further, the protein concentrates were more effective than the isolates, producing more elastic and easily spreadable oleogels. However, despite the ability of these templates to effectively absorb oil, the oil loss was quite high, with ~30–40 wt% being released upon centrifugation in an accelerated oil binding test.

In a follow-up study, Mohanan et al. expanded on this approach by incorporating a secondary crystalline structuring agent as a potential strategy to improve the OBC and rheological properties of the pulse protein oleogels [52]. High-melting monoacylglycerol (MAG) or candelilla wax was incorporated into the oil phase of cryogel-templated oleogels prepared from solutions of 5 wt% faba bean or pea protein concentrate and 0.25% XG. Different concentrations of MAG or wax (0.5–3 wt%) were dissolved in oil at 80 °C prior to introducing the oil into the aerogels. It was reported that OBC, firmness, cohesiveness, and storage moduli of the oleogels all increased with increasing concentration of either MAG or wax in the oil phase. Oil loss was reduced from ~35 wt to either 0 wt% with >2% wax or ~3–4.5 wt% with 3% MAG. This study demonstrates that combining direct and indirect oil structuring methods may provide superior functionality to single-strategy oleogelation approaches with respect to mimicking the techno-functional properties of fats.

Other ingredients tested in the foam-templated approach include rice bran protein [53], methylcellulose [46, 47], and HPMC used in combination with sodium caseinate and/or beeswax [54]. It is noted that one limiting factor in developing novel biopolymer-based foam templates is the necessity of high surface activity and the ability to form a robust, stable interfacial film at the bubble surface [53]. Moreover, some drawbacks of this approach are the reported poor oil retention when subjected to shear and the long time and high energy batch processing methods required for the dehydration of the system (usually above 48 h). Nonetheless, the foam-templated approach may prove beneficial for alternative uses, such as integrating high-value components into foods (e.g., temperature-sensitive lipids, nutraceuticals), or in drug delivery and controlled release applications.

10.2.2.2 Hydrogel-Templated Oil Absorption

The hydrogel-templated approach involves initially forming an aqueous protein network from which water is subsequently removed. The resulting structure is referred to as an aerogel, to which oil may be introduced to generate the final oleogel (Fig. 10.1c). Two methods to remove the aqueous phase before incorporating oil have thus far been reported: (i) first exchanging water with an organic solvent of intermediate polarity (commonly ethanol), followed by either freeze-drying or supercritical CO_2 drying (SCD) and (ii) directly freeze-drying the hydrogel. While the former approach requires an intermediate solvent, it differs from the solvent exchange approach (Sect. 10.3) in that the solvent is removed before introducing oil.

The use of supercritical CO_2 drying (SCD) to form aerogel templates for oleogelation was first investigated by Manzocco et al. [55] using hydrogels prepared with increasing concentrations of κ-carrageenan (0.4, 1.0, and 2.0 wt%). These were converted into alcohol gels via solvent exchange with ethanol and subsequently dried using supercritical CO_2 to obtain aerogels. The authors reported that the polymer concentration affected the aerogel structure, as higher κ-carrageenan concentrations produced a coarser structure with larger polymer aggregates; however, all the aerogels displayed a high oil absorption capacity (~80%). In a follow-up study, the same group investigated the potential of using whey protein isolate (WPI) to produce aerogels to serve as lipid-based delivery systems in foods [56]. Aerogels were obtained by either freeze-drying or SCD to evaluate their capacity for oil- and water-loading. The morphology and porosity of aerogels obtained by freeze-drying or SCD were remarkably different, with the freeze-drying procedure resulting in aerogels with larger pores and a greater void volume compared to the SCD process (Fig. 10.2). Such differences suggest it would be possible to modify the material performance by appropriate selection of the drying technique.

Another approach to producing aerogels for subsequent conversion to oleogels was carried out by Chen and Zhang [57]. Alginate/soy protein conjugates were prepared via a heat-induced Maillard reaction, and the resulting covalently linked hydrogels were freeze-dried to form the aerogel templates. These structures absorbed up to 10.9 g oil/g aerogel and displayed an oil holding capacity of ~40%.

10.2.3 Solvent Exchange

The solvent exchange approach can be divided into two main steps: (i) the production of an initial system, such as a hydrogel or collection of protein aggregates, and (ii) a sequence of solvent exchange steps, where water is first removed by

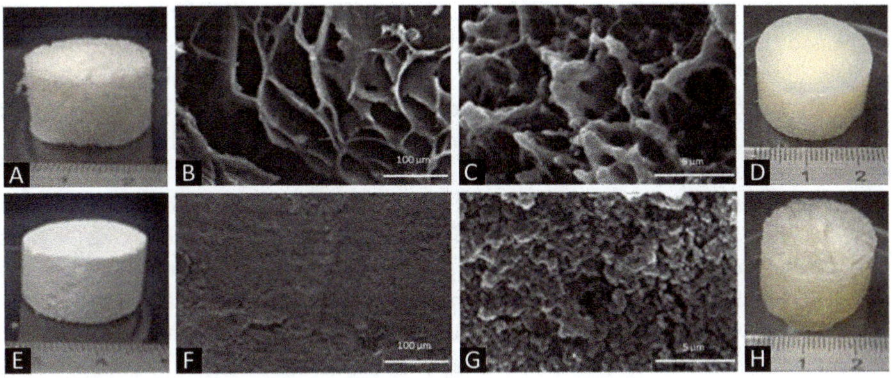

Fig. 10.2 Visual appearance and scanning electron micrographs of freeze-dried (**a**–**d**) or supercritical CO_2 dried (**e**–**h**) whey protein isolate hydrogel templates (**a**–**c**; **e**–**g**) and corresponding hydrogel-templated oleogels (**d**, **h**). Electron micrographs depict low (**b**, **f**) and high (**c**, **g**) magnification of the template structures. (Image adapted from Manzocco et al. [56] with permission from Elsevier)

introducing an intermediate solvent, and subsequent removal of the solvent by direct addition of oil. In practice, both solvent substitution steps are iterative processes in which the secondary solvent is introduced to the template (i.e., either the intermediate solvent or oil), followed by centrifugation and removal of the supernatant. The procedure is then repeated (possibly multiple times) to ensure the full removal of the initial solvent.

10.2.3.1 Solvent Exchange Via Hydrogel Templates

In the solvent exchange approach to hydrogel templates, the exchange process is performed directly to a polymeric hydrogel structure via passive diffusion [58, 59]. Intermediate solvents that have been successfully used in this approach include tetrahydrofuran and acetone [58]. Oleogels could be produced with up to 91 wt% oil trapped homogeneously in the protein matrix, with a residual water content of <1 wt%. It was shown that the OBC depended on the microstructure, polarity of the intermediate solvent, and kinetics of the solvent exchange process. Further results showed that the composition, OBC, and mechanical properties of the oleogels could be tuned by varying protein concentration. Most of the water can be removed without structural collapse due to changes in solvent polarity [6].

10.2.3.2 Solvent Exchange Via Protein Aggregates

To overcome the diffusion-limited solvent exchange process inherent to polymeric hydrogels, de Vries et al. [60] explored the use of heat-set WPI aggregates. Colloidal hydrogel aggregates (diameter ~ 200 nm) were formed by heating a dilute WPI suspension at pH 5.7 (i.e., near the protein isoelectric point) to induce protein denaturation. The resulting aggregates were then homogenized and concentrated via centrifugation, and oil was introduced via a direct solvent substitution approach (as outlined above) using acetone as an intermediate solvent. This procedure is shown schematically in Fig. 10.1d (path 1a). The resulting protein suspension was converted to an oleogel by centrifugation and removal of excess unbound oil. It was shown that using a solvent of intermediate polarity, which was both water- and oil-miscible, prevented agglomeration of the aggregates, as the particle size did not differ between the oleogel and the initial aqueous phase. While this approach is limited to proteins able to form colloidal aggregates, a wide range of globular proteins would likely be suitable for this method, including many plant-derived proteins [61].

10.2.4 Protein Aggregates

The protein aggregate approach is conceptually similar to the solvent exchange route; however, after forming the initial colloidal aggregates in an aqueous environment, they are freeze-dried to eliminate the need for intermediate solvents. The resulting aggregates may then be ground and directly dispersed in oil (Fig. 10.1d, path 2), providing a much less laborious procedure than the solvent exchange approach. De Vries et al. showed that direct freeze-drying the particles led to the formation of irreversible aggregates, which impeded dispersion in oil and inhibited effective oil binding [60]. In subsequent work, this group reported that irreversible agglomeration of WPI aggregates can be prevented by carefully controlling the drying process, thus providing a means to obtain a dried protein material that can be used directly for structuring liquid oil [59].

Even though this method shows similarities with the direct production method, there is still a need to create an initial dispersion and intermediate system (i.e., freeze-dried aggregates), and it can therefore still be considered an indirect method. In a follow-up study, de Vries et al. transferred protein aggregates from water to various solvents with different polarities to investigate the effect on agglomeration. It was shown that dried aggregates produced using solvents with a low polarity (e.g., hexane) showed good dispersibility in liquid oil compared to those dried from solvents with a high polarity [59]. A further study carried out by the same group investigated the effect of adding a small amount of water to the oleogels (0.1–0.5 g water/g protein) to modify the mechanical and rheological behavior [49]. It was shown that the water could form capillary bridges between the protein aggregates and be subsequently removed using a mild heat treatment. Despite removing the water phase, the resulting modified colloidal network displayed an increase in elastic modulus (up to three orders of magnitude), critical strain, and yield stress. This approach may provide a method of tuning the functional performance of oleogels produced using protein aggregates.

10.2.5 Other Approaches

In addition to the strategies outlined above, some further approaches to transform liquid oil into soft solid materials have also been investigated. These include emulsion-filled gels, bigels, oleofilms, and oleofoams, among others. Each of these systems is also a semi-solid structure composed mainly of oil, oleogelator, and (in some cases) water, and have great potential to be used in novel food formulations.

Emulsion-filled gels may be described as a network consisting of emulsion droplets embedded throughout a hydrogel continuous phase. These may be formed by incorporating a gelling agent into the continuous phase of an emulsion or by promoting the formation of a network of aggregated structures (e.g., protein particles or emulsion droplets) [62, 63]. Common food examples include fat-containing yogurt, cheese, and finely comminuted meats (e.g., frankfurters). Due to their ubiquity across the food industry, these systems have been extensively investigated and reviewed in depth elsewhere [62, 63]. The reader is thus referred to these works for further information. Another similar approach to emulsion-filled gels is bigels, which are viscoelastic semi-solid structures composed of two immiscible, structured phases: the oleogel and hydrogel phases. For the latter, a wide range of biopolymers can be used, with proteins and polysaccharides being the most common examples, as these polymers can form gels due to changes in intermolecular bonding [64, 65]. Bigels may be water-continuous, lipid-continuous, or bi-continuous, depending on the composition and processing conditions. For example, Cho et al. [66] produced water-continuous bigels using gelatin to structure the hydrogel phase and rice bran wax to produce a soybean oil oleogel, which was embedded in the gelatin gel as a structured emulsion. The microstructure and rheological properties of the resulting bigels were strongly influenced by the ratio of hydrogel to oleogel. The authors evaluated the bigels for 6 months under accelerated oxidation conditions to assess oxidative stability. They reported that the bigels displayed superior structural and mechanical properties to pure oleogel and hydrogel, as demonstrated by higher yield stress and reduced water and oil loss after freeze-thawing.

HIPEs were previously noted as a route to forming emulsion-templated oleogels (Sect. 10.2.1.2). However, these systems can be used directly to incorporate into food products as a fat replacer without the need for further processing steps such as drying and shearing. Other studies that have produced HIPEs using a combination of protein and polysaccharides include chickpea aquafaba [67], alginate and WPI [38], and sodium caseinate and LMP [38], among others. Studies that produced HIPEs using solely protein-based ingredients include, among others, quinoa protein [68] and lentil protein isolate [69].

HIPEs have also been explored as a means of producing dried, flexible edible oil-based films, which may be used as moisture barriers. Wijaya et al. produced oleofilms (>97 wt% edible liquid oil) by simply drying HIPEs which were stabilized by sodium caseinate and alginate with glycerol incorporated as a plasticizer for the biopolymers [70]. This work evaluated the ability of glycerol to improve the flexibility and processability of the dried products by increasing interchain spacing and reducing polymer-polymer interactions. The results of this study also demonstrated the important role of protein-polysaccharide ratio and biopolymer concentration on the stability and resistance to oil leakage of the resulting oil-based films.

As an extension of oleogels, the development of structured, aerated oils (or oleofoams) has recently gained increasing interest for food applications [71]. While these systems are often based on the assembly of crystalline platelets at the air/oil interface (i.e., via direct structuring), these gelators could potentially be used in combination with indirect structuring strategies. This concept was demonstrated by Mohanan et al., who converted pulse protein foam-templated oleogels into oleofoam [72]. In this study, oleogels were prepared at room temperature by adding oil containing MAG or candelilla wax to the freeze-dried pea or faba bean protein-stabilized foam templates. These dual-network oleogels were then whipped to create the oleofoams; however, only those oleogels containing MAG could form stable foams. As a potential shortening replacement, the oleofoams produced larger, heterogeneously distributed air cells in cake batters, but this outcome could also be improved by modifying the batter preparation procedure. This work demonstrates that aerated, structured lipids made from a combination of direct and indirect approaches could tailor the lipid profile and rheological and textural properties of food products such as baked goods.

Another approach conceptually similar to that of bigels that shows promise for developing edible materials able to mimic animal fat tissue is crosslinked structured emulsion. Dreher et al. investigated structured emulsions containing varying ratios of liquid and fully hydrogenated canola oil structured with SPI and crosslinked using transglutaminase. It was shown that the content of both protein and solid fat can be independently varied to modulate the structural and rheological properties of the crosslinked structured emulsions [73]. These crosslinked emulsions have also been incorporated as fat replacers in cured and fermented meat products or plant-based analogs with promising results [74, 75]. This approach may be particularly useful for accurately mimicking the properties of whole-muscle meats and products containing ground meats where adipose tissue contributes to the structural integrity of the lipid phase.

10.3 Polymers as Structuring Agents

Structuring edible oils using high molecular weight biopolymers such as proteins and polysaccharides has gained considerable interest due to their wide availability, food-grade status, and general consumer acceptance. Proteins are generally considered to have a nutritional benefit, while many polysaccharides may be categorized as non-digestible dietary fiber. The physicochemical properties and performance of these macromolecules depend on a variety of factors, such as their chemical composition, environmental conditions during processing (i.e., pH, ionic strength, temperature), and the presence of other ingredients. Proteins are also complex mixtures extracted from various sources, and both the content and ratio of the different proteins present, as well as extraction and processing conditions, can drastically influence their functional properties [61, 76]. Combining ingredients such as amphiphilic proteins with non-surface-active polysaccharides can also improve the desired performance and has been a popular strategy for producing oleogels by indirect methods. The general properties and applications of various polysaccharides and proteins that have been used to produce oleogels with potential food applications are outlined.

10.3.1 Polysaccharides

Polysaccharides are a chemically and structurally diverse class of macromolecules that are widely used in foods as stabilizers, structuring agents, and texture modifiers. Food polysaccharides may be derived from plant, animal, or microbial sources; they are available in high purity, have little influence on taste, and generally exhibit high functionality at relatively low concentrations [77]. Common applications include promoting hydration and water binding, viscosity enhancement, emulsification, and gelation. As oleogelators, various polysaccharides have been exploited for their surface activity, viscosity enhancement or structural support, and gelation behavior, depending on the oleogelation strategy employed. The most commonly employed polysaccharides used to produce edible oleogels include cellulose derivatives, xanthan gum, and κ-carrageenan.

10.3.1.1 Cellulose Derivatives

Cellulose is a highly abundant biopolymer that contributes to the structural integrity of plant cell walls. Chemically, cellulose is a linear (unbranched) homopolymer composed of D-glucose monomers linked through repeating β(1-4) glycosidic bonds. The orientation provided by the β-linkage results in a very linear molecule supported by intramolecular hydrogen bonding and provides high resistance to chemical and enzymatic degradation. This rigid, ordered structure also allows for extensive inter-polymer hydrogen bonds, resulting in the formation of crystalline domains known as microfibrils which assemble into highly elongated hierarchal structures with high tensile strength [78].

Cellulose ethers are commonly used as food additives and are produced using base-catalyzed substitution reactions to introduce various functional groups at the hydroxyl groups of carbons C2, C3, and C6. Both the average molecular weight and the degree of substitution will impact the functional properties of the resulting polymer. Ethylcellulose, methylcellulose (MC), hydroxypropyl methylcellulose (HPMC), and carboxymethyl cellulose have all been explored for oleogel production, either independently or in combination with other structuring agents.

Ethylcellulose is commonly used in a variety of chemical, pharmaceutical, cosmetic, and packaging applications and is currently approved for use in food applications as a flavor fixative, tablet coating, and excipient. It is produced by reacting cellulose with ethyl chloride, and the most common commercial varieties are synthesized with an average degree of substitution of ~2.5. While ethylcellulose has been extensively investigated as an oleogelator by direct dispersion in oil [79], nanoparticles of ethylcellulose produced via an antisolvent precipitation method have been used as surface-active agents to stabilize water-in-oil emulsions via a Pickering mechanism [80]. The nanoparticles were also able to stabilize water-in-oil emulsions in which the oil phase was structured by ethylcellulose via the more common direct dispersion method. This multi-phasic system was shown to have promising freeze-thaw stability, and potential as a delivery vehicle for both water-soluble (anthocyanin) and lipid-soluble (curcumin) bioactives.

MC and HPMC are the most extensively studied cellulose derivatives for indirect oil structuring. The polymers are common food additives with E-numbers E461 and E464, respectively. MC is produced by reacting alkali cellulose with methyl chloride, while HPMC can be produced through a subsequent reaction with propylene oxide to generate a mixture of methoxy and propylene glycol functional groups [81]. While these polymers are highly water-soluble and are commonly used as thickening agents and stabilizers, the hydrophobic characteristics imparted by the substituents impart several unique properties. Both MC and HPMC are effective emulsifiers, can stabilize aqueous foams, and form thermo-reversible gels above a critical temperature due to the association of their hydrophobic functional groups. As oleogelators, they have been independently used to form both foam- and emulsion-templates, either alone [32, 43] or with the addition of a secondary stabilizing hydrocolloid, such as XG [28, 82]. Due to its more hydrophilic propylene glycol functional groups, HPMC has superior foaming characteristics to MC and has been more extensively explored using the foam-templated approach [83, 84]. Limited studies have also combined HPMC with additional viscosity-enhancing hydrocolloids [48, 49], as they are not necessary to produce stable foam structures capable of withstanding freeze-drying.

Cellulose particles are another class of cellulosic derivative that have been increasingly used as natural, biodegradable functional materials across numerous fields [85]. Microcrystalline cellulose contains both crystalline and amorphous regions and exhibits a wide range of functional properties. Cellulose nanocrystals are highly crystalline domains most commonly produced by cleaving amorphous regions from cellulose fiber pulp via acid hydrolysis. They generally have a rod-shaped morphology and range in length from ~50 to 500 μm [86]. Both the source material and processing conditions will impact the morphology and functional properties of the nanoparticles. Cellulose nanocrystals have been used to stabilize oleogel-in-water emulsions containing beeswax-based oleogels via a Pickering mechanism [87]. Increasing nanocrystal concentration (0.1 up to 0.9% w/v) reduced oleogel droplet size and increased emulsion viscosity, indicating this approach could be effective for tuning the functionality of structured emulsions. Emulsion-templated oleogels have also been successfully prepared using cellulose nanoparticles as the sole emulsifying agent (0.4–4.0% w/v) using a freeze-drying procedure [88].

Cellulose particles have also been introduced into oleogels as a secondary structuring agent. Particles of varying sizes (micro- to nano-scale) have been used as fillers in oleogels structured by MAG crystal networks [89]. These fillers were reported to serve as nucleation sites for the crystal network and produced a size- and filler-content-dependent increase in the mechanical and rheological response. Oleogels have also been used to produce oil-continuous capillary suspensions of water-swollen cellulose particles (oleogels containing up to 20 wt% water) [90]. Both capillary bridging and particle clustering contributed to the rheological properties. The use of cellulose particles may thus provide a broad range of opportunities to tailor oleogel properties based on particle size, morphology, and mechanism of oil stabilization. The addition of functional groups could further expand this approach.

10.3.1.2 Xanthan Gum

XG is an anionic, branched, high molecular weight polysaccharide of bacterial origin most commonly used in foods as a viscosity enhancer. XG has a backbone identical to cellulose and a repeating branching structure at the C3 position of every second glucose monomer. These branches consist of a mannose-glucuronic acid-mannose trisaccharide, where the interior mannose is acylated, and the terminal mannose may be modified with a pyruvic acid group. XG is non-surface-active and non-gelling, but highly effective at increasing the viscosity of aqueous solutions across a wide range of pH and ionic strength [91]. The latter is caused by polymer entanglement and the formation of weak intermolecular interactions, which also make XG solutions highly susceptible to shear-thinning. These attributes are commonly utilized in food applications to stabilize suspensions at rest without imparting an overly viscous texture or slimy mouthfeel.

XG has most commonly been used in combination with surface-active hydrocolloids to structure edible oils through indirect methods. In emulsion-based oleogels, XG increases the oil/water interfacial rigidity, which can help withstand the drying processes commonly used for water removal. XG/MC [27, 28, 92], XG/HPMC [28, 35, 82, 93], and XG/gelatin [2, 94] emulsion templates have all been reported. Espert and coworkers [95] contrasted the effect of various surface-active molecules when used in combination with XG to produce emulsion-templated oleogels. Emulsions containing 20 wt% oil were stabilized with polysorbate 80, soy lecithin, whey protein, or locust bean gum (5 wt%), each thickened with XG (2 wt%) and oven-dried to produce oleogels. Although soy lecithin produced weaker gels than the other compounds (lower shear elastic modulus, G'), all four emulsifiers had similar oil-binding properties. Foam-templated oleogels incorporating xanthan gum have also been investigated using MC [48] or gelatin [18] as the surface-active agent. XG has also been used in combination with proteins to produce oleogels through several approaches, including aerogel templates with egg white protein [96], cryogel templates with pea or faba bean proteins [51, 52], emulsion templates with sodium caseinate [97], and HIPEs stabilized by pea protein [42].

10.3.1.3 κ-Carrageenan

Carrageenans are a class of anionic, unbranched polysaccharides derived from certain types of red algae. This family of polymers consists of repeating disaccharide units made up of various galactose derivatives with differing patterns of anionic sulfate substitutions. Several varieties also possess a 3,6-anhydro linkage in one of the disaccharide units. This feature promotes the formation of helical structures which contribute to their functional properties. κ-Carrageenan is one member of this family which is distinguished by having both the anhydro-linkage and a single sulfate group at the C4 position of the complementary galactose monomer. This repeat pattern forms helical structures which produce strong, rigid gels in the presence of potassium ions and has been exploited to produce aerogel-templated oleogels via supercritical CO_2 drying or freeze-drying [55, 98]. The selected drying method impacted the aerogel structure, oil loading capacity, and oil retention, with supercritical CO_2 drying causing more network shrinkage, but higher oil binding. The addition of inert fillers (particulate from lettuce homogenate) was shown to interrupt the gel structure, reducing network shrinkage and increasing oil holding capacity [98]. κ-Carrageenan has also been used in combination with soy protein isolate [29] and sodium caseinate [38] to form emulsion-templated oleogels and HIPE gels, respectively.

10.3.2 Proteins

A diverse range of proteins derived from edible sources are widely used in foods for their techno-functional properties, including water binding, gelation, emulsification, and foaming. Proteins may be viewed as naturally amphiphilic biopolymers; however, their functional behavior and interfacial activity will depend on the primary amino acid sequence, secondary- and higher-order structural features, size, the forces contributing to structural stability (e.g., hydrophobic interactions, disulfide bonds), and chemical environment (pH, ionic strength), among others. Many proteins of animal origin are highly functional, including dairy proteins (casein and whey), gelatin, and egg white proteins. Plant-derived proteins generally have less intrinsic functional attributes, but extraction and processing methods continue to be developed which may enhance their utility in promoting desirable structures and textures. Due to the sheer breadth of protein sources and parameters that can influence their functional behavior, proteins provide a promising new avenue for developing oleogels using indirect structuring methods. Here, we summarize some of the most common proteins investigated for oleogelation.

10.3.2.1 Whey Proteins

Whey is a by-product of the dairy industry which is produced during cheesemaking and casein protein production. Although they are derived from a side stream, whey proteins are highly functional and used for a broad range of applications across the food industry. They also have a high nutritional value due to their balanced amino acid composition and rapid absorption during digestion [99]. The whey protein fraction of milk can be classified as those proteins that remain dispersed in the aqueous phase at pH 4.6 (i.e., the isoelectric point of casein proteins), or after rennet-induced coagulation of casein. Whey proteins constitute ~20 wt% of the total protein content of bovine milk and are primarily composed of β-lactoglobulin (~50 wt%) and α-lactalbumin (~20 wt%), as well as immunoglobulins, serum albumin, and other minor proteins and peptides [99]. β-Lactoglobulin is a ~18 kDa globular protein that has a compact β-barrel tertiary structure with a hydrophobic core. It contains one free cysteine which contributes to the functional properties such as thermal gelation, emulsification, and foaming of commercial whey proteins [99]. α-Lactalbumin is also a compact protein (~14 kDa) with a tertiary structure dominated by α-helical and random coil motifs. The protein is highly structured due to the presence of four disulfide linkages, but contains no free cystines and thus has poor thermal gelation properties. Whey proteins will completely denature upon heating to 90 °C and are able to form fine-stranded gels at neutral pH and in the absence of salts, generally with protein concentrations exceeding ~9–10 wt% [99]. Particulate colloidal structures will form under acidic conditions or in the presence of ionic species.

As noted previously, heat-set whey protein gels prepared at varying ionic strength (pH 7) have been used to produce oleogels through the direct solvent-substitution approach using solvents with intermediate polarity [58]. As oil-structuring agents, heat-induced aggregates of whey proteins have been the most extensively characterized. These aggregates have most commonly been formed using a dilute solution (4 wt%) of whey protein isolate (>90 wt% protein) that is thermally denatured under acidic conditions (pH 5.7). The resulting weak particulate hydrogels are mechanically homogenized to produce nano-scale protein aggregates [58]. Various methods of producing oleogels with these aggregates have been explored, including direct solvent substitution [58, 60, 100], supercritical CO_2 drying after solvent substitution [101, 102], and oven drying or freeze-drying [59] (Fig. 10.1d, paths 1a, 1b, and 2, respectively). The effect of oil polarity [103] and the addition of water to the oil-suspended particles to form capillary bridges [100] have also been investigated.

Whey proteins are also known for their high interfacial activity and are effective at stabilizing oil-in-water emulsions and foams. During mechanical agitation, their low molecular weight and compact structures allow them to rapidly diffuse to the interface, where they are able to unfold and reorient to form a viscoelastic film. Upon adsorption, disulfide bonds form between β-lactoglobulin molecules, enhancing the integrity of the interfacial film, which increases the emulsion stability [104]. Despite their excellent interfacial properties, there have been limited studies using indirect emulsion- or foam-templated methods to produce oleogels stabilized by whey proteins. Purified β-lactoglobulin was the first protein used to produce emulsion-templated oleogels but required crosslinking with glutaraldehyde to stabilize the system during water removal [25]. Oleogels from WPI-stabilized emulsions have also been reported using high concentrations of xanthan gum (2 wt%) to stabilize the interface [95]. Therefore, the use of WPI-based oleogels could potentially be expanded using approaches that can enhance the integrity of the interfacial film or provide structural support between the emulsion or foam droplets.

10.3.2.2 Gelatin

Gelatin is a partially hydrolyzed form of collagen which is most commonly extracted from pork hides or beef bones as a by-product of animal processing. Gelatin is widely used for its ability to form highly elastic, thermo-reversible gels which are supported by the physical association of individual protein strands. This behavior is due to gelatin having a common repetitive amino acid sequence of (-glycine-X-Y-), where X is frequently proline and Y is the post-translationally modified hydroxyproline. The repeating motif causes individual chains to form helices, which further assemble into a characteristic three-membered triple helix supported by intramolecular hydrogen bonds. Upon heating, these structures disassemble, but will reform upon cooling, serving as junction zones that support a three-dimensional gel network. The degree of hydrolysis and resulting molecular weight thus determines the functional properties of the protein. Gelatin also has a notable fraction of charged amino acids and is therefore a surface-active protein. It has been used to produce both emulsion-templated [2] and cryogel-templated [18] oleogels in combination with XG. Emulsion-templated oleogel systems stabilized by gelatin-polyphenol complexes and added hydrocolloids have also been reported using tannic acid [105] or proanthocyanidins [94]. These complexes may provide an added benefit of reducing oxidation in the oil phase and/or serving as health-promoting compounds for the consumer [106].

10.3.2.3 Legume Proteins

Plant-derived proteins are commonly viewed as a more sustainable and environmentally friendly protein source and are increasingly being used to provide distinct textural and functional properties in a wide range of food products [107]. The performance of these proteins is strongly influenced by numerous factors, including the source, cultivar, composition, extraction procedure, environmental conditions, and processing methods employed. The most abundant class of high-purity plant protein extracts available for food applications are globular storage proteins produced by the host plant as amino acid reserves for growth and development [108]. In their native state, these proteins have more hydrophobic amino acids buried in the globular core, while the more hydrophilic and charged amino acids tend to be present at the surface and/or exposed to the surrounding environment. Heating or applying mechanical shear can induce partial denaturation of the globular structures, exposing the hydrophobic core. With sufficient denaturation, this process can induce protein-protein aggregation, leading to the formation of a network-spanning gel [109]. Applying sufficient mechanical shear in the presence of a non-polar medium such as oil or air can expose hydrophobic groups, allowing them to orient at the interface, which is a prerequisite to forming stable emulsions and foams. The stability of these systems depends on the protein's ability to unfold and orient at the interface and form a strong, cohesive layer supported by protein-protein interactions [110]. Depending on the amino acid composition, the formation of intermolecular disulfide bonds may also contribute to gel formation or interfacial stabilization (as with β-lactoglobulin). All of these factors will directly impact the performance of oleogels using plant-derived proteins through indirect methods. Among legume proteins, soy is by far the most studied, both for its techno-functional properties and its oil structuring potential, followed by pulses such as peas, faba beans, chickpeas, and lentils.

Soybeans are a legume species grown as a commodity crop. They are highly valued for their exceptionally high content of both oil and protein (~20% and ~30–40% on a dry matter basis, respectively). They are primarily cultivated as a source of vegetable oil for human consumption and food preparation. The residual high-protein meal is used mainly as animal feed, with a minor fraction being processed into protein-rich fractions for human consumption. Soy proteins are one of the most functional classes of plant-derived proteins available, due to their high solubility and excellent gelation and emulsification properties. The most abundant proteins in soy are the globular storage proteins glycinin and β-conglycinin, which comprise ~80–90% of the proteins when separated by isoelectric precipitation and thus provide the majority of the functional attributes [111]. These proteins are often referred to as the 11S (legumin) and 7S (vicilin) protein fractions, in reference to their respective sedimentation coefficients. Soy glycinin exists most commonly as a ~350 kDa hexamer, with each subunit consisting of an acidic and basic polypeptide pair linked by a single disulfide bond. Soy β-conglycinin contains three unique subunits which assemble into a ~180 kDa trimer, supported by hydrophobic and hydrogen bonding interactions [112].

Soy proteins are well known for their gelation properties, which may be induced via isoelectric precipitation, the addition of divalent cations, or heat denaturation. The structure, texture, and functional properties of these gels will differ based on the method of production; however, each of these is ultimately formed by aggregation processes of the individual globular proteins, or protein clusters [113]. As structuring agents for oleogels, aggregates are more susceptible to aggressive processing methods needed for water removal and may therefore be more useful for approaches using milder processing conditions, such as solvent substitution [1].

Due to their bulky, globular nature, glycinin and β-conglycinin are not recognized as having strong emulsifying properties. These proteins and their aggregates diffuse slowly to the interface, limiting their in-plane interactions and leading to a weak viscoelastic interfacial film [110]. Emulsions stabilized by soy proteins are thus prone to destabilization phenomena such as aggregation and coalescence and show sensitivity to environmental changes (pH or salt content) [114]. Despite these challenges, it has been shown that SPI and SPI-κ-carrageenan electrostatic complexes were able to form concentrated emulsions (60 wt% oil) under acidic conditions (pH 3), which were able to withstand water removal by oven drying to form emulsion-templated oleogels [29]. To overcome the issues associated with globular plant proteins, processing, enzymatic, or biochemical modifications may be used to improve the protein structure and/or interfacial activity [110].

An alternative mechanism by which soy and other plant proteins may exhibit emulsification activity is through the formation of nanoparticles that exhibit partial wettability. When sufficient energy input is applied, these particles can absorb at the oil/water interface with a high activation energy for removal, effectively becoming irreversibly bound. This phenomenon is referred to as Pickering stabilization and may be achieved using individual or combinations of proteins, as well as protein complexes [115]. Complexes of SPI with basil seed gum have been used to produce oleogels from Pickering emulsions [116]. Additionally, both soy [117–119] and pea proteins [120–122] have been used to form nanoparticle conjugates capable of forming Pickering emulsions. Due to the high stability of Pickering emulsion droplets, these systems may be promising candidates for more aggressive approaches to water removal for oleogel formation, such as freeze-drying, or to avoid the need for additional thickeners.

Foaming behavior represents another emerging opportunity as research on expanding the functional properties of plant-derived proteins continues to grow [123]. Pulse proteins are not as effective at stabilizing air/water interfaces as other common foaming agents, such as conventional animal-derived proteins (e.g., whey, egg white, gelatin, or casein proteins) [124]. The globular nature of plant storage proteins often provides them with lower solubility and interfacial molecular flexibility, greater pH sensitivity, reduced viscosity enhancement, and poor water retention between bubble lamella during foaming. However, there have been efforts to characterize and improve the foaming properties of proteins from several widely cultivated pulses, including soy, pea, chickpea, and faba [124].

As with emulsification behavior, foaming properties depend on the protein source, cultivar, extraction method, processing, and environmental conditions. While efforts to identify specific proteins or protein fractions that contribute to foaming have been limited, approaches to enhance the foaming properties of pulse proteins are an active area of research. Examples include thermal treatment, high-pressure processes, application of ultrasound, and enzymatic modification [124]. Further, using less concentrated protein fractions such as wet-extracted concentrates or air-classified proteins may provide some functional benefit, in addition to being more economical and sustainable than highly purified isolates. This is of particular relevance to foam-templated oleogels in which the presence of high molecular weight carbohydrates might provide added structural stability during foam production and water removal. For example, the foaming properties of commercially produced pea and faba bean concentrates and isolates (with added XG) have been characterized for use as cryogel templates [51]. It was shown that after freeze-drying, the concentrates produced smaller air cells, particularly at neutral and alkaline pH. The freeze-dried foams prepared with concentrates also had higher oil binding capacity at all pH values, but lower G′ and firmness.

10.3.2.4 Underutilized Proteins

As the food industry continues to seek out more sustainable solutions for food production, there is increasing interest in identifying novel functional ingredients from agricultural residues and low-value by-products. Novel processing and extraction methods are increasingly being explored to identify opportunities to valorize these materials and isolate compounds with functional and/or health-promoting properties [125, 126]. While the majority of oleogelator systems structured using proteins have focused on established food ingredients with well-characterized functional properties, several underutilized protein sources have also recently been investigated, including those from canola, rice bran, and bamboo.

Proteins derived from oil seeds may represent an untapped resource with suitable functionality for food applications. De-oiled canola meal is rich in two main storage proteins: a larger, globular protein (~300–350 kDa) called cruciferin and a smaller prolamin known as napin (~14 kDa). These proteins are present in varying ratios and provide a well-balanced amino acid profile and exhibit emulsifying and gelling properties [127]. It has been shown that canola protein isolate derived from cold-pressed canola meal can be used to prepare oleogels through an emulsion-templated approach without the need for thickening agents [128]. Stable emulsions (50 wt% oil) were prepared by high-pressure homogenization, followed by a thermal denaturation step. The protein-stabilized oleogels were then produced by shearing and oven drying to remove residual water.

Proteins present in by-products of food processing represent another underutilized, yet promising source of functional proteins which are being investigated with increasing frequency. Protein extracted from rice bran (a major by-product of rice milling) has been shown to have high nutritional value, low allergenicity [129], and demonstrated surface activity [130]. Rice bran protein was used to produce oleogels via a cryogel-templating approach without the need for additional hydrocolloids [53]. Oil absorption and corresponding oleogel properties depended on foaming behavior, which was strongly impacted by the pH during foaming. Aquafaba (a major by-product of legume cooking) also represents an underutilized by-product rich in proteins, starch, and surface-active small molecules [131], which can be explored as either an emulsifier or highly effective foaming agent to produce emulsion-templated and foam-templated oleogels, respectively. For instance, chickpea aquafaba-based emulsions have been evaluated as a potential substitute for palm oil in pound cake [3]. The sensory evaluation showed no statistically significant difference between cakes produced with chickpea aquafaba HIPE and palm oil for the attributes of aroma, color, texture, flavor, and overall impression. These results indicate that aquafaba-based emulsions can potentially replace palm oil in cakes, thereby reducing both total fat and saturated fatty acid content.

Combining well-characterized proteins with those from non-traditional sources may provide further opportunities to expand the functionality of both ingredients. Using this approach, oleogels have been prepared by combining soy protein isolate and proteins derived from bamboo shoots using alkaline extraction [132]. The oleogels were prepared using a freeze-dried emulsion template approach and varying the soy/bamboo shoot protein ratio. Increasing bamboo shoot protein progressively decreased the oleogel rheological viscoelastic response, firmness (by texture profile analysis), viscosity, and oil binding capacity, but increased cohesiveness. Interestingly, despite producing weaker oleogels, the free fatty acid release during in vitro static digestion decreased as the proportion of bamboo shoot protein was raised. This may suggest that oil release was correlated to product cohesiveness, i.e., the ability of the gel structure to retain its integrity after an initial deformation. Taken together, these studies underscore the promising opportunities to upcycle food manufacturing side-streams and valorize low-value products by identifying highly functional but underexploited ingredients.

10.3.3 Colloidal Complexes and Conjugates

Combining proteins and polysaccharides is a common approach to alter the functional performance of food biopolymers, including modifying gelation behavior and interfacial activity, and to achieve product stability (e.g., in protein-stabilized emulsions) [133, 134]. The interaction between biopolymers is often electrostatic in nature, and the net charge of both proteins and polysaccharides depends on the content of ionizable functional groups and environmental conditions (i.e., pH, presence of ionic species). Polysaccharides are most commonly uncharged or anionic at pH conditions relevant to foods, with chitin being a notable exception due to the presence of N-acetylglucosamine functional groups which are positively charged at neutral pH. In contrast, proteins are inherently amphiphilic and can have either a net negative or positive charge if the environmental pH is above or below the isoelectric point, respectively. The magnitude of this charge also strengthens at pH further from the isoelectric point, due to the higher concentration of protonated/deprotonated functional groups. In protein-stabilized emulsions, uncharged polysaccharides are often used in emulsions to increase solution viscosity, while anionic varieties may also provide electrostatic repulsion between emulsion droplets to reduce aggregation and coalescence [133, 135]. Appropriate selection of biopolymers, concentrations, and environmental conditions can also be used to form associative interactions that give rise to electrostatic complexes. These may take the form of mixed complexes, multilayer structures formed by sequential deposition at an interface, or Pickering emulsions using protein-polysaccharide nanoparticles [135].

Although protein-polysaccharide complexes have been studied extensively for food applications, there are relatively few reports investigating these systems as oleogelators. Electrostatic complexes of SPI and the anionic polysaccharide κ-carrageenan have been produced using an emulsion-templated approach and oven drying to remove water [29]. Concentrated emulsions (60 wt% oil) were prepared with a protein/polysaccharide ratio of 15:1 at pH 3 (based on a pre-screening evaluation). While this pH is below the isoelectric point of κ-carrageenan, it has been shown to promote the electrostatic complexation of these biopolymers [136]. The complexes formed weaker oleogels (lower G') than those prepared with SPI as the sole emulsifier, suggesting the complexes assembled at the oil/water interface did not form as extensive interactions as soy proteins alone. However, this work demonstrates the addition of polysaccharides could be an effective approach to modulate the rheological behavior of protein-based oleogels produced via emulsion templating.

Oleogels have also shown considerable promise as delivery systems of health-promoting bioactive compounds, such as sensitive lipids and fat-soluble nutraceuticals [10]. Indirect approaches to producing oleogels also provide a potential route to directly incorporate certain bioactives as structuring agents. The formation of protein-polyphenol complexes and conjugates is viewed as a promising approach to alter protein functionality and develop structures with added antioxidant activity [106, 137]. Complexes may be formed through hydrophobic and/or electrostatic interactions. An alternative approach is covalent grafting, which can be achieved using enzymatic methods, alkaline conditions, or free radicals (e.g., hydrogen peroxide/ascorbic acid redox pair) [137].

Passive electrostatic complexation has been employed to produce emulsion-templated oleogels using gelatin complexed with either proanthocyanidins [94] or tannic acid [105]. Both these systems also incorporated additional polysaccharides after forming the complexes, and templates were produced by freeze-drying or oven-drying. Tannic acid has also been combined with the protein zein, a hydrophobic prolamin storage protein derived from corn [138]. The latter complexes were used to produce colloidal particles via antisolvent precipitation, freeze-drying, and ball-milling with oil to produce a suspension of micron-sized colloidal particles. Introducing a small amount of water resulted in a self-supporting capillary suspension due to the wetting of tannic acid. This same group used an analogous approach to prepare covalent conjugates using gallic acid, (−)-epigallocatechin gallate, or tannic acid with heat-induced WPI microgels formed through the free-radical grafting approach [139]. The resulting capillary suspensions were used to improve the oxidative stability of a gelled algal oil rich in ω-3 fatty acids.

Oleogels produced using protein-polysaccharide Maillard conjugates as hydrogel templates have been reported by reacting sodium alginate with plant- or animal-derived proteins using controlled heat treatment. SPI-alginate complexes were converted to hydrogels by the addition of $CaCO_3$ to induce gelation of the polysaccharide prior to freeze-drying [57]. In contrast, aerogels prepared from gelatin-alginate conjugates were directly frozen from the solution [140]. It was reported that the heat-treated SPI-alginate conjugates produced stronger oleogels when intermediate grafting was achieved using a lower temperature and shorter time. In contrast, the gelatin-alginate conjugates had comparable oil absorption and oil holding capacity, but more extensive grafting produced oleogels with a higher elastic modulus and compressive strength, and lower free fatty acid release during simulated digestion. Therefore, both the protein source and grafting procedure will strongly impact the properties of protein-alginate complexes. Further work exploring the impact of alternative polysaccharides may provide an additional approach to tailoring the properties of oleogels structured with Maillard conjugate.

As noted previously, polysaccharides and protein-polysaccharide conjugates that have been used to strengthen the interface of both conventional emulsions and HIPEs for use as oleogel templates (Sect. 10.3.1). Examples include xanthan gum [2, 4, 28, 42], carrageenans [29, 42], gum Arabic [42], alginate [38, 42], pectin [38, 40, 42], gellan gum [42], locust bean gum [42], tara gum [13, 42], and different forms of cellulose derivatives, including MC [27], carboxymethyl cellulose [141], and HPMC [31]. In several of the above-mentioned studies, the complexation of the polysaccharides with proteins was evaluated at various pH values, concentrations, ratios protein/polysaccharide ratios, process parameters (e.g., dispersing equipment and operating speed and time), and dispersion preparation technique. One approach to improve emulsion stability is via a layer-by-layer deposition process. In this approach, the oil is first dispersed into a protein solution, forming a protein-stabilized emulsion, followed by the addition of a polysaccharide and continued emulsification, allowing the polysaccharide to associate with the protein film (generally via electrostatic interactions), forming an exterior layer. An alternate approach is to produce the emulsified system by dispersing the oil directly into a solution of the two biopolymers which form a mixed interfacial layer. It is noted that these approaches are extensively employed in the formation and stabilization of emulsions and have only recently been explored for the production of oleogels [142].

10.4 Indirect Oleogels in Food Applications

Oleogels produced by indirect methods using proteins and/or polysaccharides as structurants have been applied to many food products as a potential replacement for traditional fat sources rich in saturated or *trans*-fatty acids. Examples include sandwich cookie creams [46], cakes [27, 51, 52, 72, 83, 128, 143], muffins [84], meat patties [144–146], plant-based analogs of salami [75] and bacon [147], peanut butter [47], chocolate [33, 146, 148], and chocolate spread [93] (Table 10.1). These studies have evaluated oleogels produced by the foam or emulsion-templated approaches. Cake has been the most widely tested food product, indicating that oleogels produced by the indirect method using various proteins and polysaccharides can be a potential candidate suitable for developing functional shortening replacements. Most of these findings have shown the potential of oleogels as replacers for *trans*- and saturated fatty acids, thereby potentially improving the nutritional profile of the formulated product. Select examples of structured emulsions and emulsion-filled hydrogels developed to mimic animal adipose tissue have also been included and may represent a promising direction for future applications of

Table 10.1 Oleogels produced by indirect methods used in food applications

Oleogelator system *(a)* Gelator(s) *(b)* Oil type *(c)* Production method	**Food product** *Intended application*	**Main findings**	**Ref.**
Oleogels			
(a) MC and XG *(b)* Sunflower oil *(c)* Emulsion-templated	Cake batter *Shortening substitute*	Oleogels produced cakes with textural and rheological properties comparable to those made with commercial shortening and maintained freshness better than cakes with unstructured oil	[27]
(a) HPMC *(b)* Canola oil *(c)* Cryogel-templated	Cake batter *Shortening substitute*	Oleogels had poor structural recovery and produced batters with intermediate density relative to shortening and oil and cakes with a 'gel-like' rheological response at elevated temperatures	[83]
(a) Pea protein concentrate and XG *(b)* Canola oil *(c)* Cryogel-templated	Cake batter *Shortening substitute*	Cryogel-templated oleogels prepared at pH 9 provided the best performance as shortening substitutes; Oleogel cake batters had fewer, larger air pockets and more irregular structure than shortening; Oleogel cakes had comparable specific volume but higher hardness and chewiness than those with shortening	[51]
(a) Faba bean or pea protein concentrate and XG; MAG or CLW *(b)* Canola oil *(c)* Cryogel-templated	Cake batter *Shortening substitute*	Addition of high-melting MAG or CLW improved the oil binding and rheological properties of oleogels; MAG improved the textural properties of cakes prepared with oleogels but did not perform as well as shortening	[52]
(a) Faba bean or pea protein concentrate and XG; MAG or CLW *(b)* Canola oil *(c)* Cryogel-templated; aerated	Cake batter *Shortening substitute*	Aerated oleogels prepared with high-melting MAG or CLW produced cakes with larger, heterogeneous air cells, but provided similar textural properties to shortening-based cakes	[72]
(a) Canoa protein isolate *(b)* Canola oil *(c)* Emulsion-templated	Cake batter *Shortening substitute*	Oleogels produced cake batters with large air cells and more homogeneous fat distribution, resulting in cakes with higher volume and lower hardness relative to shortening control, but higher cohesiveness and springiness	[128]
(a) Citrus pectin/tea polyphenol-palmitate complexes *(b)* Camellia oil *(c)* Emulsion-templated	Cake batter *Butter substitute*	Oleogels (100% replacement) reduced air incorporation in cake batters, but produced similar cake volumes when using up to 3.5% citrus pectin in emulsion templates; Increasing pectin content increased product firmness and chewiness, and reduced overall acceptability	[143]

(continued)

Table 10.1 (continued)

Oleogelator system *(a)* Gelator(s) *(b)* Oil type *(c)* Production method	**Food product** *Intended application*	**Main findings**	**Ref.**
(a) HPMC; MAG *(b)* Soybean oil *(c)* Cryogel-templated	Cake batter, cookies *Butter substitute*	Increasing MAG content in HPMC-based oleogels positively correlated to hardness and chewiness, but negatively correlated to cohesiveness	[50]
(a) HPMC *(b)* Sunflower oil *(c)* Cryogel-templated	Muffin *Shortening substitute*	Oleogels produced muffin batters with larger, heterogeneously distributed air cells; Partial substitution (50%) maintained acceptable specific volume and favorable soft, chewy texture; 100% replacement caused an increase in textural hardness, springiness, and chewiness	[84]
(a) Chitosan, vanillin (crosslinker), and tween 60 *(b)* Canola oil *(c)* Emulsion-templated	Cookie *Shortening substitute*	Cookies prepared with oleogels had higher moisture loss during cooking, reduced spread, and higher browning; Oleogels with 1.5% chitosan and 1.0% vanillin (as crosslinker) produced cookies most similar to those prepared with vegetable shortening	[149]
(a) HPMC and XG/HPMC *(b)* Corn oil *(c)* Emulsion-templated/ cryogel-templated	Cookie *Shortening substitute*	Both emulsion- and cryogel-templated oleogels were firmer than commercial shortening; Cookies with emulsion templates (100% fat replacement) were equivalent in color, but both templating strategies had lower sensory ratings (appearance, flavor, crispiness, and overall appeal)	[150]
(a) HPMC or MC *(b)* Canola oil *(c)* Cryogel-templated; chopped	Sandwich cookie cream filling *Palm oil substitute*	Partial substitution (50%, 75%) produced small- and large-deformation rheological behavior similar to full-shortening control, while full replacement dramatically increased the stiffness of cream fillings	[46]
(a) HPMC or MC *(b)* Peanut oil (in situ) *(c)* Cryogel-templated; chopped	Peanut butter *Natural peanut butter stabilizer*	Templated oleogels improved the oil binding capacity of peanut butter; Peanut butter with >1 wt% HPMC or MC eliminated oil separation for a minimum of 6 months	[47]
(a) HPMC *(b)* High oleic sunflower oil *(c)* Emulsion-templated	Cocoa butter *Cocoa butter substitute with improved lipid profile*	Increasing oleogel content (50–100% fat replacement) reduced saturated fatty acid content, but progressively diminished rheological properties, melting onset temperature, and textural firmness	[33]

(continued)

Table 10.1 (continued)

Oleogelator system *(a)* Gelator(s) *(b)* Oil type *(c)* Production method	**Food product** *Intended application*	**Main findings**	**Ref.**
(a) HPMC *(b)* High oleic sunflower oil *(c)* Cryogel-templated; chopped	Chocolate *Partial cocoa butter substitute*	Increasing oleogel content reduced desirable textural properties of both cocoa butter/oleogel blends and chocolates (firmness, melting behavior); Chocolates with up to 50% replacement had similar textural and oil-binding properties to controls	[148]
(a) HPMC *(b)* High oleic sunflower oil *(c)* Emulsion-templated	Chocolate *Partial cocoa butter substitute*	Chocolate firmness increased with HPMC content (0.5–2 wt%) but was significantly weaker than control; Increasing HPMC improved the peak melting temperature, but produced a rough, sandy, and softer texture	[146]
(a) HPMC and XG *(b)* Olive oil or sunflower oil *(c)* Emulsion-templated	Chocolate spread *Partial coconut oil substitute*	50% replacement of coconut oil with oleogel produced firmer chocolate spreads, but retained a homogeneous microstructure and similar sensory attributes to the control	[93]
(a) HPMC *(b)* Canola oil *(c)* Cryogel-templated; chopped	Meat patties *Animal adipose tissue substitute*	50% replacement of beef tallow with oleogel increased sensory juiciness, tenderness, and overall acceptability relative to both control and 100% substitution	[144]
Emulsion gels/structured emulsions			
(a) Soy protein isolate; wheat fiber particles *(b)* Canola oil *(c)* Crosslinked emulsion	Dry-fermented poultry sausage *Palm fat replacement*	Oleogel hardness increased with the length of added wheat fibers; Salamis prepared with oleogels had lower hardness and increased drying time relative to those with palm fat	[74]
(a) Soy protein isolate; Sal fat *(b)* Canola oil *(c)* Crosslinked structured emulsion	Plant-based salami analog *Adipose tissue mimetic*	Transglutaminase produced crosslinked structured emulsions; Increasing solid fat content of emulsion increased firmness of both oleogels and salami, but decreased cohesiveness, springiness, and rate of weight loss during drying; Oleogels with 50% sal fat produced products with the highest sensory acceptability	[75]
(a) Soy protein isolate; shea stearin *(b)* Canola oil *(c)* Crosslinked structured emulsion	Plant-based bacon analog *Adipose tissue mimetic*	Transglutaminase, pH reduction, or divalent salts produced crosslinked structured emulsions; Covalent crosslinking was needed to provide suitable protein/fat adhesion required for practical performance	[147]

(continued)

Table 10.1 (continued)

Oleogelator system *(a)* Gelator(s) *(b)* Oil type *(c)* Production method	**Food product** *Intended application*	**Main findings**	**Ref.**
(a) Gelatin, cocoa bean shell flour *(b)* Walnut oil *(c)* Emulsion-filled gel (40% oil)	Beef burgers *Animal fat replacement*	Increasing substitution of pork backfat with emulsion-filled gel caused a progressive decrease in product hardness, cohesiveness, and chewiness, and increased oxidized lipids after cooking, but reduced cooking loss; 50% fat replacement produced sensory properties similar to full fat control	[145]

Abbreviations: *CLW* candelilla wax, *HPMC* hydroxypropyl methylcellulose, *MAG* monoglycerides, *MC* methylcellulose, *XG* xanthan gum

indirect oleogelator systems. For future studies, it will be of great value to investigate the economic feasibility, shelf life, oxidative stability, and sensory attributes to fully characterize the commercial potential of using oleogels as fat replacers in foods.

10.5 Conclusion and Future Perspectives

While the field of oleogelation has seen a rapid progression in recent years, it is still a technological challenge to develop structured oils that can effectively replace solid fats. The potential of applying oleogels produced by indirect methods at an industrial level depends not only on the sensory characteristics but also on the performance of these oleogels in the context of a food matrix, including oxidative stability, rheological behavior, and the ability to mimic the mouthfeel and texture of fats, as well as legislation, consumer acceptability, and the feasibility and scalability of the indirect production method. Yet, oleogelation using polysaccharides and proteins can play a significant role in reshaping how oil is structured for industrial applications aimed at reducing and/or replacing saturated and *trans*-fatty acids. Further research to expand the use of plant-derived oleogelators would widen the possibilities for developing novel healthier foods suitable for vegetarian and vegan diets, or cater to consumers interested in reducing their environmental impact, both of which are currently increasing trends. In their current form, scale-up remains challenging for indirect oleogelator systems, and further work on understanding the impact of extraction and processing on the functional properties of proteins will also be a critical factor. Thus, to improve the human diet by enabling nutritional reformulation with novel fat replacers such as protein- and polysaccharide-based oleogels, further research will continue to improve production methods, physical and oxidative stability, and consumer acceptability, thereby contributing to a more sustainable food system.

Acknowledgments Andrew Gravelle acknowledges financial support from the USDA National Institute of Food and Agriculture Hatch/Multi-State project. Graziele Grossi Bovi Karatay thanks the FAPESP for financial support (fellowship 2020/05074-6). Miriam Hubinger thanks CNPq research fellowship (309022/2021-5) and FAPESP research projects (2021/06863-7 and Thematic 2019/27354-3) for their financial support.

References

1. Scholten E (2019) Edible oleogels: how suitable are proteins as a structurant? Curr Opin Food Sci 27:36–42. https://doi.org/10.1016/j.cofs.2019.05.001
2. Patel AR, Rajarethinem PS, Cludts N, Lewille B, De Vos WH, Lesaffer A, Dewettinck K (2015) Biopolymer-based structuring of liquid oil into soft solids and oleogels using water-continuous emulsions as templates. Langmuir 31:2065–2073. https://doi.org/10.1021/la502829u
3. Grossi Bovi Karatay G, Rebellato AP, Joy Steel C, Dupas Hubinger M (2022) Chickpea aquafaba-based emulsions as a fat replacer in pound cake: impact on cake properties and sensory analysis. Foods 11. https://doi.org/10.3390/foods11162484
4. Vélez-Erazo EM, Bosqui K, Rabelo RS, Hubinger MD (2021) Effect of PH and pea protein: xanthan gum ratio on emulsions with high oil content and high internal phase emulsion formation. Molecules 26. https://doi.org/10.3390/molecules26185646
5. Patel AR, Dewettinck K (2016) Edible oil structuring: an overview and recent updates. Food Funct 7:20–29. https://doi.org/10.1039/C5FO01006C
6. Feichtinger A, Scholten E (2020) Preparation of protein oleogels: effect on structure and functionality. Foods 9. https://doi.org/10.3390/foods9121745
7. Limaho H, Sugiarto, Pramono R, Christiawan R (2022) The need for global green marketing for the palm oil industry in Indonesia. Sustainability 14. https://doi.org/10.3390/su14148621
8. Jasaw GS, Saito O, Gasparatos A, Shoyama K, Boafo YA, Takeuchi K (2017) Ecosystem services trade-offs from high fuelwood use for traditional shea butter processing in semi-arid Ghana. Ecosyst Serv 27:127–138. https://doi.org/10.1016/j.ecoser.2017.09.003
9. Martins AJ, Vicente AA, Pastrana LM, Cerqueira MA (2020) Oleogels for development of health-promoting food products. Food Sci Human Wellness 9:31–39. https://doi.org/10.1016/j.fshw.2019.12.001
10. Pinto TC, Martins AJ, Pastrana L, Pereira MC, Cerqueira MA (2021) Oleogel-based systems for the delivery of bioactive compounds in foods. Gels 7. https://doi.org/10.3390/gels7030086
11. Silva PM, Cerqueira MA, Martins AJ, Fasolin LH, Cunha RL, Vicente AA (2022) Oleogels and bigels as alternatives to saturated fats: a review on their application by the food industry. J Am Oil Chem Soc 99:911–923. https://doi.org/10.1002/aocs.12637
12. da Silva TLT, Arellano DB, Martini S (2019) Interactions between candelilla wax and saturated triacylglycerols in oleogels. Food Res Int 121:900–909. https://doi.org/10.1016/J.FOODRES.2019.01.018
13. Vélez-Erazo EM, Okuro PK, Gallegos-Soto A, da Cunha RL, Hubinger MD (2022) Protein-based strategies for fat replacement: approaching different protein colloidal types, structured systems and food applications. Food Res Int 156:111346. https://doi.org/10.1016/j.foodres.2022.111346
14. Wang Z, Chandrapala J, Truong T, Farahnaky A (2022) Oleogels prepared with low molecular weight gelators: texture, rheology and sensory properties, a review. Crit Rev Food Sci Nutr:1–45. https://doi.org/10.1080/10408398.2022.2027339
15. Kavya M, Udayarajan C, Fabra MJ, López-Rubio A, Nisha P (2022) Edible oleogels based on high molecular weight oleogelators and its prospects in food applications. Crit Rev Food Sci Nutr:1–24

16. Tang Q, Huang G (2022) Improving method, properties and application of polysaccharide as emulsifier. Food Chem 376:131937. https://doi.org/10.1016/j.foodchem.2021.131937
17. Shakeel A, Farooq U, Gabriele D, Marangoni AG, Lupi FR (2021) Bigels and multi-component organogels: an overview from rheological perspective. Food Hydrocoll 111: 106190. https://doi.org/10.1016/j.foodhyd.2020.106190
18. Abdollahi M, Goli SAH, Soltanizadeh N (2020) Physicochemical properties of foam-templated oleogel based on gelatin and xanthan gum. Eur J Lipid Sci Technol 122:1900196. https://doi.org/10.1002/ejlt.201900196
19. Skilling KJ, Citossi F, Bradshaw TD, Ashford M, Kellam B, Marlow M (2014) Insights into low molecular mass organic gelators: a focus on drug delivery and tissue engineering applications. Soft Matter 10:237–256. https://doi.org/10.1039/C3SM52244J
20. Sharma N, Bansal V, Sahu JK (2023) Chapter 7: Plant-based gels. In: Prakash S, Bhandari BR, Gaiani C (eds) Engineering plant-based food systems. Academic, pp 131–150. isbn:978-0-323-89842-3
21. Martins AJ, Vicente AA, Cunha RL, Cerqueira MA (2018) Edible oleogels: an opportunity for fat replacement in foods. Food Funct 9:758–773. https://doi.org/10.1039/c7fo01641g
22. Davidovich-Pinhas M (2018) Oleogels. In: Pal K, Banerjee I (eds) Polymeric gels, Woodhead publishing series in biomaterials. Woodhead Publishing, pp 231–249. isbn:978-0-08-102179-8
23. Laredo T, Barbut S, Marangoni AG (2011) Molecular interactions of polymer oleogelation. Soft Matter 7:2734–2743. https://doi.org/10.1039/C0SM00885K
24. Ashfaq A, Jahan K, Islam RU, Younis K (2022) Protein-based functional colloids and their potential applications in food: a review. LWT 154:112667. https://doi.org/10.1016/j.lwt.2021.112667
25. Romoscanu AI, Mezzenga R (2006) Emulsion-templated fully reversible protein-in-oil gels. Langmuir 22:7812–7818. https://doi.org/10.1021/la060878p
26. Patel AR (2018) Structuring edible oils with hydrocolloids: where do we stand? Food Biophys 13:113–115. https://doi.org/10.1007/s11483-018-9527-6
27. Patel AR, Cludts N, Sintang MDB, Lesaffer A, Dewettinck K (2014) Edible oleogels based on water soluble food polymers: preparation, characterization and potential application. Food Funct 5:2833–2841. https://doi.org/10.1039/c4fo00624k
28. Patel AR, Cludts N, Bin Sintang MD, Lewille B, Lesaffer A, Dewettinck K (2014) Polysaccharide-based oleogels prepared with an emulsion-templated approach. ChemPhysChem 15:3435–3439. https://doi.org/10.1002/cphc.201402473
29. Tavernier I, Patel AR, Van der Meeren P, Dewettinck K (2017) Emulsion-templated liquid oil structuring with soy protein and soy protein: κ-carrageenan complexes. Food Hydrocoll 65: 107–120. https://doi.org/10.1016/j.foodhyd.2016.11.008
30. Tavernier I, Doan CD, Van der Meeren P, Heyman B, Dewettinck K (2018) The potential of waxes to alter the microstructural properties of emulsion-templated oleogels. Eur J Lipid Sci Technol 120:1700393. https://doi.org/10.1002/ejlt.201700393
31. Meng Z, Qi K, Guo Y, Wang Y, Liu Y (2018) Macro-micro structure characterization and molecular properties of emulsion-templated polysaccharide oleogels. Food Hydrocoll 77:17–29. https://doi.org/10.1016/j.foodhyd.2017.09.006
32. Espert M, Salvador A, Sanz T (2020) Cellulose ether oleogels obtained by emulsion-templated approach without additional thickeners. Food Hydrocoll 109:106085. https://doi.org/10.1016/j.foodhyd.2020.106085
33. Alvarez MD, Cofrades S, Espert M, Salvador A, Sanz T (2021) Thermorheological characterization of healthier reduced-fat cocoa butter formulated by substitution with a hydroxypropyl methylcellulose (HPMC)-based oleogel. Foods 10. https://doi.org/10.3390/foods10040793
34. Bascuas S, Salvador A, Hernando I, Quiles A (2020) Designing hydrocolloid-based oleogels with high physical, chemical, and structural stability. Front Sustain Food Syst 4:1–8. https://doi.org/10.3389/fsufs.2020.00111

35. Bascuas S, Hernando I, Moraga G, Quiles A (2020) Structure and stability of edible oleogels prepared with different unsaturated oils and hydrocolloids. Int J Food Sci Technol 55:1458–1467. https://doi.org/10.1111/ijfs.14469
36. Meissner PM, Keppler JK, Schwarz K (2023) Lipid oxidation induced protein scission in an oleogel as a model food. Food Chem 415:135357. https://doi.org/10.1016/j.foodchem.2022.135357
37. Meissner PM, Keppler JK, Stöckmann H, Schwarz K (2020) Cooxidation of proteins and lipids in whey protein oleogels with different water amounts. Food Chem 328:127123. https://doi.org/10.1016/j.foodchem.2020.127123
38. Wijaya W, Van der Meeren P, Patel AR (2018) Oleogels from emulsion (HIPE) templates stabilized by protein–polysaccharide complexes. The Royal Society of Chemistry, pp 175–197. isbn:978-1-78262-829-3
39. Cameron NR (2005) High internal phase emulsion templating as a route to well-defined porous polymers. Polymer (Guildf) 46:1439–1449. https://doi.org/10.1016/j.polymer.2004.11.097
40. Wijaya W, der Meeren P, Wijaya CH, Patel AR (2017) High internal phase emulsions stabilized solely by whey protein isolate-low methoxyl pectin complexes: effect of PH and polymer concentration. Food Funct 8:584–594. https://doi.org/10.1039/C6FO01027J
41. Wijaya W, Sun Q-Q, Vermeir L, Dewettinck K, Patel AR, Van der Meeren P (2019) PH and protein to polysaccharide ratio control the structural properties and viscoelastic network of HIPE-templated biopolymeric oleogels. Food Struct 21:100112. https://doi.org/10.1016/j.foostr.2019.100112
42. Vélez-Erazo EM, Bosqui K, Rabelo RS, Kurozawa LE, Hubinger MD (2020) High internal phase emulsions (HIPE) using pea protein and different polysaccharides as stabilizers. Food Hydrocoll 105. https://doi.org/10.1016/j.foodhyd.2020.105775
43. Patel AR, Schatteman D, Lesaffer A, Dewettinck K (2013) A foam-templated approach for fabricating organogels using a water-soluble polymer. RSC Adv 3:22900–22903. https://doi.org/10.1039/C3RA44763D
44. Bascuas S, Morell P, Hernando I, Quiles A (2021) Recent trends in oil structuring using hydrocolloids. Food Hydrocoll 118:106612. https://doi.org/10.1016/j.foodhyd.2021.106612
45. Jiang Q, Geng M, Meng Z (2022) Enhancement effect of fat crystal network on oleogels prepared by methyl-cellulose and xanthan gum using the cryogel-templated method. J Food Process Preserv 46. https://doi.org/10.1111/jfpp.17000
46. Tanti R, Barbut S, Marangoni AG (2016) Hydroxypropyl methylcellulose and methylcellulose structured oil as a replacement for shortening in sandwich cookie creams. Food Hydrocoll 61: 329–337. https://doi.org/10.1016/j.foodhyd.2016.05.032
47. Tanti R, Barbut S, Marangoni AG (2016) Oil stabilization of natural peanut butter using food grade polymers. Food Hydrocoll 61:399–408. https://doi.org/10.1016/j.foodhyd.2016.05.034
48. Jiang Q, Du L, Li S, Liu Y, Meng Z (2021) Polysaccharide-stabilized aqueous foams to fabricate highly oil-absorbing cryogels: application and formation process for preparation of edible oleogels. Food Hydrocoll 120:106901. https://doi.org/10.1016/j.foodhyd.2021.106901
49. Jiang Q, Li P, Ji M, Du L, Li S, Liu Y, Meng Z (2022) Synergetic effects of water-soluble polysaccharides for intensifying performances of oleogels fabricated by oil-absorbing cryogels. Food Chem 372:131357. https://doi.org/10.1016/j.foodchem.2021.131357
50. Jiang Q, Yu Z, Meng Z (2022) Double network oleogels co-stabilized by hydroxypropyl methylcellulose and monoglyceride crystals: baking applications. Int J Biol Macromol 209: 180–187. https://doi.org/10.1016/j.ijbiomac.2022.04.011
51. Mohanan A, Tang YR, Nickerson MT, Ghosh S (2020) Oleogelation using pulse protein-stabilized foams and their potential as a baking ingredient. RSC Adv 10:14892–14905. https://doi.org/10.1039/C9RA07614J
52. Mohanan A, Nickerson MT, Ghosh S (2020) The effect of addition of high-melting monoacylglycerol and candelilla wax on pea and Faba bean protein foam-templated oleogelation. J Am Oil Chem Soc 97:1319–1333. https://doi.org/10.1002/aocs.12425

53. Wei F, Lv M, Li J, Xiao J, Rogers MA, Cao Y, Lan Y, Lu M, Li J, Xiao J et al (2022) Construction of foam-templated oleogels based on rice bran protein. Food Hydrocoll 124: 107245. https://doi.org/10.1016/j.foodhyd.2021.107245
54. Alizadeh L, Abdolmaleki K, Nayebzadeh K, Hosseini SM (2020) Oleogel fabrication based on sodium caseinate, hydroxypropyl methylcellulose, and beeswax: effect of concentration, oleogelation method, and their optimization. J Am Oil Chem Soc 97:485–496. https://doi.org/10.1002/aocs.12341
55. Manzocco L, Valoppi F, Calligaris S, Andreatta F, Spilimbergo S, Nicoli MC (2017) Exploitation of κ-carrageenan aerogels as template for edible oleogel preparation. Food Hydrocoll 71:68–75. https://doi.org/10.1016/j.foodhyd.2017.04.021
56. Manzocco L, Plazzotta S, Powell J, de Vries A, Rousseau D, Calligaris S (2022) Structural characterisation and sorption capability of whey protein aerogels obtained by freeze-drying or supercritical drying. Food Hydrocoll 122:107117. https://doi.org/10.1016/j.foodhyd.2021.107117
57. Chen K, Zhang H (2020) Fabrication of oleogels via a facile method by oil absorption in the aerogel templates of protein–polysaccharide conjugates. ACS Appl Mater Interfaces 12:7795–7804. https://doi.org/10.1021/acsami.9b21435
58. de Vries A, Hendriks J, van der Linden E, Scholten E (2015) Protein oleogels from protein hydrogels via a stepwise solvent exchange route. Langmuir 31:13850–13859. https://doi.org/10.1021/acs.langmuir.5b03993
59. de Vries A, Lopez Gomez Y, Jansen B, van der Linden E, Scholten E (2017) Controlling agglomeration of protein aggregates for structure formation in liquid oil: a sticky business. ACS Appl Mater Interfaces 9:10136–10147. https://doi.org/10.1021/acsami.7b00443
60. de Vries A, Wesseling A, van der Linden E, Scholten E (2017) Protein oleogels from heat-set whey protein aggregates. J Colloid Interface Sci 486:75–83
61. Akharume FU, Aluko RE, Adedeji AA (2021) Modification of plant proteins for improved functionality: a review. Compr Rev Food Sci Food Saf 20:198–224. https://doi.org/10.1111/1541-4337.12688
62. Tan C, McClements DJ (2021) Application of advanced emulsion technology in the food industry: a review and critical evaluation. Foods 10. https://doi.org/10.3390/foods10040812
63. Geremias-Andrade IM, Souki NPBG, Moraes ICF, Pinho SC (2016) Rheology of emulsion-filled gels applied to the development of food materials. Gels 2. https://doi.org/10.3390/gels2030022
64. Alves Barroso L, Grossi Bovi Karatay G, Dupas Hubinger M (2022) Effect of potato starch hydrogel:glycerol monostearate oleogel ratio on the physico-rheological properties of bigels. Gels 8. https://doi.org/10.3390/gels8110694
65. Bealer EJ, Onissema-Karimu S, Rivera-Galletti A, Francis M, Wilkowski J, la Cruz D, Hu X (2020) Protein–polysaccharide composite materials: fabrication and applications. Polymers (Basel) 12. https://doi.org/10.3390/polym12020464
66. Cho K, Tarté R, Acevedo NC (2023) Development and characterization of the freeze-thaw and oxidative stability of edible rice bran wax-gelatin biphasic gels. LWT 174:114330. https://doi.org/10.1016/j.lwt.2022.114330
67. Grossi Bovi Karatay G, Medeiros Theóphilo Galvão AM, Dupas Hubinger M (2022) Storage stability of conventional and high internal phase emulsions stabilized solely by chickpea aquafaba. Foods 11. https://doi.org/10.3390/foods11111588
68. Zuo Z, Zhang X, Li T, Zhou J, Yang Y, Bian X, Wang L (2022) High internal phase emulsions stabilized solely by sonicated quinoa protein isolate at various PH values and concentrations. Food Chem 378:132011. https://doi.org/10.1016/j.foodchem.2021.132011
69. Galvão AMMT, Vélez-Erazo EM, Karatay GGB, de Figueiredo Furtado G, Vidotto DC, Tavares GM, Hubinger MD (2022) High internal phase emulsions stabilized by the lentil protein isolate (Lens culinaris). Colloids Surf A Physicochem Eng Asp 653:129993. https://doi.org/10.1016/j.colsurfa.2022.129993

70. Wijaya W, Van der Meeren P, Dewettinck K, Patel AR (2018) High internal phase emulsion (HIPE)-templated biopolymeric oleofilms containing an ultra-high concentration of edible liquid oil. Food Funct 9:1993–1997. https://doi.org/10.1039/C7FO01945A
71. Fameau A-L, Saint-Jalmes A (2020) Recent advances in understanding and use of oleofoams. Front Sustain Food Syst 4:110
72. Mohanan A, Harrison K, Cooper DML, Nickerson MT, Ghosh S (2022) Conversion of pulse protein foam-templated oleogels into oleofoams for improved baking application. Foods 11. https://doi.org/10.3390/foods11182887
73. Dreher J, Blach C, Terjung N, Gibis M, Weiss J (2020) Formation and characterization of plant-based emulsified and crosslinked fat crystal networks to mimic animal fat tissue. J Food Sci 85:421–431. https://doi.org/10.1111/1750-3841.14993
74. Dreher J, Knorz M, Herrmann K, Terjung N, Gibis M, Weiss J (2022) Structuring oil to substitute palm fat in dry-fermented poultry sausages. Food Struct 33:100281. https://doi.org/10.1016/j.foostr.2022.100281
75. Dreher J, König M, Herrmann K, Terjung N, Gibis M, Weiss J (2021) Varying the amount of solid fat in animal fat mimetics for plant-based salami analogues influences texture, appearance and sensory characteristics. LWT 143:111140. https://doi.org/10.1016/j.lwt.2021.111140
76. Ma KK, Greis M, Lu J, Nolden AA, McClements DJ, Kinchla AJ (2022) Functional performance of plant proteins. Foods 11. https://doi.org/10.3390/foods11040594
77. Yang X, Li A, Li X, Sun L, Guo Y (2020) An overview of classifications, properties of food polysaccharides and their links to applications in improving food textures. Trends Food Sci Technol 102:1–15. https://doi.org/10.1016/j.tifs.2020.05.020
78. Klemm D, Heublein B, Fink H, Bohn A (2005) Cellulose: fascinating biopolymer and sustainable raw material. Angew Chem Int Ed 44:3358–3393
79. Gravelle AJ, Marangoni AG (2018) Ethylcellulose oleogels: structure, functionality, and food applications. In: Toldrá F (ed) Advances in food and nutrition research, vol 84. Academic, pp 1–56. ISBN 1043-4526
80. Zhang R, Yu J, Liu N, Gao Y, Mao L (2022) W/O emulsions featuring ethylcellulose structuring in the water phase, Interface and oil phase for multiple delivery. Carbohydr Polym 283:119158. https://doi.org/10.1016/j.carbpol.2022.119158
81. da Silva Júnior WF, de Oliveira Pinheiro JG, Moreira CDLFA, de Souza FJJ, de Lima ÁAN (2017) Alternative technologies to improve solubility and stability of poorly water-soluble drugs. In: Grumezescu AM (ed) Multifunctional systems for combined delivery, biosensing and diagnostics. Elsevier, pp 281–305. ISBN 9780323527255. https://doi.org/10.1016/B978-0-323-52725-5.00015-0
82. Meng Z, Qi K, Guo Y, Wang Y, Liu Y (2018) Physical properties, microstructure, intermolecular forces, and oxidation stability of soybean oil oleogels structured by different cellulose ethers. Eur J Lipid Sci Technol 120. https://doi.org/10.1002/ejlt.201700287
83. Patel AR, Dewettinck K (2015) Comparative evaluation of structured oil systems: shellac oleogel, HPMC oleogel, and HIPE gel. Eur J Lipid Sci Technol 117:1772–1781. https://doi.org/10.1002/ejlt.201400553
84. Oh IK, Lee S (2018) Utilization of foam structured hydroxypropyl methylcellulose for oleogels and their application as a solid fat replacer in muffins. Food Hydrocoll 77:796–802. https://doi.org/10.1016/j.foodhyd.2017.11.022
85. Grishkewich N, Mohammed N, Tang J, Tam KC (2017) Recent advances in the application of cellulose nanocrystals. Curr Opin Colloid Interface Sci 29:32–45. https://doi.org/10.1016/j.cocis.2017.01.005
86. Moon RJ, Martini A, Nairn J, Simonsen J, Youngblood J (2011) Cellulose nanomaterials review: structure, properties and nanocomposites. Chem Soc Rev 40:3941–3994. https://doi.org/10.1039/C0CS00108B

87. Qi W, Li T, Zhang Z, Wu T (2021) Preparation and characterization of oleogel-in-water pickering emulsions stabilized by cellulose nanocrystals. Food Hydrocoll 110:106206. https://doi.org/10.1016/j.foodhyd.2020.106206
88. Li X, Guo G, Zou Y, Luo J, Sheng J, Tian Y, Li J (2023) Development and characterization of walnut oleogels structured by cellulose nanofiber. Food Hydrocoll 142:108849. https://doi.org/10.1016/j.foodhyd.2023.108849
89. Bhattarai M, Penttilä P, Barba L, Macias-Rodriguez B, Hietala S, Mikkonen KS, Valoppi F (2022) Size-dependent filling effect of crystalline celluloses in structural engineering of composite oleogels. LWT 160:113331. https://doi.org/10.1016/j.lwt.2022.113331
90. Fuhrmann PL, Powell J, Rousseau D (2023) Structure and rheology of oil-continuous capillary suspensions containing water-swellable cellulose beads and fibres. Food Hydrocoll 139: 108503. https://doi.org/10.1016/j.foodhyd.2023.108503
91. Sworn G (2000) Handbook of hydrocolloids. Williams Woodhead Publishing Series in Food Science, Technology and Nutrition, pp 186–203
92. Meng Z, Qi K, Guo Y, Wang Y, Liu Y (2018) Effects of thickening agents on the formation and properties of edible oleogels based on hydroxypropyl methyl cellulose. Food Chem 246: 137–149. https://doi.org/10.1016/j.foodchem.2017.10.154
93. Bascuas S, Espert M, Llorca E, Quiles A, Salvador A, Hernando I (2021) Structural and sensory studies on chocolate spreads with hydrocolloid-based oleogels as a fat alternative. LWT Food Sci Technol 135:110228. https://doi.org/10.1016/j.lwt.2020.110228
94. Pan H, Xu X, Qian Z, Cheng H, Shen X, Chen S, Ye X (2021) Xanthan gum-assisted fabrication of stable emulsion-based oleogel structured with gelatin and proanthocyanidins. Food Hydrocoll 115:106596. https://doi.org/10.1016/j.foodhyd.2021.106596
95. Espert M, Hernández MJ, Sanz T, Salvador A (2022) Rheological properties of emulsion templated oleogels based on xanthan gum and different structuring agents. Curr Res Food Sci 5:564–570. https://doi.org/10.1016/j.crfs.2022.03.001
96. Jaberi R, Pedram Nia A, Naji-Tabasi S, Elhamirad AH, Shafafi Zenoozian M (2020) Rheological and structural properties of oleogel base on soluble complex of egg white protein and xanthan gum. J Texture Stud 51:925–936. https://doi.org/10.1111/jtxs.12552
97. Abdolmaleki K, Alizadeh L, Nayebzadeh K, Hosseini SM, Shahin R (2020) Oleogel production based on binary and ternary mixtures of sodium caseinate, xanthan gum, and guar gum: optimization of hydrocolloids concentration and drying method. J Texture Stud 51:290–299. https://doi.org/10.1111/jtxs.12469
98. Plazzotta S, Calligaris S, Manzocco L (2019) Structure of oleogels from κ-carrageenan templates as affected by supercritical-CO2-drying, freeze-drying and lettuce-filler addition. Food Hydrocoll 96:1–10. https://doi.org/10.1016/j.foodhyd.2019.05.008
99. Goulding DA, Fox PF, O'Mahony JA (2020) Milk proteins: an overview. In: Milk proteins, pp 21–98
100. de Vries A, Jansen D, van der Linden E, Scholten E (2018) Tuning the rheological properties of protein-based oleogels by water addition and heat treatment. Food Hydrocoll 79:100–109. https://doi.org/10.1016/j.foodhyd.2017.11.043
101. Plazzotta S, Calligaris S, Manzocco L (2020) Structural characterization of oleogels from whey protein aerogel particles. Food Res Int 132:109099. https://doi.org/10.1016/j.foodres.2020.109099
102. Plazzotta S, Jung I, Schroeter B, Subrahmanyam RP, Smirnova I, Calligaris S, Gurikov P, Manzocco L (2021) Conversion of whey protein aerogel particles into oleogels: effect of oil type on structural features. Polymers (Basel) 13:4063
103. de Vries A, Gomez YL, van der Linden E, Scholten E (2017) The effect of oil type on network formation by protein aggregates into oleogels. RSC Adv 7:11803–11812. https://doi.org/10.1039/C7RA00396J
104. Singh, H.; Li, S. Functional properties of caseins and whey proteins. 2022

105. Qiu C, Huang Y, Li A, Ma D, Wang Y (2018) Fabrication and characterization of oleogel stabilized by gelatin-polyphenol-polysaccharides nanocomplexes. J Agric Food Chem 66: 13243–13252. https://doi.org/10.1021/acs.jafc.8b02039
106. Sun X, Sarteshnizi RA, Udenigwe CC (2022) Recent advances in protein–polyphenol interactions focusing on structural properties related to antioxidant activities. Curr Opin Food Sci 45:100840. https://doi.org/10.1016/j.cofs.2022.100840
107. Yano H, Fu W (2022) Effective use of plant proteins for the development of "new" foods. Foods 11. https://doi.org/10.3390/foods11091185
108. Loveday SM (2019) Food proteins: technological, nutritional, and sustainability attributes of traditional and emerging proteins. Annu Rev Food Sci Technol 10:311–339. https://doi.org/10.1146/annurev-food-032818-121128
109. Nicolai T (2019) Gelation of food protein-protein mixtures. Adv Colloid Interf Sci 270:147–164. https://doi.org/10.1016/j.cis.2019.06.006
110. Yang J, Sagis LMC (2021) Interfacial behavior of plant proteins — novel sources and extraction methods. Curr Opin Colloid Interface Sci 56:101499. https://doi.org/10.1016/j.cocis.2021.101499
111. Nishinari K, Fang Y, Guo S, Phillips GO (2014) Soy proteins: a review on composition, agregation and emulsification. Food Hydrocoll 39:301–318. https://doi.org/10.1016/j.foodhyd.2014.01.013
112. González-Pérez S, Arellano JB (2009) 15: Vegetable protein isolates. In: Phillips GO, Williams PA (eds) Handbook of hydrocolloids, Second edn. Woodhead Publishing, pp 383–419. isbn:978-1-84569-414-2
113. Zheng L, Regenstein JM, Zhou L, Wang Z (2022) Soy protein isolates: a review of their composition, aggregation, and gelation. Compr Rev Food Sci Food Saf 21:1940–1957. https://doi.org/10.1111/1541-4337.12925
114. Kim W, Wang Y, Selomulya C (2020) Dairy and plant proteins as natural food emulsifiers. Trends Food Sci Technol 105:261–272. https://doi.org/10.1016/j.tifs.2020.09.012
115. Chen L, Ao F, Ge X, Shen W (2020) Food-grade pickering emulsions: preparation, stabilization and applications. Molecules 25. https://doi.org/10.3390/molecules25143202
116. Naji-Tabasi S, Mahdian E, Arianfar A, Naji-Tabasi S (2020) Investigation of oleogel properties prepared by pickering emulsion-templated stabilized with solid particles of basil seed gum and isolated soy protein as a fat substitute in cream. Res Innov Food Sci Technol 9:269–282. https://doi.org/10.22101/JRIFST.2020.229269.1168
117. Zhang X, Luo X, Wang Y, Li Y, Li B, Liu S (2020) Concentrated O/W pickering emulsions stabilized by soy protein/cellulose nanofibrils: influence of PH on the emulsification performance. Food Hydrocoll 108:106025. https://doi.org/10.1016/j.foodhyd.2020.106025
118. Yang H, Su Z, Meng X, Zhang X, Kennedy JF, Liu B (2020) Fabrication and characterization of pickering emulsion stabilized by soy protein isolate-chitosan nanoparticles. Carbohydr Polym 247:116712. https://doi.org/10.1016/j.carbpol.2020.116712
119. Ju M, Zhu G, Huang G, Shen X, Zhang Y, Jiang L, Sui X (2020) A novel pickering emulsion produced using soy protein-anthocyanin complex nanoparticles. Food Hydrocoll 99:105329. https://doi.org/10.1016/j.foodhyd.2019.105329
120. Feng T, Wang X, Wang X, Zhang X, Gu Y, Xia S, Huang Q (2021) High internal phase pickering emulsions stabilized by pea protein isolate-high methoxyl pectin-EGCG complex: interfacial properties and microstructure. Food Chem 350:129251. https://doi.org/10.1016/j.foodchem.2021.129251
121. Ji Y, Han C, Liu E, Li X, Meng X, Liu B (2022) Pickering emulsions stabilized by pea protein isolate-chitosan nanoparticles: fabrication, characterization and delivery EPA for digestion in vitro and in vivo. Food Chem 378:132090. https://doi.org/10.1016/j.foodchem.2022.132090
122. Yi J, Gan C, Wen Z, Fan Y, Wu X (2021) Development of pea protein and high methoxyl pectin colloidal particles stabilized high internal phase pickering emulsions for β-carotene

protection and delivery. Food Hydrocoll 113:106497. https://doi.org/10.1016/j.foodhyd.2020.106497

123. Avelar Z, Vicente AA, Saraiva JA, Rodrigues RM (2021) The role of emergent processing technologies in tailoring plant protein functionality: new insights. Trends Food Sci Technol 113:219–231. https://doi.org/10.1016/j.tifs.2021.05.004
124. Amagliani L, Silva JVC, Saffon M, Dombrowski J (2021) On the foaming properties of plant proteins: current status and future opportunities. Trends Food Sci Technol 118:261–272. https://doi.org/10.1016/j.tifs.2021.10.001
125. Andler SM, Goddard JM (2018) Transforming food waste: how immobilized enzymes can valorize waste streams into revenue streams. NPJ Sci Food 2:19. https://doi.org/10.1038/s41538-018-0028-2
126. Dey T, Bhattacharjee T, Nag P, Ritika, Ghati A, Kuila A (2021) Valorization of agro-waste into value added products for sustainable development. Bioresour Technol Rep 16:100834. https://doi.org/10.1016/j.biteb.2021.100834
127. Chmielewska A, Kozłowska M, Rachwał D, Wnukowski P, Amarowicz R, Nebesny E, Rosicka-Kaczmarek J (2021) Canola/rapeseed protein–nutritional value, functionality and food application: a review. Crit Rev Food Sci Nutr 61:3836–3856
128. Tang YR, Ghosh S (2021) Canola protein thermal denaturation improved emulsion-templated oleogelation and its cake-baking application. RSC Adv 11:25141–25157. https://doi.org/10.1039/D1RA02250D
129. Park H-Y, Lee K-W, Choi H-D (2017) Rice bran constituents: immunomodulatory and therapeutic activities. Food Funct 8:935–943. https://doi.org/10.1039/C6FO01763K
130. Chen W, Ju X, Aluko RE, Zou Y, Wang Z, Liu M, He R (2020) Rice bran protein-based nanoemulsion carrier for improving stability and bioavailability of quercetin. Food Hydrocoll 108:106042. https://doi.org/10.1016/j.foodhyd.2020.106042
131. He Y, Meda V, Reaney MJT, Mustafa R (2021) Aquafaba, a new plant-based rheological additive for food applications. Trends Food Sci Technol 111:27–42. https://doi.org/10.1016/j.tifs.2021.02.035
132. Li J, Xi Y, Wu L, Zhang H (2023) Preparation, characterization and in vitro digestion of bamboo shoot protein/soybean protein isolate based-oleogels by emulsion-templated approach. Food Hydrocoll 136:108310. https://doi.org/10.1016/j.foodhyd.2022.108310
133. Gentile L (2020) Protein–polysaccharide interactions and aggregates in food formulations. Curr Opin Colloid Interface Sci 48:18–27. https://doi.org/10.1016/j.cocis.2020.03.002
134. Wijaya W, Patel AR, Setiowati AD, Van der Meeren P (2017) Functional colloids from proteins and polysaccharides for food applications. Trends Food Sci Technol 68:56–69. https://doi.org/10.1016/j.tifs.2017.08.003
135. Rodríguez Patino JM, Pilosof AMR (2011) Protein–polysaccharide interactions at fluid interfaces. Food Hydrocoll 25:1925–1937. https://doi.org/10.1016/j.foodhyd.2011.02.023
136. Molina Ortiz SE, Puppo MC, Wagner JR (2004) Relationship between structural changes and functional properties of soy protein isolates–carrageenan systems. Food Hydrocoll 18:1045–1053. https://doi.org/10.1016/j.foodhyd.2004.04.011
137. Quan TH, Benjakul S, Sae-leaw T, Balange AK, Maqsood S (2019) Protein–polyphenol conjugates: antioxidant property, functionalities and their applications. Trends Food Sci Technol 91:507–517. https://doi.org/10.1016/j.tifs.2019.07.049
138. Wang G-S, Chen H-Y, Wang L-J, Zou Y, Wan Z-L, Yang X-Q (2022) Formation of protein oleogels via capillary attraction of engineered protein particles. Food Hydrocoll 133:107912. https://doi.org/10.1016/j.foodhyd.2022.107912
139. Wang G-S, Chen H-Y, Feng G-X, Yuan Y, Wan Z-L, Guo J, Wang J-M, Yang X-Q (2023) Polyphenol-enriched protein oleogels as potential delivery systems of omega-3 fatty acids. J Agric Food Chem 71:749–759. https://doi.org/10.1021/acs.jafc.2c06348
140. Li J, Zhang C, Li Y, Zhang H (2022) Fabrication of aerogel-templated oleogels from alginate-gelatin conjugates for in vitro digestion. Carbohydr Polym 291:119603. https://doi.org/10.1016/j.carbpol.2022.119603

141. Jiang Y, Liu L, Wang B, Sui X, Zhong Y, Zhang L, Mao Z, Xu H (2018) Cellulose-rich oleogels prepared with an emulsion-templated approach. Food Hydrocoll 77:460–464. https://doi.org/10.1016/j.foodhyd.2017.10.023
142. Santos MAS, Okuro PK, Fonseca LR, Cunha RL (2022) Protein-based colloidal structures tailoring techno- and bio-functionality of emulsions. Food Hydrocoll 125:107384. https://doi.org/10.1016/j.foodhyd.2021.107384
143. Luo SZ, Hu XF, Jia YJ, Pan LH, Zheng Z, Zhao YY, Mu DD, Zhong XY, Jiang ST (2019) Camellia oil-based oleogels structuring with tea polyphenol-palmitate particles and citrus pectin by emulsion-templated method: preparation, characterization and potential application. Food Hydrocoll 95:76–87. https://doi.org/10.1016/j.foodhyd.2019.04.016
144. Oh I, Lee JH, Lee HG, Lee S (2019) Feasibility of hydroxypropyl methylcellulose oleogel as an animal fat replacer for meat patties. Food Res Int 122:566–572. https://doi.org/10.1016/j.foodres.2019.01.012
145. Botella-Martinez C, Lucas-González R, Lorenzo JM, Santos EM, Rosmini M, Sepúlveda N, Teixeira A, Sayas-Barberá E, Pérez-Alvarez JA, Fernandez-Lopez J et al (2021) Cocoa coproducts-based and walnut oil gelled emulsion as animal fat replacer and healthy bioactive source in beef burgers. Foods 10. https://doi.org/10.3390/foods10112706
146. Espert M, Hernández MJ, Sanz T, Salvador A (2021) Reduction of saturated fat in chocolate by using sunflower oil-hydroxypropyl methylcellulose based oleogels. Food Hydrocoll 120: 106917. https://doi.org/10.1016/j.foodhyd.2021.106917
147. Herz E, Herz L, Dreher J, Gibis M, Ray J, Pibarot P, Schmitt C, Weiss J (2021) Influencing factors on the ability to assemble a complex meat analogue using a soy-protein-binder. Innovative Food Sci Emerg Technol 73:102806. https://doi.org/10.1016/j.ifset.2021.102806
148. Alvarez MD, Cofrades S, Espert M, Sanz T, Salvador A (2021) Development of chocolates with improved lipid profile by replacing cocoa butter with an oleogel. Gels 7:220. https://doi.org/10.3390/gels7040220
149. Brito GB, Peixoto VODS, Martins MT, Rosário DKA, Ract JN, Conte-Júnior CA, Torres AG, Castelo-Branco VN (2022) Development of chitosan-based oleogels via crosslinking with vanillin using an emulsion templated approach: structural characterization and their application as fat-replacer. Food Struct 32:100264. https://doi.org/10.1016/j.foostr.2022.100264
150. Li S, Wu G, Li X, Jin Q, Wang X, Zhang H (2021) Roles of gelator type and gelation technology on texture and sensory properties of cookies prepared with oleogels. Food Chem 356:129667. https://doi.org/10.1016/j.foodchem.2021.129667

Chapter 11
Ultrasound as a Tool to Taylor Oleogelation and Oleogels Physical Properties

Thais Lomonaco Teodoro da Silva and Silvana Martini

Abbreviations

BW	Beeswax
CBW	Carnauba wax
CLW	Candelilla wax
CO	Canola oil
CRO	Corn oil
EC	Ethylcellulose
G′	Storage modulus
G″	Loss modulus
HF	Hardfat
HIU	High-intensity ultrasound
HOSO	High-oleic sunflower oil
LE	Lecithin
MCTO	Medium-chain triacylglycerols oil
MG	Monoglycerides
NO	Nut oils
OBC	Oil binding capacity
OO	Olive oil
PN	Pine nut oil
PO	Peanut oil

T. L. Teodoro da Silva
Department of Nutrition, Dietetics, and Food Sciences, Utah State University, Logan, UT, USA

Food Science Department, Federal University of Lavras, Lavras, Brazil

S. Martini (✉)
Department of Nutrition, Dietetics, and Food Sciences, Utah State University, Logan, UT, USA
e-mail: silvana.martini@usu.edu

C. Andrea Palla, F. Valoppi (eds.), *Advances in Oleogel Development, Characterization, and Nutritional Aspects*,
https://doi.org/10.1007/978-3-031-46831-5_11

PPW	Propolis wax
PV	Peroxide value
RO	Rapeseed oil
SAO	Safflower oil
SBO	Soybean oil
SFA	Saturated fatty acids
SFC	Solid fat content
SFO	Sunflower oil
T	Temperature
t	Time
T_c	Crystallization temperature
TG	Triglycerides
T_{on}	Onset temperature
T_p	Peak temperature
USW	Ultrasound standing waves
WO	Walnut oil

11.1 Introduction

Oleogels are a growing area of research in lipid science and technology because they have shown great potential for use in many areas, such as cosmetics [1], pharmaceutics [2], and food [3] industries. In the food industry, the interest arises from the desire to reduce saturated fatty acids (SFA) and remove *trans* fatty acids to comply with many health recommendations [4] and normatives [5].

The first step in oleogels research was to explore viable oleogelators to structure oils without compromising cost and sensory properties. In addition, the ideal oleogelator should be food grade and readily available [6]. Finding an oleogelator that meets these expectations is not a trivial challenge; however, potential oleogelators have been identified such as vegetable waxes [7], monoglycerides (MG) [8], lecithin (LE) [9], phytosterols [10], ethylcellulose (EC) [11], fatty acids and alcohols [12], among others. Moreover, many oleogel applications related to foods have been developed for several products, such as bakery products [13, 14], margarine and spreads [15, 16], chocolate products [17, 18], ice cream [19, 20], fillings [3, 14], and meat products [21, 22]. Although there is a significant amount of research in the area, some technological properties still need improvements such as improving aeration capacity [23], hardness [14], appearance [24], oil binding capacity (OBC) [3], shelf-life [18], and sensory or consumer acceptability [15].

Many strategies have been explored to overcome these limitations, such as using multicomponent oleogels [3, 25], using alternatives routes to oleogelation such as foam-templates [26], oleofoams [27], solvent dissolution [28, 29], and ultrasound [30], among others. Ultrasound is an emerging and green technology capable of tailoring the oleogelation process [31]. Two types of ultrasound technologies have been used in oleogels, high-intensity ultrasound (HIU) and ultrasound standing waves (USW).

HIU is a high-intensity and low-frequency (20–100 kHz) ultrasound technology extensively used in lipid systems to induce and change the crystallization process [32–35]. The effect of HIU to induce crystallization can be attributed to cavitation events generated by acoustic waves that create bubbles that can either collapse or dissolve forming localized high temperature and pressure zones. In addition, bubbles can act as nuclei and induce crystallization, resulting in the structuring of the network [36]. In oleogels HIU facilitates the formation of 3D-network increasing the strength of the gel and also reducing the probability of migration of liquid oil away [25]. HIU normally is applied in temperature-controlled systems, using a horn or probe [37]. The cavitation generated by the HIU is influenced by oil/oleogelator composition and by acoustic parameters such as frequency, amplitude, power level, pulse duration, and step of the oleogelation where HIU is applied.

USW fields provide a non-invasive, contactless method for manipulating the spatial organization of objects in liquids and air [38]. A USW can be generated in a medium by producing a radiating wave using a transducer which is reflected off a surface and interferes with the wave that is traveling into the medium in the opposite direction [39]. These events generate localized pressure gradients that form zones of high-pressure (antinodes) and low-pressure (nodes) [40]. This pressure gradient promotes fields that organize the solid particles. The shape of the standing wave field gives rise to the band-like structure in oleogels, which can be tailored through manipulation of the frequency of the transmitted wave. This new oleogel structure can be achieved regardless of the type and concentration of the crystallizing structuring agent and oleogel cooling [41, 42]. USW operates on different chambers and they differ in terms of process type (batch versus continuous), chamber volume, temperature control mode (no control versus constant temperature), and type of reflecting boundary (soft such as liquid/air interface versus hard such as metal, plastic, and ceramic) [39].

The use of ultrasound in oleogels is recent with the first studies published in 2014. In this chapter, we describe new findings on the use of HIU in oleogels. We also explain how HIU can tailor the oleogelation process and what are the positive and negative effects of this technology on the physical and chemical properties of the oleogels.

11.2 Ultrasound in Oleogels

11.2.1 High-Intensity Ultrasound (HIU)

Studies that evaluate the effect of HIU on different oleogels are described below and summarized in Table 11.1.

Table 11.1 Summary of the main finding of high-intensity ultrasound (HIU) application on oleogels

Oleogel type	HIU goal	HIU conditions[a]	Main findings	References
BW[b] + SBO (CO, CRO, OO, SAO, SFO)	Reduce phase separation	2 mm probe 180 μm (10 W) 10 s In the presence of crystals	HIU reduced phase separation. The magnitude of the improvement depends on the oil type.	[43]
PPW + OO	Improve crystal structure and thermal properties	13 mm probe 100, 200 or 300 W 30 or 120 s HIU in melted oleogel	HIU induced nucleation, formed a stronger crystalline network with a higher OBC.	[44]
CLW + NO	Effect of HIU on physical properties of different NO-oleogels, stability and in vitro ingestion of β-carotene loaded sonicated oleogels.	6.36 mm probe 95 W 20 s pulses (2 s ON/2 s OFF) HIU during cooling	HIU improved viscoelasticity, OBC, and firmness The magnitude of the improvement was dependent on the oil type. HIU promoted higher stability of the β-carotene on the oleogel.	[45]
MG + HOSO	Effect of HIU on MG-oleogels physical properties using different cooling rates.	3.2 mm probe 96 W 30 s (10 s ON5s OFF) HIU in the presence of primary crystals	Small length crystals, high hardness, viscoelasticity, and OBC were found. Results highly dependent on the cooling rate and moment where HIU was applied.	[46, 47]
EC+ MCTO	Effect of HIU on EC-oleogels	20 mm probe 240 W 10% duty 10 min (40 °C) HIU used as mixer	HIU improved solubility of EC on oil without using higher T/t. Improved physical properties. Higher oxidative stability.	[31]
MG:HF + RO	Effect of HIU on physical properties of the oleogel	12.7 mm probe 50 W 10 s, 30 s or 30 s pulse (10sON/10s OFF) HIU in the presence of primary crystals	HIU reduced incompatibility of the oleogelators HIU improved all physical properties mainly in a short time (10 s).	[48].

(continued)

Table 11.1 (continued)

Oleogel type	HIU goal	HIU conditions[a]	Main findings	References
MG:HF: CLX + HOSO	Effect of HIU on physical properties of the oleogel	3.2 mm or 12.7 mm probes 120 and 216 μm (for the micro tip) and 60 and 108 μm (macro tip) 1 or 3 min HIU in melted oleogel	HIU improved hardness and reduced crystal size.	[30]
MG:HF: LE + RO	Effect of HIU on physical properties of the oleogel	12.7 mm probe 50 W 10 s HIU in the presence of primary crystals	HIU induced crystallization of HF and improved physical properties were found on binary and ternary combinations where HF was presented.	[49]

[a]All studies have used a 20 kHz HIU

[b]Abbreviations: *BW* beeswax, *SBO* soybean oil, *CO* canola oil, *CRO* corn oil, *OO* olive oil, *SAO* safflower oil, *SFO* sunflower oil, *PPW* propolis wax, *CLW* candelilla wax, *NO* nut oils, *MG* monoglycerides, *HOSO* high-oleic sunflower oil, *EC* ethylcellulose, *MCTO* medium chain triglycerides oil, *HF*hardfat, *RO* rapeseed oil, *LE* lecithin, *W* watts, *OBC* oil binding capacity, *T/t* temperature/time

11.2.1.1 Mono-Component Oleogels

Wax-Oleogels One of the first studies of HIU in wax-based oleogels was performed by Jana and Martini [43]. In this study, HIU (20 kHz, 2 mm probe, 180 μm [10 W] for 10 s) was used as a tool to change the crystallization behavior of beeswax (BW) in different liquid oils and to reduce phase separation. To achieve maximum efficiency of the ultrasound waves, HIU was applied at the onset of crystallization; therefore, HIU was applied at 300, 20, and 1.5 min for the 0.5% BW samples cooled at 0.1, 1, and 10 °C/min, respectively. These authors showed that HIU can help in the stabilization (uniform crystallization) of a crystalline network to avoid phase separation, especially when samples are processed using slow cooling rates. In faster cooling rates, although no phase separation was observed, HIU significantly increased the transmission values at the top of the tube suggesting some degree of sedimentation. In addition, the authors observed a change in crystal morphology and size, especially for the BW crystallized at slow cooling rates: big spherulites were observed in the sample crystallized without the use of HIU, and smaller crystals were observed when HIU was used. Using a higher concentration of BW (1%), and the same HIU conditions similar results were found suggesting that a significant phase separation occurs when 0.5 and 1% of BW are crystallized at the slow cooling rate (0.1 °C/min) in soybean oil and that this phase separation can be delayed using HIU. The study was further reproduced in other oils such as canola, corn, olive, safflower, and sunflower oils. Overall, the stability toward phase separation increased in the order of soybean oil < olive oil < sunflower oil < canola oil < safflower oil < and corn oil. These results suggest that (1) for the same type and amount of wax, the type of oil used affects the degree of phase separation and (2) HIU can help in the delay of phase separation to different degrees depending on the oil used. Crystal morphology does not seem to explain the differences found in phase separation as a function of the type of oil used; however, spherulites seem to be more often associated with phase separation. The reduction in phase separation due to HIU could be due to the breaking of spherulites into small needle-like crystals [43, 50]. Jana and Martini [43] have hypothesized that cavitation was responsible for forming the small needle-like crystals and that this type of crystal can interact more readily with each other, allowing for more oil entrapment and therefore delaying phase separation. Additionally, HIU resulted in a narrower melting profile as evidenced by either higher onset temperature (T_{on}) values or lower peak temperature (T_p) values in the sonicated samples. The effect of HIU in delaying phase separation is likely related to the generation of a less fractionated crystalline network where similar molecules can interact strongly among them, entrapping more oil and thus reducing or avoiding phase separation. Similar results have been previously observed in palm oil, where HIU accelerated nucleation and formed more crystals promoting the formation of a smaller and more organized crystalline network [51].

A second study in wax-oleogels evaluated propolis wax (10%) in olive oil, using different cooling rates (1.35, 1.07, and 0.62 °C/min), ultrasound power (100, 200, and 300 W) and times (30 and 120 s) using a 13 mm probe (20 kHz). In this case, HIU was applied right after melting. HIU application at 300 W ultrasonic power, 5 ° C crystallization temperature, and 120 s HIU pulse were the optimum HIU conditions to obtain maximum firmness and minimum oil migration. Results showed that HIU induced nucleation by creating small crystals that led to the formation of a strong crystalline network with high oil binding capacity. Differential scanning calorimetry analysis (DSC) showed that T_{on} and T_p of the HIU-treated sample and control were not significantly different, while T_{end} (end temperature) and enthalpy of the HIU-treated sample were higher than the non-sonicated one suggesting a higher amount of crystalline material in the sample due to sonication. Moreover, the X-ray diffraction pattern revealed no polymorphism transformation due to sonication. However, a slight increase in peroxide value (PV) from 3.7 to 4.7 meq de O_2/kg was observed for HIU-treated oleogels [44]. The authors hypothesize that cavitation and collapsing of bubbles induce primary nucleation and generate local hot spots with high temperature and pressure that can result in an increase in PV. In addition, no changes were observed in free fatty acids or fatty acid composition due to HIU [44].

Li et al. [45] studied the effects of HIU (20 kHz, 95 W, 20 s pulse -2 s ON/2 s OFF, 6.36 mm probe) on the physical properties, stability, and in vitro digestion of β-carotene enriched CLW-oleogels. Three nut oils were used (peanut [PO], pine nut [PNO], and walnut [WO] oils) with 3% of CLW. Sonicated oleogels were prepared by sonicating the samples after being taken out from 80 °C and kept at room temperature for 5 min. HIU increased the storage modulus (G′), loss modulus (G″), and firmness, and reduced oil loss of all nut oils. The magnitude of the changes in physical properties mentioned above was dependent on the type of oil and its composition. The content of stearic acid in all three oils was similar, while the concentrations of oleic and linoleic (C18:2) acids were higher in PNO and WO. As PO had the lowest content of C18:2 (33.74%) among all the three oils, the G′ value of P oleogel increased only by 22.21% after HIU treatment. The C18:2 content of PNO and WO was 45.96 and 61.65%, respectively, and their G′ values HIU treatment increased by 97.64 and 121.16%, respectively. Based on this, it was hypothesized that the high-melting fatty acids might form a stronger network that can affect the physical structure of oleogels. Moreover, the weaker network, which was formed mainly with low-melting fatty acids, could act as an additional skeleton attached to the stronger network. The effect of sonication may be more pronounced with weaker networks. This hypothesis could be visualized in the polarized light microscopy images, with a visible increase of crystals after HIU, and was attributed to the cavitation effects generated by sonication. The bubbles created by ultrasound collapsed forming smaller bubbles that could generate more sites for nucleation [45]. A similar effect was observed in low saturated fats, where HIU induced the crystallization of low-melting point triglycerides (TG) that would not crystallize without sonication [35]. Li et al. [45] also found that HIU did not change the polymorphic form of the oleogels. Regarding the β-carotene enriched oleogels, similar improvements in physical properties were found after sonication in oleogels processed with or without β-carotene. Moreover, HIU decreased significantly the degradation of β-carotene for PO and WO samples after 120 days of storage, and on PNO for 60 days. Furthermore, HIU reduced the release of β-carotene in intestinal digestion. In summary, HIU was able to improve the functional properties of wax-nut oils oleogels and their β-carotene enriched oleogels. These protective phenomena might be because HIU promoted crystal aggregation, which might cause denser networks and could improve the binding of β-carotene within the oleogel network [45].

In summary, all sonicated wax-oleogels were sonicated using HIU with 20 kHz frequency, but with different probes sizes, amplitude/power, and durations, and were applied either in the presence or absence of crystals. Sonication positively influenced the oleogelation and the physical properties of wax-oleogels when applied either in the presence or absence of crystals, which suggests the effect of cavitation either on primary or secondary nucleation. Moreover, HIU also induced crystallization of low melting point TGs of the liquid oils, which emphasized the effect of oil type. Nevertheless, the probe size does not seem to be a big player on the effect of sonication as long as the amplitude/power and duration are well designed.

Monoglycerides (MG) Oleogels Giacomozzi et al. [46, 47] investigated the effect of HIU in MG-oleogels in high oleic sunflower oil. In their first study, HIU was evaluated using different MG concentrations (3, 4.5, and 6%) and cooling rates (0.1 and 10 °C/min). HIU (20 kHz, 3.2 mm probe, 96 W, 30 s pulses- 10 s ON/5 s OFF) for both types of cooling rates was applied at the onset of crystallization. Sonication significantly increased the OBC on slowly cooled samples at higher concentrations of MG (4.5 and 6%) but was not able to improve the OBC of fast-cooled samples. Although the initial G′ values (when samples reached 25 °C) of the samples crystallized at both cooling rates were affected by HIU, no significant differences were found in the G′ values of the samples sonocrystallized after 24 hours of storage. When HIU was used for both cooling rates, no changes or lower values were observed for hardness, melting profile, and OBC. A reduction in crystal length was observed with HIU application at the final time, except in the fast-cooled MG-3% samples where an increase in crystal length was observed. It is possible that the power level of the acoustic waves used in this study generated disruption of the crystalline network that it was being formed, and the low level of supersaturation was not enough to overcome this disruption. Da Silva and Danthine [48, 49] also evaluated sonication (20 kHz, 12.7 mm probe, 50 W for 10 s, 30 s, or 10 s pulse) of MG-oleogels with 6% or 10% of MG in rapeseed oil. They found a similar negative effect of HIU on hardness and OBC, and a reduction in the amount of crystalline material formed on this type of oleogel; even when using only 10 s of sonication. This suggests that sonication disrupted the MG primary nucleation instead of inducing secondary nucleation.

The lack of differences or negative effects of HIU in MG-oleogels found in their first study made Giacomozzi et al. [47] further explore a different process approach. Similar HIU conditions, MG concentration, and cooling rate were evaluated. However, instead of HIU being applied at the onset of crystallization it was applied when the sample reached a specific crystallization temperature (T_c) determined by DSC as the temperature at which the sample achieved 48% crystallinity. In this case, HIU was able to significantly improve MG oleogels crystallization behavior, reducing crystal length and forming a stronger (higher crossover point), harder, more adhesiveness, and elastic network with higher OBC. The effects of HIU application were enhanced by cooling at 0.1 °C/min and storing at 5 °C. Neither solid fat content (SFC), polymorphism, nor thermal behavior was affected by HIU. An interesting result found in this study is that the OBC of sonicated samples showed similar or higher OBC than non-sonicated samples with a higher concentration of MG. This suggests that it would be possible to reduce the amount of MG in the oleogel by applying HIU without losing the desired physical properties [47].

In summary, sonication in MG-based oleogels depends on the moment when sonication is applied, especially on highly saturated MG studied. Thus, it is possible to improve the physical properties, and even reduce the amount of MG used if the oleogelation process and HIU are well optimized.

Ethylcellulose Oleogels (EC) Medium-chain triglycerides (MCTO) oleogels structured by EC (1–4%) prepared with ultrasound were also evaluated regarding their physical properties and oxidative stability [31]. HIU (20 kHz) was applied using a 20 mm probe, 240 W, 10% duty cycle at 40 °C for 10 min. The control sample was prepared by melting at 140 °C for approximately 35 min. The mixtures were allowed to cool to room temperature and then stored at refrigeration temperature before analysis. The most interesting result from this study was the reduction in temperature and time necessary to produce EC-oleogels. The generation of cavities that expand and collapse during sonication generates shock waves thus increasing turbulence and leading to fine mixing of different phases (EC- solid and MCTO – liquid) and thus speeding up the process of gelation. The ultrasound-assisted gelation process took

15 min (10 min for mixing and 5 min for cooling to set gel), whereas the conventional process took a total of 45 min for the formation of oleogel (35 min for mixing and 10 min for cooling). Moreover, the sonicated oleogel was produced at 40 °C, while the control sample needed 140 °C to dissolve the EC in the oil. The EC-oleogel had lower oxidative stability (6.41 h) compared to the sonicated oleogel (9.76 h). Interestingly, the MCTO alone showed an even lower stability (5.96 h) than the control oleogel, thus although the conventional process to form EC-oleogels induced some oxidation due to the higher temperature required, the conversion of MCTO into oleogel enhanced its oxidative stability thereby increasing the possibility of its applications, and the use of sonication increases this stability even more. In addition to the thermal and oxidative breakdown of the oil which negatively affects the quality and shelf-life of the oil, the heating of EC-MCTO mixture at high temperature results in the formation of bubbles at the bottom of the mixing vessel and these bubbles form a foam in the surface of this mixture hindering the process of gelation. Moreover, sonication also affected the physical properties of the EC oleogels, and sonicated oleogels showed less oil loss, and higher hardness, stickiness, and viscous and elastic modulus over 30 days of storage. The increase in the OBC was possible due to the fine mixing of the polymer promoted by cavitation favoring the formation of the more stable tridimensional network. It is assumed that the effect of ultrasound is also present in the MCTO, inducing nucleation of the oil, which also contributed to a stronger network formed [31]. The sonicated process developed by Jadhav et al. [31] to dissolve EC in oil was a groundbreak to the use of this oleogelator because previous results have faced deleterious oxidation [52–54] on this type of oleogelation due to the high time/temperature requirements. Nevertheless, when used in combination with high-temperature dissolution, HIU promoted oxidation even more on EC-oleogels [22, 55].

11.2.1.2 Multicomponent Oleogels

The use of HIU in binary and ternary oleogelators mixtures has also been evaluated. A binary combination of MG and high-melting-point-TG (hardfats:HF) was evaluated by da Silva and Danthine [48]. In this study MG:HF (0:6, 1:5, 2:4, 3:3, 4:2, 5:1, and 6:0) oleogels were produced by melting, followed by the application of HIU (20 kHz) in the presence of few crystals (primary nucleation). HIU using a 12.7 mm probe and 50% amplitude (56 W) for 10 s, 30 s, or 30 s using pulses (10sON/10sOFF) was used. The authors have reported a small increase in crystallization temperature due to cavitation (0.05 ± 0.03 °C using 10s, and 3.5 ± 1.0 °C using 30s and 10sP), which was dependent on the HIU condition used. After applying HIU for only 10 s, small needle-like crystals were observed without clusters. The other conditions also showed some improvement in reducing the number of clusters, but the results were less pronounced. In fact, in some conditions, the amount of crystalline material was reduced suggesting a possible excessive sonication time. HIU also improved all physical properties of the blends such as OBC, plasticity, hardness, and viscosity, especially if used for only 10 s. Moreover, no change in polymorphism and SFC due to sonication was observed in these oleogels. In this study, as a very high melting point lipid material was used, it is assumed that the sonication produced smaller crystals and a more organized network. HIU could have acted as (1) inducer of the secondary nucleation and with more nuclei formed crystal growth was reduced or (2) the shear generated during sonication fragmented the clusters into small needles. A very interesting result of this study is that although the blends MG:HF without HIU were very soft indicating an incompatibility between the oleogelators, HIU reduces this incompatibility, proving to be an interesting alternative for multicomponent oleogels where antagonism between the oleogelators might exist [48].

Ternary systems formed by MG:HF:CLW [30] and MG:HF:LE [49] were also evaluated using two sonication strategies: (1) the CLW system was sonicated in the absence of crystals right after completely melting the oleogelators to avoid breakage of the oleogel network due to shear [30] and (2) the LE system was sonicated in the presence of primary nuclei [49]. HIU (20 kHz) was used in both studies. On the CLW system, 2 probes (3.2 mm and 12.7 mm), two duration times (1 and 3 min), and two amplitudes 120 and 216 μm (for the micro tip) and 60 and 108 μm (for the macro tip) were used. When HIU was applied to the oleogel for 3 min using a 3.2 mm diameter tip at an amplitude of vibration of 216 μm, a reduction in crystal size and crystal area was observed with an increase in hardness and no change in G′ nor in oil loss compared to the non-sonicated oleogel was found. Other sonication conditions (lower power levels, shorter durations, and bigger tips) tested in this study reduced the hardness and elasticity of the sample and increased oil loss. No significant differences were found in the melting profiles and OBC of the sample due to HIU treatment. These results show that even though a reduction in crystal size and the crystallized area was obtained in sonicated samples, this reduction was not sufficient to significantly change the elasticity and hardness of the samples, except for the T3_A216 sample (3 min, 216 μm). Although this benefit condition was followed by a high increase in the temperature during sonication due to the localized temperature and pressure promoted by cavitation. Nevertheless, in this case, the high temperature was followed by a higher supercooling which was hypothesized to induce crystallization and as a consequence generate better physical properties [30].

In the ternary system with LE [49], in addition to the ternary system MG:HF:LE authors evaluated binary blends of MG:HF, MG:LE, and HF:LE (always totaling 10% oleogelators in equal proportions). HIU was applied for 10 s using 50 W and a 12.7 mm probe as previously optimized by da Silva and Danthine [48]. Three synergic combinations that produced 99% OBC oleogels were found: MG: HF, HF:LE, and MG:HF:LE. These combinations showed improved physical properties such as hardness, elastic modulus, and oil loss when sonicated, which was attributed to the induced secondary crystallization of the HF promoted by HIU, since this oleogelator was the common one between all blends. The oleogels formed by MG:LE were very liquid and although HIU slightly improved some physical properties, it was not sufficient to form viable oleogels. Reduction in crystal size, deagglomeration of clusters, fewer modifications on melting parameters, and no change in the polymorphism of these blends were also some common results from the sonication [49].

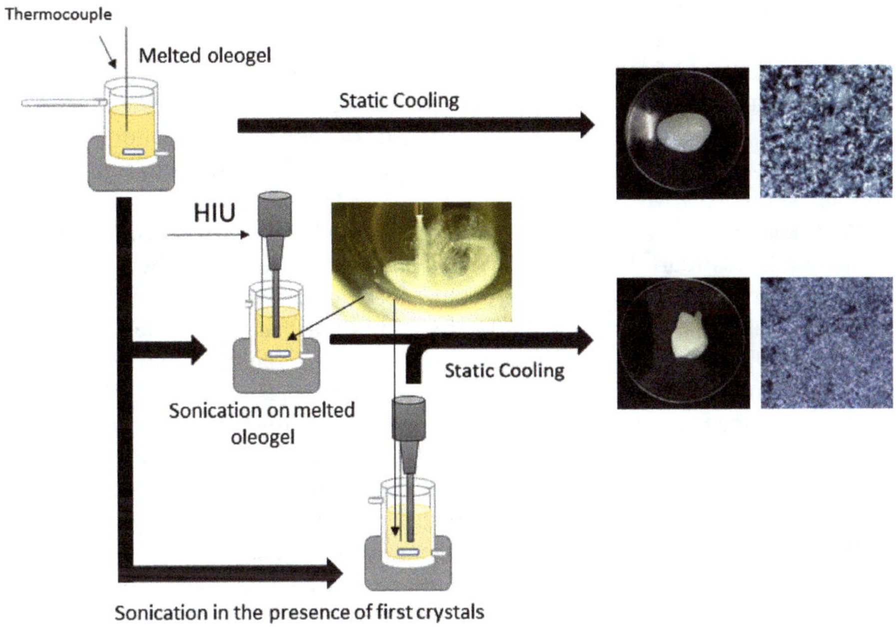

Fig. 11.1 Possible use of high-intensity ultrasound on oleogels to induce nucleation (primary or secondary) and strengthen the oleogelator network

Overall, the results discussed above suggest that contrary to what was found for single oleogelators, for binary and ternary systems the duration of sonication as well as the moment of HIU application on the oleogels process must be considered. Longer times (20 s or more) can generate some non-desired heating on the sample, which even though not negative to physical properties did not allow the same improvement on physical properties as when HIU was applied in the presence of primary nucleation and for a shorter time (10 s). This, of course, can also be a consequence of many different molecular interactions between the various oleogelators in the mixture. Figure 11.1 summarizes HIU application on oleogels and the effect of the cavitation on the macro and microstructure.

11.3 Ultrasonic Standing Waves (USW)

Due to the ability of ultrasound waves to modify gel strength [56], the use of USW has recently been of much interest for oleogels. Differently from HIU, USW oscillate in place. USW are produced when a wave is confined within boundaries. Thus when the ultrasound waves travel into a medium they reflect off a surface and they can interfere with the incident wave, creating regions of no displacement (nodes) and regions of maximum displacement (anti-nodes), which corresponds to the USW field [42]. A custom-designed ultrasound chamber was developed to direct the crystallization of MG, sunflower wax (SFW), and CLW oleogels [42]. The study evaluated the effect of frequency (0, 1, 2, and 4 MHz), cooling rate (1 and 10 °C/min), oleogelator type (MG, CLX, and SFW), and concentration (MG = 2.5%, 10%, and 20%, CLW and SFW = 2.5 and 5%) on the oleogels crystallization. Authors observed that USW fields modified the microscopic structure when the crystalline network was formed. Control samples (0 MHz) showed a very homogeneous structure; however, the ultrasound sample showed a heterogeneous structure. During crystallization of 5% MG-oleogel, MG crystals were pushed toward the nodal planes by the pressure gradient forming alternating dense and sparse bands. The number of crystalline bands (dark band) was directly proportional to the applied frequency. Moreover, applying a 10 °C /min cooling rate instead of a 1 °C/min did not statistically alter the band thickness or the number of bands with the same USW frequency. Oleogelator crystals change their dimensions (and volume) during crystallization, and the acoustic force acting on them increases during crystallization. Similar to saturated fats there is a nucleation and crystal growth step for oleogelators, eventually forming a macroscopic network. Thus, sonication forces acted on the MG crystals increasing and speeding up the translation of crystals to the nodal plane. After modeling the dynamic behavior of the crystalline material subjected to the USW, it was found that the crystals orient themselves perpendicularly to the standing wave and slowly travel toward the pressure node [42]. Control and low frequency (1 MHz) 5% MG-oleogel showed triclinic β-form; nevertheless, using higher frequencies (2 and 4 MHz) formed the less stable α-polymorphic form suggesting that USW fields with high frequencies can disturb crystallization at the nanostructural level, forming less stable polymorphic forms. Authors have tried to compare the thickness of the crystalline bands using different oleogelators and concentrations, but they were not able to calculate this parameter for all samples, mainly for waxes and higher concentrations of MG (10 and 20%) which was attributed to the tightly packed crystal network formed on those samples. This study showed for the first time that a USW field can be used as a robust method for oleogel preparation independent of oleogelator type (MG, SFW, and CLW), oleogelator concentration (between 2.5% and 20%), and cooling rate (1 °C/min and 10 °C/min) [42].

The same group has further improved the ultrasound chamber previously used [42] and Lassila et al. [41] developed three new USW chambers, with volumes from 10 to 100 mL, which were able to form a controllable standing field during MG-oleogel formation and allowed sample extraction for further analysis through optical and ultrasound imaging, uniaxial compression, and oil release. A reduction in effective stiffness was observed after USW application (2.78 ± 0.28 MPa, 1.93 ± 0.18 MPa, and 1.87 ± 0.15 MPa, corresponding to reference-, low- and medium-pressure, respectively). This implies that sonicated samples can be deformed with less force and are therefore weaker compared to reference samples. An effect was even more pronounced with higher pressures. The result was attributed to the changes observed in microstructure (more regular band-like structure only for medium-pressure sonicated samples) and to the possible heat generated by the piezoelectric ceramics during operation, which influenced the cooling rate, and structuration. The weakening of the overall oleogel structure for medium-pressure sonicated samples, as suggested by the uniaxial compression test, also seems to hasten oil release [41]. Figure 11.2 shows one of the chambers developed, where

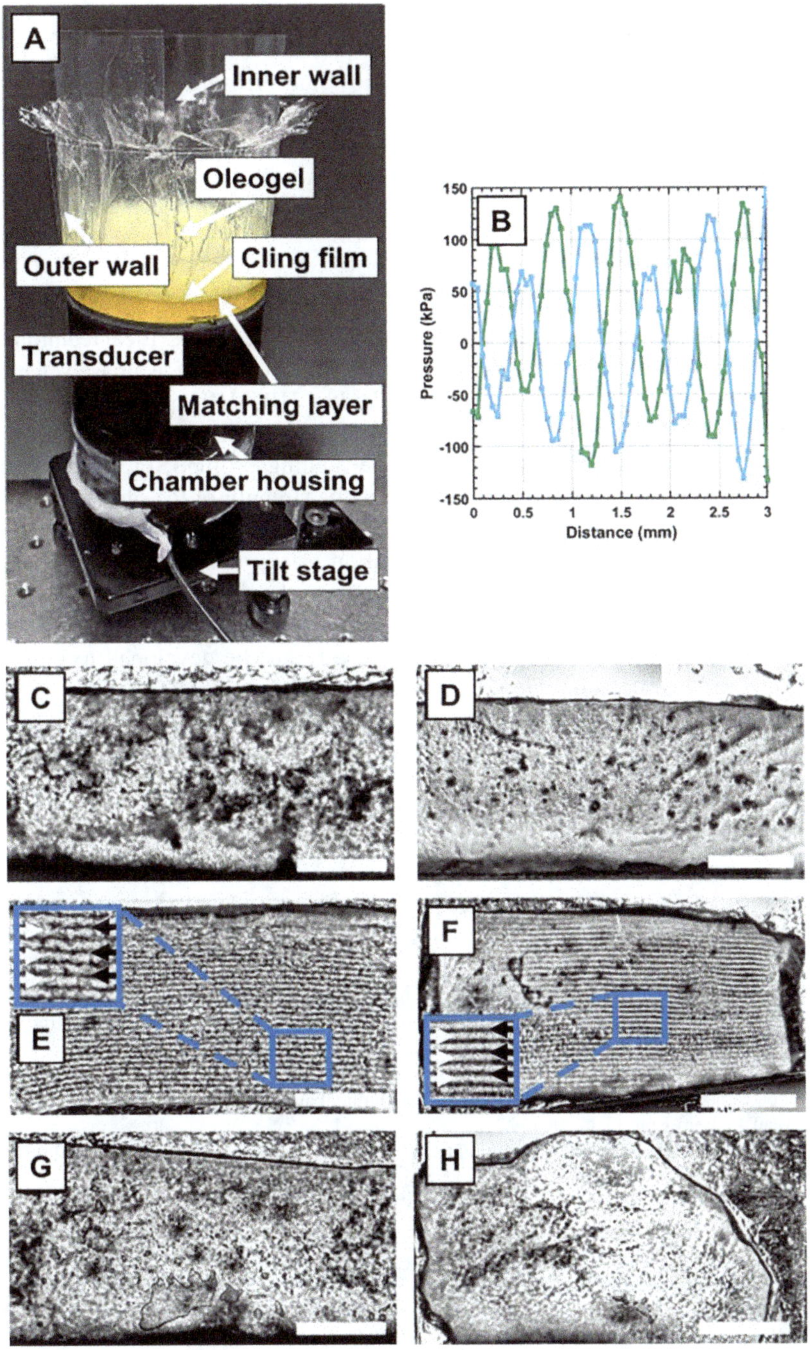

Fig. 11.2 (**a**) Chamber designed featuring a matching layer of oil between the transducer and the oleogel. (**b**) Pressure field scan recorded in oil during sonication in chamber, 2.3 MHz; x-axis represents the distance from the cling film. Optical images of (**c**) Reference sample containing 10% monoglyceride (MG), (**d**) 5% MG reference, (**e**) 10% MG medium-pressure sonicated, (**f**) 5% MG medium-pressure sonicated, (**g**) 10% MG low-pressure sonicated, and (**h**) 10% MG high-pressure sonicated sample slices. (**c–h**) Size bars equal 5 mm. (Reproduced from Lassila et al. [41], under the terms of the Creative Commons CC-BY license)

there was a sample extraction possibility, how the pressure oscillates during the ultrasound waves, and the effect of the technique on the microstructure of the oleogel.

11.4 Conclusions

The use of ultrasound in oleogels has shown promising results. Critical technological properties faced by the food industry in lipid-based foods such as phase separation, oil migration, and incompatibility are improved by HIU. HIU also seems to be able to reduce the amount of oleogelator required to structure the oleogel and favors oleogelator dissolution without changing the melting profile, SFC, and polymorphism. Moreover, HIU significantly improves the strength of the oleogel keeping a very low content of SFA. To achieve these benefits HIU must be optimized according to the oleogelator, and a combination of appropriate cooling/crystallization conditions. The use of USW also influences the microstructure of oleogels, affecting their physical properties such as strength, oil migration, and polymorphism. Nevertheless, more studies and chambers must be developed to refine this recent and promising technique.

Acknowledgments The authors are grateful for the postdoctoral fellowships and funding in Sciences, Technology, Engineering, Materials, and Agrobiotechnology (STEMA) funding OTP N° DIVE.0899-J-P, given by the ULiège University Research Council.

References

1. Botega DCZ, Nogueira C, de Moura NM, Martinez RM, Rodrigues C, Barrera-Arellano D (2021) Influence of aqueous matrices into candelilla wax organogels emulsions for topical applications. J Am Oil Chem Soc 98:317–328
2. Sagiri SS, Kasiviswanathan U, Shaw GS, Singh M, Anis A, Pal K (2016) Effect of sorbitan monostearate concentration on the thermal, mechanical and drug release properties of oleogels. Korean J Chem Eng 33:1720–1727
3. da Silva TLT, Fernandes GD, Arellano DB (2021) Development of reduced saturated fat cookie fillings using multicomponent oleogels. J Am Oil Chem Soc 48:1069–1082
4. HHS USD of H and HS, USDA USD of A (2015) 2015–2020 Dietary Guidelines for Americans. 2015–2020 Diet Guidel Am (8th Ed). https://doi.org/10.1097/NT.0b013e31826c50af
5. Food and Drug Administration (2015) Final determination regarding partially hydrogenated oils. Federal register 80(116):34650–34670
6. Co ED, Marangoni AG (2012) Organogels: an alternative edible oil-structuring method. J Am Oil Chem Soc 89:749–780
7. Martini S, Tan CY, Jana S (2015) Physical characterization of wax/oil crystalline networks. J Food Sci 80:C989–C997
8. da Pieve S, Calligaris S, Co E, Nicoli MC, Marangoni AG (2010) Shear nanostructuring of monoglyceride organogels. Food Biophys 5:211–217

9. Pernetti M, van Malssen KF, Flöter E, Bot A (2007) Structuring of edible oils by alternatives to crystalline fat. Curr Opin Colloid Interface Sci 12:221–231
10. Bot A, Den AR, Roijers EC (2008) Fibrils of γ-Oryzanol + β-Sitosterol in edible oil organogels. J Am Oil Chem Soc 85:1127–1134
11. Zetzl AK, Marangoni AG, Barbut S (2012) Mechanical properties of ethylcellulose oleogels and their potential for saturated fat reduction in frankfurters. Food Funct 3:327–337
12. Gandolfo FG, Bot A, Flöter E (2004) Structuring of edible oils by long-chain FA, fatty alcohols , and their mixtures. J Am Oil Chem Soc 81:1–6
13. Zhao M, Rao J, Chen B (2022) Effect of high oleic soybean oil oleogels on the properties of doughs and corresponding bakery products. J Am Oil Chem Soc 99:1071–1083
14. Tanti R, Barbut S, Marangoni AG (2016) Hydroxypropyl methylcellulose and methylcellulose structured oil as a replacement for shortening in sandwich cookie creams. Food Hydrocoll 61: 329–337
15. da Silva TLT, Chaves KF, Fernandes GD, Rodrigues JB, Bolini HMA, Arellano DB (2018) Sensory and technological evaluation of margarines with reduced saturated fatty acid contents using oleogel technology. J Am Oil Chem Soc 95:673–685
16. da Silva TLT, Arellano DB, Martini S (2019) Effect of water addition on physical properties of emulsion gels. Food Biophys 14:30–40
17. Patel AR, Rajarethinem P, Grędowska A, Turhan O, Lesaffer A, De Vos WH, Van de Walle D, Dewettinck K (2014) Edible applications of shellac oleogels: spreads, chocolate paste and cakes. Food Funct 5:645–652
18. Fayaz G, Amir S, Goli H, Kadivar M, Valoppi F, Barba L, Calligaris S, Cristina M (2017) Potential application of pomegranate seed oil oleogels based on monoglycerides , beeswax and propolis wax as partial substitutes of palm oil in functional chocolate spread. LWT - Food Sci Technol 86:523–529
19. Airoldi R, da Silva TLT, Ract JNR, Foguel A, Colleran HL, Ibrahim SA, da Silva RC (2022) Potential use of carnauba wax oleogel to replace saturated fat in ice cream. J Am Oil Chem Soc 99:1085–1099
20. Botega DCZ, Marangoni AG, Smith AK, Goff HD (2013) Development of formulations and processes to incorporate wax oleogels in ice cream. J Food Sci 78:1845–1851
21. Barbut S, Wood J, Marangoni A (2016) Quality effects of using organogels in breakfast sausage. Meat Sci 122:84–89
22. Gómez-Estaca J, Pintado T, Jiménez-Colmenero F, Cofrades S (2020) The effect of household storage and cooking practices on quality attributes of pork burgers formulated with PUFA- and curcumin-loaded oleogels as healthy fat substitutes. LWT - Food Sci Technol 119:108909
23. Kim JY, Lim J, Lee J, Hwang H, Lee S (2017) Utilization of oleogels as a replacement for solid fat in aerated baked goods: physicochemical , rheological , and tomographic characterization. J Food Sci 82:445–452
24. Yilmaz E, Ogutcu M (2015) Oleogels as spreadable fat and butter alternatives: sensory description and consumer perception. R Soc Chem 5:50259–50267
25. da Silva TLT, Arellano DB, Martini S (2018) Physical properties of candelilla wax, monoacylglycerols, and fully hydrogenated oil oleogels. J Am Oil Chem Soc 95:797–811
26. Patel AR, Schatteman D, Lesaffer A, Dewettinck K (2013) A foam-templated approach for fabricating organogels using a water-soluble polymer. RSC Adv 3:22900–22903
27. Liu Y, Binks BP (2021) A novel strategy to fabricate stable oil foams with sucrose ester surfactant. J Colloid Interface Sci 594:204–216
28. De Vries A, Hendriks J, Van Der Linden E, Scholten E (2015) Protein oleogels from protein hydrogels via a stepwise solvent exchange route. Langmuir 31:135850–113859
29. Lomonaco Teodoro da Silva T, Baeten V, Danthine S (2023) Modifying sucrose esters oleogels properties using different structuration routes. Food Chem 405:134927
30. da Silva TLT, Barrera DA, Martini S (2019) Use of high-intensity ultrasound to change the physical properties of oleogels and emulsion gels. J Am Oil Chem Soc 96:681–691

31. Jadhav HB, Pratap AP, Gogate PR, Annapure US (2022) Ultrasound-assisted synthesis of highly stable MCT based oleogel and evaluation of its baking performance. Appl Food Res 2:100156
32. Suzuki AH, Lee J, Padilla SG, Martini S (2010) Altering functional properties of fats using power ultrasound. Food Eng Phys Prop 75:208–214
33. Ye Y, Wagh A, Martini S (2011) Using high intensity ultrasound as a tool to change the functional properties of interesterified soybean oil. J Agric Food Chem 59:10712–10722
34. Birkin PR, Martin HL, Youngs JJ, Truscott TT, Merritt AS, Elison EJ, Martini S (2019) Cavitation clusters in lipid systems: the generation of a bifurcated streamer and the dual collapse of a bubble cluster. J Am Oil Chem Soc 96:1197–1204
35. da Silva TLT, Cooper Z, Lee J, Gibon V, Martini S (2020) Tailoring crystalline structure using high-intensity ultrasound to reduce oil migration in a low saturated fat. J Am Oil Chem Soc 97: 141–155
36. Higaki K, Ueno S, Koyano T, Sato K (2001) Effects of ultrasonic irradiation on crystallization behavior of tripalmitoylglycerol and cocoa butter. J Am Oil Chem Soc 78:513–518
37. Kadamne JV, Ifeduba EA, Akoh CC, Martini S (2017) Sonocrystallization of interesterified fats with 20 and 30% C16:0 at sn-2 position. J Am Oil Chem Soc 94:3–18
38. Drinkwater BW (2016) Dynamic-field devices for the ultrasonic manipulation of microparticles. Lab Chip 16:2360–2375
39. Valoppi F, Salmi A, Ratilainen M, Puranen T, Tommiska O, Hyvönen J, Heikkilä J, Haeggström E (2021) Ultrasonic standing wave chamber for engineering microstructures of water- and lipid-based materials. Eng Res Express 3:016002
40. Yamahira S, Hatanaka SI, Kuwabara M, Asai S (2000) Orientation of fibers in liquid by ultrasonic standing waves. Jpn J Appl Phys, Part 1 Regul Pap Short Notes Rev Pap 39:3683–3687
41. Lassila P, Valoppi F, Tommiska O, Hyvönen J, Holmström A, Hietala S, Salmi A, Haeggström E (2022) Practical scale modification of oleogels by ultrasonic standing waves. Ultrason Sonochem 85:105970
42. Valoppi F, Salmi A, Ratilainen M, Barba L, Puranen T, Tommiska O, Helander P, Heikkilä J, Haeggström E (2020) Controlling oleogel crystallization using ultrasonic standing waves. Sci Rep 10:1–13
43. Jana S, Martini S (2014) Effect of high-intensity ultrasound and cooling rate on the crystallization behavior of beeswax in edible oils. J Agric Food Chem 62:10192–10202
44. Sharifi M, Goli SAH, Fayaz G (2019) Exploitation of high-intensity ultrasound to modify the structure of olive oil organogel containing propolis wax. Int J Food Sci Technol 54:509–515
45. Li L, Taha A, Geng M, Zhang Z, Su H, Xu X, Pan S, Hu H (2021) Ultrasound-assisted gelation of β-carotene enriched oleogels based on candelilla wax-nut oils: physical properties and in-vitro digestion analysis. Ultrason Sonochem 79:105762
46. Giacomozzi AS, Palla CA, Carr E, Martini S (2019) Physical properties of monoglycerides oleogels modified by concentration , cooling rate, and high-intensity ultrasound. J Food Sci 84: 2549–2561
47. Giacomozzi A, Palla C, Carrin ME, Martini S (2020) Tailoring physical properties of monoglycerides oleogels using high- intensity ultrasound. Food Res Int 134:109231
48. da Silva TLT, Danthine S (2021) Effect of high-intensity ultrasound on the oleogelation and physical properties of high melting point monoglycerides and triglycerides oleogels. J Food Sci 86:343–356
49. Teodoro da Silva TL, Danthine S (2022) Influence of sonocrystallization on lipid crystals multicomponent oleogels structuration and physical properties. Food Res Int 154:110997
50. Blake AI, Marangoni AG (2015) The use of cooling rate to engineer the microstructure and oil binding capacity of wax crystal networks. Food Biophys 10:456–465
51. Chen F, Zhang H, Sun X, Wang X, Xu X (2013) Effects of ultrasonic parameters on the crystallization behavior of palm oil. J Am Oil Chem Soc 90:941–949

52. Gravelle AJ, Davidovich-Pinhas M, Zetzl AK, Barbut S, Marangoni AG (2016) Influence of solvent quality on the mechanical strength of ethylcellulose oleogels. Carbohydr Polym 135: 169–179
53. Gravelle AJ, Barbut S, Marangoni AG (2012) Ethylcellulose oleogels: manufacturing considerations and effects of oil oxidation. Food Res Int 48:578–583
54. Gravelle AJ, Barbut S, Quinton M, Marangoni AG (2014) Towards the development of a predictive model of the formulation-dependent mechanical behaviour of edible oil-based ethylcellulose oleogels. J Food Eng 143:114–122
55. Gómez-Estaca J, Herrero AM, Herranz B, Álvarez MD, Jiménez-Colmenero F, Cofrades S (2019) Characterization of ethyl cellulose and beeswax oleogels and their suitability as fat replacers in healthier lipid pâtés development. Food Hydrocoll 87:960–969
56. Lionetto F, Coluccia G, D'Antona P, Maffezzoli A (2007) Gelation of waxy crude oils by ultrasonic and dynamic mechanical analysis. Rheol Acta 46:601–609

Chapter 12
Designing for the Future: The Intersection of 3D Printing and Oleogels

M. Itatí De Salvo, Ivana M. Cotabarren, and Camila Palla

Abbreviations

STL	Stereolithography
3D	Three-dimensional
μ	viscosity
AC	Additional component
BG	Bigel
BW	Beeswax
CAD	Computer-aided design
CMC	Carboxymethyl-cellulose
CR	Carrageenan
CW	Candelilla wax
EC	Ethylcellulose
G''	Loss/viscous modulus
G'	Storage/elastic modulus
GG	Guar gum
HG	Hydrogel
HOASS	High oleic acid sunflower seed
HOSO	High oleic sunflower oil
HPMC	Hydroxypropyl methylcellulose
MCT	Medium chain triglyceride
ME	Melting extrusion

M. Itatí De Salvo · I. M. Cotabarren · C. Palla (✉)
Departamento de Ingeniería Química, Universidad Nacional del Sur (UNS), Bahia Blanca, Argentina

Planta Piloto de Ingeniería Química – PLAPIQUI (UNS-CONICET), Bahia Blanca, Argentina
e-mail: mdesalvo@plapiqui.edu.ar; icotabarren@plapiqui.edu.ar; cpalla@plapiqui.edu.ar

C. Andrea Palla, F. Valoppi (eds.), *Advances in Oleogel Development, Characterization, and Nutritional Aspects*,
https://doi.org/10.1007/978-3-031-46831-5_12

MG	Monoglyceride
OG/HG	Oleogel/hydrogel
OG	Oleogel
PEG40S	Polyethylene glycol monostearate
PGPR	Polyglycerol ricin alkyd ester
PS	Phytosterol
SFW	Sunflower wax
SME	Soft material extrusion
SO	Sunflower oil
SO-SW	Soybean oil-sunflower wax
SPNP	Spirulina platensis protein nanoparticle
T_g	Gel point temperature
T_P	Building platform temperature
XG	Xanthan gum
Y_s	Yield stress
$\dot{\gamma}$	Shear rate

12.1 Introduction

Food three-dimensional (3D) printing is a computerized construction process digitally controlled by a software, which can build complex 3D food products layer by layer [1]. It aims to revolutionize food manufacturing because of the advantages it presents to the food industry. One of the main benefits of food 3D printing is its ability to create customized, intricate designs that would be difficult to achieve through traditional methods [2]. This could lead to greater creativity and innovation in food design, as well as the ability to produce food with more precise nutritional content. With 3D printing, food can be tailored to the consumer's preferences in terms of shape, taste, and design. This level of customization can enhance the overall eating experience and lead to greater satisfaction for the consumer. Another benefit of food 3D printing is personalized nutrition. The technology allows to produce exact formulations with precise levels of nutrients and functional compounds essential for preventing malnutrition and diseases in an individual. This could lead to the development of personalized nutrition plans that cater to specific dietary requirements and health needs [2]. Additionally, food 3D printing has the potential to decrease both food and packaging waste and promote sustainability. In fact, it enables the production of precise portions that minimize excess material, as well as the printing of food on demand near its final consumption point, eliminating the need for secondary packaging [3]. Finally, food 3D printing has the potential to increase access to healthy food options, particularly for individuals with limited access to fresh products.

Various 3D printing methods have been evaluated for food production, such as selective laser sintering, melting/soft-material extrusion, binder jetting, and inkjet printing [4]. Among them, 3D printing based on extrusion (3DP-EXT) is the most widely adopted method. This method manufactures a 3D model by extruding fluid or semi-fluid mixtures (pastes or gels) and depositing them on a platform layer by layer. Compared to other methods, the most significant advantage is the wide variety of raw materials that can be used to form a food product [5]. The material used for 3DP-EXT should be extrudable to flow uniformly through the nozzle and possess adequate mechanical properties to maintain structural stability after printing [6]. This makes it necessary to conduct a detailed rheological and mechanical characterization of the printing materials, as well as to use specialized techniques developed specifically for evaluating the quality of printed products. Among food ingredients, lipid-based materials have shown to be essential components in the development of food 3D printing to customize them nutritionally and organoleptically [7]. For this reason, there is a growing interest in testing new lipid materials, such as oleogels, bigels, and oleogel-based emulsions, for 3D printing applications.

The quality of the final product obtained from 3D printing, both in terms of structure and nutrition, is dependent on the type of material and the printing technique employed. For this reason, it is critical to control different parameters. These include factors related to the printing process—such as extrusion speed, flow rate, nozzle size and height, and layer height—as well as those pertaining to the used food materials—such as rheological, mechanical, and thermodynamic properties. By closely controlling these parameters, it is possible to produce high-quality 3D printed food products.

In this chapter, we delve into the implications of using oleogels and bigels in 3DP-EXT. Firstly, we present a description of the most suitable 3D printing techniques for these materials. Subsequently, we examine the most important results that have emerged from all the known contributions to our understanding, which have evaluated the possibility of developing customized nutraceutical or food products based on these materials. In addition, we closely describe the tests that need to be performed on the printing inks and the printed foods to evaluate suitability of materials and quality of printed products. Finally, we summarize the opportunities, challenges, and future perspectives in this field.

12.2 Extrusion 3D Printing

The 3DP-EXT process involves a series of steps, as illustrated in Fig. 12.1, many of which are shared with other 3D printing technologies. The first step is to create a computer-aided design (CAD) file in stereolithography (STL) format that defines the

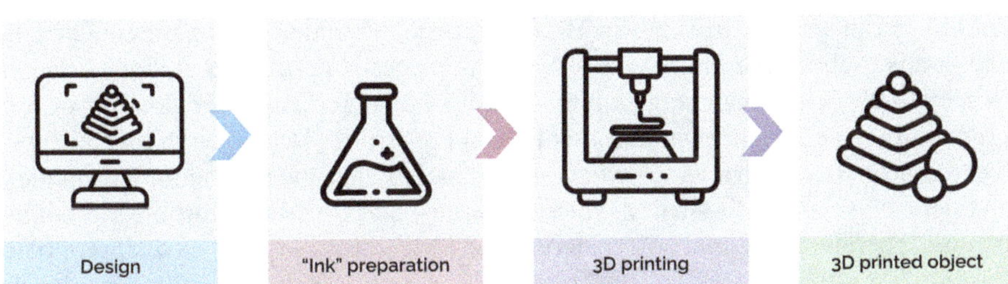

Fig. 12.1 Stages involved in extrusion 3D printing process

geometry and dimensions of the 3D object. This resulting STL file is loaded into a slicer program and the different print parameters, including layer thickness, infill, flow, layer height, and print speed, are set [8]. These parameters are critical in 3DP-EXT, as they determine the accuracy of the shape, internal microstructure, and quality of the final product [8, 9]. The "ink," formulated from food ingredients, is loaded into a syringe and then extruded through the nozzle by the force produced by a piston, screw, or air pressure. Consecutive layers are deposited by directing the syringe to points predetermined by the CAD model. Depending on the type of material used, bonding mechanisms can occur by layer stacking controlled by the rheological properties of the materials, solidification on cooling, or gel-forming extrusion.

Regarding 3DP-EXT printer development for the food sector, the first one was the Fab@home created by Cornell University academics [10]. This technology was initially used to extrude hot molten chocolate and, later, to print turkey flesh and scallops. In addition, Hao et al. [11] created The Choc Edge, the first commercially available chocolate 3D printer. Similarly, Natural Machines produced the Foodini Printer, which can print paste, chocolate, frosting, mashed potatoes, tomato, and other foods. There are also commercial printers that have been modified to allow food extrusion, such as the modified Prusa printer used by Cotabarren et al. [9] to print nutraceuticals.

12.2.1 Melting Extrusion

Melting extrusion (ME) is a process that involves extruding molten food material from a moving printhead. The material is deposited layer by layer onto the build platform and solidifies almost immediately after extrusion [12]. Temperature plays a crucial role in the melting extrusion process since the material is held at a temperature above its melting point while in the extruder. As it exits the nozzle and is deposited, it cools down (either due to the ambient temperature or with the help of a cooling system) and solidifies, fusing with the lower layers while maintaining its shape and supporting the subsequent ones [1].

12.2.2 Soft Material Extrusion

During soft material extrusion (SME), food paste is extruded from a moving nozzle and bonds to the preceding layers [12]. This process is performed at low extrusion temperatures—e.g., room temperature—and is suitable for pastes and doughs, whose printability depends only on their rheological properties [1]. Although this technology has been used with a wide variety of food ingredients, the production of complex and delicate shapes is limited, as these materials are prone to distortion and deformation. Therefore, optimal soft materials for SME must not only have appropriate mechanical strength and thixotropic recovery to maintain the produced shapes but also shear-thinning behavior to be readily extruded through the nozzle of an extrusion printer [1].

12.3 Oleogels and Bigels Used in Extrusion 3D Printing

Combining oleogels with 3DP-EXT techniques can serve two main objectives. The first one is to evaluate these materials as printing ink, focusing on their formulation and properties for use in printing geometric shapes with specific characteristics. By understanding the interaction between the materials and the printing process, it is possible to optimize the properties of the final product and achieve the desired texture, specific component distribution, and other characteristics. The second objective, which is less explored, is to use 3D printing as a process to modify the characteristics of the materials. The printing process can alter the microstructure and mechanical properties of oleogels, enabling the production of custom-designed materials with specific functionality.

Table 12.1 shows the studies that have evaluated the performance of 3DP-EXT so far, using oleogels as printing materials. As can be seen, the amount of gelator in the oleogel and the amount of oleogel in the printing material are two of the most investigated parameters due to their significant impact on the physical properties of the material, which affects both its ability to be extruded through a syringe and the quality of the printed product. A recent study investigated the 3D printing performance of mixtures composed of potato starch, whey protein, and beeswax oleogel [6]. The authors found that increasing the amount of beeswax in oleogel improved printing accuracy, with 10%, 15%, and 20% beeswax providing acceptable printing accuracy, good structural support, and extrusion performance, while 5% beeswax resulted in poor accuracy and collapse during printing. Additionally, as the amount of oleogel increased, the printing precision of the mixtures first improved and then decreased, indicating that the oleogel needs to be incorporated in an optimal amount. Another factor that influences the printing process when using oleogels is the extrusion temperature. This factor has an impact on the physical properties of the material, such as viscosity and surface tension, which consequently affect the extrusion rate, dimensions, and mechanical properties of the printed part. Some studies have shown that optimizing the extrusion temperature can significantly improve the mechanical strength, dimensional accuracy, and quality of the printed objects [6, 13]. Moreover, the extrusion temperature can also influence the adhesion between the printed layers, which is essential for ensuring the structural integrity of the part [13].

On the other hand, it is possible to introduce the oleogel into the syringe once formed and perform the extrusion of the semi-solid material by printing at temperatures below its melting point, e.g., room temperature (SME), or introduce the melted oleogel and print at temperatures above its melting point (ME), which require the use of a heated syringed. Printing with oleogels at temperatures below their melting point has advantages such as building stable structures that can maintain

Table 12.1 Oleogels used as printing materials in 3D printing-based extrusion (3DP-EXT)

Printing material formulation			Type of 3DP-EXT	Objective	Evaluated parameters	Highlighted results	Reference
Gelator (wt%)	Oil	AC					
MG (10 and 20)	HOSO	PS	ME	Evaluate printability of mixtures	PS (20–50 wt%/oleogel) and MG contents and nozzle diameter (0.83 and 0.40 mm)	A maximum of 30–40 wt% PS/oleogel and oleogels with 10–20 wt% MG resulted in stable and uniform solid forms with high weight repeatability	[9]
MG (10)	HOSO	PS	ME	Evaluate the effect of printing process variables	Flow (70–100%), extruder movement speed (1–6 mm/s), and T_P (8–20 °C)	Weight of solid forms only affected by flow. Similarity to CAD design and high self-supporting capacity at low T_P and speed (8 °C and 1 mm/s). T_P from 8 to 20 °C decreased hardness	[8]
BW (11–19)	HOSO	None	ME	Determine optimal formulation and infill of materials alternative to lard	BW content and infill level (IL) (50–100%)	Cubes of BW oleogel at optimized conditions (BW: 16 wt%, IL: 58.9%) were similar in hardness, cohesiveness, and adhesiveness to cubes printed with lard (IL: 75%)	[14]
BW (5–25)	SO	Potato starch, whey protein, and water	SME	Determine optimal formulation and printing conditions	Oleogel content (10–50 wt%), extrusion temperature (30–50 °C), printing speed (20–100 mm/s) and nozzle diameter (0.8–2 mm)	Optimum printing conditions: 20 wt% oleogel content, 30 °C of printing temperature, 40 mm/s of printing speed, and 1.2 mm of nozzle diameter	[6]

EC	MCT	PEG40S	ME and SME	Investigate the effect of addition of PEG40S on oleogel mixture printability	PEG40S content (0–5 wt%) and extrusion temperature (25–95 °C)	Optimal printability at 5 wt % surfactant and extrusion at 45 °C	[13]
Gelatin (10 wt/vol%) and gellan gum (1–1.75 wt/vol%)	MCT	Water, curcumin, and resveratrol	SME	Determine optimal formulation and printing conditions	Gellan gum concentration, nozzle diameter (0.84 and 1.2 mm), and printing speed (600–900 mm/min)	1.5 wt% of gellan gum was optimum to produce a stable 30 wt% oil-in-water emulsion for 3DP-EXT	[15]
PS and lecithin (10–50)	SO	None	SME	Evaluate the printability of oleogels	Total solid content (20–50 wt%), phytosterol-lecithin ratio, printing speed (5–45 mm/s), and number of printing layers (5–29)	PS-lecithin ratio (R) and solid content (SC) that formed self-supportive structures with shape fidelity ≥90%: 80% R and 10–30% SC or 40–60% R and 40–50% SC	[7]

AC additional component, *BW* beeswax, *EC* ethylcellulose, *HOSO* high oleic sunflower oil, *MCT* medium chain triglyceride, *ME* melting extrusion, *MG* monoglyceride, *PEG40S* polyethylene glycol monostearate, *PS* phytosterol, *SME* soft material extrusion, *SO* sunflower oil, T_P build platform temperature their shape during the printing process, reducing the risk of thermal degradation of the oleogel, which can affect the final product quality, and reducing the complexity and cost of the printing process. However, printing with semisolid materials involves working with higher viscosity materials, resulting in slower printing speeds and the potential for clogging the extrusion nozzle. Printing with molten oleogels at temperatures above their melting point has advantages, such as faster printing speeds, easier extrusion due to the liquid state of the oleogel, and higher adhesion between printed layers. Furthermore, since oleogel formation occurs during the printing process, adjusting the build platform temperature, extruded material amount, or extruder movement speed can modify the microstructure of the printed forms and, therefore, their macroscopic properties. By printing with a molten mixture composed of monoglyceride (MG) oleogel loaded with phytosterols (PSs) (20 wt% PS/oleogel), it was found that decreasing the build platform temperature (T_P) from 20 to 8 °C resulted in a 37% increase in the hardness of printed forms [8]. The authors also reported that there is a similar effect when working with the highest T_P or extrusion speed. In such cases, the first layers of printing material do not cool sufficiently to reach the semi-solid state, resulting in inadequate support for subsequent layers and eventual collapse of the structure. This finding is consistent with the material's behavior observed during the rheological sweep test, where an increase in temperature corresponded to a decrease in the elastic modulus (G'), and consequently, low self-supporting capabilities of the material [9].

The images presented in Fig. 12.2 underscore the importance of selecting an appropriate printing material formulation and printing configuration to ensure the

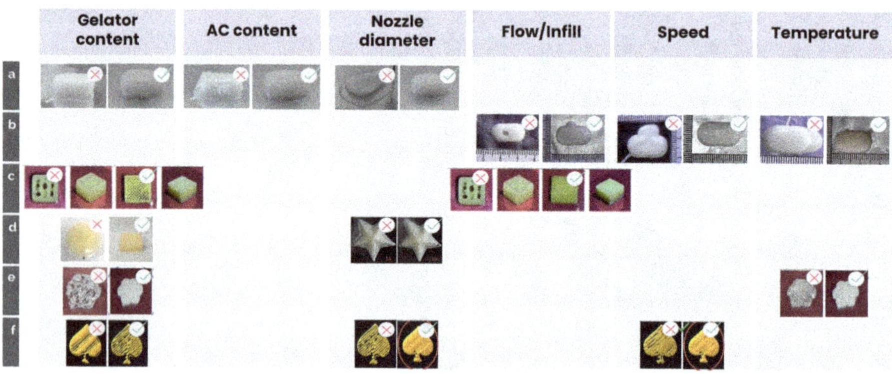

Fig. 12.2 Images of printed shapes, both failed and successful, produced with oleogels showing the effect of using different formulations and printer setting conditions (AC: Additional component): (**a**, **b**) nutraceuticals oral forms (reproduced from Cotabarren et al. [9] and De Salvo et al. [8], respectively, both with permission from Elsevier); (**c**) alternative product to animal fat (reproduced from Kang et al. [14], under the terms of the Creative Commons CC-BY license); (**d**) test parts (reproduced from Shi et al. [6], with permission from Elsevier); (**e**) test parts (reproduced from Kavimughil et al. [13], with permission from Elsevier); and (**f**) nutraceuticals forms (reproduced from Kavimughil et al. [15], with permission from Elsevier). Note: Speed in (**b**) and (**f**) refers to extruder movement and printing speed, respectively; temperature in (**b**) and (**e**) refers to build platform and extrusion temperature, respectively. Refer to Table 12.1 for more information about the formulation of the printing materials

production of high-quality printed products. As can be seen, optimizing the gelator content resulted in more precisely printed parts that were structurally stable and had an appropriate filling. The concentration of additional compounds, including bioactive additives, has also an impact on the printing process. This influence became evident in a comparison of printed solid forms containing 20–50 wt% PS/oleogel [9]. As the PS content increased, the geometry of the printed forms showed signs of deformation. Despite displaying superior stability based on rheological testing, these gels were susceptible to nozzle drag, which affected the final shape of the solid. Regarding nozzle diameters, satisfactory printings outcomes have been reported using sizes of 0.83 mm for ME and 1.2 and 1.5 mm for SME [6, 9, 13].

Bigels are a type of semisolid gel that consists of an oleogel (OG) and a hydrogel (HG), which are typically formed by mechanical mixing at specific temperatures and conditions of gel setting. Bigels present some advantages over oleogels and hydrogels as their hydrophilic and hydrophobic phases can be used to encapsulate different bioactive compounds, such as vitamins, minerals, and antioxidants, which can provide a range of health benefits. Moreover, the physicochemical properties of bigels can be tuned by adjusting the composition and amount of each phase [16]. This allows tailoring their rheological properties, making bigels potentially suitable materials for 3D printing. Table 12.2 exhibits some bigel formulations that have been tested in 3DP-EXT, all of them using SME. These studies have investigated the impact of varying the OG/HG ratio on rheological properties, printability, and supporting capacity. In general, it was observed that increasing the amount of oleogel led to improved extrusion fluidity, enhanced rheological properties, and better printability, making materials more optimal for 3D printing [17–19]. Bigels produced using oleogel from beeswax and hydrogel from carrageenan and xanthan gum exhibited an increase in viscosity and shear-thinning properties as the oleogel fraction increased. These characteristics allowed for smoother extrusion of ink from the nozzle [19]. Additionally, the bigel with 80% oleogel demonstrated the highest yield stress, complex modulus, and gel strength, indicating superior self-supporting capacity. Figure 12.3 shows images of solid forms printed by 3DP-EXT considered failed or successful based on their geometric similarity with the 3D digital model, the ability to maintain their shape after printing, and uniform surface. As noted earlier, bigels containing a higher proportion of oleogel than hydrogel produced well-formed printed shapes.

12.4 Material Printability

Along with printing settings, information about how the material behaves during and after the printing process is very important. In this sense, the term *printability* emerged, which incorporates the rheological and mechanical properties of the materials to be printed in order to evaluate: (i) the material intrinsic properties that facilitate handling and deposition by the printer (flowability through the nozzle) and (ii) the capacity to hold the structure and dimensional stability, either as the final step

Table 12.2 Bigels used as printing materials in 3D printing-based extrusion (3DP-EXT)

Printing material formulation							
Oleogel		Hydrogel					
Gelator (wt%)	Oil	Gelator (wt%)[a]	Type of 3DP-EXT	Objective	Evaluated parameters	Highlighted results	Reference
BW (10)	Soybean	HPMC (3)	SME (20 °C)	Evaluate physical properties and 3D printability of bigels	OG/HG ratios (80–50 wt% OG)	Bigels with OG $\geq$ 60 wt%: best mechanical properties and printing stability. 55 wt% OG (semi-bicontinuous): worst extrusion ability. 50 wt% OG: minimal TPA property changes pre- and post-printing	[20]
SFW (5)	Soybean	XG (1)	SME (20 °C)	Evaluate printability of nanoparticle-based bigels	OG/HG ratios (20–80 wt% OG)	High OG/HG ratios were less likely to collapse. Moisture content improved structure smoothness. 56 wt% OG: best for 3D printing. Significant reduction in mechanical properties due to nozzle shearing	[17]
EC (10) MG (2)	Soybean	Gelatin (10)	SME (20 °C)	Analyze physical properties and 3D stacking ability of bigels	OG/HG ratios (20–80 wt% OG)	60 wt% OG: optimal for 3D printing due to superior extrusion fluidity. Bigels with OG $\leq$ 40 wt%: properties similar to hydrogel, optimal for pastry decoration	[18]
BW (15)	Corn	CR and XG (1.5)	SME (30 °C)	Investigate the 3D printability, rheological properties, and microstructure of bigels	OG/HG ratios (20–80 wt% OG)	Increasing OG/HG ratio enhanced rheological properties and printability. 80 wt% OG: best material with stronger mechanical properties	[19]

CW (3)	HOASS	Fish gelatin (5)	SME (room temperature)	Study the influence of different bigel composition and emulsifiers	OG/HG ratios (0–100 wt% OG). Emulsifiers (PGPR, MGs, lecithin)	MG or lecithin best emulsifiers. Bigels of MGs with 70 wt% OG: good properties and best 3D printing quality. High HG content improved mechanical support but obstructed the printing nozzle	[21]

[a]Water as solvent. *AC* additional component, *BW* beeswax, *CMC* carboxymethyl-cellulose, *CR* carrageenan, *CW* candelilla wax, *EC* ethylcellulose, *GG* guar gum, *HOASS* high oleic acid sunflower seed, *HPMC* hydroxypropyl methylcellulose, *MG* monoglyceride, *OG/HG* oleogel/hydrogel, *PGPR* polyglycerol ricin alkyd ester, *SFW* sunflower wax, *SME* soft material extrusion, *SO-SW* soybean oil-sunflower wax, *SPNP* spirulina platensis protein nanoparticle, *XG* xanthan gum

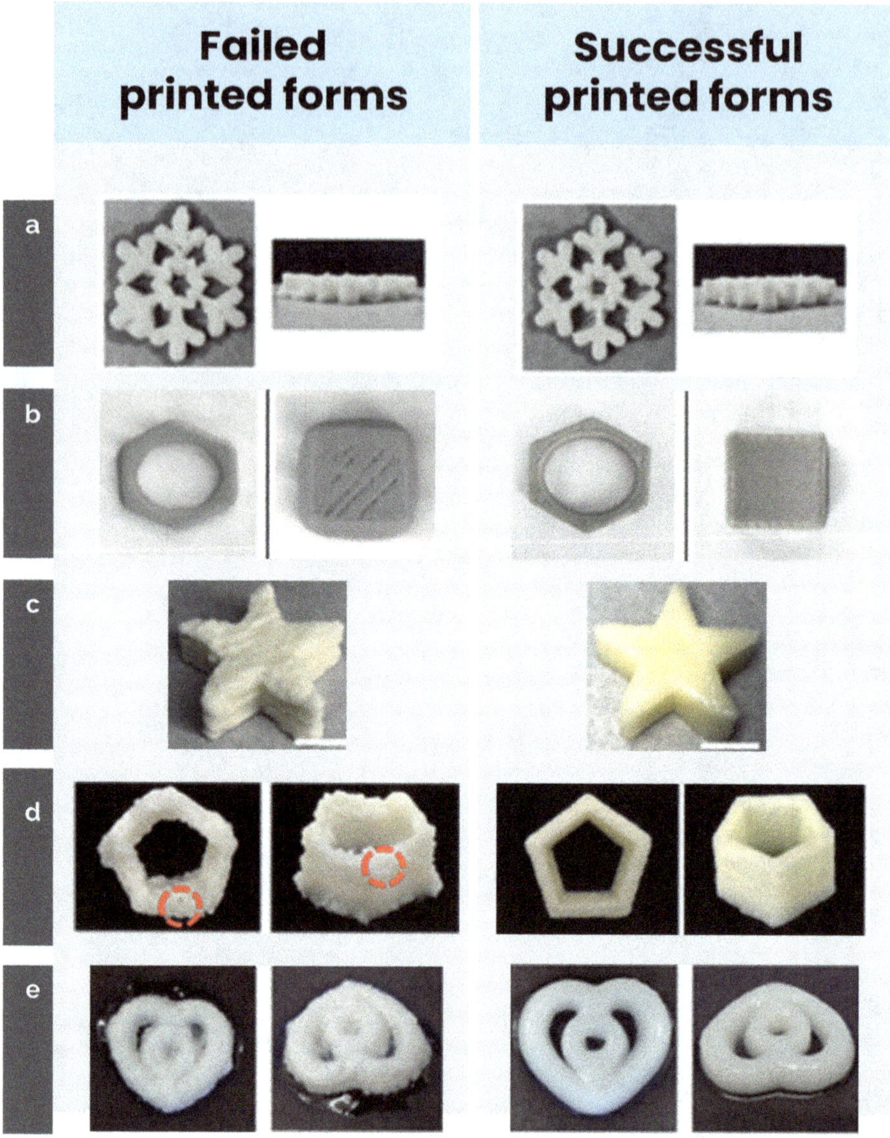

Fig. 12.3 Images of printed forms using bigels varying the oleogel/hydrogel ratio (OG/HG) or OG composition: (**a**) 55% OG (left) and 50% OG (right) (reproduced from Chen et al. [20], with permission from Elsevier); (**b**) 50% OG (left) and 56% OG (right) (reproduced from Guo et al. [17], with permission from Elsevier); (**c**) 40% OG (left) and 60% OG (right) (reproduced from Jiang et al. [18], with permission from Elsevier); (**d**) 50% OG (left) and 80% OG (right) (reproduced from Qiu et al. [19], with permission from Elsevier); and (**e**) 70% OG formulated with lecithin (left) and monoglycerides (right) (reproduced from Xie et al. [21], with permission from Elsevier)

or until post-processing [22]. To describe material printability, a deep understanding of (a) the inks' rheological characteristics, (b) the mechanical behavior of the printed solids, and (c) specifically defined printed geometric factors is required [23]. Some of these material characteristics can be assessed before the extrusion and are directly related to their printability—such as viscosity (μ), storage and loss moduli (G' and G''), and yield stress (Y_s)—while others relate to the stability and properties of the printed material, such as their shear recovery behavior or mechanical strength. Y_s, for example, is the lowest force necessary for the material to be extruded, and if this force is exceeded, the structure of the material collapses and flows. High viscosity materials may also clog the nozzle output [23]. Intimately related to the rheological properties is the inks' capability to form stable 3D structures, which can be evaluated by characterizing and comparing the 3D structure with the desired model through, for example, shape inconsistency and deformation factors. Finally, several consumer-oriented features are intimately related to the mechanical properties of the final product like texture and mouthfeel.

12.4.1 Rheological Tests

To be printed effectively, edible inks are often required to have appropriate rheology properties for ensuring a smooth printing process and guaranteeing high-quality prints. In fact, rheology has greatly driven the development of printing technology in food applications since it provides a pre-evaluation of ink printability, creating a bridge between edible inks and practical printing [24].

In this sense, the printing process can be divided into three steps: (i) *extrusion step*, where parameters such as viscosity, shear-thinning behavior, and Y_s can be used; (ii) *recovery step*, where shear recovery properties are measured; and (iii) *self-supporting step*, where parameters like G'', G', and Y_s are used as a reference [23]. All these rheological properties are highly related to ink composition and possess high relation to the smoothness of the printing procedure. G'' and G' indicate viscoelastic properties (liquid-like or solid-like behavior) of edible ink, and therefore determine structure strength. Moreover, Y_s and thixotropy could help explore status transition and structure recovery for high structure maintainability and shape retention after printing [24].

Characterization of rheology is achieved on a rheometer, using in general, a parallel plate geometry. According to Maldonado Rosas et al. [23], the following assays (Fig. 12.4) need to be performed to measure properties that are relevant to the printing process:

Amplitude Sweep Oscillation amplitude sweep measurements are made at a fixed frequency to determine the Y_s. This value is obtained as the intersection of G' and G'' curves. Y_s is an important parameter for lipid materials, as it correlates with spreadability and material stability and, in turn, results in a key indicator of the extrudability and self-supporting properties of oleogel and bigel inks. For oleogels, a

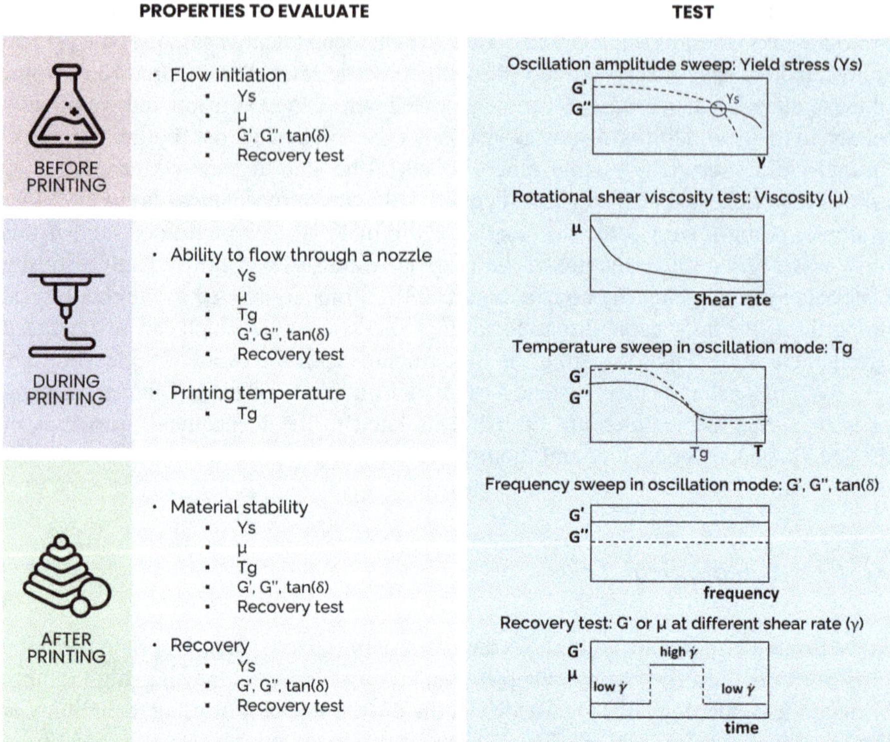

Fig. 12.4 Rheological tests used to evaluate material properties in the different stages of the extrusion 3D printing process. G': storage/elastic modulus; G'': loss/viscous modulus; T_g: gel point temperature; $\dot{\gamma}$: shear rate; tan $(\delta) = G''/G'$

significant increase in Y_s was observed with an increase in the gelator (MG content) and to a lesser extent, with the addition of PSs [9]. This was beneficial for the stability of the printed forms, while Y_s above a certain value (critical Y_s) compromised the material flowability through the nozzle. Another study reported a decrease in Y_s with the addition of surfactant to oleogel, indicating a softening and plasticizing effect that favored the printing extrusion process [13]. Similarly, for bigel inks it was found that Y_s increased with the addition of oleogel to the bigel matrix [19]. In conclusion, the larger the Y_s of the oleogel and bigel inks, as long as it is less than the critical Y_S, the better it could ensure smooth extrusion and improve its self-supporting properties. On the other hand, lower Y_s implies that the inks extrude from the nozzle smoothly, but they cannot self-support so that the product collapses after printing.

Shear-Viscosity Test Rotational shear-viscosity determinations involve measuring how the viscosity of a fluid changes with variations of the shear rate. In this sense, it was found that under the same shear rate, the increase in amount of gelator (i.e., beeswax) resulted in denser crystal networks that caused an increase in viscosity

[6]. However, when this oleogel was incorporated in the studied mixtures (potato starch, whey protein, and oleogel), an increase in its amount resulted in a decrease in both G' and viscosity as the shear rate increased. An appropriate amount of oleogel can reduce the viscosity and improve the stability of printed samples, but too much oleogel content will increase the ductility and affect the structural integrity of the printed form. In bigels, as previously mentioned, it has been found that the viscosity increases with oleogel fraction [19, 20]. Notably, it was also observed that when the oleogel addition exceeded a certain mass percentage, due to the material complex structure, the viscosity of the bigel ink no longer changes significantly. In general, bigel inks with sufficient viscosity can self-support subsequent layers for 3DP-EXT.

Temperature Sweeps in the Oscillatory Mode The sweeps performed in the oscillatory mode varying temperature at fixed frequency and strain within the linear viscoelastic region allow to determine variations in the G' and G'', which can be used to calculate the loss factor (tan $(\delta) = G''/G'$). The ink gel point temperature (T_g) is usually determined as the point where the loss factor is equal to unity. This property is especially important for setting the extrusion temperature in ME, as the ink should be in a dominant fluid state to be able to flow through the printer nozzle. In fact, Cotabarren et al. [9] proved that when oleogels are used as printing materials, extrusion temperature should be quite above T_g to avoid clogging. Interestingly, too high temperatures resulted in over-extrusion that compromised the printed geometry [13]. Therefore, careful setting should be done according to each ink gel point. Using bigels, an interesting behavior was observed as a function of the oleogel content. For samples with high oleogel content, G' showed a decreasing trend throughout the heating process, so that when the T_g was reached, the G' decreased rapidly, which demonstrated that bigels started to destabilize and behave more as liquids. In bigels with high hydrogel content, the thermal gelation of hydrogels was more important than the thermal melting of the oleogels, leading to an increase in the G' after reaching the hydrogel gelation temperature [17, 20]. Unfortunately, extrusion temperature was not a variable studied in these contributions (i.e., it was kept constant at 25 °C); therefore, it is not possible to infer the effect of this inversion in the printing process.

Frequency Sweeps The sweeps performed in the oscillatory mode varying frequency at a fixed temperature (e.g., 20 °C) and strain within the linear viscoelastic region while measuring G' and G'' response provide information about the solid- or viscous-like behavior of the inks. A material that presents $G' < G''$ will behave more like a liquid and vice-versa. For oleogel inks, in general, it is observed that the mixtures present a gel-like behavior with the predominance of G' over G'' independent of the frequency. Cotabarren et al. [9] found that G' increased with increasing gelator and PSs content. A similar result was observed by Kavimughil et al. [13] when increasing surfactant content and by Shi et al. [6] as the amount of beeswax in the oleogel increased. Regarding bigels, it was also observed that G' was higher than G'', indicating that these samples were viscoelastic semi-solids [20]. Furthermore, the addition of oleogel would significantly increase the values of G' and G'' [19]. The increase in oleogel ratio or gelator content significantly increased the mechanical strength of inks, making them able to withstand layer by layer adhesion after 3DP-EXT.

Recovery Test Elastic recovery assessments (the three-interval thixotropy test) are conducted for simulation and characterization of the shear recoverability conditions expected for the formulations during the printing process. The samples are first exposed to a low shear rate to mimic at-rest before extrusion. Then, a higher shear rate is applied to the formulations emulating the shear forces in the nozzle during printing. Finally, a low shear rate is again applied to determine the recovery of the formulations. This assay is particularly crucial for materials used in SME, such as bigels. In fact, it has been observed that bigels could recover after removing the high-speed shearing rate [19]. Notably, the oleogel-dominated bigel inks showed a very short recovery time, making these mixtures more suitable for 3DP-EXT. The quick recovery feature can enhance structural stability and offer sufficient mechanical strength, which is supportive for shape maintenance after extrusion. Similar results (i.e., viscosity or modulus recovery) have been reported for other bigels mixtures [17, 18, 25, 26].

Overall, to ensure that a mixture is technologically capable of being 3D printed—that is, it can pass through the nozzle and maintain its shape and internal structure when extruded or subjected to post-processing—rheological properties must be thoroughly investigated. This is necessary if food 3D printing is to progress toward appropriate formulations for individualized nutrition, which should be designed with optimal rheological properties to be capable of being printed or extruded.

12.4.2 Mechanical Tests

The mechanical characterization of materials is essential for 3D printing. Usually, these measurements are carried out in a texture analyzer by applying a perpendicular stress on the sample. However, it is important to be aware that when mechanical characterizing any 3D printed product, the results are not associated purely with the materials that are being used but also with the printing process parameters, such as infill percentage, extrusion and build platform temperature, and post-processing [8, 22]. This makes 3D printing a versatile fabrication technique to produce food with customized properties. Essentially, the following tests can be performed (Fig. 12.5):

Texture Profile Analysis (TPA) The TPA test can be employed to both evaluate the quality of the printed product and predict its acceptability. Hardness and adhesiveness are related to the shape fidelity of the printed part, as well as the ability of the material to be sensorially accepted or withstand storage and transportation processes. In the work of Shi et al. [6], increasing the amount of beeswax increased hardness, springiness, and chewiness of the oleogel. However, when this oleogel was combined with potato starch and whey protein to form a mixture, higher

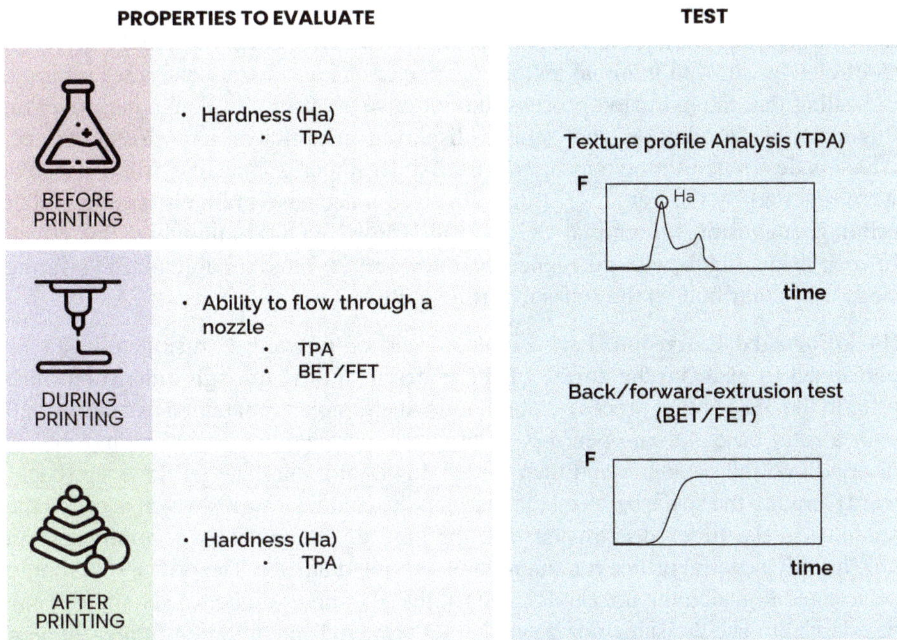

Fig. 12.5 Mechanical tests used to evaluate material properties in the different stages of the extrusion 3D printing process. *F*: force

amounts of oleogel led to a decrease in hardness and resilience, while the elasticity, chewiness, and gumminess increased first and then decreased showing a high interaction between the materials. For pure oleogel printed forms, Cotabarren et al. [9] found that hardness increased with PS content. However, the 3D printing process generated oral forms with lower hardness than those produced by manual molding, indicating that the layer by layer construction process generated some fragility. Other authors performed a TPA of oleogel printed samples right after printing and after 24 h of storage in order to understand the effect of time, solid concentration, and printing speed on texture properties [7]. Results showed that inks with the highest solid concentration at faster printing speeds produced 3D printed oleogels with high hardness values; but no significant differences were observed with low solid concentration at different speeds. Additionally, the 3D printed oleogels that were tested 24 h after printing showed hardness values almost twice higher than those obtained after printing. By analyzing gelator concentration (i.e., beeswax) and printing parameters settings, it has been observed that sample hardness increased rapidly with gelator concentration and slowly increased with increasing infill level [14]. Cohesiveness decreased with solid concentration, while a linear increase in adhesiveness was observed with increasing both concentration and infill level.

For bigels, the texture properties obtained before and after printing have been evaluated, finding that the hardness of samples with high oleogel content (>60 wt%)

decreased after 3D printing [20]. This may be due to the breakage of the crystalline network structure during the extrusion process. On the other hand, a bigel with 50 wt % of oleogel showed minimal texture parameter changes before and after printing, indicating that the extrusion process did not have an excessive damaging effect on this material. In addition, this sample displayed a lubricious and glossy surface, which is desirable in food appearance from a consumer perspective. Similar results were reported by Guo et al. [17] that observed a decrease in bigels hardness after printing, suggesting an optimal OG/HG ratio (56 wt% OG) to minimize this effect. In conclusion, bigels with the highest hardness are the most suitable for 3D printing since they contribute to the self-supporting ability.

Back/Forward-Extrusion Test Either a back or forward extrusion test can be performed to evaluate the ability of the material to flow through a narrow orifice prior to the 3D printing process. In both tests, the sample is contained in a strong cell with a solid base and an open top. In the back-extrusion test, a rod with a disc is inserted into the container and then forced down until the material flows up (backward) through the space between the disc and the container walls, which is called the annulus. In the forward extrusion test, the base of the container accommodates a nozzle with a central orifice (or annulus) of varying diameter. The orifice size should be selected to match the nozzle diameter of the 3D printer or, better yet, the cell can be adapted to use the same nozzle as the 3D printer. Then, a tightly fitting plunger, which acts almost like a piston, compresses the sample and causes forward flow through the orifice. It is important to note that these tests are appropriate for semisolid materials such as bigels or formed oleogels, since according to our laboratory work, the extrusion forces are barely detectable by the equipment when using molten oleogels. To our knowledge, there is only one contribution that uses the forward extrusion technique to assess the ability of bigels mixtures to flow through a nozzle [27]. However, this method would also be useful for testing oleogel-based inks.

12.4.3 Printing Tests

Lattice structures, 2D shapes, and different 3D geometrical figures are commonly used to assess the printing quality of soft-like printed shapes. In the case of lattice, for example, it has been proved that the distance between lines and the area within the internal squares can be correlated with the printing ability of gel-type inks [28]. In turn, different parameters that determine circularity—or are based on it—can be calculated to evaluate the printing precision. According to Maldonado Rosas et al. [23], the following tests are often performed in order to assess geometry printability (Fig. 12.6):

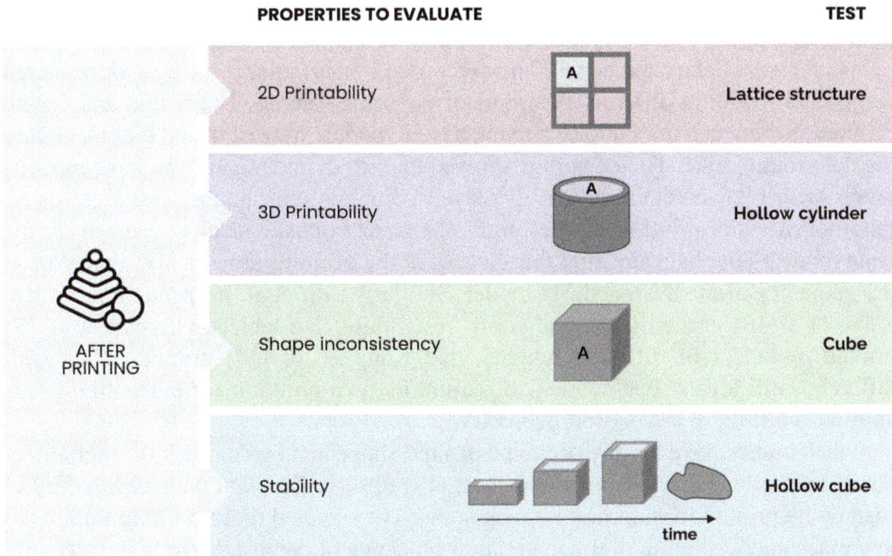

Fig. 12.6 Printing tests used to evaluate 3D printed parts properties. *A*: area

2D Printability Printability of 2D structure in the printing plane is evaluated by measuring the average area of the inner squares of the lattice structure. The calculated average areas are then compared with their corresponding designed value, and the percentage error is estimated.

3D Printability To assess the stability of the structure in the stacking direction, a hollow cylinder is printed and evaluated by measuring its inner circular area. Similar to the previous test, each measured inner circular area is compared with the corresponding designed value in order to estimate a percentage error.

Shape Inconsistency Factor and Deformation Factor A cube is printed to evaluate these two parameters. The shape inconsistency factor is calculated as the area of each face of the cube divided by its theoretical value. The average of the standard deviation of these values multiplied by 100 is defined as shape inconsistency factor. On the other hand, the deformation factor is determined as the ratio between the area of the top and the bottom faces.

Stability Determination The printing process is performed until the entire structure collapses. This parameter requires that the printing process is recorded so that the printing time before the collapse is known, as well as the corresponding sample weight. The stress parameter at the collapsing moment is obtained as the weighted sample mass multiplied by the acceleration of gravity and divided by the bottom area of the hollow cube.

Simpler models are often designed in CAD software and printed to evaluate their dimensions against the original ones. Chen et al. [20] printed models including cylinders, snowflakes and cones. They recorded images after printing and measured height and length at different positions of the same sample. The height and length differences between the samples and the design models were obtained for calculating the deformation rate. Their findings showed that for cylinder and snowflake models, bigels with high oleogel content displayed complete appearance with little or no inhomogeneity of printing. Interestingly, the exact opposite results occurred for the cone model. This indicates that the success in the printing process also depends on the geometry of the designed 3D model. Similarly, Shi et al. [6] printed cubes, De Salvo et al. [8] printed solid oral forms resembling 1 g tablets, Oliveira et al. [7] printed cylinders of different heights, and Kang et al. [14] printed cubes with different infill levels. Each case will require its own criterion for evaluating printability according to the chosen geometry.

Other studies have simply printed a desired shape and evaluated it through direct visual inspection. For instance, Qiu et al. [19] observed broken lines or inability to sustain the printed form in hollow pentagons. They related these printing failures to low mechanical strength of the material or blockage of the nozzle for high viscosity. As previously mentioned, the inks with higher OG/HG ratios have presented optimal printing performance, providing suitable consistency and fluidity to ensure smooth extrusion without choking, rapid recoverability, and outstanding mechanical properties to support printed parts. Similar studies, while using different geometries, were done by Jiang et al. [18], Shi et al. [29], Kavimughil et al. [15], Guo et al., [17], Zhai et al., [30], Kavimuguil et al., [13], and Shi et al. [25].

12.5 Opportunities, Challenges, and Future Perspectives

As mentioned throughout this chapter, 3D printing can be used to create complex matrices that may not be feasible with traditional methods. Furthermore, these matrices can be designed to improve the bioavailability of nutrients/active compounds. Thus, oleogels and bigels present excellent potential to be used for the development of materials with specific macro- and microstructures capable of protecting and/or controlling the release of these compounds. To this aim, suitable methods, such as dissolution and digestion tests, should be used or adapted to evaluate the liberation of active ingredients from the printed products. While 3D printing will not completely replace traditional methods of food production, it will allow the development of tailored products adapted for people with specific requirements [28]. Indeed, the size of these potential niche markets is today unknown.

Furthermore, several reports indicate that 3D printing technology is a part of the 4.0 industry revolution. In fact, industry 4.0 promotes decentralized production and distribution networks, whereby production is increasingly located closer to the final consumer, and everyone can become simultaneously a developer and producer. Thus, 3D printing is a perfect fit for the production methods of the future in many industries, including the production of food [28]. However, more in-depth research is required for the development of ready-to-use formulations to print food, in which oleogels could be used as healthy lipid ingredients. Furthermore, proper sanitization protocols for every step of the 3D printing process should be developed by regulatory agencies to safely develop products for human consumption. This is also a vital aspect when considering the use of sensory analysis studies to evaluate the acceptability of printed products.

On the other hand, while a huge variety of oleogels produced from different combinations of mono-, bi-, or multi-gelators with vegetable oils have been studied, only a few have been tested for 3D printing. This suggests that there is potential for further development and optimization of soft materials suitable for SMEs that better resist the forces associated with extrusion, resulting in products with improved structural integrity.

Regarding the technology challenges, 3D printing is always evolving towards new material and machinery developments. One of the most recent advances in materials is the so-called 4D printing, which combines 3D printing with stimulus-responsive materials to produce unique pieces that can switch between various configurations or appearances in response to environmental changes like heat or moisture. In some ways, food 3D printing can be thought of as 4D printing since the finished food might need to be cooked or treated before eating and because the product's design must consider the form and size changes that will occur during this post-printing. Yet, further advancements can be achieved when using oleogel-based systems. By using oleogels designed to respond to stimuli, 4D printing could, for example, create controlled-release delivery systems that can be triggered to release active compounds in response to changes in pH or temperature. Regarding machine development, the term "5D printing" has emerged. The convenience of not needing to print support structures for intricate patterns was the driving force behind the creation of five-dimensional printing. 5D printing refers to the use of a five-axis printer instead of a three-axis printer used in traditional 3D printing. In essence, a five-axis 3D printer constructs the object from various angles (X, Y, and Z with rotational and translational movements), resulting in stronger parts than with conventional 3D printing and avoiding the use of supports in many structures [28]. Even more, a combination of 4D and 5D printing has been proposed to boost the advantages of both techniques in the recently named 6D printing [31]. These technologies open up a new window for the development of even more intricate structures that can achieve specific product properties, such as targeted release of active ingredients or particular textures.

12.6 Conclusions

The use of oleogel-based materials as inks for 3D printing processes emerges as a new technological solution for food 3D printing based on solid fat replacers, with the ability of incorporating lipid-soluble active ingredients. In this chapter, a critical review of the main findings from all the contributions found in the literature to date on 3D printing of oleogels and bigels has been conducted. Interestingly, it has been found that two of the most studied and influential parameters when using oleogels and bigels as inks in 3D printing are the amount of gelator in oleogels and the amount of oleogel in bigels. Other factors, also crucial to produce successful printed forms, are the temperature of the printer nozzle and the build platform, as well as the extrusion speed. Furthermore, this chapter has provided a detailed description of the tests required for evaluating the rheological, mechanical, and printing properties of the inks and printed solid forms. It was concluded that for the material to pass through the nozzle and maintain its shape and structure when extruded or subjected to post-processing, all these properties must be thoroughly investigated. Once the materials and process conditions are fully characterized, an operation map could be defined, indicating the rheology and textures characteristics that a food mixture needs to meet to be successful for 3DP-EXT printing, independently of their composition. Finally, this chapter discussed the challenges that still need to be addressed regarding material properties, product acceptance, technology limitations, regulation aspects, among others, as well as current trends in food 3D printing, demonstrating that there is great potential for research and development in this area.

Acknowledgments The authors express their gratitude for the financial support by the Consejo Nacional de Investigaciones Científicas y Técnicas (CONICET) and Universidad Nacional del Sur (UNS) of Argentina. The authors also express their gratitude to Dis. Teresa Dutari and Eng. Diego Colaneri for their technical assistance.

References

1. Diañez I, Martínez I, Franco JM et al (2022) Advances in 3D printing of food and nutritional products. Adv Food Nutr Res 100:173–210
2. Godoi FC, Prakash S, Bhandari BR (2016) 3D printing technologies applied for food design: status and prospects. J Food Eng 179:44–54. https://doi.org/10.1016/j.jfoodeng.2016.01.025
3. Rogers H, Srivastava M, Petruzzelli AM, Frattini F (2021) Emerging sustainable supply chain models for 3D food printing. Sustainability 13:12085. https://doi.org/10.3390/SU132112085
4. Sun J, Peng Z, Zhou W et al (2015) A review on 3D printing for customized food fabrication. Procedia Manuf 1:308–319. https://doi.org/10.1016/j.promfg.2015.09.057
5. Lipton JI (2017) Printable food: the technology and its application in human health. Curr Opin Biotechnol 44:198–201. https://doi.org/10.1016/j.copbio.2016.11.015
6. Shi Y, Zhang M, Bhandari B (2021) Effect of addition of beeswax based oleogel on 3D printing of potato starch-protein system. Food Struct 27:100176. https://doi.org/10.1016/j.foostr.2021.100176
7. Oliveira SM, Martins AJ, Fuciños P et al (2023) Food additive manufacturing with lipid-based inks: evaluation of phytosterol-lecithin oleogels. J Food Eng 341:111317. https://doi.org/10.1016/J.JFOODENG.2022.111317
8. De Salvo MI, Palla CA, Cotabarren IM (2023) Effect of printing parameters on the extrusion 3D printing of oleogel-based nutraceuticals. J Food Eng 349:111459. https://doi.org/10.1016/J.JFOODENG.2023.111459

9. Cotabarren IM, Cruces S, Palla CA (2019) Extrusion 3D printing of nutraceutical oral dosage forms formulated with monoglycerides oleogels and phytosterols mixtures. Food Res Int 126: 108676. https://doi.org/10.1016/j.foodres.2019.108676
10. Periard D, Schaal N, Schaal M et al (2007) Printing food. In: 2007 International solid freeform fabrication symposium. University of Texas at Austin, Austin, pp 564–574. https://doi.org/10.26153/TSW/7242
11. Hao L, Mellor S, Seaman O et al (2010) Material characterisation and process development for chocolate additive layer manufacturing. Virtual Phys Prototyping 5:57–64. https://doi.org/10.1080/17452751003753212
12. Liu Z, Zhang M (2019) 3D food printing technologies and factors affecting printing precision. Elsevier Inc., London
13. Kavimughil M, Leena MM, Moses JA, Anandharamakrishnan C (2022) 3D printed MCT oleogel as a co-delivery carrier for curcumin and resveratrol. Biomaterials 287:121616. https://doi.org/10.1016/j.biomaterials.2022.121616
14. Kang H, Oh Y, Lee NK, Rhee JK (2022) Printing optimization of 3D structure with lard-like texture using a beeswax-based oleogels. J Microbiol Biotechnol 32:1573–1582. https://doi.org/10.4014/JMB.2209.09052
15. Kavimughil M, Leena MM, Moses JA, Anandharamakrishnan C (2022) Effect of material composition and 3D printing temperature on hot-melt extrusion of ethyl cellulose based medium chain triglyceride oleogel. J Food Eng 329:111055. https://doi.org/10.1016/j.jfoodeng.2022.111055
16. Lupi FR, Shakeel A, Greco V et al (2016) A rheological and microstructural characterisation of bigels for cosmetic and pharmaceutical uses. Mater Sci Eng C 69:358–365. https://doi.org/10.1016/J.MSEC.2016.06.098
17. Guo J, Gu X, Du L, Meng Z (2023) Spirulina platensis protein nanoparticle-based bigels: dual stabilization, phase inversion, and 3D printing. Food Hydrocoll 135:108160. https://doi.org/10.1016/J.FOODHYD.2022.108160
18. Jiang Q, Wang Y, Du L et al (2022) Catastrophic phase inversion of bigels characterized by fluorescence intensity-based 3D modeling and the formability for decorating and 3D printing. Food Hydrocoll 126:107461. https://doi.org/10.1016/J.FOODHYD.2021.107461
19. Qiu R, Wang K, Tian H et al (2022) Analysis on the printability and rheological characteristics of bigel inks: potential in 3D food printing. Food Hydrocoll 129:107675. https://doi.org/10.1016/j.foodhyd.2022.107675
20. Chen Z, Bian F, Cao X et al (2023) Novel bigels constructed from oleogels and hydrogels with contrary thermal characteristics: phase inversion and 3D printing applications. Food Hydrocoll 134:108063. https://doi.org/10.1016/J.FOODHYD.2022.108063
21. Xie D, Hu H, Huang Q, Lu X (2023) Development and characterization of food-grade bigel system for 3D printing applications: role of oleogel/hydrogel ratios and emulsifiers. Food Hydrocoll 139:108565. https://doi.org/10.1016/J.FOODHYD.2023.108565
22. Tejada-Ortigoza V, Cuan-Urquizo E (2022) Towards the development of 3D-printed food: a rheological and mechanical approach. Foods 11:1191. https://doi.org/10.3390/FOODS11091191
23. Maldonado-Rosas R, Tejada-Ortigoza V, Cuan-Urquizo E et al (2022) Evaluation of rheology and printability of 3D printing nutritious food with complex formulations. Addit Manuf 58: 103030. https://doi.org/10.1016/J.ADDMA.2022.103030
24. Cheng Y, Fu Y, Ma L et al (2022) Rheology of edible food inks from 2D/3D/4D printing, and its role in future 5D/6D printing. Food Hydrocoll 132:107855. https://doi.org/10.1016/j.foodhyd.2022.107855
25. Shi Y, Zhang M, Phuhongsung P (2022) Microwave-induced spontaneous deformation of purple potato puree and oleogel in 4D printing. J Food Eng 313:110757. https://doi.org/10.1016/J.JFOODENG.2021.110757

26. Wang Q, Rao Z, Chen Y et al (2022) Characterization of responsive zein-based oleogels with tunable properties fabricated from emulsion-templated approach. Food Hydrocoll 133:107972. https://doi.org/10.1016/J.FOODHYD.2022.107972
27. González L, Lobato A, Cotabarren IM, Palla CA (2022) Bigels inks for 3D food printing: rheological and extrusion behavior. Poster presented at the VII Congreso Internacional de Ciencia y Tecnología de Alimentos, Universidad Nacional de Córdoba, Córdoba, 4–6 October 2022
28. Godoi FC, Bhandari BR, Prakash S, Zhang M (2019) An introduction to the principles of 3D food printing. In: Fundamentals of 3D food printing and applications. Academic Press, London, pp 1–18. https://doi.org/10.1016/B978-0-12-814564-7.00001-8
29. Shi Z, Xu W, Geng M et al (2023) Oil body-based one-step multiple phases and hybrid emulsion gels stabilized by sunflower wax and CMC: application and optimization in 3D printing. Food Hydrocoll 136:108262. https://doi.org/10.1016/J.FOODHYD.2022.108262
30. Zhai X, Sun Y, Cen S et al (2022) Anthocyanins-encapsulated 3D-printable bigels: a colorimetric and leaching-resistant volatile amines sensor for intelligent food packaging. Food Hydrocoll 133:107989. https://doi.org/10.1016/J.FOODHYD.2022.107989
31. Ghazal AF, Zhang M, Mujumdar AS, Ghamry M (2023) Progress in 4D/5D/6D printing of foods: applications and R&D opportunities. Crit Rev Food Sci Nutr 63:7399–7422. https://doi.org/10.1080/10408398.2022.2045896

Chapter 13
Emulsions Containing Oleogels

Matheus Augusto Silva Santos and Rosiane Lopes da Cunha

Abbreviations

EGCG	Epigallocatechin-3-gallate
FH	Fully hydrogenated
PGPR	Polyglycerol polyricinoleate
WPI	Whey protein isolate

13.1 Introduction

Emulsions form a universe of possibilities, widely investigated in the search for the best way to stabilize the water-oil interface. A new approach for emulsion stabilization that has been highlighted in recent years is the structuring of the lipid phase from oil gelation using some ingredients known as oleogelators [1–3]. These compounds form a crystalline or fibrillar network capable of entrapping the liquid oil, producing oleogels and giving the liquid oil characteristics similar to those of solid fat, such as texture and spreadability, only from physical interactions [4–6]. The necessary replacement of trans fat, driven by concerns about human health and well-being, increases interest in this emulsified colloidal system with properties similar to solid fats [7, 8]. Thus, from the point of view of food structure design, producing emulsions with oleogels is a great advantage, as this approach can contribute to the development of more dynamic structures, capable of being incorporated more easily into food products.

M. A. S. Santos · R. Lopes da Cunha (✉)
Department of Food Engineering and Technology, School of Food Engineering, Universidade Estadual de Campinas, UNICAMP, Campinas, Brazil
e-mail: rosiane@unicamp.br

C. Andrea Palla, F. Valoppi (eds.), *Advances in Oleogel Development, Characterization, and Nutritional Aspects*,
https://doi.org/10.1007/978-3-031-46831-5_13

Another important aspect from the point of view of health is the replacement of synthetic or semi-synthetic surfactants by natural emulsifiers, mainly of vegetable origin, also considering the alignment with sustainable processes and biofunctionality [9]. In this sense, emulsions containing oleogels have shown excellent results regarding the reduction or even non-use of these components, mainly in water-in-oleogel emulsions. Furthermore, the combination of ingredients, oleogelator and emulsifier, also demonstrated synergistic effects, providing kinetic stability to these emulsions. The processes for obtaining emulsions containing oleogels have some challenges, mainly related to temperature. For oleogelators to be efficient in these systems, they need to be above the melting temperature so that they can be prepared and homogenized in the organic phase [10]. For this reason, it is necessary to maintain the emulsion production at a high temperature, generating greater technical challenges. However, some technologies have been used to reduce the size of the emulsion droplets, such as high-pressure homogenization and ultrasound [11, 12].

Combining oleogels and emulsions also opens up a new field in the delivery of bioactive compounds, as the high surface area of the droplets is associated with greater bioaccessibility [13]. Several studies have used this type of emulsion to deliver lipophilic and hydrophilic compounds, including in the same emulsified system [12, 13]. Furthermore, gelation of the oil phase adds an extra barrier to protect unstable active compounds, such as phenolic compounds [14]. This chapter presents an overview of the characteristics of emulsions containing oleogel, including the techniques applied for their production and the different structuring agents (oleogelators) used. In addition, it also addresses the potential of these emulsified systems in the delivery of bioactive compounds, modulation of lipolysis, and reduction in lipid oxidation.

13.2 Emulsions Containing Oleogel

13.2.1 Oleogel-in-Water Emulsion

Conventional oil-in-water emulsions consist of colloidal systems composed of two phases, one nonpolar oily and the other polar or aqueous. In these emulsions, the oil phase appears as spherical droplets dispersed in the aqueous phase and is present in products such as milk, mayonnaise, sauces, and other foods [15]. Different mechanisms have been used to stabilize these emulsified systems, such as changes in the composition of droplet interface or the rheology of the continuous phase. The stabilization of the interface can be divided into different strategies. The first is related to the adsorption of emulsifiers on the surface of the droplets that promote the reduction in interfacial tension and the formation of a layer capable of avoiding the phenomena of flocculation and coalescence, by steric or/and electrostatic hindrance [15]. The second is the Pickering mechanism, in which particles can be deposited at the interface, according to their wettability, promoting steric and/or mechanical hindrance between the droplets [16]. The other strategy is related to increasing viscosity of the continuous phase, causing a reduction in the mobility of the droplets and, consequently, increasing stability.

In this sense, to form stable oil-in-water emulsions, focus should be placed on the proper choice of emulsifier (proteins, polysaccharides, lipid esters, synthetic surfactants) and the physical-chemical environment (pH, salt concentration, temperature), in addition to the use of processes capable of reducing the polydispersity of the droplets. Recently, a new approach for the development of oil-in-water emulsions has been investigated; the gelation of the oil phase using compounds (oleogelators) capable of solidifying the dispersed phase only by physical changes. Oleogelators (gelling agents) are compounds that have temperature-dependent characteristics. They are liquids above the melting temperature and below it capable of forming crystals or three-dimensional (3D) networks that trap the liquid oil, ensuring rheological properties very close to solid fats [17].

Table 13.1 shows some recent studies on oleogel-in-water emulsions. A point that should be highlighted in relation to these emulsified systems is the need to use emulsifiers to promote stabilization. Up to now, only the gelation of the dispersed phase is not enough for these emulsions to be stable, being necessary to perform a combination between emulsifier and oleogelator. This subject is discussed in more detail in Sect. 13.3.

13.2.2 Water-in-Oleogel Emulsion

Water-in-oil emulsions are also colloidal systems formed by two immiscible phases; however, these emulsions are composed of water droplets dispersed in a lipid phase, such as margarine and butter [15]. These emulsions are quite challenging in terms of stability, due to the high mobility of water droplets and the restrictive range of surfactants capable of preventing flocculation, coalescence, and, consequently, the sedimentation of these droplets. The low conductivity of the continuous phase restricts the use of emulsifiers that cause electrostatic repulsion, making the main stabilization mechanism of these emulsions the steric hindrance [27]. This fact contributes to the use of some synthetic surfactants, mainly PGPR. A major challenge is to find replacement of these synthetic compounds with natural ingredients, such as lecithin [28], insoluble particles (Pickering stabilization mechanism) [29], or combinations capable of reducing the amount of synthetic compounds in food emulsions.

In this sense, continuous phase solidification has been widely used and has proven to be quite effective in stabilizing water-in-oil emulsions for a long period of time. Therefore, emulsions known as water-in-oleogel are obtained from the gelation of the continuous phase by means of oleogelators. Differently from oleogel-in-water emulsion, some studies have shown that it is possible to stabilize water-in-oleogel emulsions without emulsifiers due to the increased viscosity of the continuous (oily) phase, which reduces the mobility of the droplets. Despite this, the

Table 13.1 Ingredients used in the production of oleogel-in-water emulsions

Oil phase	Oleogelator	Emulsifier	Bioactive compound	Ref.
Fish oil (enriched in Ω-3)[a]	FH rapeseed oil	Tween 20	Curcumin	[18]
Rice brain oil	Carnauba wax	Ovalbumin and Arabic gum complexes	Astaxanthin	[19]
Essential oils (orange, lemon, and peppermint)	Phytosterols	Tea saponin	No	[11]
Camellia oil	Glycerol monolaurate	Soy protein	No	[10]
Sunflower oil	Fruit wax	Soy lecithin, WPI, Tween 80	No	[20]
Olive oil	Glycerol monostearate	Pluronic F-68 and Tween 80	Curcumin	[21]
Corn oil	Carnauba wax	Lactoferrin particles	Curcumin	[13]
Coconut oil	Beeswax, lecithin	WPI	Essential thyme oil	[22]
Sunflower oil	Candelilla wax	Ovotransferrin, carboxymethyl chitosan nanoparticles	Curcumin	[23]
Corn oil	Glycerol monostearate	Span 20 and Tween 20	β-carotene	[12]
Medium chain triglycerides oil	Stearic acid	Sodium starch octenyl succinate	No	[1]
Perilla oil	Beeswax and stearic acid	Hydroxypropyl methyl cellulose and sodium dodecyl sulphate	No	[24]
Canola oil	Stearic acid	Soy lecithin	No	[25]
Corn oil	Ethyl cellulose	Sodium caseinate	No	[7]
Corn oil	Beeswax	WPI	β-carotene	[26]
Soybean oil	Beeswax	Cellulose nanocrystals	No	[5]

Abbreviations: *FH* fully hydrogenated, *WPI* whey protein isolate
[a]Water phase with xanthan gum

use of emulsifiers is fundamental in emulsions with a high volumetric fraction of the dispersed phase [30, 31]. Some recent studies are shown in Table 13.2.

13.3 Principal Characteristics of Oleogelators in Emulsions Containing Oleogels

Tables 13.1 and 13.2 show that different oleogelators and emulsifiers are used to produce and stabilize emulsions containing oleogels. Among the oleogelators, two classes can be established, those capable of organizing themselves into crystals and those that form a fibrillar network in the oil fraction [9]. At the same time, emulsifiers

Table 13.2 Ingredients used in the production of water-in-oleogel emulsions

Oil phase	Oleogelator	Emulsifier	Bioactive compound	Ref.
Corn oil	γ-oryzanol and β-sitosterol	PGPR	β-carotene and EGCG	[32]
Rapeseed oil	Glycerol monostearate, soy lecithin, and FH rapeseed oil	No	No	[33]
Different oils[a]	Sunflower wax	Tween 20 and Span 20	No	[8]
Safflower oil (high oleic)	Ethyl cellulose	Glycerol monostearate	No	[34]
Rapeseed oil	Candelilla wax	No	No	[2]
Soybean oil	Beeswax	No	No	[3]
Corn oil	Ethyl cellulose	Ethyl cellulose particles	Curcumin and anthocyanin	[14]
Safflower oil	Candelilla wax, FH soybean oil	Glycerol monooleate, glycerol monopalmitate, PGPR	No	[35]
Rapeseed oil	Carnauba wax and beeswax	No	Iron	[36]

Abbreviations: *PGPR* polyglycerol polyricinoleate, *EGCG* epigallocatechin-3-gallate, *FH* fully hydrogenated

[a]Groundnut, sunflower, soybean, sesame, mustard, rice bran, coconut, and palm oil

range from natural polymers such as proteins and polysaccharides to synthetic ones such as PGPR in water-in-oleogel emulsions.

13.3.1 Crystal-Based Gelation

Ingredients that crystallize in the oil phase have been widely studied in the production of emulsions containing oleogels. These compounds trap the oil in a 3D network through interactions such as hydrogen bonds and *van der Waals* forces [17]. The main compounds in this class are waxes, monoglycerides, fatty acids, fatty acid esters, and fully hydrogenated fat. Most waxes have limited surface activity [20] as demonstrated for fruit wax (7% w/w), and therefore it is impossible to obtain an oleogel-in-water emulsion without the need for emulsifiers, such as proteins or low molecular weight surfactants, especially if there is no modification of the rheological behavior (increase in viscosity) of the continuous aqueous phase. On the other hand, a significant reduction in interfacial tension with candelilla wax (3% w/w) was observed when increasing the temperature [35]. The authors attribute this result to the adsorption of polar compounds, such as fatty acids, fatty alcohols, and triterpenes at the interface. These divergent results are associated with the enormous variability in wax composition. Interestingly, another oleogelators show surface activity, which can contribute to stabilizing the water-oil interface, such as mono- and diglycerides [10, 12, 34].

Another crucial factor in the stabilization of these emulsions is the crystallization profile of the compounds. For instance, beeswax and stearic acid crystals, observed by polarized light microscopy, were shorter in oleogel-in-water emulsions than in oleogel, evidencing the change in behavior with the presence of emulsifiers and water [24]. Furthermore, stearic acid crystals were observed both in the continuous and dispersed phases, possibly due to crystal growth and rupture of the interfacial layer, causing flocculation between the droplets. On the other hand, beeswax formed small crystals, present only in the dispersed phase, without exerting influence on flocculation of the droplets. The chemical composition of the oleogelators can explain these differences. Wax is formed by a mixture of compounds such as fatty acids and alcohols, hydrocarbons and wax ester, which promotes the formation of smaller crystals, while stearic acid can undergo secondary nucleation and increase the crystal size.

Wax concentration can also affect the stability of emulsions. For instance, concentrations greater than 3% w/w of beeswax caused an increase in droplet size, influencing its shape, becoming less spherical, in addition to favor a greater penetration of crystals and interface rupture [26]. The higher beeswax content also contributed to the coalescence of the droplets; upon heating the emulsion, liquid oil was released due to interface disruption. The amphiphilic nature of monoacylglycerols also favors the penetration of crystals from the oil phase into the aqueous phase, promoting network formation with neighboring droplets and allowing coalescence [12]. However, the compaction of the monoacylglycerol self-assembled structures contributed to the formation of irregular droplets, and after 72 hours, the emulsions became more rigid, influenced by crystal growth and greater packing of the oil phase. Despite this, the use of a surfactant (Tween 20) contributed to a better coating of the interface, reducing flocculation and coalescence.

13.3.2 Fibril-Based Gelation

The use of polymers or low molecular weight molecules, such as ethyl cellulose and phytosterols, respectively, has also been researched in the development of emulsions containing oleogels. For example, ethyl cellulose, above the melting temperature, can disperse in liquid oils and, after cooling, form a fibrillar network from the intra and intermolecular interaction between the polymer (hydrogen bonds), trapping the oil and forming a physical gel [37]. Similarly, phytosterols (γ-oryzanol and β-sitosterol) form tubular structures due to the interaction of hydrogen bonds between the two sterols with the entrapment of the liquid oil [38].

Ethyl cellulose was used in the production of water-in-oleogel without emulsifiers. The continuous network formed in the oil phase did not allow the formation of spherical droplets, and a defined interfacial layer was not observed [14]. The authors observed that the ethyl cellulose oleogel was honeycombs-like, with pores that trapped the liquid oil, while in the emulsions they observed a solid structure with incorporated water droplets of different sizes. The weak interaction between the polymer was not able to prevent oil release and emulsion destabilization with the application of shear forces. On the other hand, the use of glycerol monostearate combined with ethyl cellulose oleogel contributed to the stability of the water-in-oleogel emulsion [34]. The authors observed phase separation in the emulsions produced only with the monoglyceride, while the ethyl cellulose oleogel was fluid, but the combination of these ingredients in the emulsion produced a very rigid structure.

The stabilization of water-in-oleogel emulsions using a mixture of phytosterols was investigated [32], but water was incorporated with difficulty due to its tendency to bind the hydroxyl group of β-sitosterol. This prevented the formation of hydrogen bonds with the γ-oryzanol molecule, hindering oil gelation. However, the use of PGPR and a high concentration of phytosterols (15% and 20% w/w) resulted in self-standing emulsions. In addition, the high concentration allows the formation of a more compact network, enhancing the mechanical properties and resulting in a more stable emulsion. This mixture of phytosterols was also used to gel the dispersed phase of oleogel-in-water emulsions, showing the importance of an emulsifier in reducing interfacial tension and building a film between phases capable of preventing destabilization phenomena [11].

13.4 Methods for Producing Emulsions Containing Oleogels

Different methods are used to produce emulsions, categorized into high or low-energy processes. The choice depends on the ingredients and the application, such as delivery of bioactive compounds, or rheological properties modification [39]. Emulsions containing oleogel, whether oleogel-in-water or water-in-oleogel, are produced mainly by high-energy methods, that is, high shear mixers, high-pressure homogenizers, and ultrasonic processes. These methods use disruptive forces capable of promoting the formation of droplets and the breaking of these droplets into smaller units, and the final size depends on the energy introduced into the system. However, oleogel emulsions have a particularity, the need to keep the oil phase warm during emulsification since the oleogelators are solid at room temperature, and for droplet formation to be possible, the lipid phase needs to be fluid.

In this way, a standard method can be outlined and used as a common step in the production of these emulsions, as seen in Fig. 13.1. The first step involves mixing the oil and aqueous phases, in which the oleogelator is always homogenized in the lipid phase. Depending on the type of emulsion, that is, water-in-oleogel or oleogel-in-water, the emulsifier is solubilized in the oil or aqueous phase, respectively. In the second stage, the emulsification process is carried out using a rotor stator, which

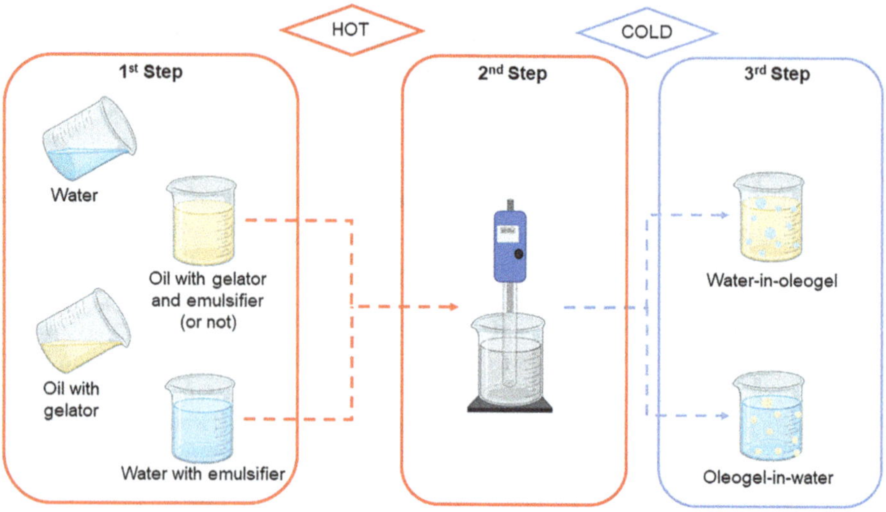

Fig. 13.1 Basic outlined production of emulsions containing oleogel

provides the necessary energy to form the water or oil droplets. These first stages are carried out at a high temperature, according to the melting temperature of each oleogelator. Finally, in the third stage, the emulsion is cooled, causing the formation of crystals or three-dimensional networks in the oil phase (oleogel), immobilizing the liquid oil.

To reduce droplet size, decrease polydispersity and, consequently, increase the stability of these emulsions, the following high-energy processes or combinations between them are used.

13.4.1 High-Pressure Homogenization

The high-pressure homogenization process consists of passing the emulsion through equipment containing a narrow passage between the valves responsible for causing the rupture of larger droplets in an emulsion to droplets with a very small diameter, also reducing polydispersity [15]. However, to use this technology to produce emulsions with oleogel, the process must occur quickly and at temperatures higher than the melting temperatures of the oleogelators. Several studies have used high-pressure homogenization as the final step to produce these emulsified systems, using different oleogelators, such as beeswax and stearic acid [24], candelilla wax and fully hydrogenated soybean oil [35], monoglycerides [12], ethyl cellulose and glyceryl monostearate [34] and fruit wax [20].

As an example of the application of this technology, an oleogel-in-water emulsion was produced using beeswax and a mixture of emulsifiers (hydroxypropyl methylcellulose and sodium dodecyl sulfate) by a combined approach [24]. In the first step, a coarse emulsion (rotor-stator) was produced before forming a fine emulsion in a high-pressure homogenizer. The final emulsion had a droplet size between 200 and 300 nm. In another study, droplets between 150 and 200 nm were obtained in emulsions containing monoglyceride (as oleogelators) and surfactants (as emulsifier) using the same technique [12]. For comparative purposes, an emulsion produced only with a rotor-stator, with the oil phase gelled with candelilla wax, had a droplet size between 50 and 60μm [23]. Therefore, a positive effect of the combination of high-energy processes can be observed in the production of these emulsions, providing higher kinetic stability due to reduction in the droplet size.

13.4.2 Ultrasound

The ultrasound process is based on the breakup of droplets by cavitation and turbulent flow effects. The equipment generates high-intensity ultrasonic waves, causing the formation of high pressure and shear gradients, capable of reducing the size of droplets to the nanometric scale [40]. The production of emulsions containing oleogels has been extensively investigated using this technique, which can be used alone or in combination with the rotor-stator or even after magnetic stirring of the two phases.

In addition to the possibility of obtaining emulsions with a more uniform droplet size, ultrasound has been presented as an interesting technology in the crystallization of oleogelators in the lipid phase. Ultrasound improved formation of monoglyceride crystals in the oil phase of water-in-oleogel emulsions, in addition to causing greater entrapment of water droplets in the continuous phase network formed after cooling the emulsion [33]. Another advantage of using this technique is related to the heating of the system caused by cavitation, which is very useful as it causes the melting of the oleogel and, consequently, facilitates the production of stable oleogel-in-water emulsions [13]. Any energy contribution can lead to dissipation in the form of heat, decreasing the viscosity of the oleogel and greatly facilitating droplet size reduction [18].

Another parameter that should also be considered is the ultrasound power used, which can contribute to the decrease in the size of the dispersed droplets. However, in an emulsified colloidal system produced with olive oil, monoglycerides and surfactants (Tween 80 and Pluronic F-68) [21], non-significant effect on droplet size reduction was observed at high power levels (>280 W). This result was associated with loss of surfactant functionality, or a higher rate of droplets collisions caused by a highly turbulent flow. Furthermore, ultrasound time can cause the same phenomenon, although longer times have less influence on size reduction.

13.5 Potential Application of Emulsions Containing Oleogels

Oleogels are widely reported to replace saturated and trans fats; however, their application in aqueous food matrices can be challenging [32]. Therefore, the development of emulsified systems containing oleogels has been widely investigated because they can be more easily integrated into food products. Furthermore, these emulsions can also act as delivery systems for bioactive compounds [18] and minerals [36], increase stability during freezing and thawing [19] and retard lipid oxidation [24, 25]. A summary of emulsions containing oleogels is presented in Fig. 13.2.

Many bioactive compounds have been carried both by oleogel-in-water and water-in-oleogel emulsions. Some examples are curcumin [13], epigallocatechin gallate [32], β-carotene [12], astaxanthin [19], anthocyanins [14] among others. For instance, the delivery of curcumin in oleogel-in-water emulsion has been shown to be more efficient than in oleogel alone [23], due to lower exposure of the bioactive compound in the emulsion caused by the emulsifier barrier at the interface in synergy with gelation of the oily matrix. Furthermore, these emulsions also have great potential to provide a suitable condition for the delivery of multiple active compounds (lipophilic and hydrophilic), such as curcumin (gelled oil phase) and anthocyanins (aqueous phase) in water-in-oleogel emulsions [14].

These emulsified systems also play an important role in the oxidative stability of liquid oils rich in polyunsaturated fatty acids. Lipid oxidation is a complex process, dependent on several factors, such as the type of oil and its concentration in the emulsion, the emulsifier and oleogelator used, the presence of pro-oxidant agents,

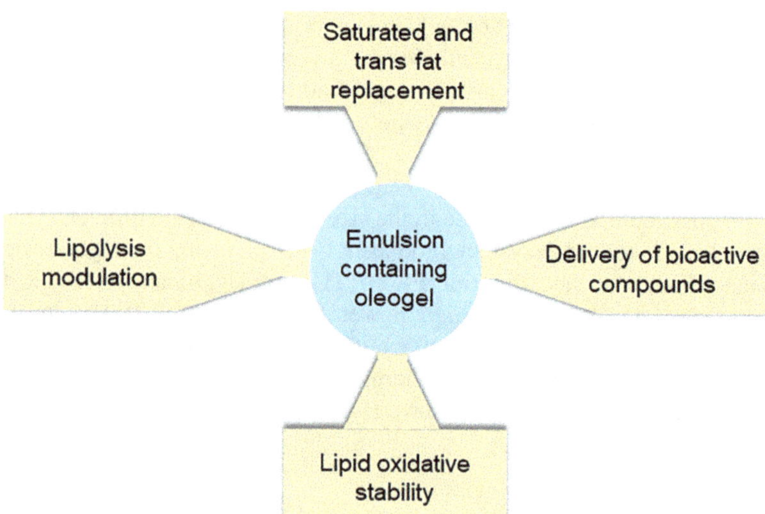

Fig. 13.2 Summary of applications of emulsions containing oleogel

the surface area of the droplets, among others [18]. However, oil gelation has been shown to be quite positive in reducing lipid oxidation, decreasing mass transfer and, consequently, the production of free radicals and oxidation compounds, such as hydroperoxides [36].

Among the oleogelators used, different behaviors regarding the oxidative stability of the oil can also be observed. For instance, the use of beeswax was more efficient in reducing lipid oxidation compared to the use of stearic acid in the formulation of emulsion containing oleogel [24] under the same temperature conditions. The authors attributed the results to the distribution of crystals in the oily matrix. That is, the three-dimensional structure of the oleogelator can act as a physical barrier, for example, against O_2 dissolved in the continuous phase or cause the disruption of the interfacial layer, impairing the stability of the emulsion.

A new approach that has been investigated is the digestibility of these colloidal systems, assessing lipolysis and bioaccessibility of bioactive compounds. Both lipolysis and bioaccessibility of curcumin were superior in emulsions with carnauba wax oleogel than oleogel, explained by the greater exposure of lipids to the action of lipase due to the greater surface area of the gelled droplets [13]. In the same sense, the lipolysis of the emulsion containing the oily fraction gelled with candelilla wax was superior to the oleogel [23]. The dispersion of oleogel in gastrointestinal fluids is limited but can be facilitated when present in an emulsion.

13.6 Conclusions

Most studies are still focused on exploring the ingredients or combinations capable of stabilizing these emulsions and little on the application of these systems in foods. The advance in the use of natural ingredients is noticeable, exploring the synergism of compounds in multicomponent systems as a strategy for reducing synthetic or semi-synthetic emulsifiers. Furthermore, the improvement of production techniques for these emulsions has extrapolated mechanical agitation only to incorporate the dispersed phase in more efficient processes, such as high-pressure homogenization and ultrasound. These high-energy processes have played a crucial role in reducing droplet size and changing the three-dimensional network in the oil fraction, contributing to stabilizing emulsions for longer.

Considering all the advantages, these systems have great application possibilities. However, the large amount of oleogelator used in these emulsions, sometimes greater than necessary to carry out the structuring of the non-emulsified oil, is still a challenge. In this sense, multicomponent systems can be a way to achieve self-standing structures with potential application in food. These gelled emulsions can also act as excellent carriers of bioactive compounds, mainly lipophilic, which are trapped in the oily network, ensuring greater protection and delivery, as can be seen by the greater bioaccessibility when compared to oleogels. Furthermore, studies have shown that lipolysis during the digestion process can also be modulated in oleogel-in-water emulsions, since the greater surface area of the droplets can improve lipase access to the lipid fraction. Another positive aspect of these emulsions is related to greater lipid oxidative stability, even when using polyunsaturated oils. Therefore, the development of emulsions containing oleogels proves to be a very competitive and attractive field in the sense of delivering products with reduced lipid content to the consumer, with a more functional appeal, contributing to the improvement of health and well-being.

Acknowledgments The authors are grateful to the financial agencies: Coordenação de Aperfeiçoamento de Pessoal de Nível Superior – Brazil (CAPES) – Finance Code 001. National Council for Scientific and Technological Development – CNPq (grant #307094/2021-9), and #2019/27354-3, FAPESP.

References

1. Wu Y, Chen F, Zhang C et al (2021) Improve the physical and oxidative stability of O/W emulsions by moderate solidification of the oil phase by stearic acid. LWT 151:112120. https://doi.org/10.1016/j.lwt.2021.112120
2. Hong X, Zhao Q, Chen J et al (2022) Fabrication and characterization of oleogels and temperature-responsive water-in-oil emulsions based on candelilla (Euphorbia cerifera) wax. Food Chem 397:133677. https://doi.org/10.1016/j.foodchem.2022.133677
3. Gao Y, Lei Y, Wu Y et al (2021) Beeswax: a potential self-emulsifying agent for the construction of thermal-sensitive food W/O emulsion. Food Chem 349:129203. https://doi.org/10.1016/j.foodchem.2021.129203
4. Urbánková L, Sedláček T, Kašpárková V, Bordes R (2021) Formation of oleogels based on emulsions stabilized with cellulose nanocrystals and sodium caseinate. J Colloid Interface Sci 596:245–256. https://doi.org/10.1016/j.jcis.2021.02.104
5. Qi W, Li T, Zhang Z, Wu T (2021) Preparation and characterization of oleogel-in-water Pickering emulsions stabilized by cellulose nanocrystals. Food Hydrocoll 110:106206. https://doi.org/10.1016/j.foodhyd.2020.106206
6. Abdullah LL, Javed HU, Xiao J (2022) Engineering emulsion gels as functional colloids emphasizing food applications: a review. Front Nutr 9:890188. https://doi.org/10.3389/fnut.2022.890188
7. Zhang Y, Wang Y, Zhang R et al (2022) Tuning the rheological and tribological properties to simulate oral processing of novel high internal phase oleogel-in-water emulsions. Food Hydrocoll 131:107757. https://doi.org/10.1016/j.foodhyd.2022.107757
8. Holey SA, Sekhar KPC, Mishra SS et al (2021) Sunflower wax-based oleogel emulsions: physicochemical characterizations and food application. ACS Food Sci Technol 1:152–164. https://doi.org/10.1021/acsfoodscitech.0c00050
9. Wijarnprecha K, de Vries A, Sonwai S, Rousseau D (2021) Water-in-oleogel emulsions—from structure design to functionality. Front Sustain Food Syst 4:566445. https://doi.org/10.3389/fsufs.2020.566445
10. Pan J, Tang L, Dong Q et al (2021) Effect of oleogelation on physical properties and oxidative stability of camellia oil-based oleogels and oleogel emulsions. Food Res Int 140:110057. https://doi.org/10.1016/j.foodres.2020.110057
11. Chen XW, Yin WJ, Yang DX et al (2021) One-pot ultrasonic cavitational emulsification of phytosterols oleogel-based flavor emulsions and oil powder stabilized by natural saponin. Food Res Int 150:110757. https://doi.org/10.1016/j.foodres.2021.110757

12. Zhang R, Cui M, Ye J et al (2022) Physicochemical stability of oleogel-in-water emulsions loaded with β-carotene against environmental stresses. LWT 155:112965. https://doi.org/10.1016/j.lwt.2021.112965
13. Xia T, Gao Y, Liu Y et al (2022) Lactoferrin particles assembled via transglutaminase-induced crosslinking: utilization in oleogel-based Pickering emulsions with improved curcumin bioaccessibility. Food Chem 374:131779. https://doi.org/10.1016/j.foodchem.2021.131779
14. Zhang R, JingjingYu LN et al (2022) W/O emulsions featuring ethylcellulose structuring in the water phase, interface and oil phase for multiple delivery. Carbohydr Polym 283:119158. https://doi.org/10.1016/j.carbpol.2022.119158
15. McClements DJ (2004) Food emulsions: principles, practices, and techniques. CRC Press, Boca Raton. https://doi.org/10.1201/9781420039436
16. Guida C, Aguiar AC, Cunha RL (2021) Green techniques for starch modification to stabilize Pickering emulsions: a current review and future perspectives. Curr Opin Food Sci 38:52–61. https://doi.org/10.1016/j.cofs.2020.10.017
17. Doan CD, Tavernier I, Okuro PK, Dewettinck K (2018) Internal and external factors affecting the crystallization, gelation and applicability of wax-based oleogels in food industry. Innov Food Sci Emerg Technol 45:42–52. https://doi.org/10.1016/j.ifset.2017.09.023
18. Vellido-Perez JA, Ochando-Pulido JM, Brito-de la Fuente E, Martinez-Ferez A (2021) Effect of operating parameters on the physical and chemical stability of an oil gelled-in-water emulsified curcumin delivery system. J Sci Food Agric 101:6395–6406. https://doi.org/10.1002/jsfa.11310
19. Gao Y, Wang Z, Xue C, Wei Z (2022) Modulation of fabrication and nutraceutical delivery performance of ovalbumin-stabilized oleogel-based nanoemulsions via complexation with gum arabic. Foods 11:1859. https://doi.org/10.3390/foods11131859
20. Okuro PK, Gomes A, Cunha RL (2020) Hybrid oil-in-water emulsions applying wax(lecithin)-based structured oils: tailoring interface properties. Food Res Int 138:109798. https://doi.org/10.1016/j.foodres.2020.109798
21. Palla CA, Aguilera-Garrido A, Carrín ME et al (2022) Preparation of highly stable oleogel-based nanoemulsions for encapsulation and controlled release of curcumin. Food Chem 378:132132. https://doi.org/10.1016/j.foodchem.2022.132132
22. Gong W, Guo X-l, Huang H-b et al (2021) Structural characterization of modified whey protein isolates using cold plasma treatment and its applications in emulsion oleogels. Food Chem 356:129703. https://doi.org/10.1016/j.foodchem.2021.129703
23. Dong Y, Wei Z, Wang Y et al (2022) Oleogel-based Pickering emulsions stabilized by ovotransferrin–carboxymethyl chitosan nanoparticles for delivery of curcumin. LWT 157:113121. https://doi.org/10.1016/j.lwt.2022.113121
24. Wei F, Miao J, Tan H et al (2021) Oleogel-structured emulsion for enhanced oxidative stability of perilla oil: influence of crystal morphology and cooling temperature. LWT 139:110560. https://doi.org/10.1016/j.lwt.2020.110560
25. Zhuang X, Gaudino N, Clark S, Acevedo NC (2021) Novel lecithin-based oleogels and oleogel emulsions delay lipid oxidation and extend probiotic bacteria survival. LWT 136:110353. https://doi.org/10.1016/j.lwt.2020.110353
26. Zhang Y, Lu Y, Zhang R et al (2021) Novel high internal phase emulsions with gelled oil phase: preparation, characterization and stability evaluation. Food Hydrocoll 121:106995. https://doi.org/10.1016/j.foodhyd.2021.106995
27. Ushikubo FY, Cunha RL (2014) Stability mechanisms of liquid water-in-oil emulsions. Food Hydrocoll 34:145–153. https://doi.org/10.1016/j.foodhyd.2012.11.016
28. Balcaen M, Steyls J, Schoeppe A et al (2021) Phosphatidylcholine-depleted lecithin: a clean-label low-HLB emulsifier to replace PGPR in w/o and w/o/w emulsions. J Colloid Interface Sci 581:836–846. https://doi.org/10.1016/j.jcis.2020.07.149
29. Lefroy KS, Murray BS, Ries ME (2022) Relationship between size and cellulose content of cellulose microgels (CMGs) and their water-in-oil emulsifying capacity. Colloids Surf A Physicochem Eng Asp 647:128926. https://doi.org/10.1016/j.colsurfa.2022.128926

30. Wu X, Xu N, Cheng C et al (2022) Encapsulation of hydrophobic capsaicin within the aqueous phase of water-in-oil high internal phase emulsions: controlled release, reduced irritation, and enhanced bioaccessibility. Food Hydrocoll 123:107184. https://doi.org/10.1016/j.foodhyd.2021.107184
31. Liu Y, Lee WJ, Tan CP et al (2022) W/O high internal phase emulsion featuring by interfacial crystallization of diacylglycerol and different internal compositions. Food Chem 372:131305. https://doi.org/10.1016/j.foodchem.2021.131305
32. Pinto TC, Martins AJ, Pastrana L et al (2022) Water-in-oleogel emulsion based on γ-oryzanol and phytosterol mixtures: challenges and its potential use for the delivery of bioactives. J Am Oil Chem Soc 99:1045–1053. https://doi.org/10.1002/aocs.12636
33. da Silva TLT, Danthine S (2022) High-intensity ultrasound as a tool to form water in oleogels emulsions structured by lipids oleogelators. Food Biophys 17:361–374. https://doi.org/10.1007/s11483-022-09728-9
34. García-Ortega ML, Toro-Vazquez JF, Ghosh S (2021) Development and characterization of structured water-in-oil emulsions with ethyl cellulose oleogels. Food Res Int 150:110763. https://doi.org/10.1016/j.foodres.2021.110763
35. García-González DO, Yánez-Soto B, Dibildox-Alvarado E et al (2021) The effect of interfacial interactions on the rheology of water in oil emulsions oleogelled by candelilla wax and saturated triacylglycerols. LWT 146:111405. https://doi.org/10.1016/j.lwt.2021.111405
36. Naktinienė M, Eisinaitė V, Keršienė M et al (2021) Emulsification and gelation as a tool for iron encapsulation in food-grade systems. LWT 149:111895. https://doi.org/10.1016/j.lwt.2021.111895
37. Gravelle AJ, Davidovich-Pinhas M, Zetzl AK et al (2016) Influence of solvent quality on the mechanical strength of ethylcellulose oleogels. Carbohydr Polym 135:169–179. https://doi.org/10.1016/j.carbpol.2015.08.050
38. Martins AJ, Cerqueira MA, Pastrana LM et al (2019) Sterol-based oleogels' characterization envisioning food applications. J Sci Food Agric 99:3318–3325. https://doi.org/10.1002/jsfa.9546
39. McClements DJ, Jafari SM (2018) Improving emulsion formation, stability and performance using mixed emulsifiers: a review. Adv Colloid Interface Sci 251:55–79. https://doi.org/10.1016/j.cis.2017.12.001
40. Leong T, Kentish S (2011) The fundamentals of power ultrasound: a review. Acoust Aust 39:54–63

Chapter 14
Bigels: An Innovative Hybrid of Hydrogels/Oleogels for Food Applications

Somali Dhal, Miguel A. Cerqueira, Doman Kim, and Kunal Pal

Abbreviations

AV	p-Anisidine value
DMA	Dynamic mechanical analysis
dp-CNC	Date palm-derived cellulose nanocrystals
DSC	Differential scanning calorimetry
DSD	Droplet size distribution
EC	Ethylcellulose
EGCG	Epigallocatechin gallate
FAME	Fatty acid methyl ester
FTIR	Fourier transform infrared
GCP	Graft copolymerization
GG	Guar gum
GMS	Glycerol monostearate
H/O	Hydrogel dispersed in oleogel
HLB	Hydrophilic-lipophilic balance
HMWOs	High-molecular-weight oleogelators

S. Dhal · K. Pal (✉)
Department of Biotechnology and Medical Engineering, National Institute of Technology Rourkela, Rourkela, India
e-mail: palk@nitrkl.ac.in

M. A. Cerqueira
International Iberian Nanotechnology Laboratory, Braga, Portugal
e-mail: miguel.cerqueira@inl.int

D. Kim
Department of International Agricultural Technology & Institute of Green BioScience and Technology, Seoul National University, Seoul, Gwangwon-do, South Korea
e-mail: kimdm@snu.ac.kr

C. Andrea Palla, F. Valoppi (eds.), *Advances in Oleogel Development, Characterization, and Nutritional Aspects*,
https://doi.org/10.1007/978-3-031-46831-5_14

LMWOs	Low-molecular-weight oleogelators
O/H	Oleogel dispersed in hydrogel
PV	Peroxide value
PVA	Polyvinyl alcohol
PVP	Polyvinyl pyrrolidone
SEs	Sucrose esters
SR	Stress relaxation
TGA	Thermogravimetric analysis
XRD	X-ray diffraction

14.1 Introduction

In comparison to earlier gel formulations, bigels are a relatively modern formulation. Bigel is a biphasic semisolid system derived from two different gel phases, that is, oleogel as the hydrophobic phase and hydrogel as the hydrophilic phase [1]. These gels are the result of combining oleogel and hydrogel in a certain ratio and homogenizing the mixture at a high shear rate. The composition of bigels contributes to their physicochemical and functional properties. Bigels combine the best features of both hydrogels and oleogels, allowing them to circumvent their core issues. As a result, these biphasic systems outperform the two gels individually. One of the biggest benefits of bigels is that they may be used to transport both lipophilic and hydrophilic bioactive compounds. When compared to other biphasic systems (emulsions and emulgels), which rely on surfactants for stability, these gels have superior thermodynamic and kinetic stability due to the semisolid structure of both phases [2]. Bigels have greater physicochemical stability because the mobile phases are trapped in the three-dimensional network of the oleogel and the hydrogel, resulting in a more finely dispersed structure. As a result, they do not suffer phase separation even when maintained at room temperature for up to 6–12 months. Such gels have moisturizing and cooling properties, enhanced spreadability, improved drug permeability, and the ability to modify the drug release rate [3]. Bigel systems are particularly desirable to the pharmaceutical, cosmetic, and food industries because of these traits. Also, there is a possibility to change the physical properties of bigels by adjusting the quality of any single phase [4]. Depending on the requirement and the commercial applications, various materials can be used to synthesize bigels. The ultimate characteristics of bigels are determined by the nature of the hydrophobic and hydrophilic phases [5]. As a result, the distinct phases are crucial in selecting suitable materials for these phases. The aqueous phase (hydrogel) is formulated mainly using synthetic or natural hydrophilic polymers (e.g., agar, sodium alginate, gelatin, carrageenan, and pectin) that forms a colloidal network of polymer chains in the water. The hydrophobic phase (oleogel) for bigel matrices has been developed using fatty alcohols and fatty acids, plant-derived wax, glyceryl stearate, and policosanol as oleogelators [6, 7] and various vegetable oils as the oil phase

[8]. Based on the component used for individual phases, the characteristics of bigels vary, and these changes can be characterized in terms of physical, mechanical, electrical, structural, and rheological properties [9]. This chapter will deal with different types of bigels, materials used, factors affecting bigels, characterization techniques, and food/nutrition-based applications of bigels.

14.1.1 Types of Bigels

The formulation and physicochemical properties of various bigels have been the subject of an in-depth study conducted by scientists over the last decade. Bigels can be differentiated into four categories according to the spatial organization and the nature of oleogels and hydrogels. The following are the different categories of bigels with a distinct set of qualities (Fig. 14.1) [10]:

- *Oleogel dispersed in hydrogel (O/H):* The formation of this kind of bigel involves dispersing the oleogel inside the architecture of the hydrogel. Numerous researchers have looked at this kind of bigels for several applications. For instance, Zhu et al. [11] fabricated bigels by combining glycerol monostearate-beeswax-based oleogel with high acyl gellan gum hydrogel in varying quantities. The microstructure of bigels revealed the scattered presence of oleogel. The incorporation of the spherical oil droplets into the continuous hydrogel matrix indicates the establishment of an oleogel-in-hydrogel structural configuration.
- *Bicontinuous bigel:* These bigels are formulations in which both phases appear to be continuous. Formation of such bigels involves an oleogel fraction of ~60% when the oil phase turns into a continuous phase. The hydrogel phase does not form droplets in these situations and gets thoroughly mixed within the oleogel phase, creating bicontinuous structures. Kodela et al. [12] formulated bigels by varying the ratios of agar hydrogel and stearyl alcohol oleogel. The microscopic examination of bigels revealed that, at greater proportions of oleogel, the interior phase was not spherical but had irregular-shaped features. This demonstrated that when the oleogel fraction was raised over a critical limit, bicontinuous bigels were formed.
- *Hydrogel dispersed in oleogel (H/O):* In this type of bigel, the hydrogel is present as the dispersed phase within the continuous matrix of oleogel. For delivering lipophilic bioactive chemicals, Zheng et al. [13] created bigels from carrageenan hydrogel and monoglyceride oleogels. H/O bigel has the same structure as water-in-oil emulsions, but both the water and oil phases are in the gel state.
- *Complex bigel:* These systems are prepared by homogenizing an emulsion gel with the oleogel phase in different fractions. To create a complex matrix-in-matrix type at the highest oleogel fraction, Lupi et al. [10] used extra virgin olive oil as the oil phase, a combination of policosanol and glyceryl monostearate as the oleogelator, and low methoxyl pectin as the hydrogelator. The highest oleogel percent yielded this matrix-in-matrix structure.

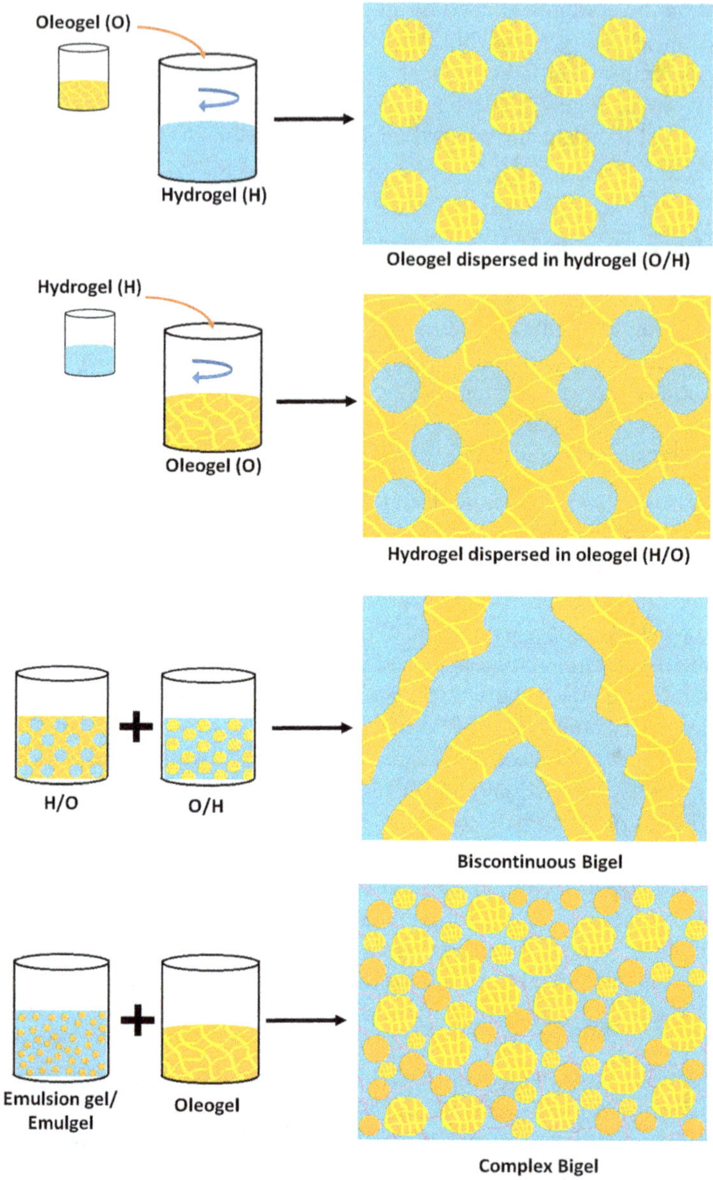

Fig. 14.1 Schematic representation of different types of bigels. (Adapted from [14], under the terms of the Creative Commons CC-BY license, and [15], with permission from Elsevier)

Bigel systems are further classified as conventional or unconventional based on whether they are made using monocomponent oleogel and hydrogel phases or multicomponent oleogel/hydrogel phases [15]. Conventional bigels are made by adjusting the concentration of a single component, such as oleogel or hydrogel. In order to optimize the rheological features of the resulting bigel systems, conventional bigel systems have been extensively investigated in the literature as a function of several aspects, such as oleogel/hydrogel ratio and hydrogelator type. Certain studies have shown that combining atypical bigel systems with normal ones is possible. These innovative solutions resulted in better technical and functional results. Unconventional bigels include multicomponent oleogel/hydrogel-based bigels, emulsion gel-based bigels, and colloidal particles-based bigels. Such bigels bring up new alternatives for upgrading the technological features of conventional bigels [15].

14.2 Materials Used in the Preparation of Bigels

Several compounds are utilized for the formation of hydrogels and oleogels. The application of the formed bigels heavily influences the choice of these compounds. Hydrogels can be obtained using natural and/or synthetic polymers. It is possible to describe hydrogels as jelly-like systems in which an appropriate hydrogelator, either naturally occurring or synthetically produced, creates a three-dimensional (3D) network to entrap the polar or aqueous phase. Natural polymers such as polysaccharides (e.g., starch, alginate, and agarose) and proteins (such as collagen and gelatin) have been widely utilized. Natural polymers may be modified using synthetic monomers via graft copolymerization, allowing them to be used to make superabsorbent hydrogels. Graft copolymerization (GCP) of gum polysaccharides with various vinyl monomers is one of the most researched and reliable methods for altering natural polymers. In recent years, interest has risen in synthesizing GCP-based polysaccharides, which permit combining the favorable characteristics of synthetic and natural polymers. Grafting may be accomplished by the use of standard redox grafting procedures, microwave irradiation, irradiation, or electron beams [16]. Water-soluble linear polymers such as polyvinyl alcohol (PVA) and polyvinyl pyrrolidone (PVP) have also been employed among synthetic polymers [17, 18]. The formation of hydrogels from these polymers can be accomplished via a variety of crosslinking processes, including (a) chemical reaction between the polymer chains, (b) ionizing radiation, and (c) physical interactions, such as entanglements [19].

Like hydrogels, several oleogel systems are also required to fabricate bigels. An oleogel is a semisolid substance composed of liquid oil trapped inside a network of structuring molecules called an oleogelator. Gelator molecules typically self-assemble to produce a 3D network stabilizing the oil phase. This is accomplished by developing physical connections through Van der Waals and hydrogen bonding, which encourage crystallization, crosslinking, and stacking of molecules. The liquid phase found in the formulation of oleogels is usually a vegetable oil derived from plants. Examples of such oils include sunflower oil [20, 21], extra virgin olive oil [22], rice bran oil [23], sesame oil [24], linseed oil [25], soybean oil [26], almond oil

[27], jojoba oil [28], canola oil [29], and tea tree oil [28]. Some biocompatible systems also use medium-chain triglycerides and isopropyl myristate as the oil phase [9]. Materials such as monoglycerides, phytosterols, and waxes show good oil-structuring abilities. The excellent gelation capability of oleogelators allows them to meet application-specific features. Based on molecular weight, oleogelators can be classified as low-molecular-weight oleogelators (LMWOs) and high-molecular-weight oleogelators (HMWOs) (Fig. 14.2). The LMWOs used in the oleogel formulation include monoglycerides, glyceryl fatty acid esters, ceramides, natural waxes, hydroxylated fatty acids, wax esters, lecithin, phytosterols, oligopeptides, glyceryl monostearate, and sorbitan monoesters [17, 26, 30, 31]. Furthermore, lecithin is a well-known natural oleogelator that comprises biological systems' richest group of phospholipids. It is obtained chiefly from sunflower, soybean, and egg yolk. Lecithin permits bioactive molecules to be incorporated into the aqueous and organic phases. Other LMWOs used in bigels include 12-hydroxystearic acid, stearyl alcohol, cetyl alcohol, and fumed silica [32]. Polymers are used as HMWOs and they can be categorized into proteins and polysaccharides (Fig. 14.2). Some examples of these polysaccharides are ethylcellulose (EC), hydroxypropyl methylcellulose, pectin, chitin, xanthan gum, guar gum (GG), kappa-carrageenan, alginate, hyaluronic acid, and chitosan, while examples of these proteins include gelatin, caseinate, soy protein isolate, and alpha-lactoglobulin [33]. But as of now, the only direct method for making oleogels made of polymers is to employ EC, a chemically altered version of the naturally occurring polymer cellulose [34].

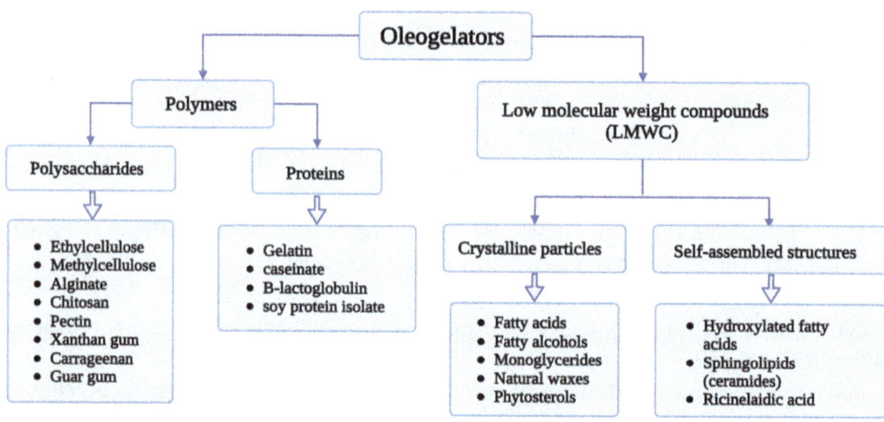

Fig. 14.2 Widely available oleogelators used for bigel formulation. (Reproduced from [35], under the terms of the Creative Commons CC-BY license)

14.3 Factors Affecting the Formation of Bigels

Usually, bigels are of O/H type when the oleogel concentration is less than 50%. Then, they change into a bicontinuous type when the oleogel concentration is ≥60%. The H/O type of bigels is formed after raising the oleogel concentration to ≥75%. Therefore, various structural morphologies may be generated by varying the oleogel: hydrogel ratio, which can considerably impact the rheological behavior of bigels [11]. However, if the gel network is damaged by the shearing action, the phase separation of bigels may happen. The most significant element dictating the rheological behavior of bigels appears to be the mass or volume ratios of oleogels and hydrogels. According to several studies, higher oleogel percentages improve the hardness, smooth texture, cohesion, stickiness, viscosity, larger yield stress values, and stability of bigels. After a specific oleogel:hydrogel ratio, the system might start to become hydrogel-dominated, which may limit the good influence of oleogel fraction on bigels characteristics.

Once the hydrogel content exceeds the oleogel fraction, the rheological and textural properties depend on the hydrogel fractions. For instance, it has been shown that a higher percentage of hydrogel results in harder bigels. The spreadability, work of shear, hardness, stickiness, work of adhesion, and percent stress relaxation of bigel systems are all considerably affected by the concentration of the hydrogelator [4]. In addition to hydrogelator concentration, the physicochemical attributes of bigels may be influenced by the composition of the hydrogelator. Compared to the system containing linear polysaccharides, bigels using branched polysaccharides as hydrogelators usually have superior gel strength and enhanced deformation resistance but inferior stress relaxation capabilities. Hence, the physicochemical properties of the bigel might change eventually [23].

In a recent publication, Lu et al. [36] developed an O/H type of bigel system to deliver nutraceuticals such as curcumin and epigallocatechin gallate (EGCG). They utilized glycerol monostearate (GMS) as an oleogelator and gelatin as a hydrogelator. Curcumin and EGCG were dissolved in the oleogel and hydrogel phases, respectively. Investigation of such a system revealed that the hardness and viscosity of the bigels increased with the GMS content. The bigels containing higher GMS content resulted in a stable crystal network structure, which enhanced their mechanical and thermal stability (Fig. 14.3). This study also revealed that the release of the mentioned nutraceutical also depended on the amount of the gelator. Such findings proved that the physicochemical properties of bigels can be altered by varying the gelator concentration. Also, bigels can codeliver nutritional compounds depending on their polarities and can be utilized to develop functional foods.

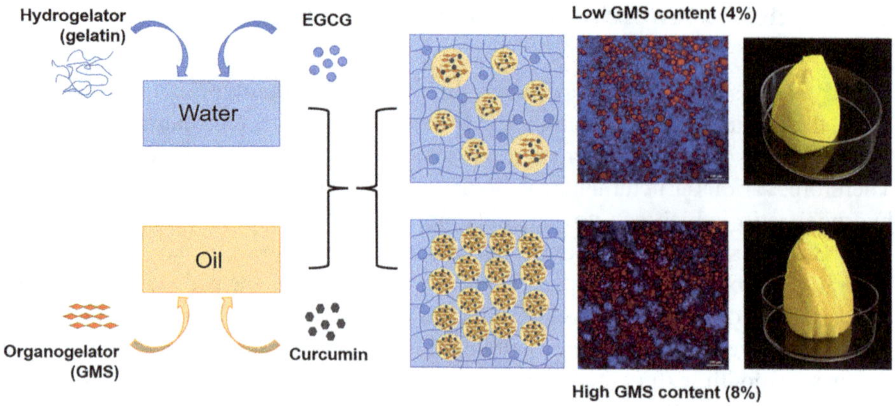

Fig. 14.3 Schematic for the formulation of an O/H type of bigel system. (Reproduced from [36], with permission from American Chemical Society)

14.4 Characterization Techniques of Bigels

As materials and the preparation methods vary, the characteristics of bigels differ accordingly. In general, preliminary characterization is carried out to confirm the formation of bigels through the tube inversion method [37, 38] and determine whether the material flows under its weight. A proper bigel is deemed to have formed if the sample does not flow. After bigels have been formed, several characteristics, including color, consistency, smoothness, transparency, viscosity, and pH, are assessed [39]. Bigels are frequently referred to as milky white systems, a characteristic linked to light dispersion from interfaces of both phases. They often have a smooth feel and are opaque [20]. Researchers have further characterized bigels based on mechanical strength, electrical properties, thermal stability, rheological, antimicrobial efficiency, and bioactive agent release rates [30, 38, 40]. Researchers have looked at the feasibility of using some food-grade bigels as fat replacers by studying their mechanical properties, rheology, and thermal stability. In this section, we shall discuss the different techniques that are utilized for the characterization of bigels.

14.4.1 Compositional Analysis

Fatty acid profiling largely determines vegetable oil's properties, composition, and uses in the oleogel phase. Characterization of the lipid component of foods by fatty acid analysis provides crucial data on feed and food quality. FAME (fatty acid methyl ester) analysis is regarded as an important method for characterizing fats and oils and estimating the total fat content in food products. FAME is the end product of the transesterification of fats with alcohol. This technique facilitates the determination of the presence or absence of *trans*-fatty acids in the formulation of bigels when analyzed using a gas chromatography instrument [29]. Following the quantification of total FAME in the derived sample, the total quantity of fat in the form of triglycerides is determined using the original sample weight.

Lipid oxidation is a chain of processes negatively affecting oil-based food products' quality and shelf life. Such a process of lipid oxidation involves the reaction between unsaturated fatty acids and oxygen to generate lipid hydroperoxides. The generated hydroperoxides are then broken down into small, volatile compounds such as aldehydes and ketones. These minute molecules generate off-flavors and aromas, reducing the acceptability of lipid-containing products, such as oils and meat. Therefore, examining lipid oxidation factors before using bigels in food items is essential.

According to previous research, the peroxide value (PV) and the p-anisidine value (AV) are often used to measure the amount of primary and secondary oxidation, respectively. The American Oil Chemists' Society (AOCS)-recommended official iodometric technique "Cd 8-53" is used to determine PV. The iodine generated by potassium iodide given as a reducing agent to the oxidized sample dissolved in a chloroform-acetic acid combination is measured by titration. The released iodine is titrated with conventional sodium thiosulfate ($Na_2S_2O_3 \cdot 5H_2O$) to a starch endpoint. AOCS "Cd 18-90" is used to determine the AV of samples spectrophotometrically [29]. Conventionally, the AV value is defined as 100 times the optical density measured at 350 nm in a cuvette of a solution containing 1.00 g of oil in 100 mL of a solvent/reagent combination. This technique detects the quantity of aldehydes (mostly 2-alkenals and 2,4-dienals) in samples, including vegetable fats and oils, by reacting the aldehydic compounds in the oil and p-anisidine in an acetic acid solution. With the use of the following formulae, the PV and AV values are determined:

$$\mathrm{PV} = \frac{(S - B) * M * 1000}{W} \tag{14.1}$$

where S is the volume of $Na_2S_2O_3 \cdot 5H_2O$ in the sample titration (mL), B is the volume of $Na_2S_2O_3 \cdot 5H_2O$ in the blank titration (mL), M is the molarity of the thiosulfate solution, and W is the weight of the sample (g).

$$\mathrm{AV} = 25 * \frac{1.25 * (A_\mathrm{C} - A_\mathrm{S})}{W} \tag{14.2}$$

where A_C is the absorbance of isooctane with p-anisidine solution, A_S is the absorbance of sample solution, and W is the weight of the sample (g).

14.4.2 Mechanical Characterization

Textural characteristics are most likely the most researched aspect of bigels. Bigels may be evaluated via texture analysis by assessing their mechanical resistance to stress. Various equipment, including viscometers [9], rheometers [13], and mechanical testers [17], are used to study the mechanical characteristics of bigels, such as viscosity and stress relaxation. Different characteristics (such as sample surface, sample shape, sample mass, and test speed) significantly impact the results obtained by the abovementioned instruments. In a basic gel strength measurement, a 0.5-inch-diameter cylinder probe is penetrated into the gel system at a constant pace. The gel strength is expressed as the maximum force necessary to travel a certain distance before irreversible deformation. However, by probing further into the gel, the rupture force and elasticity/brittleness of the gel may be determined [1].

The viscoelastic properties of bigels are reflected in the results of a textural investigation called stress relaxation (SR). An SR analysis involves inserting a probe into the formulation until a certain distance. The result is a force equal to its theoretical maximum, indicated by the symbol F_0. Then the strain is held constant for a specific period (known as the relaxation period). During this period, the measured force diminishes rapidly and approaches a near-constant asymptotic value. This forge value is known as residual force (F_r), which marks the resistance faced by the retained elastic portion of bigels. Overall, the bigel stress relaxation profiles are similar to the viscoelastic solid stress relaxation profiles. The percent SR (%SR) is calculated using the following equation:

$$\%\mathrm{SR} = \left(\frac{F_0 - F_r}{F_0}\right) \times 100 \tag{14.3}$$

where F_r is the residual force and F_0 is the maximum force.

A recent study by Shaikh and co-authors (2022) tailored the properties of GG-hydrogel and candelilla wax/sesame oil oleogel by incorporating date palm-derived cellulose nanocrystals (dp-CNC). The SR profiles of bigels were obtained using a mechanical tester, and data were used to evaluate the %SR (Fig. 14.4). The F_0 value of the control sample (BG0) was reported to be lower than the dp-CNC-containing bigels (BG1-BG4). Subsequent addition of dp-CNC significantly increased F_0 values of the formulations, and F_{60} values followed a trend comparable to F_0 values. The calculated parameter, %SR, provides a rationale for the fluidity of bigel formulations. A higher %SR number corresponded to a more fluid-like formulation. Bigels with lower dp-CNC content had more elastic components, while bigels with greater dp-CNC content had more fluidic components [41]. Consequently, via SR investigations, the viscoelastic nature of bigel formulations may be identified, which is useful for customizing their characteristics.

Non-Newtonian pseudoplastic flow and shear-thinning behavior of bigels have also been established using rheological experiments. According to these findings, a higher amount of hydrogel strengthens the hydrogen bonds in the polymeric network

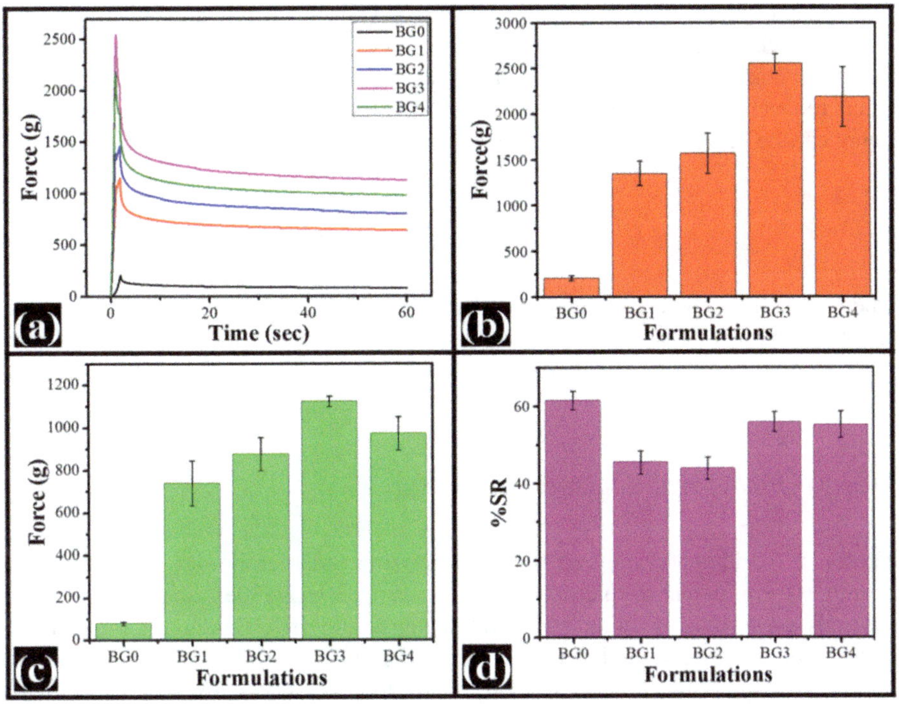

Fig. 14.4 Stress relaxation study of bigels with increasing concentration of dp-CNC*. (**a**) SR profile, (**b**) F_0, (**c**) F_{60}, and (**d**) %SR profiles. (Reproduced from [41], under the terms of the Creative Commons CC-BY license). (*BG0, BG1, BG2, BG3, and BG4 contain 0-, 5-, 10-, 15-, and 20-mg dp-CNC, respectively)

of bigels. As a result, the stiffness and viscosity of bigels increase. The viscosity of oleogels is measured using a cone and plate viscometer. Typically, measurements are taken with a shear rate between 20 and 100 s^{-1} at room temperature. Non-Newtonian fluid viscosity profiles can be defined by fitting the data to an equation based on the Ostwald-de Waele power law:

$$\tau = K.\gamma^n \quad (14.4)$$

where τ is the shear stress, K is the flow consistency coefficient, γ is the shear rate, and n is the power law index, which measures the shear-thinning property.

The viscoelastic nature of bigels can also be studied through creep experiments. A reduction in the viscosity of bigels was seen as the number of creep cycles increased [1]. Bigel with a lower oleogel concentration had a higher % creep recovery than bigel with a higher oleogel content. The creep experiment data can be fitted into the four-element Burgers mathematical model for viscoelasticity (Eq. 14.5). The Burgers model is a standard linear viscoelastic model that is used to characterize the viscoelastic behavior of polymeric systems. This model demonstrates all of the fundamental characteristics of viscoelasticity: when a load is applied, the model undergoes elastic deformation, followed by creep, which leads to permanent deformation. It comprises an immediate elastic recovery followed by a decreasing rate of creep recovery with permanent deformation. In other words, once the tension is removed, the model assumes that the sample partly recovers from the deformation.

$$J_c(t) = J_0 + J_1\left[1 - \exp\left(\frac{-t}{t_1}\right)\right] + \frac{t}{\eta_0} \tag{14.5}$$

where η_0 is the pure viscosity of the material, t_1 is the retardation time, $J_c(t)$ is the creep compliance at any time t, J_1 is the retarded creep compliance, and J_0 is the instantaneous creep compliance.

14.4.3 Structural Characterization

The most fundamental method for characterizing the morphology of bigels is to examine them under a microscope. Many studies have described microscopic methods for assessing bigel structural features and investigating the reciprocal distribution of organic and aqueous phases within bigels (Table 14.1). Different optical parameters such as droplet size distribution (DSD) or polydispersity of the dispersed phase can also be extracted from the microphotographs using different software. Such types of software include Image J software [42], Microtrac software [27], NI Vision Assistant-2010 software [21], and DHS Image database software [10]. DSD is calculated by computing the breadth of the droplet size distribution, which yields a dimensionless number known as the SPAN factor [23]. The SPAN factor is calculated as follows:

$$\text{SPAN} = \frac{(D_{90} - D_{10})}{D_{50}} \tag{14.6}$$

where D_{10}, D_{50}, and D_{90} represent the cumulative particle size distribution of 10, 50, and 90%, particles, respectively.

Fourier transform infrared (FTIR) spectroscopy determines the functional groups and chemical interactions present in bigel formulations [46]. Many bigel systems have been studied using an FTIR spectrophotometer in the attenuated total reflectance (ATR) mode in the wavenumber range of 4000–500 cm^{-1} [38, 47]. Both the polymer composition in the hydrogel and the proportion of gels affect the FTIR spectrum of bigels. FTIR is also employed to determine the involvement of hydrogen bonding in the bigel structure. Bigels often show a wide peak in their FTIR spectra in the wavenumber region of 3300–3200 cm^{-1} [48]. This peak arises primarily due to the hydrogel's intramolecular or intermolecular hydrogen bonding.

Table 14.1 Different microscopic techniques used to study bigel systems

Microscopic technique	Bigel system studied	Observations	References
Fluorescence microscopy + 3D digital imaging	Proteins/sunflower oil-based bigels Synthetic polymers/ sunflower oil-based bigels Carbopol/sesame oil-based bigels	Observation of droplet size	[21, 44]
Bright field microscopy	Natural gums/sunflower oil-based bigels Gelatin-agar/soybean oil-based bigels	Information on the dispersion of the dispersed phase and observation of the globular phases within a continuum phase	[17, 44]
Polarized optical microscope	Polymers/fish oil-based bigels Potato starch hydrogel and GMS oleogel	Fat crystals show birefringence property and hydrogel phase appears darker when viewed under polarized light	[45]
Confocal laser scanning microscopy	Guar gum/sesame oil-based bigels Polysaccharides/sunflower oil/fumed silica-based bigels	Fluorescent dyes identify the distribution of both oil and aqueous phases	[24, 32]
Cryo-scanning electron microscope	Fumed silica/based bigels	Observation of the distribution pattern of aqueous and organic phases	[32]
Phase contrast microscopy	Pectin/olive oil-based bigels Cosmetic system/olive oil-based bigels	Observation of particle size	[10, 22]
Field emission scanning electron microscopy	Soybean oil oleogel and gelatin-agar hydrogel-based bigels	Observation of the surface microstructure of dried and freeze-dried bigels	[44]

The FTIR spectroscopy method has also been used to evaluate interactions or compatibility between bioactive chemicals and bigels [49].

X-ray diffraction (XRD) analysis of bigel systems is performed to comprehend the amorphous or crystalline nature of bigels [6, 13, 36]. This study may be used to examine crystal characteristics. XRD analysis provides information on the crystalline organization, crystal shape, size, and molecular packing inside crystals. The shape of the crystals is determined using small-angle and wide-angle X-ray diffraction. Co-K has been usually reported as the radiation source ($\lambda = 1.79$ Å). The X-ray diffractometer is usually operated at 35 kV and 25 mA during the analysis of the samples. Evaluation of the samples is being performed within the diffraction angles from 5° to 50° (2θ) at a rate of 5° 2θ/min. Diffraction profiles were often deconvoluted to obtain accurate peak positioning and calculate other dependent parameters. These parameters include d-spacing (d), crystallite size (D), lattice strain (ε), and dislocation density (δ). By using Brag's law (Eq. 14.7), we can determine the

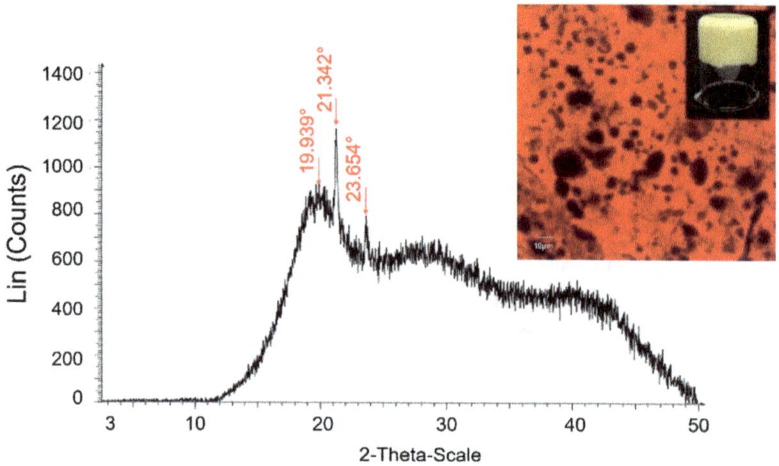

Fig. 14.5 X-ray pattern, confocal image, and structural behavior for beeswax-alginate bigel. (Reproduced from [29], under the terms of the Creative Commons CC-BY license)

d-spacing, and using the Debye-Scherrer equation (Eq. 14.8), we can calculate the crystallite size [50]. The lattice strain and dislocation density can be calculated using Eqs. 14.9 and 14.10. Quilaqueo et al. [29] recently investigated the function of hydrogel type in the production of bigels (beeswax/canola oil-based oleogel and alginate hydrogel) for fat substitutes in cookies. The X-ray diffraction pattern of bigels revealed semicrystalline solid with crystalline and amorphous components (Fig. 14.5).

$$\lambda n = 2d \sin \theta \tag{14.7}$$

where λ is 1.79 Å, n is an integer value, and θ is the diffraction angle.

$$D = \frac{\lambda\, k}{\beta \cos \theta} \tag{14.8}$$

where k is the Scherrer constant, β (radian) is the full width at half maxima (FWHM) at 2θ.

$$\varepsilon = \frac{\beta}{4 \tan \theta} \tag{14.9}$$

where ε is the lattice strain, β (radian) is the FWHM at 2θ, and θ is the diffraction angle.

$$\delta = \frac{1}{D^2} \qquad (14.10)$$

where δ is the dislocation density and D is the crystallite size.

14.4.4 Thermal Characterization

Thermal analysis refers to a set of procedures that measure a physical property of a substance and its reaction product(s) as a function of temperature. As the bigel is heated, its physical characteristics are often measured, including its temperature, thermal expansion, or thermodynamic parameters such as heat capacity, enthalpy, and entropy. Differential scanning calorimetry (DSC) is the most common method for the thermal investigation of bigels. Thermogravimetric analysis (TGA) and dynamic mechanical analysis (DMA) are also being utilized for such purposes. The choice of thermal analysis technique depends on the type of gel and the properties that need to be studied. DSC measures the difference in the heat needed to raise the sample and reference temperature as a function of temperature. In general, the temperature of the sample holder rises linearly as a function of time in DSC analysis. A DSC experiment yields a curve of heat flow versus temperature or time. The area under the curve is proportional to the heat flow. By integrating the peak corresponding to a certain transition, this curve may be used to derive transition enthalpies. DSC may also detect more subtle physical changes such as glass transitions. For thermal analysis of bigels, a small amount of bigel samples can be used. The sample is usually put in an aluminum crucible sealed with a pierced aluminum lid. The reference is taken as a sealed empty pan. After that, heating and cooling thermal scanning are performed for each sample. Melting profiles are captured across a specific temperature range. The analysis is performed under an N_2 environment maintained at a 40-mL/min flow rate. The heating and cooling DSC profiles can be obtained by scanning at a rate of 3.0 °C/min in the required temperature range [51].

Figure 14.6 represents a DSC thermogram of bigels with sucrose esters (SEs) with different hydrophilic-lipophilic balance (HLB) values. The study was carried out by Golodnizky et al. [51] to explore the effect of different SEs on the structure and properties of bigels (glycerol/gelatin water phase and GMS/Lecithin/SE oil phase). DSC analysis facilitated the determination of the oleogel's melting temperature and the hydrogel's evaporation temperature. The first peak, at ~59 °C, was present only in the bigel thermograms but absent in the thermogram of the hydrogel. As a result, this peak was associated with the melting temperature of GMS in oil. The second and third peaks were ascribed to the evaporation of free and bound water molecules from the bigel system, respectively. DSC analysis makes it possible to predict the thermal behavior of bigel systems prepared with different components.

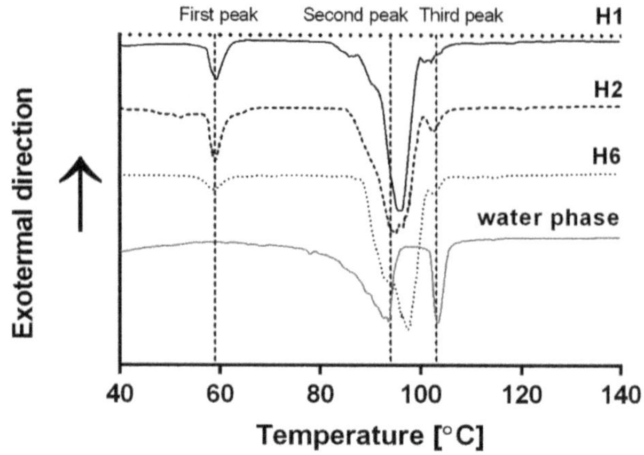

Fig. 14.6 DSC thermogram of H1 (black curve, bigel with SE of HLB1), H2 (dashed black curve, bigel with SE of HL21), H6 (dotted gray curve, bigel with SE of HLB6), and the water phase (gray curve). (Reproduced from [51], under the terms of the Creative Commons CC-BY license)

14.4.5 Electrical Properties

Recent interest has focused on instrumental approaches for characterizing samples because of their quick measurement times and lack of destructive effects. The electrical characteristics of a product or substance are measured to determine its quality. Bigels' electrical characteristics may be analyzed using an impedance analysis. The mobility of the polymer chains in a bigel sample causes resistance to the passage of current (resistivity) and the capacity to store energy in an electric field (capacitance) when an alternating current is applied. This method, specifically, keeps track of how the impedance (Z) of the sample changes as a function of the angular frequency of an applied voltage signal. In a wide frequency range, the impedance spectrum contains all the information about the electric characteristics of the bigel system. The frequency of the alternating current used in the measurement can affect the results, as the polymer chains would respond differently at different frequencies. At low frequencies, the movement of the polymer chains is dominated by viscous behavior, and the gel would have a relatively high resistivity and low capacitance. At high frequencies, the movement of the polymer chains is dominated by elastic behavior, and the gel would have a relatively low resistivity and high capacitance. This information can be used to study the gel's microstructure, including the polymer chains' size and distribution and the material's viscoelastic properties. Additionally, the resulting data can be used to determine the gel's storage modulus, loss modulus, and tan delta, which can indicate the elasticity and viscosity of the gel. Impedance analysis can also be used to monitor changes in gels with respect to time.

Figure 14.7 represents the impedance profile of bigels with an increasing concentration of dp-CNC in GG-hydrogel and sesame oil/candelilla wax oleogel, as reported by Shaikh et al. [41]. The impedance profile suggests the presence of typical capacitive behavior, where the impedance is high in the low-frequency range and reduces at high-frequency ranges. The authors correlated the observed impedance with the microstructure of bigels. They stated that smaller and homogenous droplets

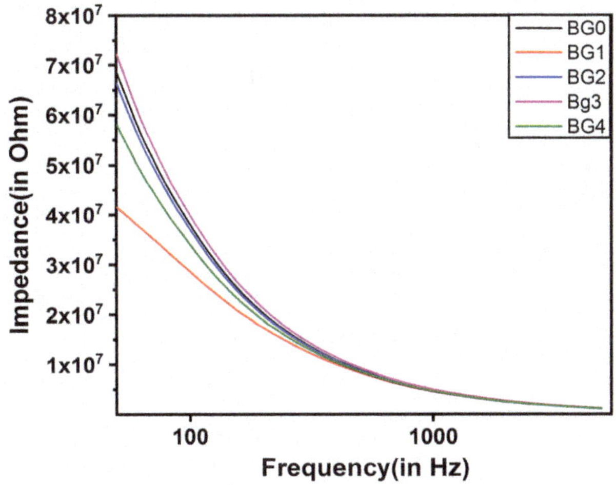

Fig. 14.7 Impedance spectrum of bigels. (Reproduced from [41], under the terms of the Creative Commons CC-BY license)

in hydrogels acted as numerous small capacitive elements, which increased the impedance of bigels.

14.5 Food-Based Applications of Bigels

Bigels are used in a variety of food applications. Different combinations of hydrogels and oleogels have been used to formulate bigels. Recent studies show that they have been used as thickening agents in food products to give them the desired texture and fine-tune their properties [29, 52]. As a fat replacer in low-fat and fat-free products, they can provide a creamy texture and mouthfeel similar to that of full-fat products. The water retention properties of bigels have been helpful in meat products. Recent works of literature suggested that applying bigels in sausages and deli meats can help retain moisture and improve texture and juiciness [53]. Bigels can encapsulate flavors, nutrients, and other food ingredients and release them in a controlled manner over time. Such systems are versatile and valuable tools for food scientists and manufacturers to improve the texture and taste of food. Using bigels in food is a relatively new development, and scientists are now working on modifying bigels' characteristics, so they may be used as fat replacements. There has been the suggestion that bigels may be used instead of fat in foods such as butter and cream cheese. Ghiasi et al. [54] were the first to show that low-fat beef patties prepared using bigels had better cooking properties than animal-fat burgers, proving bigels as a fat alternative in meals. In a recent article, researchers have tried to develop bigels consisting of gelatin hydrogels and GMS-enhanced EC oleogels using the component-induced catastrophic phase inversion method and studied their 3D-printing application in food [55]. Since the continuous phase consisted of EC oleogels, these bigels exhibited remarkable fluidity, making them more appropriate as 3D

Table 14.2 Application of bigels in different food products and nutrient delivery

Oil phase	Oleogelator	Hydrogelator	Potential application	References
Sunflower oil + olive pomace oil	Carnauba wax	Gelatin and collagen	Developing healthier fat-based product	[56]
Sunflower oil	Ethylcellulose	Wheat starch	Partial animal fat replacement	[54]
Canola oil	Beeswax	Sodium alginate and carboxymethylcellulose	Replacement of saturated and trans fats in cookies	[29]
Vegetable oil	Monoglycerides	Gelatin and κ-carrageenan	Replacement of animal fat in fermented sausages	[57]
Soybean oil	Rice bran wax	Gelatin	Novel edible bigel for food applications	[49]
Sunflower oil	Candelilla wax and 12-hydroxystearic acid	Sodium polyacrylate	A delivery system for vitamin E	[58]
Soybean oil	Glycerol monostearate and beeswax	Gellan gum	Delivery of nutraceuticals, i.e., lycopene	[11]

printing ink. The suggested procedure of catastrophic phase inversion of bigels gives the means to regulate the type of bigels by modifying the composition to acquire the desired physical attributes. This kind of adjustable bigels may readily replace conventional solid fat. Another group of researchers evaluated the role of bigels as fat replacers in cookies [29]. The oleogel phase comprised of beeswax and canola oil. On the other hand, the hydrogel phase consisted of sodium alginate and carboxymethylcellulose. The cookies obtained using the developed oleogels had a hardness similar to those with original shortening. This application showcased the huge potential of bigels as a fat substitute. Table 14.2 lists some of the recent applications of bigels as a fat replacer in food products.

14.6 Conclusion

Researchers have been experimenting with semisolid systems such as bigels, created by altering conventional gels such as hydrogels and oleogels, for potential usage in the food and nutrition industry. Recent years have seen the development and refinement of a variety of bigel formulations to meet the requirements of various applications, especially in the food industry. These bigels have been regarded as fat mimetics due to their potential to replace solid fats, for instance, butter. When it comes to substituting solid fats in food items, bigel is a particularly tempting alternative since they have successfully proven its functioning in multiple applications. Bigels may be used to improve the fatty acid profile of a product and can also be utilized to deliver lipid-soluble bioactives, both of which are advantages over traditional methods of achieving these goals. However, when such gel-based products are utilized as the fat phase in margarine and spread, issues have been encountered in producing the right mouthfeel, hardness, and long-term stability. This opens the door to further investigation and modification of such fat systems to make them acceptable for the necessary food application. This chapter focuses on different types of bigel systems, the materials used in their synthesis, the variables that affect the characteristics, the methods that have been developed to characterize them, and their application in the food industry. The properties of bigels can be tailored by modifying their composition and structure, making them a promising area of research for scientists and engineers. Although research has been done on such systems, no commercially viable solutions exist. Therefore, further research is needed to fully understand and exploit the capabilities of these derived systems.

Acknowledgments Ms. Somali Dhal would like to express her gratitude to NIT Rourkela for the fellowship that she was awarded for completing her master's thesis.

References

1. Shakeel A, Lupi FR, Gabriele D, Baldino N, De Cindio B (2018) Bigels: a unique class of materials for drug delivery applications. Soft Mater 16:77–93
2. Sharma C, Agrawal D, Goyal R, Sharma AK, Khandelwal M (2021) Organogel: a new approach in topical drug delivery system. Eur J Pharm Med Res 8:304–307
3. Singh VK, Qureshi D, Nayak SK, Pal K (2018) Polymeric gels. Elsevier, pp 265–282
4. Shakeel A, Farooq U, Iqbal T, Yasin S, Lupi FR, Gabriele D (2019) Key characteristics and modelling of bigels systems: a review. Mater Sci Eng C 97:932–953
5. Raytthatha N, Vyas J, Shah I, Upadhyay U (2022) Bigels: a newer system–an opportunity for topical application. Hamdan Med J 15:113
6. Yang J, Zheng H, Mo Y, Gao Y, Mao L (2022) Structural characterization of hydrogel-oleogel biphasic systems as affected by oleogelators. Food Res Int 158:111536
7. Fayaz G, Calligaris S, Nicoli MC (2020) Comparative study on the ability of different oleogelators to structure sunflower oil. Food Biophys 15:42–49
8. Pang M, Shi Z, Lei Z, Ge Y, Jiang S, Cao L (2020) Structure and thermal properties of beeswax-based oleogels with different types of vegetable oil. Grasas Aceites 71:e380–e380
9. Martín-Illana A, Notario-Pérez F, Cazorla-Luna R, Ruiz-Caro R, Bonferoni MC, Tamayo A, Veiga MD (2021) Bigels as drug delivery systems: from their components to their applications. Drug Discov Today 27:1008
10. Lupi FR, Shakeel A, Greco V, Rossi CO, Baldino N, Gabriele D (2016) A rheological and microstructural characterisation of bigels for cosmetic and pharmaceutical uses. Mater Sci Eng C 69:358–365
11. Zhu Q, Gao J, Han L, Han K, Wei W, Wu T, Li J, Zhang M (2021) Development and characterization of novel bigels based on monoglyceride-beeswax oleogel and high acyl gellan gum hydrogel for lycopene delivery. Food Chem 365:130419

12. Kodela SP, Pandey PM, Nayak SK, Uvanesh K, Anis A, Pal K (2017) Novel agar–stearyl alcohol oleogel-based bigels as structured delivery vehicles. Int J Polym Mater Polym 66:669–678
13. Zheng H, Mao L, Cui M, Liu J, Gao Y (2020) Development of food-grade bigels based on κ-carrageenan hydrogel and monoglyceride oleogels as carriers for β-carotene: roles of oleogel fraction. Food Hydrocoll 105:105855
14. Singh VK, Qureshi D, Nayak SK, Pal K (2018) In: Pal K, Banerjee I (eds) Polymeric gels. Woodhead Publishing, pp 265–282
15. Shakeel A, Farooq U, Gabriele D, Marangoni AG, Lupi FR (2021) Bigels and multi-component organogels: an overview from rheological perspective. Food Hydrocoll 111:106190
16. Mittal H, Ray SS, Okamoto M (2016) Recent progress on the design and applications of polysaccharide-based graft copolymer hydrogels as adsorbents for wastewater purification. Macromol Mater Eng 301:496–522
17. Sahoo S, Singh VK, Uvanesh K, Biswal D, Anis A, Rana UA, Al-Zahrani SM, Pal K (2015) Development of ionic and non-ionic natural gum-based bigels: prospects for drug delivery application. J Appl Polym Sci 132(38) 42561. https://doi.org/10.1002/app.42561
18. Behera B, Dey S, Sharma V, Pal K (2015) Rheological and viscoelastic properties of novel sunflower oil-span 40-biopolymer–based bigels and their role as a functional material in the delivery of antimicrobial agents. Adv Polym Tech 34
19. Ahmed EM (2015) Hydrogel: preparation, characterization, and applications: a review. J Adv Res 6:105–121
20. Bollom MA, Clark S, Acevedo NC (2020) Development and characterization of a novel soy lecithin-stearic acid and whey protein concentrate bigel system for potential edible applications. Food Hydrocoll 101:105570
21. Behera B, Sagiri SS, Pal K, Pramanik K, Rana UA, Shakir I, Anis A (2015) Sunflower oil and protein-based novel bigels as matrices for drug delivery applications—characterization and in vitro antimicrobial efficiency. PPTEn 54:837–850
22. Lupi F, Gentile L, Gabriele D, Mazzulla S, Baldino N, De Cindio B (2015) Olive oil and hyperthermal water bigels for cosmetic uses. J Colloid Interface Sci 459:70–78
23. Paul SR, Qureshi D, Yogalakshmi Y, Nayak SK, Singh VK, Syed I, Sarkar P, Pal K (2018) Development of bigels based on stearic acid–rice bran oil oleogels and tamarind gum hydrogels for controlled delivery applications. J Surfactant Deterg 21:17–29
24. Singh VK, Banerjee I, Agarwal T, Pramanik K, Bhattacharya MK, Pal K (2014) Guar gum and sesame oil based novel bigels for controlled drug delivery. Colloids Surf B Biointerfaces 123: 582–592
25. Wróblewska M, Szymańska E, Szekalska M, Winnicka K (2020) Different types of gel carriers as metronidazole delivery systems to the oral mucosa. Polymers 12:680
26. Mukherjee S, Ash D, Majee SB, Biswas GR (2020) Studies on Span based Soy-bigels with HPMC. Res J Pharm Technol 13:353–360
27. Andonova VY, Peneva PT, Apostolova EG, Dimcheva TD, Peychev ZL, Kassarova MI (2017) Carbopol hydrogel/sorbitan monostearate-almond oil based organogel biphasic formulations: preparation and characterization of the bigels. Trop J Pharm Res 16:1455–1463
28. Kanoujia J, Parashar P, Singh N, Saraf SA (2021) Tea tree and jojoba oils enriched bigel loaded with isotretinoin for effective management of acne. Indian J Nat Prod Resour 12(1):158–163
29. Quilaqueo M, Iturra N, Contardo I, Millao S, Morales E, Rubilar M (2022) Food-grade Bigels with potential to replace saturated and trans fats in cookies. Gels 8:445
30. Maji R, Omolo CA, Jaglal Y, Singh S, Devnarain N, Mocktar C, Govender T (2021) A transferosome-loaded bigel for enhanced transdermal delivery and antibacterial activity of vancomycin hydrochloride. Int J Pharm 607:120990
31. Scholten E (2018) in Edible oleogels. pp 285–305 Elsevier
32. Patel A, Mankoč B, Sintang MB, Lesaffer A, Dewettinck K (2015) Fumed silica-based organogels and 'aqueous-organic'bigels. RSC Adv 5:9703–9708

33. Feichtinger A, Scholten E (2020) Preparation of protein oleogels: effect on structure and functionality. Foods 9:1745
34. Gravelle AJ, Davidovich-Pinhas M, Zetzl AK, Barbut S, Marangoni AG (2016) Influence of solvent quality on the mechanical strength of ethylcellulose oleogels. Carbohydr Polym 135: 169–179
35. Pascuta MS, Varvara R-A, Teleky B-E, Szabo K, Plamada D, Nemeş S-A, Mitrea L, Martău GA, Ciont C, Călinoiu LF, Barta G, Vodnar DC (2022) Polysaccharide-based edible gels as functional ingredients: characterization, applicability, and human health benefits. Gels 8:524
36. Lu Y, Zhong Y, Guo X, Zhang J, Gao Y, Mao L (2022) Structural modification of O/W Bigels by glycerol Monostearate for improved co-delivery of curcumin and epigallocatechin Gallate. ACS Food Sci Technol 6:975
37. Soni K, Gour V, Agrawal P, Haider T, Kanwar IL, Bakshi A, Soni V (2021) Carbopol-olive oil-based bigel drug delivery system of doxycycline hyclate for the treatment of acne. Drug Dev Ind Pharm 47:954–962
38. Vergara D, Loza-Rodríguez N, Acevedo F, Bustamante M, López O (2022) Povidone-iodine loaded bigels: characterization and effect as a hand antiseptic agent. J Drug Deliv Sci Technol 72:103427
39. Andonova V, Peneva P, Georgiev GS, Toncheva VT, Apostolova E, Peychev Z, Dimitrova S, Katsarova M, Petrova N, Kassarova M (2017) Ketoprofen-loaded polymer carriers in bigel formulation: an approach to enhancing drug photostability in topical application forms. Int J Nanomedicine 12:6221
40. Corredor-Chaparro MY, Vargas-Riveros D, Mora-Huertas CE (2022) Hypromellose–Collagen hydrogels/sesame oil organogel based bigels as controlled drug delivery systems. J Drug Deliv Sci Technol 75:103637
41. Shaikh HM, Anis A, Poulose AM, Madhar NA, Al-Zahrani SM (2022) Development of Bigels based on date palm-derived cellulose nanocrystal-reinforced Guar Gum hydrogel and sesame oil/Candelilla Wax Oleogel as delivery vehicles for Moxifloxacin. Gels 8:330
42. Hamed R, Mahmoud NN, Alnadi SH, Alkilani AZ, Hussein G (2020) Diclofenac diethylamine nanosystems-loaded bigels for topical delivery: development, rheological characterization, and release studies. Drug Dev Ind Pharm 46:1705–1715
43. Singh VK, Anis A, Banerjee I, Pramanik K, Bhattacharya MK, Pal K (2014) Preparation and characterization of novel carbopol based bigels for topical delivery of metronidazole for the treatment of bacterial vaginosis. Mater Sci Eng C 44:151–158
44. Wakhet S, Singh VK, Sahoo S, Sagiri SS, Kulanthaivel S, Bhattacharya MK, Kumar N, Banerjee I, Pal K (2015) Characterization of gelatin–agar based phase separated hydrogel, emulgel and bigel: a comparative study. J Mater Sci Mater Med 26:1–13
45. Rehman K, Amin MCIM & Zulfakar MH (2014) Development and physical characterization of polymer-fish oil bigel (hydrogel/oleogel) system as a transdermal drug delivery vehicle. J Oleo Sci:ess14101
46. Singh VK, Anis A, Al-Zahrani S, Pradhan DK, Pal K (2014) FTIR, electrochemical impedance and iontophoretic delivery analysis of guar gum and sesame oil based bigels. Int J Electrochem Sci 9:5640–5650
47. Patel ND, Kanaki NS (2021) Quality evaluation of the Tugaksheeree samples by ATR-FTIR spectroscopy using multicomponent analysis. J Nat Prod 11:377–382
48. Barragán-Martínez LP, Molina-Rodríguez A, Román-Guerrero A, Vernon-Carter EJ, Alvarez-Ramirez J (2022) Effect of starch gelatinization on the morphology, viscoelasticity, and water structure of candelilla wax–canola oil–starch hybrid gels. J Food Process Preserv 46:e16520
49. Saffold AC, Acevedo NC (2021) Development of novel rice Bran Wax/Gelatin-based biphasic edible gels and characterization of their microstructural, thermal, and mechanical properties. Food Bioprocess Technol 14:2219–2230
50. Fasolin LH, Martins AJ, Cerqueira M, Vicente A (2021) Modulating process parameters to change physical properties of bigels for food applications. Food Struct 28:100173

51. Golodnizky D, Davidovich-Pinhas M (2020) The effect of the HLB value of sucrose ester on physiochemical properties of Bigel systems. Foods 9:1857
52. Dong Y, Wei Z, Xue C (2021) Recent advances in carrageenan-based delivery systems for bioactive ingredients: a review. Trends Food Sci Technol 112:348–361
53. Guo J, Cui L, Meng Z (2022) Oleogels/emulsion gels as novel saturated fat replacers in meat products: a review. Food Hydrocoll 108313
54. Ghiasi F, Golmakani M-T (2022) Fabrication and characterization of a novel biphasic system based on starch and ethylcellulose as an alternative fat replacer in a model food system. Innov Food Sci Emerg Technol 78:103028
55. Jiang Q, Wang Y, Du L, Li S, Liu Y, Meng Z (2022) Catastrophic phase inversion of bigels characterized by fluorescence intensity-based 3D modeling and the formability for decorating and 3D printing. Food Hydrocoll 126:107461
56. Baltuonytė G, Eisinaitė V, Kazernavičiūtė R, Vinauskienė R, Jasutienė I, Leskauskaitė D (2022) Novel formulation of Bigel-based vegetable oil spreads enriched with Lingonberry Pomace. Foods 11:2213
57. Zampouni K, Siachou C, Katsanidis E (2022) Design of biphasic structures for replacing saturated fats in food systems. Pub Health Tox 2(Supplement 1):A107. https://doi.org/10.18332/pht/149613
58. Martinez RM, Magalhaes WV, Da Silva SB, Padovani G, Nazato LIS, Velasco MVR, Da Silva Lannes SC, Baby AR (2021) Vitamin E-loaded bigels and emulsions: physicochemical characterization and potential biological application. Colloids Surf B Biointerfaces 201:111651

Chapter 15
Oleofoams: Formulation Rules and New Characterization Methods Based on X-Rays and Neutrons to Advance Current Understanding

Anne-Laure Fameau and Elliot Paul Gilbert

Abbreviations

DSC	Differential scanning calorimetry
NMR	Nuclear magnetic resonance
SANS	Small-angle neutron scattering
SAS	Small-angle scattering
SAXS	Small-angle X-ray scattering
USANS	Ultra-small-angle neutron scattering
USAXS	Ultra-small angle X-ray scattering
WAXS	Wide-angle X-ray scattering

15.1 Introduction

Oleogelation of edible oils is a good alternative for the replacement of saturated fat in food formulations. To achieve oleogelation, gelator molecules are added to an edible oil, which convert the liquid oil into a gel (see previous chapters on oleogels). While

A.-L. Fameau (✉)
Université Lille, CNRS, INRAE, Centrale Lille, UMET, Lille, France
e-mail: anne-laure.fameau@inrae.fr

E. P. Gilbert
Centre for Nutrition and Food Sciences, and Australian Institute for Bioengineering and Nanotechnology, The University of Queensland, Brisbane, QLD, Australia

Australian Centre for Neutron Scattering, ANSTO, Lucas Heights, NSW, Australia
e-mail: elliot.gilbert@ansto.gov.au

C. Andrea Palla, F. Valoppi (eds.), *Advances in Oleogel Development, Characterization, and Nutritional Aspects*,
https://doi.org/10.1007/978-3-031-46831-5_15

this strategy may significantly improve the nutritional profile of full-fat products, due to the use of large amounts of oils within the formulation, the calorific content remains similar to the conventional solid fat product. Oleofoams (oil foams) are based on the incorporation of air bubbles inside an oleogel [1]. Oleofoams are a promising option for food products combining both a reduced fat content and a new appealing texture due to the presence of air bubbles [2]. Oleofoams have been described in several patents to partially replace butter in food products such as cakes, biscuits, laminated pastry, and mayonnaise [3–6]. Oleofoams were shown to decrease the fat content while still retaining a similar taste to the original products [3–6]. In the cosmetic and pharmaceutical industries, oleofoams could also be a good alternative for oily products, since oleofoams are easier for consumers to apply on the skin and hair due to their rheological properties [7–9]. Oleofoams more broadly exhibit further advantages such as long-term stability (for months even above room temperature), reduction in microbial spoilage (as water is absent and so preservatives are not needed). In addition, only one or even no additive is required to produce them; therefore, oleofoams could enter in the category of "clean label" products in various industries such as in food, cosmetics, and pharmaceuticals. However, the literature and understanding of oleofoam systems is still scarce in comparison to the aqueous foams despite their wide potential [10]. The main reason is that oleofoams are much more difficult to stabilize as the choice of foam stabilizer is very limited compared to their aqueous counterparts, explaining also why they are less studied in the literature [9].

The main difference between aqueous foams and oleofoams comes from the relatively large difference in the surface tension of the solvents. Water has a very high surface tension of 71.9 $mN{\cdot}m^{-1}$ at 25 °C, whereas most of the edible oils have surface tensions relatively lower between ca. 15 and 35 $mN{\cdot}m^{-1}$. In aqueous foams, a large variety of hydrocarbon-based surfactants, proteins, polymers, and particles can adsorb at the interface decreasing the air-water surface tension. For systems based on edible oils, the low surface tension between air and oil makes the adsorption of the most common hydrocarbon-based surfactants used to produce aqueous foams energetically unfavorable, leading to no adsorption at the air-oil surface [11]. Most of the suitable oil-soluble surfactants need to form crystalline particles to adsorb at the air-oil surface and allow foam formation and stabilization [11]. Indeed, most of the research on oleofoams has focused on finding a suitable foam stabilizer based on the oleogel literature and on trying to understand the relationship between crystalline properties inside the oleogels and the resulting foaming properties (foamability and foam stability) [12]. However, in common with aqueous foams, oleofoams are multiscale materials with relevant structural length scales spanning from a few nanometers up to millimeters. Different characterization methods need to be combined to obtain a comprehensive understanding of the oleofoam structure, which is the prerequisite to link formulation, processing, and the resulting macroscopic properties. This understanding is essential to control and adjust the properties of oleofoams in terms of rheological properties and their stability with respect to time and temperature, and improve the process for their manufacture [13]. As for aqueous foams, the characterization of oleofoams is challenging since they are turbid systems; the direct optical observation of bubbles and foam structure is limited to a small quantity of sample and thus with limited numbers of bubbles leading to a restricted understanding of some three-dimensional features. Moreover, oleofoams deform under low levels of shear and are highly sensitive to temperature fluctuations. In oleofoams, the measure of the liquid fraction is also much more challenging in comparison to aqueous foams. For aqueous foams, electrical conductivity is widely used to measure the liquid fraction in the foam, but for nonconducting fluids, measurements are much more complicated [14].

Powerful techniques based on the X-ray and neutron could be used to better understand the structure of oleofoams at the different relevant scales of these systems as it is already the case for aqueous foams [15–18]. In small-angle scattering (SAS) techniques, the intensity of scattered radiation versus the magnitude of the scattering vector (q) is measured. These techniques are commonplace in the study for soft matter, since they are noninvasive and without the need for special sample preparation methods (e.g., staining, exposure to vacuum) leading to a minimization of sample artifacts in contrary to microscopy techniques [16]. However, the reader needs to keep in mind that SAS yields nonvisual information unlike microscopy and provides structural information in the so-called reciprocal space. Thus, to extract information with SAS, the scattering data have to be either inverted back to real direct space or fitted with reciprocal space models available [19]. Since neutrons are an isotopically sensitive probe of structure, one advantage of small-angle neutron scattering (SANS) in the case of oleofoams is the potential for contrast variation by using deuterated components [15]. The aim of this chapter is to highlight the techniques and the corresponding structural information, which could be obtained by X-ray (SAXS) and neutron (SANS) scattering as well as by tomography on oleofoams in order to stimulate the interest of the readers to apply them in the future. That is why we also recommend the readers to have a look to reviews on SAS applied to food science if they want to go further on the understanding of these techniques and how to apply them in details [16, 19, 20].

15.2 Oleofoams: Formulation and Process

15.2.1 Formulation of Oleofoams

Oleofoams are composed of air bubbles dispersed in a continuous edible oil phase (Fig. 15.1). The process to produce oleofoams is based on two distinct steps. The first is to produce an oleogel; that is to say that an edible vegetable oil is used as the continuous phase into which a high-melting edible component is added. This mixture needs to be heated above the solubility boundary of the mixture to dissolve the high-melting edible component. By cooling the mixture below the solubility boundary, the high-melting edible component crystallizes leading to an oleogel. In the second step, the mixture is whipped to introduce air bubbles leading to oleofoams (Fig. 15.1a). Crystalline particles from various systems have been used to obtain

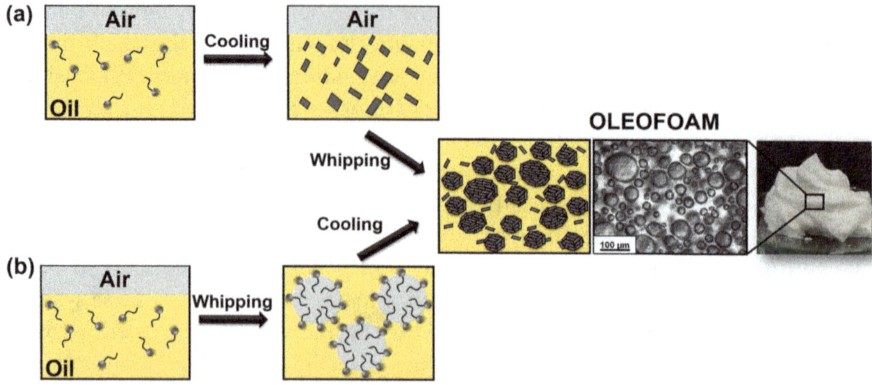

Fig. 15.1 Schematic showing the two oleofoam production strategies. (**a**) Crystalline particles need first to be formed to stabilize air bubbles produced by whipping. (**b**) Surfactant molecules can stabilize air bubbles produced by whipping but, for long-term stabilization, cooling is required to produce crystalline particles. The resulting oleofoam is stabilized by both crystalline particles at the bubbles surface and in the oil continuous phase as observed by optical light microscopy

oleofoams: mono-, di- and/or triglycerides, fatty alcohols, fatty acids, mixtures of phytosterols and mono- or diglycerides, and cocoa butter [8, 21–27]. Another approach to produce oleofoams is to use an oil or fat containing a high proportion of saturated medium-length fatty acid chains in order to have additive-free edible oleofoams such as coconut oil or shea butter [28, 29]. By controlling the temperature, the triglycerides with a high melting point crystallize while the other triglycerides with a lower melting point remain in a liquid state.

15.2.2 Emerging Systems to Produce Oleofoams

Recently, a new approach to produce oleofoams has been described [30]. Oleofoams can be stabilized directly by the addition of molecular surfactants before crystallization has occurred by using sucrose ester surfactant or sorbitan monoester [30, 31] (Fig. 15.1b). Oleofoams can be produced from the one-phase molecular solution because the triglycerides molecules form H-bonds with the hydroxyl groups of the surfactant [30]. In this case, the aeration process is first performed in the one-phase molecular region above the solubility boundary of the mixture. The prerequisite is that the oil-soluble surfactant should contain hydroxyl groups and that the alkyl chain lengths of the triglycerides and surfactant should be comparable to promote interfacial crystallization on cooling. These molecular complexes can adsorb at the air-oil surface, rendering efficient foam formation upon whipping above the solubility boundary. However, an efficient oleofoam stabilization is obtained only after rapid cooling of the foam to induce crystal formation in situ. In comparison to all the previous studies on oleofoams described above, this foaming strategy has the advantage of achieving much greater overrun (foam quantity).

15.2.3 Production Methods of Oleofoams

For all the systems cited above, the most popular foaming process is the whipping technique (electric beater, rotor-stator homogenizer, or kitchen food mixer). The whipping can take place during or after the cooling process, when the foam is stabilized only by crystalline particles (Fig. 15.1a). When foams are first stabilized by the surfactant layer at the air-oil surface, the whipping need to be first applied before cooling to increase the quantity of foam produced in the crystallized state (Fig. 15.1b). In whipping, the production of foam is based on air entrainment and progressive bubble breakup under shear [32]. During shearing, the gas fraction increases and the average bubble size decreases. Both the rheological properties of the oil dispersion and the whipping speed control the characteristic gas fraction and bubble size of the foaming mixture. To produce edible oleofoams on an industrial scale, high shear rotor-stator mixers such as Mondomix VL (Bühler company) are frequently used in the food industry for aqueous foam as well as oleofoam production [26]. It is important to point out that whatever the foaming process for these systems, the foam formation and stabilization is only possible when crystals are present in solution. The melting and crystallization temperatures of crystalline particles are linked to the solubility boundary, and temperature is therefore a crucial control parameter [9]. It is important to notice that oleofoams have also been produced by another foaming technique using depressurization but, thus far, only for cosmetic applications [33].

15.3 Structure and Properties of Oleofoams: State-of-the-Art of the Current Knowledge

15.3.1 Oleofoam Stabilization Mechanisms

For oleofoams, the presence of crystals is crucial for long-term stabilization. Each foam bubble is covered by adsorbed crystalline particles alone or below a surfactant monolayer, which prevents the relaxation of the bubble to a spherical shape [9]. This very high foam stability is due to the dense layer of adsorbed crystals at bubble surfaces that considerably reduce both coarsening and coalescence [9]. Moreover, the fraction of nonadsorbed crystals, which increases with crystal concentration, leads to gel network formation in the continuous oil phase reducing buoyancy-driven creaming of bubbles within the foam. The aging process is thus arrested leading to long-term stability. The presence of crystals both in bulk and at the interfaces is easily seen by using microscopy techniques such as polarized light microscopy, confocal Raman microscopy, or cryo-scanning electron microscopy (cryo-SEM) (Fig. 15.1).

15.3.2 Crystalline Particle Properties Leading to Oleofoams

The key criterion for crystals to adsorb at the air/oil interface is to exhibit a suitable contact angle below 90° [34]. An elegant study by Mishima and coworkers used synchrotron radiation microbeam X-ray diffraction (XRD) to understand the arrangement of triacylglycerol crystals around the air bubbles within an oleofoam [35]. They demonstrated that the lamellar planes of crystals near the air-oil surface are arranged almost parallel to the surface, that is, the lamellar planes composed of methyl end groups are facing the air phase, whereas the lateral planes composed of glycerol groups are connected to each other through crystals adsorbed at the air-oil surface [35].

For oleofoams stabilized by crystalline particles, both the foamability and foam stability depend on the size, shape, and concentration of particles. Needle-like and platelet crystals lead to efficient adsorption, whereas spherulitic particles seem less favorable for the adsorption [26, 36]. For oleogels, it is well known that the simplest way to tune the size and shape of crystalline particles is to use different cooling and shearing rates and also introduce tempering (cooling followed by slight warming) [37]. A clear link has been drawn between the properties of oleogels and those of foams in the case of oleogels based on fatty alcohol and fatty acid in sunflower oil forming platelet-shaped crystals [38]. Oleogel properties such as hardness and stability against oil loss are correlated with their resulting foaming properties in terms of foamability, foam firmness, and stability [7, 8]. For example, the foamability and foam stability of oleofoams based on fatty alcohol or fatty acid in sunflower oil were increased by increasing the cooling rate during the oleogel formation before whipping. Such enhancement in terms of oleofoam properties was shown to be linked to an increase of crystal quantity and a decrease in size inside the oleogel [7, 8]. However, there was no direct measurement on the crystal quantity and their size inside the foam after the oleogel whipping.

Another easy way to tune the quantity of crystals and their size without tuning the cooling rate or the shearing rate described in the oleogel field is to use a mixture of fatty alcohols and fatty acids with the same hydrocarbon chain length but at a specific molar ratio [39]. The optimal molar ratio of these two components is 3:1 of fatty alcohol:fatty acid for which co-crystallization occurs forming mixed crystals of small size and in high quantity [8, 40]. However, for other systems in which tempering was applied, the effect of polymorph type was not so clear and was not always observed on the foam properties [26, 35, 41]. These different conclusions could result from the fact that the tempering process not only influences the type of polymorph but also affects the quantity of crystals and the bulk rheological properties, which have a strong impact on the oleofoam properties [9].

15.3.3 Rheological Properties of Oleofoams

The main requirement to produce oleofoams from oleogels, from rheological perspective, is the need to behave as viscous dispersions once the whipping or mixing process starts in order to allow air incorporation; this implies specific rheological properties (such as yielding) for the base oleogels [42]. The rheological properties of oleofoams are still far from understood, since it results from both the bubbles coated by crystals and the rheology of the gelled continuous phase made of crystals [42, 43]. However, the rheological properties also strongly depend on the amount of incorporated air bubbles and their size. All these intermediate-scales parameters then control the macroscopic stability and rheology of the produced oleofoams. For the industrial applications of oleofoams in terms of the scale-up, it is very important that the oleofoam can be easily pumped through pipes and mixed with other components and it is directly linked to the rheological properties [2].

15.4 Multiscale Characterization of Oleofoams

15.4.1 Foam Structure and Rheological Properties

The structure of oleofoams is usually studied by using microscopy techniques or by indirect measurements such as oil drainage and texture profile analysis [8]. The main problem with microscopy techniques is sample preparation leading to artifacts in the microstructure, and which provide only two-dimensional (2D) information and a limited understanding of the three-dimensional (3D) foam structure [13]. X-ray tomography is a much better technique than microscopy to study the structure of oleofoams in three dimensions since, in contrast to visible light, X-rays are much less scattered by foams. By using X-ray tomography, the 3D oleofoam structure can be obtained in a matter of seconds with micrometric resolution [44]. Only one recent study used X-ray tomography to study oleofoam based on cocoa butter [13]. By using this technique, the authors quantitatively determine the number, size, and shape of the bubbles, as well as the thickness of the continuous phase. As a result, they could compare the effect of the process and time on the oleofoam structure (Figs. 15.2 and 15.3). For example, it is possible by looking at slices of the reconstructed volumes in a 2D plane to see a clear distinction between the gas phase (dark gray pixels) and the continuous phase (light gray pixels) enabling the thickness of the oleogel between the bubbles to be calculated and compared. By using X-ray tomography, it is possible to determine the thickness of oleogels present between the bubbles (around 10μm) and the remaining large domains of unwhipped oleogels as highlighted by the red narrow in Fig. 15.2.

The dispersed gas phase can be described in terms of bubble size and shape distribution as shown in Fig. 15.3. The size distribution of the air bubbles inside oleofoams can be estimated and compared for different processes and oleofoam

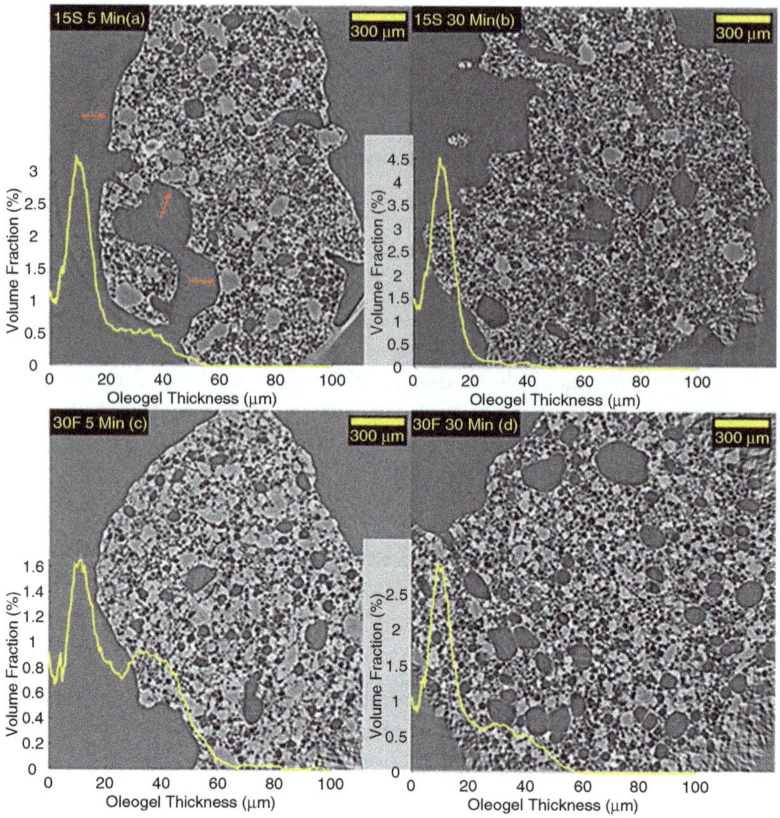

Fig. 15.2 Tomographic slice of an oleofoam based on 15 wt.% of cocoa butter crystals (15S) in a high oleic sunflower oil after (**a**) 5 or (**b**) 30 minutes of aeration by whipping the oleogel. The distribution of the oleogel thickness in yellow is overlaid on the respective images. The large oleogel fragments are highlighted with a red arrow. (Reproduced from [13] with permission from the American Chemical Society. This article is available under the terms of the creative commons attribution license (http://creativecommons.org/licenses/by/4.0/))

formulations as shown in Fig. 15.3. Another advantage of this technique, in contrast to microscopy techniques, is the examination of the shape of the bubbles in 3D, that is, the extent of bubble sphericity can be analyzed (Fig. 15.3); indeed, the nonspherical shape of the bubbles in oleofoams is clearly seen in Fig. 15.3e–f.

Small-angle scattering techniques, and indeed its lower q analog ultra-small-angle scattering (USAXS and USANS), are also valuable characterization tools for aqueous liquid foams, but unfortunately still not used for oleofoams. Specific setups have been designed to study aqueous foams produced by bubbling by SANS [18, 45]. Aqueous foams produced by other techniques such as hand-shaking and double-syringes techniques can also be studied by SANS without using specific measurement devices [46, 47]. SAXS and SANS have the advantages of being both sensitive to the foam structure but also to the foam stabilizer structure (crystalline

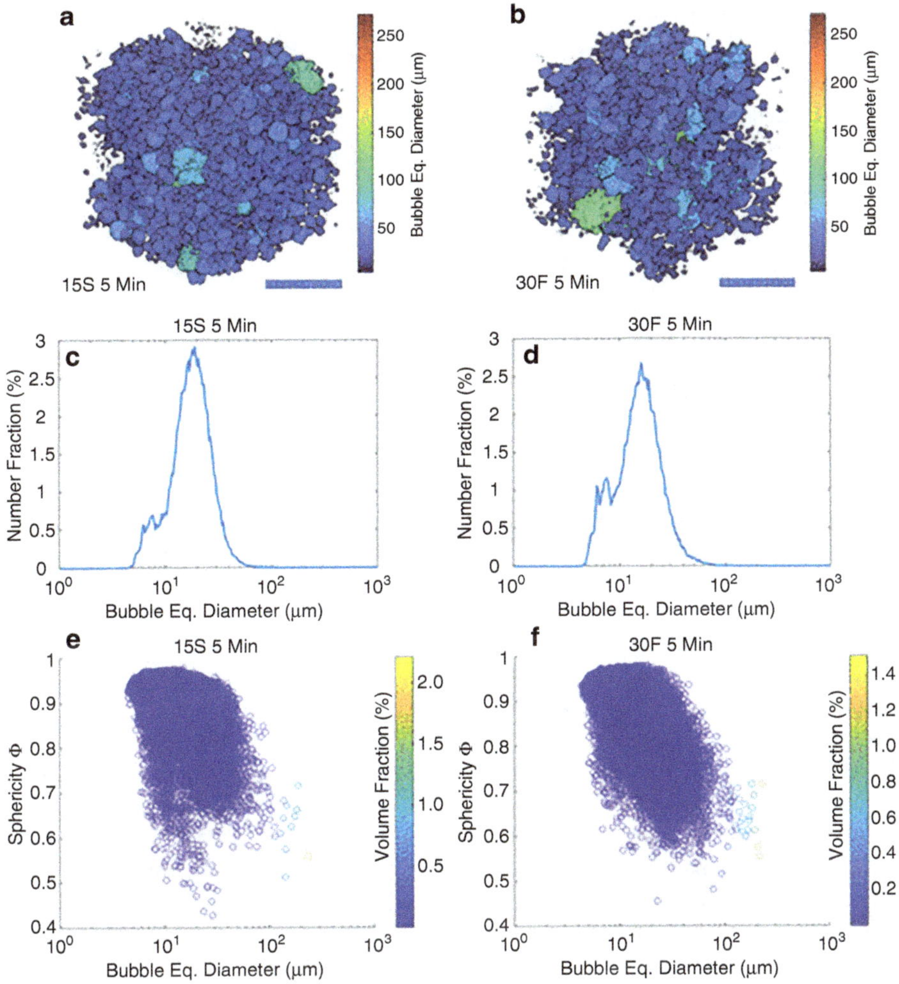

Fig. 15.3 3D renderings of representative volume of interest of two oleofoam samples: (**a**) 15 wt.% of cocoa butter (15S) and (**b**) 30 wt.% of cocoa butter (30 S) in high oleic sunflower oil after 5 minutes of aeration by whipping. The equivalent bubble diameter distribution for both samples is represented: (**c**) 15 wt.% of cocoa butter (15S) and (**d**) 30 wt.% of cocoa butter (30S). The corresponding scatter plots with bubble size and sphericity (**e**, **f**). The color bar shows the volume fraction occupied by each bubble. (Reproduced from [13] with permission from the American Chemical Society. This article is available under the terms of the creative common attribution license (http://creativecommons.org/licenses/by/4.0/))

particles in the case of oleofoams) [15]. Neutrons offer the opportunity to take advantage of isotopic sensitivity enabling different components within oleofoams to be specifically studied via the so-called contrast variation methods. There are few studies on the use of SANS for aqueous foams [18, 45–48]. However, emulsions have been studied extensively by this method [16, 49]. From the use of SANS, one

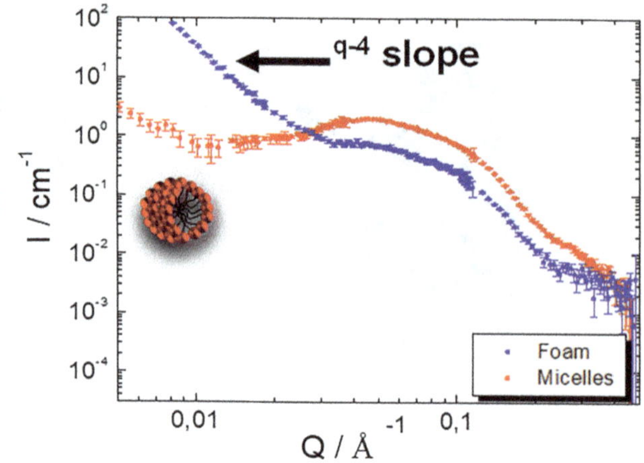

Fig. 15.4 SANS from micelles in aqueous solution (red) and from the associated foam formed from this micellar solution (blue). At low q, the q^{-4} slope from the Porod law can be used to extract the mean bubble size

could expect to extract different information on the oleofoam structure. Most of the oleofoam systems described in the literature are wet foams with high liquid fraction (above 36%). In oleofoams most of the bubbles are thus not in contact and thus the presence of the so-called thin films is limited. By knowing the liquid fraction, it would be possible to estimate a mean bubble size quantitatively with a high statistical precision inside the oleofoam due to the scattering occurring at the air/oil surface without the need of any sample preparation. The size of bubbles would be determined directly by using the q^{-4} decay at low q as described by the Porod law [15] (Fig. 15.4). For more information on how to extract the bubble size from SANS measurements, the reader is directed to the review by Mikhailovskaya et al. [15]. Temperature control is routinely conducted during SANS experiments, thus providing the potential to dynamically follow changes of bubbles size at different temperatures close to the melting temperatures of crystals. Rheo-SAXS and rheo-SANS could also give important information by measuring simultaneously both foam structure and rheological properties, and better characterize what happens for the foam structure under deformation [50]. Rheo-SAXS has already been used successfully to characterize the behavior of fat during crystallization and shear and the resulting network formation [51].

15.4.2 Characterization Methods of the Crystalline Phase Within the Foam

Oleofoams are specific foam systems since crystalline particles are present both at the interface and in bulk, and they are sensitive to temperature. Based on the literature, what is currently missing is the characterization in situ of crystals inside oleofoams in the bulk oleogel phase between the air bubbles. Information on the size, shape, polymorph type, and quantity is needed to better understand the

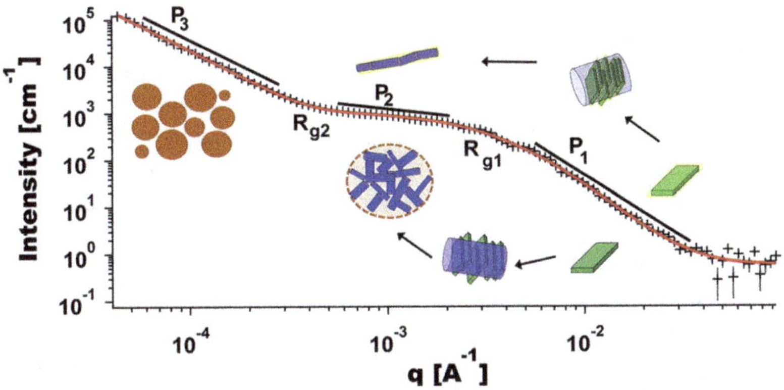

Fig. 15.5 USAXS and SAXS data for a system made from tristearin as TAG into triolein as liquid oil. The red line shows the fitting using the unified fit model while solid black lines show the different slopes obtained [55]. At the smallest length scale, the morphology of the surface of the crystalline nanoplatelets is obtained from the slope P_1 in the Porod region, while the size is obtained from the radius of gyration (R_{g1}) in the Guinier region. At intermediate q, from the Guinier-Porod model (P_2 and R_{g2}), the presence of cylinders was observed. The crystalline nanoplatelets aggregate into cylinders due to the stacking of the platelets as shown in the figure. P_2 and mass fractal values can be determined in the intermediate q region. In the smallest q region, the Porod exponent P_3 and R_{g3} can be also determined giving information on the surface fractal dimension. (Reproduced from [55] with permission from Elsevier)

oleofoam properties. Only wide-angle X-ray scattering (WAXS) characterization has been performed until now giving rise to information on the polymorph inside the foam [23, 24]. However, different techniques need to be combined to study these crystals in the continuous phase of the oleofoam. The best way to gain a better understanding of these foams at different length scales is to use and adapt the characterization techniques established to study classical aqueous liquid foams, emulsions, and/or oleogels. For example, low-field solid-state nuclear magnetic resonance (NMR) and differential scanning calorimetry (DSC) are best suited to determine the quantity of crystals. The size, aggregation state, shape, and polymorph of crystalline particles would require the combination of different microscopy techniques with scattering techniques such as SAXS and/or SANS as already used to study oleogels and aqueous foams stabilized by solid particles [15, 52, 53]. For example, SAXS and ultra-small-angle scattering (USAXS) have been shown to be very important techniques to describe the triglyceride (TAG) structure in liquid oils from the type of polymorph, to the packing of the lamellae, and also the shape and size of the TAG platelet as shown in Fig. 15.5 [54, 55]. SANS has already been used also to characterize oleogels [56–58]. For example, the self-assembled structures giving rise to oleogel formation in the mixtures of β-sitosterol and γ-oryzanol have been revealed by using SANS combined with solvent contrast variation [58]. By using decane as solvent in its hydrogenated form and deuterated form in different proportions, and keeping the oleogelator components in their hydrogenated forms (β-sitosterol and γ-oryzanol), it was possible to precisely determine the structure at the nanoscale at the origin of the gelling behavior. The oleogel structure is formed by both β-sitosterol/γ-oryzanol tubules, forming larger structures both trough intertubule aggregation and interconnection via junction zones leading to a space-filling network of fibrils [58]. This study is a good example to describe the value of SANS in comparison to SAXS, since the use of manipulation of contrast variation helps to test different structural models in order to be sure about the fitting of the scattering curves. SANS and USANS have also been applied to understand the gelling behavior of oleogels based on oleic acid and sodium oleate [56, 57]. In this case the solvent contrast variation method was used with two solvents: decane and hexadecane, which are easy to buy in deuterated forms. Inverse micellar structures and lamellar phases were measured before the formation of a crystalline space-filling network leading to oleogel formation. At low q by USANS, a mass fractal-like behavior was observed. SANS could also be used with the contrast variation method by using deuterated triglycerides oil, deuterated fatty acids, deuterated fatty alcohols, etc. The use of SANS for oleogel characterization is just at the beginning. The same approach could be used to characterize the structure, shape, and aggregation of crystalline particles within the oleofoam by using SAS and USAS techniques. There may also be opportunities to conduct simultaneous SAS and DSC studies [59, 60].

15.4.3 *Crystals at the Interface*

In terms of interfacial characterization of crystals, not only is the contact angle important but also the structure of crystals at the interface inside oleofoams. One aim in these foam systems is to separate the structure and the interfacial effects of crystals from their effects in the bulk continuous phase. Experiments on planar model surfaces are needed. For example, by using interfacial rheology, Mishra et al. studied the crystal layer formation of tripalmitin and monopalmitin at the planar air-oil surface to gain insight into the adsorption of crystals at the interface [36]. They studied crystallization in situ by cooling the vessel below the crystallization temperature; they showed that monopalmitin initially adsorbs molecularly and, upon crystallization, an elastic network is formed rapidly. Tripalmitin however formed β crystals in bulk, which adsorbed to the surface forming a plastic deformable network of high absolute strength. Tripalmitin crystals stabilize the air-oil surface by a Pickering-type mechanism, whereas monopalmitin exists as both molecules and subsequent crystals [36]. However, there are significant opportunities to study model planar surfaces, to determine the kinetics of adsorption of species and the resulting layer thickness and structure using neutron and X-ray reflectivity techniques. Such techniques could be helpful in gaining further insight into the structure of the layer during cooling and crystallization but also to capture changes during heating. X-ray and neutron reflectivity have already applied to characterize a wide range of complex structures at the air/water interface but also at the oil/water surface [49, 61, 62].

15.5 Conclusions

Oleofoams are very promising for the food industry since the benefits include long-term stability above room temperature, healthier food products with low fat content, and new sensorial properties. In recent years, the main building blocks explaining oleofoam stability have been identified, allowing a first picture to be developed based on contributions from interfaces and bulk. While crystals govern the entire foam stability, a deeper understanding on the stabilization mechanisms and the foam structure is still required; this, in turn, offers the possibility to design new food products as well as oil foams for cosmetic and pharmaceutical applications. In the future, it will be important to continue the recent works to establish the detailed link between the foam structure and their resulting rheological properties depending on the crystal concentration, their size, and shape, both in bulk between the air bubbles and at the interfaces. The main problem remains to separate bulk and interfacial effects, which could be discriminated by combining techniques at different length scales such as NMR, microscopy techniques, and especially neutron and X-ray scattering techniques [15–17, 52, 63, 64]. For the readers interested in the use of SAS techniques after reading this chapter to characterize oleofoams but also oleogels, a review by Gilbert [16] describes the method as well as how it has been applied across a wide range of food-relevant colloidal systems. For general information on SAS, the SAS portal is an excellent resource for with comprehensive tutorials, software, books, and articles [19]. Similarly, for the application of neutron methods more broadly, including reflectivity, to food materials, the reader is directed to the review by Lopez-Rubio and Gilbert [20].

References

1. Fameau A-L, Cousin F, Saint-Jalmes A (2017) Morphological transition in fatty acid self-assemblies: a process driven by the interplay between the chain-melting and surface-melting process of the hydrogen bonds. Langmuir 33. https://doi.org/10.1021/acs.langmuir.7b02651
2. Heymans R, Tavernier I, Dewettinck K, Van der Meeren P (2017) Crystal stabilization of edible oil foams. Trends Food Sci Technol 69:13–24
3. Gunes ZD, Schafer O, Chisholm H, Deyber H, Pelloux C, Binks BP (2018) Lipid based foam. WO 2016/150978
4. Gehin-Delval C, Chisholm H, Gunes ZD, Deyber H, Pelloux C, Schafer O, Nouzille CA, Destribats MJ (2019) Food composition comprising gas bubbles. US 10 , 383 , 352 B2
5. Chisholm H, Gunes ZD, Gehin-Delval C, Nouzille CA, Garvey E, Destribats MJ, Chandrasekaran SN, Vieira JB, German J, Binks BP (2018) Aerated confectionery material. US 2018 / 0064127 A1
6. Gehin-Delval C, Chisholm H, Chung W, Deyber H, Destribats MJ, Gunes ZD, Pelloux C (2019) Method for forming a laminated pastry. US 2019 / 0200625 A1
7. Callau M, Jenkins N, Sow-Kébé K, Levivier C, Fameau A-L (2021) The effect of vegetable oil composition on the structural properties of oleogels based on behenyl alcohol/behenic acid oleogelator system. J Cosmet Sci 72:399–405

8. Callau M, Sow-Kébé K, Jenkins N, Fameau A-L (2020) Effect of the ratio between fatty alcohol and fatty acid on foaming properties of whipped oleogels. Food Chem 333:127403. https://doi.org/10.1016/j.foodchem.2020.127403
9. Fameau A-L, Binks BP (2021) Aqueous and oil foams stabilized by surfactant crystals: new concepts and perspectives. Langmuir 37:4411–4418. https://doi.org/10.1021/acs.langmuir.1c00410
10. Dickinson E (2020) Advances in food emulsions and foams: reflections on research in the neo-Pickering era. Curr Opin Food Sci 33:52–60
11. Binks BP, Vishal B (2021) Particle-stabilized oil foams. Adv Colloid Interf Sci 291:102404. https://doi.org/10.1016/j.cis.2021.102404
12. Fameau A-L, Saint-Jalmes A (2020) Recent advances in understanding and use of oleofoams. Front Sustain Food Syst 4. https://doi.org/10.3389/fsufs.2020.00110
13. Metilli L, Storm M, Marathe S, Lazidis A, Marty-Terrade S, Simone E (2022) Application of X-ray microcomputed tomography for the static and dynamic characterization of the microstructure of Oleofoams. Langmuir 38:1638–1650
14. Salonen A, Lhermerout R, Rio E, Langevin D, Saint-Jalmes A (2012) Dual gas and oil dispersions in water: production and stability of foamulsion. Soft Matter 8:699–706
15. Mikhailovskaya A, Zhang L, Cousin F, Boue F, Yazhgur P, Muller F, Gay C, Salonen A (2017) Probing foam with neutrons. Adv Colloid Interf Sci 247:444–453
16. Gilbert EP (2019) Small-angle X-Ray and neutron scattering in food colloids. Curr Opin Colloid Interface Sci 42:55–72
17. Metilli L, Francis M, Povey M, Lazidis A, Marty-Terrade S, Ray J, Simone E (2020) Latest advances in imaging techniques for characterizing soft, multiphasic food materials. Adv Colloid Interf Sci 279:102154
18. Lamolinairie J, Dollet B, Bridot J-L, Bauduin P, Diat O, Chiappisi L (2022) Probing foams from the nanometer to the millimeter scale by coupling small-angle neutron scattering, imaging, and electrical conductivity measurements. Soft Matter 18:8733
19. SAS portal (n.d.). http://smallangle.org/. Accessed 29 Nov 2022
20. Lopez-Rubio A, Gilbert EP (2009) Neutron scattering: a natural tool for food science and technology research. Trends Food Sci Technol 20:576–586
21. Qiu C, Lei M, Lee WJ, Zhang N, Wang Y (2021) Fabrication and characterization of stable oleofoam based on medium-long chain diacylglycerol and β-sitosterol. Food Chem 350:129275
22. Metilli L, Lazidis A, Francis M, Marty-Terrade S, Ray J, Simone E (2021) The effect of crystallization conditions on the structural properties of oleofoams made of cocoa butter crystals and high oleic sunflower oil. Cryst Growth Des 21:1562–1575
23. Fameau A-L, Lam S, Arnould A, Gaillard C, Velev OD, Saint-Jalmes A (2015) Smart nonaqueous foams from lipid-based Oleogel. Langmuir 31:13501–13510
24. Binks BP, Garvey EJ, Vieira J (2016) Whipped oil stabilised by surfactant crystals. Chem Sci 7: 2621–2632. https://doi.org/10.1039/C6SC00046K
25. Shrestha RG, Shrestha LK, Solans C, Gonzalez C, Aramaki K (2010) Nonaqueous foam with outstanding stability in diglycerol monomyristate/olive oil system. Colloids Surfaces A Physicochem Eng Asp 353:157–165
26. Gunes DZ, Murith M, Godefroid J, Pelloux C, Deyber H, Schafer O, Breton O (2017) Oleofoams: properties of crystal-coated bubbles from whipped Oleogels-evidence for pickering stabilization. Langmuir 33:1563–1575
27. Brun M, Delample M, Harte E, Lecomte S, Leal-Calderon F (2015) Stabilization of air bubbles in oil by surfactant crystals: a route to produce air-in-oil foams and air-in-oil-in-water emulsions. Food Res Int 67:366–375
28. Binks BP, Marinopoulos I (2017) Ultra-stable self-foaming oils. Food Res Int 95:28–37
29. Liu Y, Binks BP (2021) Foams of vegetable oils containing long-chain triglycerides. J Colloid Interface Sci 583:522–534
30. Liu Y, Binks BP (2021) A novel strategy to fabricate stable oil foams with sucrose ester surfactant. J Colloid Interface Sci 594:204–216

31. Liu Y, Binks BP (2022) Fabrication of stable Oleofoams with Sorbitan Ester Surfactants. Langmuir 38:14779–14788
32. Drenckhan W, Saint-Jalmes A (2015) The science of foaming. Adv Colloid Interf Sci 222:228–259
33. Shrestha LK, Aramaki K, Kato H, Takase Y, Kunieda H (2006) Foaming properties of monoglycerol fatty acid esters in nonpolar oil systems. Langmuir 22:8337–8345
34. Fameau A-L, Saint-Jalmes A (2017) Non-aqueous foams: current understanding on the formation and stability mechanisms. Adv Colloid Interf Sci 247:454–464. https://doi.org/10.1016/j.cis.2017.02.007
35. Mishima S, Suzuki A, Sato K, Ueno S (2016) Formation and microstructures of whipped oils composed of vegetable oils and high-melting fat crystals. J Am Oil Chem Soc 93:1453–1466. https://doi.org/10.1007/s11746-016-2888-4
36. Mishra K, Bergfreund J, Bertsch P, Fischer P, Windhab EJ (2020) Crystallization induced network formation of tri-and monopalmitin at the MCT oil/air interface. Langmuir 36:7566–7572
37. Marangoni AG (2004) Fat crystal networks. CRC Press
38. Callau M, Sow-Kébé K, Nicolas-Morgantini L, Fameau AL (2020) Effect of the ratio between behenyl alcohol and behenic acid on the oleogel properties. J Colloid Interface Sci 560:874–884
39. Blach C, Gravelle AJ, Peyronel F, Weiss J, Barbut S, Marangoni AG (2016) Revisiting the crystallization behavior of stearyl alcohol: stearic acid (SO: SA) mixtures in edible oil. RSC Adv 6:81151–81163. https://doi.org/10.1039/C6RA15142F
40. Shah DO (1971) Significance of the 1: 3 molecular ratio in mixed surfactant systems. J Colloid Interface Sci 37:744–752
41. Heymans R, Tavernier I, Danthine S, Rimaux T, Van der Meeren P, Dewettinck K (2018) Food-grade monoglyceride oil foams: the effect of tempering on foamability, foam stability and rheological properties. Food Funct 9:3143–3154. https://doi.org/10.1039/C8FO00536B
42. Mishra K, Dufour D, Windhab EJ (2020) Yield stress dependent foaming of edible crystal-melt suspensions. Cryst Growth Des 20:1292–1301
43. Saha S, Saint-Michel B, Leynes V, Binks BP, Garbin V (2020) Stability of bubbles in wax-based oleofoams: decoupling the effects of bulk oleogel rheology and interfacial rheology. Rheol Acta 59:1–12
44. Meagher AJ, Mukherjee M, Weaire D, Hutzler S, Banhart J, Garcia-Moreno F (2011) Analysis of the internal structure of monodisperse liquid foams by X-ray tomography. Soft Matter 7: 9881–9885
45. Kühnhammer M, Widmann T, Kreuzer LP, Schmid AJ, Wiehemeier L, Frielinghaus H, Jaksch S, Bögershausen T, Barron P, Schneider H (2021) Flexible sample environments for the investigation of soft matter at the European Spallation Source: part III—the macroscopic foam cell. Appl Sci 11:5116
46. Fameau A-L, Saint-Jalmes A, Cousin F, Houinsou Houssou B, Novales B, Navailles L, Nallet F, Gaillard C, Boué F, Douliez J-P (2011) Smart foams: switching reversibly between ultrastable and unstable foams. Angew Chemie Int Ed 50:8264–8269. https://doi.org/10.1002/anie.201102115
47. Arnould A, Cousin F, Salonen A, Saint-Jalmes A, Perez A, Fameau A-L (2018) Controlling foam stability with the ratio of Myristic acid to choline hydroxide. Langmuir 34:11076–11085. https://doi.org/10.1021/acs.langmuir.8b02261
48. Zhang L, Mikhailovskaya A, Yazhgur P, Muller F, Cousin F, Langevin D, Wang N, Salonen A (2015) Precipitating sodium dodecyl sulfate to create ultrastable and stimulable foams. Angew Chemie Int Ed 54:9533–9536
49. Penfold J, Thomas RK (2014) Neutron reflectivity and small angle neutron scattering: an introduction and perspective on recent progress. Curr Opin Colloid Interface Sci 19:198–206
50. Eberle APR, Porcar L (2012) Flow-SANS and Rheo-SANS applied to soft matter. Curr Opin Colloid Interface Sci 17:33–43

51. Nikolaeva T, Den Adel R, Velichko E, Bouwman WG, Hermida-Merino D, Van As H, Voda A, Van Duynhoven J (2018) Networks of micronized fat crystals grown under static conditions. Food Funct 9:2102–2111
52. Fameau A-L, Salonen A (2014) Effect of particles and aggregated structures on the foam stability and aging. Comptes Rendus Phys 15:748–760
53. Marangoni AG, Van Duynhoven JPM, Acevedo NC, Nicholson RA, Patel AR (2020) Advances in our understanding of the structure and functionality of edible fats and fat mimetics. Soft Matter 16:289–306
54. Ramel PR, Co ED, Acevedo NC, Marangoni AG (2016) Structure and functionality of nanostructured triacylglycerol crystal networks. Prog Lipid Res 64:231–242
55. Peyronel F, Pink DA, Marangoni AG (2014) Triglyceride nanocrystal aggregation into polycrystalline colloidal networks: ultra-small angle X-ray scattering, models and computer simulation. Curr Opin Colloid Interface Sci 19:459–470
56. Nikiforidis CV, Gilbert EP, Scholten E (2015) Organogel formation via supramolecular assembly of oleic acid and sodium oleate. RSC Adv 5:47466–47475. https://doi.org/10.1039/c5ra05336f
57. Cornet S, de Campo L, Martínez-Sanz M, Scholten E, Gilbert EP (2021) Small-angle neutron scattering reveals basis for composition dependence of gel behaviour in oleic acid-sodium oleate oleogels. Innov Food Sci Emerg Technol 73:102763
58. Gilbert EP (2022) Building blocks of β-sitosterol-γ-oryzanol gels revealed by small-angle neutron scattering and real space modelling. Food Funct 13:7123–7131
59. Pullen SA, Booth N, Olsen SR, Day B, Franceschini F, Mannicke D, Gilbert EP (2014) Design and implementation of a differential scanning calorimeter for the simultaneous measurement of small angle neutron scattering. Meas Sci Technol 25:55606
60. Gilbert EP, Nelson A, Sutton D, Terrill N, Martin C, Lal J, Lang E (2005) Phase separation in the organic solid state: the influence of quenching protocol in unstable n-alkane blends. Mol Cryst Liq Cryst 440:93–105
61. Campbell RA (2018) Recent advances in resolving kinetic and dynamic processes at the air/water interface using specular neutron reflectometry. Curr Opin Colloid Interface Sci 37: 49–60
62. Skoda MWA (2019) Recent developments in the application of X-ray and neutron reflectivity to soft-matter systems. Curr Opin Colloid Interface Sci 42:41–54
63. Marangoni AG, Garti N (2018) Edible oleogels: structure and health implications. AOC Press, San Diego
64. Low LE, Siva SP, Ho YK, Chan ES, Tey BT (2020) Recent advances of characterization techniques for the formation, physical properties and stability of Pickering emulsion. Adv Colloid Interf Sci 277:102–117

Chapter 16
Physical and Oxidative Stability of Oleogels During Storage

Hong-Sik Hwang and Jill K. Winkler-Moser

Abbreviations

AV	Anisidine value
CDV	Conjugated diene value
DSC	Differential scanning calorimetry
FAME	Fatty acid methyl ester
GC	Gas chromatography
MDA	Malondialdehyde
NMR	Nuclear magnetic resonance
PUFAs	Polyunsaturated fatty acids
PV	Peroxide value
RH	Relative humidity
SPME	Solid-phase micro-extraction
TBA	Thiobarbituric acid
TBARS	2-Thiobarbituric acid-reactive substances
TOTOX	Total oxidation values

H.-S. Hwang (✉) · J. K. Winkler-Moser
U.S. Department of Agriculture, Agricultural Research Service, National Center for Agricultural Utilization Research, Functional Food Research, Peoria, IL, USA
e-mail: hongsik.hwang@usda.gov; jill.moser@usda.gov

C. Andrea Palla, F. Valoppi (eds.), *Advances in Oleogel Development, Characterization, and Nutritional Aspects*,
https://doi.org/10.1007/978-3-031-46831-5_16

16.1 Introduction

Oleogels (or organogels) have received great interest as healthy alternatives to conventional solid fats containing high saturated fats. Oleogels are typically prepared by dissolving a small amount of oleogelator in liquid oils such as vegetable oil and fish oil with heat followed by cooling to room temperature to produce semi-solid materials. Initial studies focused on the development of oleogels with physical properties that were similar to conventional solid fats. With increasing success in achieving the desired physical properties of oleogels, attaining satisfactory physical and oxidative stability during storage and transportation has become an important property to examine. Special attention should be taken for their oxidative stability because oleogels are often prepared with vegetable oil or fish oil containing high contents of unsaturated fatty acids, which are much more prone to oxidation than saturated fatty acids.

Conventional solid fats can undergo physical changes during short- or long-term storage, including transitions from less stable to more stable polymorphs, the formation of new crystals, sintering, and Ostwald ripening [1, 2]. Oleogels can also undergo changes in physical properties owing to the formation of new crystals and changes in the crystal network during storage [1]. Sintering can occur to form solid bridges in narrow gaps between fat crystals via non-covalent bonds, which can result in increased firmness [1, 2]. It is not surprising that oleogels undergo oxidation during storage since they are made of oils with high contents of mono- and polyunsaturated fatty acids, which can quickly oxidize. The physical and oxidative stabilities of oleogels depend on the types of oleogelator and oil, the storage conditions, and other factors.

This chapter will summarize the physical and oxidative stabilities of different kinds of oleogels based on recent studies. The information in this chapter may help direct future studies for the practical application of oleogels in food products. Although this chapter focuses on oleogels prepared by direct methods, in which oleogelators are directly mixed or dissolved in oils, and methods to evaluate their stability, it should be noted that these methods can also be applied to oleogels obtained through indirect methods such as foam, emulsion or aerogel templating methods, and solvent exchange processes. In addition, this chapter focuses on understanding the stability of oleogels alone, while there are many other studies on the stability of oleogel-containing materials such as emulsions and other food products. It is expected that the stability of an oleogel may change when it is incorporated in an emulsion or a food product due to the interactions with other ingredients used.

16.2 Physical Stability

The term "physical stability" has also been used to express the stability during centrifugation, texture analysis, or rheological analysis of freshly made oleogels. However, in this chapter, these properties are regarded as initial gel properties, and only the physical stability determined by periodic examinations for changes in the physical properties of oleogels during storage is discussed.

16.2.1 Methods to Measure Physical Stability of Oleogels

Table 16.1 summarizes the methods that have been used to examine the physical stability of oleogels and some example studies where these methods were used. The physical stability of oleogels can be examined during storage at different temperatures, but one must be cautious during accelerated stability testing to ensure that the temperature is kept below the melting or phase transition temperature of the oleogel; otherwise, the results may not correlate with stability at actual storage temperatures [3].

Texture analysis is one of the most widely used methods to evaluate the physical stability of oleogels. This test is typically conducted by the penetration test, where a probe penetrates through a sample at a controlled speed to a desired depth. From the resultant force–distance curve, the firmness (or hardness), cohesiveness (or work of shear), adhesiveness (or work of adhesion), and stickiness (or adhesive force) can be determined from the maximum force, the work required during the downward movement of the probe, the work required during retreat of the probe, and the negative peak force, respectively [4, 5]. Firmness and cohesiveness are generally well correlated. Adhesiveness often does not correlate well with firmness and cohesiveness, especially when oleogels have solid-like properties [6].

Oil binding capacity is a measure of the separation of oil from the oleogel. It has been defined and measured in several different ways, but the most common method to measure and report oil binding capacity is by subjecting the oleogel to centrifugation and measuring the weight of oil lost per weight of oleogel [7]. Blake et al. [8] described the liquid oil in oleogels as either physically entrapped by the oleogel network, or strongly bound, where the oil may interact with or be adsorbed to the surface of oleogelator solid particles. Thus, oleogelator type and its interactions with the oil as well as gelator concentration and preparation techniques that influence the oleogel microstructure will influence the oil binding capacity. Changes in oil binding capacity during storage has been used to evaluate the physical stability of oleogels. Phase separation can occur even without a force such as centrifugation due to changes in the crystal network and crystal aggregation during storage. Therefore, changes in the physical appearance, such as syneresis, have also be used to evaluate physical stability of oleogels during storage [3].

Table 16.1 Methods used to measure physical stability of oleogels

Method	Typical parameters	Oleogelator	Oil	References
Texture analysis	Travel depth: 5, 10, 12, or 23 mm Penetration speed: 0.5, 1, or 3 mm/s Pulling out speed: 5, 10 mm/s Probes: cylindrical or conic probes Probe diameter: 6, 10, 12.5, or 13, mm	Sorbitan monostearate	Sweet almond oil	[3]
		Beeswax	Fish oil	[22]
		Carnauba wax, monoglycerides	Virgin olive oil	[19]
		Sunflower wax, carnauba wax	Cod liver oil	[20]
		Monoglycerides	Olive oil	[11]
		Berry wax, sunflower wax	Rice bran oil	[12]
		Monoglycerides	High oleic sunflower oil	[15]
		Monoglycerides, mixture of monoglycerides and phytosterols	Olive oil	[14]
Oil binding capacity	Centrifuge conditions: 4500 rpm for 30 min, 9000 rpm for 15 min, 10,000 rpm for 15 min, or 13,000 rpm for 15 min	Ethylcellulose, stearic acid, monoglycerides, octadecanol, rice bran wax, sunflower wax, beeswax, mixture of γ-oryzanol and β-sitosterol	Sunflower oil	[18]
		Monoglycerides	High oleic sunflower oil	[26]
		Monoglycerides	High oleic sunflower oil	[15]
		Monoglycerides	Rapeseed oil	[28]
Macroscopic appearance	N/A	Sorbitan monostearate	Sweet almond oil	[3]
		Mixture of starch and lecithin	Blend of rice bran oil and sesame oil	[78]
		12-Hydroxystearic acid, beeswax	Castor oil	[34]
		Beeswax, sorbitan monostearate (Span 60)	Fish oil	[21]
Rheological analysis	Clearance set: 0.052, 1, or 2 mm Shear frequency: 0.1, 1, or 10 Hz Shear rate: 0.01–200 s^{-1} Shear strain: 0.01–100, or 0.001–10%.	Ethylcellulose, stearic acid, monoglycerides, octadecanol, rice bran wax, sunflower wax, beeswax, mixture of γ-oryzanol and β-sitosterol	Sunflower oil	[18]
		Mixture of β-sitosterol and lecithin	Peanut oil, soybean oil	[9]
		Monoglycerides	High oleic sunflower oil	[26]

(continued)

Table 16.1 (continued)

Method	Typical parameters	Oleogelator	Oil	References
		Monoglycerides	High oleic sunflower oil	[15]
		Monoglycerides, rice wax, ethylcellulose, mixture of oryzanol and ß-sitosterol	Extra virgin olive oil	[30]
Viscosity	Gap: 0.52 mm Shear rate: 0–200 s^{-1}	Mixtures of β-sitosterol and lecithin	Sunflower oil	[10]
Melting property (DSC)	Cooling and heating rates: 5 or 10 °C/min	Berry wax, sunflower wax	Rice bran oil	[12]
		Mixture of candelilla wax and fully hydrogenated soybean oil	High oleic safflower oil	[13]
		Monoglycerides, mixture of monoglycerides and phytosterols	Olive oil	[14]
		Monoglycerides	High oleic sunflower oil	[15]
X-ray diffraction	Sample-to-detector distance: 55 mm, 360 mm Scan range: 1–2.6°, 15–30°, or 10–40°	Berry wax, sunflower wax	Rice bran oil	[12]
		Monoglycerides	Rapeseed oil	[28]
		Mixture of candelilla wax and fully hydrogenated soybean oil	High oleic safflower oil	[13]

Rheology has also been used to assess the stability of oleogels during storage over time. For example, changes to the storage (G′) and loss (G″) moduli were used to monitor the physical stability of oleogels formed with β-sitosterol-lecithin in soybean oil or peanut oil during storage [9]. Apparent viscosity (Pa·s) was used to monitor the physical property change of oleogels of β-sitosterol and lecithin in sunflower oil during storage at 15 °C for 60 days [10]. In many studies, two or more methods were used to understand the physical stability of oleogels.

X-ray diffraction, microscopy, and melting profiles have been used to further understand the changes in physical properties. Microscopic methods such as polarized light microscopy are often used to explain observations of changes in the physical properties by examining the changes in the number and shape of oleogelator crystals during storage. For example, Ojijo et al. [11] used the results from polarized light microscopy to explain the increased firmness of an oleogel. This study observed that the firmness of the oleogel sample prepared with 7% monoglycerides in olive oil followed by shearing at 8000 rpm for 60 s gradually increased during storage at 25 °C for 8 weeks. Polarized light microscopy showed aggregation of monoglyceride crystals and increases in crystallinity, which was thought to explain the increased firmness observed in the oleogel during storage. Doan et al. [12] used X-ray diffraction, polarized light microscopy, and differential scanning calorimetry (DSC) to understand the changes in the physical properties of rice bran oil oleogels with berry wax and sunflower wax. In this study, 5% berry wax oleogels, of which the gel strength increased (increased G′) during storage, showed growth and clustering of wax crystals, development of rosette spherulites, polymorphic transition, and Ostwald ripening during storage. Melting profiles and X-ray diffraction patterns of this berry wax oleogel also significantly changed during storage, which indicated a transition of berry wax crystals from α to β morphology. Other examples of studies that used DSC to monitor changes in melting profiles as a supplementary method to explain the physical property changes of oleogels include studies by Ramírez-Gómez et al. [13] and Doan et al. [2]. However, some other studies reported that melting properties of oleogels were not significantly affected by storage while there were some changes in gel strength [14]. The study by Giacomozzi et al. [15] also reported no specific changes in the melting behavior although there were significant changes in hardness, cohesiveness, elasticity, and crystal length of monoglycerides-high oleic sunflower oil oleogels stored for 8 weeks at 5 °C.

16.2.2 Physical Stability of Wax-Based Oleogels

Table 16.2 summarizes studies of the physical stability of oleogels made with waxes such as sunflower wax, carnauba wax, beeswax, rice bran wax, and berry wax in different oils. The concentrations of wax in these studies ranged from 3% to 11% in weight. Storage temperature and time ranged from 4 to 25 °C and from 4 weeks to 6 months, respectively.

Sunflower wax oleogels All three studies with sunflower wax oleogels in Table 16.2 found that these oleogels had stable physical properties. Hazelnut oil oleogels made with 3, 7, and 10% sunflower wax showed steady adhesiveness during storage at 4 and 20 °C for 90 days [16]. Firmness of oleogels with 3, 7, and 10% sunflower wax in fish oil did not change after 90 days at 4 and 20 °C [17]. The firmness of oleogels with 3 and 5% sunflower wax was also steady while that of 10% sunflower wax slightly increased during the 90 days storage. Oleogels made with 5% sunflower wax in rice bran oil showed stable gel strength during 1-month storage at 5 °C, which was monitored by rheology properties [12]. Oil binding capacity and rheological properties of 10% sunflower wax oleogel were analyzed during a 30-day storage period at 20 °C [18]. Oil binding capacity of the sunflower wax oleogel remained steady, while storage (G′) and loss (G″) moduli increased during storage. It seemed that sunflower wax crystals underwent sintering between crystals via non-covalent bonds during storage which caused the increases in storage (G′) and loss (G″) moduli.

Table 16.2 Physical stability of wax-based oleogels during storage

Wax	Amount	Oil	Conditions	Method	Stability	References
Sunflower wax	3, 7, and 10%	Hazelnut oil	4 and 20 °C, 90 days	Firmness, adhesiveness	Stable, Slightly increased firmness with 10% sunflower wax	[16]
	3, 7, and 10%	Fish oil	4 and 20 °C, 90 days	Firmness	Stable	[17]
	5%	Rice bran oil	5 °C, 4 weeks	Rheological properties	Stable	[12]
	10%	Sunflower oil	20 °C, 30 days	Oil binding capacity, rheological properties	Steady oil binding capacity; increased G′ and G″	[18]
Carnauba wax	3, 7, and 10%	Virgin olive oil	4 and 20 °C, 90 days	Firmness, adhesiveness	Stable	[19]
	3, 7, and 10%	Hazelnut oil	4 and 20 °C, 90 days	Firmness, adhesiveness	Stable	[16]
	3, 7, and 10%	Cod liver oil	4 and 20 °C, 90 days	Firmness, adhesiveness	Stable, slight increase with 10% wax at 20 °C	[20]
Beeswax	5, 10, 15, and 20%	Fish oil	25 °C/60% RH, 6 months	Appearance	Stable up to 6 months	[21]
	5, 10, 15, and 20%	Fish oil	40 °C/75% RH, 6 months	Appearance	Stable up to 6 months for 10–20% wax; syneresis at 3 months for 5% wax	[21]
	5%	Fish oil	4 °C, 90 days	Firmness, cohesiveness, stickiness, adhesiveness	Increased gel strength at 90 days	[22]
	3, 7, and 10%	Cod liver oil	4 and 20 °C, 90 days	Firmness, adhesiveness	Stable	[20]
	10%	Sunflower oil	20 °C, 30 days	Oil binding capacity, rheological properties	Steady oil binding capacity; slightly decreased G′ and G″	[18]

(continued)

Table 16.2 (continued)

Wax	Amount	Oil	Conditions	Method	Stability	References
Rice bran wax	10%	Sunflower oil	20 °C, 30 days	Oil binding capacity, rheological properties	Slightly decreased oil binding capacity; steady G′ and G″	[18]
	7, 9, and 11% (crude)	Rice bran oil	20 °C, 30 days	Firmness	Steady or slightly increased firmness	[23]
Berry wax	5%	Rice bran oil	5 °C, 4 weeks	Rheological properties	Increased G′ and G″	[12]

% = wt.%

Carnauba wax oleogels Based on the three studies on carnauba wax oleogels in Table 16.2, these oleogels also seem to have stable physical properties when stored at refrigeration and room temperatures. Virgin olive oil oleogels with 3, 7, and 10% carnauba wax, which were examined with firmness and adhesiveness, were found to be stable during storage up to 90 days at 4 and 20 °C [19]. The firmness and adhesiveness of oleogels made with 3, 7, and 10% carnauba wax in hazelnut oil or cod liver oil were also found be stable during a storage period of 90 days at 4 and 20 °C [16, 20].

Beeswax oleogels Fish oil oleogels prepared with beeswax were examined for their syneresis during storage at 25 °C/60% relative humidity (RH) or at 40 °C/75% RH [21]. Fish oil oleogels with 10, 15, and 20% beeswax in plastic containers did not show syneresis for up to 6 months at 25 °C/60% RH and at 40 °C/75% RH. The fish oil oleogel with 5% beeswax was also stable up to 6 months at 25 °C/60%RH, but it exhibited syneresis at 3 months at 40 °C/75% RH. Since beeswax oleogels have relatively low melting points compared to other wax oleogels, and the lower concentration of wax results in lower melting point, it was thought that crystal network of the 5% beeswax oleogel might have been weakened at 40 °C. Oleogel with 5% beeswax in fish oil seemed to be slightly softer after 45 days at 4 °C, but it showed increased gel strength in firmness, cohesiveness, adhesiveness, and stickiness after 90 days [22]. Cod liver oil oleogels with 3, 7, and 10% beeswax had stable firmness and adhesiveness during storage at 4 and 20 °C for 90 days [20]. In a different study, 10% beeswax in sunflower oil oleogels had no change in oil binding capacity, but slightly decreased storage (G′) and loss (G″) moduli over 30 days at 20 °C [18].

Rice bran wax oleogels Oil binding capacity of oleogels made with 10% rice bran wax in sunflower oil decreased by 6% during a 30-day storage at 20 °C, while sunflower wax and beeswax oleogels did not have reduced oil binding capacity [18]. However, storage (G′) and loss (G″) moduli of the rice bran wax oleogels did not change. In another study [23], firmness of rice bran oil oleogels with 7, 9, and 11% crude rice bran wax was unchanged or slightly increased during storage at 20 ° C for 30 days.

Berry wax oleogels Berry wax is primarily composed of free fatty acids of which >80% are palmitic acid (C16) [24]. Gel strength assessed by rheological properties of 5% berry wax in rice bran oil oleogels increased during 1-month storage at 5 °C [12]. It was found that the microcrystalline berry wax particles aligned, reorganized, and the crystals also underwent polymorphic transition and Ostwald ripening during storage.

In general, most studies found that physical properties of wax-based oleogels were stable during storage with some exceptions with beeswax oleogels, of which the gel strength slightly decreased at 20 °C and showed syneresis at 40 °C/75% RH in 3 months. Some studies reported increased gel strength (increased firmness or rheology parameters) with sunflower, rice bran, and berry waxes after storage, while one study with rice bran wax oleogels reported their slightly decreased oil binding capacity. The increase in gel strength of wax-based oleogels was thought to be due to the reorganization of wax crystals [25] and the sintering between crystals via non-covalent bonds [12]. The increased gel strength could be problematic for the development of food products with these oleogels [1, 12]. However, this problem may be more easily solved and less problematic than the decreased gel strength and syneresis. For example, one possible solution is to facilitate sintering and reorganization of wax crystals by storing the oleogel at a slightly higher temperature than the storage temperature while oil oxidation is carefully prevented, which may lead to a faster stabilization of gel strength.

16.2.3 Physical Stability of Monoglyceride Oleogels

Table 16.3 summarizes studies of the physical stability of monoglyceride-based oleogels. Firmness and adhesiveness (stickiness) of oleogels made with 3, 7, and 10% monoglycerides in virgin olive oil decreased during storage at 4 and 20 °C over 90 days [19]. In another study, the oil binding capacity, firmness, and storage modulus (G') of the oleogel made with 6% monoglycerides in high oleic sunflower oil significantly decreased during storage for 8 weeks at 5 °C [15]. A study by Li et al. [26] also found that the firmness and oil binding capacity of high oleic sunflower oil oleogels formed with 7% glycerol monostearate, glycerol monobehenate, and their mixtures decreased during storage at room temperature

Table 16.3 Physical stability of monoglyceride oleogels during storage

Monoglycerides concentration	Oil	Conditions	Method	Stability	References
3, 7, and 10%	Virgin olive oil	4 and 20 °C, 90 days	Firmness, adhesiveness	Decreased firmness and adhesiveness	[19]
6.6%	High oleic sunflower oil	5 °C, 8 weeks	Oil binding capacity, rheological properties, textural properties	Significantly decreased firmness, G′, and G″; slightly decreased oil binding capacity	[15]
7%	High oleic sunflower oil	Room temp., 30 days	Firmness, oil binding capacity	Decreased firmness and oil binding capacity	[26]
10%	Sunflower oil	20 °C, 30 days	Oil binding capacity, rheological measurements	Steady oil binding capacity; decreased G′ and G″	[18]
1, 3, and 6%	Soybean oil	5 and 25 °C, 6 weeks	Firmness, adhesiveness	Stable	[27]
6%	Rapeseed oil	20 °C, 8 weeks	Oil binding capacity, rheological properties	Steady oil binding capacity and G′	[28]
3, 7, and 10%	Fish oil	4 and 20 °C, 90 days	Firmness	Stable	[17]
10, 15, and 20%	Olive oil	−20, 5, and 25 °C, 14 days	Firmness	All steady except for 15 and 20% oleogels at 5 and 25 °C	[14]

% = wt.%

for 30 days. Firmness of 7% glycerol monolaurate in high oleic sunflower oil oleogel also decreased from day 3 to 30, while the oil binding capacity was steady. Oil binding capacity of oleogels made with 10% monoglycerides in sunflower oil did not change after 30 days at 20 °C [18]. However, storage (G′) and loss (G″) moduli significantly decreased after 15 days storage.

The decrease in the strength of monoglyceride oleogels during storage is believed to be either due to the polymorphic transformation of monoglyceride crystals from the sub-α or α-form to β-polymorph [18, 26], or due to a decrease in the strength of the bonds that stabilize the microstructure of the crystalline network [15]. Giacomozzi et al. [15] also explained the decrease in gel strength of monoglyceride oleogels during storage by the decrease in the fractal dimension (D). Since the D-value indicates the homogeneity of the crystalline mass distribution, the decrease in D-value during storage indicated that monoglyceride crystals had more disordered molecular organization with less evenly distributed (or clustered) mass after storage.

On the other hand, some studies reported no changes to the physical properties of monoglyceride oleogels during storage. For example, the firmness and adhesiveness of soybean oil oleogels prepared with 1, 3, and 6% monoglycerides during storage at 5 and 25 °C for a period of 6 weeks did not significantly change [27]. Oil binding capacity and G′ of 6% monoglycerides-rapeseed oil oleogel were steady during storage at 20 °C for 8 weeks [28]. Firmness of oleogels with 3, 7, and 10% monoglycerides in fish oil was also steady for 90 days at 4 and 20 °C [17]. Zampouni et al. [14] observed that stability was affected by the concentration of monoglycerides as well as the storage temperature. The firmness of olive oil oleogels with 10, 15, and 20% monoglycerides was steady at −20 °C for 14 days. At 5 °C, the firmness of 15 and 20% monoglycerides oleogels decreased while that of 10% monoglycerides oleogel was unchanged, while at 25 °C, the firmness of these oleogels increased in 1–3 days, and then gradually decreased as the storage time progressed.

Studies to date have thus reported mixed results for the physical stability of monoglyceride oleogels. Commercial monoglycerides are made from different fats/oils and have different fatty acids. They often contain significant amounts of di- and triglycerides as well as free fatty acids since they are prepared by glycerolysis of fats/oils or by esterification of glycerin with fatty acids. For example, when three different commercial glyceryl monostearates were analyzed with proton nuclear magnetic resonance (NMR), it was found that all three samples contained triglycerides and two samples had a significant amount of diglycerides [29]. The study by Choi et al. [29] further showed that the differences in the commercial glyceryl monostearates had a significant impact on their gelling properties. Therefore, the mixed results in the physical stability of monoglyceride oleogels may be due to the different compositions of monoglycerides used. Since the gel strength of monoglyceride oleogels significantly decreased in several studies, the long-term physical stability should be carefully examined, and solutions for the trend toward decreased gel strength must be provided.

16.2.4 Physical Stability of Oleogels Made with Other Gelators

Physical stabilities of oleogels made with oleogelators other than wax or monoglycerides are summarized in Table 16.4.

Oleogels with ethylcellulose Oleogels formed with 10% ethylcellulose in extra virgin olive oil were examined under three different storage conditions, at 20 °C for 120 days, at 30 °C for 75 days, and at 40 °C for 35 days [30]. Oil binding capacity significantly decreased after 120 days at 20 °C and 75 days at 30 °C, while it was relatively stable at 40 °C for 35 days. Firmness also dramatically decreased after 120 days at 20 °C and 75 days at 30 °C, while it rather increased after 35 days at 40 °

Table 16.4 Physical stability of other oleogels during storage

Oleogelator (%)	Oil	Conditions	Method	Stability	References
Ethylcellulose (10%)	Extra virgin olive oil	20 °C, 120 days, 30 °C, 75 days, or 40 °C, 35 days	Oil binding capacity, firmness, and rheology	In general, the gel strength decreased except for the firmness at 40 °C, which was increased after 35 days.	[30]
Fatty alcohols (5%)	Peanut oil	20 °C, 14 and 30 days	Oil released, and back extrusion	Shorter alcohol: decreased gel strength Longer alcohols: relatively stable	[31]
Mixture of β-sitosterol and lecithin (16%)	Sunflower oil	15 °C, 60 days	Apparent viscosity	Increased viscosity	[10]
Mixture of β-sitosterol and lecithin (16%)	Soybean oil	Temperature not specified, 10 days	Rheological properties	Damaged gel network, became fluid-like	[9]
Mixture of β-sitosterol and lecithin (16%)	Peanut oil	Temperature not specified, 10 days	Rheological properties	Damaged gel network, but remained as gel-like solid	[9]
Mixture of β-sitosterol and γ-oryzanol (10%)	Sunflower oil	20 °C, 30 days	Oil binding capacity	Steady oil binding capacity	[18]
Sorbitan fatty acid ester (15, 20%)	Fish oil	25 °C/60% RH or 40 °C/75% RH, 6 months	Appearance	Stable at 25 °C/60% RH, Syneresis from 3 months at 40 °C/75% RH	[21]
Sorbitan monostearate (19%)	Sweet almond oil	20 °C, 40 °C, 3 months	Firmness, adhesiveness	Steady (or slight increase) firmness and adhesiveness at 20 °C; Markedly decreased firmness and adhesiveness at 40 °C	[3]

(continued)

Table 16.4 (continued)

Oleogelator (%)	Oil	Conditions	Method	Stability	References
Mixture of sorbitan fatty acid ester and cetearyl alcohol fatty acid (4–10%)	Virgin olive oil	25 °C, 90 days	Firmness, cohesiveness (work of shear)	Gradually decreased firmness and cohesiveness	[32]
Mixture of sorbitan tristearate and lecithin (6, 8, 10%)	Soybean oil	5 and 25 °C, 6 weeks	Firmness and adhesiveness	Steady at 25 °C; slightly increased firmness and adhesiveness at 5 °C	[27]

% = wt.%

C. Rheological parameters (G′ and G″) significantly decreased during the three different storage conditions indicating that, in general, the gel strength of ethylcellulose oleogels decreased over time.

Oleogels with fatty alcohols Oil release and gel strength were measured with back extrusion for oleogels of 5% fatty alcohol in peanut oil during storage at 20 °C for 14 and 30 days [31]. While longer fatty alcohols such as $C_{20}OH$ and $C_{22}OH$ showed negligible oil release, shorter fatty alcohols including $C_{16}OH$ and $C_{18}OH$ had significant oil release during storage. Samples prepared under a faster cooling rate showed higher oil release. The gel strength also showed similar trends that the strength of oleogels made with $C_{16}OH$ and $C_{18}OH$ decreased during the storage, while those with $C_{20}OH$ and $C_{22}OH$ relatively stable and that the increased cooling rate resulted in a faster reduction of gel strength.

Oleogels with the mixture of β-sitosterol and lecithin Apparent viscosity was used to evaluate the physical stability of the oleogel made with 16% β-sitosterol and lecithin in sunflower oil during storage at 15 °C for 60 days [10]. It was found that the viscosity increased during storage, which was thought to be due to the formation of junction zones which link the microplatelet units into a three-dimensional network. However, another study with 16% β-sitosterol-lecithin in soybean oil oleogel stored for 10 days reported that the gel network was damaged (G″ became close to G′), and the gels had fluid-like properties after storage [9]. The same study found that 16% β-sitosterol-lecithin oleogel in peanut oil also showed the damaged internal network structure (G′ and G″ decreased), but remained as gel-like solid after the storage.

Oleogels with the mixture of γ-oryzanol and β-sitosterol The only study of storage stability of oleogels made with a mixture of γ-oryzanol and β-sitosterol (10%) showed that the oil binding capacity of sunflower oil oleogels did not change after 30 days at 20 °C, indicating that this oleogel maintained physical stability under these storage conditions [18].

Sorbitan fatty acid ester oleogels Fish oil oleogels made with 15 or 20% sorbitan monostearate (Span 60) did not show syneresis after 6 months of storage at 25 °C/ 60% RH, but they were less stable at 40 °C/75% RH, showing syneresis starting after 3 months [21]. Sweet almond oil oleogel with 19% sorbitan monostearate showed no significant change in firmness and adhesiveness during storage at 20 °C for 3 months, but these values markedly decreased during the storage at 40 °C [3]. Cohesiveness (or work of shear) and firmness of virgin olive oil oleogels prepared with 4–10% mixture (50:50) of sorbitan esters of olive oil fatty acids and cetearyl alcohol esters with olive oil fatty acids gradually decreased during storage for 90 days at 25 °C [32].

Oleogels with the mixture of sorbitan tristearate and lecithin The firmness and adhesiveness of soybean oil oleogels prepared with 6, 8, and 10% mixture of sorbitan tristearate and lecithin (50:50) did not significantly change during storage at 25 °C for a period of 6 weeks, but these values slightly increased after 6 weeks at 5 °C [27].

16.2.5 Summary and Additional Information on Physical Stability of Oleogels

Wax-based oleogels had relatively stable physical properties during storage, while some showed increased gel strength. For monoglyceride-based oleogels, while some studies observed steady gel strength during storage, more studies found decreased gel strength. It is likely that the differences in the compositions of monoglycerides used are responsible for the different results. Other oleogels such as those made with a mixture of β-sitosterol and lecithin also provided mixed results. Discrepant results of these oleogels were thought to be due to different storage conditions, oleogelators, oils, sources of materials determining the exact composition, and preparation parameters such as cooling rate, shear rate, and temperature [2, 33–35].

As additional information, antioxidants that increase oxidative stability can also increase physical stability. For example, when an antioxidant, resveratrol, was added in oleogels made with the mixture of β-sitosterol and lecithin in peanut oil and soybean oil oleogels, the network structure of these oleogels was more stable than those without resveratrol during 10 days storage [9]. Some additives may positively affect the oleogel crystal network to improve physical stability. For examples, when 5% phytosterol was added in addition to 15% monoglycerides in olive oil oleogel, it was found that this oleogel had higher initial firmness than 20% monoglycerides oleogel and showed potential to have greater physical stability during storage at −20, 5, and 25 °C for 14 days [14]. It was postulated that the phytosterols prevented aggregation of the monoglycerides crystals, which allowed for an extended crystal network resulting in improved textural properties.

16.3 Oxidative Stability

Oxidation of oil and oil-containing food products may occur when they are in contact with air during manufacturing, transportation, cooking, and storage. Heat, light, prooxidants, and transition metals accelerate oil oxidation, and often antioxidants are added in addition to inherent antioxidants in oil to protect the nutritional value and maintain the quality of oil and oil-containing foods until they are consumed [36]. Many studies have demonstrated the potential use of oleogels with vegetable oils and omega-3 oils as healthy alternatives to highly saturated conventional fats in foods based on their physical properties. However, it is well known that these healthy oils rich in unsaturated fatty acids are much more prone to the oxidation than saturated fats. Oxidation of these fatty acids may produce off-flavors and odors, as well as destroy essential fatty acid and other sensitive bioactive lipids and nutrients. In addition, co-oxidation of non-lipid components such as proteins, carbohydrates, and vitamins can impact the texture, flavor, and overall quality of foods. Therefore, the oxidative stability of oleogels is an important property to examine.

As will be discussed in detail in later sections and summarized in Table 16.5, many studies have reported that oleogels had higher oxidative stability than the unstructured bulk oils used in their preparation. The mechanisms explaining the protective effects have not been fully elucidated, as most of the studies to date were designed to compare oxidative stability but not to determine underlying mechanisms of protection. However, in one study, Hwang et al. [37] demonstrated that linear diffusion of dye-containing oil through 3% wax oleogels held at 35 °C was in the order of beeswax ≈ rice bran wax > candelilla wax ≈ sunflower wax oleogels, and that higher diffusion rates correlated with higher rates of oxidation in oleogels at this temperature. In addition, 3% sunflower wax oleogels prepared at higher cooling rates (11 and 25 °C/min) were shown to oxidize more slowly than those prepared at lower cooling rates (5 °C/min). Since faster cooling rates were previously shown to result in sunflower wax oleogels with smaller crystal size, and greater crystal density and oleogel firmness [38], these results together suggest that denser wax crystalline networks may be providing a protective barrier to the immobilized oil droplets. The dye diffusion experiments demonstrated differences in how the wax network slowed the diffusion and mixing of oil, so in some oleogels, oxidized oil close to the surface of the oleogel, which will have prooxidant activity, is slower to mix with oil droplets further in the interior. It is possible that the barrier could slow the diffusion of oxygen and light as well, but this has not been tested. Thus, oleogelation may be protecting oil from oxidation in a similar way that encapsulation and nanoencapsulation are used, so the properties of the oleogelator could also have an impact on this protective effect, similar to the properties of encapsulation wall materials [39, 40].

However, in some cases, oleogels were less stable than the corresponding oil. For example, in the study by Hwang et al. [37] where all 3% rice bran wax, sunflower wax, beeswax, and candelilla wax oleogels slowed oxidation of fish oil at 35 °C, the

Table 16.5 Oxidative stability of oleogels during storage

Oleogelator	Amount	Oil	Conditions	Method	Stability	References
Beeswax	10%	Black cumin oil	35 °C, 8 weeks	PC, CDV	More stable than oil	[53]
	3%	Fish oil	35 and 50 °C, up to 56 days	PV, CDV, FAME	More stable than oil at 35 °C; Less stable than oil at 50 °C	[37]
	5%	Fish oil	4 °C, 90 days	PV	Less stable than oil	[22]
	3 and 6%	Sunflower oil	35 °C, 6 days	PV	Less stable than oil	[41]
	3, 7, and 10%	Cod liver oil	4 and 20 °C, 90 days	PV	Similar to oil	[20]
	3, 7, and 10%	Hazelnut oil	4 and 25 °C, 90 days	PV	Stable (PV lower than 0.7 mEq O_2/kg)	[65]
	3–7%	Camellia oil, soybean oil, sunflower oil, flaxseed oil	5 and 25 °C, 90 days	PV	At 5 °C: Slight increase in camellia and soybean oil oleogels; no increase with sunflower and flaxseed oil oleogels. At 25 °C: increased for all the samples	[66]
Rice bran wax	3%	Fish oil	35 and 50 °C, up to 56 days	PV, CDV, FAME	More stable than oil at 35 and 50 °C	[37]
	7, 9, and 11%	Rice bran oil	4 and 20 °C, 30 days	PV	Not compared with oil. More stable with more wax	[23]
Carnauba wax	7%	Black cumin oil	35 °C, 8 weeks	PC, CVD	More stable than oil	[53]
	3, 7, and 10%	Virgin olive oil	4 and 20 °C, 90 days	PV	Stable (slightly faster increase than commercial margarine)	[19]
	3, 7, and 10%	Hazelnut oil	4 and 20 °C, 90 days	PV	Similar to a commercial shortening	[16]

Sunflower wax	7%	Black cumin oil	35 °C, 8 weeks	PC, CDV	More stable than oil	[53]
	3, 7, and 10%	Hazelnut oil	4 and 20 °C, 90 days	PV	More stable than a commercial shortening	[16]
	3%	Fish oil	35 and 50 ° C, up to 56 days	PV, CDV, FAME	More stable than oil at 35 and 50 °C	[37]
Candelilla wax	3%	Fish oil	35 and 50 ° C, up to 56 days	PV, CDV, FAME	More stable than oil at 35 and 50 °C	[37]
Apple wax, orange wax	0.5–5%	Linseed oil	25 °C, 3 days	isothermal calorimetry	More stable than oil with >1% wax	[64]
Berry wax	6%	Flaxseed oil	25 °C, 30 days	PV	Similar to oil	[67]
Mixtures of rice bran wax, sunflower wax and beeswax	5%	Pumpkin seed oil, sunflower oil	Room temperature, 8 months	FAME, Volatiles	Similar to oil	[60]
Monoglycerides	7, 9, 10, and 12%	Amaranth oil	4 and 25 °C, 2 months	PV	More stable than oil at 25 °C; Similar to oil at 4 ° C	[69]
	6.6%	High oleic sunflower oil	5 °C, 8 weeks	PV	More stable than oil	[15]
	1, 3, 5, and 10%	Camellia oil	40 °C, 14 days	PV, TBARS	More stable than oil	[57]
	6%	Flaxseed oil	25 °C, 30 days	PV	Similar to oil	[67]

(continued)

Table 16.5 (continued)

Oleogelator	Amount	Oil	Conditions	Method	Stability	References
	3, 7, and 10%	Hazelnut oil	4 and 25 °C, 90 days	PV	Stable (PV lower than 0.7 mEq O_2/kg)	[65]
	3, 7, and 10%	Virgin olive oil	4 and 20 °C, 90 days	PV	Stable (PV lower than 2 mEq O_2/kg)	[19]
Ethylcellulose	2, 4, 6, and 8%	Soybean oil	45 °C, 5 days	PV	More stable than oil	[51]
	12%	Corn oil	37 °C, 12 days	PV	Less stable than oil	[70]
Mixture of sorbitan olivate and cetearyl olivate	4–10%	Olive oil	25 °C, 90 days	PV	More stable than oil	[32]

% = wt.%

same molten oleogels oxidized faster than fish oil at 90 °C, indicating that the waxes or minor components in the waxes had prooxidant activity when the wax oleogels were melted. While it seemed that the positive effect by oleogelation was in competition with the negative effect by the prooxidant activity of waxes, since these oleogels had higher oxidative stability than the bulk fish oil at 35 °C, the protection effect prevailed over the prooxidant effect of the oleogelators at this temperature. Another study reported that oleogels prepared with 3 and 6% beeswax in sunflower oil oxidized faster than sunflower oil even at 35 °C [41]. The study also examined the oxidation of molten oleogels at 90 °C with Oxitest method and found that molten oleogels oxidized faster than sunflower oil. Beeswax was reported to contain wax esters (58.0%), hydrocarbon (26.8%), free fatty acids (8.8%), and long chain alcohols (6.4%) [24]. One or more components in beeswax may have prooxidant activity. For example, it is well known that free fatty acids have prooxidant activity in oil [42, 43]. Sobolev et al. [44] separated wax component fractions of beeswax, and found that oleogels prepared with fractions containing monoesters and hydrocarbons were the most oxidatively stable at 35 °C, while the pure BW was the least stable, followed by fractions containing di- and tri-esters, fatty alcohols, and free fatty acids.

Another factor that may influence the outcome of oxidative stability studies is that many oleogel types require high heat to melt and disperse the oleogelator prior to cooling and gel formation. If care is not taken, this extended heating period can rapidly induce lipid oxidation, especially in sensitive oils such as those high in polyunsaturated fatty acids (PUFAs). To protect the oleogel quality, and especially for oxidative stability studies, heating and stirring steps should be conducted with an inert gas such as nitrogen or argon in the headspace to eliminate air, and under limited light exposure to prevent photooxidation. In addition, all treatment comparisons including the bulk oils that are often used as the "control" should undergo the same conditions. Many studies omit this information about sample treatment and preparation, which is critical to evaluating the quality of an oxidative stability study and for interpreting the results.

16.3.1 Analytical Methods to Examine the Oxidative Stability of Oleogels

The same analytical methods developed for the assessment of lipid oxidation in bulk oils and food products can also be used for oleogels, but some caution must be taken in interpreting the results of some accelerated methods where the temperature is higher than the melting point of oleogel. For example, the experimental temperatures of the standard methods such as Rancimat test (AOCS Method Cd 12–57) and Oxitest (AOCS Method Cd-12c-16) may be higher than the melting point of the oleogel to be tested. While 3% beeswax-fish oil oleogel showed greater oxidative stability than oil at 35 °C, it oxidized faster than oil at 50 °C [37]. The peak melting point of the beeswax oleogel was 50.1 °C, and it seemed some beeswax crystals in the oleogel melted at 50 °C to negate the protection effect by oleogelation, while the prooxidant activity of beeswax (or its components) prevailed. Although it may take a long time, it is ideal to use the actual storage temperatures rather than accelerated experiments unless the study is designed to understand the oxidation of a molten oleogel. Since the major purpose of this chapter is to compare the oxidative stabilities of oleogels in solid-like state, the studies conducted at a temperature higher than the melting point of oleogels are not included in this chapter. In addition, oxidative stability testing should be carried out using the principles of testing oxidative stability tests in bulk oils, emulsions, and other food products in that the starting oil and oleogelator should be as fresh as possible, should not contain high levels of lipid oxidation products or prooxidants, and using standardized temperatures, sampling times, and analytical procedures to accurately compare oxidation induction times and rates of oxidation in early and late stages [45] (AOCS Recommended Practice Cg 5-97; AOCS Analytical Guidelines Cg 3-91).

Peroxide value (PV) The complex oxidation reactions of oil start with the formation of the major primary oxidation products, hydroperoxides, which further decompose and undergo other reactions to produce many different secondary products such as aldehydes, ketones, alcohols, dimers, and polymers. Therefore, the determination of PV is the most widely used method for the oxidative stability of oleogels [44]. The PV is typically measured by titration to determine the amount of iodine, which is formed by the reaction of hydroperoxide with the iodide ion (AOCS method, Ja 8-87) [46] or by spectrometric measurement of the iron-thiocyanate complex that is formed when ferrous ion (Fe^{2+}) is oxidized to ferric ions (Fe^{3+}) by hydroperoxide [47]. However, it should be noted that hydroperoxides are unstable and subject to decomposition to secondary oxidation products, and one of the problems of the PV method is that the value initially increases and then decreases with time as shown Fig. 16.1 [48].

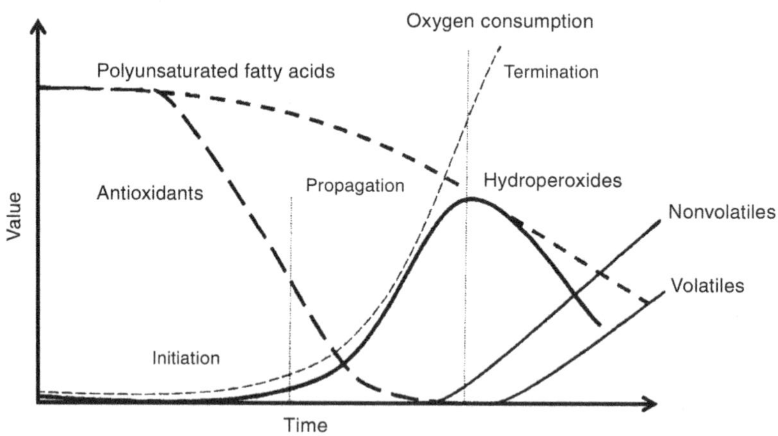

Fig. 16.1 Changes in concentrations of oxidation products and antioxidants by time. (Reproduced from [48], under the terms of the Creative Commons CC BY NC DD 4.0 license)

Therefore, it is recommended to use the PV to assess the early stages of lipid oxidation, not to use the value beyond the peak, and not to use the measurement of PV at one time point to assess the oxidative stability. Furthermore, since the oil oxidation comprises of many different reactions leading to many different oxidation products, it is highly recommended to use at least two analytical methods, preferably, both primary and secondary oxidation products to accurately evaluate the oxidative stability [48, 49]. For this reason, recently, more and more studies have used two or more analytical methods. For example, Park et al. [50] measured secondary lipid oxidation products by the 2-thiobarbituric acid-reactive substances (TBARS) assay and volatile oxidation products such as hexanal and nonanal. Fu et al. [51] used anisidine value (AV) and TBARS assay, and Hwang et al. [37] used conjugated diene value (CDV) and changes in fatty acid profiles in addition to PV.

Conjugated diene value (CDV) CDV is another measure of primary oxidation products, which can be determined spectrophotometrically at 233–235 nm [52]. This method is also used for the determination of oleogel oxidation [53].

Total oxidation values (TOTOX) The TOTOX value is obtained from PV and anisidine value (AV) (TOTOX = 2PV + AV). AV is determined by the concentration of the reaction products between the reagent, *p*-anisidine, and aldehydes formed by oxidation of oil, which is determined spectrophotometrically at 350 nm. TOTOX also has been used for the determination of the oxidative stability of oleogels [51].

2-Thiobarbituric acid-reactive substances (TBARS) Malondialdehyde (MDA) is an oxidation product of oil and reacts with the reagent, thiobarbituric acid (TBA), forming a pink chromogen, which is spectrophotometrically measured at 532–535 nm [54]. This assay has been used to determine the oxidation of oleogels such as oleogels formed with beeswax in high oleic sunflower oil oleogels [55], those of ethylcellulose and beeswax in a mixture of oils [56], and glycerol monolaurate camellia oil-based oleogels [57].

Headspace oxygen consumption Oxygen is consumed as the oxidation proceeds. The consumption of headspace oxygen during oxidation of oil, which is determined using a gas chromatography, has been used to determine oxidation of oils and oleogels [51].

^{1}H NMR The proton NMR method is a relatively new method for the assessment of lipid oxidation, which has been proven to be reliable and convenient. One disadvantage is that compounds at very low concentrations cannot be detected. Signals of oxidation products such as conjugated dienes and aldehydes are shown in the ^{1}H NMR spectrum at 5–7 and 9–10 ppm, respectively, of which the intensities increase as the oxidation proceeds [58]. This method was also used for ethylcellulose-soybean oil oleogels [51].

Headspace volatiles A variety of volatiles are formed in the headspace of an oil sample as oxidation progresses, especially in secondary lipid oxidation. Potent odorants associated with lipid oxidation such as 4-heptenal, 3-methyl-1-butanol, 2-hexenal, and 1-penten-3-ol [59] as well as some other aldehydes such as pentanal, hexanal, *t*-2-pentenal, *t*-2-heptenal, and *t*-2-octenal [60, 61] are used to monitor the oxidation of oil. These volatiles are often determined by gas chromatography (GC). Solid-phase micro-extraction (SPME) fiber and an auto-sampler are often used with GC to make the analysis faster and easier. Hexanal levels in pumpkin seed oil and sunflower oil oleogels formed with ternary mixtures of rice bran wax, sunflower wax, and beeswax were used to evaluate the oxidative stability of these oleogels [60].

Sensory analysis Sensory evaluation is useful for correlating changes in chemical markers of oil oxidation, especially secondary lipid oxidation products, with the formation of off-flavors and odors (AOCS Recommended Practice Cg 2-83). Sensory evaluation is especially important in the evaluation of food products, because small but statistically significant differences in chemical markers are not as important to food quality as whether a consumer can detect off-flavors, odors, or changes in appearance or texture [62]. None of the studies with pure oleogels summarized in this chapter have included sensory evaluation. However, there are additional studies in the literature, not covered in this chapter, where oleogels are incorporated into food products and sensory evaluation was included as part of the storage analysis, such as described by da Silva et al. [63].

Fatty acid methyl ester (FAME) analysis Oil oxidation causes the decrease in the concentration of unsaturated fatty acids resulting in changes in the fatty acids composition of oil, which can be monitored by the FAME analysis. In the FAME analysis, first, the oil sample (triglycerides) is converted to FAMEs by the interesterification with a methylation agent, and then the concentrations of FAMEs are determined by GC. Oxidation of pumpkin seed oil and sunflower oil oleogels prepared with ternary mixtures of rice bran wax, sunflower wax, and beeswax was evaluated with the changes in fatty acid composition during the storage at room temperature for 8 months [60].

Isothermal calorimetry Lipid oxidation is an exothermic reaction, and therefore, isothermal calorimetry was found to be a convenient method to measure oxidation of oleogels [64]. In this method, induction times were measured to determine the oxidative stability of oleogels.

Table 16.5 summarizes studies of the oxidative stability of oleogels made with many kinds of waxes and their mixtures, monoglycerides, ethylcellulose, and the mixture of sorbitan ester and fatty acid ester. As shown in Table 16.5, all of these studies except for two studies used PV as the primary method to determine the oxidative stability, and many studies used PV alone, while some used one or more other analytical methods such as CDV, FAME analysis, headspace volatiles, and TBARS in addition to PV. Storage temperatures used were 4 or 5 °C for refrigeration storage, 20–25 °C as room temperature, and 35–50 °C to mimic the processing temperatures and high-end storage temperatures. The storage time ranged from 5 to 90 days.

16.3.2 Oxidative Stability of Wax-Based Oleogels

Beeswax oleogels Black cumin oil oleogels prepared with 10% beeswax had lower PV and CDV compared to the control, black cumin oil, during storage at 35 °C for 8 weeks [53]. Fish oil oleogels with 3% beeswax also had lower PV, CDV, and decomposition of fatty acids during storage at 35 °C [37]. However, fish oil oleogels with 5% beeswax had higher PV than bulk fish oil after 90 days at 4 °C [22]. In another study, oleogels with 3 and 6% beeswax in sunflower oil increased in PV faster than the bulk sunflower oil during storage at 35 °C for 6 days [41]. PV of oleogels with 3, 7, and 10% beeswax in hazelnut oil, a high oleic acid oil, after 90 days at 4 °C and 20 °C was slightly higher than a commercial spread subjected to the same storage conditions, but the PVs of all of the oleogel samples were less than 0.7 mEq O_2/kg after 90 days [65]. This study underscored the importance of using oils with high oxidative stability to prepare stable oleogels. The PVs of beeswax oleogels with different types of vegetable oil, including camellia oil, soybean oil, sunflower oil, or flaxseed oil, were evaluated during storage at 5 and 25 °C for 90 days [66]. This study also found that the oxidative stability of an oleogel was dependent on the oil used. At 5 °C, PV of camellia oil and soybean oil oleogels slightly increased, while PV of sunflower and flaxseed oil oleogels did not increase at 5 °C. However, PV increased in all the samples at 25 °C.

Rice bran wax oleogels Hwang et al. [37] found that 3% rice bran wax-fish oil oleogels had higher stability than bulk fish oil at 35 and 50 °C when monitored with PV, CDV, and FAME. PV of rice bran oil oleogels with 7, 9, and 11% crude or refined rice bran wax gradually increased during storage at 4 and 20 °C for 30 days [23]. It was found that the increase in PV was slower when the concentration of rice bran wax in oleogels increased from 7 to 11%. Although this study did not directly compare the oxidative stability of oleogels with that of oil, the increased oxidative stability with increased amount of oleogelator indirectly indicated the protective effect of oleogelation.

Carnauba wax oleogels Black cumin oil oleogel with 7% carnauba wax was found to be more stable than the bulk oil when the PV and CDV were compared during storage at 35 °C for 8 weeks [53]. PV of oleogels made with 3, 7, and 10% carnauba wax and virgin olive oil gradually increased for 90 days at 4 and 20 °C, which was shown to be similar or slightly faster than that of a commercial margarine [19]. PV of hazelnut oil oleogels prepared with 3, 7, and 10% carnauba wax wase not significantly different from that of the bulk hazelnut oil during storage at 4 and 20 °C in the dark for 90 days [16].

Sunflower wax oleogels PV and CDV of oleogel with 7% sunflower wax in black cumin oil were lower than those of oil during storage at 35 °C for 8 weeks [53]. PV of hazelnut oil oleogels prepared with 3, 7, and 10% sunflower wax was lower than that of a commercial shortening during storage at 4 and 20 °C in dark for 90 days [16]. Oleogels of 3% sunflower wax in fish oil had lower PV, CDV, and decreases in omega-3 PUFAs content at 35 and 50 °C compared to the bulk fish oil [37].

Other wax oleogels Hwang et al. [37] found that 3% candelilla wax-fish oil oleogel was more stable than bulk fish oil at 35 and 50 °C. Waxes (1–5%) extracted from apple and orange peels by using supercritical carbon dioxide also retarded oxidation of linseed oil at 25 °C [64]. Oleogel prepared with 6% berry wax in flaxseed oil had similar PV to oil at 25 °C for 30 days [67]. Oleogels formed with 5% ternary mixture of rice bran wax, sunflower wax, and beeswax in pumpkin seed oil and sunflower oil were found to have similar stability to oil at room temperature for 8 months when they were examined with FAME analysis and volatiles [60].

Comparative studies between wax-based oleogels In the study with black cumin oil oleogels prepared with 10% beeswax, 7% sunflower wax, and 7% carnauba wax, which were stored at 35 °C for up to 8 weeks [53], the order of the oxidative stability determined by PV was 7% sunflower wax oleogel > 7% carnauba wax oleogel > 10% beeswax oleogel > oil. The oxidative stability order by CDV was 10% beeswax oleogel > 7% sunflower oleogel > 7% carnauba wax oleogel > oil. In another study with fish oil oleogels containing 3% sunflower, rice bran, candelilla, and beeswaxes [37], rice bran wax oleogel had higher PV than the other three oleogels after storage at 35 °C for 7 days, as well as faster oxidation of omega-3 PUFAs up to 56 days at the same temperature. Meanwhile, candelilla wax oleogel had lower CDV than the other three oleogels over the storage period. All of the oleogels had better oxidative stability than the bulk oil under these conditions indicating that oleogelation effectively protected oil from oxidation. As shown in these studies, results on the oxidative stability of wax-based oleogels can vary with waxes, storage temperatures, storage times, and analytical methods. Since the different analytical methods showed different results, it is important to use at least two analytical methods not to lead to wrong conclusions on the oxidative stability of oleogels.

The overall results with wax-based oleogels indicated that these oleogels had higher oxidative stability than the corresponding oil as long as the temperature was not close to the melting point of oleogels. The higher oxidative stability of wax oleogels than oil might be attributed to immobilization of oil in the three-dimensional network of wax crystals which may provide protection from oxygen, light, and diffusion of prooxidant compounds or oxidized oil from the surface of oleogels. Some wax-based oleogel were reported to have higher or similar oxidative stability to commercial shortenings [16], while some had slightly faster oxidation than commercial margarine [19]. In addition, chicken samples fried in 5% carnauba wax-canola oil oleogel had significantly slower oxidation compared to chicken samples fried in canola oil during the storage for 8 days [68].

16.3.3 Oxidative Stability of Monoglyceride Oleogels

The oxidative stability of oleogels with 7, 9, 10, and 12% monoglycerides in amaranth oil was measured with PV during 2 months at 4 and 25 °C [69]. The results showed that all of the oleogels had higher oxidative stability than amaranth oil at 25 °C and similar stability with amaranth oil at 4 °C. The increase in PV of the oleogel with 6.6% monoglycerides in high oleic sunflower was significantly slower than that of oil during storage at 5 °C for 8 weeks [15]. PV and TBARS of camellia oil oleogels with 1, 3, 5, and 10% glycerol monolaurate were lower than oil during storage at 40 °C of 14 days [57]. A study with 6% glycerol monostearate in flaxseed oil oleogel found that PV of this oleogel was similar to that of oil during a 30-day storage at 25 °C [67].

Some studies also reported satisfactory oxidative stability of monoglyceride oleogels, although the direct comparison with oil was not made. For examples, Öğütcü and Yilmaz [19] found that PV of the oleogels prepared with 3, 7, and 10% carnauba wax in virgin olive oil stayed below 2 mEq O_2/kg during storage at 4 and 20 °C for 90 days indicating the good oxidative stability of these oleogels, although the increase in PV of these oleogels is somewhat faster than that of commercial margarine. Another study reported that PV of 3, 7, and 10% monoglycerides oleogels in hazelnut oil was slightly higher than that of a commercial spread sample during storage at 4 °C and 20 °C for 90 days, but all the oleogel samples had the PV lower than 0.7 mEq O_2/kg after 90 days [65]. Overall results indicated that monoglyceride-based oleogels also had higher stability than the corresponding oil indicating that the structure of monoglyceride oleogel is beneficial to oxidative stability of oil.

16.3.4 Oxidative Stability of Other Oleogels

Ethylcellulose oleogels Ethylcellulose oleogels showed mixed results. Soybean oil oleogels with 2, 4, 6, and 8% ethylcellulose had slower increase in PV than soybean oil during storage at 45 °C for 5 days [51]. However, in a study with 12% ethylcellulose in corn oil oleogel, it was found that the oleogel had faster increase in PV than oil [70]. On the other hand, emulsions prepared with ethylcellulose-corn oil oleogel had significantly slower oxidation than control emulsions made with corn oil during storage at 30 °C for 20 days [71]. The contradictory results may be attributed to different preparation conditions. Gravelle et al. [72] studied the effect of preparation conditions on the properties of oleogel. The study found that longer holding time at 140 °C during the preparation of ethylcellulose-canola oil oleogels produced significantly more primary and secondary oxidation products.

Sorbitan fatty acid ester oleogels Virgin olive oil oleogels formed with 4–10% 50: 50 mixture of sorbitan olivate (sorbitan esters with olive oil fatty acids) and cetearyl olivate (esters of cetearyl alcohol and olive oil fatty acids) had lower PV than the control, olive oil, during storage for 90 days at 25 °C [32].

16.3.5 Summary and Additional Information on the Oxidative Stability of Oleogels

In general, oleogels seems to have higher oxidative stability than bulk oil due to immobilization of oil within the network of oleogelators. In a few cases, oleogels oxidized faster than the bulk oil, which was thought to be due to the prooxidant activity of oleogelator or its minor components. However, most of the studies thus far have evaluated the oxidative stability of oleogels in comparison to the bulk oil, but there are fewer studies comparing oleogel stability to the saturated fats that they are meant to replace. For example, in one such study, oleogels with 2, 4, and 6% hydroxypropyl methylcellulose in canola oil were more stable than bulk canola oil during storage, but they had much lower oxidative stability than beef tallow [73]. In another study, beef heart patties manufactured with 10% beeswax rapeseed oil oleogel were less stable than patties made with beef fat during cold storage, as determined by PV, TBARS, and carbonyl content [74]. Beef burgers where 25 and 50% animal fat were replaced with 10% ethylcellulose in sesame oil oleogels had higher TBARS than beef patties prepared with animal fat during 3 months frozen storage [75]. Special attention should be made for the oxidative stability of oleogels containing PUFAs. Antioxidants can be used to increase the oxidative stability of oleogels [76], and increasing cooling rate during the preparation of oleogels can also increase the firmness and oxidative stability [37]. In addition, depending on the type of oleogels, constituents in the target food can affect the oxidative stability of oleogel. For example, it was reported that whey protein isolate-safflower oil oleogels had faster oxidation when the water content increased to 1, 2, and 3% when monitored with CDV [77].

16.4 Conclusions

In general, wax-based oleogels had satisfactory physical stability during storage and had better oxidative stability than the bulk oils used in their preparation. Monoglyceride oleogels seem to have higher oxidative stability than oil, but they decreased in gel strength during storage in many cases. For other types of oleogelators, discrepant results have been reported for their physical and oxidative stability, and therefore more studies are needed. In addition, there is a lack of sensory evaluation of changes to oleogels or foods containing oleogels during storage, and this deficit needs to be addressed to determine the effect of changes to physical and oxidative stability on the sensory characteristics and acceptability. Since the physical and oxidative stabilities of a specific oleogel are affected by many factors, better understanding of these factors will enable the development of production conditions that optimize for physical and oxidative stability of each oleogel type. In addition, researchers should continue to search for oleogelators and oils to produce oleogels with better stability during storage. Test conditions are recommended to be similar to actual storage conditions that are expected to be encountered in its intended application. For oxidative stability studies, it is recommended to use more than one analytical method for more accurate assessment of oxidation. Efforts should be made to improve the oxidative stability of oleogel-containing foods by optimizing antioxidants and oleogel preparation to achieve higher stability.

References

1. Johansson D, Bergenståhl B (1995) Sintering of fat crystal networks in oil during post-crystallization processes. J Am Oil Chem Soc 72(8):911–920. https://doi.org/10.1007/BF02542069
2. Doan CD, Tavernier I, Okuro PK et al (2018) Internal and external factors affecting the crystallization, gelation and applicability of wax-based oleogels in food industry. Innov Food Sci Emerg Technol 45:42–52. https://doi.org/10.1016/j.ifset.2017.09.023
3. Almeida IF, Bahia MF (2006) Evaluation of the physical stability of two oleogels. Int J Pharm 327(1):73–77. https://doi.org/10.1016/j.ijpharm.2006.07.036
4. Hurler J, Engesland A, Poorahmary Kermany B et al (2012) Improved texture analysis for hydrogel characterization: gel cohesiveness, adhesiveness, and hardness. J Appl Polym Sci 125(1):180–188. https://doi.org/10.1002/app.35414
5. Bourne MC, Kenny JF, Barnard J (1978) Computer-assisted readout of data from texture profile analysis curve. J Texture Stud 9(4):481–494. https://doi.org/10.1111/j.1745-4603.1978.tb01219.x
6. Hwang H-S, Kim S, Winkler-Moser JK et al (2022) Feasibility of hemp seed oil oleogels structured with natural wax as solid fat replacement in margarine. J Am Oil Chem Soc 99(11):1055–1070. https://doi.org/10.1002/aocs.12619
7. Da Pieve S, Calligaris S, Co E et al (2010) Shear nanostructuring of monoglyceride organogels. Food Biophys 5(3):211–217. https://doi.org/10.1007/s11483-010-9162-3
8. Blake AI, Co ED, Marangoni AG (2014) Structure and physical properties of plant wax crystal networks and their relationship to oil binding capacity. J Am Oil Chem Soc 91(6):885–903. https://doi.org/10.1007/s11746-014-2435-0
9. Qu K, Qiu H, Zhang H et al (2022) Characterization of physically stable oleogels transporting active substances rich in resveratrol. Food Biosci 49:101830. https://doi.org/10.1016/j.fbio.2022.101830
10. Wan WB, Han LJ, Liu GQ et al (2014) Effect of storage conditions on apparent viscosity of oleogel developed by β-sitosterol and lecithin with sunflower oil. In: Advanced materials research, pp 903–907. https://doi.org/10.4028/www.scientific.net/AMR.1004-1005.903
11. Ojijo NKO, Kesselman E, Shuster V et al (2004) Changes in microstructural, thermal, and rheological properties of olive oil/monoglyceride networks during storage. Food Res Int 37(4):385–393. https://doi.org/10.1016/j.foodres.2004.02.003
12. Doan CD, Tavernier I, Sintang MDB et al (2017) Crystallization and gelation behavior of low- and high melting waxes in rice bran oil: a case-study on berry wax and sunflower wax. Food Biophys 12(1):97–108. https://doi.org/10.1007/s11483-016-9467-y
13. Ramírez-Gómez NO, Acevedo NC, Toro-Vázquez JF et al (2016) Phase behavior, structure and rheology of candelilla wax/fully hydrogenated soybean oil mixtures with and without vegetable oil. Food Res Int 89:828–837. https://doi.org/10.1016/j.foodres.2016.10.025
14. Zampouni K, Soniadis A, Moschakis T et al (2022) Crystalline microstructure and physico-chemical properties of olive oil oleogels formulated with monoglycerides and phytosterols. LWT 154:112815. https://doi.org/10.1016/j.lwt.2021.112815

15. Giacomozzi AS, Carrín ME, Palla CA (2021) Storage stability of oleogels made from monoglycerides and high oleic sunflower oil. Food Biophys 16(3):306–316. https://doi.org/10.1007/s11483-020-09661-9
16. Ögütcü M, Yilmaz E (2015) Characterization of hazelnut oil oleogels prepared with sunflower and carnauba waxes. Int J Food Prop 18(8):1741–1755. https://doi.org/10.1080/10942912.2014.933352
17. Öğütcü M, Temizkan R, Arifoğlu N et al (2015) Structure and stability of fish oil organogels prepared with sunflower wax and monoglyceride. J Oleo Sci 64(7):713–720. https://doi.org/10.5650/jos.ess15053
18. Fayaz G, Calligaris S, Nicoli MC (2020) Comparative study on the ability of different oleogelators to structure sunflower oil. Food Biophys 15(1):42–49. https://doi.org/10.1007/s11483-019-09597-9
19. Öğütcü M, Yilmaz E (2014) Oleogels of virgin olive oil with carnauba wax and monoglyceride as spreadable products. Grasas Aceites 65(3):e040. https://doi.org/10.3989/gya.0349141
20. Öğütcü M, Arifoğlu N, Yilmaz E (2015) Storage stability of cod liver oil organogels formed with beeswax and carnauba wax. Int J Food Sci Technol 50(2):404–412. https://doi.org/10.1111/ijfs.12612
21. Daman Huri MF, Ng SF, Zulfakar MH (2013) Fish oil-based oleogels: Physicochemicals characterisation and in vitro release of betamethasone dipropionate. Int J Pharm Pharm 5(3): 458–467
22. Yılmaz E, Öğütcü M, Arifoglu N (2015) Assessment of thermal and textural characteristics and consumer preferences of lemon and strawberry flavored fish oil organogels. J Oleo Sci 64(10): 1049–1056. https://doi.org/10.5650/jos.ess15113
23. Wang N, Chen J, Zhou Q et al (2021) Crude wax extracted from rice bran oil improves oleogel properties and oxidative stability. Eur J Lipid Sci Technol 123(6):2000091. https://doi.org/10.1002/ejlt.202000091
24. Doan CD, To CM, De Vrieze M et al (2017) Chemical profiling of the major components in natural waxes to elucidate their role in liquid oil structuring. Food Chem 214:717–725. https://doi.org/10.1016/j.foodchem.2016.07.123
25. Fayaz G, Goli SAH, Kadivar M et al (2017) Potential application of pomegranate seed oil oleogels based on monoglycerides, beeswax and propolis wax as partial substitutes of palm oil in functional chocolate spread. LWT 86:523–529. https://doi.org/10.1016/j.lwt.2017.08.036
26. Li J, Guo R, Bi Y et al (2021) Comprehensive evaluation of saturated monoglycerides for the forming of oleogels. LWT 151:112061. https://doi.org/10.1016/j.lwt.2021.112061
27. Si H, Cheong LZ, Huang J et al (2016) Physical properties of soybean oleogels and oil migration evaluation in model praline system. J Am Oil Chem Soc 93(8):1075–1084. https://doi.org/10.1007/s11746-016-2846-1
28. Rondou K, De Witte F, Rimaux T et al (2022) Multiscale analysis of monoglyceride oleogels during storage. J Am Oil Chem Soc 99:1019–1031. https://doi.org/10.1002/aocs.12645
29. Choi KO, Hwang HS, Jeong S et al (2020) The thermal, rheological, and structural characterization of grapeseed oil oleogels structured with binary blends of oleogelator. J Food Sci 85(10): 3432–3441. https://doi.org/10.1111/1750-3841.15442
30. Alongi M, Lucci P, Clodoveo ML et al (2022) Oleogelation of extra virgin olive oil by different oleogelators affects the physical properties and the stability of bioactive compounds. Food Chem 368:130779. https://doi.org/10.1016/j.foodchem.2021.130779
31. Valoppi F, Calligaris S, Marangoni AG (2017) Structure and physical properties of oleogels containing peanut oil and saturated fatty alcohols. Eur J Lipid Sci Technol 119(5):1600252. https://doi.org/10.1002/ejlt.201600252
32. Erinç H, Okur I (2021) Determination of physical and chemical properties of oleogels prepared with olive oil and olive-based emulsifier. J Food Process Preserv 45(6):e15545. https://doi.org/10.1111/jfpp.15545
33. Trujillo-Ramírez D, Lobato-Calleros C, Jaime Vernon-Carter E et al (2019) Cooling rate, sorbitan and glyceryl monostearate gelators elicit different microstructural, viscoelastic and

textural properties in chia seed oleogels. Food Res Int 119:829–838. https://doi.org/10.1016/j.foodres.2018.10.066
34. Pasquali RC, Sacco N, Bregni C (2010) Stability of lipogels with low molecular mass gelators and emollient oils. J Dispers Sci Technol 31(4):482–487. https://doi.org/10.1080/01932690903212263
35. Palla CA, Dominguez M, Carrín ME (2022) An overview of structure engineering to tailor the functionality of monoglyceride oleogels. Compr Rev Food Sci Food Saf 21(3):2587–2614. https://doi.org/10.1111/1541-4337.12930
36. Choe E, Min DB (2006) Mechanisms and factors for edible oil oxidation. Compr Rev Food Sci Food Saf 5(4):169–186. https://doi.org/10.1111/j.1541-4337.2006.00009.x
37. Hwang HS, Fhaner M, Winkler-Moser JK et al (2018) Oxidation of fish oil oleogels formed by natural waxes in comparison with bulk oil. Eur J Lipid Sci Technol 120(5):1700378. https://doi.org/10.1002/ejlt.201700378
38. Hwang HS, Kim S, Evans KO et al (2015) Morphology and networks of sunflower wax crystals in soybean oil organogel. Food Struc 5:10–20. https://doi.org/10.1016/j.foostr.2015.04.002
39. Jurić S, Jurić M, Siddique MAB et al (2022) Vegetable oils rich in polyunsaturated fatty acids: Nanoencapsulation methods and stability enhancement. Food Rev Int 38(1):32–69. https://doi.org/10.1080/87559129.2020.1717524
40. Machado M, Rodriguez-Alcalá LM, Gomes AM et al (2022) Vegetable oils oxidation: mechanisms, consequences and protective strategies. Food Rev Int. In press. https://doi.org/10.1080/87559129.2022.2026378
41. Frolova YV, Sobolev RV, Sarkisyan VA et al (2021) Approaches to study the oxidative stability of oleogels. IOP Conf Ser: Earth Environ Sci 677(3):032045. https://doi.org/10.1088/1755-1315/677/3/032045
42. Miyashita K, Takagi T (1986) Study on the oxidative rate and prooxidant activity of free fatty acids. J Am Oil Chem Soc 63(10):1380–1384. https://doi.org/10.1007/BF02679607
43. Frega N, Mozzon M, Lercker G (1999) Effects of free fatty acids on oxidative stability of vegetable oil. J Am Oil Chem Soc 76(3):325–329. https://doi.org/10.1007/s11746-999-0239-4
44. Sobolev R, Frolova Y, Sarkisyan V et al (2022) Effect of beeswax and combinations of its fractions on the oxidative stability of oleogels. Food Biosci 48:101744. https://doi.org/10.1016/j.fbio.2022.101744
45. Decker EA, Warner K, Richards MP et al (2005) Measuring antioxidant effectiveness in food. J Agric Food Chem 53(10):4303–4310. https://doi.org/10.1021/jf058012x
46. Ozogul Y, Yuvka İ, Ucar Y et al (2017) Evaluation of effects of nanoemulsion based on herb essential oils (rosemary, laurel, thyme and sage) on sensory, chemical and microbiological quality of rainbow trout (Oncorhynchus mykiss) fillets during ice storage. LWT 75:677–684. https://doi.org/10.1016/j.lwt.2016.10.009
47. Shantha NC, Decker EA (1994) Rapid, sensitive, iron-based spectrophotometric methods for determination of peroxide values of food lipids. J AOAC Int 77(2):421–424. https://doi.org/10.1093/jaoac/77.2.421
48. Pignitter M, Somoza V (2012) Critical evaluation of methods for the measurement of oxidative rancidity in vegetable oils. J Food Drug Anal 20(4):772–777. https://doi.org/10.38212/2224-6614.2024
49. Hwang H-S (2020) A critical review on structures, health effects, oxidative stability, and sensory properties of oleogels. Biocatal Agric Biotechnol 26:101657. https://doi.org/10.1016/j.bcab.2020.101657
50. Park C, Bemer HL, Maleky F (2018) Oxidative stability of rice bran wax oleogels and an oleogel cream cheese product. J Am Oil Chem Soc 95(10):1267–1275. https://doi.org/10.1002/aocs.12095
51. Fu H, Lo YM, Yan M et al (2020) Characterization of thermo-oxidative behavior of ethylcellulose oleogels. Food Chem 305:125470. https://doi.org/10.1016/j.foodchem.2019.125470

52. Corongiu FP, Milia A (1983) An improved and simple method for determining diene conjugation in autoxidized polyunsaturated fatty acids. Chem Biol Interact 44(3):289–297. https://doi.org/10.1016/0009-2797(83)90056-X
53. Orhan NO, Eroglu Z (2022) Structural characterization and oxidative stability of black cumin oil oleogels prepared with natural waxes. J Food Process Preserv In press. https://doi.org/10.1111/jfpp.17211
54. Kumar S, Krishna Chaitanya R, Preedy VR (2018) Chapter 20 - Assessment of antioxidant potential of dietary components. In: Preedy VR, Watson RR (eds) HIV/AIDS. Academic Press, pp 239–253. https://doi.org/10.1016/B978-0-12-809853-0.00020-1
55. Martins AJ, Cerqueira MA, Cunha RL et al (2017) Fortified beeswax oleogels: effect of β-carotene on the gel structure and oxidative stability. Food Funct 8(11):4241–4250. https://doi.org/10.1039/c7fo00953d
56. Gómez-Estaca J, Pintado T, Jiménez-Colmenero F et al (2019) Assessment of a healthy oil combination structured in ethyl cellulose and beeswax oleogels as animal fat replacers in low-fat, PUFA-enriched pork burgers. Food Bioprocess Technol 12(6):1068–1081. https://doi.org/10.1007/s11947-019-02281-3
57. Pan J, Tang L, Dong Q et al (2021) Effect of oleogelation on physical properties and oxidative stability of camellia oil-based oleogels and oleogel emulsions. Food Res Int 140:110057. https://doi.org/10.1016/j.foodres.2020.110057
58. Goicoechea E, Guillen MD (2010) Analysis of hydroperoxides, aldehydes and epoxides by 1H nuclear magnetic resonance in sunflower oil oxidized at 70 and 100 °C. J Agric Food Chem 58(10):6234–6245. https://doi.org/10.1021/jf1005337
59. Singh P, Singh TP, Gandhi N (2018) Prevention of lipid oxidation in muscle foods by milk proteins and peptides: a review. Food Rev Int 34(3):226–247. https://doi.org/10.1080/87559129.2016.1261297
60. Puşcaş A, Mureşan A, Socaci S et al (2022) Cold pressed pumpkin seed oil fatty acids, carotenoids, volatile compounds profiles and infrared fingerprints as affected by storage time and wax-based oleogelation. J Sci Food Agric 103(2):680–691. https://doi.org/10.1002/jsfa.12180
61. Hwang H-S, Winkler-Moser JK, Tisserat B et al (2021) Antioxidant activity of Osage orange extract in soybean oil and fish oil during storage. J Am Oil Chem Soc 98(1):73–87. https://doi.org/10.1002/aocs.12458
62. Malcolmson LJ, Winkler-Moser JK (2020) Flavor and sensory aspects. In: Bailey's industrial oil and fat products, pp 1–17. https://doi.org/10.1002/047167849X.bio032.pub2
63. da Silva TLT, Chaves KF, Fernandes GD et al (2018) Sensory and technological evaluation of margarines with reduced saturated fatty acid contents using oleogel technology. J Am Oil Chem Soc 95(6):673–685. https://doi.org/10.1002/aocs.12074
64. Valoppi F, Haman N, Ferrentino G et al (2020) Inhibition of lipid autoxidation by vegetable waxes. Food Funct 11(7):6215–6225. https://doi.org/10.1039/D0FO01022G
65. Yılmaz E, Öğütcü M (2014) Properties and stability of hazelnut oil organogels with beeswax and monoglyceride. J Am Oil Chem Soc 91(6):1007–1017. https://doi.org/10.1007/s11746-014-2434-1
66. Pang M, Shi Z, Lei Z et al (2020) Structure and thermal properties of beeswax-based oleogels with different types of vegetable oil. Grasas Aceites 71(4):e380. https://doi.org/10.3989/GYA.0806192
67. Barroso NG, Okuro PK, Ribeiro APB et al (2020) Tailoring properties of mixed-component oleogels: wax and monoglyceride interactions towards flaxseed oil structuring. Gels 6(1):5. https://doi.org/10.3390/gels6010005
68. Adrah K, Adegoke SC, Nowlin K et al (2022) Study of oleogel as a frying medium for deep-fried chicken. J Food Meas Charact 16(2):1114–1123. https://doi.org/10.1007/s11694-021-01237-6

69. Kamali E, Sahari MA, Barzegar M et al (2019) Novel oleogel formulation based on amaranth oil: physicochemical characterization. Food Sci Nutr 7(6):1986–1996. https://doi.org/10.1002/fsn3.1018
70. Liu N, Lu Y, Zhang Y et al (2020) Surfactant addition to modify the structures of ethylcellulose oleogels for higher solubility and stability of curcumin. Int J Biol Macromol 165:2286–2294. https://doi.org/10.1016/j.ijbiomac.2020.10.115
71. Wei F, Miao J, Tan H et al (2021) Oleogel-structured emulsion for enhanced oxidative stability of perilla oil: influence of crystal morphology and cooling temperature. LWT 139:110560. https://doi.org/10.1016/j.lwt.2020.110560
72. Gravelle AJ, Barbut S, Marangoni AG (2012) Ethylcellulose oleogels: manufacturing considerations and effects of oil oxidation. Food Res Int 48(2):578–583. https://doi.org/10.1016/j.foodres.2012.05.020
73. Oh I, Lee J, Lee HG et al (2019) Feasibility of hydroxypropyl methylcellulose oleogel as an animal fat replacer for meat patties. Food Res Int 122:566–572. https://doi.org/10.1016/j.foodres.2019.01.012
74. Gao Y, Li M, Zhang L et al (2021) Preparation of rapeseed oil oleogels based on beeswax and its application in beef heart patties to replace animal fat. LWT 149:111986. https://doi.org/10.1016/j.lwt.2021.111986
75. Moghtadaei M, Soltanizadeh N, Goli SAH et al (2021) Physicochemical properties of beef burger after partial incorporation of ethylcellulose oleogel instead of animal fat. J Food Sci Technol 58(12):4775–4784. https://doi.org/10.1007/s13197-021-04970-4
76. Jeong S, Lee S, Oh I (2021) Development of antioxidant-fortified oleogel and its application as a solid fat replacer to muffin. Foods 10(12):3059. https://doi.org/10.3390/foods10123059
77. Meissner PM, Keppler JK, Stöckmann H et al (2019) Influence of water addition on lipid oxidation in protein oleogels. Eur J Lipid Sci Technol 121(9):1800479. https://doi.org/10.1002/ejlt.201800479
78. Sengupta A, Chattopadhyay S, Ghosh M (2022) Innovative formulation of oleogels using bioactive compounds and sal starch and characterization of their products. Future Food: J Food Agric Soc 10(1):1–9. https://doi.org/10.17170/kobra-202110144899

Chapter 17
Oleogels for Delivery and Protection of Bioactive Molecules

Artur J. Martins, Buse N. Gürbüz, Mahnoor Ayub, Rui C. Pereira, Lorenzo M. Pastrana, and Miguel A. Cerqueira

Abbreviations

ALA	Alpha-linolenic acid
ALA	Eicosapentaenoic acid
CAGR	Compound annual growth rate
CHD	Coronary heart diseases
CP	Citrus pectin (CP)
CVD	Cardiovascular disease
DGLA	Dihomo-gamma-linolenic
DHA	Docosahexaenoic acid
GLA	Gamma-linolenic acid
GMS	Glycerol monostearate
HDL	High-density lipoprotein
HIU	High-intensity ultrasound
HPMC	Hydroxypropyl methylcellulose
LA	Linolenic acid
LDL	Low-density lipoprotein
MAG	Monoacylglycerol (MAGs)
OPE	Oleogel-in-water Pickering emulsion
RBX	Rice bran wax
TFA	Trans-fatty acids

Artur J. Martins, Buse N. Gürbüz, Mahnoor Ayub and Rui C. Pereira contributed equally with all other contributors.

A. J. Martins · B. N. Gürbüz · M. Ayub · R. C. Pereira · L. M. Pastrana · M. A. Cerqueira (✉)
International Iberian Nanotechnology Laboratory, Braga, Portugal
e-mail: artur.martins@inl.int; buse.gurbuz@inl.int; mahnoor.ayub@inl.int; rui.pereira@inl.int; lorenzo.pastrana@inl.int; miguel.cerqueira@inl.int

C. Andrea Palla, F. Valoppi (eds.), *Advances in Oleogel Development, Characterization, and Nutritional Aspects*,
https://doi.org/10.1007/978-3-031-46831-5_17

Tp-palmitate Tea polyphenol-palmitate
WHO World Health Organization

17.1 Introduction

In the last years, the consumption pattern of foods has been in constant shift due to environmental and societal particularities. There is a devastating cumulative incidence of non-communicable diet-related chronic diseases (e.g., obesity, diabetes, and cardiovascular disease) affecting the world population and to address this; it is necessary to implement strategies that foster a wide range of healthier eating patterns [1]. An abundant number of vitamins, phytochemicals, and peptides are present in a multitude of different food matrices and are pointed as being capable of reducing the incidence of said chronic diseases and malnutrition [2, 3]. Using natural compounds as ingredients for the preparation of functional foods is seen as a safe, feasible approach that can improve both health-related and organoleptic properties, thus increasing the consumer acceptance of said food products. Also, an increasing number of foods are now labelled and advertised (e.g., in the European Union) as bearing nutrition and health claims [4]. However, that can only come to reality if nutrients and bioactives are ingested in sufficient quantities during extended consumption periods [5]. To fulfil the goal of producing fortified foods, it is required that the amount of bioactive compounds is enough to accomplish its purpose. Several natural matrices have been studied as a source of bioactives within the development of functional foods [6]. All of this has established an inevitable challenge for the food industrial players to come up with the elaboration of novel fortified food products that could actually fit the consumer's demands on taste and functionality.

Worldwide data demonstrates a significant growth in the fortified or functional food market. In detail, the global functional food market size was valued at USD 305.4 billion in 2022, with the cardio-health sector representing the largest value share for 2022 and carotenoids being the ingredient with the higher expansion (CAGR—compound annual growth rate of 7.4%). Regarding the end product, the dairy segment led the race in terms of revenue in the year 2022. The prediction is for the global market to hit around USD 597.1 billion in 10 years, growing at a registered CAGR of 6.93% during this same period. Nowadays, the Asia-Pacific market is the largest one, with a revenue share of over 35.9% in 2022, followed by the North American region with a 27% revenue [7].

Aspects like sustainability, technological processing, and bioactive tailored bioavailability (e.g., personalized nutrition) are now shaping the evolving process of functional food design, towards the improvement in life quality and well-being [8]. Since food products have distinct bioactive compounds in their constitution (which nurse their own assorted biological properties) when specific foods are consumed together, there is a real possibility for nutrient interaction at the molecular level (i.e., phytosterols binding ability towards proteins and peptides). This could lead to inhibitory effects which would cause a different impact if certain foodstuffs were to be consumed exclusively [2]. With all that in mind, the dispute is aligned with the fabrication of food products with an enriched nutritive background, which is able to fit diverse food matrices while still being cost-competitive.

Some of the most interesting bioactive molecules that can be added to foods (and have been recently studied regarding their in-vitro bioaccessibility in different food sources) are vitamins D [9], curcumin [10], carotenoids [11–14], sterols [15], calcium [16], fatty acids [16, 17], polyphenols [18], and peptides [19–22]. Technologies targeting the formulation of foods, considering the bioactives incorporation, their stability, and delivery, are also essential to allow the highest efficacy when they are ingested by the consumer. Therefore, the chemical nature of the bioactives and the food matrix composition must always be considered [23]. Regarding bioactives' fate, there are several endogenous factors like oral bioavailability, solubilization of bioactives in gastrointestinal fluids, nutrient absorption at the epithelial level, and intra-cellular chemical and biochemical transformations. The nutrient bioavailability is also affected by exogenous aspects that compromise the bioactives' physicochemical properties such as food processing environment, storage conditions, and the food matrix [24].

In recent years, oleogels and oleogel-derived structures have emerged as a new and effective strategy for the production of soft-mater systems based on oil structuring techniques. A set of exploratory examples regarding oleogel application in novel food products, with scalability potential, has increased the overall global interest in these promising edible materials. Oleogel-based systems are being endorsed not only due to structural/mechanical functionalities but also in terms of their capacity to deliver bioactive molecules. They can be explored at the macro-, micro-, and nanoscales, stemming from diverse release behaviors and likewise distinctive applications. Besides their capability to incorporate lipophilic bioactives via oil phase affinity, certain oleogelators can deliver significant intrinsic bioactivity (e.g., phytosterols) like cholesterol-lowering properties [25]. Oleogel structures are commonly produced through direct dispersion procedures, where the oleogelator molecules are typically dissolved (in adequate amounts) in the lipid/vegetable solvent under specific conditions (i.e., above melting point) followed by a cooling step that will trigger the formation of a gelling network, responsible for bearing the formation of a solid-like assembly [26]. In the past few years, there has been an increased interest regarding the production and applicability of bigels or hybrid gels. These studies have explored the use of indirect routes towards the production of oleogel systems (e.g., recurring to the use of biopolymers) to induce the gelation of the single or both components (oil and aqueous phases) [26–29]. These bi-continuous systems are grounded on the physical oil entrapment that is prompted by the presence of an oil gelator or oleogelator, or even by the combination of different oleogelators that form a network with enough strength to induce solid-like properties to the developed systems [30, 31]. Therefore, different systems can be produced through distinct oleogelation routes. Employing the often called emulsification template method, it is possible to develop systems that simultaneously incorporate and deliver lipophilic and hydrophilic bioactive compounds. This chapter aims to shed light on the possibilities concerning the use of oleogels and oleogel-derived systems for the delivery and protection of bioactive molecules.

17.2 Nutritional and Health Aspects of Ingredients

17.2.1 Fatty Acids and Oils

Fatty acids are considered the building blocks of lipids giving diversity and specificity to fats and oils [32]. Fatty acids are defined as carboxylic acids obtained from the hydrolysis of triglycerides with long linear aliphatic chains which are either straight, branched, saturated, or unsaturated [33]. Each fatty acid varies in its chain length, the position of a double bond, and the degree of saturation where classification is based on the presence and absence of a double bond or the number of carbon atoms present in the alkyl chain length [34]. Depending on the number of carbon atoms in the alkyl chain length, fatty acids are divided into three categories: short-chain (2–4 carbon atoms), medium-chain (6–10 carbon atoms), and long-chain fatty acids (12–26 carbon atoms). They are classified as saturated (no double bond), monounsaturated (one double bond), and polyunsaturated (two or more double bonds) depending on the type of bond they have in their chains which are important for metabolic, energetic, and structural activities [35, 36].

Saturated fatty acids are usually solid at room temperature because of their tightly packaged structure and have a higher melting point than unsaturated fatty acids. The melting point of fatty acids depends on the length and degree of unsaturation of hydrocarbon chains. Therefore, saturated fatty acids ranging between 12 and 24 have a waxy consistency at ambient temperature. Contrary to this, unsaturated fatty acids are liquid at room temperature and possess a lower melting point. The difference lies in the different degrees of packing among fatty acid molecules. Palmitic acid, behenic acid, and stearic acid are examples of saturated fatty acids, whereas linolenic acid, linolenic acid, and oleic acid are examples of unsaturated fatty acids [34]. The chemical structures of saturated and unsaturated fatty acids are presented in Table 17.1.

Fatty acids can further be divided into two main categories, namely essential and non-essential fatty acids. Essential fatty acids are lipids that the body needs for the proper functioning of many physiological systems. These are called essential because of humans' inability to synthesize these fatty acids naturally; therefore, essential fatty acids should be included in the diet and thus meet biological demands [39]. Essential fatty acids can be divided into two main families, namely omega-3 and omega-6 fatty acids with lipid components present within each of its groups [40]. Omega-3 fatty acids include docosahexaenoic acid (DHA), alpha-linolenic acid (ALA), and eicosapentaenoic acid (EPA). The main sources to obtain omega-3 fatty acids include salmon, tuna, mackerel, sardines, and herring. However, omega-6 essential fatty acids include linolenic acid (LA), gamma-linolenic acid (GLA), dihomo-gamma-linolenic (DGLA), and arachidonic acid. Omega-6 dietary sources include palm oil, grapeseed oil, sunflower oil, soybean oil, poultry, nuts, and cereals [41].

Dietary fats (fats and oils) are an essential component of the human's diet and are a major source of energy. Dietary fats provide 9 kcal of energy in 1 g of fat which is

Table 17.1 Chemical structure and melting point of fatty acids

Fatty acid	Structure		Melting point (°C)
Saturated			
Lauric	$C_{12:0}$	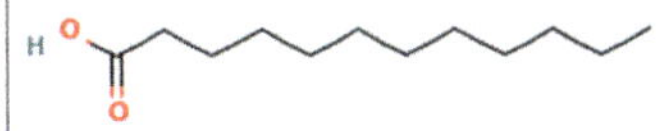	44
Palmitic	$C_{16:0}$	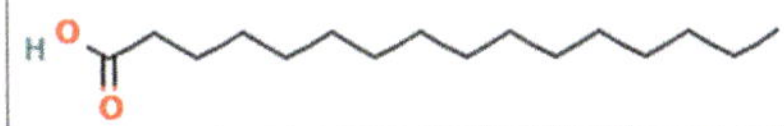	63
Stearic	$C_{18:0}$	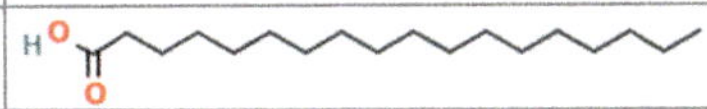	69
Behenic	$C_{22:0}$	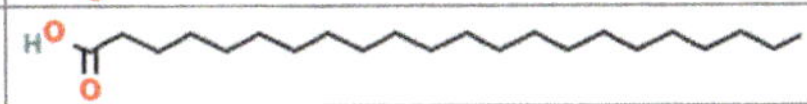	80
Unsaturated			
Oleic	$C_{18:1}$	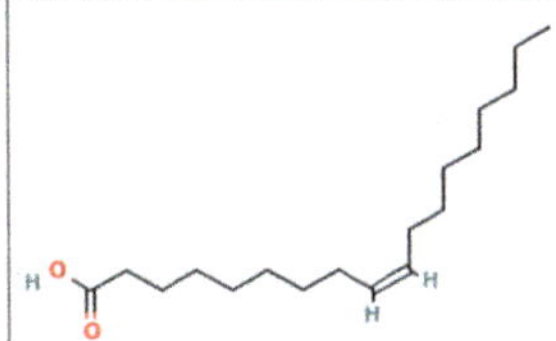	11
Linolenic	$C_{18:2}$	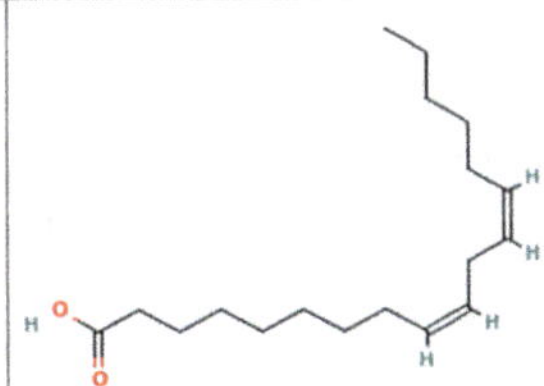	−5
Linolenic	$C_{18:3}$	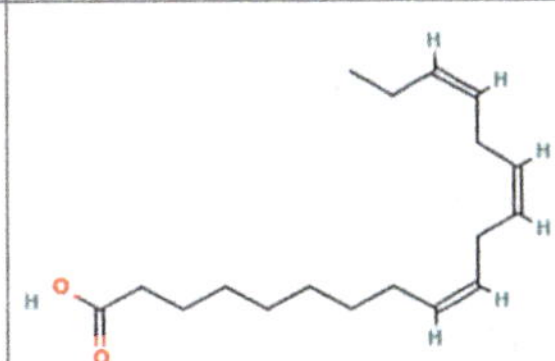	−11
Arachidonic	$C_{20:4}$	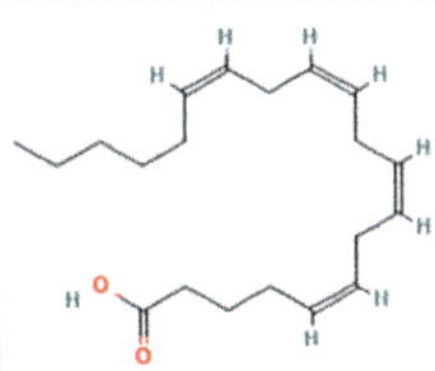	−50
Eicosapentenoic	$C_{20:5}$	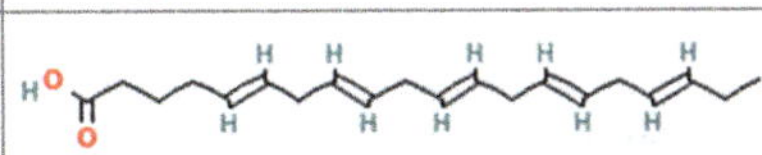	−54

(continued)

Table 17.1 (continued)

Fatty acid	Structure		Melting point (°C)
Docosahexenoic	$C_{22:6}$	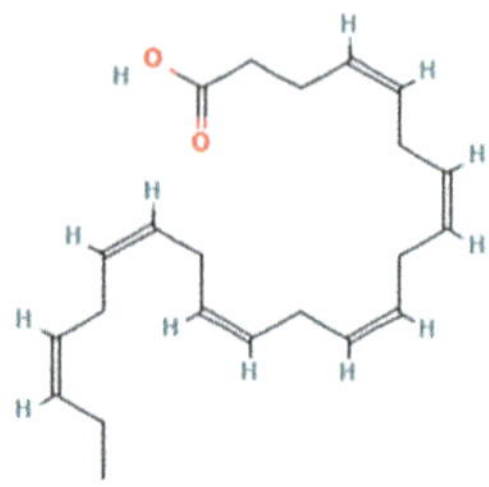	−44[a]

Sources: Trinick and Duly [37]

[a]Reference for DHA melting point is taken from Calder [38]. Chemical structure images obtained from https://pubchem.ncbi.nlm.nih.gov/

twice more than that contributed by carbohydrates and protein (4 kcal per gram) [42]. Dietary fats and oils are known for their good and bad characteristics. One of the main advantages of dietary fat is that it acts as a carrier of fat-soluble vitamins (A, D, E, and K). It also enhances the bioavailability of fat-soluble bioactive substances, e.g., the absorption of fat-soluble micronutrients. Fat is a crucial structural part of cell membranes and lipoprotein particles and supports cell function in the human body. It also provides insulation to the body against extreme temperature changes and pressure, which protects the organs [43]. Fat and oils also play a vital role in food by enhancing the texture of food and organoleptic properties [44]. In contrast, fats are also known for several disadvantages for the human body. In 2003, the WHO [45] confirmed the adverse effects of dietary *trans*-fatty acids (TFA) resulting in coronary heart diseases (CHD). The risk of suffering CHD is favored by an increase in LDL (low-density lipoprotein) and a reduction in HDL (high-density lipoprotein) which could promote inflammation and endothelial dysfunction. These series of events could then possibly affect the insulin resistance and displacement of essential fatty acids from the membranes which disturb other key membrane-related functions. It is further reported that TFA increases the risk of multiple CVDs (cardiovascular diseases) and heart-related issues [46].

17.2.2 Gelators

In general, the term "gel" refers to the state of matter in between solids and liquids; viscous than liquid and more elastic than solids. Hence, gelators illustrate a molecule that is either a polymer or a low molecular weight molecule forming a three-dimensional gel network in the solvent via a covalent bond or noncovalent interactions [47]. These gelators are widely used in food applications to transform liquid oils into solid fats without forming TFA and allow unsaturated fatty acids to solidify at mild temperatures. There are two types of oleogelators; low molecular weight oleogelators and high molecular weight oleogelators. Low molecular weight oleogelators consist of compounds having low molecular mass and the ability to self-assemble using Van der Waals forces, hydrophobic, and hydrogen bonds depending on shear forces and temperature. However, high molecular weight oleogelators form a three-dimensional structure to entrap oil by hydrogen bonding [48].

The low molecular weight gelators are widely used in the manufacturing of oleogels which includes waxes, β-Sitosterol, γ-oryzanol, ceramides, lecithin, fatty acids and alcohol, mono-acylglycerol [49], lecithin/phytosterols [50], and waxes/mono-acylglycerols [48, 51]. These gelators also could impart health benefits and structural stability to the food system. For example, phytosterols (β-sitosterol, and γ-oryzanol) are famous for their ability to lower blood cholesterol in the body [52]. Lecithin, on the other hand, can interact with other gelators, e.g., waxes as a co-gelator and can form direct dispersion improving the rheological characteristics of food [53].

Other compounds like antioxidants are also used to improve the quality of gels. Tocopherol (vitamin E) is an example of such a compound which also contributes against cardiovascular diseases [54]. Nikiforidis and Scholten [55] in their study used alpha-tocopherol and lecithin to alter the structural geometry of organogelation to supramolecular structures. They concluded that by combining alpha-tocopherol and lecithin, the organogel reached a thermo-reversible temperature of around 35 °C which makes the gel desirable to be used as crystalline-free edible lipid. Luo et al. [56] developed a camellia oil-based oleogel structure using tea polyphenol-palmitate (Tp-palmitate) at 2.5% (m/v) and citrus pectin at 1.5, 2.5, 3.5, and 4.5% (m/v) concentration using an emulsion method. The results depicted that polyphenol-rich oleogel imparts a high antioxidant activity with citrus pectin enhancing the oleogel strength and oil binding capacity of the gel.

Oleogels are now being widely used in many areas of the food industry such as meat, dairy, bakery, and confectionery as a healthier substitute to fats and oils to reduce or cut down saturated and *trans*-fats [48]. Therefore, different gelators are used for various food applications. For instance, various natural waxes have been used as oleogelators in pastries, ice cream, and meat-based products which include candelilla wax, carnauba wax, sunflower wax, rice bran wax, and beeswax [26, 28, 29, 48, 53]. Phytosterols-based gelators (β-sitosterol, γ-oryzanol) were reported to be used in the development of ice creams and meat products, shellac, and ethylcellulose for chocolates, and β-sitosterol, γ-oryzanol, and methylcellulose are also used in meat products as oleogelators [28]. With the help of oleogelators and their widespread use in food applications, solid fats can be replaced without compromising the food quality and organoleptic properties [48].

17.2.3 Other Ingredients with Health and Nutritional Claims

Oleogels can be used not only to mimic fats but are also a good matrix for the delivery of various bioactive compounds, with higher functionality, including vitamins, antioxidants, and peptides [53]. Bioactive compounds are usually sensitive to external factors, including heat, light, pH, and physiological conditions where oleogels can provide an interesting structured lipid system to safely protect and transport these bioactive compounds [57].

In this segment, curcuminoids have been studied greatly in oleogels [53]. Vellido-Pérez et al. [57] in their study worked on curcumin which is a natural polyphenolic compound having therapeutic properties such as anticancer, antibacterial, antioxidant, and anti-inflammatory actions. However, curcumin, being water insoluble and sensitive to external factors like light and heat, presents low bioavailability. Using an oleogel structure with curcumin, they were able to not only maximize the loaded curcumin content but also retarded the oil oxidation. Ferulic acid is known for its antioxidant and anti-inflammatory properties and is widely used in anti-ageing creams. In 2013, ferulic acid was incorporated in oleogels using olive oil as the medium to get its functional benefits [58]. Ferulic acid was introduced to an acidic ambient to mimic stomach conditions where oleogel was able to protect the nutraceutical functions of ferulic acid and also depicted great rheological properties. The results showed that oleogels can help to preserve and protect the functional properties of functional ingredients and the ingredients can be delivered to provide health benefits.

β-carotene is a type of carotenoid that is responsible for giving strong red-orange color to plants, acts as an antioxidant, and is also a precursor of vitamin A. Therefore, fortifying food products with β-carotene could be a viable substitute for vitamin A and its ability to be converted to vitamin A can expand its health benefits including improvement in the immune system, eye health, and cognitive performance [59]. Various studies reported the use of β-carotene, while using various types of oils with different types of gelators which resulted in different conclusions and uses. β-carotene with canola oil and ethylcellulose was reported to improve the stability of β-carotene and protect it against oxidation [60]. Corn oil with monoacylglycerols (MAGs) oleogel enhanced the light, heat stability, and solubility of β-carotene [61]. β-carotene is also reported to improve oleogel structure by improving the strength of the gel and its oil binding capacity [62].

Bei et al. [63] used nisin and D-limonene in an oleogel-based emulsion which effectively improved the antimicrobial properties for the production and supported the preservation of foods. D-limonene is hydrophobic in nature and prone to oxidative degradation which causes it to lose the lemon-like flavor. Therefore, incorporating such bioactive compounds in oleogel-based systems is suitable for the protection of the compound and could also provide a substitute as a natural food preservative [53].

Several other bioactive compounds are being used in oleogels for either strengthening the structure of oleogel or enhancing the bio-delivery of bioactive compounds.

For instance, tea polyphenols are reported to enhance the storage stability of oleogels and the antioxidant properties of edible oil [64]. Betulin, quercetin, hesperidin, capsaicin, lutein ester, phytosterols, and other volatile aromas are also used in various studies for different purposes in oleogel-based systems for the delivery of bioactive compounds [53].

17.3 Oleogels as a Delivery System

17.3.1 Strategies for the Incorporation of Bioactive Compounds

Bioactive compounds are characterized by being unstable under critical conditions, such as physiological or food processing conditions, so they must be incorporated into matrices that can protect and maintain their integrity [29]. Oleogels, due to their structure, constitute a suitable matrix for the delivery of bioactive compounds, by simultaneously protecting the bioactive molecules against oxidation and loss of functionality and, on the other hand, controlling their release at the intestinal level [53]. Chemical interactions between different minor compounds in oil can be complex and influence the stability of the oleogel and the bioactive compounds over time. In this line, according to the hydrophilic or hydrophobic nature of the bioactive compound of interest, the production methods and the type and quality of the oil can be adjusted to improve the stability of the molecule [29, 65].

Most of the studies rely on the integration of liposoluble compounds in edible oleogels, so the incorporation of hydrosoluble compounds in oil-structuring-based systems remains unexplored [66]. Oils rich in polyunsaturated fatty acid (PUFA) or organic solvents with shorter carbon-chain backbones should be selected to provide an important impact on lipid droplets or bioactive compound absorption at the intestinal level. Based on this, several methods can be used to impart a gel-like structure to edible oils, through different oleogelators, direct and indirect methods of production, and versatile emulsion and biphasic strategies (Fig. 17.1).

In the direct dispersion method, the oleogelator is dispersed into the oil in the liquid state at temperatures above the melting point (or glass transition temperature), followed by a cooling phase, to promote the development of the oleogelator network. Due to the hydrophobic medium, higher amounts of lipophilic bioactive compounds can be loaded onto oleogels [67]. Also, the gel network can improve the solubility of crystalline bioactives by altering their crystallization [68, 69]. Cui et al. [61] observed that the retention of β-carotene was higher in corn oil-based oleogels than in corn oil, mainly in higher concentrations of monoglyceride, when subjected to heat and UV light conditions. Likewise, similar findings were accessed by the development of curcumin-loaded oleogels structured by β-sitosterol and lecithin. Curcumin did not affect the mechanical strength and crystalline network of the oleogel. This crystalline network coupled with the antioxidant properties of

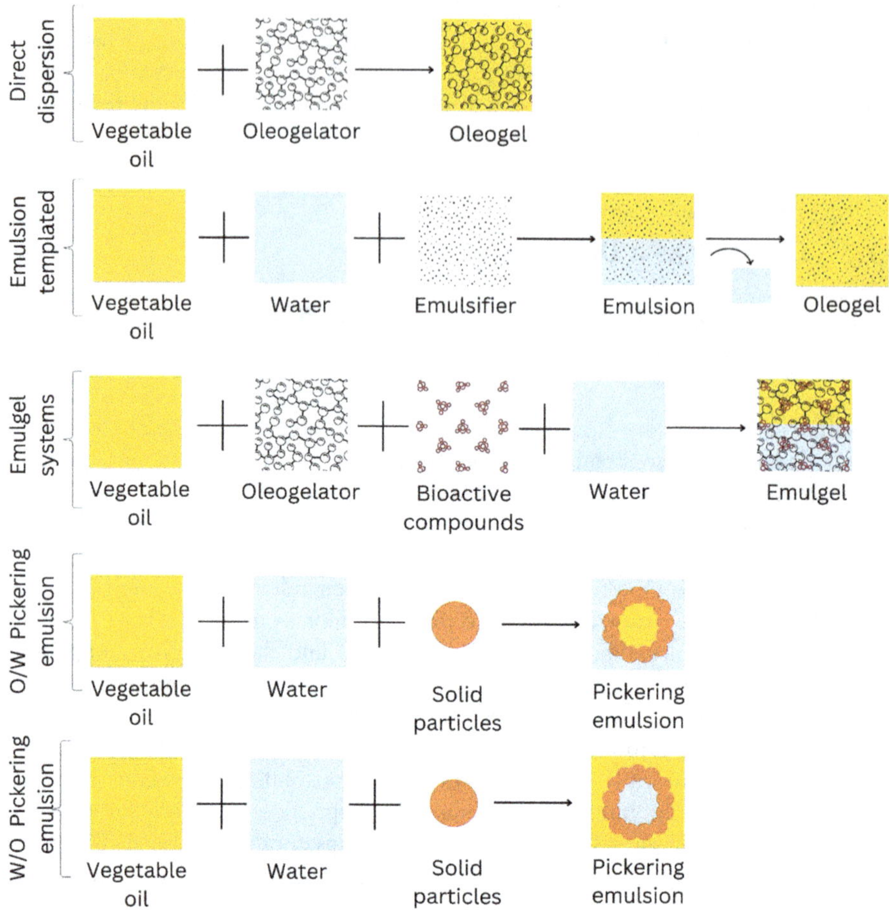

Fig. 17.1 Strategies for the development of lipid-based systems for the incorporation of bioactive compounds

curcumin reduced the development of oxidation products, showing higher oxidative stability of curcumin oleogel compared to curcumin-free oleogel [70].

To overcome the low solubility of hydrophobic compounds in the oil and the compatibility of oleogels with water-based food products, some emulsification strategies can be applied, by using low or high-energy methods such as ultrasonic treatment, high-pressure homogenization, and high-shearing [71, 72]. Indirect approaches to oil structuring led to the production of more complex structures, namely oleogel-based emulsions or Pickering emulsions, with the great advantage of including different types of oleogelators, such as proteins, polysaccharides, and polymers. Emulgel systems are a promising method for using both hydrophilic gelators and bioactive compounds that cannot be directly dispersed in the oil phase to achieve the necessary gel strength and structure. Moreover, in emulgels, the incorporated ingredients and compounds have limited mobility because of the compact gel network, which acts as a barrier against severe conditions. The main challenge in emulgel development is to protect the bioactive compounds against degradation when the gel structure starts to collapse during storage [53, 68]. Several studies have reported that the incorporation of bioactive compounds within emulsion gels improved their stability. A study showed the effect of ultrasonication processing parameters to produce oleogel-based nanoemulsions for encapsulation and release of curcumin. Curcumin was added to oleogel before the mixture of both phases. The nanoemulsions with the lowest oleogel/aqueous phase (5/95) were stable for 10 months, with higher encapsulation efficiency than the formulation without monoglyceride [73]. Another work reported that the chemical stability of β-carotene in oleogel-in-water Pickering emulsion (OPE) was higher than that in conventional Pickering emulsion (oil-in-water). The presence of cellulose nanocrystals improved the freezing-thawing and physical stability of OPE, which was able to maintain the β-carotene concentration for 15 days at room temperature, equally increasing its bioaccessibility [74]. Tea polyphenols are bioactive compounds with antioxidant properties and free-radical scavenging and exhibit low solubility in oils, so their use in lipid base food products or fat mimic systems is committed. Luo et al. [56] developed camellia oil-based oleogels by an emulsion-templated method. Camellia oil was structured with tea polyphenols and palmitate as oleogelators with increasing concentrations of citrus fiber. The emulsion was prepared using a high-shearing mechanical method, followed by a drying step. The tea polyphenol-based oleogels showed high anti-oxidative activity, where higher content in citrus pectin led to higher stability and viscoelasticity of the emulsions as well as strength of oleogels.

17.3.2 Oleogels as Delivery Systems

Delivery systems for food applications are used to improve bioavailability, and controlled release of bioactive components, while they show high compatibility with the food matrix without affecting any properties of the food product; likewise, they protect the selected compounds against critical conditions that can emerge during several steps of the delivery mechanism. There are two major groups for delivery systems in food products, (i) biopolymer-based delivery systems (e.g., hydrogels, protein-polysaccharide complex and polymeric micelle) and (ii) lipid-based delivery systems (e.g., emulsions, liposomes, solid lipid particles, nanostructured lipid carriers, self-dispersing lipid formulations) [75]. The selection of the delivery systems is dependent on several factors, such as the solubility of the bioactive compound. For example, lipid-based delivery systems are reported to have better functionality in the encapsulation of liposoluble active compounds. As they present low solubility in aqueous environments (typical of most foods and the human digestion process), lipid-based delivery systems have become a suitable option to increase the bioavailability of these components [76, 77].

The interest in using oleogels as delivery systems to regulate lipid digestion and delivery of nutrients and bioactive molecules has arisen likewise the increased number of studies for digestibility of the oleogel in the digestive tract. Slow release of lipids into the bloodstream can be achieved through oil-water-monoglyceride gels; therefore, post-prandial triglyceride, free fatty acid, and insulin levels can be reduced [78]. More recently, it was reported that the modification of gel strength of emulsified oil due to oleogelation could control lipid digestion improving the delivery of lipid-soluble molecules, considering their manageable structures and food-grade composition. Also, oleogels can increase the bioaccessibility of lipophilic molecules through the delivery of the molecules in soluble form and can provide additional protection benefits as well as the ability of controlled release [69]. Similarly, oleogels have been reported to increase the controlled release of lipid-soluble nutraceuticals while reducing the post-prandial plasma triglycerides and leading to a decrease in the adipose tissue accumulation [79].

Oleogels are complex systems that several important characteristics, such as gel strength and microstructure, can be changed mainly due to the different types of oleogelation methods (e.g., direct and indirect methods), origin and unsaturation level of vegetable oils, and the characteristics of the oleogelators (e.g., high and low molecular weight oleogelators). Furthermore, other structural properties of oleogels such as stiffness, porosity, permeability, water holding capacity, and swelling ratio should be considered while selecting the formulations and methods to obtain oleogels to ensure improved protection and controlled release of incorporated components. Due to the recent improvements in the food area, numerous formulations have been developed using a variety of oils and oleogelators. For instance, canola oil oleogels were prepared with different oleogelators (distilled monoglycerides (E471), β-sitosterol + γ-oryzanol, ethyl cellulose 20 cP, ethyl cellulose 45 cP) in different concentrations (8, 10, and 15%) and different oleogelation mechanisms (crystalline and polymer network). The authors explored the relation of mechanical, rheological properties, and rate of lipolysis under simulated intestinal conditions. The study concluded that there is a relation between rheological properties and different oleogelators and oleogelation mechanisms where the increase of oleogelator content led to stronger gels and storage modulus at gel state. Also, it was reported that increased concentrations of β-sitosterol + γ-oryzanol mixture in the oleogel showed a decreased rate of the free fatty acid release; however, other oleogels did not show significant differences in the rate of free fatty acid release for all concentrations [80]. For instance, in vitro digestive lipolysis of canola oil oleogels was explored to determine the impact of different oleogelators and oleogelation mechanism, and the rate of lipolysis and release of potential hydrophobic nutraceutical due to matrix breakdown can be affected by the nature of the structuring network, as well as the type and the concentration of the oleogelator. Likewise, Guo et al. [81] prepared oleogels with soybean oil with different concentrations of rice bran wax (RBX) as an oleogelator to determine the impact on in vitro intestinal lipid digestion and reported that the intestinal digestion of lipids can be delayed independently of the RBX concentration. However, the influence of the oleogelator should not be underestimated, since different quantities and types can affect the structure of the oleogel. For instance, the different concentrations of the glycerol monostearate (GMS) can influence the gel network, where the increased GMS concentration allows to produce of stronger gels, as Cui et al. [61] reported that increasing GMS concentrations also increased the retention ratio of β-carotene; thus, this oleogel provides chemical stability and improved solubility. Likewise, Calligaris et al. [82] prepared sunflower oil oleogels with different oleogelators to determine the effect on lipolysis and bioaccessibility of the curcuminoids by in vitro digestion and reported that the bioaccessibility of the curcuminoids is affected by oleogelator selected, as well as the structure and strength of the oleogel influence the lipid digestion and extended lipolysis besides bioaccessibility of the lipophilic bioactive compounds. Additionally, the nature and the size of the crystals of the oleogels are also affected by the different ratios and types of the oleogelators. The lower lipid digestion can occur due to self-assembled microstructures of oleogelators, where they can act as physical barriers and, consequently, improve extended digestion time and controlled release in the human gastrointestinal tract [83]. Considering the preparation methods of oleogels, high-intensity ultrasound (HIU) is becoming an important homogenizing technology and it can reduce the release of bioactive compounds in vitro intestinal digestion. A study confirmed that candelilla wax and nut oils (peanut, pine nut and walnut oil) oleogels enriched with β-carotene and treated with HIU showed reduced release of β-carotene in intestinal digestion [84]. This reduction may be caused due to the presence of crystals that provide a denser structure to the oleogel network; thus, these crystals can inhibit the oil release and the action of the enzyme on the oil network. Additionally, the novel technology of 3D printing was used to obtain oleogel as a co-delivery carrier for bioactive compounds of curcumin and resveratrol, through emulsion templated method using medium-chain oil and gelation and gellan gum [85]. Moreover, Kavimughil et al. [85] reported that oleogelators can play a significant role in lipase activity when considering the free fatty acids release profile, and the oleogel can increase the protection of the bioactive compounds under oral and gastric conditions due to in vitro results, thus they can be delivered to intestinal target safely beside the increased bioaccessibility of these bioactive compounds. Yet, Wang et al. [86] determined the stability and in vitro bioaccessibility of astaxanthin through the influence of self-assembled structures, using oleogels prepared with GMS and reported that improved stability of the astaxanthin is achieved by oleogels due to their compact network systems acting as physical barriers against critical environmental stress, as well as enhanced bioaccessibility.

17.3.3 Incorporation in Food Products

The increased concern for health leads to healthy dietary habits such as consuming less saturated fat since the excess consumption of them can cause the formation of atherosclerosis as well as increased coronary heart diseases [87]. However, fats are important components of a food product as they can affect the texture and flavor;

Table 17.2 Food products with oleogels with bioactive compounds and the corresponding oleogelator and oil type

Food product	Oleogelator and oil type	Bioactive compound	References
Pastries	GMS and candelilla wax; sunflower oil	β-carotene	[95]
	Zein; soybean oil	β-carotene	[99]
	Tea polyphenol-palmitate; Camellia-oil	Tea polyphenol-palmitate	[56]
	Carnauba wax, beeswax, β-sitosterol, GMS and soy lecithin; sunflower oil	β-sitosterol	[93]
	Chitosan, Tween®60 and vanillin; canola oil	Vanillin	[94]
	γ-oryzanol, β-sitosterol and sucrose stearate/ascorbyl palmitate; menhaden oil blended with caprylic acid or stearic acid	Sterols	[100]
Meat-based	Ethylcellulose or beeswax; olive, linseed and fish oils	Curcumin	[90]
	Cocoa bean shell flour; walnut oil	Polyphenolic compounds	[101]
	γ-oryzanol and β-sitosterol; linseed oil	Sterols	[91]
	Phytosterol and γ-oryzanol; sunflower oil	Sterols	[102]
	Monoglycerides and plant sterol mixture; sunflower oil	Sterols	[103]
	γ-oryzanol and β-sitosterol; linseed oil	Sterols	[104]
Ice cream	Phytosterols and γ-oryzanol; sunflower oil	Sterols	[96]
Dairy	Monolaurin, Schizochytrium spp. algal oil	Tocopherols, 1-*o*-galloylglycerol and gallic acid	[98]
Chocolate	β-sitosterol and lecithin, corn oil	β-sitosterol	[97]

GMS glycerol monostearate

therefore, oleogels are considered a promising saturated fat replacement in food products such as baked goods, dairy, meat products, and ice cream [88]. In Table 17.2, examples of saturated fat replacement with the addition of antioxidants of these food categories are presented.

Meat and meat products are consumed worldwide and the interest in replacing animal fat with healthier fats is gaining more importance [89]. For instance, Gómez-Estaca et al. [90] developed burgers with a replacement of saturated fat with oleogels prepared using polyunsaturated fatty acids (PUFA) (a mixture of olive oil, linseed and fish oils) and different oleogelators (ethyl cellulose, ethyl cellulose with curcumin, beeswax, and beeswax with curcumin) to determine the effect of household practices (chilled storage and cooking) on quality attributes of these developed burgers. They reported that lipid oxidation was reduced due to curcumin, in both household practices. However, curcumin-added samples obtained reduced sensory scores for the appearance. Moreover, burgers prepared with only beeswax or beeswax with curcumin showed acceptable technological properties, as well as good overall sensory characteristics when compared to burgers prepared with ethyl cellulose or ethyl cellulose with curcumin. Similarly, Martins et al. [91] produced pork patties replacing the saturated fat in different proportions (25 and 75% of replacement) with oleogels prepared using linseed oil and sterols. They reported that both proportions of the replacement showed no differences between the patties with and without sterols oleogels regarding textural properties (hardness, cohesiveness, and chewiness). However, the same study also demonstrated that the 25% replacement showed feasible results in the acceptance and preferences tests.

Bakery products such as bread, cakes, cookies, and pastries contain solid fats mainly to improve flavor and textural properties [92]. For instance, in the study of Tanislav et al. [93], the replacement of saturated fat with oleogels prepared with sunflower oil and different oleogelators (carnauba wax, β-sitosterol:beeswax, β-sitosterol:lecithin, and glycerol monostearate) in doughs was evaluated. They reported that oleogel prepared with β-sitosterol:beeswax showed the highest oil retain capacity, and slightly higher hardness than the reference margarine. Also, doughs prepared with the incorporation of β-sitosterol:lecithin oleogel showed the lowest values for both elastic and viscous modulus in frequency sweep tests. Additionally, the same study concluded that samples prepared with β-sitosterol mixtures can be a good option to decrease the saturated fat content in doughs, as they showed feasible textural and rheological properties. Also, in another study, the chitosan-based canola oil oleogels were developed via crosslinking with vanillin in different concentrations (1 or 3%) using the emulsion-template method. The study reported that oleogels prepared with both 1 and 3% of vanillin can create a network and this network was able to entrap the oil via Schiff bases formation with chitosan. However, the chitosan-based oleogel prepared with 3% of vanillin showed a more open microstructure, lower oil binding capacity, low mechanical resistance, and weaker gel (lower G') in rheology assessments than chitosan-based oleogel prepared with 1% vanillin. Thus, they concluded that the chitosan-based oleogel contains 1% vanillin presents the best properties to be used as fat replacers in cookies [94]. In another study, muffins with β-carotene oleogels showed the highest oxidative stability, indicating that oleogels produced with candelilla wax, GMS, and sunflower oil positively affected the storage stability of the muffin, without changing the specific volume and total porosity when compared to control muffin [95]. A study proposed an efficient way to structure camellia oil-based oleogels using the emulsion-templated method with 2.5% Tp-palmitate particles and different concentrations of citrus pectin (CP). The coordination of Tp-palmitate particles with citrus pectin proved to be effective in stabilizing O/W emulsions and corresponding oleogels, as evidenced by droplet size measurements, rheological tests, and microscopic analysis (Fig. 17.2a). These camellia oil-based oleogels were incorporated in cakes instead of butter, which were shown to be slightly firmer and exhibiting lower hedonic scores than those made with butter (Fig. 17.2b). Nonetheless, substituting butter with camellia oil-based oleogels can be a promising way to decrease *trans* fatty acid and saturated fat levels, leading to a healthier nutritional profile [56].

Ice cream is the most popular frozen dairy product, with good potential to be a functional food, through the incorporation of certain nutrients, plant-based

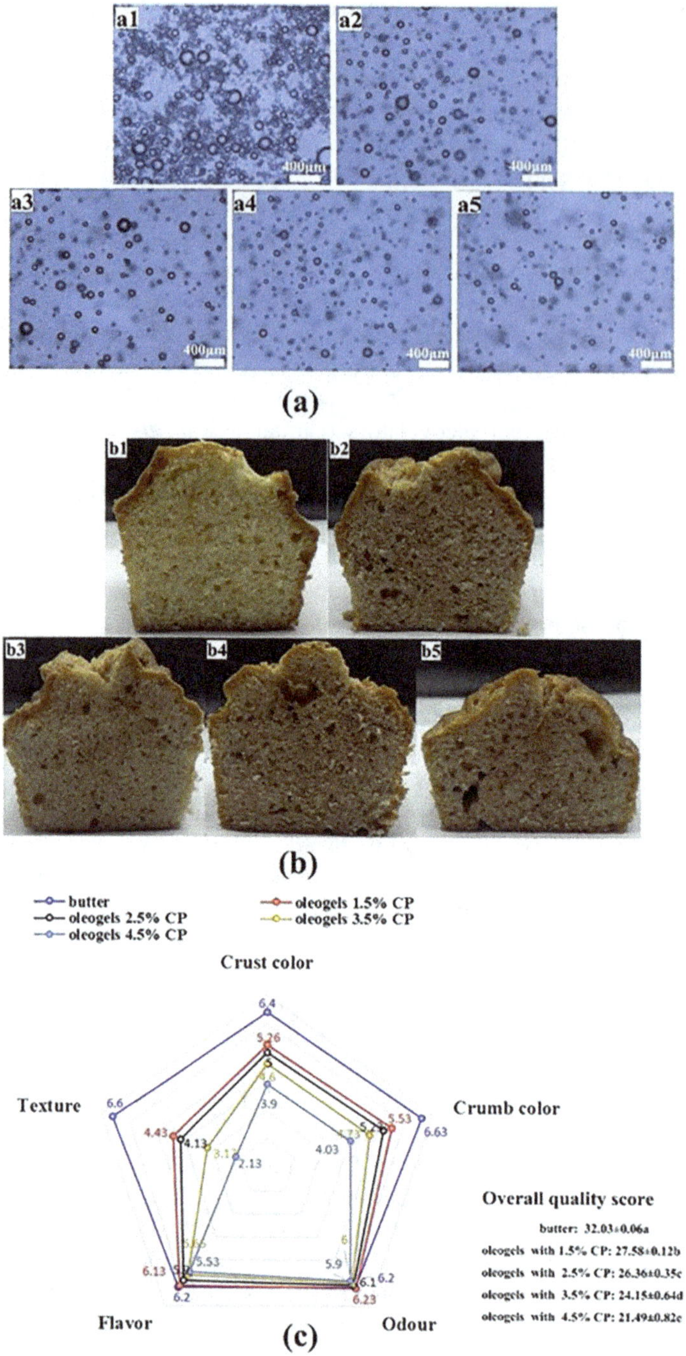

Fig. 17.2 (**a**) Light microscopy images of cake batters prepared using butter (a1), oleogels with 1.5 (a2), 2.5, (a3), 3.5 (a4), and 4.5% (a5) (w/v) CP, respectively. (**b**) Photographs of cake products prepared using butter (b1), oleogels with 1.5 (b2), 2.5, (b3), 3.5 (b4), and 4.5% (b5) (w/v) CP, respectively. (**c**) Comparative sensorial parameters of cakes prepared using butter, oleogels with 1.5, 2.5, 3.5, and 4.5% (w/v) CP, respectively. (Reprinted from [56] with permission from Elsevier)

unsaturated fats, and bioactive compounds. For instance, oleogels were produced using structured sunflower oil with different oleogelators (phytosterols and γ-oryzanol) to replace the saturated fat in artisanal ice creams. The study reported that ice creams produced with oleogels and 12% of gelators showed higher firmness values, higher overrun, higher melting starting time, and similar melting rate with the reference ice cream. Thus, the study concluded that in ice creams, replacement of saturated fat with unsaturated fat and increased nutritional properties due to sterols was possible, as ice creams produced with oleogels prepared with 12% of gelators showed similar or better quality characteristics when compared to ice creams of reference [96].

The chocolate structure consists of two phases; the continuous phase is a blend of several fats, including cocoa butter, milk fat and other vegetable fats, and the dispersed phase is composed of cocoa and milk powders and sugar. Cocoa butter is responsible for texture and flavor release, so its substitution or reduction should be carefully analyzed. Li and Liu [97]developed chocolate reduced in saturated fat using oleogels. The oleogel was produced using corn oil and three different oleogelators (monoglyceric stearate, β-sitosterol/lecithin and ethyl cellulose) individually. They reported that oleogel prepared with monoglyceric stearate was able to form solid-like appearance chocolate with 100% replacement of cocoa butter. Also, they reported that partial (50%) replacement of cocoa butter with ethyl cellulose oleogels showed higher hardness and yielding stress than chocolates prepared with β-sitosterol/lecithin and monoglyceric stearate. Also, the same study showed that each chocolate produced with 50% of oleogels showed a different crystalline structure. However, they reported that oleogel can be a feasible option to replace the saturated fat by unsaturated fat in chocolates.

Dairy products, especially yoghurt, contain high amounts of saturated fat; however, it is an important part of the adult diet. Hyatt et al. [98] developed oleogels to decrease the saturated fat content in yoghurts and increase the antioxidant content. Thus, oleogels were produced using monolaurin¸ *Schizochytrium* algal oil, tocopherols, 1-o-galloylglycerol, and gallic acid. Four different types of yoghurts were produced with algal oil with antioxidants, oleogel with antioxidants, oleogel with microencapsulated antioxidants, and microencapsulated oleogel with microencapsulated antioxidants. They reported that yoghurts produced with a combination of oleogelation and microencapsulation showed similar rheological and thermal properties and increased oxidative stability compared to the reference yoghurts. Thus, they concluded that microencapsulated oleogel and microencapsulated antioxidants is a feasible option for low saturated and high oxidative stabile yoghurts and combination of oleogelation and microencapsulation can be used in foods, cosmetics and pharmaceuticals.

17.4 Conclusion and Future Perspectives

The development of functional foods that could bring additional health benefits to consumers is a commercial strategy for several companies; however, the technological and economic challenges are, in some cases, limiting their entrance and acceptance in the market. Therefore, the development of new healthy ingredients or carriers of bioactive molecules that can be then easily used by companies are needed. In this regard, oleogels and oleogelation strategies have appeared in the last years to deliver functionality to food products, not only by replacing "bad" fats from foods but also by being used as a carrier of bioactive compounds with nutritional and health benefits.

The research work in the field is increasing, and it is possible to find several examples, with most of the work being focused on the use of oil-soluble compounds. It is expected that examples of the co-delivery of oil and water-soluble compounds in the same system increase with the use of emulgels and bigels. One of the challenges is to show the advantages of using oleogels as a delivery system, where the use of in vitro systems mimicking the human gut, have been used. However, more in vivo studies need to be performed to clearly show how these structures behave and could improve the bioavailability of the compounds when compared with traditional strategies.

The future in the area can pass by controlling the structure breakdown, the lipolysis rate control, and at the same time, the bioavailability of bioactive compounds by using tailor-made oleogel systems. Another strategy to increase the bioavailability of the compounds can be the incorporation of loaded micro and nanocapsules into the oleogels, in a way that they can easily be absorbed by the human gut.

Acknowledgements NUTRI4ICECARE: Ice cream with incorporation of compounds of nutritional interest (NORTE-01-0247-FEDER-045380), cofinanced by Norte 2020 and the European Union, through the European Regional Development Fund (ERDF).

References

1. Kris-Etherton PM, Etherton TD, Fleming J (2014) Human Nutrition | Cardiovascular and obesity health concerns. In: Dikeman M, Devine CBT-E of MS (Second E (eds). Academic, Oxford, pp 105–110
2. Pérez-Gregorio R, Soares S, Mateus N, de Freitas V (2020) Bioactive peptides and dietary polyphenols: two sides of the same coin. Molecules 25. https://doi.org/10.3390/molecules25153443
3. Shi M, Gu J, Wu H et al (2022) Phytochemicals, nutrition, metabolism, bioavailability, and health benefits in lettuce–a comprehensive review. Antioxidants 11. https://doi.org/10.3390/antiox11061158
4. European Parliament & Council (2006) Regulation (EC) 1924/2006 on nutrition and health claims made on foods. Off J Eur Communities L404:1–15

5. Wu P, Chen XD (2021) Validation of in vitro bioaccessibility assays — a key aspect in the rational design of functional foods towards tailored bioavailability. Curr Opin Food Sci 39: 160–170. https://doi.org/10.1016/j.cofs.2021.03.002
6. Acevedo-Fani A, Dave A, Singh H (2020) Nature-assembled structures for delivery of bioactive compounds and their potential in functional foods. Front Chem 8:1–22. https://doi.org/10.3389/fchem.2020.564021
7. Precedence Research Functional Food Market - Global Market Size, Trends Analysis, Segment Forecasts, Regional Outlook 2023–2032. https://www.precedenceresearch.com/functional-food-market. Accessed 29 Mar 2023
8. Jiménez-Munoz LM, Tavares GM, Corredig M (2021) Design future foods using plant protein blends for best nutritional and technological functionality. Trends Food Sci Technol 113:139–150. https://doi.org/10.1016/j.tifs.2021.04.049
9. Tan Y, Li R, Liu C et al (2020) Chitosan reduces vitamin D bioaccessibility in food emulsions by binding to mixed micelles. Food Funct 11:187–199. https://doi.org/10.1039/C9FO02164G
10. Lu X, Zhu J, Pan Y, Huang Q (2019) Assessment of dynamic bioaccessibility of curcumin encapsulated in milled starch particle stabilized Pickering emulsions using TNO's gastrointestinal model. Food Funct 10:2583–2594. https://doi.org/10.1039/C8FO02495B
11. Liu W, Gao H, McClements DJ et al (2019) Stability, rheology, and β-carotene bioaccessibility of high internal phase emulsion gels. Food Hydrocoll 88:210–217. https://doi.org/10.1016/j.foodhyd.2018.10.012
12. Nimalaratne C, Savard P, Gauthier SF et al (2015) Bioaccessibility and digestive stability of carotenoids in cooked eggs studied using a dynamic in vitro gastrointestinal model. J Agric Food Chem 63:2956–2962. https://doi.org/10.1021/jf505615w
13. Reboul E, Richelle M, Perrot E et al (2006) Bioaccessibility of carotenoids and vitamin E from their Main dietary sources. J Agric Food Chem 54:8749–8755. https://doi.org/10.1021/jf061818s
14. Tan Y, Zhang Z, Zhou H et al (2020) Factors impacting lipid digestion and β-carotene bioaccessibility assessed by standardized gastrointestinal model (INFOGEST): oil droplet concentration. Food Funct 11:7126–7137. https://doi.org/10.1039/D0FO01506G
15. Blanco-Morales V, López-García G, Cilla A et al (2018) The impact of galactooligosaccharides on the bioaccessibility of sterols in a plant sterol-enriched beverage: adaptation of the harmonized INFOGEST digestion method. Food Funct 9:2080–2089. https://doi.org/10.1039/C8FO00155C
16. Swackhamer C, Zhang Z, Taha AY, Bornhorst GM (2019) Fatty acid bioaccessibility and structural breakdown from in vitro digestion of almond particles. Food Funct 10:5174–5187. https://doi.org/10.1039/C9FO00789J
17. Paulo MC, Marques J, Cardoso C et al (2020) The development of a novel functional food: bioactive lipids in yogurts enriched with Aurantiochytrium sp. biomass. Food Funct 11:9721–9728. https://doi.org/10.1039/D0FO01884H
18. Acevedo-Fani A, Ochoa-Grimaldo A, Loveday SM, Singh H (2021) Digestive dynamics of yoghurt structure impacting the release and bioaccessibility of the flavonoid rutin. Food Hydrocoll 111:106215. https://doi.org/10.1016/j.foodhyd.2020.106215
19. Baptista DP, Salgaço MK, Sivieri K, Gigante ML (2020) Use of static and dynamic in vitro models to simulate Prato cheese gastrointestinal digestion: effect of lactobacillus helveticus LH-B02 addition on peptides bioaccessibility. LWT 134:110229. https://doi.org/10.1016/j.lwt.2020.110229
20. Corstens MN, Berton-Carabin CC, Schroën K et al (2018) Emulsion encapsulation in calcium-alginate beads delays lipolysis during dynamic in vitro digestion. J Funct Foods 46:394–402. https://doi.org/10.1016/j.jff.2018.05.011
21. Gómez-Mascaraque LG, Miralles B, Recio I, López-Rubio A (2016) Microencapsulation of a whey protein hydrolysate within micro-hydrogels: impact on gastrointestinal stability and potential for functional yoghurt development. J Funct Foods 26:290–300. https://doi.org/10.1016/j.jff.2016.08.006

22. Tagliazucchi D, Helal A, Verzelloni E et al (2016) Composition and properties of peptides that survive standardised in vitro gastro-pancreatic digestion of bovine milk. Int Dairy J 61:196–204. https://doi.org/10.1016/j.idairyj.2016.06.002
23. Battino M, Belwal T, Prieto MA (2023) Valorization of food products using natural functional compounds for improving organoleptic and functional chemistry. Food Chem 403. https://doi.org/10.1016/j.foodchem.2022.134181
24. Dima C, Assadpour E, Dima S, Jafari SM (2020) Bioavailability of nutraceuticals: role of the food matrix, processing conditions, the gastrointestinal tract, and nanodelivery systems. Compr Rev Food Sci Food Saf 19:954–994. https://doi.org/10.1111/1541-4337.12547
25. Martins AJ, Cerqueira F, Vicente AA et al (2022) Gelation behavior and stability of multicomponent sterol-based Oleogels. Gels 8:37. https://doi.org/10.3390/gels8010037
26. Martins AJ, Vicente AA, Cunha RL, Cerqueira MA (2018) Edible oleogels: an opportunity for fat replacement in foods. Food Funct 9:758–773. https://doi.org/10.1039/c7fo01641g
27. Huang Z, Guo B, Gong D, Zhang G (2023) Oleogel-structured emulsions: a review of formation, physicochemical properties and applications. Food Chem 404:134553. https://doi.org/10.1016/j.foodchem.2022.134553
28. Martins AJ, Vicente AA, Pastrana LM, Cerqueira MA (2020) Food science and human wellness Oleogels for development of health-promoting food products. Food Sci Human Wellness 9:31–39. https://doi.org/10.1016/j.fshw.2019.12.001
29. Silva PM, Cerqueira MA, Martins AJ et al (2022) Oleogels and bigels as alternatives to saturated fats: a review on their application by the food industry. JAOCS. J Am Oil Chem Soc
30. Patel AR (2017) A colloidal gel perspective for understanding oleogelation. Curr Opin Food Sci 15:1–7. https://doi.org/10.1016/j.cofs.2017.02.013
31. Shakeel A, Farooq U, Gabriele D et al (2021) Bigels and multi-component organogels: an overview from rheological perspective. Food Hydrocoll 111:106190. https://doi.org/10.1016/j.foodhyd.2020.106190
32. Glatz JFC (2011) Challenges in fatty acid and lipid physiology. Front Physiol 2:1–3. https://doi.org/10.3389/fphys.2011.00045
33. Wang HMD, Li XC, Lee DJ, Chang JS (2017) Potential biomedical applications of marine algae. Bioresour Technol 244:1407–1415. https://doi.org/10.1016/j.biortech.2017.05.198
34. Siram K, Habibur Rahman SM, Balakumar K et al (2019) Chapter 4: Pharmaceutical nanotechnology: brief perspective on lipid drug delivery and its current scenario. In: Grumezescu AMBT-BA of N (ed) . William Andrew Publishing, pp 91–115
35. Lobb K, Chow CK (2008) Fatty acid classification and nomenclature. In: Chow CK (ed) Fatty acids in foods and their health implications, Third edn. CRC Press Taylor & Francis, Boca Raton, pp 1–16
36. Puglisi M (2019) Chapter 47: Dietary fat and sports performance. In: Bagchi D, Nair S, Sen CKBT-N and ESP (Second E (eds). Academic, pp 555–569
37. Trinick TR, Duly EB (2013) Hyperlipidemia: overview. In: Caballero BBT-E of HN (Third E (ed). Academic, Waltham, pp 442–452
38. Calder PC (2016) The DHA content of a cell membrane can have docosahexaenoic acid. 69:8–21. https://doi.org/10.1159/000448262
39. Wilson BA, Pollard RD, Ferguson DS (2014) Nutriential hazards: macronutrients: essential fatty acids. In: Motarjemi YBT-E of FS (ed) . Academic Press, Waltham, pp 95–102
40. Kapalka GM (2010) Chapter 4 - substances involved in neurotransmission. In: Kapalka GMBT-N and HT for C and A (ed) Practical resources for the mental health professional. Academic, San Diego, pp 71–99
41. Macsai M, Mojica G (2013) 34 - medical Management of Ocular Surface Disease. In: Holland EJ, Mannis MJ, Conjunctiva L, Tear Film WBBT-OSDC (eds) . W.B. Saunders, London, pp 271–281
42. Lichtenstein AH (2005) Hyperlipidemia | Nutritional management. In: Caballero BBT-E of HN (Second E (ed) Encyclopedia of human nutrition. Elsevier, Oxford, pp 491–499
43. Lichtenstein AHBT-RM in FS (2021) Lipids (fats and oils). Elsevier

44. Trugo NMF, Torres AG (2003) FATS | Requirements. In: Caballero BBT-E of FS and N (Second E (ed). Academic, Oxford, pp 2279–2284
45. Organization WH (2003) Diet, nutrition, and the prevention of chronic diseases: report of a joint WHO/FAO expert consultation. World Health Organization
46. Uauy R, Aro A, Clarke R, et al (2009) REVIEW WHO Scientific Update on trans fatty acids : summary and conclusions. pp 68–75. https://doi.org/10.1038/ejcn.2009.15
47. Vemula PK, Campbell NR, Zhao F et al (2011) 4.421 - self-assembled prodrugs. In: Ducheyne PBT-CB (ed) . Elsevier, Oxford, pp 339–355
48. Manzoor S, Masoodi FA, Naqash F, Rashid R (2022) Oleogels: promising alternatives to solid fats for food applications. Food Hydrocoll Heal 2:100058. https://doi.org/10.1016/j.fhfh.2022.100058
49. Schaink HM, van Malssen KF, Morgado-Alves S et al (2007) Crystal network for edible oil organogels: possibilities and limitations of the fatty acid and fatty alcohol systems. Food Res Int 40:1185–1193. https://doi.org/10.1016/j.foodres.2007.06.013
50. Okuro PK, Malfatti-Gasperini AA, Vicente AA, Cunha RL (2018) Lecithin and phytosterols-based mixtures as hybrid structuring agents in different organic phases. Food Res Int 111:168–177. https://doi.org/10.1016/j.foodres.2018.05.022
51. da Silva TLT, Arellano DB, Martini S (2018) Physical properties of Candelilla wax, Monoacylglycerols, and fully hydrogenated oil Oleogels. JAOCS, J Am Oil Chem Soc 95: 797–811. https://doi.org/10.1002/aocs.12096
52. Matheson A, Dalkas G, Clegg PS, Euston SR (2018) Phytosterol-based edible oleogels: a novel way of replacing saturated fat in food. Nutr Bull 43:189–194. https://doi.org/10.1111/nbu.12325
53. Pinto TC, Martins AJ, Pastrana L, Pereira MC (2021) Oleogel-based Systems for the Delivery of bioactive compounds in foods. 1–24
54. Traber MG (2006) Modern nutrition in health and disease. In: Shils ME, Shike M, Ross AC, Caballero B, Cousins R (eds)
55. Nikiforidis C V, Scholten E (2014) RSC Advances. 2466–2473. https://doi.org/10.1039/c3ra46584e
56. Luo SZ, Hu XF, Jia YJ et al (2019) Camellia oil-based oleogels structuring with tea polyphenol-palmitate particles and citrus pectin by emulsion-templated method: preparation, characterization and potential application. Food Hydrocoll 95:76–87. https://doi.org/10.1016/J.FOODHYD.2019.04.016
57. Vellido-pérez JA, Rodríguez-remacho C, Rodríguez- J (2019) Optimization of Oleogel formulation for curcumin Vehiculization and lipid oxidation stability by multi-response surface methodology. 75:427–432. https://doi.org/10.3303/CET1975072
58. Calabria U, Gabriele D (2013) Food & Function. https://doi.org/10.1039/c3fo60259a
59. Eggersdorfer M, Wyss A (2018) Carotenoids in human nutrition and health. Arch Biochem Biophys 652:18–26. https://doi.org/10.1016/j.abb.2018.06.001
60. Chloe MO, Davidovich-Pinhas M, Wright AJ et al (2017) Ethylcellulose oleogels for lipophilic bioactive delivery--effect of oleogelation on in vitro bioaccessibility and stability of beta-carotene. Food & Funct 8:1438–1451
61. Cui M, Mao L, Lu Y et al (2019) Effect of monoglyceride content on the solubility and chemical stability of β-carotene in organogels. LWT 106:83–91. https://doi.org/10.1016/J.LWT.2019.02.042
62. Martins AJ, Cerqueira MA, Cunha RL, Vicente AA (2017) Fortified beeswax oleogels: effect of \$β\$-carotene on the gel structure and oxidative stability. Food & Funct 8:4241–4250
63. Bei W, Zhou Y, Xing X et al (2015) Organogel-nanoemulsion containing nisin and D-limonene and its antimicrobial activity. Front Microbiol 6:1010
64. Shi R, Zhang Q, Vriesekoop F, et al (2014) Preparation of Organogel with tea polyphenols complex for enhancing the Antioxidation properties of edible oil

65. Hwang HS (2020) A critical review on structures, health effects, oxidative stability, and sensory properties of oleogels. Biocatal Agric Biotechnol 26:101657. https://doi.org/10.1016/J.BCAB.2020.101657
66. Okuro PK, Martins AJ, Vicente AA, Cunha RL (2020) Perspective on oleogelator mixtures, structure design and behaviour towards digestibility of oleogels. Curr Opin Food Sci 35:27–35. https://doi.org/10.1016/J.COFS.2020.01.001
67. Vintiloiu A, Leroux JC (2008) Organogels and their use in drug delivery — a review. J Control Release 125:179–192. https://doi.org/10.1016/J.JCONREL.2007.09.014
68. Mao L, Lu Y, Cui M et al (2020) Design of gel structures in water and oil phases for improved delivery of bioactive food ingredients. Crit Rev Food Sci Nutr 60:1651–1666. https://doi.org/10.1080/10408398.2019.1587737
69. O'Sullivan CM, Barbut S, Marangoni AG (2016) Edible oleogels for the oral delivery of lipid soluble molecules: composition and structural design considerations. Trends Food Sci Technol 57:59–73. https://doi.org/10.1016/J.TIFS.2016.08.018
70. Li L, Wan W, Cheng W et al (2019) Oxidatively stable curcumin-loaded oleogels structured by β-sitosterol and lecithin: physical characteristics and release behaviour in vitro. Int J Food Sci Technol 54:2502–2510. https://doi.org/10.1111/IJFS.14208
71. Cheng Y, Donkor PO, Ren X et al (2019) Effect of ultrasound pretreatment with mono-frequency and simultaneous dual frequency on the mechanical properties and microstructure of whey protein emulsion gels. Food Hydrocoll 89:434–442. https://doi.org/10.1016/J.FOODHYD.2018.11.007
72. Mao L, Lu Y, Cui M et al (2019) Design of gel structures in water and oil phases for improved delivery of bioactive food ingredients. 101080/1040839820191587737. 60:1651–1666. https://doi.org/10.1080/10408398.2019.1587737
73. Palla CA, Aguilera-Garrido A, Carrín ME et al (2022) Preparation of highly stable oleogel-based nanoemulsions for encapsulation and controlled release of curcumin. Food Chem 378:132132. https://doi.org/10.1016/J.FOODCHEM.2022.132132
74. Qi W, Zhang Z, Wu T (2020) Encapsulation of β-carotene in oleogel-in-water Pickering emulsion with improved stability and bioaccessibility. Int J Biol Macromol 164:1432–1442. https://doi.org/10.1016/J.IJBIOMAC.2020.07.227
75. Gonçalves RFS, Martins JT, Duarte CMM et al (2018) Advances in nutraceutical delivery systems: from formulation design for bioavailability enhancement to efficacy and safety evaluation. Trends Food Sci Technol 78:270–291. https://doi.org/10.1016/J.TIFS.2018.06.011
76. Fathi M, Mozafari MR, Mohebbi M (2012) Nanoencapsulation of food ingredients using lipid based delivery systems. Trends Food Sci Technol 23:13–27. https://doi.org/10.1016/J.TIFS.2011.08.003
77. Pool H, Mendoza S, Xiao H, McClements DJ (2013) Encapsulation and release of hydrophobic bioactive components in nanoemulsion-based delivery systems: impact of physical form on quercetin bioaccessibility. Food Funct 4:162–174. https://doi.org/10.1039/C2FO30042G
78. Wright AJ, Pietrangelo C, MacNaughton A (2008) Influence of simulated upper intestinal parameters on the efficiency of beta carotene micellarisation using an in vitro model of digestion. Food Chem 107:1253–1260. https://doi.org/10.1016/J.FOODCHEM.2007.09.063
79. Tan SY, Wan-Yi Peh E, Marangoni AG, Henry CJ (2017) Effects of liquid oil vs. oleogel co-ingested with a carbohydrate-rich meal on human blood triglycerides, glucose, insulin and appetite. Food Funct 8:241–249. https://doi.org/10.1039/C6FO01274D
80. Ashkar A, Laufer S, Rosen-Kligvasser J et al (2019) Impact of different oil gelators and oleogelation mechanisms on digestive lipolysis of canola oil oleogels. Food Hydrocoll 97:105218. https://doi.org/10.1016/J.FOODHYD.2019.105218
81. Guo Q, Wijarnprecha K, Sonwai S, Rousseau D (2019) Oleogelation of emulsified oil delays in vitro intestinal lipid digestion. Food Res Int 119:805–812. https://doi.org/10.1016/j.foodres.2018.10.063

82. Calligaris S, Alongi M, Lucci P, Anese M (2020) Effect of different oleogelators on lipolysis and curcuminoid bioaccessibility upon in vitro digestion of sunflower oil oleogels. Food Chem 314:126146. https://doi.org/10.1016/J.FOODCHEM.2019.126146
83. Dong L, Lv M, Gao X et al (2020) In vitro gastrointestinal digestibility of phytosterol oleogels: influence of self-assembled microstructures on emulsification efficiency and lipase activity. Food Funct 11:9503–9513. https://doi.org/10.1039/D0FO01642J
84. Li L, Taha A, Geng M et al (2021) Ultrasound-assisted gelation of β-carotene enriched oleogels based on candelilla wax-nut oils: physical properties and in-vitro digestion analysis. Ultrason Sonochem 79:105762. https://doi.org/10.1016/J.ULTSONCH.2021.105762
85. Kavimughil M, Leena MM, Moses JA, Anandharamakrishnan C (2022) 3D printed MCT oleogel as a co-delivery carrier for curcumin and resveratrol. Biomaterials 287:121616. https://doi.org/10.1016/J.BIOMATERIALS.2022.121616
86. Wang S, Chen K, Liu G (2022) Monoglyceride oleogels for lipophilic bioactive delivery – influence of self-assembled structures on stability and in vitro bioaccessibility of astaxanthin. Food Chem 375. https://doi.org/10.1016/j.foodchem.2021.131880
87. Wang Q, Afshin A, Yakoob MY et al (2016) Impact of nonoptimal intakes of saturated, polyunsaturated, and trans fat on global burdens of coronary heart disease. J Am Heart Assoc 5. https://doi.org/10.1161/JAHA.115.002891
88. Singh A, Auzanneau FI, Rogers MA (2017) Advances in edible oleogel technologies – a decade in review. Food Res Int 97:307–317. https://doi.org/10.1016/J.FOODRES.2017.04.022
89. Jimenez-Colmenero F, Salcedo-Sandoval L, Bou R et al (2015) Novel applications of oil-structuring methods as a strategy to improve the fat content of meat products. Trends Food Sci Technol 44:177–188. https://doi.org/10.1016/J.TIFS.2015.04.011
90. Gómez-Estaca J, Pintado T, Jiménez-Colmenero F, Cofrades S (2020) The effect of household storage and cooking practices on quality attributes of pork burgers formulated with PUFA- and curcumin-loaded oleogels as healthy fat substitutes. LWT 119:108909. https://doi.org/10.1016/J.LWT.2019.108909
91. Martins AJ, Lorenzo JM, Franco D et al (2019) Omega-3 and polyunsaturated fatty acids-enriched hamburgers using sterol-based Oleogels. Eur J Lipid Sci Technol 121:1900111. https://doi.org/10.1002/EJLT.201900111
92. Demirkesen I, Mert B (2019) Recent developments of oleogel utilizations in bakery products. 101080/1040839820191649243 60:2460–2479. https://doi.org/10.1080/10408398.2019.1649243
93. Tanislav AE, Puşcaş A, Păucean A et al (2022) Evaluation of structural behavior in the process dynamics of Oleogel-based tender dough products. Gels 8:317. https://doi.org/10.3390/GELS8050317
94. Brito GB, Peixoto VODS, Martins MT et al (2022) Development of chitosan-based oleogels via crosslinking with vanillin using an emulsion templated approach: structural characterization and their application as fat-replacer. Food Struct 32:100264. https://doi.org/10.1016/J.FOOSTR.2022.100264
95. Jeong S, Lee S, Oh I (2021) Development of antioxidant-fortified Oleogel and its application as a solid fat replacer to muffin. Foods 10:3059. https://doi.org/10.3390/FOODS10123059
96. Moriano ME, Alamprese C (2017) Organogels as novel ingredients for low saturated fat ice creams. LWT 86:371–376. https://doi.org/10.1016/J.LWT.2017.07.034
97. Li L, Liu G (2019) Corn oil-based oleogels with different gelation mechanisms as novel cocoa butter alternatives in dark chocolate. J Food Eng 263:114–122. https://doi.org/10.1016/J.JFOODENG.2019.06.001
98. Hyatt JR, Zhang S, Akoh CC (2023) Combining antioxidants and processing techniques to improve oxidative stability of a Schizochytrium algal oil ingredient with application in yogurt. Food Chem 417:135835. https://doi.org/10.1016/J.FOODCHEM.2023.135835

99. Chen XW, Fu SY, Hou JJ et al (2016) Zein based oil-in-glycerol emulgels enriched with β-carotene as margarine alternatives. Food Chem 211:836–844. https://doi.org/10.1016/J.FOODCHEM.2016.05.133
100. Willett SA, Akoh CC (2019) Physicochemical characterization of yellow cake prepared with structured lipid Oleogels. J Food Sci 84:1390–1399. https://doi.org/10.1111/1750-3841.14624
101. Botella-Martinez C, Lucas-González R, Lorenzo JM et al (2021) Cocoa coproducts-based and walnut oil gelled emulsion as animal fat replacer and healthy bioactive source in beef burgers. Foods 10:2706. https://doi.org/10.3390/FOODS10112706
102. Panagiotopoulou E, Moschakis T, Katsanidis E (2016) Sunflower oil organogels and organogel-in-water emulsions (part II): implementation in frankfurter sausages. LWT 73:351–356. https://doi.org/10.1016/J.LWT.2016.06.006
103. Kouzounis D, Lazaridou A, Katsanidis E (2017) Partial replacement of animal fat by oleogels structured with monoglycerides and phytosterols in frankfurter sausages. Meat Sci 130:38–46. https://doi.org/10.1016/J.MEATSCI.2017.04.004
104. Agregán R, Barba FJ, Gavahian M et al (2019) Fucus vesiculosus extracts as natural antioxidants for improvement of physicochemical properties and shelf life of pork patties formulated with oleogels. J Sci Food Agric 99:4561–4570. https://doi.org/10.1002/JSFA.9694

Chapter 18
Oleogel Characterization: Physical, Physicochemical, and Chemical Techniques

Fernanda Peyronel and Elena Dibildox-Alvarado

Abbreviations

AED	Atomic emission detector
AFM	Atomic force spectroscopy
AOCS	American Oil Chemists Society
APCI	Atmospheric chemical ionization
BFM	Bright field microscope
C	Circularity of aggregates
CAN	Crystalline aggregates number
Cryo	Cryogenic
DAPI	4′,6-Diamidino-2-phenylindole
DLS	Dynamic light scattering
DSC	Differential scanning calorimeter
ECD	Electron capture detector
ED	Equivalent diameter of aggregates
ELSD	Evaporative light scattering detection
EM	Electron microscopy
FA	Fatty acid
FID	Free Induction Decay
FTIC	Fluorescein-5-isothiocyanate
FW	Fruit waxes GC gas chromatography

F. Peyronel (✉)
Food Science Department, University of Guelph, Guelph, ON, Canada
e-mail: fsvaikau@uoguelph.ca

E. Dibildox-Alvarado
Facultad de Ciencias Químicas, Laboratorio de Biopolímeros Alimentarios, Universidad Autónoma de San Luis Potosí, San Luis Potosi, Mexico
e-mail: dibildox@uaslp.mx

C. Andrea Palla, F. Valoppi (eds.), *Advances in Oleogel Development, Characterization, and Nutritional Aspects*,
https://doi.org/10.1007/978-3-031-46831-5_18

GLC	Gas-liquid chromatography
GSC	Gas-solid chromatography
HF	Heat-Flux
HPLC	High-Pressure Liquid Chromatography
L	Length of aggregates
LEC	Soya lecithin
MDGC	Multidimensional gas chromatography
MPL	Metal-phosphate luminophores
MRS	Magnetic resonance spectroscopy
NMR	Nuclear magnetic resonance spectroscopy
OM	Weight of solid
OOO	Triolein
OSI	Oxidative stability index
PC	Power-compensated
PLM	Polarized light microscopy
pNMR	Pulsed nuclear magnetic resonance
PP	Penalty point
PV	Peroxide value
PV-LED	Peroxide value light emitting diode
SEM	Scanning electron microscopy
SFA	Saturated fatty acid
SFC	Solid fat content
SSS	Tristearin
SW	Weight of oil
TAGs	Triacylglyceride
TCD	Thermal conductivity detector
TEM	Transmission electron microscopy
Tg	Transition temperature
TPA	Texture profile analyzer
TW	Weight of oleogel
UV/VIS	Ultraviolet/visible

18.1 Introduction

Various approaches can be employed to structure oleogels, with the help of molecules called oleogelators. The structures that the oleogelators can form are confined within oily matrices that are typically free of *trans* fats and low in saturated fatty acids. Oleogels can be produced by utilizing one or multiple oleogelators within a singular oil or a blend of oils and can also be structured using water to form different types of emulsions, such as simple or bigels (w/o, o/w).

In order to design effective strategies for studying and evaluating these structures, as well as the behavior and stability of oleogels during shelf life, it is crucial to select

Table 18.1 Techniques and methods to be discussed in this chapter that can be used to study oleogels

Structure analysis (mm, μm, nm, atomic scale)		Chemical characterization	Thermal characterization
18.2.1 Cone penetration test	18.2.6 DLS	18.3.1 Raman	18.4.1 DSC
18.2.2 Compressed test	18.2.7 Cryo SEM and Cryo TEM	18.3.2 FTIR	
18.2.3 Oil migration test	18.2.8 Confocal microscopy	18.3.3 GC	
18.2.4 Oil flowing test	18.2.9 AFM	18.3.4 HPLC	
18.2.5 PLM/BF microscopy	18.2.10 Pulsed NMR	18.3.5 Oxidative stability	

appropriate methodologies. Predictions based only on the oleogelator and oil to use are frequently insufficient in achieving an accurate diagnosis. Therefore, it is important to choose or adapt analytical techniques based on the desired final structure and matrix of the oleogel, as well as the parameters to be evaluated.

This chapter describes the basics of many analytical techniques and some methods that can be applied to the characterization of oleogels and is focused on the results obtained from them, highlighting their use and how they have been implemented in the study of oleogels. Each technique or method brings a different insight into the oleogel structure; hence, the choice of which one to use depends on the final objective. Usually, the desired application of the oleogel dictates the parameters needed to be evaluated; hence, different techniques might be required. Table 18.1 lists the techniques that will be discussed in this chapter.

Additionally, this chapter will provide a detailed examination of the penalty points (PPs) associated with each technique, along with strategies to minimize these points using the Eco-Scale—an analytical methodology devised by Gałuszka and colleagues [1, 2]. The Eco-Scale employs the principles of "Green Chemistry" [3, 4] to rate a process in terms of its environmental and human health impact. PPs are utilized by the Eco-Scale to quantify the harm caused by reagents, solvents, and equipment based on their environmental and health impact. Methods with higher PPs are deemed to be more hazardous. Therefore, the aim is to select reagents, solvents, and equipment that result in the least amount of PPs.

To mitigate the PPs resulting from reagents and solvents, several strategies can be adopted, including the elimination or substitution of toxic compounds, or the reduction in their quantity. The energy consumption of equipment can also be curbed by utilizing devices that are more efficient or limiting their operation duration. Furthermore, the principles espouse the adoption of multi-analyte or multi-parameter methods, sample number and size reduction, and avoidance of derivatization. These guidelines not only serve to decrease the impact of the analytical process on the environment and human health but also ensure the safety of operators. The Eco-Scale also takes into account the PPs associated with waste disposal during the analytical process. However, the focus of this chapter is on the PPs resulting from by the use of reagents, solvents, and energy consumption by the equipment. The quantification of the PPs of each chemical (reagent or solvent) [1, 2] will be calculated as follows:

$$\text{PPs} = \text{PP reagent/solvent amount} \times [\text{Pictograms PP} + \text{Hazardous PP}] \quad (18.1)$$

PP reagent/solvent amount is a number (0, 1, 2, 3) based on the amount of reagent/solvent used as described in the Eco-Scale. Pictogram PP refers to 0, 1, 2, etc., according to the number of WHMIS-regulated pictograms associated with the reagent or solvent. Hazardous PP refers to 0, 1, or 2 depending if the label of the chemical has no warning word (0), or if the word "warning" (1) or "hazard" (2) is displayed on the label. The amount of energy consumed by the equipment will also be estimated based on the penalty points assigned by the Eco-Scale [1, 2]. The total PPs will be determined by adding the PPs associated with all the chemicals used, including the PPs attributed to the energy consumption resulting from the operation of the equipment.

Each section expands on the principles of the technique or method, including recommendations on its application to oleogels, and presents data obtained via selected technique or method. In order to provide a concise summary of each technique or method, a four-column table has been constructed and placed at the end of each section. The four-column table details the spatial length scale that the technique can investigate, the equipment operating cost (condensed as "$" for low, "$ $" for medium, and "$$$" for high), the parameters that can be measured using the technique, and the corresponding PPs.

18.2 Structure Analysis Techniques

18.2.1 Cone Penetration Test

The cone penetration test is a widely used technique to evaluate the consistency or hardness of soft materials such as butter or animal/vegetable spreads and semisolid fats [5–7]. Traditionally, a metal cone is dropped into the material, and an analog dial reports the results. While the manual cone penetrometer instrument is still used by many research groups, digital versions are now being manufactured.

A cone penetrometer consists of a stand to which a metal shaft is attached, with either a dial or a digital screen connected to the shaft. Different probes can be attached to the end of the shaft, with the cone being the most popular. The choice of probe depends on the anticipated range of firmness of the material to be analyzed. Harder samples require probes with a smaller diameter or a smaller cone angle. The probe is positioned directly above the surface of the sample with penetration occurring under the probe's dead weight. Attention is needed to ensure that the specimen's surface is smooth and flat.

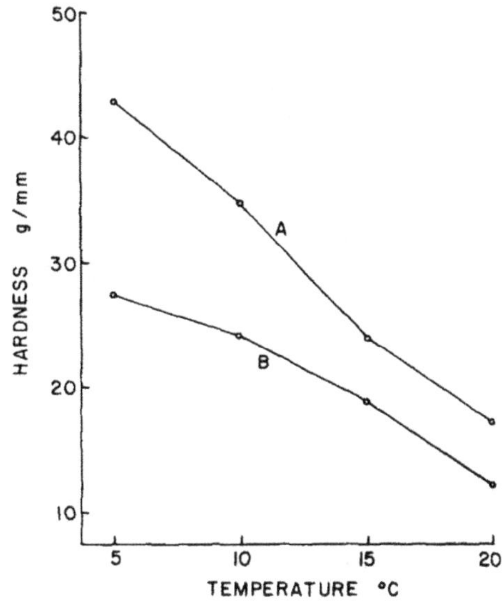

Fig. 18.1 Changes in hardness with increase of the temperature for butter (A) and margarine (B). (Reproduced from [5] with permission from AOCS Press)

There are two modes of operation: one involves releasing the cone for a certain length of time, while the other involves letting go of the cone and waiting until it comes to a stop. This allows for a rapid empirical evaluation of the material's properties. Oleogels, which fall into this category of soft materials, can also be analyzed using a cone penetrometer.

The results of cone penetration are typically interpreted in terms of the depth of penetration, regardless of the method used to stop the movement of the cone. In order to maintain consistent conditions, various recipes are tested with constant cone angle, mass of the dropping assembly, and temperature. de Man [5, 8] demonstrated a correlation between penetration depth and hardness in butter, due to thixotropic changes, when the cone was allowed to move for 5 s. For example, Fig. 18.1 illustrates the changes in hardness with temperature for butter and margarine, where penetrometer depths were converted to hardness using the equation: hardness = (mass of the cone assembly)/(depth of penetration in mm obtained from [5]).

The analysis of the yield value of a material is influenced when employing the approach where the probe is allowed to come to rest. During this process, the applied stress decreases until it reaches the yield value, at which point the material's deformation ceases. In 1959, Haighton [6] demonstrated that the logarithm of the penetration is inversely proportional to the logarithm of the yield value, and proposed a relationship between yield and depth of penetration for fats and margarines. However, it should be noted that some oleogels might be too soft to be probed with a heavy cone. Furthermore, as this instrument does not provide absolute results, selecting an alternative cone for comparison is not a concern unless a standardized method, such as method Cc 16–60 from AOCS [9], is being followed [9].

Penalties Points

The absence of solvents to run the equipment means that there are no PP created by chemical use. Additionally, the manual version of the cone penetrometer does not generate PP from electricity, while the digital version is used for such a brief time, that there is no PP created due to electricity. Therefore, the technique's overall PP value is zero.

Relevant Information Regarding the Technique

Spatial length scale	Operation of equipment ($, $$, $$$)	Parameters to measure	Penalty points
Millimeters	$	Spreadability Firmness	0

18.2.2 *Compressed Test*

A texture profile analyzer (TPA) is a widely used instrument for conducting compression tests on food materials. Various designs are available in the market, featuring either one or two vertical arms with a horizontal section to which the probe is attached. The probe is lowered and raised with great precision using a control system. During a TPA test, the sample is placed on the platform beneath the probe, and the horizontal arm with the attached probe is lowered to penetrate the sample. The test measures the strain, stress, and force required to compress, break, or penetrate the sample, which are registered during the test. One type of measurement is to control the rate of probe penetration until a specific final depth is reached (back extrusion or strength test) while recording the force exerted by the probe [10, 11]. Another type of measurement involves performing two consecutive compression cycles to simulate "mastication." This test generates rapid results and provides information on the sample's main textural parameters, such as firmness, adhesiveness, springiness, cohesiveness, elasticity, gumminess, and chewability. Regardless of the method used, force is the key parameter, and force versus time or distance (as deformation or displacement) is typically recorded.

A typical TPA double compression test measurement is illustrated in Fig. 18.2, where the Szczesniak mastication profile was used to identify parameters. The most commonly used variables in such tests include hardness, the maximum force required to compress the sample which is the height of the first peak; adhesiveness, which is the first negative area; cohesiveness, or the extent to which the sample could be deformed prior to rupture which is the second positive area divided by the first positive area; springiness, or the ability of the sample to recover its original form after deformation is removed, which is the time of second bite divided by the time of first bite; gumminess, or the force needed to disintegrate a semisolid sample to a steady state of swallowing which is hardness multiplied by cohesiveness; chewiness,

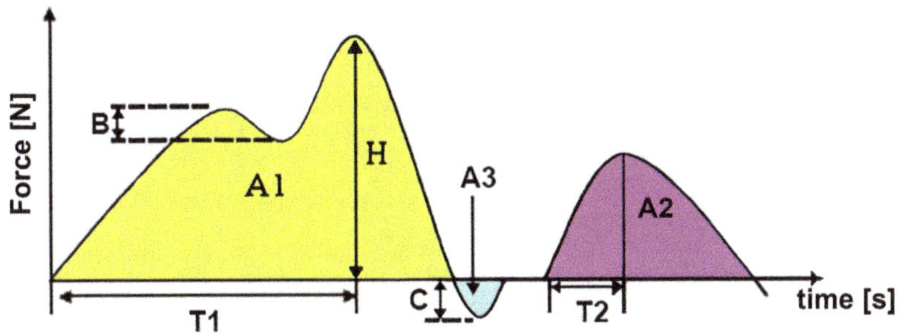

Fig. 18.2 Texture analyzer profile using Szczesniak profile. B = brittleness, H = hardness, C = cohesion strength, T1 = indentation, A3 = adhesiveness, A2/A1 = cohesiveness, T2/T1 = elastic quality, H × A2/A1 × T2/T1 = masticability, H × A2/A1 = gumminess. (Reproduced from [12] with permission from Elsevier)

Fig. 18.3 Back extrusion results on an oleogel made using canola oil and 15% ethyl cellulose. (Reproduced from [10] with permission from Elsevier)

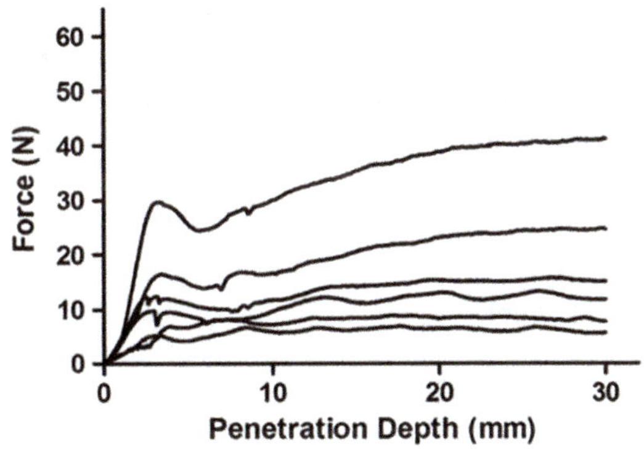

or the work needed to chew a solid sample to a steady state of swallowing which is gumminess multiplied by springiness. Brittleness refers to the fracture behavior where the more force indicates less brittle; indentation is the force needed to keep a permanent deformation.

Each area in Fig. 18.2 provides a number that can be compared between samples. However, the lack of "standards" makes it difficult to carry out quantitative measurements with this instrument. Nevertheless, using a TPA removes the subjectivity of panelists and allows for sample comparisons, although it is important to conduct comparisons under the same experimental conditions. This technology is particularly useful when comparing oleogels made with different amounts or types of oleogelators. A control product that consumers accept as optimal can be used "as a standard".

When the samples are too soft, the mastication profile may not be adequate, and in such cases, the TPA can be used for extrusion. This is applicable for oleogels, as shown in Fig. 18.3, which displays the result of a back extrusion measurement using a probe that was moved at 1.5 mm/s for an oleogel made with 15 wt% ethyl cellulose in canola oil, as reported by Gravelle and colleagues [13].

Table 18.2 Results of a TPA test on oleogels made using various olegelators in rapeseed oil

Oleogelator	Maximum force of deformation (N)	Hardness (N)	Work of shear (N. mm)	Stickiness (N)
CAN (2%)	5.52 ± 0.33	12.68 ± 1.06	35.57 ± 3.88	−10.06 ± 0.83
MAG (5%)	0.77 ± 0.15	6.47 ± 0.46	18.29 ± 1.34	−7.50 ± 0.31
BW (3%)	1.60 ± 0.31	5.60 ± 0.42	13.00 ± 1.05	−6.22 ± 0.39
BY (3%)	0.73 ± 0.15	4.39 ± 0.42	9.74 ± 1.44	−4.87 ± 0.41
SUN (1%)	3.33 ± 0.42	15.78 ± 3.12	46.62 ± 8.60	−20.37 ± 4.53

Adapted from [11]
Work of shear was determined as the area given by A1 in Fig. 18.2. Stickiness was determined as the cohesion strength (C) in Fig. 18.2

Kupiec and colleagues [11] compared different oleogels made with various olegelators in rapeseed oil. Those authors used different oleogelators, such as candelilla wax (CAN 2%), monoglycerides (MAG 5%), yellow beeswax (BW 3%), white beeswax (BY 3%), and sunflower beeswax (SUN 1%), with each number in the brackets representing the weight percentage. Table 18.2 summarizes the results obtained by the group. For statistics, please refer to the original paper.

By looking at the results on Table 18.2, conclusions about different characteristics of the oleogels can be easily made.

Penalty Points

This test does not require any reagents or solvents, resulting in a PP of 0 for chemical usage. In addition, the electricity consumption is minimal, using less than 0.1 kWh, resulting in a PP of 0 for electricity usage.

Relevant Information Regarding the Technique

Spatial length scale	Operation of equipment ($, $$, $$$)	Parameters to measure	Penalty points
Millimeters	$	Firmness Adhesiveness Springiness Cohesiveness Elasticity Gumminess Chewability Spreadability	1

18.2.3 Oil Migration Test

Oil migration is a diffusion-based movement [14] of liquid oil from an oil-rich phase to a lipid-poor or lipid-solid phase in a food product. This phenomenon is observed in various products, including cream-filled cookies, chocolate-coated goods, multi-layer filled chocolate products, and many composite products that are coated with chocolate. The oil migration can result in various quality issues, including changes in texture and flavor, reduced shelf life, and decreased product appeal. Factors that affect oil migration include the composition and structure of fats and oils, processing techniques, and temperature and storage conditions. Fats with high melting points and high solid content tend to have lower oil migration rates. Products containing emulsifiers can enhance structural stability and prevent oil migration. Failure to control oil migration can lead to quality issues such as fat bloom, softening of chocolate, and hardening of the filling. The phenomenon of oil migration is closely related to the structural properties of the lipid's crystal lattice, which determines its oil-retaining ability [15].

Various techniques can be utilized to investigate oil migration, including fluorescence recovery after photobleaching [16], magnetic resonance imaging [17], compression test, differential scanning calorimetry (DSC), X-ray diffraction, and many different microscopy techniques. All those techniques require specialized equipment. However, there are two methods commonly employed that do not necessitate sophisticated instrumentation. They are the time-dependent monitoring of stained oil using a dye, such as Nile red [18], and the gravimetric method [15].

The stained-oil method involves adding Nile red to the material, which binds to the lipids and allows for visual observation of the stained oil. The measurement is performed using a flatbed scanner, with the pixel intensity used to determine the amount of migrated oil [18]. In contrast, the gravimetric method involves forming the processed sample into discs, which are stored at specific temperatures for a specified period before analysis. Multilayered discs can also be utilized to facilitate or enhance analysis. The disc is placed on top of a filter paper to allow the oil to flow toward the paper using only gravity as the driving force. The method enables the determination of oil loss from the fat crystal network by weighing the filter papers (wt. paper) and the sample disc (wt. disc) before (t=0) and after storage (t) [15]. Oil loss is calculated using the following formula:

$$\text{Oil loss} = \frac{\text{wt.paper}\ (t) - \text{wt.paper}\ (t=0)}{\text{wt.disc}\ (t=0)} \times 100\%$$

However, there are no official standards set in the food industry for either of these two methods. For instance, the amount of Nile red added is arbitrary, and the size and weight of the sample discs are also arbitrary. Thus, each research group can choose which absorbent paper to use, the number of hours or days the sample rests on the paper, and the storage temperature. It should be emphasized that these two inexpensive and straightforward methods are useful for sample comparison only. For example, Cho and colleagues [19] recently employed the gravimetric method to quantify oil and water losses in frozen/thaw edible rice bran wax-gelatin biphasic gels. The authors found that after one freeze-thaw cycle, the biphasic gels had less total liquid loss by approximately 50% compared to the oleogels.

Penalty Points

When considering the gravimetric method, there are no penalty points since no reagents or solvents are involved, and no equipment that consumes electricity is required. In the case of time-dependent monitoring of stained oil using a dye, Nile red carries with it 0 penalty points when the amount is less than 10 mL or 1 PP when it is between 10 and 100 mL. The energy consumption for the scanner can be between 1 and 2, depending on penalties points (when energy is below 1.5 kWh) and up to 2 PPs. Hence, the maximum for this method could be 2 PPs.

Relevant Information Regarding the Method

Spatial length scale	Operation of equipment ($, $$, $$$)	Parameters to measure	Penalty points
Nanometers to micrometers	$	Weight of oil displaced Color intensity	0–2

18.2.4 Oleogel/Oil Flowing Test

The flowing test is a simple method used to determine the minimum concentration of an oleogelator required to produce a material that does not flow under its weight. The material to be studied is placed inside a vial and either inverted or centrifuged after a certain time. If the material does not flow after inversion or if there is no separation when using the centrifuge, then the material can be called an oleogel.

When using the direct method to make oleogels, the oleogelator is dissolved in the oil phase at sufficiently high temperatures to generate a solution, and then stirred for some minutes before it gets cool down, usually to ambient conditions. The time needed for the oleogel to set can range from hours to weeks. Once the desired period has elapsed, vials containing the samples are turned upside down or inserted in a centrifuge. The critical concentration is determined by visually comparing the vials. The vial that uses the lowest amount of oleogelator and does not flow or separate is identified as the one with the "critical concentration." Vials are typically prepared using increments of 1% of the oleogelator. Critical concentrations for specific oleogelators in specific oils can be found in the literature. However, changing the oil or oleogelators may require repeating the flowing test.

Centrifugation can be used to determine the weight of the non-gelated oil. The method employed during centrifugation is referred to as the "oil binding capacity," which determines the quantity of oil released from the sample following the centrifugal process. To measure how much oil is liberated, the weight of the oleogel placed in the tube for centrifugation is measured (TW). After centrifugation, the tube is inverted, and the non-bound oil is collected and weighed (OW). The weight of the solid remaining in the tube can be calculated as SW = TW − OM. Sometimes, the percentage of oil liberated (or not trapped in the solid matric) is calculated using the formula (SW − OW)/(TW − SW) × 100. This allows for a quick comparison between samples. However, there are no standards for the weight of the oleogel or the parameters to use during centrifugation. Each research group must exercise their discretion in determining the weight of the sample to centrifugate and the parameters to use in the centrifuge (e.g., speed of rotation or temperature).

Penalty Points

There are no penalty points associated with the oleogel/oil flowing test (or inversion method), as there are no reagents or solvents involved and no consumption of electricity. When the centrifuge is used, the energy required to run the centrifuge (electricity ≤1.5 kWh) can generate 1 penalty point.

Relevant Information Regarding the Method

Spatial length scale	Operation of equipment ($, $$, $$$)	Parameters to measure	Penalty points
Millimeters	$	Visual flow of material	0–1

18.2.5 Polarized Light and Bright Field Microscopy (PLM/BFM)

18.2.5.1 Polarized Light Microscopy

Polarized light microscopy (PLM) is a robust contrast-enhancing technique that exploits the unique optical properties of anisotropy to reveal valuable information on the structural and compositional properties of diverse materials, such as oleogelators in oleogels. A PLM is a bright field microscope in which the sample is illuminated with oriented single plane waves, by positioning a polarizer between the light source and the specimen, and an analyzer between the objective rear aperture and the observation tubes or camera port. This technique utilizes the optical properties of anisotropy to reveal information about the structure and composition of materials [20], enabling the differentiation of isotropic and anisotropic materials. PLM has proven to be a useful technique to identify solid crystals in edible fats [21, 22] because solid fat crystals are birefringent, hence appear bright under polarized light. PLM can be used for both qualitative and quantitative analysis to provide insights into the morphology and organization of the oleogelator in the sample.

The solid structure of an oleogel can be determined by placing approximately 10 μL (a drop) of the sample onto a microscope slide and covering it with a cover slide (Fig. 18.4). It is important not to add too much sample as it must be squeezed with the cover slide. Alternatively, some oleogels that are manufactured using the direct method can be formed directly on the microscope stage. In this case, the oil and oleogelator are pre-mixed and a drop of the mixture at room temperature is deposited on the microscope sample holder. The pre-mixed sample is then subjected to a temperature cycle, which may include thermal memory elimination, continuous or intermittent cooling ramps, or isothermal structuring periods. Sometimes, a shear cycle may also be applied. These techniques can be performed using a microscope that has the necessary capabilities (e.g., cold stage using liquid nitrogen (Fig. 18.4)

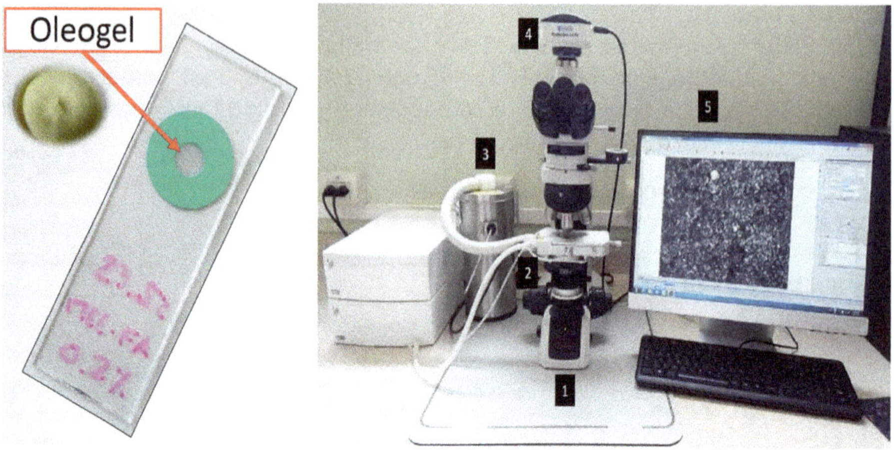

Fig. 18.4 Location of an oleogel in the microscope slide and the equipment used to capture PLM images: (1) Body of the microscope. (2) Temperature control stage. (3) Liquid nitrogen reservoir. (4) Camera. (5) Image analysis software for managing the information collected during the measurement. (*Source*: Courtesy of Food Biopolymer Lab, Faculty of Chemical Sciences of the Autonomous University of San Luis Potosí)

or shear cell). A note to make here is that one needs to be aware that the material is "not free" to expand in three dimensions, but it is rather confined to the small space between the microscope slide and its cover slip.

Most microscopes are equipped with objectives of different magnifications, ranging from low (~4 or 5X) to the highest of all, usually 100X. The selection of the objective depends on the size of the structures to be observed. Images can be recorded as a function of the structuring oleogelator and/or storage time of the oleogel, or as a function of the temperature. The shapes and sizes of the structures formed can be evaluated by image analysis. An optical microscope is limited in the resolution and the smallest particles that can be measured, usually limited to micrometer sizes. Free software for image analysis is accessible on the web, or commercial software can be used for more measurements that are specialized. To perform these measurements, it is necessary to convert the acquired grayscale images to binary images [23] (as shown in Fig. 18.5 for the case when the temperature was 23.5 °C). It is advisable to take measurements from at least four different micrographs of the oleogel sample. The data is typically reported as the mean and its standard deviation.

A grayscale image is normally obtained by capturing the light reflected from the sample onto the sensor of the camera. To obtain a binary image, a threshold must be applied to the grayscale image. The threshold is a value chosen to separate the foreground (the structures of interest) from the background (the rest of the image). The threshold is needed to identify where the particle of interest has its outer limit. Pixels with a value above the threshold are set to white and those below the threshold are set to black. This process creates a binary image where the structures of

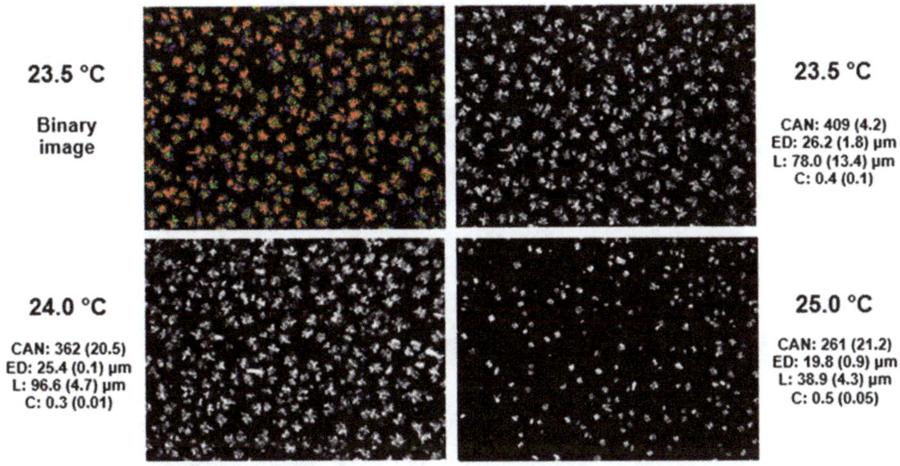

Fig. 18.5 PLM microphotographs showing the water-in-palm oil emulsions crystallized at three different isothermal temperatures using microcrystalline cellulose as the oleogelator. (CAN: Crystalline aggregates number. ED: Equivalent diameter of aggregates. L: Length of aggregates. C: Circularity of aggregates). (*Source*: Courtesy of Food Biopolymer Lab, Faculty of Chemical Sciences of the Autonomous University of San Luis Potosí)

interest are highlighted in white, and the background is in black. This binary image can then be used for image analysis to obtain parameters such as crystal number, area, equivalent diameter, perimeter, length, width, circularity, elongation, intensity, among others. A binary image can easily be colored with the right software. Figure 18.5 displays four parameters of a water-in-palm oil emulsion using microcrystalline cellulose as structuring agent and prepared at three different temperatures. These parameters can aid in the interpretation of the observed changes in crystalline structure over time.

18.2.5.2 Bright Field Microscopy

A bright field microscope (BFM), also referred to as a compound light microscope, is an optical microscope that uses light rays to produce a dark image against a bright background [24]. While some BFM are designed exclusively for bright field mode, this instrument can be used as a polarizing light microscope (PLM) by adding a polarizer and an analyzer in the light path. Both qualitative and quantitative analyses can be performed using a BFM, similar to the PLM. Qualitative analysis is relatively straightforward when there are noticeable differences in the morphology, size, and number of structures present in the images. However, quantitative analysis requires a mathematical approach that involves measuring specific features. The selection of relevant features is a crucial step, which necessitates some expertise on the operator. Nowadays, software packages that accompany microscopes, or freely accessible

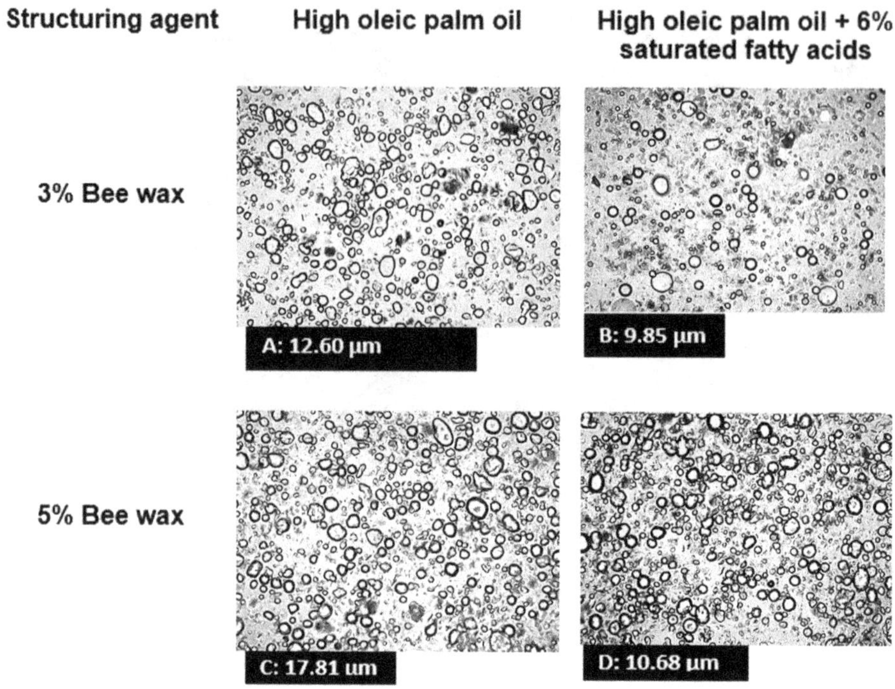

Fig. 18.6 Microphotographs taken in bright field mode using a 20× objective of oleogelled emulsions containing 3% (**a**, **b**) and 5% (**c**, **d**) beeswax, and high oleic palm oil with different content of saturated fatty acids: 32% (**a**, **c**), 38% (**b**, **d**). The number indicates the average equivalent diameter of the water droplets. (*Source*: Courtesy of Food Biopolymer Lab, Faculty of Chemical Sciences of the Autonomous University of San Luis Potosí)

ones, can be used for quantification. Quantitative analysis enables accurate comparison of samples, and in the case of gelled and bigellated emulsions, bright field microscopy provides a straightforward way to calculate droplet size and distribution in the emulsion.

Figure 18.6 demonstrates the influence of varying concentrations of beeswax gelling agent (3% and 5% of beeswax) and the quantity of saturated fatty acid (SFA) present in high oleic palm oil (32% and 38% SFA) on the water drop size of a gelled water-in-oil emulsion. The impact of differing concentrations of gelling agent and solids in the oleogel is visually apparent from the increased number of water droplets of varying sizes. However, a more comprehensive analysis of the effect can be achieved by determining the equivalent diameter using image analysis techniques, or better still, obtaining a histogram or curves that provide a clear representation of the droplet size distribution present in the emulsion.

Penalties Points

The energy consumption of both BFM and PLM instruments is less than 1.5 kWh, resulting in a PP = 1. However, if the analysis of the oleogel requires temperatures below ambient, liquid nitrogen is necessary, and the cost of the nitrogen must be considered. As liquid nitrogen is classified as a hazardous material with a warning symbol, it carries a PP of $2 \times (1 + 1) = 4$.

Relevant Information Regarding the Technique

Spatial length scale	Operation of equipment ($, $$, $$$)	Parameters to measure	Penalty points
Micrometers to millimeters	$$$	Crystal sizes Crystal number Crystal morphologies and dimensions	1–4

18.2.6 *Dynamic Light Scattering*

Dynamic light scattering (DLS) is a non-invasive technique commonly used to characterize particles, whether they are in a solution or as powders. Dynamic means that there is movement, which can be created by Brownian motion or because the instrument moves the particles to be measured. Light scattering refers to the use of "light", usually a laser and the phenomena of scattering. We are here discussing an instrument that "moves" the particles to be analyzed, like a Mastersizer. This method does not require calibration standards, but it does necessitate the measurement of an empty sample holder before the sample measurement, similar to other scattering techniques. The empty sample holder is automatically measured by the instrument prior to collecting data for the sample, and the software automatically removes the background from the sample before showing the results. However, the user can prevent this from happening by consulting the relevant manuals.

In principle, this measurement technique involves unperturbed laser light shining on particles, which scatter the light. The instrument's detectors register the intensities as a function of the scattered angle. At the molecular level, this can be understood as incoming photons interacting with the electron density of the atoms in the sample. The interaction between photons and electrons results in the emission of photons at various angles, which the detectors capture. The scattering theory employs the scattering q-vector as the variable, and the raw data collected by the instruments represent the magnitude of the scattering q-vector versus intensity, which is the number of photons counted at each detector.

The software utilized for data analysis employs the widely accepted Mie theory, which reports data as a function of either volume density or surface density of particles with respect to particle size, rather than intensity with respect to the magnitude of the q-vector. Alternatively, the Fraunhofer approximation can be used in lieu of the Mie theory. The Mie theory uses a spherical or non-spherical particle, chosen by the user, to model the collected data. Therefore, when the data is displayed on the screen, it is reported as the volume density on the *Y*-axis versus the particle size on the *X*-axis. Have in mind, that this data have already been processed, and is not the raw data. The idea of the manufacturer is to simplify the user's work by showing results of the correct model that fits the data best. The "model" requires particle and media related parameters such as refractive index, absorption indexes for the equipment's two lasers, and densities. The user also selects if the particles are spherical, non-spherical or opaque in order for the model to accurately compute the surface distribution or the volume distribution.

In a liquid measurement, the sample is dispersed in a medium (dispersant) and kept in constant motion to prevent flocculation while simultaneously diluting the sample. Dilution is crucial to prevent multiple scattering, which arises when many particles are present. Prior to analysis, the appropriate dispersant is added to the dispersant unit, followed by a few drops of the sample. The degree of dilution is controlled by a variable called "obscuration," which can be monitored in real time as the sample is added to the dispersant, and after a short stabilization period for the sample and laser. Obscuration values may vary between 0.5% and 20%, depending on the sample, dilution, and the speed at which the dispersant is moved by the stirrer. As a result, it is imperative to develop a unique method for each sample to be analyzed, and new method development must be carried out when the sample's composition changes. Some users have the idea that on this instrument, parameters like obscuration or rpm should be kept constant regardless of the samples created (which migth vary on the amount of emulsifiers or the oils used for the emulsions). This is a mistake, as the theory used by the instrument has nothing to do with those parameters to report the correct size distribution.

Figure 18.7. A shows the result of a liquid measurement using an obscuration of ~10%. Note that the *X*-axis is in a logarithmic scale, making it easy to visualize many length scales simultaneously, while the *Y*-axis is the volume density %. Figure 18.7b is a histogram that emphasizes that this instrument measures size distributions, with the *Y*-axis now representing volume %.

In Fig. 18.7a, a wide distribution ranging from 0.1 to 1100 μm is observed, with three distinct modes in the data. The largest peak among the three modes is observed at ~30 μm. However, it is important to note that this does not mean that the majority of the particles are of this size, as the volume density is only around 5% and not 100%. To further characterize the particle size distribution, DLS users commonly use parameters such as D [3,2] (Surface Weighted Mean), D [4,3] (Volume Weighted Mean), or Dv10, Dv50, and Dv90. These last three parameters indicate the size below which 10 or 50 or 90% of the total particles exist, where the DvXX value represents the size. Interested readers can refer to the Mastersizer 3000 User Manual and User Guide for more detailed information.

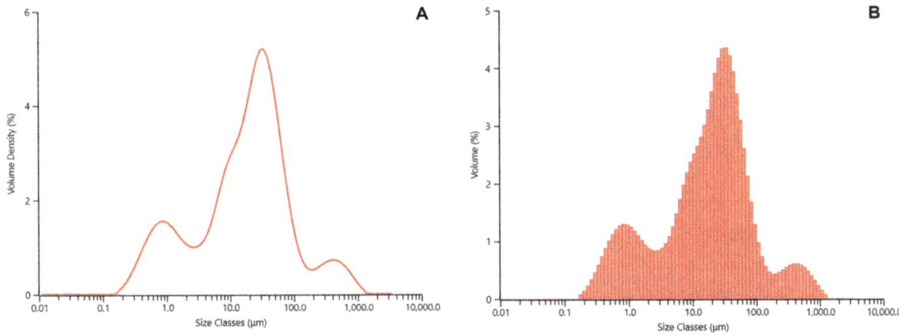

Fig. 18.7 Example of a particle size distribution obtained using a Malvern Mastersizer 3000. (**a**) Volume percentage as a function of particle size (size classes). (**b**) Histogram displaying volume percentage as a function of particle size. (*Source*: Courtesy of the Core Lab at the Food Science Department, University of Guelph, Canada)

Precisely quantifying oleogels using dynamic light scattering (DLS) systems poses challenges due to the inherent risk of disrupting the desired particulate structure during dispersion in the dispersant, potentially leading to their dissolution. As mentioned before, the DLS equipment necessitates the placement of droplets of the oleogel within a suitable dispersant, such as oil or water. However, this approach may compromise the accuracy of the measurements due to dilution effects of the oleogel under study. It is evident that DLS can serve as a suitable technique for quantifying oleogel-based emulsions or nano-emulsions, rather than directly assessing the properties of oleogels themselves.

Modern DLS instruments have the capability to measure powders, but the parameters required for introducing the sample into the equipment differ from those used for measuring liquids. To measure liquids, users must input values such as obscuration and RPM of the stirrer in the dispersant unit. However, when measuring powders, users must select the air pressure, tray feeding rate, obscuration, and hoper gap size.

Penalty Points

The DLS technique may require the use of an organic solvent for liquid measurements, which can result in penalty points (PP). When water is used, no PP is assigned for the chemical used. However, when ethanol is used, the PP is calculated as $2 \times (2 + 2) = 8$, as the dispersant unit can hold 100 mL of liquid, and ethanol has two pictograms and the “hazard” word. Nevertheless, since the measurement can be completed in a few minutes, the amount of electricity used is likely to be less than 1.5 kWh, resulting in a PP of 1. However, when using the powder technique, there are no penalties point due to reagents or chemicals. In this case, the only PPs are due to run the equipment and the vacuum machine that removes the powder, resulting in PP of 1.

Relevant Information Regarding the Technique

Spatial length scale	Operation of equipment ($, $$, $$$)	Parameters to measure	Penalty points
Nanometers to micrometers	$$	Particle size distribution q-vector vs. intensity	1–9 for liquids 1 for powders

18.2.7 Scanning and Transmission Electron Microscopy

Electron microscopy (EM), encompassing both scanning electron microscopy (SEM) and transmission electron microscopy (TEM), represents a powerful analytical technique that utilizes beams of electrons to illuminate samples for visualization and sizing of exceedingly small particles. Notably, at the nanoscale, a particle represents a collection of molecules, whereas at the micrometer scale, it may comprise an aggregation of particles of nanometer dimensions.

EM focuses an electron beam onto the sample, controlled by coils and apertures instead of glass lenses as used in light microscopy. The SEM scans the electron beam across the surface of the sample from top to bottom and from right to left, while the TEM transmits the beam through the sample. Both types of microscopies utilize an electron gun with an electron source, a Wehnelt cylinder, and an anode to form and accelerate the beam. When the electron beam interacts with the sample, it can be absorbed or diffracted by the atoms in the sample, creating secondary and backscattered electrons, X-rays, visible light, and heat. The SEM utilizes secondary and backscattered electrons for imaging, which show morphology and topography and illustrate contrasts in composition, respectively.

A cryogenic (Cryo) system used in conjunction with SEM improves the imaging when contrast is lacking at room temperature. The system preserves the internal structure of the sample by rapidly freezing it at liquid-nitrogen temperatures. The frozen sample is then transferred to a preparation chamber under vacuum conditions, where it undergoes manipulations such as cold fracturing, sublimation of surface ice, and nanometer-thick cadmium/gold coating. The fractured and coated sample is transferred to the SEM chamber for further analysis under cryogenic conditions. The constant −170 °C temperature is guaranteed by the Cryo system attached to the gun of the SEM. Overall, EM and its types provide powerful imaging techniques for visualizing and sizing particles at small scales, and the use of a Cryo system with SEM enhances imaging capabilities by preserving the internal structure of the sample.

The transmission electron microscope works by directing a beam of electrons through a thin film of the sample before capturing the image on a fluorescent screen for observation. It is critical to have a thin layer (around 100 nm) to prevent excessive absorption of the electron beam by the material, which would prevent it from reaching the screen. This equipment takes advantage of the high speed of electrons; the faster they move, the shorter the wavelength and the greater the quality and detail of the image. The fluorescent screen shows light and dark areas, with lighter regions indicating where more electrons passed through the sample and darker areas representing denser regions. These differences provide valuable information on the sample's structure, texture, shape, and size. For certain samples, it is necessary to freeze them with liquid nitrogen to "fix" them and prevent changes over time.

What sets EM apart from light microscopy is its superior resolution. Resolution refers to the ability to distinguish two closely spaced points (particles) as two separate entities. EM can achieve resolutions between 0.2 and 0.05 nm, with the accelerating voltage being a key factor. In a scanning electron microscope, the accelerating voltage can range from 500 to 30,000 eV. Electrons accelerated by a potential of 30 keV have shorter wavelengths than those accelerated by 5 keV, resulting in better resolution for the 30 keV beam. TEM can handle voltages between 40 and 100 keV, offering even higher resolution than SEM.

Visualizing oleogels with SEM is challenging because the contrast between the oleogelator and the oil is not very distinct as can be seen in Fig. 18.8a. However, TEM offers higher resolution and allows for a clear identification of the features as seen in Fig. 18.8b. Proper sample preparation is crucial for successful observation using either technique. De-oiling is a useful preparation technique for visualizing oleogels. It involves removing the oil from the sample while keeping it at refrigeration temperatures to prevent melting of the desired components during observation. De-oiling can be accomplished using organic solvents [25–27] or soaps [28–30]. However, care must be taken during de-oiling, as it can lead to various issues that may affect the matrix of interest.

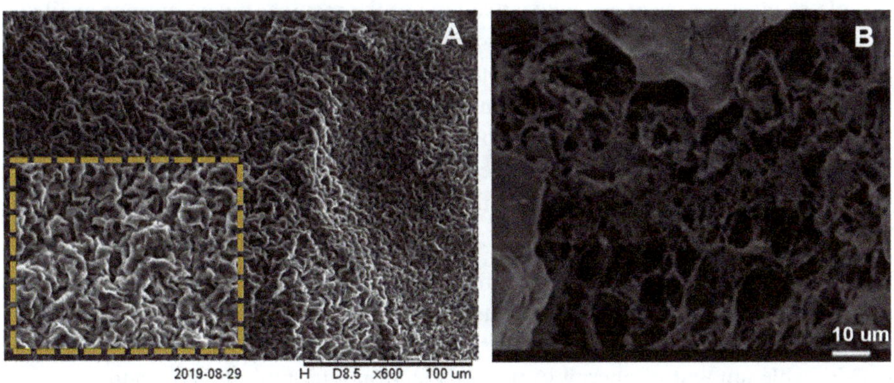

Fig. 18.8 Electron microscopy imagines of two oleogels. (**a**) SEM of an oleogel made using 10% berry wax in flaxseed oil. (Reproduced with permission from [31] under the terms and conditions of the Creative Commons Attribution License (CC BY)). (**b**) Cryo TEM of an oleogel made using 7% oleogelator (79% of fruit wax, Myrica cerifera with 21% soya lecithin) in sunflower oil. (Reproduced from [32], with permission from The Royal Society of Chemistry)

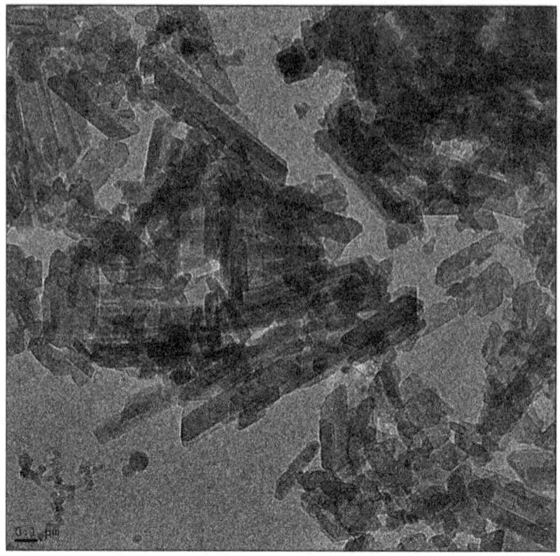

Fig. 18.9 Oleogelator particles. The oleogel was prepared using 20% tristearin in triolein. The cryo TEM image was obtained after de-oiling the oleogel to reveal only the olegelator basic unit. (Reproduced from Peyronel et al. [35] with permission from AIP Publishing)

TEM has been utilized in various systems, such as those where a hard fat is dispersed in a soft one [30, 33, 34]. For instance, Fig. 18.9 illustrates the case where nanoplatelets of tristearin (SSS) were de-oiled from triolein (OOO), the dispersant used in the oleogel. The preparation of the sample to obtain this image was crucial, as the OOO had to be entirely removed while preserving the desired particles to be observed from being destroyed in the de-oiling process [33, 35].

Over the past decade, advances in technology for data acquisition and software analysis have facilitated users to collect and process data without much consideration for the entire process. Generally, the data obtained from both SEM and TEM are ready for analysis. The software saves the scale with the captured photo, and users can measure the desired crystals if needed. The microscope software enables sample analysis, or users can take advantage of the many software options available on the market. One such popular free software is Image J. When performing crystal measurements, it is essential to define the crystal's boundary, which can be achieved by using a threshold that employs contrasting colors. A threshold is typically used when the user wants the software to measure and report automatically. Manual measurement may not be ideal, as the user must decide on the boundary for each crystal, which is challenging to do with the naked eye.

Penalties Points

Using SEM and TEM instruments incurs penalties points due to their high costs and energy consumption. These instruments are typically located in a central facility, and users pay to access and use them. Operating an electron microscope (EM) requires a considerable amount of energy to run the electron gun, mobile parts, and pumps to produce a vacuum. The energy consumption is typically greater than 1.5 kWh, resulting in a penalty point (PP) of 2. If Cryo is also used, there is an additional cost associated with the liquid nitrogen used, which is never less than 100 m. As liquid nitrogen has a warning label, it incurs an additional PP of 2, resulting in a total of 4 PPs. Special glues, such as glutaraldehyde, may be required for SEM, which falls below 10 mL and incurs no penalty points.

Additionally, SEM requires argon and either cadmium or gold for sputter coating. Argon is dispensed at a low flow rate for 10–30 min, resulting in a reagent PP of 1 (between 10 and 100 mL). Argon has only one pictogram and no hazardous working in its label, resulting in a total PP of 1. Cadmium is used in very small amounts (less than 10 mL) and incurs no penalty points. Therefore, the PP for operating this instrument can range from 2 PPs to 7 PPs (2 + 0 + 4 + 1).

Relevant Information Regarding the Technique

Spatial length scale	Operation of equipment ($, $$, $$$)	Parameters to measure	Penalty points
Nanometers to micrometers	$$$	Crystal sizes Crystal morphologies	2–7

18.2.8 Confocal Microscopy

A confocal microscope uses a laser instead of electrons or visible light to create images. The laser is scanned across the sample to produce the image, with a narrower depth field than a traditional "wide field" light microscope. The name "confocal" comes from the fact that only information from the plane of focus is detected, achieved using a pinhole that cuts off signals that are out of focus. In addition to its narrow depth of field, the confocal microscope has several advantages, including its ability to resolve smaller details than a conventional light microscope (~100 nm) and its rapid image collection, which can be previewed before saving. The confocal microscope is particularly useful for obtaining three-dimensional images, as the focus plane can be repositioned, and the images stacked together.

To be analyzed with a confocal microscope, samples need to be fluorescent or reflective. The microscope is typically operated in epi-fluorescence mode, where the illuminated and emitted light travels through the same objective lens. This mode is particularly useful for imaging thick samples (over 10 μm deep). Fluorescence allows researchers to include stains that make molecules of interest fluoresce, allowing for the viewing of specific features in the sample. Multiple dyes can be used simultaneously to differentiate components in the sample. Most confocal microscopes have one or more lasers that can excite most common fluorescent probes using a mercury lamp (e.g., fluorescein-5-isothiocyanate (FTIC), rhodamines, and cyanine dyes), with UV lasers also available to excite probes such as 4′,6-diamidino-2-phenylindole or DAPI or Fura-2.

The reflected mode is used to image surfaces or reflective stains within samples, while the transmitted mode functions similarly to a conventional light microscope, allowing samples to be viewed by bright field, darkfield, phase contrast, or differential interference contrast. Some confocal manufacturers now support full color transmitted light imaging.

Results obtained from a confocal microscope can be interpreted similarly to any other microscope, including the size and shape of crystals, as well as the density of the crystals. Software is recommended for analysis, as the human eye may struggle to consistently distinguish between the crystal and background.

One group that took advantage of the reflection of the laser on needle-like crystals is Holey and colleagues [36]. They did not use any dyes and did not create 3D images by scanning at different focal points, resulting in a micrograph like what can be obtained from bright field. On the other hand, the gel made by Gu and coworkers [37] was a simple emulsion using 50% distilled water, 43.8%, soybean oil, 2.2% PGPR and 5% rice bran wax. The stirring was done at 90 °C before it was quickly cooled to 4 °C. Figure 18.10 shows confocal images of the above-mentioned gel. The authors had used Nile blue to stain the wax and Nile Red to stain the oil phase. It can be seen in Fig. 18.10 how easily one can distinguish one phase from another by using a different wavelength of the laser.

Penalties Points

There are some considerations to keep in mind when using a confocal microscope regarding the number of PP. While fluorescence typically requires staining, the amount needed is usually less than 10 mL, resulting in no PP. However, operating lasers does require a significant amount of power, with the Eco-scale indicating that the PPs = 2 for consumption over 1.5 kWh.

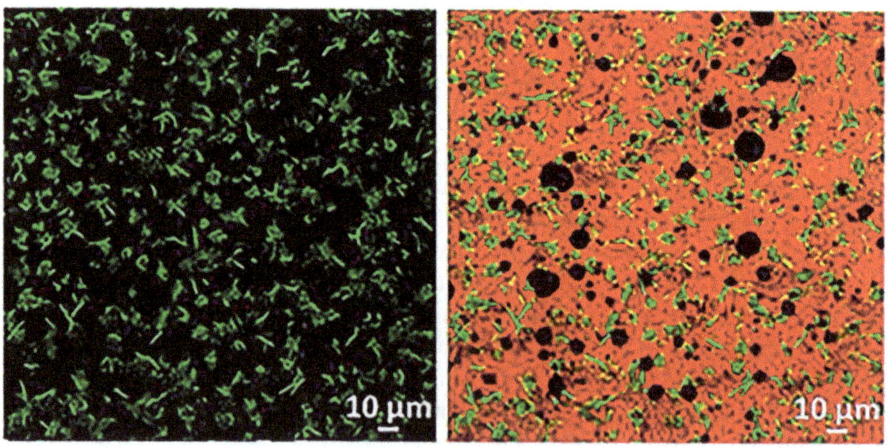

Fig. 18.10 Confocal images of a simple emulsion made using water, rice bran wax (green) and soybean oil (red droplets). The black droplets are believed to be water. (**a**) Fluorescence image of the wax crystals, (**b**) overlay fluorescence image of the oil phase and wax crystal using two different stains. (Reproduced from [37] with permission from Elsevier)

Relevant Information Regarding the Technique

Spatial length scale	Operation of equipment ($, $$, $$$)	Parameters to measure	Penalty points
Nanometers to micrometers	$$$	Crystal sizes Crystal morphologies	2

18.2.9 Atomic Force Microscopy

Atomic force spectroscopy (AFM) is a powerful tool used to detect structures of different stiffness buried within a sample, using stiffness tomography. Typically, an AFM instrument consists of an arm to which a cantilever is attached, with the cantilever chosen according to the material to be analyzed. The dimensions of the cantilever are typically in the scale of micrometers, while the radius of the tip is usually on the scale of a few nanometers to a few tens of nanometers. The atom at the apex of the tip "senses" individual atoms on the sample surface when it forms incipient chemical bonds with each atom. The system measures the forces between the chemical bonds obtained in an accurate and precise manner for each type of atom expected in the sample. A spatial map of the interactions is made by measuring the deflection at many points on a 2D surface. The beam-deflection method is popular due to its high sensitivity and simple operation, and because cantilevers can be fabricated relatively cheaply with sharp integrated tips, without requiring electrical contacts or other special treatments.

In a study conducted by Zetzl and coworkers [38], ethyl cellulose was utilized at two distinct viscosities in conjunction with either soya bean oil or canola oil, revealing the formation of a spongy structure through atomic force microscopy (AFM) analysis. However, the application of this technique posed certain challenges, primarily involving the effective removal of the oil without destroying the solid matrix to image, and the appropriate setup of the AFM instrument. To address these issues, the authors submerged the oleogel in isobutanol for a duration of 24 h, allowing for the subsequent collection of AFM images while the oleogel matrix remained immersed in water.

Penalties Points

It should be noted that if organic solvents are used in the sample preparation, there may be some penalties points (PP) associated with the technique. Equation (18.1) could be as simple as PP = 0, if the amount of solvent used is less than 10 mL, or it could become PP = 4 if between 10 and 100 mL of 2-propanol is used (2 pictograms and the word "hazard"). Additionally, an atomic force microscope is a sophisticated instrument with many different parts that require electricity to operate. Its consumption might be less than 1.5 kWh per hour or over it, hence the PP due to electricity consumption can be either 1 or 2. We have not included the penalties points (PP) associated with cantilever preparation, but it should be noted that the fabrication process involves the use of chemicals, which may contribute to PP.

Relevant Information Regarding the Technique

Spatial length scale	Operation of equipment ($, $$, $$$)	Parameters to measure	Penalty points
Nanometers to micrometers	$$	Particle sizes	1–6

18.2.10 Pulse Nuclear Magnetic Resonance (NMR)

Nuclear magnetic resonance spectroscopy (NMR spectroscopy) or magnetic resonance spectroscopy (MRS) is a widely employed analytical technique utilized to investigate the local magnetic fields surrounding atomic nuclei. The technique relies on measuring the absorption of electromagnetic radiation in the radio frequency range spanning from ~4 to 900 MHz. The sample of interest is positioned within a magnetic field, and the NMR signal is generated by exciting the nuclei with radio waves, ultimately inducing nuclear magnetic resonance that is detected by highly sensitive radio receivers. The intramolecular magnetic field surrounding an atom in a molecule alters the resonance frequency, thereby enabling the elucidation of the electronic structure of a molecule and its individual functional groups. While a frequency of 900 MHz is suitable for exciting hydrogen nuclei, the nuclear magnetic moments of different nuclei vary, leading to distinct resonance frequencies that are proportional to the field strength [39].

NMR spectroscopy is a versatile analytical tool that comprises three fundamental steps: first, nuclear spins are aligned within a magnetic field; second, a radiofrequency (RF) pulse perturbs the nuclear spins; and third, the resulting NMR signal is detected either during or after the RF pulse. For the analysis of oleogels, pulsed nuclear magnetic resonance (pNMR) represents a preferred technique. Unlike traditional NMR instruments that require substantial equipment and infrastructure, pNMR is a benchtop instrument that is readily accessible to a wide range of researchers. While pNMR has long been utilized to measure the solid fat content (SFC) of edible fats [40], it also holds promise for the simultaneous determination of fat and water content [41] and for the analysis of gelled and bigelled emulsions, provided that the water concentration does not exceed 15% [42]. Notably, pNMR represents a non-destructive analytical technique that is well suited for the investigation of complex systems [43].

In the context of oleogel analysis, pNMR is commonly employed for the measurement of solid fat content (SFC). During pNMR analysis, the oleogel is subjected to electromagnetic excitation, which leads to interactions between liquid-state nuclei and the surrounding nuclei. The pNMR is a dedicated instrument that only induces protons, or hydrogens atom to go into an excited state and since uses pulses, as soon as the pulse is removed, the hydrogen nuclei relax from their excited state. Solid-state hydrogen nuclei relax faster than the liquid-state hydrogen nuclei. The interaction of the nuclei is recorded as they relax in a curve called the Free Induction Decay (FID). The data from the FID enables the calculation of the number of solids-proton nuclei in the sample. This technique is particularly useful in analyzing oleogels due to the high number of hydrogen atoms present in almost all the molecules. A compound present in the liquid phase will retain its full diffusive mobility, while a constituent of a crystal or polymeric network will be substantially immobilized within the gel matrix. The percentage of solid matter in the system, referred to as the solids content, is commonly used to express the amount of solid content in the sample. By employing this method, researchers can accurately determine the solids content and assess the performance of oleogels in various applications.

Over the years, various standardized methods have been established to measure the SFC, including the AOCS direct method Cd 16b-93 [44], AOCS indirect method Cd 16–81 [45], as well as international standards such as ISO 8292 and IUPAC 2.150. These methods provide users with reliable and reproducible means of measuring SFC in order to ensure consistency and accuracy in the assessment of edible fats. These measurements are typically conducted using benchtop equipment, such as the one illustrated in Fig. 18.11a, which must be calibrated prior to use. This assessment is performed after oleogels have been formed and involves a storage period at a predetermined temperature to evaluate the effect of the gelator on structuring development. The oleogel can be directly placed in the sample container for measurement or molten oleogel can be poured into the container and subjected to a preset program that may include agitation and temperature control, as shown in Fig. 18.11b–e. To obtain precise and reproducible measurements, it is essential to comply with the recommended amount or height of the sample in the container specified by the equipment supplier. To ensure accuracy, it is advisable to allow a minimum of 24 h at the structuring temperature before taking measurements and to stabilize the oleogel for at least 1 h at the temperature designated for the measurement of SFC.

To facilitate thermal treatment of the oil sample for the purpose of structuring it prior to performing the SFC measurement, a variety of heating media can be employed, as depicted in Fig. 18.11b–e, contingent upon the specific experimental configuration. The reading on the equipment is obtained by directly inserting the tube containing the oleogel into the compartment of the resonance equipment. For this analysis, it is recommended to perform three to five measurements per sample to ensure accuracy and reproducibility.

Figure 18.12a displays the SFC as a function of temperature for palm olein, palm stearin, three of their mixtures and beeswax, which are materials commonly used to make oleogels. On the other hand, Fig. 18.12b shows that the amount of beeswax has an important impact on the SFC profile. The more beeswax used, the higher the SFC, no matter which blend of olein/stearin was studied.

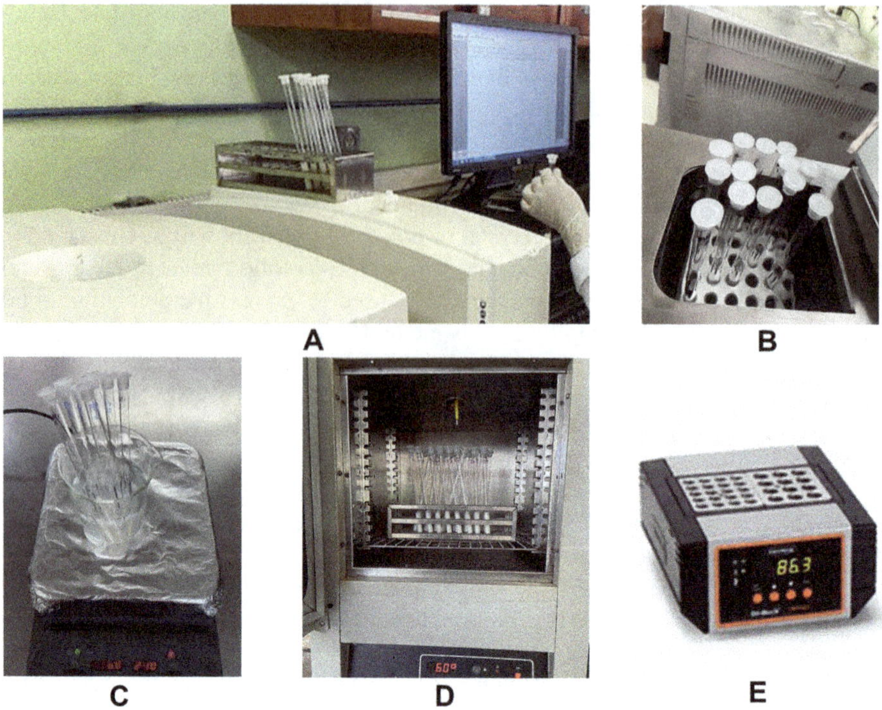

Fig. 18.11 Pulse nuclear magnetic resonance analyzer unit (**a**). Glass tubes in: a water bath (**b**), hot plate (**c**), electrical oven (**d**) and a dry heater (**e**). (*Source*: Courtesy of Food Biopolymer Lab, Faculty of Chemical Sciences of the Autonomous University of San Luis Potosí, except (**e**) (Techne Dri-block))

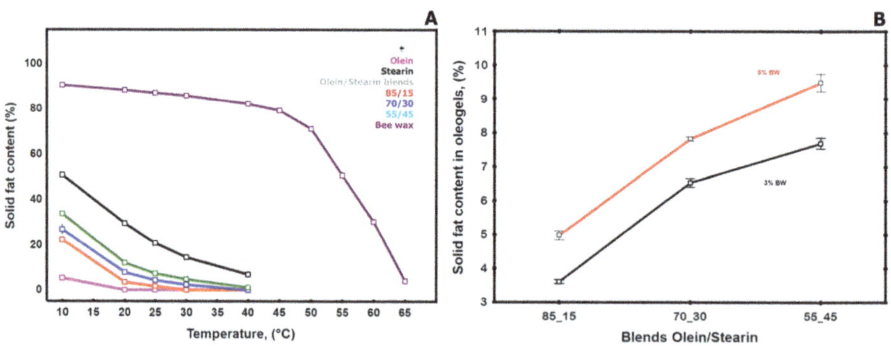

Fig. 18.12 Solid fat content of individual and mixed oils and beeswax. (**a**) Palm oil olein, palm stearin, their mixtures (85/15, 70/30, 55/45 – olein/stearin-) and beeswax. (**b**) 3% and 5% beeswax added to mixtures of olein/stearin (85/15, 70/30, 55/45). (*Source*: Courtesy of Food Biopolymer Lab, Faculty of Chemical Sciences of the Autonomous University of San Luis Potosí)

Penalties Points

This technique uses pNMR equipment to measure the SFC and does not require the use of reagents or solvents. There are two sources of electricity consumption: the first comes from the temperature profile that the sample must undergo, and the second comes from the use of the equipment required for measuring the SFC. The pNMR equipment only requires energy to generate resonance and to apply it in the oleogel. Typically, a pNMR equipment consumes less than 1.5 kWh of energy, hence resulting in PP = 1. When carrying out the AOCS protocol for the SFC measurement, 5 water baths are needed, the PP for each of them is 0, but since there are five, it could be considered a PP = 1. Thus, the total PPs required for this process could be as much as 2.

Relevant Information Regarding the Technique

Spatial Length Scale	Operation of Equipment ($, $$, $$$)	Parameters to Measure	Penalty Points
Millimeters	$$$	Solid fat content Fat solids content curves Total fat and water content in oleogel emulsion	1–2

18.3 Chemical Characterization

18.3.1 Raman Spectroscopy

Raman spectroscopy involves shining a monochromatic laser beam of light on a sample, which can be absorbed, transmitted, reflected, or scattered. This technique is concerned with inelastic scattering, which deals with different frequencies associated with the vibrational frequencies of the molecules in the sample. Although Raman scattering is about a million times less intense than elastic (Rayleigh) scattering, it provides valuable information about the sample. The Raman shift is the energy difference between the incident light and the Raman scattered light, and the resulting spectrum shows the intensity of the scattered light on the vertical axis and the Raman shift on the horizontal axis. The Raman shift can be either above the incident laser frequency (Stokes scattering) or below (anti-Stokes scattering). Additionally, Rayleigh scattering is a sharp, strong peak located between the Stokes and anti-Stokes peaks. In Raman spectroscopy, the intensity of the scattered light depends on the degree of polarizability (electron volume) for vibration modes for bonds such as S–S, C–C, and total carbon number (CN). This technique is also sensitive to homonuclear molecular bonds such as C–C, C=C, and C≡C bonds.

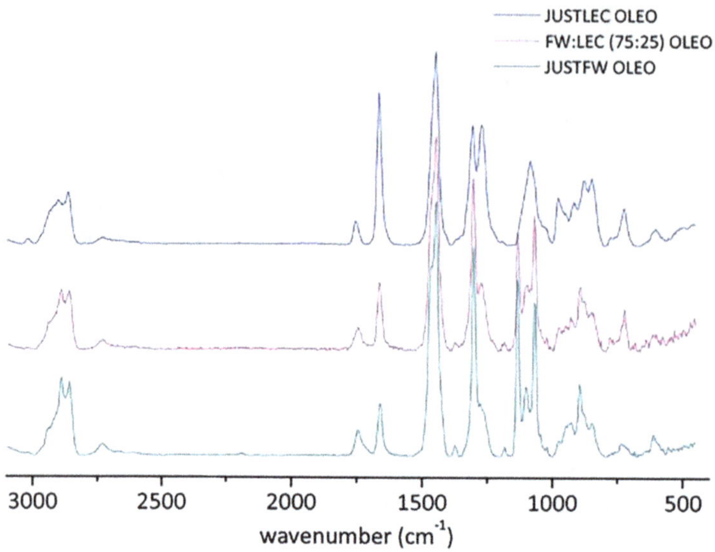

Fig. 18.13 Raman spectrum of lecithin (JUSTLEC OLEO), the mixtures of fruit wax and lecithin (FW:LEC 75:25 OLEO) and fruit wax (JUST FW OLEO). (Reproduced from [32], with permission from The Royal Society of Chemistry)

For example, in a study by Okuro and colleagues [32] Raman spectra of two oleogelators used in sunflower oil were analyzed. The oleogelators were fruit waxes (FW) and soya lecithin (LEC), and changes in the shape and frequency of the molecular vibrations' bands were monitored to explain possible interactions. The authors observed peaks at 3021 and 720 cm^{-1} in lecithin oleogels, which correspond to the asymmetric stretching vibrations of the choline head group and the C-N stretching, respectively. The asymmetric CH_2 stretching mode was 2899 and 2889 cm^{-1} for LEC and FW, respectively. When LEC and FW were mixed, the authors observed a peak at 2886 cm^{-1}. They also found that the symmetric CH_2 stretching mode was the same for both single-component oleogels (2858 cm^{-1}) but shifted to 2860 cm^{-1} in the mixture (Fig. 18.13).

Penalty Points

Although there is no sample preparation for this technique, the equipment itself requires large amounts of electrical power (more than 1.5 kWh), resulting in two penalty points.

Relevant Information Regarding the Technique

Spatial length scale	Operation of equipment ($, $$, $$$)	Parameters to measure	Penalty points
Atomic scale	$$$	Bonds between atoms	2

18.3.2 Fourier Transform Infrared (FTIR) Spectroscopy

Fourier transform spectroscopy is a powerful analytical technique that enables the simultaneous detection of all wavelengths of infrared light transmitted or reflected by a sample. Rather than scattering, the technique relies on the detection of interference. When the sample is irradiated with infrared radiation, the covalent bonds within its molecules get excited, producing vibrations that include stretching and bending modes. The absorbed wavelength of light is a function of the energy difference between the at-rest and the excited vibrational states. Fourier transform infrared (FTIR) spectroscopy uses an interferometer to modulate the wavelength of light from a broadband infrared source. The signal detected by the instrument is an interferogram, which is then analyzed using Fourier transforms to obtain a single-beam infrared spectrum. The resulting FTIR spectra are usually presented as plots of intensity versus wavenumber (cm^{-1}), where wavenumber is the reciprocal of the wavelength. The intensity can be plotted as either the percentage of light transmittance or absorbance at each wavenumber.

FTIR analysis provides valuable information for material identification by identifying molecular components and structures. The absorption peaks in the fingerprint region (Region 2, <1000 cm^{-1}) are unique to each molecule and can be compared to known standards for identification. However, the fingerprint region is complex, and assigning peaks can be challenging. The functional group region (Region 1, 1000–4000 cm^{-1}) is the most informative and contains peaks that correspond to specific functional groups. Polar covalent bonds are IR active, indicating that they will absorb in the IR spectrum. IR spectroscopy is sensitive to heteronuclear functional group vibrations and polar bonds, such as OH stretching in water. The intensity of the spectral absorption depends on the size of the dipole moment for vibration modes for bonds such as C=O and O–H. Some functional groups can be "viewed" as combinations of different bond types. For example, an ester (CO_2R) contains both C=O and C–O bonds, which are typically seen in an IR spectrum of an ester.

FTIR analysis can be a qualitative tool for material identification, but it can also be used as a quantitative tool to quantify specific functional groups when standard reference materials and a good understanding of the chemistry are available. The intensity of the absorbance correlates with the quantity of "functional groups" present in the sample. However, the presence of other chemical and matrix factors can cause the peak position to shift. Lookup tables for IR spectra can be found in the literature, but they are limited in scope and depth compared to the millions of industrial chemicals available. As a result, researchers often rely on previous research in the same area to identify bands.

Figure 18.14 shows an FTIR spectrum of an oleogel made with a mixture of β-sitosterol and beeswax as the oleogelator in sunflower oil [46]. The fingerprint region of the spectrum reveals significant differences in the position of peaks when the β-sitosterol and beeswax are separated. The peak at 1460 cm^{-1} corresponds to the –OH bending, while the peak at 1170 cm^{-1} is associated with the –CO

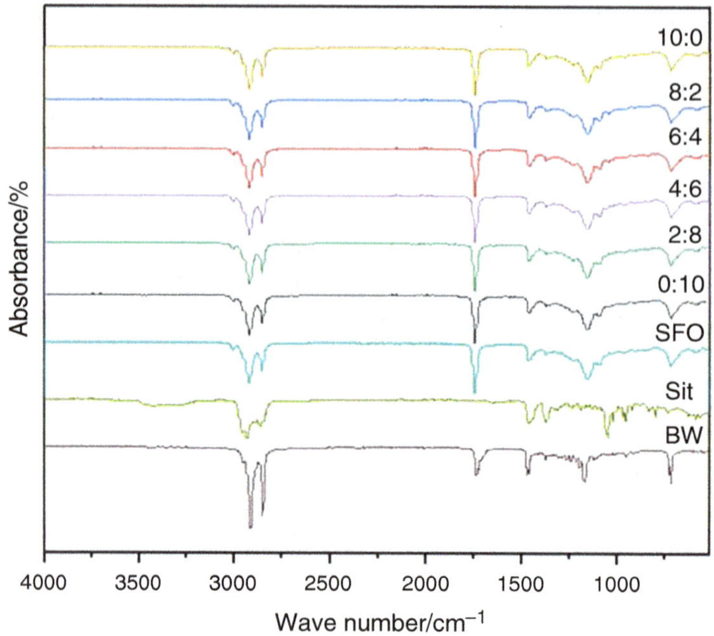

Fig. 18.14 FTIR spectra of two oleogelators (Beeswax, BW, and β-sitosterol, SIT), the oil (sunflower SFO) used to make the oleogel and six different mixtures of oleogelators (SIT:BW). The oleogels contained 10% of the mix of oleogelators, (0:10, 2:8, 4:6, 6:4, 8:2, 10:0). (Reproduced from [46] with permission from John Wiley and Sons)

stretching. In contrast, there is little difference among the oleogels in the functional group region. Both β-sitosterol and beeswax show peaks in the C−H stretching region (2800–3000 cm^{-1}). The FTIR results shown in Fig. 18.14 suggest that the oleogels made with mixtures of β-sitosterol and beeswax were not formed by creating new covalent bonds.

Penalty Points

It should be noted that the FTIR instrument used for this analysis requires a significant amount of electrical power, as it uses a laser in the microscope. As a result, it is considered to be a high-power instrument, with PPs = 2.

Relevant Information Regarding the Technique

Spatial length scale	Operation of equipment ($, $$, $$$)	Parameters to measure	Penalty points
Atomic scale	$$$	Characterize elements and functional groups	2

18.3.3 Gas Chromatography

Gas chromatography (GC) is a powerful analytical separation technique used for the separating volatile compounds present in a gas phase. The technique can be applied to samples that are either in the gas phase or require prior derivatization and pyrolysis. A gaseous sample is injected onto the head of a chromatographic column, which is typically several meters long and filled with a stationary phase. Elution of the analytes in the samples is brought about by the flow of an inert gaseous mobile phase such as helium, argon, nitrogen, carbon dioxide, and hydrogen. The stationary phase can either be a solid adsorbent (GSC, gas-solid chromatography) or a liquid immobilized on an inert support (GLC, gas-liquid chromatography), with the latter being the more commonly used technique. Separation of analytes is achieved based on their relative vapor pressures and affinities to the stationary phase [47]. Hence, the choice of column is critical, as it significantly impacts the separation of analytes. Samples are usually not suitable for direct introduction into the GC instrument. It is crucial to refer to the literature for optimal sample preparation protocols.

Despite its origins in the separation of amino acids in 1950, gas chromatography has since become a widely utilized technique due to its high sensitivity, precision, and accuracy, as well as its ability to operate at fast rates and with small sample sizes (about few milligrams). To apply this technique, the sample must be volatile and heat stable [48] which permits the separation and analysis of the vegetable oils that comprise oleogels through GC, allowing for the determination of their respective fatty acid and triacylglycerides profile.

In order to determine the fatty acid (FA) profile in the oily phase of an oleogel (pure or blended vegetable oil), it is imperative to first derivatize the oily phase to separate the FAs from the glycerol molecule [49], wherein the FA will dissolve in the organic solvent employed. On the other hand, to quantify the triacylglyceride (TAG) profile of the oleogel oily phase, the sample should be dissolved solely in a solvent compatible with the column to be used [50]. The prepared sample is introduced into the chromatographer via the injection port, using a microsyringe to dispense the desired amount into the vaporization chamber. The majority of commercial gas chromatographs support injection in two modalities: split or divided injections, which utilize a sample divider to direct excess sample to waste and splitless or without division injections. The vaporization chamber is typically heated to 50 °C above the lowest boiling point of the analytes present in the sample. The vaporized sample is mixed with the carrier gas at the head of the column and transported through the column.

To ensure accuracy and reproducibility in the determination, process, precise control of the column temperature is essential. The column temperature is controlled using an equipment oven, which can operate in either isothermal programming where the column temperature is maintained constant throughout the separation, or temperature programming, where the column temperature is increased continuously or in steps as the separation progresses. The rates of temperature increase during temperature ramps can vary depending on the elution interval of the fatty acids in the column utilized [47]. Additionally, multidimensional gas chromatography (MDGC/GCxGC) is now available as a chromatography method for separating complex samples with components that have similar retention times. This technique involves passing the eluent through two or more columns, as opposed to the traditional single column approach.

The outlet of the GC column is the detector, which captures data as intensity over time. Several types of detectors are commonly used in GC, including the flame ionization detector, mass spectrometer (MS), electron capture detector (ECD), thermal conductivity detector (TCD), and atomic emission detector (AED), among others. The flame ionization detector and mass spectrometer are the most employed detectors for separating fatty acids. The GC software records the raw data as a chromatogram, which can be further processed by the user. The chromatogram displays peaks at specific elution times, corresponding to the retention times of the analytes present in the sample. For instance, In the case of FAs in the oily phase of an oleogel, the retention time is dependent on the length of the hydrocarbon chain and degree of unsaturation of each FA. In a well-resolved sample, the number of analytes is equivalent to the number of peaks observed, and the relative abundance of each component is represented by the area under the curve of each peak relative to the total area [47]. If standards are used, absolute quantification can be performed for each of peaks corresponding to each analyte. To compliment this methodology, the official methods of the AOCS can be consulted [51, 52], or published works on the quantification of short-chain, unsaturated and *trans* fatty acids [53, 54].

Oleogels are commonly produced using unsaturated oils as solvent [55]. To demonstrate the application of this technique, we present the FA chromatogram of avocado oil in Fig. 18.15. Chromatography was performed using a flame ionization detector, with the injector temperature maintained between 220 and 250 °C and the

FATTY ACID (FA)	COMPOSITION IN AVOCADO OIL (%, w/w)
Palmitic, C16:0	14.96 (0.24)*
Palmitoleic, C16:1	5.52 (0.04)
Stearic, C18:0	0.72 (0.03)
Oleic, C18:1c	59.58 (0.76)
Oleic, C18:1c	5.03 (0.12)
Linoleic, C18:2	13.24 (0.45)
Linolenic, C18:3	0.95 (0.01)
FA Saturated	15.68 (0.24)
FA Insaturated	84.32 (0.76)
-Monounsaturated	70.13 (0.76)
-Polyunsaturated	14.19 (0.45)
-*Trans*	0.00 (0.00)
* Mean and (standard deviation), n=4	

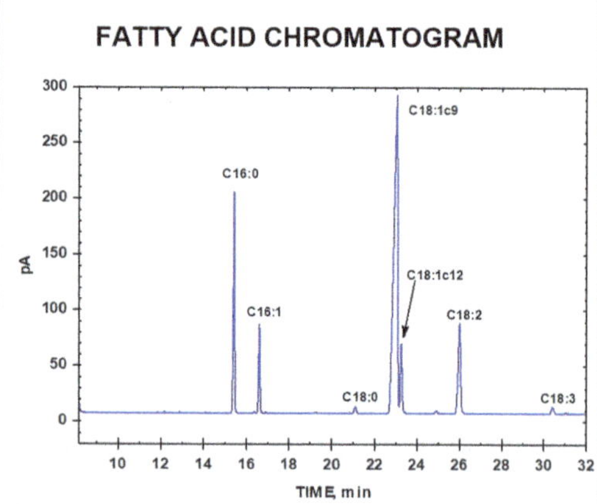

Fig. 18.15 Fatty acid profile and chromatogram obtained by gas chromatography of avocado oil used in an oleogel formulation. (*Source*: Courtesy of Food Biopolymer Lab, Faculty of Chemical Sciences of the Autonomous University of San Luis Potosí)

PRINCIPAL TAGs	COMPOSITION IN AVOCADO OIL (%, w/w)
POP	2.89 (0.00)*
POPo	7.14 (0.00)
PLPo	2.42 (0.06)
POO	18.24 (0.06)
POL	14.77 (0.02)
PLL	3.78 (0.00)
SOO	1.33 (0.05)
OOO	33.96 (0.32)
OLO/OOL	8.63 (0.12)
OLL/LOL	1.60 (0.12)

*Mean and (Standard Deviation), n=4.

P: Palmitic, Po: Palmitoleic, S: Stearic, O: Oleic, L: Linoleic

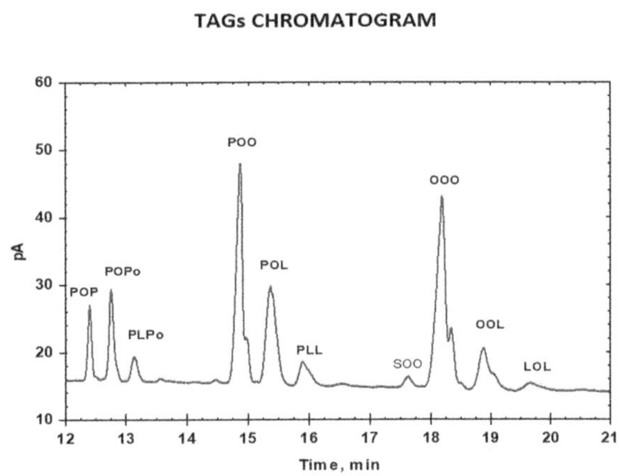

Fig. 18.16 Main triacylglycerides (TAGs) in avocado oil, which were analyzed by gas chromatography and the corresponding chromatogram. (*Source*: Courtesy of Food Biopolymer Lab, Faculty of Chemical Sciences of the Autonomous University of San Luis Potosí)

detector temperature around 250–270 °C. The AOCS official method [49] was utilized for derivatization, and separation was conducted using capillary columns. Recently, long-length capillary columns (90 at 100 m) have become available on the market, allowing for the separation of omega-3 and -6 FAs, as well as *cis* and *trans* isomers of FA methyl ester, which are typically found in oleogels. The use of high efficiency column may significantly reduce the analysis time (≈10 min vs. ≈50 min). The chromatograph for avocado oil was obtained using a cyanopropyl siloxane-based stationary phases with a recommended temperature of 180 °C. One or more temperature ramps can be employed to increase the temperature to values of 210 °C or higher, aiding in separation. A split mode injection of 100:1 or 50:1 ratio with helium as the carrier gas provided good resolution; we used a 100:1 ratio and a 1 μL injected sample volume. FA methyl esters were identified by comparing their retention times to those of FA standards analyzed under the same conditions as the sample and were quantified as a percentage of the area of the corresponding peak with respect to the sum of the areas of all peaks. Additionally, calibration curves may be created for specific analytes.

The physical structure of oleogels can be influenced by the composition of TAGs and their interaction with the oleogelator, as noted by Van Bockstaele and colleagues [56] TAGs have a more complex chemical structure than FAs, necessitating higher temperatures for their volatilization. The composition of TAGs can be determined via GC using the AOCS Ce 5–86 method [57]. Briefly, this technique involves a phenyl-methyl-polysiloxane-based capillary column with a length of 30 m, internal diameter of 0.32 mm, and a stationary phase thickness of 0.15 μm. The column temperature starts at 200 °C and is increased in two heating ramps to 300 °C at 25 °

C/min and to 350 °C at 30 °C/min. Helium is utilized as a carrier gas with a flow rate of 1.36 mL/min, while the injector and detector temperatures are set to 360 and 370 °C, respectively. The samples are dissolved in chloroform (10 mg/mL) and injected in split mode (50:1), with 1 μL of this solution.

The manufacturer of the chromatographer (e.g., ChemStation MSD from Agilent Technology) provides software that is utilized for data analysis. The peaks are identified based on the CN, excluding glycerol carbons, and their confirmation is done by analyzing commercial standards under identical conditions as the sample. The retention time of each peak directly corresponds to the CN, as shown in the chromatogram. In cases where there is the presence of TAG with the same CN, the carbon equivalent number can also be calculated [58]. Figure 18.16 presents the TAGs profile found in the avocado oil used to obtain the FA profile. The predominant TAGs consist of unsaturated hydrocarbon chains, which aligns with the FA findings. GC chromatography can also be employed to investigate the chemical profiles of waxes and gums, as noted by Doan and colleagues [59]. This method therefore offers a comprehensive understanding of the composition of oleogels, waxes, and gums, which can be valuable in various applications.

Penalties Points

The GC technique involves the use of solvents and reagents for sample derivatization, which can result in the imposition of penalty points (PPs). The PPs associated with the organic solvents and reagents used range from 4 to 13. Furthermore, the maintenance of the GC detector flame necessitates the continuous flow of nitrogen, helium, hydrogen, and air gases, resulting in an additional 2 PPs for gas consumption. The electricity consumption of a GC instrument, which exceeds 1.5 kWh, contributes an extra 2 PPs. The environmental impact of these PPs must be considered while assessing the sustainability of the GC technique.

Relevant Information Regarding the Technique

Spatial length scale	Operation of equipment ($, $$, $$$)	Parameters to measure	Penalty points
Atomic scale	$$ to $$$	Molecules identification	8–17

18.3.4 High-Pressure Liquid Chromatography (HPLC)

The High-Pressure Liquid Chromatography (HPLC) is a powerful tool for separating substances that are in suspension. The HPLC system works by utilizing the polarity differences of the eluent fluids (mobile phase) and the packing material (stationary phase) in the column to separate the compounds (analytes) in the sample. The equipment mixes the eluent with the analyte and pumps it under pressure through the column. The specific intermolecular interaction between the analyte and the stationary phase causes different molecules in the analyte to be retained inside the column for different lengths of time. This results in the different components of the analyte being eluted at different times and shown on the signal captured by the detector (chromatogram).

A signal versus time chromatogram shows characteristic peaks that correspond to the different analytes present in the sample. The peaks should be separated from each other, and this is achieved by selecting the right mobile and stationary phase as well as the eluents and the conditions of operation of the HPLC. Depending on the composition of the mobile phase, two different modes are generally applicable. If the makeup of the mobile phase remains constant during the separation process, the HPLC system is defined as an isocratic elution system. When the composition of the mobile phase is changed during the process, the HPLC system is defined as a gradient elution system.

The column is the heart of any HPLC system and is designed specifically to separate certain compounds. The column's efficiency correlates with its inner diameter, length, and type and particle size of the packing material. Different packing materials support different separation mechanisms, including normal-phase, reversed-phase, size exclusion, ion exchange, affinity, chiral, or hydrophilic interaction HPLC.

As the analyte is eluted, the detector registers the time and amount of a substance that is eluted from the column. The detector perceives the change in the composition of the eluent and converts this information into an electrical signal, which is evaluated by the aid of a computer. Different detector units, such as refractometric, UV/VIS, electrochemical, and fluorescence detectors, have been used to identify different compounds. Each individual peak detected provides qualitative and quantitative information for the analyte. The area of a peak is proportional to the concentration of the substance.

The chromatography data management software can calculate the concentration of the sample by integration after standards are run. This requires a calibration curve, which nowadays, the data management software can handle easily and efficiently. Quantitative information can then be obtained. Ideally, the peaks are recorded as a Gaussian bell-shaped curve.

The HPLC technique is highly useful for analyzing oleogels in order to determine the compounds that constitute the oil and oleogelator. Proper sample preparation is of utmost importance and requires the sample to be in solution. Edible oils contain TAGs as their main components, but minor components must also be considered to ensure accurate results. When analyzing complex systems like oleogels using HPLC, it is crucial to select an appropriate eluent for each molecule (oleogelators vs. oil phase) to avoid misinterpretation of the results. Reversed-phase HPLC is commonly used for analyzing TAGs in oils and fats [60]. Various detectors have been utilized for TAGs identification, such as ultraviolet (UV) refractive index [61], evaporative light scattering detector [62–64] and mass spectrometry [65, 66]. Regio-isomers of TAGs have been examined in vegetable oils using atmospheric chemical ionization (APCI) mass spectrometry [60, 63, 65, 67]. The positional distribution of fatty acyl chains of TAG has been analyzed in palm oil, cocoa butter, beef, pork, and chicken

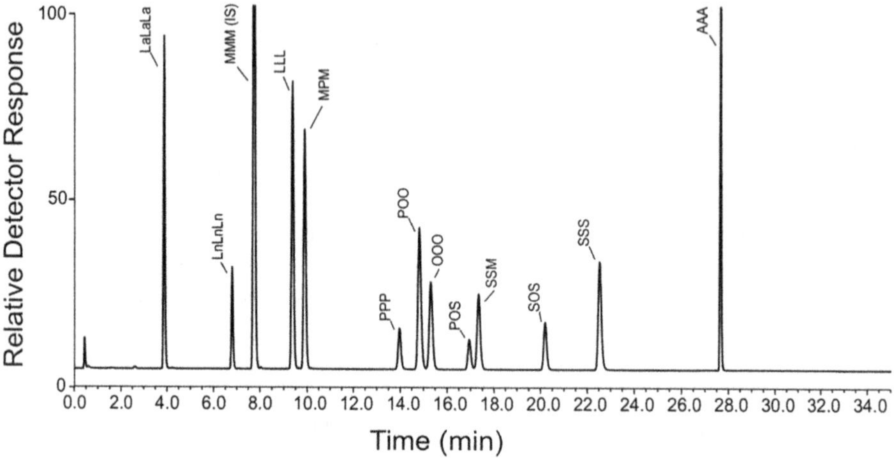

Fig. 18.17 UPLC-ELSD chromatogram of a mix of triacylglyceride (TAG) standards. Where the peaks correspond to the following TAGs: Triolein (OOO), trilaurin (LaLaLa), trilinolenin (LnLnLn), trilinolein (LLL), tripalmitin (PPP), tristearin (SSS), triarachidion (AAA), trimyristin (MMM), 1-pal-mito, 2,3-olein (POO), 1-palmito, 2-oleo, 3-stearin (POS), 1,2-stearo, 3-myristin (SSM), and 1,3-myristo, 2-palmitin (MPM). (Reproduced from [68] with permission from John Wiley and Sons)

fats using HPLC-APCI-MS. Individual regio-isomers were identified and quantified by linear calibration against the standards [65].

Figure 18.17 displays the peaks of some TAGs run on a reversed-phase liquid chromatography coupled with an evaporative light scattering detection (ELSD) by Ross and colleagues [68]. The authors utilized an Agilent Zorbax C18-XDB (100 mm, 62.1 mm, 1.8 μm) C18 column heated to 55 °C with a flow rate of 0.5 mL/min and an injection volume of 1 μL. Methanol and acetone were used as mobile phases, and a gradient was employed to elute the prepared standards using a sample that took 35 min in total. The molecules utilized for standards (samples) were prepared in dilutions of OOO in CH_3Cl_2 and $MeCl_2$ containing approximately 0.7 mg/mL of trimyristin (MMM) as the internal standard. Samples were prepared at a concentration of approximately 5 mg/mL by dissolving oils or fats in the CH_2Cl_2 internal standard solution.

It is important to note that Fig. 18.17 is just one example of how TAGs can be detected, and the resolution and ability to detect separate peaks depend not only on the HPLC but also on the detector. Mass spectrometry coupled with HPLC is a powerful technique that can help identify compounds that are difficult to separate into different peaks.

Penalties Points

The use of an HPLC incurs penalties points due to the solvents required for sample preparation and elution into the column. The Eco-Scale categorizes solvent use into three categories: less than 10 mL, 10–100 mL, and more than 100 mL. However, running an HPLC requires equilibration and continuous elution of the analyte, so the amount of solvent used is typically over 100 mL. Organic solvents are not considered "good solvents," with two to four pictograms and either the "warning" or "hazard" word. Therefore, penalties points from one organic solvent can range from 2 to 6 PPs in the lower bracket and can increase to 6–18 PPs when a combination of solvents is used. Additionally, the electricity usage of an HPLC is greater than 1.5 kWh, resulting in the addition of 2 PPs due to electricity usage.

Relevant Information Regarding the Technique

Spatial length scale	Operation of equipment ($, $$, $$$)	Parameters to measure	Penalty points
Atomic scale	$$$	Molecules identification	8–20

18.3.5 Oxidative Stability

The oxidation of unsaturated fats is a well-established phenomenon that contributes to the deteriorations of oleogels. Fu and collaborators [69] have demonstrated that oleogels can undergo oxidative degradation due to their high content of unsaturated FAs from oils. This oxidative process can lead to undesirable changes in the aroma, color, and flavor of the product, as well as the loss of certain nutrients. Moreover, the formation of potentially harmful substances can further compromise the safety of the oleogels, ultimately resulting in a shortened shelf life. The economic implications of this phenomenon are significant, as it renders the affected oleogels unsuitable for consumption. Therefore, the scientific community must focus on developing effective strategies to prevent and mitigate this type of degradation. The economic implications of this phenomenon are significant, as it renders the affected oleogels unsuitable for consumption. Therefore, the scientific community must focus on developing effective strategies to prevent and mitigate this type of degradation.

Lipolysis or hydrolytic rancidity can occur in lipids due to the activity of lipases present in oils. However, the primary mode of lipid oxidation is initiated through a free radical chain propagation reaction, which involves the formation of peroxides and hydroperoxides from fatty acids and oxygen, known as auto-oxidation [70]. These compounds are highly unstable and can readily undergo breakdown, generating additional free radicals and perpetuating the oxidation process through a chain mechanism involving three distinct phases.

The induction phase is initiated by external factors such as light, heat, and trace heavy metals, which trigger the formation of peroxides and radical species. In the propagation phase, the oxidation of free radicals results in the formation of additional hydroperoxides and free radicals, which in turn participate in the oxidation chain. Ultimately, in the end phase, the concentration of reactive compounds reaches a critical threshold, triggering their interaction with one another and leading to the formation of deteriorated products. The concentration of peroxide radicals diminishes as the formation of degraded products reaches a state of equilibrium, ultimately leading to a significant reduction in the quality and shelf life of lipids.

The oxidative stability of oleogels can be assessed by evaluating their oxidation level as a function of temperature and storage time. Chemical measurement of oxidation involves the quantification of compounds generated during the oxidation process, including peroxides, aldehydes, and ketones. Several methods are available for this purpose, including thiobarbituric acid, p-anisidine value, the amount of free fatty acid or conjugated dienes, and analysis of volatile oxidation products. Among these methods, peroxide quantification is commonly used to assess oxidation in oleogels [19, 69]. The peroxide value (PV) is expressed in milliequivalents or mmol of active oxygen per kilogram of lipid and measures the initial oxidation state of an oil. The American Oil Chemists' Society (AOCS) and the International Union of Pure and Applied Chemistry (IUPAC) offer standard procedures for this assessment.

Recently, VerraGlo LLC and the University of Kentucky collaborated [71] to develop patented foundational technology for metal-phosphate luminophores (MPL) sensors with peroxidase-like activity. These sensors emit luminescence proportional to the hydroperoxide content of edible fats and oils, offering a novel technique for peroxide value measurement that does not use or generate toxic solvents. This technique is abbreviated as PV-LED (Peroxide value light emitting diode) and has several advantages over traditional methods of peroxide quantification.

Regarding PV, Cho and colleagues [19] monitored the oxidation state of edible rice bran wax-gelatin biphasic gels for 175 days and found no significant changes in PV (between 0.17 and 2.62 mEq/kg), even in biphasic gels added with Cu2+ as a pro-oxidant agent. Giacomozzi and colleagues [72] also measured PV to study the storage stability of oleogels made from monoglycerides and high oleic sunflower oil, and similarly did not find significant changes in PV. However, Fu and colleagues [69] studied the thermo-oxidative behavior of ethyl cellulose oleogels and found that the presence of the gelling agent ethyl cellulose interfered with the reagents used in the determination of PV (KOH and $Na_2S_2O_3$), making this test inefficient and lowering the p-Anisidine value for evaluating the oxidation of oleogels. In a critical review of oleogels including oxidation, Park and Maleky [73] concluded that the oxidative stability of oleogels requires further research, particularly concerning the lack of information on the oxidative stability of foods formulated with oleogel and the effect of antioxidants in oleogel foods.

When evaluating the oxidation behavior of oleogels during shelf life, it is not only important to quantify the oxidation products present but also to predict the oxidative stability as a function of storage time through accelerated tests. There are various types of lipid analyses to predict this stability, with the most common being the active oxygen method (AOM) and the oxidative stability index (OSI) or Rancimat method [74], which expose the sample to increased temperature and elevated oxygen pressure. The AOM method, which was introduced in 1933, was the most widely used method for several years. O'Keefe and colleagues [75] used this method to compare the oxidative stability of peanut oil with high and normal oleic acid content. The AOCS officially adopted the AOM method under reference Cd 12–57 [76]. This

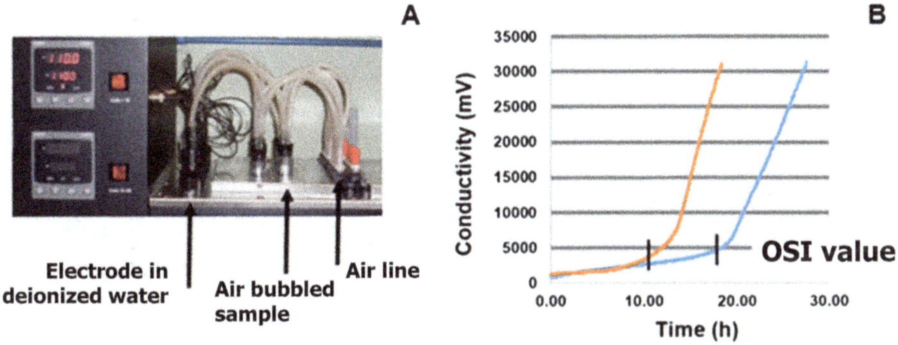

Fig. 18.18 Oxidative stability: (**a**) arrangement of elements in the instrument and (**b**) conductivity results as shown by the instrument for two vegetable oil samples indicating the oxidative stability index (OSI) value. (*Source*: Courtesy of Food Biopolymer Lab, Faculty of Chemical Sciences of the Autonomous University of San Luis Potosí)

method involves monitoring the peroxide index evolution in a sample subjected to the passage of an air current at a temperature of 97.8 °C. The stability of the sample is determined as the time required for the fat to reach a peroxide value (PV) of 100 mEq oxygen/kg lipid or, in some cases, to reach a PV of 10 mEq oxygen/kg lipid, which is the value at which the rancidity of the product is perceived. The OSI or Rancimat method is a variation of the AOM method that uses higher temperatures. The AOCS officially adopted the OSI method under reference Cd 12b-92 [77], with 110 °C as the standard value, although temperatures up to 140 °C are also used. This method involves bubbling purified air through the sample of fat or oil in a glass tube placed in a thermostatic aluminum heating block (Fig. 18.18a). As oxidation proceeds, the air leaving the sample passes into a container with deionized water, whose conductivity is monitored with a special electrode. The induction period or OSI value, in hours, is calculated by the equipment software as the point where a significant change in the increase in conductivity occurs (Fig. 18.18b), corresponding to the maximum of the second derivative of this with respect to time. de Man and colleagues [78] reported that the OSI value reached was mainly due to the presence of formic organic acid (more than 50%), followed by the presence of acetic, propionic, butyric, valeric, and caproic acids, as lipid oxidation products, for different types of lipids.

At present, a quick and convenient small-scale oxidation test endorsed by the AOCS (Cd-12c-16 [79] and the American Society for Testing and Materials (D 8206 ASTM International) is available. This method requires less time and handling than traditional oxidative stability techniques, making it ideal for solid, doughy, and liquid samples. Furthermore, this method boasts excellent reproducibility. Frolova and colleagues [80] and Li and colleagues [81] have both investigated the oxidative stability of oleogels using rapid testing methods. Frolova's research group conducted a study examining beeswax oleogels in sunflower oil (SO) at concentrations of 3% and 6%. The processed oleogels, as well as pure SO, were subjected to a 30-min heating period at 90 °C. To evaluate the impact of the oleogel technology on oxidative stability, an untreated SO sample was included as a reference. Subsequently, all the samples were stored at 35 °C, and their peroxide value (PV) was assessed over a duration ranging from 0 to 144 h. The findings indicated that the oleogel production method had minimal influence on the oxidative stability of sunflower oil. However, the incorporation of beeswax as a structuring agent resulted in an elevation of the peroxide value during storage, indicating a pro-oxidant effect of beeswax dependent on its concentration in the oleogel. Therefore, after 144 h of storage, the PV (measured in mEq/kg) for the samples were approximately 2.5, 3.2, 12.8, and 14.5, respectively, corresponding to SO, SOox, oleogel 3%, and oleogel 6%.

In the international market, there are different brands of automated equipment for the determination of oxidative stability during storage.

DSC is another approach to evaluate oxidative stability as a function of temperature. In Sect. 18.3, general guidelines for this technique are provided with two differences to note. First, the calorimeter should have an alternate line to supply air (or pure oxygen if the oxidation study requires it, or nitrogen for non-oxidation control tests) to the internal cell. Second, the sample must be exposed to the flow of air or oxygen (i.e., the crucible should not be sealed). The measurement can be carried out at a constant temperature, or the sample can be subjected to a temperature sweep. The equipment software should be programmed with the desired conditions for the oxidation study. Typically, oxidative stability is evaluated at a constant temperature. In this case, the baseline response in the thermogram undergoes a significant change in its slope at the onset of oxidation, which corresponds to the oxidation induction time for the sample analyzed. In this case, the heat flux, expressed as the horizontal baseline in the thermogram, undergoes a significant change when oxidation starts, manifesting this as a deviation. In general, the crossing point of the tangents before and after the change in heat flux indicates the oxidation induction time of the analyzed sample (Fig. 18.19).

In 2002, Tan and colleagues [82] conducted a comparative investigation on the oxidative stability of various commercial vegetable oils, evaluated by both the Rancimat method and DSC at a constant temperature of 110 °C. The results of the study revealed that the DSC method produced a faster oxidative induction time compared to the Rancimat method. Depending on the type of edible oil tested, the oxidation induction time values obtained via DSC were 7–260% lower than those obtained by the Rancimat method. Given its shorter testing time, the DSC method could potentially be applied to oleogel systems.

Penalties Points

The oxidative stability technique evaluated by chemical methods involves the use of solvents and reagents for titration, which can result in the imposition of penalty points (PPs). The PPs associated with reagents used range from 2 to 6. The maintenance of the DSC necessitates the continuous flow of nitrogen, but its use does not generate PPs. The electricity consumption of a DSC, OSI, or Rancimat equipment, which exceeds 1.5 kWh in each case, contributes an extra 2 PPs. The environmental

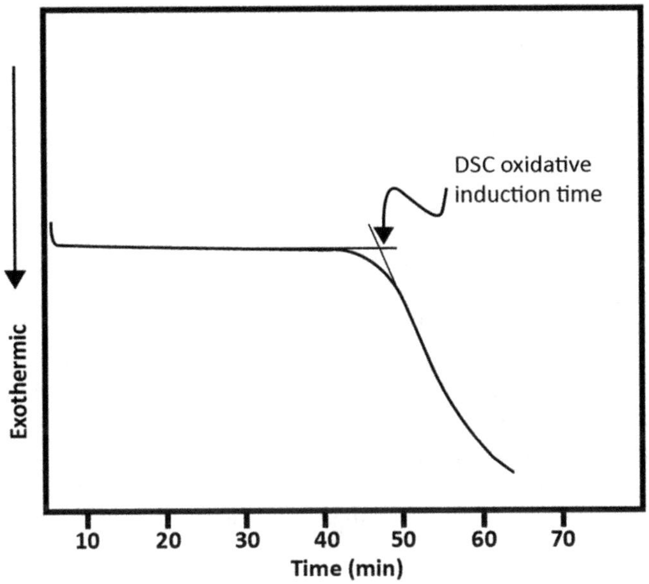

Fig. 18.19 Typical differential scanning calorimetry oxidation curve at constant temperature in an edible vegetable oil

impact of these PPs must be considered while assessing the sustainability of the oxidation technique.

Relevant Information Regarding the Technique

Spatial length scale	Cost ($, $$, $$$)	Application	Penalty points
Atomic scale	$–$$	Quantification of oxidation products	6 when using chemical methods 8 non-automatic AOM 2 OSI-Rancimat

18.4 Thermal Characterization

18.4.1 Differential Scanning Calorimetry Technique

A differential scanning calorimeter detects endothermic and exothermic transitions due to the heat that is either provided or removed from the sample. The word "differential" is used to indicate that a sample and a reference are used and that the difference between these two is the signal reported. Although this technique does not provide direct information about its chemical composition under given experimental conditions, it does provide useful information on the nature of the thermodynamic changes associated with the transformation of oils and fats from one physical state to another. These thermodynamic characteristics are sensitive to their general chemical composition and, therefore, can be used in qualitative and quantitative ways in their identification [83].

The complexity of the thermal profiles of fats and oils is due to the great variety of TAGs as their main constituents [84]. This is why both fats and oils do not have specific melting and crystallization temperatures, but show melting and crystallization profiles [83].

DSC devices are built according to one of two basic measuring principles: the heat-flux (HF) principle and the power-compensated DSC principle. The HF DSC allows heat to flow separately into the sample and the reference. This flow is measured while the sample temperature changes at a constant rate. For this type of DSC analysis, a crucible (also called pan) for the sample and a crucible for the reference are placed on the same platform. Integrated temperature sensors on the platform are used to measure the temperature of the crucibles. The platform is located inside one temperature-controlled furnace. Inert gas is passed through the cell at a constant flow rate.

There exist different designs of HF DSC. The one mentioned above falls in the category of platform or disk type. Another one uses cylinders instead of a platform, and the third kind is a turret type [85]. The principle is the same [85, 86] for all of them as well as the range of temperatures that they covered (≈ -50 to ≈ 500 °C).

A micro differential DSC is a modified HF DSC that has high sensitivity with a narrower temperature range (-20 to ≈ 120 °C). This type of device is ideal to study crystallization because the cooling and heating rates can be lower than 0.001 °C/min which is suitable for determining phase transitions like intermediate phases between solid and liquid, only detectable when the heating/cooling rate is very slow. When the HF DSC is used under modulation (MD) or under pressure (P), the measurement becomes very precise. Both MD and P allow the separation of endo- or exo-peaks that might overlap in the DSC scans. In MD mode, the typically linear heating ramp is overlaid with the sinusoidal function defined by a frequency and amplitude to produce a sine wave shape temperature versus time function.

In a power-compensated (PC) DSC, the temperatures of the sample crucible and the reference crucible are kept the same while both temperatures are increased or decreased linearly. This device measures the power needed to maintain the sample crucible temperature equal to the reference crucible temperature. To achieve this, the design of the DSC calls for two independent heating units or two furnaces.

The latest power compensating DSC is called the hyper PC-DSC. This is a fast scan DSC that can perform valid heat flow measurements with fast linear controlled rates (up to 500 K/min) especially by cooling, where the rates are higher than with the classical PC DSC. Standard DSC operates under 10 K/min. The benefits of such devices are increased sensitivity at higher rates providing the best results for an analysis of melting and crystallization of metals or detection of glass transition temperature (Tg) in medications.

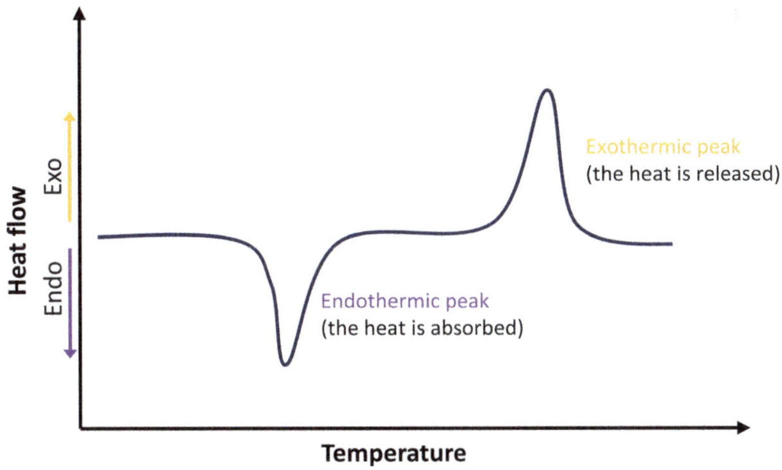

Fig. 18.20 Schematic diagram of a thermogram where heat flow was plotted as a function of the temperature (which can also be replaced by time)

Figure 18.20 shows the typical information obtained from a DSC when the temperature is either increased or decreased at a constant rate. An endothermic peak and an exothermic peak indicate the range of melting or crystallization.

Figure 18.21 illustrates the crystallization behavior of the material as measured by DSC. The data was collected by cooling the sample from 80 to 20 °C at a controlled rate, and the heat flow generated by the sample was recorded as a function of temperature. The graph clearly shows the exothermic heat flow peak associated with the crystallization process, indicating the onset and completion of the solidification process. Researchers have presented varying viewpoints regarding the characterization of the temperature at which a material undergoes crystallization. Some researchers refer to the highest point on the crystallization peak as the temperature associated with crystallization, while others consider the temperature at which the process initiates as the defining point. However, upon observing Fig. 18.20, it becomes evident that crystallization occurs within a temperature range.

The area under the peak, marked as H_c in Fig. 18.21, corresponds to the enthalpy of the phase transition (liquid to solid). If the system is under constant pressure and when the only work done in the system is due to heat, then one can equate the heat removed from the sample to the enthalpy of crystallization. Units for enthalpy are Joules (J) or J/g.

Any of the calorimeters described above can be used to study the structuring of oleogels at a constant temperature, which is an area that has been little explored until now [87]. This approach enables the determination of the induction time to structure the oleogels, as well as the enthalpy involved in the process (Fig. 18.22). For materials such as oleogels, which consist of a liquid phase and a structuring agent, the induction time is closely related to the structuring agent's concentration and its chemical structure. A change in the structuring agent concentration or a change in its chemical structure can lead to a larger or smaller induction time, indicating a larger or slower gelation process.

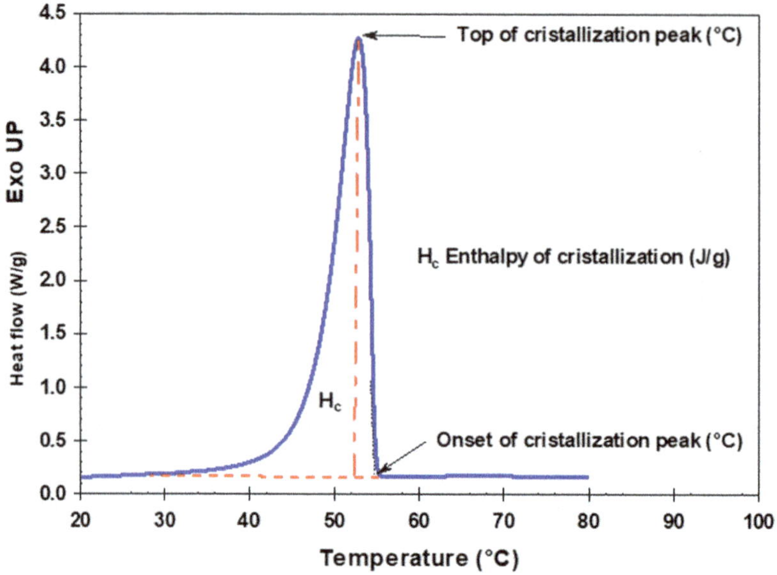

Fig. 18.21 Heat flow versus temperature showing crystallization of the material (the DSC was run by lowering the temperature from 80 to 20 °C)

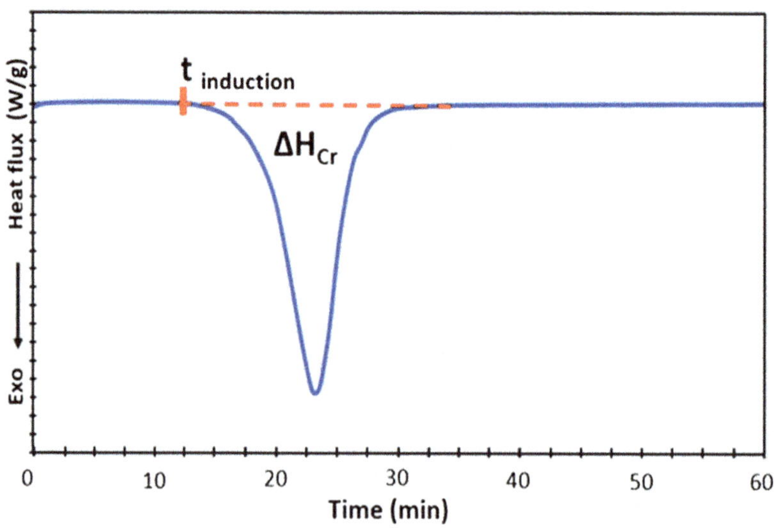

Fig. 18.22 Scheme of a DSC isothermal structuring thermogram

Penalties Points for the DSC Test

The equipment does not require any reagent or solvents to run, and the samples do not need to be prepared in any way. Hence, the PP due to the chemical used = 0. However, the equipment requires electricity to run. Runs in a DSC are never shot, requiring more than 15 min. Hence, the kWh is typically over 1.5, hence PP = 2.

Relevant Information Regarding the Technique

Spatial length scale	Operation of equipment ($, $$, $$$)	Parameters to measure	Penalty points
Millimeters micrometers	$$	Melting point Glass transition Heat capacity Enthalpy of phase transition	2

18.5 Conclusions

The objective of this chapter is to provide a thorough and all-encompassing overview of 16 techniques and methods that can be employed to investigate the structure, chemical composition, and thermal behavior of oleogels. The selection of the appropriate technique or method for investigating oleogels is dependent on the specific objectives of the study. Therefore, this chapter presents a detailed summary of each technique or method, including the length scale that it can probe and the parameters that can be evaluated. Furthermore, the Eco-Scale methodology is introduced to evaluate the environmental and health impacts of each technique by quantifying the number of reagents and solvents used and the energy consumption associated with operating the equipment using a penalty point system. It is worth mentioning that some techniques have low penalties (2 or less) as they require cheap materials such as in the oil migration studies that use filter paper or, in the consistency studies that use a manual cone penetrometer. Conversely, other techniques may have the same low penalty points but demand the use of expensive equipment, such as Raman spectroscopy. However, all techniques with more than 6 penalty points require expensive equipment and the use of reagents or solvents.

This chapter aims to raise awareness of the potential trade-offs between the cost, environmental impact, and information generated by each technique. Ultimately, minimizing the penalties points incurred during the study of oleogels can promote safe operations, reduce environmental harm, and achieve scientifically rigorous results. The information provided in this chapter can serve as a guide for researchers and professionals to select the most appropriate technique for their study of oleogels, while minimizing the impact on the environment and human health.

References

1. Gałuszka A, Migaszewski Z, Namieśnik J (2013) The 12 principles of green analytical chemistry and the SIGNIFICANCE mnemonic of green analytical practices. Trends Anal Chem 50:78–84. https://doi.org/10.1016/j.trac.2013.04.010
2. Gałuszka A, Migaszewski ZM, Konieczka P, Namieśnik J (2012) Analytical Eco-Scale for assessing the greenness of analytical procedures. Trends Anal Chem 37:61–72. https://doi.org/10.1016/j.trac.2012.03.013
3. Anastas P, Warner J (1998) Green chemistry: theory and practice. Oxford University Press, Oxford/New York
4. Anastas PT (2003) Meeting the challenges to sustainability through green chemistry. Green Chem 2:G29
5. De Man JM, Gupta S, Kloek M, Timbers GE (1984) Instrumentation for thermopenetrometry of fats. J Am Oil Chem Soc 61:1569–1570
6. Haighton AJ (1959) The measurement of the hardness of margarine and fats with cone penetrometers. J Am Oil Chem Soc 36:345–348
7. Dixon BD, Parekh JV (1979) Use of the cone penetrometer for testing the firmness of butter. J Texture Stud 10:421–434
8. de Man JM (1983) Consistency of fats: a review. J Am Oil Chem Soc 60:82–87
9. AOCS (2017) AOCS Cc 16–60: consistency, penetrometer method. In: Official methods and recommended practice of the AOCS, 7th edn. AOCS Press, Urbana
10. Gravelle AJ, Barbut S, Marangoni AG (2012) Ethylcellulose oleogels: manufacturing considerations and effects of oil oxidation. Food Res Int 48:578–583. https://doi.org/10.1016/j.foodres.2012.05.020
11. Kupiec M, Zbikowska A, Marciniak-Lukasiak K, Kowalska M (2020) Rapeseed oil in new application: assessment of structure of oleogels based on their physicochemical properties and microscopic observations. Agriculture 10:1–11. https://doi.org/10.3390/agriculture10060211
12. Chen L, Opara UL (2013) Approaches to analysis and modeling texture in fresh and processed foods – a review. J Food Eng 119:497–507
13. Gravelle AJ, Barbut S, Marangoni AG (2013) Fractionation of ethylcellulose oleogels during setting. Food Funct 4:153–161. https://doi.org/10.1039/c2fo30227f
14. Galdámez JR, Szlachetka K, Duda JL, Ziegler GR (2009) Oil migration in chocolate: a case of non-Fickian diffusion. J Food Eng 92:261–268. https://doi.org/10.1016/j.jfoodeng.2008.11.003
15. Dibildox-Alvarado E, Rodrigues JN, Gioielli LA, Toro-Vazquez JF, Marangoni AG (2004) Effects of crystalline microstructure on oil migration in a semisolid fat matrix. Cryst Growth Des 4:731–736
16. Maleky F (2012) Oil migration through fats – quantification and its relationship to structure. In: Structure-function analysis of edible fats. AOCS Press, Urbana, pp 207–230
17. Deka K, MacMillan B, Ziegler GR et al (2006) Spatial mapping of solid and liquid lipid in confectionery products using a 1D centric SPRITE MRI technique. Food Res Int 39:365–371. https://doi.org/10.1016/j.foodres.2005.08.009
18. Marty SA, Baker KB, Dibildox-Alvarado EC et al (2005) Monitoring and quantifying of oil migration in cocoa butter using a flatbed scanner and fluorescence light microscopy. Food Res Int 38:1189–1197
19. Cho K, Tarté R, Acevedo NC (2023) Development and characterization of the freeze-thaw and oxidative stability of edible rice bran wax-gelatin biphasic gels. LWT 174:114330. https://doi.org/10.1016/j.lwt.2022.114330
20. Scharf T (2007) Polarized light. In: Polarized light in liquid crystals and polymers. Wiley, Hoboken, pp 1–18
21. Rousseau D, Hill AR, Marangoni AG (1996) Restructuring butterfat through blending and chemical interesterification. 2. Microstructure and polymorphism. J Am Oil Chem Soc 73:973–981. https://doi.org/10.1007/BF02523404

22. Shi Y, Liang B, Hartel RW (2005) Crystal morphology, and textural properties of model lipid systems. J Am Oil Chem Soc 82:399–408
23. Campos R (2013) Experimental methodology. In: Structure and properties of fat crystal networks, 2nd edn. CRC Press, Taylor & Francis Group, Boca Raton
24. Mokobi F (2022) Compound light microscope- definition, principle, parts. Brightfield Microscope. https://microbenotes.com/brightfield-microscope/. Accessed 4 Apr 2023
25. Acevedo NC (2012) Characterization of the nanostructure of triacylglycerol crystal networks. In: Structure-function analysis of edible fats. AOCS Press, Urbana, pp 5–24
26. Heertje L, Leunis M, Heertje I (1997) Measurement of shape and size of fat crystals by electron microscopy. LWT Food Sci Technol 30:141–146
27. Chawla P, de Man JM (1990) Measurement of the size distribution of fat crystals using a laser particle counter. J Am Oil Chem Soc 67:329–332
28. Poot C, Dijkshoorn W, Haighton AJ, Verburg CC (1975) Laboratory separation of crystals from plastic fats using detergent solution. J Am Oil Chem Soc 52:69–72
29. Jewell GG, Meara ML (1970) A new and rapid method for the electron microscopic examination of fats. J Am Oil Chem Soc 47:535–538
30. Maleky F, Smith AK, Marangoni A (2011) Laminar shear effects on crystalline alignments and nanostructure of a triacylglycerol crystal network. Cryst Growth Des 11:2335–2345
31. Barroso NG, Okuro PK, Ribeiro APB, Cunha RL (2020) Tailoring properties of mixed-component oleogels: wax and monoglyceride interactions towards flaxseed oil structuring. Gels 6:5. https://doi.org/10.3390/gels6010005
32. Okuro PK, Tavernier I, Bin Sintang MD et al (2018) Synergistic interactions between lecithin and fruit wax in oleogel formation. Food Funct 9:1755–1767. https://doi.org/10.1039/c7fo01775h
33. Acevedo NC, Marangoni AG (2010) Characterization of the nanoscale in triacylglycerol crystal networks. Cryst Growth Des 10:3327–3333. https://doi.org/10.1021/cg100468e
34. Pink DA, Quinn B, Peyronel F, Marangoni AG (2013) Edible oil structures at low and intermediate concentrations. I. Modeling, computer simulation, and predictions for X ray scattering. J Appl Phys 114:234901. https://doi.org/10.1063/1.4847996
35. Peyronel F, Ilavsky J, Mazzanti G et al (2013) Edible oil structures at low and intermediate concentrations: II. Ultra-small angle X-ray scattering of in situ tristearin solids in triolein. J Appl Phys 114:234902. https://doi.org/10.1063/1.4847997
36. Holey SA, Sekhar KPC, Mishra SS et al (2021) Effect of oil unsaturation and wax composition on stability, properties and food applicability of oleogels. J Am Oil Chem Soc 98:1189–1203. https://doi.org/10.1002/aocs.12536
37. Gu X, Du L, Meng Z (2023) Comparative study of natural wax-based W/O emulsion gels: microstructure and macroscopic properties. Food Res Int 165:112509. https://doi.org/10.1016/j.foodres.2023.112509
38. Zetzl AK, Gravelle AJ, Kurylowicz M et al (2014) Microstructure of ethylcellulose oleogels and its relationship to mechanical properties. Food Struct 2:27–40. https://doi.org/10.1016/j.foostr.2014.07.002
39. Faller R (2021) UCD biophysics 241: membrane biology. LibreTexts, University of California, Davis
40. Sivakanthan S, Fawzia S, Madhujith T, Karim A (2022) Synergistic effects of oleogelators in tailoring the properties of oleogels: a review. Compr Rev Food Sci Food Saf 21:3507–3539. https://doi.org/10.1111/1541-4337.12966
41. Castell-Palou A, Rosselló C, Femenia A, Simal S (2013) Simultaneous quantification of fat and water content in cheese by TD-NMR. Food Bioprocess Technol 6:2685–2694. https://doi.org/10.1007/s11947-012-0912-8
42. Todt H, Guthausen G, Burk W et al (2006) Water/moisture and fat analysis by time-domain NMR. Food Chem 96:436–440
43. Peyronel F, Co ED, Marangoni AG (2017) Physical characterization of fats and oils. In: Akoh CC (ed) Food lipids, 4th edn. CRC Press, Taylor & Francis Group, Boca Raton

44. AOCS (2009) AOCS official method AOCS Cd 16b–93: solid fat content (SFC) by low-resolution nuclear magnetic resonance, direct method. In: Official methods and recommended practice of the AOCS, 6th edn. AOCS Press, Urbana
45. AOCS (2009) AOCS official method AOCS Cd 16–81: solid fat content (SFC) by low-resolution nuclear magnetic resonance, indirect method. In: Official methods and recommended practice of the AOCS, 6th edn. AOCS Press, Urbana
46. Pang M, Wang X, Cao L et al (2020) Structure and thermal properties of β-sitosterol-beeswax-sunflower oleogels. Int J Food Sci Technol 55:1900–1908. https://doi.org/10.1111/ijfs.14370
47. McNair HM, Miller JM, Snow NH (2019) Basic gas chromatography, 3rd edn. Wiley, Hoboken. E-Book 978-1-119-45078-8
48. Kaur G, Sharma S (2018) Gas chromatography-a brief review. Int J Inf Comput Sci 5:125–131
49. AOCS (2020) AOCS Ce 2–66: preparation of methyl esters of fatty acids. In: Official methods and recommended practice of the AOCS, 7th edn. AOCS Press, Urbana
50. Flores Ruedas RJ, Dibildox-Alvarado E, Pérez Martínez JD, Murillo Hernández NI (2020) Enzymatically interesterified hybrid palm stearin as an alternative to conventional palm stearin. CYTA J Food 18:1–10. https://doi.org/10.1080/19476337.2019.1699168
51. AOCS (2020) AOCS Ce 1h–05: cis-,trans, saturated, monounsaturated and polyunsaturated fatty acids in vegetable or non-ruminant animal oils and fats by capillary GLC. In: Official methods and recommended practice of the AOCS, 7th edn. AOCS Press, Urbana
52. AOCS (2021) AOCS Ce 1i–07: saturated, cis-monounsaturated, and cis-polyunsaturated fatty acids in marine and other oils containing long chain polyunsaturated fatty acids (PUFAs) by capillary GLC. In: Official methods and recommended practice of the AOCS, 7th edn. AOCS Press, Urbana
53. Delmonte P, Milani A, Kramer JKG (2021) Tutorial for the characterization of fatty acid methyl esters by gas chromatography with highly polar capillary columns. J AOAC Int 104:288–299
54. Golay PA, Moulin J, Alewijn M et al (2016) Determination of labeled fatty acids content in milk products, infant formula, and adult/pediatric nutritional formula by capillary gas chromatography: collaborative study, final action 2012.13. J AOAC Int 99:210–222. https://doi.org/10.5740/jaoacint.15-0140
55. Rumayor-Escobar A, de la Peña MM, de la Rosa-Millán J et al (2023) Effect of high intensity ultrasound on soybean and avocado oleogels' structure and stability. Food Struct 36:100315. https://doi.org/10.1016/j.foostr.2023.100315
56. Van Bockstaele F, Romanus M, Penagos IA, Dewettinck K (2022) Functionality of natural waxes in hybrid fat crystal networks. In: Development of trans-free lipid systems and their use in food products. Royal Society of Chemistry, Cryodon
57. AOCS (2009) AOCS Ce 5–86: triglycerides by gas chromatography. In: Official methods and recommended practice of the AOCS, 6th edn. AOCS Press, Urbana
58. Cordova-Barragan M, Marangoni AG, Peyronel F, Dibildox-Alvarado E (2021) Crystallization enhancement by a high behenic acid stabilizer in a palm oil-based model fat blend and its corresponding fat-reduced water-in-oil emulsion. J Am Oil Chem Soc 98:413–424. https://doi.org/10.1002/aocs.12461
59. Doan CD, To CM, De Vrieze M et al (2017) Chemical profiling of the major components in natural waxes to elucidate their role in liquid oil structuring. Food Chem 214:717–725. https://doi.org/10.1016/j.foodchem.2016.07.123
60. Castilho PC, Costa MDC, Rodrigues A et al (2004) Characterization of triacylglycerols in madeira laurel oil by HPLC-atmospheric pressure chemical ionization-MS. J Am Oil Chem Soc 81:913–919. https://doi.org/10.1007/s11746-004-1001-9
61. Jham GN, Muller HV, Cecon P (2008) Triacylglycerol molecular species variation in stored coffee beans determined by reverse-high-performance liquid chromatography/refractive index detector. J Stored Prod Res 44:82–89. https://doi.org/10.1016/j.jspr.2007.05.003
62. Amaral JS, Cunha SC, Alves MR et al (2004) Triacylglycerol composition of walnut (Juglans regia L.) cultivars: characterization by HPLC-ELSD and chemometrics. J Agric Food Chem 52: 7964–7969. https://doi.org/10.1021/jf048918n

63. van der Klift EJC, Vivó-Truyols G, Claassen FW et al (2008) Comprehensive two-dimensional liquid chromatography with ultraviolet, evaporative light scattering and mass spectrometric detection of triacylglycerols in corn oil. J Chromatogr A 1178:43–55. https://doi.org/10.1016/j.chroma.2007.11.039
64. Foubert I, Dewettinck K, Van de Walle D et al (2007) Physical properties: structural and physical characteristics. In: The lipid handbook, 3rd edn. CRC Press, Taylor & Francis Group, Boca Raton, pp 535–590
65. Fauconnot L, Hau J, Aeschlimann JM et al (2004) Quantitative analysis of triacylglycerol regioisomers in fats and oils using reversed-phase high-performance liquid chromatography and atmospheric pressure chemical ionization mass spectrometry. Rapid Commun Mass Spectrom 18:218–224. https://doi.org/10.1002/rcm.1317
66. Byrdwell WC, Neff WE (1999) Non-volatile products of triolein produced at frying temperatures characterized using liquid chromatography with online mass spectrometric detection. J Chromatogr A 852(2):417–432
67. Mottram HR, Woodbury SE, Evershed RP (1997) Identification of triacylglycerol positional isomers present in vegetable oils by high performance liquid chromatography/atmospheric pressure chemical ionization mass spectrometry. Rapid Commun Mass Spectrom 11:1240–1252. https://doi.org/10.1002/(SICI)1097-0231(199708)11:12<1240::AID-RCM990>3.0.CO;2-5
68. Ross KL, Hansen SL, Tu T (2011) Reversed-phase analysis of triacylglycerols by ultra performance liquid chromatography-evaporative light scattering detection (UPLC-ELSD). Lipid Technol 23:14–16. https://doi.org/10.1002/lite.201100083
69. Fu H, Lo YM, Yan M et al (2020) Characterization of thermo-oxidative behavior of ethylcellulose oleogels. Food Chem 305:125470. https://doi.org/10.1016/j.foodchem.2019.125470
70. Kamal-Eldin A, Pokorný J (2005) Lipid oxidation products and methods used for their analysis. In: Analysis of lipid oxidation, 1st edn. AOCS Press, Urbana
71. Boatright W (2020) Commercial assay kit using novel compound semiconductor materials for measuring peroxide value in edible fats & oils. National Institute of Food and Agriculture. https://www.sbir.gov/node/1906399
72. Giacomozzi AS, Carrín ME, Palla CA (2021) Storage stability of oleogels made from monoglycerides and high oleic sunflower oil. Food Biophys 16:306–316. https://doi.org/10.1007/s11483-020-09661-9/Published
73. Park C, Maleky F (2020) A critical review of the last 10 years of oleogels in food. Front Sustain Food Syst 4:139. https://doi.org/10.3389/fsufs.2020.00139
74. Verleyen T, Van Dyck S, Adams CA (2005) Accelerated stability test. In: Analysis of lipid oxidation, 1st edn. AOCS Press, Urbana
75. O'Keefe SF, Wiley VA, Knauft DA (1993) Comparison of oxidative stability of high- and normal-oleic peanut oils. J Am Oil Chem Soc 70:489–492
76. AOCS (1993). Surplus. AOCS Cd 12–57: fat stability, active oxygen method. In: Official methods and recommended practice of the AOCS, 7th edn. AOCS Press, Urbana
77. AOCS (2022) AOCS Cd 12b–92: oil stability index. In: Official methods and recommended practice of the AOCS, 7th edn. AOCS Press, Urbana
78. de Man JM, Tie F, Deman L (1987) Formation of short chain volatile organic acids in the automated AOM method. J Am Oil Chem Soc 64:993–996
79. AOCS (2017) AOCS Cd 12c–16: accelerated oxidation test for the determination of the oxidation stability of foods, oils and fats using the oxitest oxidation test reactor. In: Official methods and recommended practice of the AOCS, 7th edn. AOCS Press, Urbana
80. Frolova YV, Sobolev RV, Sarkisyan VA, Kochetkova A (2021) Approaches to study the oxidative stability of oleogels. IOP Conf Ser Earth Environ Sci 677(3):032045
81. Li L, Wan W, Cheng W et al (2019) Oxidatively stable curcumin-loaded oleogels structured by β-sitosterol and lecithin: physical characteristics and release behaviour in vitro. Int J Food Sci Technol 54:2502–2510. https://doi.org/10.1111/ijfs.14208

82. Tan CP, Man YBC, Selamat J, Yusoff A (2002) Analytical, Nutritional and Clinical Methods Section. Comparative studies of oxidative stability of edible oils by differential scanning calorimetry and oxidative stability index methods. Food Chem 76:385–389
83. Tan CP, Che Man YB (2000) Differential scanning calorimetric analysis of edible oils: comparison of thermal properties and chemical composition. J Am Oil Chem Soc 77:143–155. https://doi.org/10.1007/s11746-000-0024-6
84. Ali AR, Dimick PS (1994) Thermal analysis of palm mid fraction, cocoa butter and milk fat blends by differential scanning calorimetry. J Am Oil Chem Soc 71:299–302
85. Kodre K, Attarde S, Yendhe P et al (2014) Differential scanning calorimetry: a review. Res Rev J Pharm Anal 3:11–22
86. Peyronel F, Campos R (2012) Methods used in the study of the physical properties of fats. In: Structure-function analysis of edible fats. AOCS Press, Urbana, pp 231–294
87. Wang X, Ma D, Liu Y et al (2022) Physical properties of oleogels fabricated by the combination of diacylglycerols and monoacylglycerols. J Am Oil Chem Soc 99:1007–1018. https://doi.org/10.1002/aocs.12622

Chapter 19
Rheology-Based Techniques

Braulio Macias-Rodriguez

Abbreviations

LAOS	Large amplitude oscillatory shear
LVR	Linear viscoelastic region
SAOS	Small amplitude oscillatory shear

19.1 Introduction

"Rheology" is a branch of physics concerned with deformation and flow behavior experienced by complex fluids or soft materials such as foods, when acted on by forces. Such forces may be "naturally" exerted or deliberately applied during processing, use, and consumption. For example, at equilibrium, the fat microstructure is held by van der Waals-London forces, whereas under dynamic conditions, the microscopic ensemble is dictated by shear causing plastic flow. Rheological properties are also related to the application, use, and sensorial properties of fats and encode the effects of formulation and processing imposed on the fat microstructure. These insights in turn are used to establish rheology-structure models linking shear modulus and microstructure, rheology-texture relationships such as describing firmness in terms of shear compliance, and rheology-formulation-processing relationships such as describing the effect of molecular composition, shear/cooling fields on firmness. All these aspects are of paramount importance in designing alternative lipid-structuring systems.

B. Macias-Rodriguez (✉)
Deparment of Food Science and Human Nutrition, University of Illinois Urbana-Champaign, Urbana, IL, USA

C. Andrea Palla, F. Valoppi (eds.), *Advances in Oleogel Development, Characterization, and Nutritional Aspects*,
https://doi.org/10.1007/978-3-031-46831-5_19

In this chapter, I touch upon the abovementioned aspects, but focusing largely on the rheological and mechanical characterization of lipid-based materials, also applicable to oleogels. It is important to highlight that rheological and mechanical properties are material properties intimately related to textural attributes. While in some cases, it is possible to draw clear relationships among them; in some other cases, rheological measurements need to be supplemented by textural measurements. Part of this challenge arises due to the highly heterogenous and multiscale structure of lipids and the complexity of texture as a multisensorial property [1]. I briefly describe the foundations of the main methods to measure rheological, mechanical properties, and textural attributes, and detail the underlying equations describing such measures. A larger section is devoted to large amplitude oscillatory shear. Finally, experimental caveats and study cases relevant to lipid-based materials and oleogels are also provided.

19.2 Rheological and Textural Methods

19.2.1 Empirical Methods

19.2.1.1 Penetrometry

Penetrometry has been by far the most popular methods to evaluate the texture of edible fats. This method offers simple and inexpensive characterization which is useful for investigating the effect of processing conditions on consistency and making correlations with sensory panels. The method is based on the resistance of a material to be penetrated or indented by a test body: rod, cone, sphere, needle, etc. The resistance is measured as load, depths, or rates of indentation/penetration of the testing body. To date, cones remain the most common geometries to measure consistency [2]. A schematic representation of a typical cone is shown in Fig. 19.1.

Conical configurations varying in loads and angles cover a range of textures. Most studies include constant load experiments or to a less extent constant rate experiments, both of which show good agreement with one another [4, 5]. The general rule for cone angle selection is the harder (more solid) the material, the smaller the cone angle should be. The most common methods include the AOCS method Cc16-60 and the method proposed by Haighton [4]. Results are reported as penetration depths or converted into yield values, hardness, or spreadability indexes, using various equations dependent on the testing body and test conditions [4]. Considering that most of the force is used to overcome the yield point, the stress at which deformations are a combination of elastic "reversible" and plastic "irreversible" deformations, and provided that the motion is slow, an empirical yield value C is defined as the force load per unit cross-sectional area of the cone as given by,

$$C = \frac{K'W}{p^n} \tag{19.1}$$

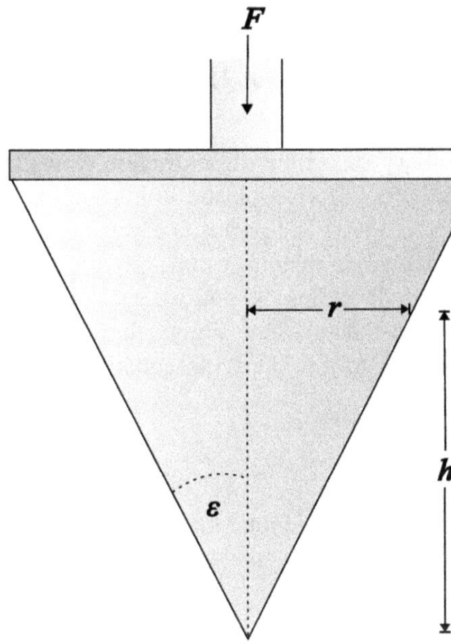

Fig. 19.1 Schematic representation of a conical probe. Applied force F, cone radius r, arbitrary penetration height h. (Reproduced from [3], with permission from Springer Nature)

Table 19.1 Examples for yield point values for margarine [4]

C-values		
<100	200–800	>1500
Very soft, not shape retaining	Best consistency, easy to spread	Too hard, no more spreadable

where K' is a constant dependent on the cone angle, W is the weight of the cone (g), p is the penetration depth raised to a fractional power $1.4 \leq n \leq 2$ as found empirically. The values of these empirical coefficients appear to depend on hardness and type of fat [4, 6] suggesting a complex relationship between the cone and yield value (e.g., for butter $n \approx 1.6$). The yield value may be affected by frictional forces between the fat and the tested material, though these are negligible for truncated cones [2]. Table 19.1 shows exemplary C values calculated according to Eq. 19.1 to help classify the consistency of margarine:

Additional definitions of apparent yield stress, hardness, and spreadability have been proposed [2]. Despite the advantage of cone penetrometry, some of the main arguments made against its use include poor reproducibility for firm butters, use of arbitrary testing conditions (e.g., penetration time), and ill-defined measures of yield value [2]. These assertions are supported by studies reporting the inability of cone penetrometry to differentiate among textural attributes of butter, limited degree of correlation of yield values with spreadability, and lack of universal material measures [4, 7, 8].

19.3 Fundamental Methods

19.3.1 Pressure-Driven Flow: Capillary Rheometry

Capillary or orifice die extrusion flows are useful for determining viscosities and yield stress of viscoplastic fluids such as edible fats under controlled extrusion-like processes [9, 10]. For this, a capillary rheometer is used which consists of a driving a tool namely a piston or ram, moving at a linear speed v to force material through a barrel with radius D_B and a die of radius D and length L (Fig. 19.2) [11].

From the pressure signal, the apparent die wall shear stress σ_w^{app} is calculated according to the following equation:

$$\sigma_w^{app} = \frac{\Delta P_{tot}}{2} \frac{R}{L} \tag{19.2}$$

where ΔP_{tot} is the total pressure drop with respect to the die exit, R is the die radius, and L is the die length. Apparent die wall shear rates $\dot{\gamma}_w^{app}$ are related to ram velocities v according to:

$$v = \frac{\dot{\gamma}_w^{app} R^3}{4R_B^2} \tag{19.3}$$

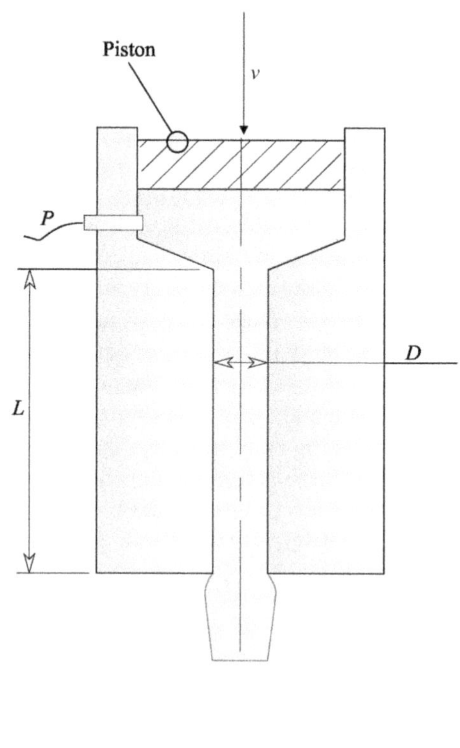

Fig. 19.2 Schematic view of a material being extruded through an orifice die with diameter D and length L in a capillary rheometer. Pressure P is applied by a piston with velocity v and measured by a manometer during the test. (Reproduced from [3], with permission from Springer Nature)

where R^3 is the die radius and R_B^2 is the barrel radius. To interpret the results in terms of true material measures, corrections including Bagley, Mooney, and Weissenberg-Rabinowitsch to account for entrance pressure losses due to contractional flow, wall slip, and non-Newtonian flow profile must be applied, respectively.

Bagley Correction
The overall pressure loss ΔP_{tot} for fluid flow through a die may be expressed as:

$$\triangle P_{\text{tot}} = \triangle P_{\text{visc}} + \triangle P_{\text{Bag}} \tag{19.4}$$

where $\triangle P_{\text{visc}}$ is the viscous pressure drop in the die and ΔP_{Bag} the entrance flow pressure drop component, called Bagley pressure. The Bagley correction consists in obtaining pressures drops at least at two shear rates using capillaries with the same D but varying lengths L ratios and extrapolation to a die of zero length. The true die wall shear stress can be calculated as follows:

$$\sigma_w = \frac{\triangle P_{\text{tot}} - \triangle P_{\text{Bag}}}{2} \frac{R}{L} \tag{19.5}$$

Mooney Correction
From the expression of the apparent die wall shear rate $\dot{\gamma}_w^{\text{app}}$:

$$\dot{\gamma}_w^{\text{app}} = \frac{4\dot{Q}}{\pi R^3} = \frac{4v}{R} \tag{19.6}$$

We can decompose the velocity into the sum of the slip ν_{slip} and true ν_{true} velocity contributions:

$$\dot{\gamma}_w^{\text{app}} = \frac{4v}{R} = \frac{1}{R}\left(4v_{\text{true}} + 4v_{\text{slip}}\right) = \dot{\gamma}_w + \frac{4\nu_{\text{slip}}}{R} \tag{19.7}$$

The slope of the apparent die wall shear rate as function of the reciprocal radius $1/R$ for a given die wall shear stress yields $4v_{\text{slip}}$. The slip contribution is determined using a set of dies with the same L/D but different diameters D. Equation 19.7 is used to estimate the wall slip velocity, which then is fitted to a discrete function such as a power-law or polynomial based on a best-fit approach, and then the wall shear rate for Newtonian fluids $\dot{\gamma}_w$ is obtained.

Weissenberg-Rabinowitsch Correction
The true non-Newtonian die shear rate $\dot{\gamma}$ can be expressed as:

$$\dot{\gamma} = \dot{\gamma}_w \left[\frac{1}{4}\left(3 + \frac{d \ln \dot{\gamma}_w}{d \ln \sigma_w}\right)\right] \tag{19.8}$$

The dependency of $\ln(\dot{\gamma}_w) = f(\ln(\sigma_w))$ may not always be linear. Therefore, higher-degree polynomial functions are fitted onto the data to yield a continuously differentiable function. Then, the first derivative of the fitted polynomial function can be used to calculate the true non-Newtonian die shear rate $\dot{\gamma}$.

Using the stress σ_w and the corrected non-Newtonian shear rate $\dot{\gamma}_w$ at the wall, the shear viscosity η can be estimated:

$$\eta = \frac{\sigma_w}{\dot{\gamma}_w} \tag{19.9}$$

The entrance pressure can also be used to estimate the extensional viscosity μ. Pressure drops can also be utilized to calculate the yield stress [9]. Considering constant volume, zero length orifice die, plastic behavior, and rate dependence of pressure, the following relationship has been proposed:

$$P = 2(\sigma_0 + \alpha V^n) \ln \frac{D_0}{D} \tag{19.10}$$

where P denotes the pressure to deform a material from its original diameter D_0 to a final diameter D, σ_0, α, and n can be considered material constants independent of die geometry and extrusion rate. The extensional yield stress σ_0 can be converted into a shear stress according to Von Mises yield criterion,

$$\sigma_0 = \frac{\sigma_y}{\sqrt{3}} \tag{19.11}$$

Castro et al. [9] recently applied this method to the characterization of the yield stress of soft solids and found good correlation with results obtained from rotational steady shear measurements.

19.3.2 Compression

Compression tests are one of the most popular tests for determining fundamental rheological, fracture properties, and empirical textural attributes due to their practicability and easiness of interpretation (Fig. 19.3). Compression tests consist in deforming a specimen of known dimensions such as a cylinder at constant force (creep) or at constant crosshead speed (axial compression) for a determined time.

For creep compression, deformation is measured as a function of time, whereas for axial compression, force is recorded as a function of time. Most literature involves uniaxial compression, in which, force-time curves are converted into

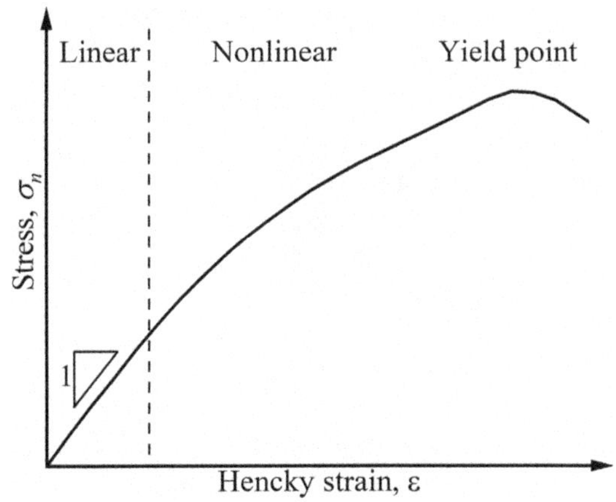

Fig. 19.3 Schematic illustration of a soft material being compressed at constant speed rate. Stress-strain curves depict linear and nonlinear regimes from which an apparent elastic modulus, yield stress, and strain can be determined

normal stress σ_n versus Hencky strain ε or strain rate $\dot{\varepsilon}$ curves. Treating fats and oleogels as incompressible materials, σ_n, ε and $\dot{\varepsilon}$ can be calculated as follows [12]:

$$\sigma_n = \frac{F}{A} \tag{19.12}$$

$$\varepsilon = \ln\left(\frac{h_t}{h_0}\right) \tag{19.13}$$

$$\dot{\varepsilon} = \left(\frac{d\varepsilon}{dt}\right) = \frac{\dot{h}}{h} \tag{19.14}$$

where A is the circular area of a cylinder with diameter D, h_0, and h_t are the specimen heights at the beginning and during the test. From the uniaxial compression curves, apparent rheological measures including the Young's modulus E_{app}, the yield point $\sigma_{\text{y\_app}}$, and the viscosity η_{app} can be calculated. The E_{app} is measured as the first derivative $d\sigma_n/d\varepsilon$ of the beginning of the compression curve where $\varepsilon \to 0$, the $\sigma_{\text{y\_app}}$ is defined as the maximum stress and the η_{app} is determined from the slope $d\sigma_n/d\dot{\varepsilon}$. Some limitations of compressive test include their limited range of accuracy since a true elastic modulus cannot be measured for yield strains below ~1% which is within the yielding region of edible fats and oleogels, manifestation of cracks or strain localization due to large strains, buckling or bulging of the sample specimens, and strong frictional effects at the boundary of the sample. Adequate specimen sizes $h_0/D \sim 1$–1.5 and lubricated plates (e.g., coated with oil or Teflon) can be used to circumvent the last two effects [12].

19.3.3 Drag Flow: Shear Rheology

Shear rheometry can be performed under rotational or oscillatory modes. The rotational mode, consisting of a steady shear input, enables the determination of compliance, shear-thinning flow behavior, and time dependence. The oscillatory mode comprises periodic shear inputs made of small or large strains or stresses at fixed or varying frequency.

19.3.3.1 Steady Shear Flow

Steady shear flow measurements are useful to determine the static or dynamic yield stress σ_y, depending on how the shear-rate ramps are performed, and shear-thinning behavior (Fig. 19.4a). "Slurry type" samples having a $\sigma_y \lesssim 10^0$–10^2 Pa are more adequate for steady-shear rates than stiffer samples, which are more prone to flow instabilities such as edge fracture, slip. When conducting steady-shear flow experiments, dependence on gap size can indicate confined flow. For the global flow behavior, the data can be described by the Herschel-Bulkley model:

$$\sigma = \sigma_y + K\dot{\gamma}^n \tag{19.15}$$

where σ_y denotes the dynamic or static yield stress for descending or ascending shear, respectively, K is a proportionality constant, and n is a power law index.

19.3.3.2 Step Shear: Creep and Recovery

Creep and recovery tests consist in applying a step stress input $\sigma(t) = \sigma_0 H(t)$, where $H(t)$ is a Heaviside step function and following the evolution of the strain $\gamma(t)$ in the material over time, to obtain the compliance $J(t) \equiv \gamma(t)/\sigma_0$ (Fig. 19.4). Based on these

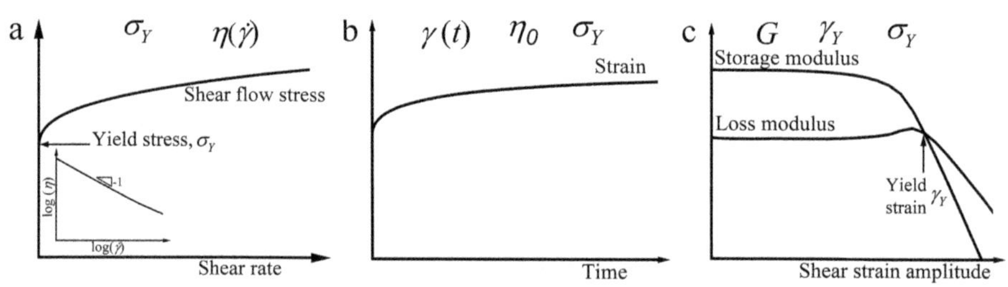

Fig. 19.4 Schematic illustration of different shear modes: (**a**) steady shear, (**b**) step shear: creep phase, and (**c**) oscillatory shear. Key rheological properties are also depicted, including (**a**) yield and viscosity from a steady shear test, (**b**) strain, zero-shear viscosity, and yield stress from a step shear creep experiment, and (**c**) viscoelastic moduli, yield strain, and yield stress from an oscillatory amplitude shear experiment

material measures, rheological definitions for textural attributes such as firmness, springiness, and rubberiness have been proposed for lipid-filled gels such as cheese.

Firmness

Firmness can be judged while deforming a piece of cheese with the mouth (tongue and palate, incisors, front teeth or molars) or by hand. Past studies suggest that firmness is a linear viscoelastic property. Instrumentally, it can be defined as a resistance to creep, and can be expressed by the inverse of the maximum compliance $J(t)$ measured at the end of the creep phase. The time at which a texture attribute is measured is called the time of observation. For firmness, this time is denoted as t_f. Thus, firmness F is mathematically defined as:

$$F \equiv \frac{1}{\max\{J(t)\}} = \frac{1}{J(t_f)} \tag{19.16}$$

and has units of Pa.

Springiness

Springiness deals with sudden responses that are evaluated over a short period of time and thus the use of a rate is appropriate. Springiness is defined as the absolute secant rate of recovery just after the stress is released and is judged at a time of observation $t_s = t_f + \Delta t_s$. Thus, springiness S is mathematically defined as:

$$S \equiv \frac{|J(t_s) - J(t_f)|}{t_s - t_f} = \frac{|J(t_f + \Delta t_s) - J(t_f)|}{\Delta t_s} \tag{19.17}$$

and has units of $(\text{Pa s})^{-1}$, which is equal to the inverse of the units of viscosity. In practice one judges S of a material such as cheese, by observing the instantaneous response when the stress releases. It is thus logical to take $\Delta t_s \ll t_f$, which is the minimum response time of modern rheometers $\Delta t_s = 0.1$ s.

Rubberiness

Rubberiness is related to the extent of strain recovery during a measurement time interval (t_f, t_r), where t_r is the time for measuring rubberiness R. If the strain is fully recovered at the time $t = t_r$, then $R = 1$. If there is no strain recovery at $t = t_r$, then $R = 0$. Thus, rubberiness is mathematically defined as:

$$R \equiv \frac{J(t_f) - J(t_r)}{J(t_f)} = 1 - \frac{J(t_f + \Delta t_r)}{J(t_f)} = 1 - F(t_f)J(t_f + \Delta t_r) \tag{19.18}$$

which is a dimensionless quantity, where Δt_r is the elapsed time of recovery for measuring rubberiness. Similar measures of F and R were proposed based on oscillatory measures of the complex modulus G^*, frequency ω_f, and the yield strain amplitude γ_0.

The time dependence of creep tests makes them particularly suitable for evaluating mechanical responses both for short time (material compliance and firmness) and for long times (such as creeping of stacked butter in store). The load and the duration of the time should be high and long enough, respectively, to induce sufficient creep motion but not too high or long to trigger formation of cracks [13].

Another use of creep tests is for the determination of the yield stress σ_y, where for $\sigma_0 < \sigma_y$ results in γ(t) curves characterized by an increase in strain and then a plateau or saturation (a hallmark of the solid regime of pastes), and $\sigma_0 > \sigma_y$ the same curves tend to a straight line with slope 1 in logarithmic scale, indicating infinite deformation at constant rate (a hallmark of the liquid regime of pastes) [14]. Experimentally, for stresses $\sigma_0 < \sigma_y$, creep compliance curves overlap or nearly overlap onto each other, whereas for stresses $\sigma_0 > \sigma_y$, deviations from this behavior occur and indicate the onset of nonlinear behavior. However, calculation of σ_y by this method may prove inefficient and unfeasible in stiff pastes due to heterogenous flow imposed by sudden stress jumps and higher sensitivity of this method to structural changes over long periods at constant stress [15].

19.3.3.3 Oscillatory Shear

During oscillation, a sinusoidal input function (strain or stress) is applied, and the associated response is measured (Fig. 19.4). The response comprises in-phase (stored energy or elastic modulus) and out-of-phase (loss energy or loss modulus) components with respect to the input function. Depending on the amplitude of the input function, oscillatory shear can be divided into two main regimes: small amplitude and large amplitude, which probe linear and nonlinear viscoelastic regions, respectively. Compared to other fundamental tests, oscillatory shear allows simultaneous characterization of elastic and viscous properties in a broad spectrum of flow conditions.

Small Amplitude Oscillatory Shear (SAOS)

Small amplitude oscillatory shear (SAOS) tests are useful to measure viscoelastic properties of the underlying microstructure. SAOS imposes relatively small strains or stresses in the linear viscoelastic region (LVR) typical of materials interacting via short-range van der Waals forces such as lipids and oleogels [16]. During SAOS tests, strain and stress maintain their linear proportionality and the crystal network exhibits viscoelastic solid-like behavior ($G' > G''$) characterized by high moduli G', $G'' \approx 10^4$–10^6 Pa, and weak frequency (ω) dependence [16–18]. For a sinusoidal strain excitation $\gamma(t) = \gamma_0 \sin(\omega t)$, a sinusoidal stress response $s(t) = g_0 \sin(\omega t + \delta)$ is obtained at the same input frequency ω, and with phase angle δ. The response can be decomposed as

$$s(t) = \gamma_0 G'(\omega)\sin(\omega t) + \gamma_0 G''(\omega)\cos(\omega t) \tag{19.19}$$

In which G', the in-phase elastic modulus or stored energy represents the real component, and G'', the out-of-phase viscous modulus or dissipated energy represents the imaginary component of the complex modulus $G^*(\omega) = G'(\omega) + iG''(\omega)$ at a given frequency ω [11]. Linear viscoelastic properties are useful for developing the structure-property relations such as linking interaction forces holding a network to its linear moduli. The magnitude of the elastic modulus G' of fats depends in a complex manner on at least three factors: volume fraction (SFC/100), crystal microstructure, and crystalline interactions [17]. The elastic modulus G' and G'' provides a combined measure of primary crystallization, aggregation, and network formation, as well as of material hardness or firmness [17]. A Fourier analysis, which converts the time function into frequency domain, of the stress response reveals in the LVR region that only the first or fundamental harmonic ($n = 1$) associated with ω occurs.

Large Amplitude Oscillatory Shear (SAOS)

The LAOS regime occurs above a yielding strain or stress, where materials plastically deform, and their response is no longer sinusoidal. The yield stress σ_y is an important property linked to sensory attributes such as spreadability. From oscillatory experiments, σ_y can be defined in several ways namely the stress: (i) at which the elastic G' starts to decay by certain percentage, (ii) where power-law fits of stress-strain curves intersect, and (iii) where G' and G'' crossover. The first two methods give the lowest estimates of σ_y and γ_y, whereas the latter provides the highest values of σ_y and γ_y since the material experiences substantial yielding. For fats and oleogels, the onset of nonlinear behavior occurs at $\gamma_y \approx 0.1\%$. There are several approaches to interpret the nonlinear LAOS data, including examining the behavior of the first-harmonic moduli, time-domain stress waveforms $\sigma(t)$ or two-coordinate axes figures referred as to Lissajous-Bowditch curves, and analyzing the raw waveforms via FT rheology, Chebyshev stress decomposition, sequence of physical processes, etc. [19, 20]. The first-harmonic or average viscoelastic moduli G' and G'' denote the global ("full" or "intercycle") stress response. Figure 19.3 shows an example of a typical edible fat: butter. The material undergoes average elastic softening coupled with increase dissipation or thinning due to disruption of the crystal network as strain increases. Time-domain raw signals $\sigma(t)$ and Lissajous-Bowditch curves qualitatively distinguish among material response and capture the onset of nonlinear behavior. Lissajous-Bowditch curves are closed loop plots of γ_0 on the abscissa and $\sigma(t)$ on the ordinate (elastic representation) or $\dot{\gamma}_0$ on the abscissa and $\sigma(t)$ on the ordinate (viscous representation). Figure 19.5 shows raw elastic Lissajous-Bowditch plots of butter within and outside the LVR for butter at $T = 18$ °C.

Within the LVR region (γ_0 <0.01%), the plots mirror nearly "perfect" ellipses, where the tangent slope corresponds to G' and the area enclosed by the ellipse represents G''. Beyond amplitudes where yielding begins (γ_0 >0.01%), the plots become gradually distorted and acquire square-like shapes enclosing increasingly larger areas [19]. Typical features, e.g., global strain softening, and additional local features, e.g., intracycle stiffening, masked by the average viscoelastic moduli can be

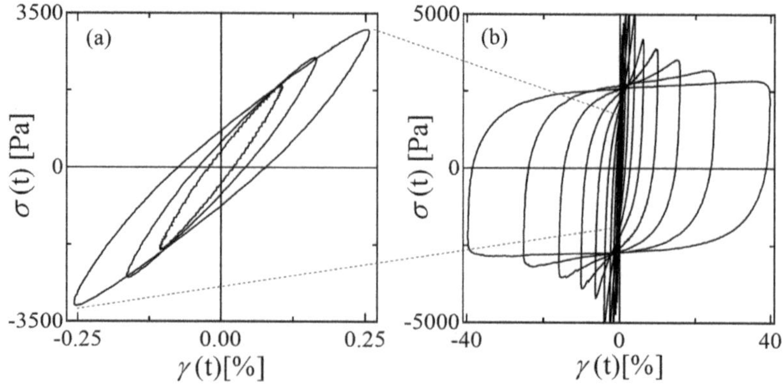

Fig. 19.5 Raw Lissajous-Bowditch plots (elastic perspective) of butter (T = 18 °C) obtained from an oscillatory shear test (ω = 3.6 rad/s) (**a**) linear to mildly nonlinear transition (SAOS to LAOS), (**b**) fully nonlinear response (LAOS). (Reproduced from [3], with permission from Springer Nature)

visualized. Global or average elastic softening is manifested as intercycle clockwise rotation in the slope of the stress-strain curve at strain minima $\gamma_0 = 0$ (i.e., at the "origin" where strain rate $\dot{\gamma}$ is at maxima) toward the strain-axis. Local strain stiffening is clearly visible as the intracycle upturn of the shear stress at strain maxima $\gamma_0 = \max$ (i.e., at the "extreme" where $\dot{\gamma} = 0$). Stress overshoots, akin to those observed during a start-up shear test, appear in the upper left quadrant of the Lissajous-Bowditch and indicate yielding. The reversibility of the observed behavior during flow reversal evokes microstructure "healing" or thixotropy [21]. While the first-harmonic viscoelastic moduli are the common output of most rheometers, locally defined measures are calculated using frameworks developed to analyze the nonlinear response such as FT, Chebyshev stress decomposition method, and sequence of physical processes. The FT analysis converts time-domain periodic signals into frequency-domain signals that make up a spectrum encompassing first and higher-order odd harmonics (n = 1, 3, 5...) in the case of nonlinear response. Despite the quantitative nature of this approach, FT analysis provides no physical insights. Other frameworks such as the Chebyshev stress decomposition and sequence of physical processes overcome this limitation. The Chebyshev stress decomposition method separates the overall stress into elastic and viscous components fitted by Chebyshev polynomials of the first kind, which are interrelated to Fourier coefficients. The Chebyshev polynomials are reduced to the following nonlinear measures [19]:

$$G'_{\mathrm{M}} \equiv \frac{d\tau}{d\gamma} = \sum_{n:\mathrm{odd}} nG'_n = e_1 - 3e_3 + \ldots, \tag{19.20}$$

$$G'_{\mathrm{L}} \equiv \frac{\tau}{\gamma} = \sum_{n:\mathrm{odd}} G'_n(-1)^{\frac{n-1}{2}} = e_1 + e_3 + \ldots, \tag{19.21}$$

$$\eta'_{\mathrm{M}} \equiv \frac{d\tau}{d\dot{\gamma}} = \frac{1}{\omega}\sum_{n:\mathrm{odd}} nG''_n(-1)^{\frac{n-1}{2}} = \upsilon_1 - 3\upsilon_3 + \ldots, \tag{19.22}$$

$$\eta'_{\mathrm{L}} \equiv \frac{\tau}{\dot{\gamma}} = \frac{1}{\omega}\sum_{n:\mathrm{odd}} G''_n = \upsilon_1 + \upsilon_3 + \ldots, \tag{19.23}$$

where G'_{M} is the minimum-strain or tangent modulus at $\gamma(t) = 0$ and G'_{L} is the large-strain or secant modulus at $\gamma(t) = \gamma_{\mathrm{max}}$. Likewise, h'_{M} is the minimum-rate viscosity and h'_{L} is the large-rate viscosity. The letters e_n and v_n refer to elastic and viscous Chebyshev coefficients of nth order fitting the data. All these material functions reduce to G' and G'' ($\eta' = G''/\omega$) in the LVR region. Comprehensive descriptions on the fundamentals and use of the Chebyshev polynomials framework are available in the literature [19, 22, 23]. Later, the "Sequence of Physical Processes" (SPP) method based on the Frenet-Serret theorem was developed [20]. This method has some important advantages over the Chebyshev decomposition approach because it describes the nonlinear moduli continuously in time and space rather than just in specific strain and strain rates of the oscillatory cycle. The Frenet-Serret theorem defines each point on the Lissajous curve by three vectors: one tangent to the response curve (T) (e.g., stress if strain is the independent excitation variable) one pointing to the center of the curve (N), and one being the cross product of the first two vectors (B). The projections of the binomial vectors (B) on the stress-strain curve define the transient elastic (G'_t) moduli, while its projection on the stress and strain rate curve defines the transient viscous (G''_t) moduli:

$$G'_t = -\frac{B_\gamma}{B_\sigma} = \frac{d\sigma}{d\gamma} \tag{19.24}$$

$$G''_t = -\frac{B_{\frac{\dot{\gamma}}{\omega}}}{B_\sigma} = \frac{d\sigma}{d\left(\frac{\dot{\gamma}}{\omega}\right)} \tag{19.25}$$

where B_γ, $B_{\dot{\gamma}/w}$, B_σ are the projections of binormal vector $B(t)$ along with the strain, strain rate, and stress axis, respectively. Plotting of the instantaneous G''_t versus G'_t plot (Cole-Cole plot) is a powerful way to evaluate the transient response of the material within an oscillation cycle. More details on the SPP approach can be found in the literature [20, 24].

Experiment LAOS Type and Data Collection

Major experimental considerations to keep in mind when performing LAOS experiments include selection of the most suitable experiment (i.e., LAOStrain vs. LAOStress), the form of data collection, rheometric artifacts prior to processing, and analysis of data. Due to the narrow linear viscoelastic regime of lipids and oleogels, careful handling of the sample is required to minimize disruption of the original network. This can be achieved by loading cylindrical pre-formed samples with controlled normal force, loading with geometries that minimize sample disturbance (e.g., vane) or by crystallization in-situ [18, 25]. Prior to data processing and analysis, it is important to be aware with possible artifacts arising during rheological and textural characterization as described in Sect. 19.4.

Selection of Experiment Type

The selection of LAOStrain versus LAOStress techniques and the measurement of corresponding material functions depend on three main factors: instrument specifications, material under investigation and its application, and structure-rheology models. Instrument specifications include rheometer design, performance, and sensitivity. Depending on the rheometer design, stress-control or strain-control may be more suitable one over the other. Additional intrinsic nonlinearities may arise such as for instruments that employ active deformation control to measure strain [26]. Microstructure "sensitivity" to deformation or loads also influences the selection of the protocol. For lipid-based materials and oleogels, LAOStrain allows better input control and deeper probing into the nonlinear regime than LAOStress that causes large strain jumps in the nonlinear region [27, 28]. Material application also defines properties of interest, e.g., butter, spreads, and shortenings undergo nonlinear shear rates during processing or usability, namely during lamination and spreading, and thus LAOStrain reflects better their use. Textural properties can also be described in terms of rheological material functions from LAOStrain with consistent measuring units in contrast to LAOStress [27]. Finally, selection of the input function also depends on the sensitivity of structure-rheology models to specific material functions, which help infer molecular-, nano-, or microscale structures [19].

Data Collection, Processing, and Analysis

LAOS requires acquiring raw oscillatory waveforms, which typically consists in collecting strain-stress raw signals using standard capabilities included in commercial rheometers software (e.g., TA Orchestrator software, Raw data LAOS waveform) [29, 30]. Beyond the LVR region, in the LAOStrain regime, the raw stress begins to show transient decays in amplitude. Any LAOS analysis using FT or stress decomposition Chebyshev representation incorporates periodic steady stress waveforms. Achievement of steady state flow can be verified by observing overlapping of the Lissajous-Bowditch loops, an equilibrium of the first-harmonic viscoelastic moduli G' and G'' or a plateau value of the shear stress over time (e.g., in a time sweep) at certain set of $\{\omega, \gamma_0\}$. In practice, however, a steady state may never be reached and hence, the number of applied cycles at a specific set of $\{\omega, \gamma_0\}$ is merely based on experimental observations. Data processing and analysis of LAOStrain is performed with a custom-written freeware MATLAB routine [31]. The software requires time series signals of strain and stress along with user-specified parameters for analysis. Data processing of the signal requires inputting multiple steady oscillatory cycles to increase the signal to noise ratio. The software smooths the total stress signal by filtering out even and no-integer harmonics associated with random noise and broken shear symmetry. A cutoff frequency, typically $I_n/I_1 < 0.05$ are not considered, limits noise attributed to higher frequency components masking the real signal. The output of the software includes Fourier and Chebyshev spectrum of the stress response, intercycle, and intracycle viscoelastic material measures and time-series signals of elastic and viscous stresses, $\sigma^{e}(\gamma(t))$ and $\sigma^{e}(\dot{\gamma}(t))$, respectively [31]. A similar routine developed for LAOS analysis based on the sequence of physical processes is thoroughly described in Le et al. [32].

19.4 Experimental Artifacts

In general, rheological and textural measurements can be affected by several experimental artifacts, which "at best" change slightly the absolute values of material functions, and "at worst" communicate false mechanical behavior. For example, for cone penetrometry, samples need to be properly conditioned prior to experimentation and the use of standard methods such as those described by the American Oil Chemists' Society Cc1606 are encouraged to avoid arbitrary testing conditions. For compression tests, it is of upmost importance to prepare specimens as flat as possible due to their limited range of accuracy at $\lesssim 1\%$ which is within the yielding region of fats and oleogels. Buckling or bulging of the sample specimens and strong frictional effects at the boundaries are other artifacts that can emerge and be avoided by preparing adequate specimen sizes $h_0/D \sim 1$–1.5 and lubricated plates (e.g., coated with oil or Teflon) [12]. Treatises compiling experimental challenges to avoid "bad data" collection in shear rheology in soft materials are available in the literature [33, 34]. It is important that the practitioner is aware of these issues to mitigate/minimize their occurrence and select an appropriate experimental window during shear measurements. For lipids and oleogels, the most important sources of artifact include slip, edge fracture, and gap under-filling during "in situ" crystallization. Given the self-lubricated nature of lipids, it would be rather rare that slip does not occur. Key indications of slip include (1) "free motion" of the sample between the contacting boundaries and even migration outside the gap when slip is strongly present, (2) gap dependence of the apparent material response, (3) reduced flow stress, (4) irregular fluctuations in the transient response of the stress-strain signal in the Lissajous-Bowditch curves, (5) asymmetric and open intracycle Lissajous-Bowditch loops, (6) significant signal contribution from even harmonics $I_2/I_1 > 0.1$, and (7) secondary loops in Lissajous-Bowditch curves also indicate slip in some cases. Approaches to check for and minimize slip include looking for any of the signatures abovementioned, collecting data at different gaps, performing marker tests, testing rheological behavior with different measuring geometries, selecting an

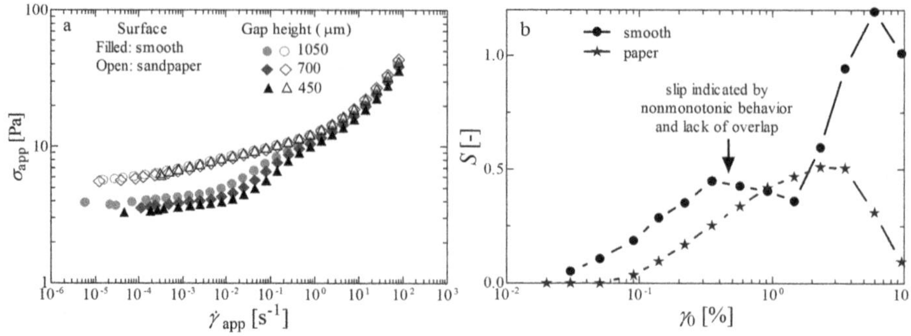

Fig. 19.6 (**a**) Wall slip on a smooth surface and elimination of wall slip on a sandpapered surface, respectively, for a water-in-oil emulsion (Nivea Lotion) tested using parallel plates of diameter $D = 40$ mm at multiple gaps $h = 450$–1050 mm. (Adapted from Ewoldt et al. [33]). (**b**) Wall slip and edge failure on smooth surface and absence of wall slip on a surface to which a cellulose filter paper has been attached, in a cake shortening using parallel plates of diameter $D = 20$ mm at a fixed gap $h = 1300$ mm and fixed frequency $\omega = 3.6$ rad/s. (Reproduced from [23], with permission from Taylor and Francis)

appropriate experimental window: define frequencies and amplitudes that prevent/minimize slip, applying a normal force control: NFC $\approx$ 0–0.5 N during measurements to maximize sample-geometry contact during measurement and used of surfaces, e.g., sandblasted, crosshatched, that facilitate contact between the sample and measurement geometries. Edge failure is another potential artifact arising in stiff and brittle samples and can be detected by visual observation of the edge of the sample, marker tests, reduction in the apparent stress due to decrease in effective contact area, and gap dependence of the data. Sample under-filling occurs due to volume contraction upon crystallization or sol-to-gel transition in the rheometer and manifests as the development of a large negative normal force. Viable ways to eliminate underfilling include applying a zero normal force control or using internal gap adjustment procedures to compensate for sample contraction [35]. Overall, any of these artifacts can contribute to erroneous estimation of the viscoelastic moduli, waveforms distortion, and premature edge failure. Figure 19.6 illustrates examples where wall slip and edge failure manifest in a water-in oil emulsion and a cake shortening. In Fig. 19.6a, wall slip manifests as a lack of overlap of the material response as a function of gap height h, and a reduced stress response. In Fig. 19.6b, wall slip and edge failure cause premature yielding and a nonmonotonic behavior.

19.5 Case Studies: Bulk Fats, Oleogels, and Particle-Filled Gels

In the following sections, several case studies relevant to edible fats and oleogels are presented. In these, structure-rheology-texture relationships are established via rheology.

19.5.1 *Fat Crystal Networks (Roll-In and All-Purpose Shortening)*

A roll-in or laminating shortening is a soft fat material with high toughness. Toughness indicates the damage-tolerance of a material and quantifies its ability to dissipate energy under large loadings. A roll-in fat is used in the manufacture of croissants, Danish, and puff pastry where it enables uniform lamination, good gas retention, and lift volume, flakiness, and good mouthfeel [34]. During processing, a roll-in fat is extruded, squeezed, and shaped into micron-width films without breaking catastrophically. By contrast, an all-purpose shortening is used in multiple bakery applications mainly cake and icing but not in laminates, since it performs poorly. The baker is well familiar with this material performance and discriminates a "good" from a "bad" laminating fat by tactile perception. From scientific and health perspectives, it is desirable to pinpoint key rheological functions that describe the functionality of these materials as to enable inverse design of roll-in fats with no *trans* and lower saturates. Compression tests mimicking tactile perceptions were performed on each sample illustrate strikingly different macroscopic behavior (Fig. 19.7). On the one hand, a roll-in fat behaves as a ductile soft solid; on the other hand, an all-purpose fat behaves as a brittle soft solid material.

To elucidate material measures underlying the macroscopic behavior, LAOStrain deformations were applied, and stress responses analyzed and interpreted using Lissajous-Bowditch curves (not shown) and nonlinear measures (Fig. 19.7). In both material classes, linear viscoelasticity dominates the stress response at $\gamma_0 \approx 0.05\%$ (i.e., tight elliptical Lissajous-Bowditch loops). Above strains $\gamma_0 \approx 0.09\%$, Lissajous-Bowditch distort because of periodic nonlinear variations in the stress response particularly visible in local points of the stress response. In general, both samples display similar qualitative features, e.g., stress upturns at γ_0 (i.e., lower elasticity or energy storage at $\gamma = 0$ or $\dot{\gamma}_0$) in the elastic perspective and "bends" at $\dot{\gamma}_0$ (i.e., lower viscosity or energy dissipation at $\dot{\gamma} = 0$ or γ_0) in the viscous perspective indicating intracycle strain stiffening and intracycle shear thinning, respectively. At high strains $\gamma_0 \geq 6\%$, self-intersections and secondary loops occur at $\dot{\gamma}_0$ in the viscous curves attributed to stronger and quicker unloading of instantaneous elastic stresses, competition between network destruction and formation at high shear rates and even slippage [21]. Within an elastic LAOS cycle, e.g., at $\gamma_0 \approx 6\%$, the peak in stress demarks two regions: a nearly linear region preceding

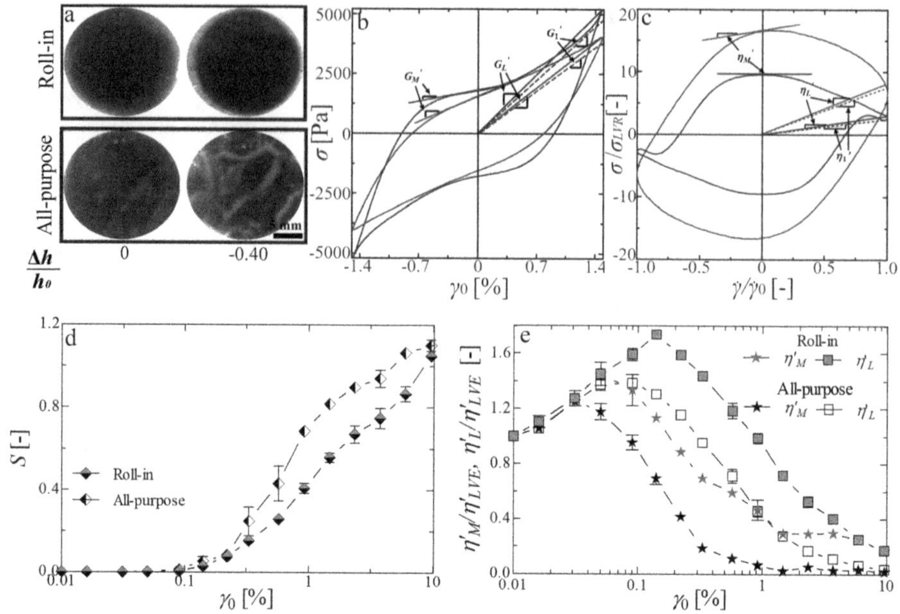

Fig. 19.7 (**a**) Views from below a transparent bottom plate supporting different mechanical behavior of fat shortenings. A roll-in shortening behaves as a ductile-like solid, whereas an all-purpose shortening resembles a brittle-like solid under compression. (**b**) Lissajous-Bowditch curve (elastic perspective) showing strain γ_0 versus stress σ for a selected data point $\gamma_0 = 1.47\%$, displaying graphical representation of local and global LAOS elastic measures: minimum-strain modulus G'_{M}, maximum-strain modulus G'_{L}, and average first-harmonic moduli G'_1 for fat shortenings. An all-purpose shortening displays more prominent stress upturns with higher stress maximum $\sigma \approx 5000$ Pa than a laminating shortening $\sigma \approx 4000$ Pa. (**c**) Lissajous-Bowditch curve (viscous perspective) showing shear rate normalized by shear rate input $\dot{\gamma}/\dot{\gamma}_0$ versus stress normalized by stress $\sigma/\sigma_{\mathrm{LVR}}$ within the LVR at $\gamma_0 = 0.01\%$. Curves are obtained from a selected strain input $\gamma_0 = 1.47\%$ ($\dot{\gamma}_0 = 0.05\ \mathrm{s}^{-1}$) and display graphical representation of local and global LAOS viscous measures: minimum-rate viscosity h'_{M}, maximum-rate viscosity h'_{L}, and average first-harmonic viscosity η'_1. An all-purpose shortening displays a Lissajous-Bowditch curve with stress upturns that enclose a smaller area than a roll-in shortening. Nonlinear local (**d**) elastic and (**e**) viscous measures for roll-in and all-purpose shortenings calculated from LAOS data at $\omega = 3.6$ rad/s. (d) Local elastic measures are integrated in the adimensional strain stiffening ratio $S = \left(G'_{\mathrm{L}} - G'_{\mathrm{M}}\right)/G'_{\mathrm{L}}$. (**e**) Viscous measures are parametrized by the linear dynamic viscosity η'_{LVE} at $\gamma_0 = 0.01\%$. (Adapted and reproduced from [34], with permission from Springer)

the overshoot that extends roughly from the lower reversal point to the stress overshoot and a "flow" region following the overshoot. This linear stress region is associated with residual elasticities. In the "flow" region, as shear rate increases (or strain decreases), the stress decreases reaching a minimum. Subsequently, as the strain increases (or shear rate decreases), the stress increases again indicating thixotropy or network restructuring [18]. This progresses until the end of the half-cycle ($\gamma_0 = 0$ or $\dot{\gamma} = \dot{\gamma}_0$), and then the sequence is repeated during flow reversal.

Larger areas enclosed by the elastic Lissajous-Bowditch curves at high strains ($\gamma_0 \geq 6\%$), indicate increased plastic response, more evident in a roll-in shortening. Local LAOS measures indicate that a roll-in shortening and all-purpose store energy in a similar manner (i.e., comparable $S = \left(G'_L - G'_M\right)/G'_L$); however, they dissipate energy in a strikingly different way (i.e., different h'_M/η'_{LVE} and h'_L/η'_{LVE}). A roll-in shortening dissipates ~10× energy of an all-purpose shortening. To link the observed nonlinear mechanical response to nano-to-micro scale structures, ultra-small angle x-ray scattering was performed. Data fitted to the Unified Fit model revealed two major structural differences on: (1) the number of levels making up the fat hierarchy, (2) the morphology and size of nanoplatelets. A roll-in shortening had three structural levels, while an all-purpose shortening only two structural levels. A roll-in shortening had nanoplatelets characterized by "smooth" surfaces and average sizes about 10-fold smaller: ~50 nm, whereas an all-purpose shortening had nanoplatelets with "rough" surfaces and average size of ~500 nm. Based on rheology and the structural insight, it was suggested that a roll-in shortening dissipates more effectively shear deformations due to: (1) an "extra" hierarchical level that allows better energy allocation, (2) sliding of microscopic layer-like crystal aggregates in which the liquid oil could serve as a lubricant. Furthermore, the fact that several compositions share a unique rheological "fingerprint": increased viscous dissipation, indicates that the structure-function framework rather than exact bulk composition determines performance [34].

19.5.2 *Particle-Filled Dispersions (High-Fat and Low-Fat Semi-Hard Cheeses and Spreads)*

19.5.2.1 Soft Fillers: Viscoelastic Lipid Droplets Filling a Protein Matrix

From a material perspective, cheese can be considered an emulsion-filled gel. Rheology and texture are largely governed by the fat filler, its volume fraction, and its physicochemical properties: melting. Fat acts as a perfect filler as it tailors textural attributes such as firmness, springiness, rubberiness, and influences cheese processability. Any reformulation effort to produce low-fat cheese requires careful consideration of rheology and texture to anticipate quality and sensory trade-offs. In this case study, relationships between rheology and structure are presented for Gouda cheeses formulated with zero ($\phi_{fat} = 0$) and full fat content ($\phi_{fat} = 0.3$) [27]. By using LAOStrain, the effect of the fat filler on the "firm to fluid" or yielding transition of cheese is elucidated. Intercycle damage progression and failure of full-fat cheese differs from that of zero-fat cheese (Fig. 19.8). The roman numbers denote the various stages of stress response for both cheeses; I: elastic (i.e., linear grow-rate of the intercycle stress and recoverable strain), I-II: elastoviscoplastic (irrecoverable strain due to initiation of microcrack formation), II: global stress $\sigma'_{max}(\gamma_0)$ or elastic failure stress, III: decay of elastic stress (due to percolation of crack into larger fractures). The main differences are that a full-fat cheese shows larger initial

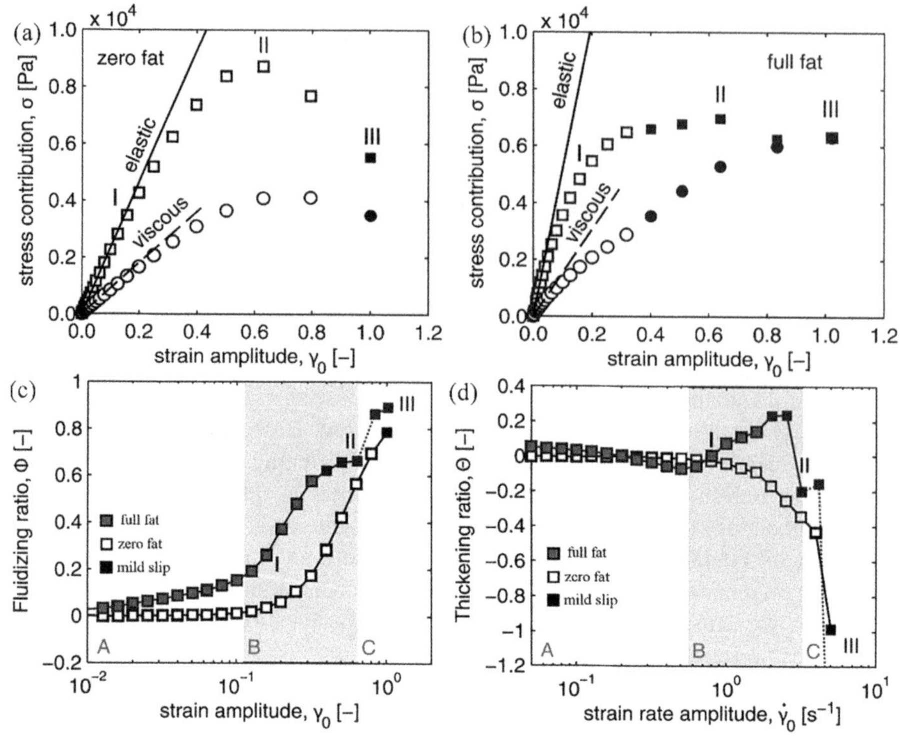

Fig. 19.8 Top panels (**a**, **b**): Evolution of the intracycle maxima of the elastic stress $\sigma'_{\max}(\gamma_0)$ (hollow squares) and viscous stress $\sigma''_{\max}(\gamma_0)$ (hollow circles) as a function of the strain amplitude γ_0 for (**a**) zero-fat and (**b**) full-fat cheese measured at $T = 25$ °C and a frequency $\omega = 5$ rad/s. The continuous and dashed lines represent the predictions of the linear viscoelastic constitutive model. Both zero-fat and full-fat cheese display an intercycle maximum of the elastic stress σ'_f (indicated by the numeral II) at a failure strain amplitude $\gamma_f \approx 0.7$. The intercycle maximum is defined as the failure criterion for the food gel. The full-fat cheese curve (**b**) displays a broad plateau in the maximum elastic stress $\sigma'_{\max}$ and a small decrease beyond the failure point. By contrast, the zero-fat cheese, shown in (**a**), displays a more clearly pronounced peak in the $\sigma'_{\max}$ max curve. The filled symbols indicate the strain amplitudes at which mild slip was observed. Bottom panels (**c**, **d**): Strain sweeps showing the evolution in (**c**) the fluidizing ratio ϕ and (**d**) the thickening ratio Θ, of zero-fat cheese and full-fat cheese measured at a temperature $T = 25$ °C and angular frequency $\omega = 5$ rad/s. (**c**) Both cheese formulations show comparable ultimate magnitudes of the fluidizing ratio; however, the rise of ϕ of full-fat cheese is more gradual and sets in at lower strains. (**d**) The non-Newtonian fluid properties of full-fat and zero-fat cheese are characterized by the evolution of the thickening ratio, which reveals three flow regimes A, B, and C. Zero-fat cheese displays continuous intercycle thinning, whereas full-fat cheese shows some initial thinning, followed by thickening and thinning. Beyond cycle II, the sample of full-fat cheese is no longer homogeneous, indicated with a dotted line in (**c**) and (**d**). The strain and strain-rate amplitudes at which mild slip is observed are indicated using filled symbols in each figure. (Reproduced from [27], with permission from Elsevier)

growth-rate (steeper slope) and larger elastoviscoplastic response (i.e., plastic response toward lower strain amplitudes I-II, and broad plateau prior to failure III) than a zero-fat cheese, since fat plasticizes the cheese matrix [27]. The maximum stress response $\sigma'_{max}(\gamma_0)$ (its magnitude and rate of decline) is a strong indicator of "brittleness" (i.e., "the tendency to break under the condition of minimal previous plastic deformation"), a similar observation made in the first case study. To quantify these differences, Lissajous-Bowditch curves, and locally defined measures are inspected, the latter depicted in Fig. 19.8c, d.

Quantitatively, the local metrics G'_M and G'_K measure the onset of plastic flow and accumulation of damage in the elastic network and η'_M and η'_L, the associated viscous responses. The level of "fluidization" or the extent of "solid-to-fluid" transition in elastic and viscous perspectives are defined as:

$$\Phi \equiv \frac{G'_K - G'_M}{G'_K} = \frac{12e_3 + 20e^5 + \ldots}{e_1 + 9e_3 + 25e_5 + \ldots} \quad \text{Elastic} \tag{19.26}$$

$$\Theta \equiv \frac{\eta'_K - \eta'_M}{\eta'_K} = \frac{12\upsilon_3 + 20\upsilon^5 + \ldots}{\upsilon_1 + 9\upsilon_3 + 25\upsilon_5 + \ldots} \quad \text{Viscous} \tag{19.27}$$

These definitions are analogous to "stiffening" and "thickening" ratios described in Ewoldt et al. [30], with the only difference being that both G'_K and η'_K are tangential (not secant) measures to highlight modulus and viscosity at maximum strain/strain-rate, respectively [27, 30]. Figure 19.8c, d shows Φ and Θ for zero-fat and high-fat cheeses where three regimes are distinguished: (A) elastic response without plastic flow $\Theta \approx 0$ for zero-fat cheese versus initial elastic response followed by mild fluidization $\Phi > 0$ and thinning $\Theta < 0$ due to strain localization in the emulsified fat component, (B) start of fluidization/thinning $\Phi > 0$ and $\Theta > 0$ due to microcrack nucleation and propagation for zero-fat cheese versus continued fluidization $\Phi > 0$ and with shear "thickening" $\Theta > 0$ akin to rubber toughening for full-fat cheese, (C) extreme fluidization/thinning attributed to failure, which is driven by crack propagation and sample-spanning fractures for both zero-fat and high-fat cheeses.

A similar study for cheese spreads encompassing the interpretation of material responses based on the sequence of physical processes using the Frenet-Serret theorem was conducted [32]. In all cases, the Cole-Cole plots are positioned higher than the $G'_t = G''_t$ line, suggesting a predominant elastic behavior. In the linear viscoelastic region, the Cole-Cole plots appeared as very small deltoids between 0 and 4% strain. As the material enters the nonlinear region (>4%), the deltoids gradually increase in size due to material displacement and microstructural changes during an oscillatory cycle. In Fig. 19.9, the strain range between 10% and 50% induced the most displacement for both samples, although the deltoids of low-fat cheese spreads are bigger than those of high-fat cheese spreads within this strain range. These findings were qualitatively consistent with G'_L and G'_M obtained from

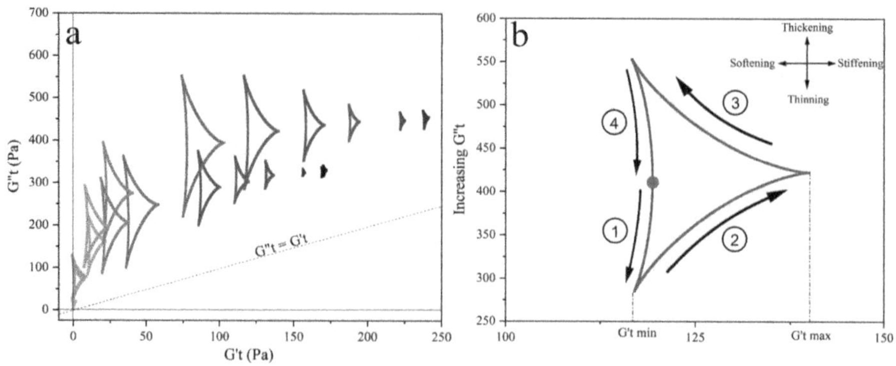

Fig. 19.9 (**a**) Cole-Cole plots of processed cheese spreads at increasing amplitudes. Low-fat cheese spreads lie higher than those of high-fat cheese spreads. (**b**) Zoom-in of a single Cole-Cole plot for a low-fat cheese spread demonstrating their general behavior during one oscillatory cycle. The dot denotes the starting point of the oscillatory cycle. (Reproduced from [32], with permission from Elsevier)

the Chebyshev decomposition method (not shown). Beyond 50% strain, the size of the deltoids in both samples decrease and become progressively comparable in size. This suggests that both cheeses have experienced similar irreversible network decay at higher deformations (80–1000%). Additionally, the shape of all deltoids for both cheeses experience a similar process of physical changes in each oscillatory cycle, described in 4 stages. These stages include (1) shear thinning and strain softening, (2) shear thickening and strain stiffening, (3) shear thickening and strain softening, (4) shear thinning and strain softening. These transitions coincided with those reported by Chebyshev parameters, with the additional transition from shear thinning (region 1) to shear thickening (stage 2) at the beginning of the oscillation cycle.

19.5.2.2 Hard Fillers: Rigid Particles Dispersed in a Lipid Gel Network

Many lipid-based suspensions such as chocolate, icing, and other confectionery contain rigid granular particles such as sugar and bulking agents, dispersed in a continuous phase. Filler addition is often sought to modulate stiffness, processability, and sensory perception. Successful replacement of the lipid phase by an oleogel phase requires partly that the suspensions exhibit similar rheological response. Recent studies suggest that in both cases, the mechanics of the composite material is akin to those hard fillers dispersed in a yield stress fluid. In what follows, the case of starch dispersed in a tripalmitin lipid gel network is presented. Starch particles were treated as noninteracting hard spheres based on preliminary findings. The elastic modulus, yield stress, and yield strain were described and interrelated through simple scaling laws from a micromechanical homogenization analysis of hard spheres isotropically distributed in yield stress fluids. For all models, the maximum packing fraction ϕ_m was equal to the random close packing of polydisperse hard

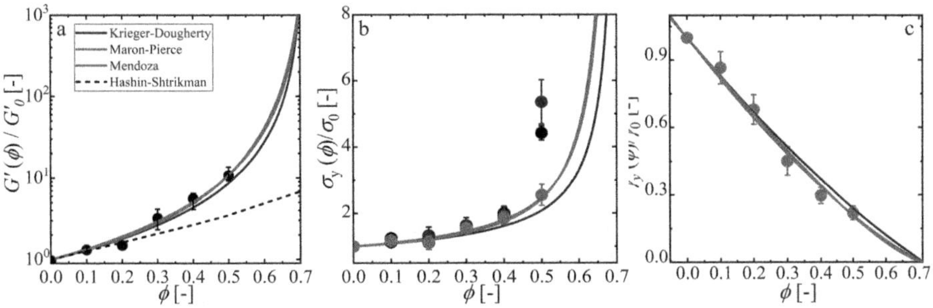

Fig. 19.10 Dimensionless rheological measures: (**a**) elastic modulus, (**b**) yield stress, and (**c**) yield strain, all as function of particle volume fraction, dispersed in a lipid gel. Experimental data is well described by empirical models: Krieger-Dougherty, Maron-Pierce, and Mendoza models, forced-fitted with a theoretical maximum packing volume fraction of 0.72 ± 0.01. (Reproduced from [36], with permission from Elsevier under the terms of the Creative Commons Attribution 4.0 International (CC BY 4.0) license)

sphere ϕ_{RCP} calculated based on a geometrical theory and using the log normal distribution of starch particles dispersed in oil as input. Figure 19.10 shows the dimensionless elastic moduli, yield stress, and yield strain as a function of volume fraction of starch. For all cases, the rheology is appropriately described using Krieger-Dougherty, Maron-Pierce, and Mendoza models derived from granular suspensions.

To illustrate, the forms of the Krieger Dougherty model for the elastic moduli, yield stress, and yield strain adequately described the rheology:

$$\frac{G'(\phi)}{G'_0} = \frac{1}{(1-\phi/\phi_m)^{2.5\phi_m}} \tag{19.28}$$

$$\frac{\sigma_y(\phi)}{\sigma_0} = \sqrt{\frac{1-\phi}{(1-\phi/\phi_m)^{2.5\phi_m}}} \tag{19.29}$$

$$\frac{\gamma_y(\phi)}{\gamma_0} = \sqrt{(1-\phi)(1-\phi/\phi_m)^{2.5\phi}} \tag{19.30}$$

where G'_0, σ_0, and γ_0 are the plateau modulus, the yield stress, and the yield strain of the continuous lipid phase, respectively, and ϕ_m is the maximum volume packing. For the use and applicability of all models, readers are referred to the original study [36].

19.6 Conclusions

Rheological and mechanical measures encode key information about the static and dynamic mesostructures responsible for macroscopic performance, textural attributes, and processing of fats and oleogels. Several tests are available to probe the mechanics and rheology of fats and oleogels ranging from traditional and simple methods such as penetrometry to elaborate and advanced methods such as large amplitude oscillatory shear. Penetrometry tests can be useful to discriminate among different fat and oleogels. However, only controlled flows such as large amplitude oscillatory shear closely resemble nonlinear and transient flows encountered during processing and product use. By coupling large amplitude flows with other nonlinear flows such as extrusion-like flows, it is possible to access a broad map of viscoelasticity as a function of timescales and deformations. Such efforts will guide the inverse design and substitution of fats by oleogels. Connecting structure with rheology via phenomenological models and coupling rheology with structural probes can also iluminate the dynamic microscopic ensemble underlying the rheological response.

References

1. Szczesniak AS (2002) Texture is a sensory property. Food Qual 13:215–225. https://doi.org/10.1016/s0950-3293(01)00039-8
2. Wright AJ, Scanlon MG, Hartel RW, Marangoni AG (2008) Rheological properties of milkfat and butter. J Food Sci 66:1056–1071. https://doi.org/10.1111/j.1365-2621.2001.tb16082.x
3. Macias-Rodriguez BA, Marangoni AG (2020) Rheology and texture of cream, milk fat, butter and dairy fat spreads. In: Truong T, Lopez C, Bhandari B, Prakash S (eds) Dairy fat products and functionality: fundamental science and technology. Springer International Publishing, Cham, pp 245–275
4. Haighton AJ (1959) The measurement of the hardness of margarine and fats with cone penetrometers. J Am Oil Chem Soc 36:345–348. https://doi.org/10.1007/bf02640051
5. Hayakawa M, deMan JM (1982) Consistency of fractionated milk fat as measured by two penetration methods. J Dairy Sci 65:1095–1101. https://doi.org/10.3168/jds.S0022-0302(82)82317-5
6. Tanaka M, Man JMDE, Voisey PW (1971) Measurement of textural properties of foods with a constant speed cone penetrometer. J Texture Stud 2:306–315. https://doi.org/10.1111/j.1745-4603.1971.tb01007.x
7. Haighton AJ (1965) Worksoftening of margarine and shortening. J Am Oil Chem Soc 42:27–30. https://doi.org/10.1007/BF02558248
8. Shama F, Sherman P (1970) The influence of work softening on the viscoelastic properties of butter and margarine. J Texture Stud 1:196–205. https://doi.org/10.1111/j.1745-4603.1970.tb00723.x
9. Castro M, Giles DW, Macosko CW, Moaddel T (2010) Comparison of methods to measure yield stress of soft solids. J Rheol 54:81–94. https://doi.org/10.1122/1.3248001
10. Mishra K, Grob L, Kohler L, Zimmermann S, Gstöhl S, Fischer P, Windhab EJ (2021) Entrance flow of unfoamed and foamed Herschel–Bulkley fluids. J Rheol 65:1155–1168. https://doi.org/10.1122/8.0000286

11. Macosko CW (1994) Rheology: principles, measurements, and applications. Wiley-VCR, Hoboken
12. Kloek W, van Vliet T, Walstra P (2005) Mechanical properties of fat dispersions prepared in a mechanical crystallizer. J Texture Stud 36:544–568. https://doi.org/10.1111/j.1745-4603.2005.00031.x
13. Deman JM, Beers AM (1987) Fat crystal networks: structure and rheological properties. J Texture Stud 18:303–318. https://doi.org/10.1111/j.1745-4603.1987.tb00908.x
14. Coussot P, Tabuteau H, Chateau X, Tocquer L, Ovarlez G (2006) Aging and solid or liquid behavior in pastes. J Rheol 50:975–994. https://doi.org/10.1122/1.2337259
15. Dinkgreve M, Paredes J, Denn MM, Bonn D (2016) On different ways of measuring "the" yield stress. J Nonnewton Fluid Mech 238:233–241. https://doi.org/10.1016/j.jnnfm.2016.11.001
16. Van den Tempel M (1961) Mechanical properties of plastic-disperse systems at very small deformations. J Colloid Sci 16:284–296. https://doi.org/10.1016/0095-8522(61)90005-8
17. Narine SS, Marangoni AG (1999) Mechanical and structural model of fractal networks of fat crystals at low deformations. Phys Rev E 60:6991–7000. https://doi.org/10.1103/physreve.60.6991
18. Macias-Rodriguez B, Marangoni AG (2016) Rheological characterization of triglyceride shortenings. Rheol Acta 1–13. https://doi.org/10.1007/s00397-016-0951-6
19. Ewoldt RH, Bharadwaj NA (2013) Low-dimensional intrinsic material functions for nonlinear viscoelasticity. Rheol Acta 52:201–219. https://doi.org/10.1007/s00397-013-0686-6
20. Rogers SA (2012) A sequence of physical processes determined and quantified in Laos: an instantaneous local 2d/3d approach. J Rheol 56:1129–1151. https://doi.org/10.1122/1.4726083
21. Ewoldt RH, McKinley GH (2010) On secondary loops in Laos via self-intersection of lissajous–bowditch curves. Rheol Acta 49:213–219. https://doi.org/10.1007/s00397-009-0408-2
22. Macias Rodriguez BA (2019) Nonlinear rheology of fats using large amplitude oscillatory shear tests. In: Marangoni AG (ed) Structure-function analysis of edible fats, 12th edn. AOCS Press, pp 169–195
23. Macias-Rodriguez BA, Marangoni AA (2018) Linear and nonlinear rheological behavior of fat crystal networks. Crit Rev Food Sci Nutr 58:2398–2415. https://doi.org/10.1080/10408398.2017.1325835
24. Rogers SA, Lettinga MP (2012) A sequence of physical processes determined and quantified in large-amplitude oscillatory shear (Laos): application to theoretical nonlinear models. J Rheol 56:1–25. https://doi.org/10.1122/1.3662962
25. Thareja P, Street CB, Wagner NJ, Vethamuthu MS, Hermanson KD, Ananthapadmanabhan KP (2011) Development of an in situ rheological method to characterize fatty acid crystallization in complex fluids. Colloids Surf A Physicochem Eng Asp 388:12–20. https://doi.org/10.1016/j.colsurfa.2011.07.038
26. Merger D, Wilhelm M (2014) Intrinsic nonlinearity from laostrain—experiments on various strain- and stress-controlled rheometers: a quantitative comparison. Rheol Acta 53:621–634. https://doi.org/10.1007/s00397-014-0781-3
27. Faber TJ, Van Breemen LCA, McKinley GH (2017) From firm to fluid – structure-texture relations of filled gels probed under large amplitude oscillatory shear. J Food Eng 210:1–18. https://doi.org/10.1016/j.jfoodeng.2017.03.028
28. Ramamirtham S, Shahin A, Basavaraj MG, Deshpande AP (2017) Controlling the yield behavior of fat-oil mixtures using cooling rate. Rheol Acta 56:971–982. https://doi.org/10.1007/s00397-017-1048-6
29. Läuger J, Stettin H (2010) Differences between stress and strain control in the non-linear behavior of complex fluids. Rheol Acta 49:909–930. https://doi.org/10.1007/s00397-010-0450-0
30. Ewoldt RH, Hosoi AE, McKinley GH (2008) New measures for characterizing nonlinear viscoelasticity in large amplitude oscillatory shear. J Rheol 52:1427–1458. https://doi.org/10.1122/1.2970095

31. Ewoldt RH, Hosoi AE, McKinley GH (2009) Nonlinear viscoelastic biomaterials: meaningful characterization and engineering inspiration. Integr Comp Biol 49:40–50. https://doi.org/10.1093/icb/icp010
32. Le AM, Erturk MY, Kokini J (2023) Effect of fat on non-linear rheological behavior of processed cheese spreads using coupled amplitude-frequency sweeps, fourier transform-chebyshev polynomials method, sequence of physical processes, and quantitative network analysis. J Food Eng 336:111193. https://doi.org/10.1016/j.jfoodeng.2022.111193
33. Ewoldt RH, Johnston MT, Caretta LM (2015) Experimental challenges of shear rheology: how to avoid bad data. In: Spagnolie SE (ed) Complex fluids in biological systems: experiment, theory, and computation. Springer, New York, pp 207–241
34. Macias-Rodriguez BA, Ewoldt RH, Marangoni AG (2018) Nonlinear viscoelasticity of fat crystal networks. Rheol Acta 57:251–266. https://doi.org/10.1007/s00397-018-1072-1
35. Mao B, Divoux T, Snabre P (2016) Normal force controlled rheology applied to agar gelation. J Rheol 60:473–489. https://doi.org/10.1122/1.4944994
36. Macias-Rodriguez BA, Velikov KP (2022) Elastic reinforcement and yielding of starch-filled lipid gels. Food Struct 32:100257. https://doi.org/10.1016/j.foostr.2022.100257

Chapter 20
Image Analysis for Oleogel and Oleogel-Based System Characterization

Camila Palla and Fabio Valoppi

Abbreviations

AFM	Atomic force microscopy
CLSM	Confocal scanning laser microscopy
CRM	Confocal Raman microscopy
Cryo-SEM	Cryogenic scanning electron microscopy
Cryo-TEM	Cryogenic transmission electron microscopy
CTI	X-ray computed tomography imaging
D	Fractal dimension
DIP	Digital image processing
FFT	Fast Fourier transform

Camila Palla and Fabio Valoppi contributed equally

C. Palla (✉)
Departamento de Ingeniería Química, Universidad Nacional del Sur (UNS), Bahía Blanca, Argentina

Planta Piloto de Ingeniería Química - PLAPIQUI (UNS-CONICET), Bahía Blanca, Argentina
e-mail: cpalla@plapiqui.edu.ar

F. Valoppi
Electronics Research Laboratory, Department of Physics, University of Helsinki, Helsinki, Finland

Department of Food and Nutrition, University of Helsinki, Helsinki, Finland

Helsinki Institute of Sustainability Science, Faculty of Agriculture and Forestry, University of Helsinki, Helsinki, Finland

Helsinki Institute of Life Science, University of Helsinki, Helsinki, Finland
e-mail: fabio.valoppi@helsinki.fi

C. Andrea Palla, F. Valoppi (eds.), *Advances in Oleogel Development, Characterization, and Nutritional Aspects*,
https://doi.org/10.1007/978-3-031-46831-5_20

MG	Monoglyceride
ML	Machine learning
MRI	Magnetic resonance imaging
PAF	Pore area fraction
PLM	Polarized light microscopy
RM	Rheo-microscope
SEM	Scanning electron microscopy
TEM	Transmission electron microscopy
UV	Ultraviolet
XRD	X-ray diffraction

20.1 Introduction

Imaging is a qualitative tool commonly employed to visually support numerical data. Besides giving tangible and concrete support to data, images contain embedded information that can be extracted and quantified using image analysis. Thus, image analysis can be a powerful tool able to enrich the pool of techniques available to characterize and explain material properties. In general, we can refer to image analysis as the set of actions needed to extract specific information from digital images. Data obtained through image analysis are then interpreted based on the equipment used to generate the original images. Usually, we can extract information about the structure and behavior of materials at macro-, micro-, and nanoscale. The type of information obtained is based on the intensity and distribution of pixels (in 2D images) or voxels (in 3D images), which are the smallest elements forming an image. Image analysis can be performed using freeware or licensed software. These programs often embed most of the necessary commands and plugins for image editing and processing, as well as the possibility to automate the process through macros and scripts.

In the field of oleogels, images are usually obtained through digital cameras, light microscopy (such as brightfield and polarized light), scanning or transmission electron microscopy (SEM or TEM), and their cryogenic versions (Cryo-SEM or Cryo-TEM). Other advanced or more recent imaging techniques applied to oleogels include magnetic resonance imagining, Raman microscopy, and atomic force microscopy. From these images, it is possible to calculate multiple parameters to characterize crystal/polymer networks in oleogels, as well as the spatial distribution of oleogels in food products when used as ingredients. When images are taken by varying time (during storage), temperature (during heating and cooling at defined rates), or shear (controlled deformation through the application of stress), we can determine the influence of such external forces/stimuli on oleogels and calculate—depending on the modified variable—kinetics of crystallization and network formation, polymorphic transformation, melting and network breakup, or oil release kinetics, to name a few. Thus, image analysis is not only useful for understanding the overall structure and behavior of oleogels but also for predicting their potential stability and applicability in real food products.

In this chapter, we summarize and explain the most common procedures and software used in the literature to analyze images of oleogels and similar lipid-based materials under static or dynamic conditions with the aim of obtaining relevant macro-, micro-, and nanostructural information. We have also included some examples of applications.

20.2 The Use of Image Analysis in Oleogel Characterization

20.2.1 Image Acquisition Techniques

Various imaging techniques have been used to acquire images of oleogels or oleogel-based systems, as shown in Table 20.1. Well-established 2D and 3D imaging techniques, commonly used for visualizing food microstructures, are mainly employed to capture and characterize micron and submicron-sized features of oleogels. Advanced imaging techniques are also employed to extract more specific information. In this section, we provide a brief overview of these techniques, their main characteristics, and their applications in the field of oleogels.

Optical (polarized light and brightfield) microscopy has been mainly adopted to qualitatively observe crystalline particles and their networks (Fig. 20.1a, b). A major advantage of optical microscopy is that it provides real-time, full-color images of samples in a non-destructive way [1]. Furthermore, by coupling a microscope with a stage that controls temperature, pressure, or shear, a variety of dynamic conditions can be applied to observe changes in the appearance of the sample. As an example, the rheo-microscope (RM) module allows optical images of internal structures of gels to be recorded simultaneously with rheological measurements, making it possible to visualize the effects of shear and deformation on sample structures.

Electron microscopy is used to image nanostructures and employ an electron beam instead of light which allows to increase the resolution limits from around 400–700 nm to 1–0.05 nm [2, 3]. Both scanning electron microscopy (SEM), which explores the surface topology of a sample, and transmission electron microscopy (TEM), which aims to visualize the internal structure of thinly sliced samples, have been used to analyze the morphological characteristics of oleogel networks (Fig. 20.1c, d). Cryogenic electron microscopy (Cryo-SEM and Cryo-TEM) is especially relevant for use with oleogel samples where the structure can be affected by high temperatures. With Cryo-SEM/TEM the fixation is performed by freezing the sample in an ultrafast cooling stage, protecting the material from direct damage from electron beam irradiation.

Confocal scanning laser microscopy (CSLM) uses a laser beam as a light source. A typical confocal microscope contains four laser lines covering the range from ultraviolet (UV) to infrared. The mode of operation involves focusing the light beam on small volumes, thereby imaging very fine focal planes that provide highly detailed optical sections through the sample [3, 4]. CSLM also allows multiple sections to be scanned at different depths to reconstruct a final 3D image of the

Table 20.1 Techniques used to capture images of oleogels, oleogel-based systems, and related lipid systems, along with the information extracted from image analysis

Technique	Measured features/analysis	System	References
Bright field microscopy	Crystal shape	Stearic acid/monoglyceride oleogels	[18]
	Crystal size	Wax and hydrolyzed wax oleogels	[19]
	Angles between crystal faces	Wax oleogels	[20]
	Bubble size kinetics	Wax-based oleofoam	[21]
Polarized light microscopy (PLM)	Crystal size Crystal size distribution Crystal shape Crystalline area/pore area Crystal fractal dimension	Wax oleogel Monoglyceride oleogel	[22] [23] [24]
Rheo-microscope	Crystal size Crystal size distribution Crystal shape Crystalline area/pore area Crystal fractal dimension Crystallization kinetics	Monoglyceride oleogel	[25] [23]
Scanning electron microscopy (SEM) and Cryo-SEM	Polymer network Pore diameter	Ethylcellulose oleogel	[6]
	Crystal shape	Fruit wax and lecithin oleogel	[26]
	Crystal network	Wax and hydrolyzed wax oleogels	[19]
Transmission electron microscopy (TEM) and Cryo-TEM	Crystal shape	Stearic acid/monoglyceride oleogels	[18]
	Lamellar dimension	Wax oleogel Wax oleogelsmonoglyceride Monoglyceride (MG)/stearic acid/ stearyl alcohol/ 2-hydroxyethyl stearate/octadecane oleogel	[27] [28]
Confocal laser scanning microscopy (CLSM)	Crystal size Crystal size distribution Crystal shape	Monoglyceride oleogel	[23]

	Crystalline area/pore area Crystal fractal dimension		
	Interfacial morphology	Beeswax oleogel-in-water Pickering emulsion	[29]
	Oleogel and hydrogel phase distribution	Bigel	[30]
Atomic force microscopy (AFM)	Polymer network Pore diameter Pore depth	Ethylcellulose oleogel	[6]
	Fibril shape Fibril growth mechanism	Sitosterol-oryzanol oleogel	[7]
Confocal Raman microscopy (CRM)	Size and porosity of fat crystal flocs Fractal dimension	Micronized fat crystals in oil	[10]
	Crystallinity index Polymorphic form identification	Porcine fat	[11]
	Microstructure Maps of component distribution	Ethylcellulose and anhydrous milk fat oleogel	[12]
	Stability against phase separation	Oleogel-based oleofoams	[5]
Magnetic resonance imaging (MRI)	Mobility of lipid protons	Sweet bread containing monoglyceride oleogel	[13]
X-ray computed tomography imaging (CTI)	Object volume Structure thickness Porosity	Oleogel-based filling cream	[15]
	Porosity	Oleogel-based muffins	[31]
Digital camera (macroscopic images)	Oil release kinetics	Monoglyceride oleogel	[17]
	Crumb grain features (air cells size and number)	Oleogel-based muffin	[16]

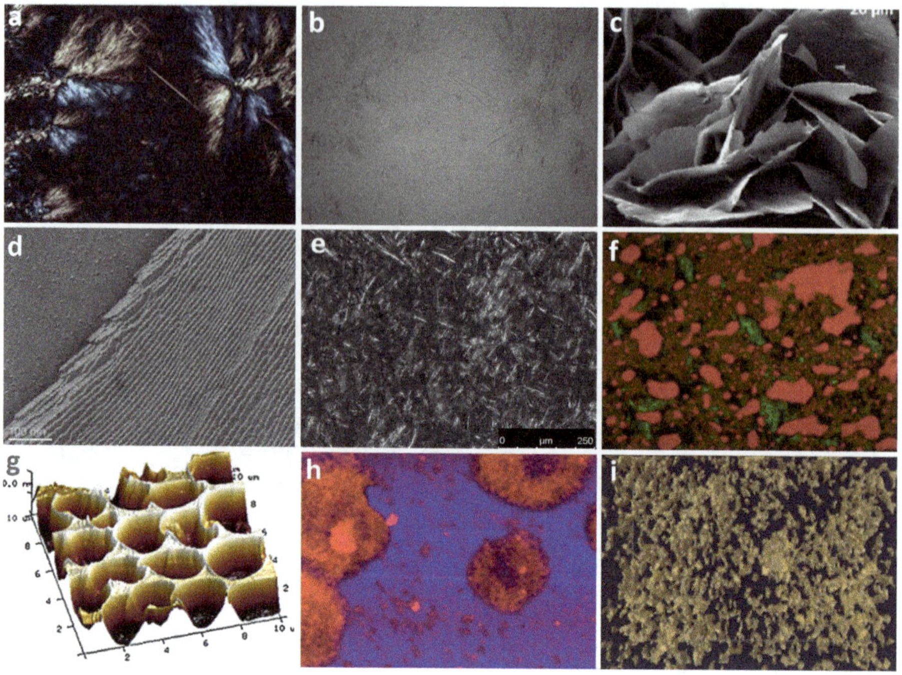

Fig. 20.1 Images acquired with different imagining techniques: (**a**) PLM, sample: monoglyceride (MG): phytosterol oleogel. (Reproduced from [32], with permission from John Wiley & Sons); (**b**) Brightfield microscopy, sample: MG: phytosterol oleogel. (Reproduced from [32], with permission from John Wiley & Sons); (**c**) Cryo-SEM, sample: sunflower wax oleogel. (Reproduced from [19], with permission from John Wiley & Sons, under the terms of the Creative Commons CC-BY-NC-ND 4.0 license); (**d**) TEM, sample: stearic acid oleogel. (Reproduced from [18], with permission from Elsevier); (**e**) CLSM, sample: MG oleogel; (**f**) CLSM, sample: 50:50 bigel (dyed hydrogel fluoresces green and dyed oleogel fluoresces red). (Reproduced from [30], with permission from Elsevier, under the terms of the Creative Commons CC-BY-NC-ND 4.0 license); (**g**) AFM, sample: ethylcellulose oleogel. (Reproduced from [6], with permission from Elsevier); (**h**) CRM, sample: ethylcellulose and anhydrous milk fat oleogel. (Reproduced from [12], with permission from Elsevier); (**i**) CTI, sample: filling creams produced with MG: wax oleogel. (Reproduced from [15], with permission from Elsevier, under the terms of the Creative Commons CC-BY-NC-ND 4.0 license). See Table 20.1 for the meaning of the abbreviations

sample. CLSM has been used to study the crystal network characteristics in oleogels (Fig. 20.1e), but also the interfacial morphology in two-phase and multiphase systems derived from oleogels, such as simple emulsions, Pickering emulsions, bigels (Fig. 20.1f), and oleofoams. Recently, CLSM was used to analyze fluorescence recovery after photobleaching as a method to determine the stability of oleofoams against possible phase separation phenomena [5].

Another advanced imaging technique is atomic force microscopy (AFM), which generates 3D topographical images of samples, with nanometer or even atomic resolution, by measuring the force interactions between a nanometer-sized probe and the material surface [4]. AFM has been used to study morphological characteristics of crystal and polymeric networks (Fig. 20.1g), geometrical characteristics of crystal surfaces, and the growth mechanism of tubular gelators [6–8].

Confocal Raman microscopy (CRM) combines Raman spectroscopy and confocal microscopy to obtain chemical composition and structural information over a small defined sample area. The identification of compounds present in the sample surface is based on the so-called Raman effect, which is the inelastic scattering of photons produced when monochromatic light interacts with sample molecules [9]. CRM compositional maps have been used to determine the size and porosity of floc crystal networks and fractal dimension in dispersions of micronized fat crystals in oil [10]. The degree of crystallinity and the type and amount of major crystal polymorphs were imaged simultaneously by CRM in porcine adipose tissue, which was used as a model fat system [11]. These methods can also be implemented to study oleogels due to the similarity between the systems. Recently, CRM was applied to visualize the structural distribution of components in oleogels from ethylcellulose and milk fat, as well as to map lipids with different degrees of mobility (liquid vs. solid) within oleogels (Fig. 20.1h) [12].

On the other hand, imaging techniques developed for medical use have been applied to study oleogel-based food products. Nuclear magnetic resonance imaging (MRI) was used to study the distribution of the lipid fraction in oleogel-based sweet bread [13]. On the other hand, X-ray computed tomography imaging (CTI) can be employed to investigate the internal structure of materials in a three-dimensional way. As X-rays pass through the sample, they attenuate according to material density and atomic number, and the intensity of the emerging X-ray is recorded in a series of two-dimensional images, which can be processed to produce cross-sectional images. These multiple images taken from different angles of the sample are then processed using computer software to reconstruct the sample 3D model [3, 14]. This makes CTI an ideal tool for observing the microstructure of porous foods. In the field of oleogels, CRI has been applied to determine morphological parameters of aerated structures of filling creams (Fig. 20.1i) [15]. This method can also be used to examine the structures of other oleogel-based products, including aerated chocolate, bread, cakes, and frozen foods [14].

Images from digital cameras can be suitable to determine macrostructure and macroscopic behavior of specific foods, in which the size of the features is large enough to be captured using low magnifying tools. Processed digital images have been used to determine surface characteristics, porosity, and crumbly grain characteristics in cookies and bakery products, as well as to visualize the consistency of oleogel-based filling creams [13, 16]. Recently, a novel method to determine the oil release kinetics of oleogels during storage was proposed based on automated image analysis [17].

20.2.2 Sample Preparation

Image analysis is a tool that can be applied to any type of images, regardless of sample preparation. For this reason, it is critical that users prepare samples for visualization using an appropriate protocol that minimizes material modification to ensure that the images they obtain are accurate representations of the original sample. Images with artifacts due to improper sample handling will still give results which, however, can be of questionable significance. For example, if a monoglyceride oleogel is imaged using an optical microscope, but the sample is first spread on the glass slide and then a glass coverslip is firmly pressed on top of the sample, the obtained images will likely show a broken and loose monoglyceride network which does not represent the original structure within the oleogel. If these images are analyzed, one can still obtain information, which would lead to improper conclusions. On the other hand, sample treatment is also necessary for certain imaging techniques because of the interference that, for example, oil might cause to the gelator network. In this case, deoiling needs to be applied before imaging. However, the type and amount of solvent, as well as the contact time, and application of fixating agents are factors that greatly influence the outcome [6, 33, 34]. Thus, the deoiling process needs to be optimized to be effective in removing the oil from the oleogel without damaging the network structure.

20.2.3 Commonly Used Software for Image Analysis

To analyze images, they should be in digital form or digitalized after acquisition. Working with digital images allows editing and processing of images before extracting information. Indeed, images need to be transformed to highlight features or objects to be analyzed, increasing their contrast with respect to the background, or masking parts of the image that are not important or that can affect the output of the transformation. More information on image processing is given in the following paragraph. In general, the reader should know at this point that there are two steps in image analysis: processing and information extraction. Dedicated software used in image analysis can perform both operations. However, there are some software developed as image editors that can easily perform the first step but possess limited or no functions to execute the second step. Although there are various software and apps for computers, tablets, and smartphones, here we report the most used and promising ones for image analysis of lipid-based materials and oleogels.

ImageJ and its different versions/evolutions, such as Fiji and ImageJ2, are freeware/open-source software (free of cost) that have been under continuous development since Rasband's pioneering work on the first image analysis software in 1987 [35]. This family of software (https://imagej.net/learn/flavors) has all the necessary basic functions to carry out image analysis for oleogel research. In addition, there are many publicly available plugins developed by the online community and users that extend the capabilities of these software, such as the FracLac plugin, which allows for the calculation of fractal dimensions of oleogel crystal networks. This family of software has a user-friendly interface that resembles many other software operating in Windows or MacOS.

Image Pro Plus is a licensed software sold by Media Cybernetics that works on 2D images and can be extended to 3D and 4D (time-lapse) images with an additional module. Like ImageJ and its derived software, Image Pro Plus has a user-friendly interface and provide all the necessary functions to perform image analysis for oleogel research.

Photoshop is a licensed software sold by Adobe, which was developed as a raster graphics editor allowing to create and edit images. This software can be used to process and transform images before analysis.

Besides dedicated software for image analysis, programming languages and computing environments like Python (freeware) and MATLAB (from MathWorks, licensed) can also be used for this purpose. Due to the versatility of these languages, it is possible to directly write a code/script that uses in-built functions to process and extract information from images. Although these programming languages are versatile, users need to learn the syntax and find the appropriate command or set of commands to carry out image analysis. To facilitate this operation, libraries and toolboxes are available for both platforms. An example of an image analysis library developed for Python is Scikit-image, where the scientist/programmer community has documented the implementations of common algorithms for image analysis with the aim to facilitate teaching [36]. Another example is the Image Processing Toolbox available for MATLAB, which includes standard algorithms for image analysis and the application of related workflows.

20.2.4 Useful Commands for Image Analysis

In this section, we present the most useful image processing commands that facilitate information extraction, as well as commands that allow to determine basic parameters in image analysis.

Processing is necessary since it optimizes images for extracting information. The main purpose of this step is to visually enhance those features to be analyzed. There are four commands that are mostly used in image processing: *grayscale 8-bit* conversion, *contrast enhancement*, *masking*, and *thresholding*. Besides these four commands, *background subtraction* and *segmentation* are also used in lipid-based materials image processing. The processed images can then be analyzed by *counting/measuring* tools and *Fast Fourier Transform (FFT)*.

Digital images are usually obtained using the RGB color model. This means that each color can be obtained by summing red, green, and blue colors at different intensities and ratios. A digital image is then the result of the overlap of three distinct images having intensities ranging from the darkest to lightest of red, green, and blue colors. Although color can carry information, as seen in CLSM and PLM micrographs, most of the analysis carried out on oleogel images only requires grayscale images. Therefore, it is common to convert images from RGB to *8-bit grayscale* format. The latter is a compressed image format where the values for each color at each pixel location are averaged (simple average or weighted average, depending on the type of function used), obtaining as a result an image that shows different tonalities of gray. Another commonly used command in image processing is *contrast enhancement*, which helps to identify objects and their features by maximizing their intensity difference with respect to other objects and the background. When illumination is not uniform throughout an image, it is sometimes necessary to *subtract* the *background* before modifying the contrast. In case parts of the image are not of interest, the most convenient action is to create a *mask* to eliminate them. This action creates customizable shapes that are overlapped to the image on specific regions selected by the user. Masks can then be colored to avoid interfering with further image processing. The last widely used command in oleogel image analysis is *thresholding*. This command allows the conversion of 8-bit grayscale images into binary (black and white) images by setting a cutoff value (threshold) of luminosity. Pixels above the threshold are converted to white pixels (usually part of the object of interest ,e.g., crystals in PLM images), while pixels below the threshold are converted to black color, which becomes part of the background. This operation reduces the complexity of images and simplifies the operations during information extraction. Sometimes *thresholding* leads to the opposite result: the objects of interest become black, and the background becomes white. This outcome does not interfere with the analysis as long as the user is aware of the color assigned to the objects of interest. *Thresholding* leads to better results when carried out after all other commands described before. This is because enhancing the difference between the background and the objects of interest improves the thresholding result since this command assumes that images have a bimodal luminosity distribution (light objects + dark background, or vice versa) [37]. If thresholding is carried out on the entire image, it is called *global thresholding*. On the other hand, if thresholding is carried out on each pixel and the cutoff value is based on the neighbor pixels within a restricted area, it is called *local thresholding*. In general, *global thresholding* is more common in oleogel image analysis. It may happen that images show objects that are touching each other, for example, adjacent crystals. If a user applies any of the commands described above, the objects will permanently merge, leading to incorrect results. To avoid this problem, another useful command is *segmentation*, which separates or divides two or more adjacent objects.

After image processing, one can proceed with the extraction of information from images. There are mainly two types of commands used for this scope when analyzing images of oleogel samples: *counting/measuring* and *FFT*. *Counting/measuring* command allows to count, calculate area, dimensions, etc., of objects (e.g., crystals and oil stains) shown in any image. It is also possible to use embedded commands like the *line profile* or *plot profile*, which allow to calculate the intensity of gray value over distance in a line or in a region of interest. In the latter example, the analysis should be carried out on an 8-bit grayscale image. The second type of command,

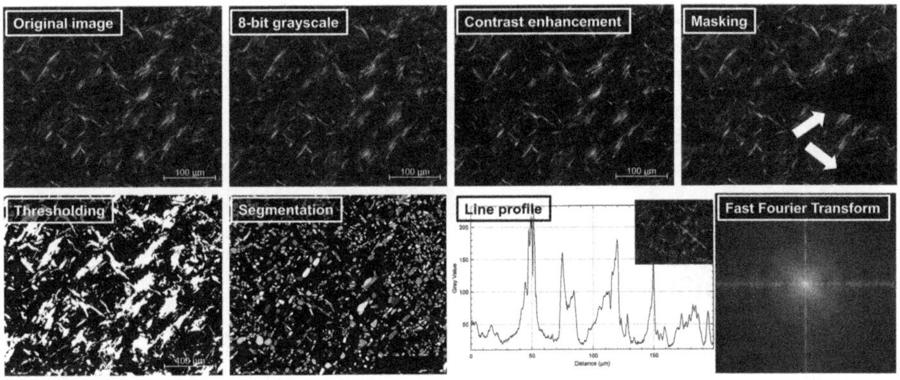

Fig. 20.2 Example of commands used in image analysis. Original image is a polarized light micrograph of a 2.5% w/w sunflower wax oleogel. (Images were processed using Fiji)

FFT, is a mathematical transformation of the image from the spatial to the frequency domain, returning as a result a power spectrum. From the power spectrum, it is possible to obtain information about the sine and cosine components of the signal of the original image in the spatial domain, where each point represents a particular frequency. If repeating patterns are present, they manifest in the power spectrum as spots that propagate in the opposite direction of the repetition in the original pattern. The spots contain information about the repeating distance of the pattern, which is useful in determining, for example, lamellar structures imaged using TEM. It is also worth mentioning that *FFT* can also be used to process images before analyzing them. Indeed, once the image is transformed, it is possible to mask some of the signals in the power spectrum and then apply the *inverse Fourier Transform* to obtain a modified version of the original image (for example, smoothing defects of the original image). An example of all commands applied to an oleogel PLM micrograph described in this section is shown in Fig. 20.2.

As a final note, there may be cases where a series of commands need to be applied to large sets of images. For example, *spatial calibration*, *contrast enhancement*, *masking*, *thresholding*, and *object counting* might have to be executed on hundreds of images. In this case, repeated actions can be automated in image analysis procedures. Depending on the program/programming language used, this can be implemented by recording *macros* (which record the actions performed on the mouse or keyboard and repeat them in sequence) or by embedding *dedicated functions* in the developed code. For more information on other commands available for image analysis, the reader can refer to the help/guide menu of the selected image analysis software as well as to the extensive book of Russ [38] on image analysis of food microstructure.

Practical Recommendations for Image Analysis

- Make sure the images obtained are accurate representations of the original sample by using a sample preparation protocol that minimizes material modification.

- Remember to carry out spatial calibration before processing and analyzing your images. You can use either the scale bar in your image or a reference object of known dimensions. This is the first action that should be performed in the software of your choice after opening or importing the image you want to analyze, before using any of the commands below.

20.3 What Information Can We Extract?

In this section, we describe qualitative and quantitative parameters related to oleogels that can be obtained from image analysis by applying the image processing commands discussed above and other specific commands.

20.3.1 Crystal Shape and Network Morphologies

Oleogels can exhibit different colloidal structures depending on the gelator they are formed from, such as polymeric, crystalline, and self-assembled fibrillar networks [39]. The morphological features of crystalline networks can be defined by the shape, size, and state of agglomeration of crystal particles. These characteristic parameters are usually determined through image analysis of PLM micrographs.

Via PLM observation, the crystal morphologies of lipid gelators, such as MG, waxes, and fatty acids, have been suggested to be needle-like, which are organized into spherulites/rosettes or irregularly shaped clusters [23, 25, 40]. Other gelators, like phytosterols, have shown multiple morphologies with aggregates consisting of fibrous, spherulite, and plate-like crystals [32]. In general, crystalline network morphology can be easily identified using oleogel systems with low gelator concentrations to avoid large-scale agglomerates that could lead to misinterpretation. Furthermore, the use of more than one crystal visualization method is highly recommended [41]. This was evidenced in the study of sunflower wax oleogels, which showed a needle-like crystalline morphology when observed by PLM, while platelets were seen by SEM examination [33, 42]. It was suggested that the needle-like morphology might be a view of a protruding edge of the platelets [33]. PLM micrographs have also been used to evaluate qualitative and quantitative changes in the crystalline network morphology of oleogels from waxes and MG during aging storage [43, 44]. On the other hand, polymeric networks such as ethylcellulose oleogels cannot be visualized under PLM due to the absence of birefringence, requiring alternative imaging techniques with higher resolution like AFM and Cryo-SEM [6].

20.3.2 Crystal Size and Crystal Size Distribution

By measuring the size or length of crystals, it is possible to characterize crystals quantitatively. Using this approach, the effects of composition or processing variables on crystal structure can be systematically and accurately identified. When oleogels are composed of uniform-shaped crystals, the average crystal size, or the average crystal length in the case of needle-like particles, and the standard deviation are suitable parameters to characterize the crystalline network. A PLM, CLSM, and RM micrograph can show dozens of crystals that can be measured using a measuring tool from suitable software (e.g., ImageJ) with a previously calibrated scale. Crystal length measurements should be made on individual crystals, even in the case of cluster formations [44]. However, a particle size distribution analysis is most appropriate when oleogels are composed of crystals of irregular sizes. This approach involves using the total of size measurements (recommended value >100) to generate a frequency histogram; for example, using the Histogram tool from the Analysis ToolPak of Excel (Microsoft Office). The size distribution is then described in terms of D_{50}, the median, and Span, the width. Span is calculated as $(D_{90} - D_{10})/D_{50}$, with D_{10}, D_{50}, and D_{90} being the particle size where 10, 50, and 90% of the population, respectively, lies below these values [23].

20.3.3 Pore Area Fraction and Crystalline Area

The determination of the area unoccupied by crystals, known as the pore area fraction (PAF), is a simple and useful method for characterizing the crystalline microstructure of oleogels. To calculate PAF (%), images are converted to *8-bit type* and processed using *global auto thresholding* to transform the original image into the form of white objects on a black background. The area (%) of white pixels in each image is representative of solid crystalline material, quantifiable by processing software. Therefore, the PAF can be calculated as the remaining area [23, 45]. In the case of images that cannot be processed by the automatic thresholding tool due to the color similarity between the crystals and the oil, a command that segments the color space present in the images can be used to divide the areas corresponding to crystals and oil. For example, the *Color Inspector 3D plugin* from ImageJ, which allows the generation of color histograms by segmentation of the color space presented in the images. With this, different colors of the image are identified—using RGB color model—and their absolute and relative frequencies are listed. To reduce the number of segments, images can be converted into *32-bit gray scale* images, resulting in histograms of 32 colors. From the histogram, the percentage of area corresponding to each segment is obtained. By visual inspection, each segment can be classified as either crystalline or oil structure, and then the areas corresponding to each type are summed up [23].

Previous work has pointed out the importance of imaging techniques and sample preparation in obtaining representative PAF values. When PLM, CLSM, and RM images of MG oleogels produced in a rheometer were used to determine PAF, different results were obtained among techniques [23]. The authors attributed this behavior to the fact that PLM and CLSM images were acquired from oleogel samples that were removed from the rheometer plate and placed between a slide and a coverslip. This procedure would keep the crystal structure intact, but probably not the oil distribution, which would be affected by the pressure exerted on the coverslip to achieve a thin enough sample for microscopy. In contrast, the RM images were captured directly *in situ* during oleogel formation, without subsequent manipulation, resulting in more accurate PAF results. Similarly, special care should be taken when samples are prepared by depositing a drop of molten oleogel on a microscope slide and covering it with a coverslip. A standardized procedure should be followed to avoid introducing variations between samples through manipulation.

20.3.4 The Box-Counting Fractal Dimension of Crystalline Area

The fractal dimension (D) is an index of space-filling properties used to describe surfaces whose irregular nature cannot be explained by traditional geometric terms. This parameter can be used as an indicator of the crystal mass distribution in oleogel networks. The higher D, the more homogeneously distributed mass and/or uniformly filled space [45]. Different computing methods can be applied to obtain D, with *box counting* (D_{bc}) being the most widely used. For this purpose, micrographs of oleogels are converted into *8-bit binary* images. Using the tool for D_{bc} calculation of an image analysis software, this parameter is determined automatically. In the case of ImageJ, D_{bc} can be determined by the *FracLac plugin* (option: BC). For more information on *FracLac plugin*, the reader can refer to its online manual (https://imagej.net/ij/plugins/fraclac/FLHelp/BoxCountingOptions.htm).

The D_{bc} algorithm consists of systematically laying a series of grids of decreasing length size r (the boxes), being $N(r)$ the number of boxes required to cover the mass area present in the image. For each successive length (scale), the N value is recorded. Then D_{bc} is calculated by evaluating how N varies with decreasing r [46]:

$$D_{bc} = \lim_{r \to 0} \frac{\log N(r)}{\log (1/r)} \tag{20.1}$$

In practice, D_{bc} is calculated as the slope of the line obtained by plotting log $(1/r)$ vs. log $(N(r))$. On the other hand, if three-dimensional gel features are to be represented, an additional dimension must be added to obtain D ($D = D_{bc} + 1$) [47].

D_{bc} has been used as a microstructural parameter to analyze the effects of the type of gelator and its concentration and cooling temperature during oleogel production

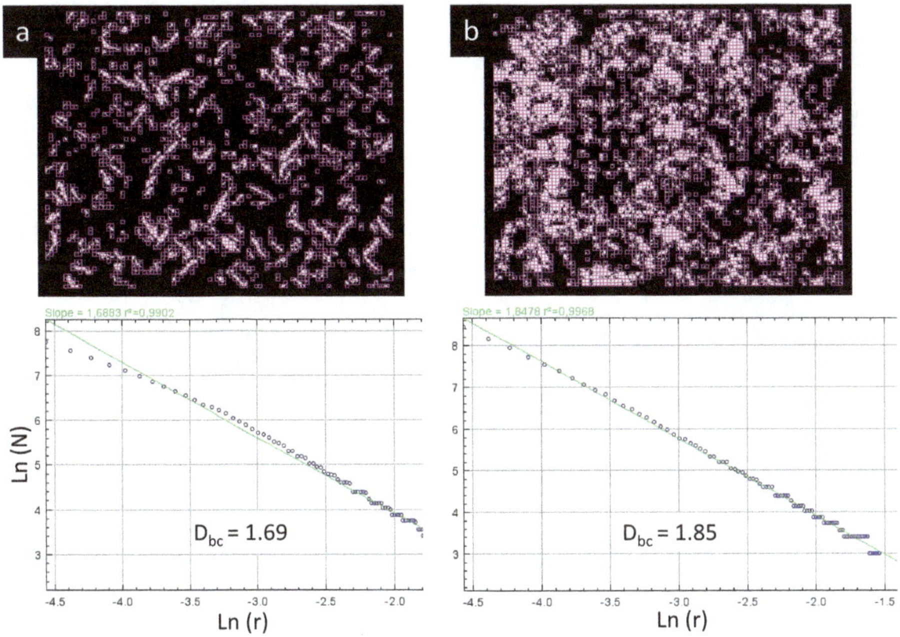

Fig. 20.3 PLM images of oleogel-based emulsions from (**a**) commercial and (**b**) recovered sunflower waxes processed by *FracLac* using box-counting analysis. The graphs show the linear regression generated by the program. The parameters in *FracLac* were set as follows: Image type = "use binary" and "locked black background" (it means that measurement is made over white part of figure, which correspond to crystalline area), number of positions for grids G = 1, maximum sizes = 45% of the total image area. Grids are indicated in violet color. Wax oleogels were produced at the same wax concentration (2% w/w). (Images adapted from [48], with permission from Wiley)

[23, 45, 48]. For example, oleogel-based emulsions formulated with recovered sunflower waxes showed higher D_{bc} values than those formulated with commercial ones (Fig. 20.3), which was consistent with a network formed by smaller crystals [42, 48]. Furthermore, the relationship between D_{bc} and solid lipid content has been studied in systems composed of MG or saturated triacylglycerols (tripalmitin and tristearin) and liquid triacylglycerols (medium chain triacylglycerols) [24]. The authors related the minimum solid lipid content able to reach a plateau value of D_{bc} to the network-forming ability and the type of crystals developed (e.g., platelet-like crystals and spherulites).

20.3.5 Kinetics of Crystallization

Image analysis can be used to determine the type of nucleation and dimensionality of crystal growth, both critical factors that determine microstructure characteristics and,

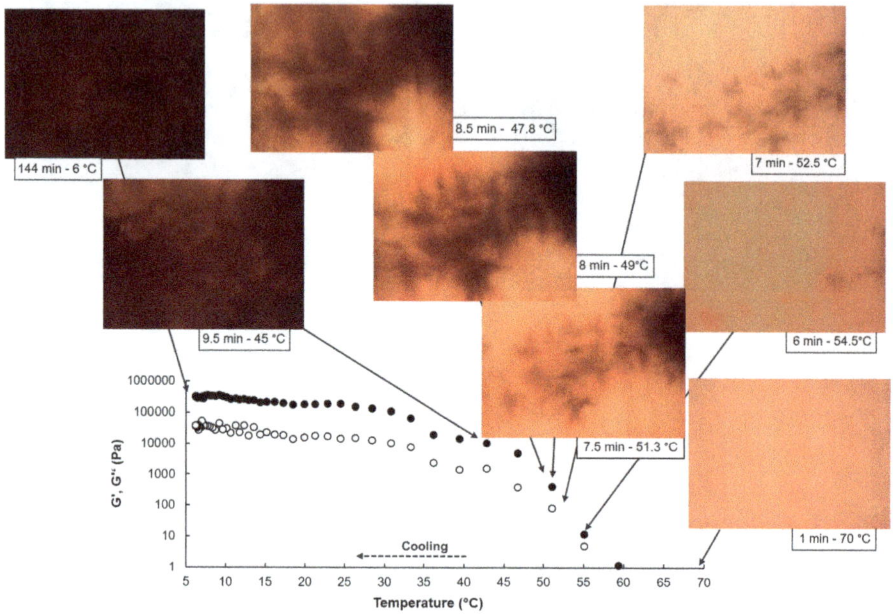

Fig. 20.4 Rheo-microscope images acquired during the formation process of a monoglyceride oleogel in a sweep temperature test (G′ [•] and G″ [○]). All acquired images were processed as indicated in Sect. 20.3.5 to model the kinetic of crystallization. (Adapted from [23], with permission from Elsevier)

consequently, the macroscopic behavior of oleogels. For this purpose, micrographs of the sample are taken during the cooling process. At first, only a homogeneous phase corresponding to the molten oleogel can be imaged. In the following stages, as the temperature decreases, the appearance of the first crystal nuclei and the subsequent crystal growth and branching is recorded (Fig. 20.4) [23]. After that, the images are processed to obtain a parameter (Y_s) that represents the crystal phase volume formed until a given time [49].

Non-isothermal crystallization kinetics of crystal particles can be appropriately described by a modified Avrami model, which is an adaptation of the original equation from isothermal cooling to non-isothermal cooling conditions [49, 50]:

$$\frac{Y_s}{Y_{max}} = 1 - e^{-k_{app}(t - t_0)^n} \tag{20.2}$$

where Y_{max} is the maximal crystal phase volume, k_{app} is the apparent crystallization rate constant, t is the time, t_0 is the induction time, and n is the Avrami index, which indicates both the nucleation mode (sporadic or instantaneous) and the dimensionality of growth (1D, 2D, 3D) [49]. k_{app}, t_0, and n parameters can be obtained by nonlinear curve fitting of experimental data of Y_s/Y_{max} versus t.

This approach was applied by Rogers and Marangoni [50] to analyze the nucleation and fiber growth kinetics of 12-hydroxystearic acid in canola oil. The authors used brightfield images acquired during cooling to first identify the type of nucleation process and then to determine the crystal fiber growth rate using Eq. 20.2, with Y_s being the average fiber length measured by ImageJ. Similarly, Avrami indices were accurately calculated from data extracted from PLM micrographs during crystallization of 12-hydroxystearic acid, stearic acid, and trihydroxystearin, with Y_s determined as crystal length, platelet area, and sphere area, respectively (Ho Lam & Rogers, 2011). More recently, Palla et al. [23] studied the crystallization of MG oleogels at different cooling profiles using RM images acquired during temperature sweep rheological tests and processed using the Color Inspector 3D plugin of ImageJ (see Sect. 20.3.3). The authors followed the crystallization process by quantifying the amount of crystalline solid material over time, that is, the percentage of crystal area in each RM image. Thus, Y_s/Y_{max} was defined as %crystal area (t)/%maximal crystal area. An Avrami index close to 2 was found for all cooling profiles tested, which was associated with an MG crystallization characterized by sporadic nucleation and crystal growth in needle-like form (1D). Interestingly, the authors also found that the RM images could be used to identify the temperature range at which the MG crystals formed, thereby determining a more restricted time domain in which Eq. 20.2 could also be used to calculate the Avrami index but based on the evolution of the complex viscosity.

20.3.6 Polymorphism and Lamellar Dimension

Oleogels obtained by structuring oil with self-assembly crystalline gelators can contain repeating stacked lamellar structures at nanoscale. To the best of our knowledge, these nanostructures have been observed in TEM images of oleogels derived from waxes, stearic acid, stearyl alcohol, and MG [18, 27, 28], where they resembled those present in crystalline nanoplatelets composed of saturated triacylglycerols [51]. In this case, image analysis can be used to determine the dimensions of these lamellar structures. For example, by applying FFT to selected areas of TEM images where the lamellar structures can be observed, it is possible to obtain the power spectrum of the image and calculate interplanar distances. This method was described by Sánchez-Becerril et al. [27] on candelilla wax-based oleogels, where the authors determined the presence of CW β' polymorphic form, further confirmed using XRD. Through the power spectrum of TEM images, lamellar dimensions of triacylglycerols in crystal nanoplatelets were also determined [52]. In general, characteristic dimensions of any repeating patterns in lipid-based materials can be calculated by applying FFT to TEM images [53], complementing the image analysis with XRD analysis.

20.3.7 Oil Mobility

Oil mobility can be determined by calculating the relaxation time of protons associated with lipids by using nuclear magnetic resonance (NMR) [54, 55]. However, to determine the location of mobile protons in a matrix/sample, it is necessary to use magnetic resonance imaging (MRI). In this case, the grayscale images obtained show bright areas that can be attributed to mobile protons belonging to oil or water. In a work by Calligaris et al. [13], oil mobility in sweet breads containing oleogels, other fat replacers, and palm oil and sunflower oil (controls) was calculated on magnetic resonance images using image analysis. The authors first demonstrated that bright areas on images could be attributed to oil (by suppressing the signal generated by lipids using a dedicated Spin-Echo mode). They then acquired images of sweet bread slices and calculated the total area of bright pixels (above a set threshold) with respect to the total area of the slice, concluding that lipid structuring plays a critical role on lipid distribution and mobility in their samples.

20.3.8 Oil Release

An important parameter for evaluating oleogel structure and stability is oil release. The oil contained in oleogels is mobile and can migrate out from the oleogel structure. This phenomenon, known as oil release or oil leakage, can occur at different magnitudes and times depending on the trapping ability of the gelator network and possible molecular rearrangements of gelators during storage [55–57]. Recently, Valoppi et al. [17] developed a novel automated method to determine oil release from oleogels using image analysis. The authors designed an inexpensive system featuring a tripod with a circular opening, filter paper, a digital camera, and a computer. Oleogels placed on a filter paper produce an oil stain that radiates out from the sample. By acquiring images from the bottom of the tripod over time, it is possible to obtain information on the evolution of the area of the oil stain and calculate the oil release kinetics. The authors automated image acquisition, processing and analysis using custom-made scripts for ImageJ. The result of the analysis is shown in Fig. 20.5. This method has also been used to evaluate the oil release of oleogels modified using ultrasonic standing waves [58].

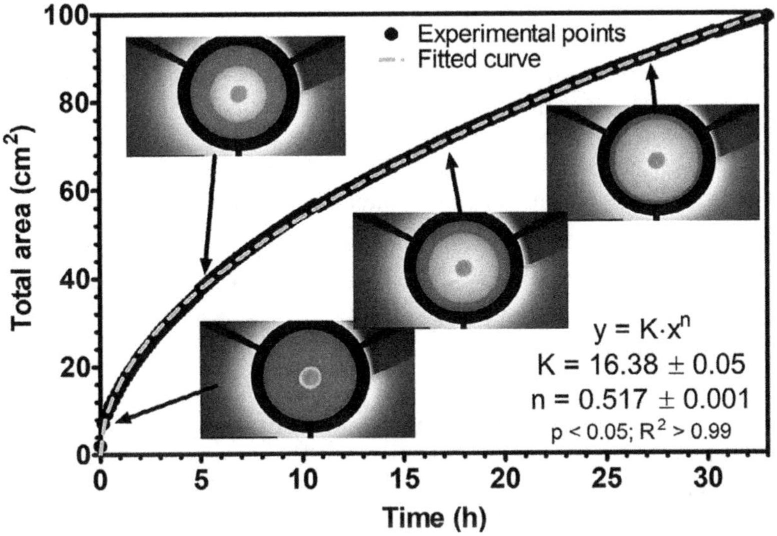

Fig. 20.5 Evolution over time of an oil stain area formed by the radial diffusion of oil from an oleogel placed on a filter paper. Data were generated using a set of 200 images captured over 33 h. Images reported in the figure show the evolution of the oil stain (annular white area) at selected time points. Fitted curve and regression analysis results are also shown. (Reproduced from [17], under the terms of the Creative Commons CC-BY license)

20.4 The (Potential) Use of Machine Learning in Image Analysis of Oleogels

Machine learning (ML) is a branch of artificial intelligence (AI), in which a computer learns from given datasets without being directly programmed, and builds mathematical models that are later used to perform specific tasks. Computers are usually trained on large datasets (labeled in case of supervised ML, or unlabeled in case of unsupervised ML) that are used to build and refine the models developed. Since ML is an iterative process, the models developed during learning are constantly improved based on new imputed datasets. ML is a versatile technique that can handle large and complex datasets, such as images, and it has been applied in image analysis of various materials [59]. Indeed, images can be considered as matrices of numbers where each pixel is identified by coordinate values (e.g., x and y in 2D images) associated with color values expressed using the RGB color model (as discussed in Sect. 20.2.4). Therefore, images can be treated as mathematical functions and be used as datasets for building models through ML. In this case, images can be analyzed using computational methods like digital image processing (DIP) by employing programming languages like Python. However, it is worth mentioning that usually the images acquired by the techniques described above (Sect. 20.2.1) represent small datasets for ML training. To overcome this problem, some solutions are available. In particular, dataset augmentation, which includes zooming, flipping, and contrast change on the same image (one image can be used multiple times), or transfer learning, where a certain problem is solved using a pre-trained ML on similar datasets, can be used to enlarge datasets for ML [59]. Although ML is a powerful tool for image analysis, to the best of our knowledge, there are no articles mentioning the use of this technique or other forms of AI for the analysis of oleogel images. However, due to the continuing advances in ML application and oleogel area, we might expect further research in this field.

20.5 Conclusions

Image analysis is a powerful tool to investigate the nano, micro, and macro characteristics of oleogels and oleogel-based systems. To obtain reliable results, images need to be representative of the sample and it is vital to avoid artifacts caused by improper sample preparation. Moreover, using more than one visualization technique is highly recommended. Throughout this chapter, the main techniques available for acquiring oleogel images have been presented, as well as the most used software and commands for image processing and analysis. The emergence of advanced microscopy techniques and the development of powerful software have made it possible to extract a wide variety of information from oleogel images. Image analysis finds a wide field of application in oleogels formed by a crystalline network, in which not only qualitative features can be determined but also quantitative parameters such as crystal size, porosity and network crystalline area, fractal dimension, kinetics of crystallization, and even the type of polymorphic form present and the lamellar structure. On the other hand, macroscopic images can be especially useful to evaluate the oil stability behavior of oleogels, as well as the macroscopic characteristics of food products made with oleogels. It is expected that in the coming years image analysis will be applied to the study of more complex oleogel-based systems, and that novel tools such as machine learning and other forms of artificial intelligence may also bring new methodologies to the study of oleogels.

Acknowledgments Camila Palla acknowledges the Consejo Nacional de Investigaciones Científicas y Técnicas (CONICET) of Argentina. Fabio Valoppi acknowledges the European Union's Horizon 2020 research and innovation program funding under the Marie Skłodowska-Curie grant agreement No. 836071.

References

1. Shi X, Zhao S, Wang F, Jiang Q, Zhan C, Li R, Zhang R (2021) Optical visualization and imaging of nanomaterials. Nanoscale Adv 3:889–903. https://doi.org/10.1039/d0na00945h
2. Ebnesajjad S (2011) Surface and material characterization techniques. In: Ebnesajjad S (ed) Handbook of adhesives and surface preparation. William Andrew Publishing, Norwich, pp 31–48
3. Verboven P, Defraeye T, Nicolai B (2018) Measurement and visualization of food microstructure. In: Devahastin S (ed) Food microstructure and its relationship with quality and stability. Woodhead Publishing, Sawston, pp 3–28
4. Metilli L, Francis M, Povey M, Lazidis A, Marty-Terrade S, Ray J, Simone E (2020) Latest advances in imaging techniques for characterizing soft, multiphasic food materials. Adv Colloid Interface Sci 279:102154. https://doi.org/10.1016/j.cis.2020.102154
5. Andriotis EG, Monou PK, Komis G, Bouropoulos N, Ritzoulis C, Delis G, Kiosis E, Arsenos G, Fatouros DG (2022) Effect of glyceryl monoolein addition on the foaming properties and stability of whipped oleogels. Gels 8:705. https://doi.org/10.3390/gels8110705
6. Zetzl AK, Gravelle AJ, Kurylowicz M, Dutcher J, Barbut S, Marangoni AG (2014) Microstructure of ethylcellulose oleogels and its relationship to mechanical properties. Food Struct 2: 27–40. https://doi.org/10.1016/j.foostr.2014.07.002
7. Matheson AB, Koutsos V, Dalkas G, Euston S, Clegg P (2017) Microstructure of beta-sitosterol: Gamma-oryzanol edible organogels. Langmuir 33:4537–4542. https://doi.org/10.1021/acs.langmuir.7b00040
8. Lupi FR, De Santo MP, Ciuchi F, Baldino N, Gabriele D (2018) The role of edible oils in low molecular weight organogels rheology and structure. Food Res Int 111:399–407. https://doi.org/10.1016/j.foodres.2018.05.050
9. Hassen SM, Mofdal MEE (2018) Using Raman spectroscopy to identify unknown materials. Int J Appl Phys 5:29–32. https://doi.org/10.14445/23500301/IJAP-V5I3P105
10. Martens KJA, van Dalen G, Heussen PCM, Voda MA, Nikolaeva T, van Duynhoven JPM (2018) Quantitative structural analysis of fat crystal networks by means of raman confocal imaging. J Am Oil Chem Soc 95:259–265. https://doi.org/10.1002/aocs.12035
11. Motoyama M, Ando M, Sasaki K, Nakajima I, Chikuni K, Aikawa K, Hamaguchi HO (2016) Simultaneous imaging of fat crystallinity and crystal polymorphic types by raman microspectroscopy. Food Chem 196:411–417. https://doi.org/10.1016/j.foodchem.2015.09.043
12. Gómez-Mascaraque LG, Tran C, O'Callaghan T, Hogan SA (2021) Use of confocal Raman imaging to understand the microstructure of anhydrous milk fat-based oleogels. Food Struct 30: 100228. https://doi.org/10.1016/j.foostr.2021.100228
13. Calligaris S, Manzocco L, Valoppi F, Nicoli MC (2013) Effect of palm oil replacement with monoglyceride organogel and hydrogel on sweet bread properties. Food Res Int 51:596–602. https://doi.org/10.1016/j.foodres.2013.01.007
14. Cornish A (2019) X-ray ct imaging of food. Food Sci Technol 33:30–33. https://doi.org/10.1002/fsat.3303_8.x
15. Kim M, Hwang H-S, Jeong S, Lee S (2022) Utilization of oleogels with binary oleogelator blends for filling creams low in saturated fat. LWT- Food Sci Technol 155:112972. https://doi.org/10.1016/j.lwt.2021.112972
16. Giacomozzi AS, Carrin ME, Palla CA (2018) Muffins elaborated with optimized monoglycerides oleogels: from solid fat replacer obtention to product quality evaluation. J Food Sci 83: 1505–1515. https://doi.org/10.1111/1750-3841.14174
17. Valoppi F, Lassila P, Salmi A, Haeggström E (2021) Automated image analysis method for oil-release test of lipid-based materials. MethodsX 8:101447. https://doi.org/10.1016/j.mex.2021.101447
18. Rosen-Kligvasser J, Davidovich-Pinhas M (2021) The role of hydrogen bonds in tag derivative-based oleogel structure and properties. Food Chem 334:127585. https://doi.org/10.1016/j.foodchem.2020.127585

19. Wettlaufer T, Hetzer B, Flöter E (2021) Characterization of oleogels based on waxes and their hydrolyzates. Eur J Lipid Sci Technol 123:2000345. https://doi.org/10.1002/ejlt.202000345
20. Brykczynski H, Hetzer B, Flöter E (2022) An attempt to relate oleogel properties to wax ester chemical structures. Gels 8:579. https://doi.org/10.3390/gels8090579
21. Saha S, Saint-Michel B, Leynes V, Binks BP, Garbin V (2020) Stability of bubbles in wax-based oleofoams: decoupling the effects of bulk oleogel rheology and interfacial rheology. Rheol Acta 59:255–266. https://doi.org/10.1007/s00397-020-01192-x
22. Blake AI, Marangoni AG (2015) The use of cooling rate to engineer the microstructure and oil binding capacity of wax crystal networks. Food Biophys 10:456–465. https://doi.org/10.1007/s11483-015-9409-0
23. Palla C, de Vicente J, Carrin ME, Galvez Ruiz MJ (2019) Effects of cooling temperature profiles on the monoglycerides oleogel properties: a rheo-microscopy study. Food Res Int 125:108613. https://doi.org/10.1016/j.foodres.2019.108613
24. Calligaris S, Valoppi F, Barba L, Anese M, Nicoli MC (2018) B-carotene degradation kinetics as affected by fat crystal network and solid/liquid ratio. Food Res Int 105:599–604. https://doi.org/10.1016/j.foodres.2017.11.062
25. Lupi FR, Greco V, Baldino N, de Cindio B, Fischer P, Gabriele D (2016) The effects of intermolecular interactions on the physical properties of organogels in edible oils. J Colloid Interface Sci 483:154–164. https://doi.org/10.1016/j.jcis.2016.08.009
26. Okuro PK, Tavernier I, Bin Sintang MD, Skirtach AG, Vicente AA, Dewettinck K, Cunha RL (2018) Synergistic interactions between lecithin and fruit wax in oleogel formation. Food Funct 9:1755–1767. https://doi.org/10.1039/c7fo01775h
27. Sánchez-Becerril M, Marangoni AG, Perea-Flores MJ, Cayetano-Castro N, Martínez-Gutiérrez H, Andraca-Adame JA, Pérez-Martínez JD (2018) Characterization of the micro and nanostructure of the candelilla wax organogels crystal networks. Food Struct 16:1–7. https://doi.org/10.1016/j.foostr.2018.02.001
28. Golodnizky D, Rosen-Kligvasser J, Davidovich-Pinhas M (2021) The role of the polar head group and aliphatic tail in the self-assembly of low molecular weight molecules in oil. Food Struct-Neth 30:100240. https://doi.org/10.1016/j.foostr.2021.100240
29. Qi WH, Li T, Zhang ZS, Wu T (2021) Preparation and characterization of oleogel-in-water pickering emulsions stabilized by cellulose nanocrystals. Food Hydrocoll 110:106206. https://doi.org/10.1016/j.foodhyd.2020.106206
30. Cho K, Tarté R, Acevedo NC (2023) Development and characterization of the freeze-thaw and oxidative stability of edible rice bran wax-gelatin biphasic gels. LWT- Food Sci Technol 174:114330. https://doi.org/10.1016/j.lwt.2022.114330
31. Jeong S, Lee S, Oh I (2021) Development of antioxidant-fortified oleogel and its application as a solid fat replacer to muffin. Foods 10:3059. https://doi.org/10.3390/foods10123059
32. Bin Sintang MD, Rimaux T, Van de Walle D, Dewettinck K, Patel AR (2016) Oil structuring properties of monoglycerides and phytosterols mixtures. Eur J Lipid Sci Technol. https://doi.org/10.1002/ejlt.201500517
33. Blake AI, Marangoni AG (2015) Plant wax crystals display platelet-like morphology. Food Struct 3:30–34. https://doi.org/10.1016/j.foostr.2015.01.001
34. Rogers MA, Smith AK, Wright AJ, Marangoni AG (2007) A novel cryo-sem technique for imaging vegetable oil based organogels. J Am Oil Chem Soc 84:899–906. https://doi.org/10.1007/s11746-007-1122-9
35. Schneider CA, Rasband WS, Eliceiri KW (2012) Nih image to imagej: 25 years of image analysis. Nat Methods 9:671–675. https://doi.org/10.1038/nmeth.2089
36. van der Walt S, Schonberger JL, Nunez-Iglesias J, Boulogne F, Warner JD, Yager N, Gouillart E, Yu T, scikit-image, c. (2014) Scikit-image: image processing in python. PeerJ 2:e453. https://doi.org/10.7717/peerj.453
37. Bankman IN (2009) Handbook of medical image processing and analysis. Elsevier, Amsterdam
38. Russ JC (2005) Image analysis of food microstructure. CRC Press, Boca Raton

39. Co ED, Marangoni AG (2018) Chapter 1 - Oleogels: an introduction. In: Marangoni AG, Garti N (eds) Edible oleogels, 2nd edn. AOCS Press, Urbana, pp 1–29
40. Sagiri SS, Singh VK, Pal K, Banerjee I, Basak P (2015) Stearic acid based oleogels: a study on the molecular, thermal and mechanical properties. Mater Sci Eng C 48:688–699. https://doi.org/10.1016/j.msec.2014.12.018
41. Flöter E, Wettlaufer T, Conty V, Scharfe M (2021) Oleogels—their applicability and methods of characterization. Molecules 26:1673. https://doi.org/10.3390/molecules26061673
42. Merchán Sandoval J, Carelli A, Palla C, Baumler E (2020) Preparation and characterization of oleogel emulsions: a comparative study between the use of recovered and commercial sunflower waxes as structuring agent. J Food Sci 85:2866–2878. https://doi.org/10.1111/1750-3841.15361
43. Doan CD, Tavernier I, Bin Sintang MD, Danthine S, Van de Walle D, Rimaux T, Dewettinck K (2017) Crystallization and gelation behavior of low- and high melting waxes in rice bran oil: a case-study on berry wax and sunflower wax. Food Biophys 12:97–108. https://doi.org/10.1007/s11483-016-9467-y
44. Giacomozzi AS, Carrin ME, Palla CA (2021) Storage stability of oleogels made from monoglycerides and high oleic sunflower oil. Food Biophys 16:306–316. https://doi.org/10.1007/s11483-020-09661-9
45. Blake AI, Marangoni AG (2015) The effect of shear on the microstructure and oil binding capacity of wax crystal networks. Food Biophys 10:403–415. https://doi.org/10.1007/s11483-015-9398-z
46. Torre IG, Heck RJ, Tarquis AM (2020) Multifrac: an imagej plugin for multiscale characterization of 2d and 3d stack images. SoftwareX 12:100574. https://doi.org/10.1016/j.softx.2020.100574
47. Dàvila E, Parés D (2007) Structure of heat-induced plasma protein gels studied by fractal and lacunarity analysis. Food Hydrocoll 21:147–153. https://doi.org/10.1016/j.foodhyd.2006.02.004
48. Merchán Sandoval J, Carelli A, Baümler E, Palla CA (2020) Relationship between microstructure, rheological behavior and kinetic stability of oleogel emulsions produced with recovered and commercial sunflower waxes. In: International conference of production research (ICPR), Bahía Blanca, Argentina
49. Ho Lam RS, Rogers MA (2011) Experimental validation of the modified Avrami model for non-isothermal crystallization conditions. CrstEngComm 13:866–875. https://doi.org/10.1039/c0ce00523a
50. Rogers MA, Marangoni AG (2008) Non-isothermal nucleation and crystallization of 12-hydroxystearic acid in vegetable oils. Cryst Growth Des 8:4596–4601. https://doi.org/10.1021/cg8008927
51. Acevedo NC, Marangoni AG (2010) Characterization of the nanoscale in triacylglycerol crystal networks. Cryst Growth Des 10:3327–3333. https://doi.org/10.1021/cg100468e
52. Ramel PR, Co ED, Acevedo NC, Marangoni AG (2016) Structure and functionality of nanostructured triacylglycerol crystal networks. Prog Lipid Res 64:231–242. https://doi.org/10.1016/j.plipres.2016.09.004
53. Bunjes H, Steiniger F, Richter W (2007) Visualizing the structure of triglyceride nanoparticles in different crystal modifications. Langmuir 23:4005–4011. https://doi.org/10.1021/la062904p
54. Gravelle AJ, Davidovich-Pinhas M, Zetzl AK, Barbut S, Marangoni AG (2016) Influence of solvent quality on the mechanical strength of ethylcellulose oleogels. Carbohydr Polym 135:169–179. https://doi.org/10.1016/j.carbpol.2015.08.050
55. Rogers MA, Wright AJ, Marangoni AG (2008) Engineering the oil binding capacity and crystallinity of self-assembled fibrillar networks of 12-hydroxystearic acid in edible oils. Soft Matter 4:1483–1490. https://doi.org/10.1039/b803299h
56. Valoppi F, Calligaris S, Marangoni AG (2017) Structure and physical properties of oleogels containing peanut oil and saturated fatty alcohols. Eur J Lipid Sci Technol 119:1600252. https://doi.org/10.1002/ejlt.201600252

57. Blake AI, Co ED, Marangoni AG (2014) Structure and physical properties of plant wax crystal networks and their relationship to oil binding capacity. J Am Oil Chem Soc 91:885–903. https://doi.org/10.1007/s11746-014-2435-0
58. Lassila P, Valoppi F, Tommiska O, Hyvönen J, Holmström A, Hietala S, Salmi A, Haeggström E (2022) Practical scale modification of oleogels by ultrasonic standing waves. Ultrason Sonochem 85:105970. https://doi.org/10.1016/j.ultsonch.2022.105970
59. Pratap A, Sardana N (2022) Machine learning-based image processing in materials science and engineering: a review. Mater Today Proc 62:7341–7347. https://doi.org/10.1016/j.matpr.2022.01.200

Chapter 21
Synchrotron-Based Analysis

Luisa Barba and Fernanda Peyronel

Abbreviations

GIXD	Grazing incidence X-ray diffraction
MG	Monoglyceride
PXRD	Powder X-ray diffraction
SAXD	Small angle X-ray diffraction
SAXS	Small angle X-ray scattering
SCXRD	Single crystal X-ray diffraction
USAXS	Ultra-small angle X-ray scattering
USSW	Ultrasonic standing wave
WAXD	Wide angle X-ray diffraction
WPI	Whey protein isolate
XRD	X-ray diffraction
XRDT	Temperature/time dependent powder X-ray diffraction
μXRD	Micro-X-ray diffraction

L. Barba
CNR – Institute of Crystallography, Basovizza, Italy
e-mail: luisa.barba@ic.cnr.it

F. Peyronel (✉)
Food Science Department, University of Guelph, Guelph, ON, Canada
e-mail: fsvaikau@uoguelph.ca

C. Andrea Palla, F. Valoppi (eds.), *Advances in Oleogel Development, Characterization, and Nutritional Aspects*,
https://doi.org/10.1007/978-3-031-46831-5_21

21.1 Oleogels Characterization by X-Ray Diffraction (XRD) from Synchrotron Light Source

21.1.1 Synchrotron Light Sources

The quality of the radiation source plays a critical role in all radiation-matter interactions. Synchrotron light facilities are advanced scientific facilities that utilize closed-trajectory particle accelerators, which also function as optimal X-ray sources as they are designed to maximize the emission of synchrotron radiation (Fig. 21.1). X-ray radiation is produced by accelerating charged particles (electrons) traveling at close to the speed of light. The electron beam is deflected and accelerated in a close orbit using devices known as benders. The electron beam emits coherent photons that can be tuned to a desired energy range through devices called wigglers and undulators.

The most relevant properties of synchrotron sources for X-ray diffraction experiments are tunable wavelength and brilliance. The former enables selection of the most suitable photon energy for a given sample's chemical composition and structural properties by modulating the magnetic field of the device or introducing a monochromator downstream. Furthermore, it allows balancing the need for the best angular resolution (angular separation of diffracted beams) with that for the

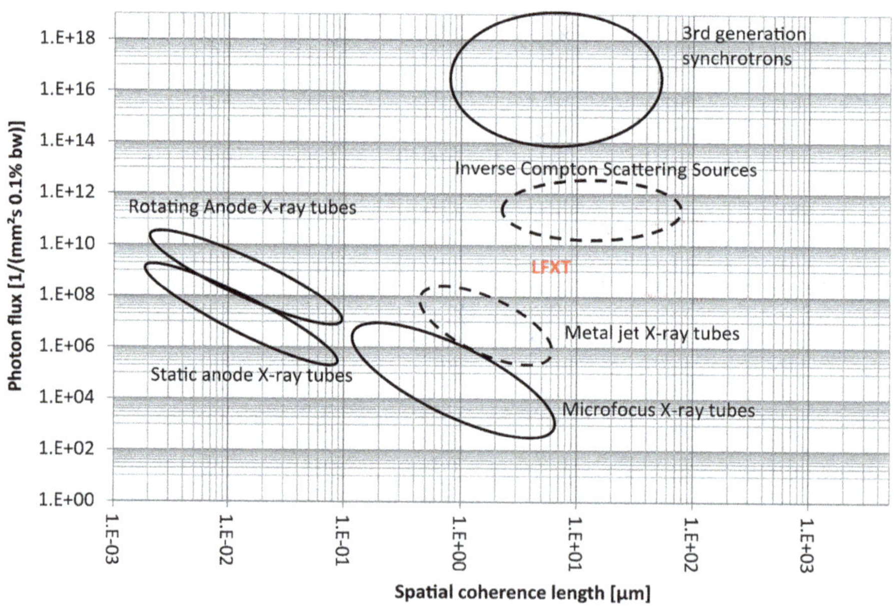

Fig. 21.1 Performance comparison of various X-ray sources. Spatial coherence length is a byproduct of source brilliance. (Reproduced from [1] under the terms of the Creative Commons Attribution 3.0 license)

completeness of a diffraction pattern, which is the ability to detect the lowest or highest attainable diffraction angles, due to the dependence of the diffraction angle on beam energy. Brilliance, which is the flux normalized to the solid angle of the radiation cone and focal spot dimensions, is maximized by minimizing beam size and divergence while maximizing photon flux. Brilliance is critical for data collection sensitivity, peak separation, and speed. Sensitivity characterizes patterns with a good signal-to-noise ratio, even for weak features. Peak separation allows precise evaluation of individual peak intensities; beam divergence is one of the principal factors (among those depending on the experimental setting) of peak width increase and consequent overlapping. Fast data collections are essential especially for materials suffering from radiation damage, and for high-throughput sample screening.

Photon transmission from the source to the experimental setup occurs within a vacuum, facilitated by an elaborate system of devices comprising a beamline. Within this beamline, filters and choppers are employed to eliminate undesirable components of the beam, while slits function to constrain its dimensions. A monochromator is utilized to select a specific energy for the beam, and mirrors are used to alter its direction and divergence. Upon reaching the sample, a portion of the photons undergoes scattering and diffraction, subsequently being captured by a detector. Conversely, photons that continue along the original trajectory (i.e., the "direct" beam) are eventually intercepted by a beamstopper situated immediately after the sample, thus preventing damage to the detector. The size of the beamstopper must match that of the direct beam, thereby determining the angular lower resolution limit of a diffraction pattern.

21.1.2 *Diffraction Techniques: SCXRD, PXRD, μXRD, GIXD*

Diffraction takes place when X-rays are elastically scattered by atoms, molecules, or groups of molecules, regularly arranged in space over long ranges, referred to also as "correlation lengths." Usually, monochromatic radiation is employed. If the radiation wavelength is comparable to or smaller than the dimensions of the elementary repeating unit, the regular arrangement of scatterers determines interferential effects. Because of the interplay of constructive and destructive interference due to reticular symmetries, the diffracted pattern consists of a symmetrical distribution of X-ray photon density elastically scattered in specific directions of space. The reciprocal lattice is the set of all wavevectors of the diffracted waves. Geometry and symmetries of a diffracted pattern depend on those of the ordered molecules in real space, and the intensity of every spot, diffracted by a family of crystallographic planes, is related to the scattering power of the atoms lying on that same plane. The heavier the atoms, the bigger the scattering factor and the more intense the diffracted peak. More precisely, the intensity distribution of diffracted photons is proportional to the square modulus of the Fourier transform of the ordered electron density in the sample. A combination of experimental and computational techniques can reconstruct the lattice in real space, hence getting a map of the average positions of the atoms in the molecule, their bond lengths, and the angles between them [2].

The most effective technique to achieve molecular structural determination is single crystal X-ray diffraction (SCXRD). Nonetheless, the hard X-rays diffraction technique has reached wider application fields. Powder diffraction (PXRD) can attain structural determination also, especially relying on synchrotron light X-ray sources; when this is not possible, pattern analysis, with particular emphasis on peaks profile parameters, can at least provide information on crystallite size in nanocrystalline materials, and on the presence of stacking faults, microstrain, and other defects undermining long-range order of molecules. Preferential orientation of micro- and nanocrystals is an unwanted sample feature in this case, and much pain is taken to prepare isotropic powder samples of the proper granulometry to obtain irreproachable data [3, 4].

On the other hand, the characterization of intermediate states between isotropically distributed micro- or nanocrystalline samples and well-oriented ones in one (fibrils) two (lamellae) or three (crystals) directions, while still far from a complete quantitative interpretation, offers useful insights on these aggregation states. To get to know local orientations of ordered molecules or groups of molecules in samples, micro-X-ray diffraction (μXRD) can be employed. Micro-beam X-ray diffraction measurements can unveil local orientation (spatially averaged within the irradiated volume in the sample). This technique is well suited to perform crystallization mapping on extended samples, characterizing the presence of ordered structures, their mass percentage with respect to the amorphous components, and their preferential orientation along specific directions [5]. Finally, insoluble monolayers of molecules like fatty acids, fatty alcohols, and phospholipids, prone to self-assembling in ordered thin films on liquid surfaces, can be characterized by means of grazing incidence X-ray diffraction (GIXD), which reduces penetration in the substrate as much as possible delivering information on the self-assembled layers symmetry and unit cell [6].

21.1.3 Steps in the Diffractometric Analysis of an Oleogel XRD Pattern

Oleogel diffraction patterns are analyzed following the same methodology employed in powder diffraction analysis, even if their composition is only in part crystalline (Fig. 21.2). The separation of the contribution of diffracted and amorphous scattering signals in a diffraction pattern is a necessary preliminary step in the interpretation [7]. Synchrotron light sources help in as much as they provide well-collimated, intense, and bright beam, and this minimizes direct beam contribution to scattered light. Depending on the wide-angle X-ray scattering contribution to the background, a baseline can be inferred by visual inspection of the X-ray pattern; its

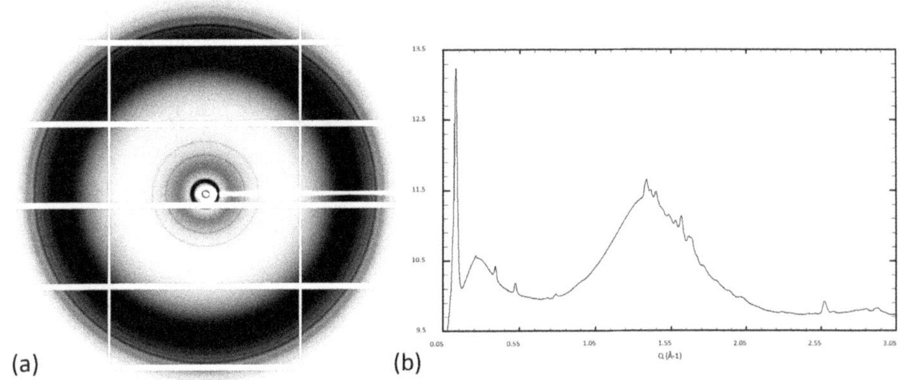

Fig. 21.2 XRD 2D pattern of a monoglyceride oleogel (**a**) and its integrated intensity (**b**) represented in logarithmic scale as a function of reticular vector q

scattered and diffracted components can be modeled by a variety of free-share and proprietary programs [8–10], which can identify the repetition unit cell and therefore assign the correct Miller indexes to the pattern peaks. In the process, we also obtain estimated values of peaks parameters: angular position, integrated intensity, full width at half maximum, along with their errors. In addition, these parameters enable the estimation of the crystallinity index and average dimensions of diffracting crystallites using a variety of models (e.g., Warren-Averbach, paracrystal model) and software [11–14].

The ordering of long-chain molecules as lipids, triglycerides, polysaccharides, waxes, can generate diffraction in oleogels. Their behavior is analog to that of lipidic aggregated phases: due to intermolecular forces (Van der Waals, hydrogen bonds, π-π stacking) molecules align preferentially along their longest dimension, with different interdigitation and inclination, generating lamellae. The in-plane molecular arrangement in lamellae varies, mainly in response to temperature, determining a number of polymorphic phases. Diffraction at lower angles (small angle X-ray diffraction or scattering, SAXD or SAXS) allows the calculation of interplanar distances in lamellae. The angular positions of diffraction peaks generated from a family of crystallographic planes whose interplanar spacing is d, follow Bragg law (Eq. 21.1).

$$n\lambda = 2d \sin\theta \tag{21.1}$$

where n is the order of the reflection, λ the X-rays wavelength, d the interplanar distance, and θ is half of the angle comprised between the diffracted beam and the incident X-ray beam direction. Hence, lower angle peaks due to lamellar organization will be found in a diffraction pattern at angular positions given by Eq. 21.2.

Table 21.1 Characteristic *d*-spacings associated with different polymorphs in lipids

Polymorph	Characteristic short spacings (Å)	Chain-packing subcell
α	One dominating line in the interval 4.12–4.20	Hexagonal
β′	3.80 and 4.20 or 4.27, 3.97, and 3.71	Orthorhombic
β	4.6 plus other weak lines between 3.8 and 5.4	Triclinic

$$2\theta = 2\arcsin\left(\frac{n\lambda}{2d}\right) \tag{21.2}$$

The longer the interplanar distance, the lower the angle of the first order diffraction peak, while the successive will be found at higher angles. There is a relationship between these parameters and the modulus q of the reciprocal lattice vector, which is shown in Eq. 21.3.

$$q_n = \frac{4\pi}{\lambda}\sin\theta = n\frac{2\pi}{d} \tag{21.3}$$

Representing diffraction patterns as a function of reciprocal lattice spacing has the property that successive order of peaks coming from the same crystallographic family of planes are equidistant and easier to visualize. As a rule of thumb, SAXD takes place at values of $q < 1\ \text{Å}^{-1}$ in oleogels.

The detection of the contribution of different chemical species to the diffracted pattern can also take place at this stage, as the presence of different lamellar organization at the same temperature can be attributed to the diffraction of different moieties, with different chain lengths and number of saturated bonds.

The number of detectable orders in SAXD depends on the long-range order in sample crystallites, and increases the precision of the calculation of *d*. Once again, a higher number of diffracted orders is easier to detect if the incident beam is collimated and intense, as in Synchrotron light source facility.

The long spacings are associated with molecular lengths, so monolayer, bilayer, or multilayer repetition unit can be inferred by the comparison of interplanar distances and molecular chain lengths. Nevertheless, interdigitation and inclined configurations can complicate the attribution, and only single crystal diffraction experiments, where feasible, can elucidate it [15, 16].

Characteristic Bragg peaks of the WAXD region allow recognizing the polymorphic forms present and their dependence on temperature or, alternatively, to the gelation path followed during sample preparation. As an example, the three principal polymorphic forms of lipids are labeled α, β′, and β in order of increasing close-packing and stability [17] (Table 21.1).

The recognition of the chain-packing subcell from the WAXD region is about as far as XRD can get to recognize crystalline phases of oleogels. The prevalence of one or the other of the polymorph of an oleogel component correlates with some of its more important structural and thermodynamic qualities, as stability and melting point.

21.1.4 Examples

Some examples of the results obtained with data collected at the beamline XRD1 at the Elettra Synchrotron in Trieste, Italy, [18] are presented. Since 2004, joint characterization of polymorphism and phase transitions in oils by means of calorimetric and temperature/time dependent powder X-ray diffraction (XRDT) was an objective of the research group [19]. In a few years, the focus shifted from oils to oleogels and structured emulsions. The first of such investigations regarded aroma partition properties of saturated monoglyceride–oil–water gels where monoglycerides (MG) self-assembled to form walls surrounding oil droplets. Gels with different MG/oil ratio, investigated by PXRD, showed no significant differences in the *d* spacing of the wide-angle region; they had all the same in-plane (lateral) molecular ordering. On the contrary, the small angle region of the diffractograms was consistent with two different longitudinal organization in bilayers of different widths, showing one or the other or both depending on MG/oil ratio. It was then possible to explain the correlation between aroma partition and MG/oil ratio on account of the capability of different bilayer structure to entrap lipophilic compounds [20].

Another study focused on MG incorporated in O/W emulsions, along with two different stabilizers, that is Tween 20 and whey protein isolate (WPI). The objective was to elucidate the crystallizing behavior of MG in the emulsions. In this case, variable temperature X-ray diffraction experiments (XRDT) were performed. Samples were heated from 25 to 90 °C and then cooled to 0 at 5 °C/min providing thermograms showing that at room temperature, in the samples stabilized with Tween 20, MG crystals (β form) were stacking in a well-ordered lamellar style with a bilayer thickness of 49.5 Å. On heating, at 44 °C, the peak shifted to 46.2 Å and a new series of peaks, characteristics of the orthorhombic unit cell packing of α and sub-α forms, emerged. The XRD pattern of WPI structured emulsions showed a clear diffraction peak at 49.01 Å, indicating the presence of a lamellar organization in the system. Few other peaks could be detected (22.98, 16.24, 4.55 Å) at a lower interplanar distance. The peaks were mostly of very low intensity, signaling the low level of lamellar organization of MG in WPI emulsion. This information allowed a more complete interpretation of calorimetric data, along with unequivocal crystallographic phase recognition [20].

XRD mapping technique allowed testing ultrasonic standing wave (USSW) fields' capability to modify oleogel structure. In USSW chambers, during crystallization, the growing crystals moved toward the USSW nodal planes. Homogeneous, horizontal bands of microcrystals formed independently of oleogelator type, concentration, and cooling rate. The thickness of these bands was proportional to the USSW wavelength. The XYZ stage of the experimental station diffractometer lodged USSW chambers developed ad hoc to expose the oleogel to temperature control, ultrasonic field and offer a window of about 1 cm^2 for incident and diffracted X-rays. Chambers were moved stepwise vertically and horizontally in a plane perpendicular to the X-ray beam. This latter was collimated by a double set of slits to a spot size of 50 × 50 μm^2 collecting a diffractogram at every step, hence producing a diffraction map of the sample. The analysis of the diffraction maps showed that microcrystal bands in correspondence of standing waves nodal planes were richer in monoglyceride crystals compared to the sparse bands, following the intensity trend of the monoglyceride lamellar structure 001 reflection during vertical scans. Moreover, the dense crystalline bands had a constant abundance of crystalline monoglycerides in the horizontal direction. All treated samples showed the formation of microcrystals, revealed by the presence of spots on the diffraction rings. In contrast, statically crystallized samples showed homogeneous diffraction rings indicating that no microcrystals formed. The nanostructural organization of monoglycerides was a triclinic β-form with a lamellar thickness of 48 –49 Å for oleogels statically crystallized. Similar results were obtained for oleogels crystallized at 1 MHz, while increasing the USSW field frequency to 2 and 4 MHz, monoglycerides tended to crystallize in the less thermodynamically stable hexagonal α-form with a lamellar thickness of 51.5 Å. USSW fields with higher frequencies can disturb monoglyceride crystallization at the nanostructural level, favoring less stable polymorphic forms. Therefore, USSW fields can affect both micro- and nanostructural levels depending on the selected frequency [21].

21.2 Ultra-Small Angle X-Ray Scattering (USAXS) Using Synchrotron Radiation on Oleogels

This section focuses on USAXS, which was carried out at beamline 9ID located at the Advanced Photon Source (APS) in the Argonne National Laboratory, under the supervision of Dr. Ilavsky. Dr. Ilavsky's setup [22] allows for rapid measurements in both the SAXS and WAXS regions to complement the USAXS data. For further details, readers can refer to the latest publication by Ilavsky and coworkers [23]. The beamline also provides software for data reduction [24] and analysis, including a semi-automatic desmear protocol [25]. The Irena software allows users to take advantage of either the Unified Fit model [26, 27] or the Hammouda model [28] to analyze USAXS data. The selection of the model to use comes from additional information which might include measurements using other techniques or it could be computer simulations using mathematical models [29, 30]. The Irena software utilizes a nonlinear regression analysis to determine the optimal parameters for each model, using multiple scattering intensity "levels" associated with user-selected q-regions. However, selecting the correct starting and ending points for each level is crucial and often requires multiple attempts as the software may fail to converge. Users typically rely on prior knowledge of the sample obtained through other techniques to select the appropriate q-regions. Each level represents a structural

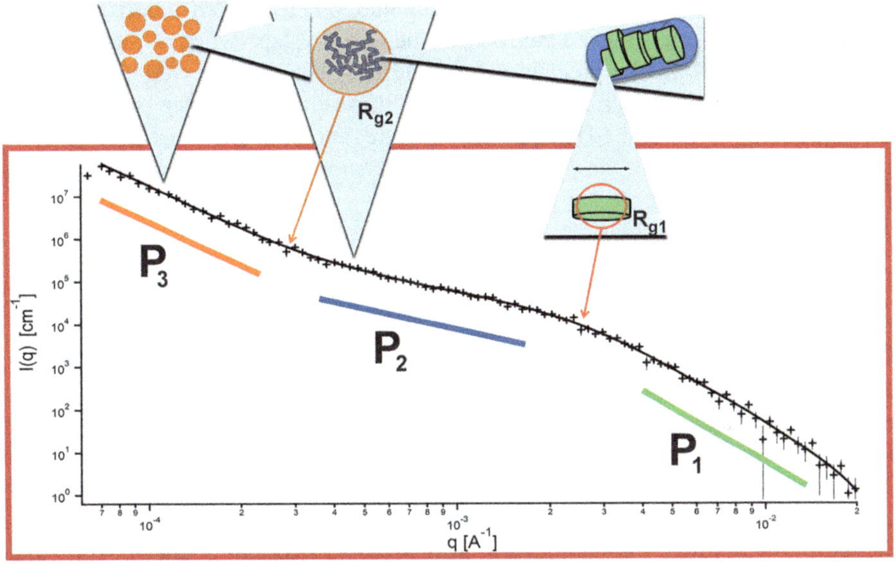

Fig. 21.3 Data obtained at beamline 9ID at APS for a system with a 20% crystalline triacylglycerol molecule (tristearin) in a liquid oil (triolein). The data has been fitted using three levels in the Unified Fit model. The best fit was obtained when the levels were linked. The straight lines (visually observed when the data is plotted as log-log) indicate the Porod regions (P), while the two curvatures indicate the Guinier regions (R_g). (Picture adapted from [36] with the permission of AIP Publishing)

level at a specific length scale range or the scattering from a particular component present in the sample. The levels can be independent of each other or linked, with the understanding that the objects observed at the smaller level are involved in the aggregation that manifests at a higher level. A structural level is characterized by a Guinier function [31] at the smallest q-values, including a "knee," and the associated power-law regime or Porod function [32] for larger q-values.

Pioneering work by Peyronel et al. [34–37] involved the analysis of edible fats in the USAXS region, using mathematical models and simulation predictions [29, 30, 33–35]. That research paved the way for further analysis in other areas, such as oleogels.

A typical output for reduced and desmeared data collected in the USAXS region for a system with 20% solids and 80% liquid oil can be seen in Fig. 21.3.

Figure 21.3 should be understood starting at the larger q-values and moving to the smallest one, as the larger q-values correspond to the smallest spatial length scales. Briefly, P_1 and R_{g1} belong to level 1, P_2 and R_{g2} constitute level 2, and from level 3, one can only see P_3. When the fit is given a fractal interpretation, then the surface morphology of the nanocrystal is observed. The surface fractal dimension is related to the $|P_1|$ value, and the average size of the nanocrystal is associated with the R_{g1}

values. Nanocrystals are believed to aggregate into long rods (TAGwoods) when the cooling is fast [37]. It is those long rods or, the nanocrystals (when crystallization is performed slowly), that aggregate to form clusters that are observed at smaller q-values (level 2). The mass fractal dimension of those clusters is given by $|P_2|$ and the average cluster size is related to R_{g2}. Further aggregation of the clusters is manifested in the last level and either the surface fractal dimension or the mass fractal dimension is obtained from $|P_3|$.

21.2.1 Examples

Rondou et al. [38] studied oleogels created using monoglycerides at 6% in rapeseed oil. USAXS was one of the techniques that they used. These authors used the Unified fit model to unveil 1 level or 2 levels in the q-region covering 0.0002 to 0.1 $Å^{-1}$. Those authors looked at the average nanocrystal size (using the equivalent diameter) after 4 and 8 weeks of storage. Not surprisingly, they found that the oleogel that was crystalized at a lower rate, 66 °C/min gave larger equivalent diameters compared with the ones crystallized at 78 °C/min. The difference in the equivalent diameter between 4 and 8 weeks of storage was ~35%, although the number could be much smaller if only the authors had reported error bars. The authors did not observe clusters of nanocrystals for the samples made with the slowest cooling rate of 66 °C/min; however, they observed a second Guinier for the oleogel made using a faster cooling rate of 78 °C/min. It makes sense that they observed cluster under this cooling rate regime because they observed that the average crystalline thickness was ~35% smaller for the fast-cooling rate oleogel than the slow cooling oleogel. The results presented by those authors can be related to the observations by Peyronel et al. [37] where they show differences between a crystallization done at a fast rate and one done at a slower rate. Peyronel et al. claimed the formation of TAGwoods something not discussed in the Rondou et al. [38] paper.

Peyronel et al. [39] studied three molecules (Triacontane, TC, stearic acid, SA, and behenyl lignocerate, BL) to be used as oleogelators using the USAXS technique. They only did one colling rate to form the oleogel as their goal was to identify possible synergetic effects when using 2 pair of molecules or the 3 molecules together. Those authors showed no mixed crystal formation when pair of molecules were melted and cooled to form the oleogels. However, the data in the USAXS region suggested that, when using all three molecules to make an oleogel, a more compacted fractal structure was obtained. The authors did not check how much oil can one fractal structure trapped compared with another fractal structure.

Tsung and coworkers [40] characterized studied the structural properties of zein-based oleogels using USAXS. The results showed the presence of at least two structural levels in the oleogels. The Guinier approximation revealed that the zein molecule was the primary unit or basic building block of 2D and 3D structures observed in oleogels. Short-range 2D structures were formed by zein or zein/oleic acid, while 3D structures were formed by protein associations or the network structure. The data suggested that there is an optimal amount of ethanol-water for a given system, where zein molecules have enough mobility to develop a supramolecular structure. The Kratky plots showed that the shape of the curve indicated a rod-like shape of the zein molecule at high.

Valoppi and coworkers [41] studied oleogels made using binary systems containing different alcohols in peanut oil. They found two structural levels. The first structural level showed that all systems studied had diffuse surfaces and not well-defined edges between solid and liquid phase. The radii of the spherical objects formed by C22OH and C20OH were similar but smaller than those formed by C18OH and C16OH. The second structural level showed that the spherical objects aggregated via reaction-limited cluster-cluster aggregation to form a disordered distribution of mass embedded in a 3D space.

21.3 Conclusions

X-ray diffraction is a powerful technique used to determine the structure of molecules by analyzing the elastic scattering of X-rays by atoms or groups of atoms arranged in space. It is employed in various fields like single crystal X-ray diffraction, powder diffraction, micro-X-ray diffraction, and grazing incidence X-ray diffraction. Oleogels' diffraction patterns can be analyzed using the same methodology as powder diffraction analysis. Separating diffracted and amorphous scattering signals is necessary for correct interpretation of results. SAXD is used to calculate interplanar distances in lamellae, and peak orders can be visualized.

Ultra small angle X-ray scattering (USAXS) allows the characterization of the nano- to micro-structures in materials like oleogels. The software used for data analysis employs nonlinear regression analysis to determine the best parameters based on observed scattering intensity levels associated with different q-regions. USAXS can provide information regarding the surface and mass fractal dimensions, as well as the average size of nanocrystals and clusters. Studies have shown that USAXS can identify possible synergetic effects when using multiple molecules as oleogelators and can provide insights into the oleogel's structure and properties. In summary, the USAXS technique allows for the determination of structural properties, such as the shape, size, and characteristic intermolecular distances of macromolecules or colloidal aggregates, as well as pore sizes and other relevant information.

Acknowledgments F.P likes to thank the Food Science Department at the University of Guelph for supporting her research. She also wants to thank Dr. Ilavsky and his team at APS. Research at this facility uses resources from the U.S. Department of Energy (DOE) Office of Science User Facility operated for the DOE Office of Science by Argonne National Laboratory under Contract No. DE-AC02-06CH1135.

References

1. Bartzsch S, Oelfke U (2017) Line focus x-ray tubes—a new concept to produce high brilliance x-rays. Phys Med Biology Instit Phys Eng Med IOP 62:8600–8615
2. Giacovazzo C, Monaco HL, Artioli G et al (2011) Fundamentals of crystallography. Oxford University Press, Oxford
3. David WIF, Shankland K, McCusker LB, Bärlocher C (2006) Structure determination from powder diffraction data. Oxford University Press, Oxford
4. Idziak SHJ (2018) Powder X ray diffraction of triglycerides in the study of polymorphism. In: Marangoni AG (ed) Structure function analysis of edible fats, 2 nd edn. AOCS Press/Elsevier Inc., pp 73–99
5. Riekel C, Burghammer M, Davies R (2010) Progress in micro and nano diffraction at the ESRF ID13 beamline. IOP Conf Ser Mater Sci Eng 14(1):012013
6. Fuller GG, Vermant J (2012) Complex fluid-fluid interfaces: rheology and structure. Ann Rev Chem Biomol Eng 3:519–543. https://doi.org/10.1146/annurev-chembioeng-061010-114202
7. Bates S (2010) Amorphous solid forms: the use of X ray Powder Diffraction (XRPD). In: PPXRD 9 pharmaceutical powder X ray diffraction symposium (sponsored by the International Centre for Diffraction Data)
8. Roisnel T, Rodriguez-Carvajal J (2001) WinPLOTR: a windows tool for powder diffraction pattern analysis. Mater Sci Forum 378–381:118–123
9. Coelho AA (2018) TOPAS and TOPAS-academic : an optimization program integrating computer algebra and crystallographic objects written in C++. J Appl Crystallogr 51:210–218. https://doi.org/10.1107/S1600576718000183
10. Altomare A, Burla MC, Giacovazzo C et al (2001) *Quanto* : a Rietveld program for quantitative phase analysis of polycrystalline mixtures. J Appl Crystallogr 34:392–397. https://doi.org/10.1107/S0021889801002904
11. Enzo S, Fagherazzi G, Benedetti A, Polizzi S (1988) A profile-fitting procedure for analysis of broadened X-ray diffraction peaks I. Methodology. J Appl Crystallogr 21:536–542. https://doi.org/10.1107/S0021889888006612
12. Hindeleh AM, Hosemann R (1991) Microparacrystals: the intermediate stage between crystalline and amorphous. J Mater Sci 26:5127–5133. https://doi.org/10.1007/BF01143202
13. Toby BH, von Dreele RB (2013) *GSAS-II* : the genesis of a modern open-source all purpose crystallography software package. J Appl Crystallogr 46:544–549. https://doi.org/10.1107/S0021889813003531
14. Rodriguez-Carvajal J (1993) Recent advances in magnetic structure determination by neutron powder diffraction. Phys B Cond Matter 192:55–69
15. Garti N, Sato K (2001) Crystallization processes in fats and lipid systems. Marcel Dekker, New York
16. Garti N, Sato K (1988) Crystallization and polymorphism of fats and fatty acids. Marcel Dekker, New York
17. Ramel PR, Co ED, Acevedo NC, Marangoni AG (2016) Structure and functionality of nanostructured triacylglycerol crystal networks. Prog Lipid Res 64:231–242
18. Lausi A, Polentarutti M, Onesti S et al (2015) Status of the crystallography beamlines at Elettra. Eur Phys J Plus 130. https://doi.org/10.1140/epjp/i2015-15043-3
19. Calligaris S, Arrighetti G, Barba L, Nicoli MC (2008) Phase transition of sunflower oil as affected by the oxidation level. J Am Oil Chem Soc 85:591–598. https://doi.org/10.1007/s11746-008-1241-y20
20. Mao L, Calligaris S, Barba L, Miao S (2014) Monoglyceride self-assembled structure in O/W emulsion: formation, characterization and its effect on emulsion properties. Food Re Int 54:81–88
21. Valoppi F, Salmi A, Ratilainen M et al (2020) Controlling oleogel crystallization using ultrasonic standing waves. Sci Rep 10:14448. https://doi.org/10.1038/s41598-020-71177-6

22. Ilavsky J, Jemian PR, Allen AJ et al (2009) Ultra-small-angle X-ray scattering at the advanced photon source. J Appl Crystallogr 42:469–479
23. Ilavsky J, Zhang F, Andrews RN et al (2018) Development of combined microstructure and structure characterization facility for in situ and operando studies at the advanced photon source. J Appl Crystallogr 51:867–882. https://doi.org/10.1107/S160057671800643X
24. Ilavsky J (2012) Nika : software for two-dimensional data reduction. J Appl Crystallogr 45: 324–328. https://doi.org/10.1107/S0021889812004037
25. Ilavsky J, Jemian PR (2009) Irena: tool suite for modeling and analysis of small-angle scattering. J Appl Crystallogr 42:347–353
26. Beaucage G (1995) Approximations leading to a unified exponential/power-law approach to small-angle scattering. J Appl Crystallogr 28:717–728
27. Beaucage G (1996) Small-angle scattering from polymeric mass fractals of arbitrary mass-fractal dimension. J Appl Crystallogr 29:134–146
28. Hammouda B (2010) A new Guinier-Porod model. J Appl Crystallogr 43:716–719. https://doi.org/10.1107/S0021889810015773
29. Pink DA, Quinn B, Peyronel F, Marangoni AG (2013) Edible oil structures at low and intermediate concentrations. I. Modeling, computer simulation, and predictions for X ray scattering. J Appl Phys 114:234901. https://doi.org/10.1063/1.4847996
30. Quinn B, Peyronel F, Gordon T et al (2014) Aggregation in complex triacylglycerol oils: coarse-grained models, Nanophase separation, and predicted X-ray intensities. J Phys Condens Matter 26:464108. https://doi.org/10.1088/0953-8984/26/46/464108
31. Guinier A, Fournet G (1955) Small-angle scattering of X-rays. Wiley, New York
32. Porod G (1951) Die Röntgenkleinwinkelstreuung von dichtgepackten kolloiden Systemen. Kolloid-Z 124:83–114
33. Pink DA, Peyronel F, Quinn B et al (2015) Condensation versus diffusion. A spatial-scale-independent theory of aggregate structures in edible oils: applications to model systems and commercial shortenings studied via rheology and USAXS. J Phys D Appl Phys 48:384003. https://doi.org/10.1088/0022-3727/48/38/384003
34. Razul MSG, MacDougall CJ, Hanna CB et al (2014) Oil binding capacities of triacylglycerol crystalline Nanoplatelets: nanoscale models of Tristearin solids in liquid Triolein. Food Funct 5: 2501–2508. https://doi.org/10.1039/c3fo60654f
35. MacDougall CJ, Razul MSG, Papp-Szabo E et al (2012) Nanoscale characteristics of triacylglycerol oils: phase separation and binding energies of two-component oils to crystalline nanoplatelets. Faraday Disc 158:425–433
36. Peyronel F, Ilavsky J, Mazzanti G et al (2013) Edible oil structures at low and intermediate concentrations: II. Ultra-small angle X-ray scattering of in situ Tristearin solids in Triolein. J Appl Phys 114:234902. https://doi.org/10.1063/1.4847997
37. Peyronel F, Pink DA, Marangoni AG (2014) Triglyceride nanocrystal aggregation into poly-crystalline colloidal networks: ultra-small angle X-ray scattering, models and computer simulation. Curr Opin Colloid Interface Sci 19:459–470. https://doi.org/10.1016/j.cocis.2014.07.001
38. Rondou K, De Witte F, Rimaux T et al (2022) Multiscale analysis of monoglyceride oleogels during storage. J Am Oil Chem 99:1019. https://doi.org/10.1002/aocs.12645
39. Peyronel F, Cooney J, Papp-Szabo E, Martini S, Pink D (2023) X-ray characterization of three possible edible oleogelators: experiment and theory. J Am Oil Chem. 1–4. https://doi.org/10.1002/aocs.12732
40. Tsung KL, Ilavsky J, Padua GW (2020) Formation and characterization of Zein-based Oleogels. J Agric Food Chem 68:13276–13281. https://doi.org/10.1021/acs.jafc.0c00184
41. Valoppi F, Calligaris S, Marangoni AG (2016) Phase transition and polymorphic behavior of binary systems containing fatty alcohols and peanut oil. Cryst Growth Des 16:4209–4215. https://doi.org/10.1021/acs.cgd.6b00145

Chapter 22
Computer Simulations: Molecular Dynamics Simulations

George Dalkas, Andrew B. Matheson, Paul Clegg, and Stephen R. Euston

Abbreviations

CH_3	Methyl group
CHEMS	Cholesteryl hemisuccinate
FTIR	Fourier transform infra-red spectroscopy
H-bond	Hydrogen bond
MD	Molecular dynamics
NMR	Nuclear magnetic resonance
R_g	Radius of gyration
RMSD	Root-mean-square displacement
SANS	Small angle neutron scattering

G. Dalkas
Institute of Mechanical, Process and Energy Engineering, Heriot-Watt University, Edinburgh, UK

Q2 Solutions, The Alba Campus, Rosebank, Livingston, UK
e-mail: dalkas@upatras.gr

A. B. Matheson · P. Clegg
School of Physics and Astronomy, University of Edinburgh, Edinburgh, UK
e-mail: a.matheson@ed.ac.uk; paul.clegg@ed.ac.uk

S. R. Euston (✉)
Institute of Biological Chemistry, Biophysics and Bioengineering, Heriot-Watt University, Edinburgh, UK

Department of Physics, Toronto Metropolitan University, Toronto, ON, Canada
e-mail: s.r.euston@hw.ac.uk

C. Andrea Palla, F. Valoppi (eds.), *Advances in Oleogel Development, Characterization, and Nutritional Aspects*,
https://doi.org/10.1007/978-3-031-46831-5_22

22.1 Introduction

It is generally accepted that saturated fats are a risk factor for cardiovascular disease as they increase the level of cholesterol in the blood stream [1]. With consumers becoming more aware of how their diet affects their health, demand for healthier foods has led to various strategies for the replacement of saturated fat in foods [2]. In large part, these involve attempts to replace fat with carbohydrate or protein-based fat mimetics. Despite the substantial research that has accompanied these attempts to replace fat, many of the products are not rated highly by consumers as it has proven difficult to fully reproduce the texture that solid fats impart on food structure. Simply replacing saturated, solid fats with polyunsaturated oils is not feasible due to undesirable textural changes. One way to facilitate the use of polyunsaturated oils in place of solid fats is to use oleogelation [3]. In this, the liquid oils are solidified by the addition of an oleogelling molecule. One widely studied oleogelating system is a combination of a plant phytosterol (often β-sitosterol) and the sterol ester γ-oryzanol. In this system, self-assembly of the sterol and sterol ester into tubules leads to the formation of a network gel that solidifies the liquid oil. Despite extensive research efforts, there are still questions over the mechanism of self-association, the role of the individual sterol components, inter-tubule interactions that control longer-range gel structure, and the reasons why the tubule structure is sensitive to water. We have used molecular dynamics (MD) simulation to investigate these questions, and this chapter is a summary of the progress of our work. We concentrate here on reporting the results of the studies, with details of the methodology omitted. Any reader who is interested in the details of the molecular docking and molecular dynamics methods that we use is referred to our previous review [4] or the individual studies described below.

22.2 Molecular Dynamics Simulation of Phytosterol Oleogels

22.2.1 *Self-Association of Sterols and Sterol Esters*

The ability of β-sitosterol and γ-oryzanol to self-associate in oils as an oleogelating system has been known for some time [5]. Various experimental techniques have given clues to the structure of the self-associated molecules, with these believed to form hollow core-shell tubules. Bot et al. [5, 6] believe that self-association progresses through the formation of a sitosterol–oryzanol dimer. In this, the flat sterane cores partially overlap and interact through dispersion forces, adopting a wedge shape. The dimer is believed to be stabilized by a single hydrogen bond between a hydroxyl group on the sterol and the carbonyl of the ester group in sterol ester. As further self-association of dimers progresses, the wedge shape causes the dimers to stack into a tubule that adopts a helical twist as the tubule grows [5–7]. To

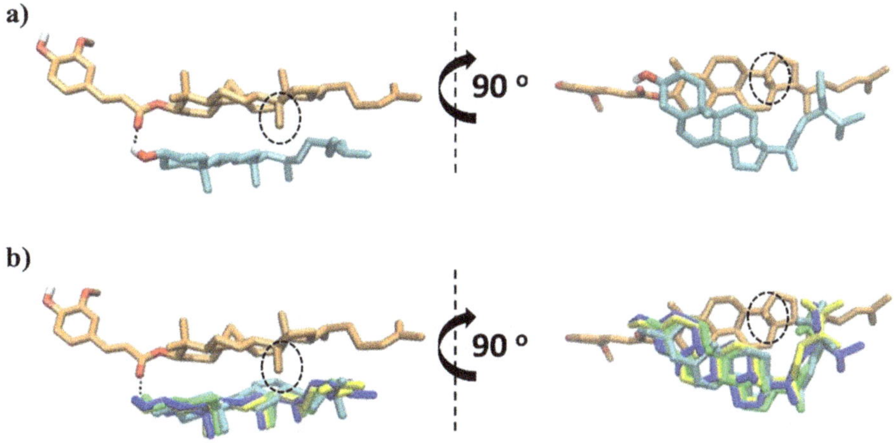

Fig. 22.1 Views differing by 90° of the lowest energy conformation for the molecular models of (**a**) sitosterol (cyan carbons) bound to cycloartenyl ferulate (orange carbons) and (**b**) phytosterols (sitosterol in cyan, cholestanol in yellow, cholesterol in green, stigmasterol in blue) bound to cycloartenyl ferulate (orange carbons). Oxygens of cycloartenyl ferulate are colored with red and hydrogens with white. Potential hydrogen bond is indicated with dashed line, and the cycloartenyl ferulate's methyl group that prevents parallel stacking of the sterane groups is indicated with dashed circle. (Reprinted/Adapted from [8], with permission from American Chemical Society)

investigate this mechanism, we carried out simulated docking experiments [8] using a modified version of AUTODOCK [9].

Figure 22.1 shows the lowest energy docking simulation for a dimer of cycloartenyl ferulate (one of the four structurally similar components of commercial γ-oryzanol) docked to one of four sterols (sitosterol, cholestanol, cholesterol, and stigmasterol). The docking simulation confirms the general theory for dimer structure, which showed a wedge shape with a putative hydrogen bond between a hydroxyl group on the sterol and the carbonyl group of the ester bond between the sterane core and ferulate of the oryzanol stabilizing the lowest energy conformation. In addition, we noted that the presence of a methyl group (C30) on oryzanol, attached to C14 of the steroidal core, and orthogonal to it [8]. This introduces steric hindrance preventing the complete overlap of the two steroidal cores of the dimer. This would seem to be the origin of the wedge shape for the dimer. This orthogonal methyl group is not present in β-sitosterol, nor any of the other sterols tested for gelling ability with oryzanol shown in Fig. 22.1b. This raises the question of how sensitive the self-association and gelation is to changes to the steroidal core of the sterol and sterol ester. This is an important question as to date although many sterols have been shown to form gels with oryzanol, no other sterol ester has yet been found to gel with sterols. Knowing how changes to the steroidal core of both the sterol and oryzanol affect gelling may help us to design and synthesize more efficient gelling systems.

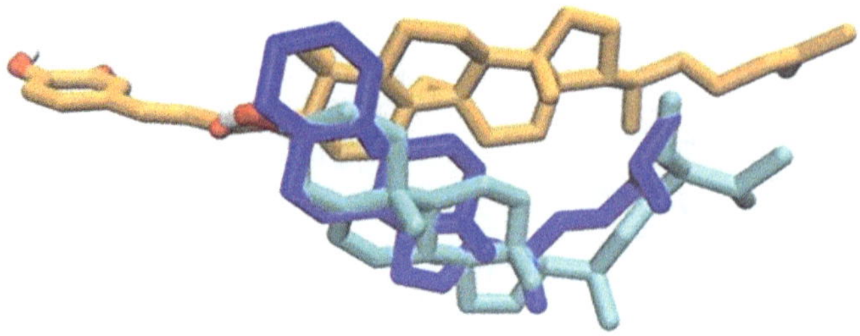

Fig. 22.2 Molecular models of sitosterol (in cyan) and cholestane (blue) bound to cycloartenyl ferulate (orange carbons). Oxygens are colored with red and hydrogens with white. (Reprinted/Adapted from [8], with permission from American Chemical Society)

To look at the effect of having an orthogonal CH_3 at C14 on the sterol as well, we tested lanosterol and cycloartenol (produced by saponification of oryzanol) for their gelling ability with oryzanol [10]. Molecular docking reveals a lowest energy dimer for these systems with a stabilizing hydrogen bond directing the sterols toward a parallel head-to-head/tail-to-tail conformation, the same as observed for sitosterol + oryzanol (Fig. 22.1a). The fact that the gels from both systems were weaker than seen for sitosterol and oryzanol [10] suggests that quite subtle changes to the sterols will change gelling ability. Interestingly, lanosterol was able to form weak gels with itself, whereas sitosterol on its own did not gel. Docking models showed that whereas two lanosterol molecules were able to form a wedge-shaped parallel dimer, the most energetically favored dimer for two sitosterols in the absence of the orthogonal CH_3 was a non-wedge-shaped anti-parallel head-to-tail dimer. This highlights the importance of the orthogonal CH_3 to the gel formation in single sterol systems.

To illustrate the importance of the hydrogen bond on dimer formation, we compared the behavior of various sterols and cholestane. Cholestane, a trisaturated triterpene, has a modified steroidal core that lacks the hydroxyl group at C3, that is, involved in H-bonding in the sitosterol + oryzanol dimer, and the double bond between C6 and C7 that is found for cholesterol, the equivalent sterol. Cholestane has been reported to not form a gel with oryzanol while cholesterol (the sterol most closely related to cholestane) does [11]. Cholestane adopts a slightly different conformation to the sterols when docked with oryzanol [8] (Fig. 22.2) since it cannot form the intermolecular hydrogen bond in the dimer. This suggests that this H-bond is important to tubule and gel formation and that small changes to sterol structure (and hence dimer structure with oryzanol) can have large effects on gelling ability.

The H-bond between the sterol and sterol ester is not the only structural feature that appears to be important in dimer and tubule formation. Substituting alternative sterol esters for oryzanol shows that also the ferulate and CH_3 groups at position C3 and C14 play a role. Cholesteryl hemisuccinate (CHEMS), where the ferulate moiety of oryzanol is replaced by succinate, does not form a gel with sitosterol. Fourier transform infra-red spectroscopy (FTIR), which has been used to detect the presence of an H-bond in sitosterol + oryzanol oleogels [6], does not indicate the presence of

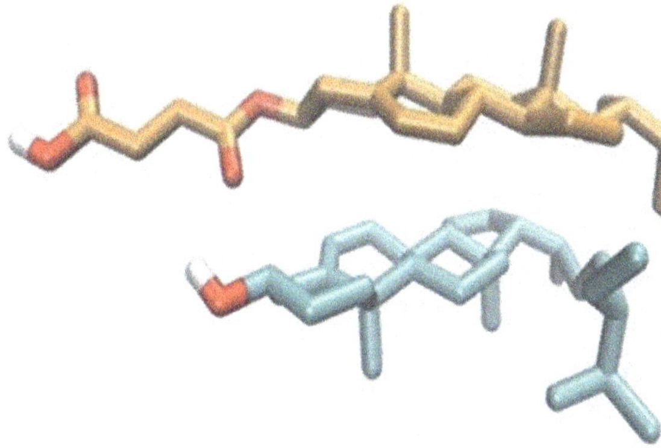

Fig. 22.3 Views of the lowest energy conformation for the molecular models of sitosterol (in cyan) bound to cholesteryl hemisuccinate (CHEMS) (orange carbons). Oxygens are colored with red and hydrogens with white. (Reprinted/Adapted from [8], with permission from American Chemical Society)

this bond in sitosterol + CHEMS systems [8]. This is confirmed in docking studies, where although a parallel dimer is formed (Fig. 22.3), the steroid cores are more overlapped and do not form the wedge-shaped dimer. This moves the molecules into a position where the H-bond cannot form. To see how the succinate may also contribute to this, we docked sitosterol-ferulate with sitosterol. This sterol ester has the ferulate moiety in common with oryzanol but lacks the C4 and C14 methyl groups. Sitosterol-ferulate also did not form a wedge-shaped dimer, but it did form the H-bond with sitosterol. We interpreted this as indicating that the key to the formation of tubules by oryzanol and sitosterol was both the presence of the ferulate and C4 and C14 methyl's, while the H-bond functions principally to select for parallel conformations.

To test the stability of the dimer, molecular dynamics simulations of the lowest energy dimer for both sitosterol + oryzanol and sitosterol + CHEMS were carried out [8]. Figure 22.4 shows that the sitosterol + oryzanol dimer was stable over a 10 ns simulation, while the sitosterol + CHEMS dimer started to break apart, as evidenced by the greater increase in the root-mean-square displacement (RMSD; Fig. 22.4c).

22.2.2 Formation and Stability of Tubules

Formation of a dimer is the first stage in the formation of a sterol–sterol ester tubule. It has been hypothesized from small angle neutron scattering (SANS) studies that the tubules are formed of a hollow core with one or two distinct sterol shells [12, 13]. The overall diameter of the tubule is around 10 nm. Although we have evidence that the dimer is stabilized by dispersion forces between the steroid cores and an H-bond, how a tubule is stabilized is less clear. To elucidate this, we constructed an in silico tubule fragment and simulated this using molecular dynamics simulation (Fig. 22.5).

Decane was used as the organic solvent phase for the simulations, as this is easier to simulate than a triglyceride. SANS experiments have shown that the tubule

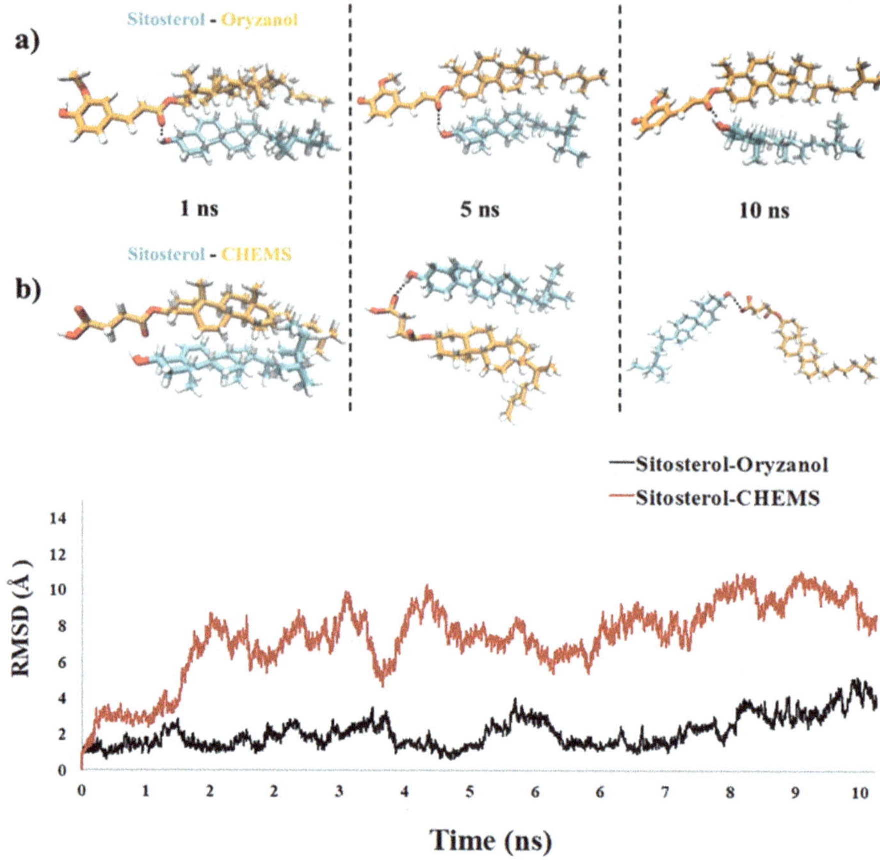

Fig. 22.4 Snapshots of the (**a**) sitosterol–oryzanol and (**b**) sitosterol–CHEMS dimers during the MD simulation. Hydrogen bonds are indicated with dashed line. (**c**) Time-dependent RMSD from the starting conformations for the sitosterol–oryzanol and sitosterol–CHEMS dimers. (Reprinted/Adapted from [8], with permission from American Chemical Society)

structure is conserved in triglyceride and alkane solvents [12]. The sitosterol–oryzanol tubule (Fig. 22.5a) is stable over the 50 ns of the simulation, showing a relatively low RMSD change (Fig. 22.5c) and a relatively high proportion of hydrogen bonds (Fig. 22.5d) from both dimer and other H-bonds [8]. We also noted that there was a significant contribution to interaction within the tubule from π–π stacking of the aromatic rings of ferulate groups of the oryzanol, which further stabilized the tubule [8]. Contrasting this to the sitosterol + CHEMS tubule (Fig. 22.5b), it is noticeable that a sitosterol–CHEMS tubule is more disordered after 50 ns of simulation, with a larger deviation in RMSD (Fig. 22.5c), and has significantly lower numbers of hydrogen bonds (Fig. 22.5d) [8]. Also, CHEMS lacks the ferulate group, so π–π interactions are not able to further stabilize the tubule.

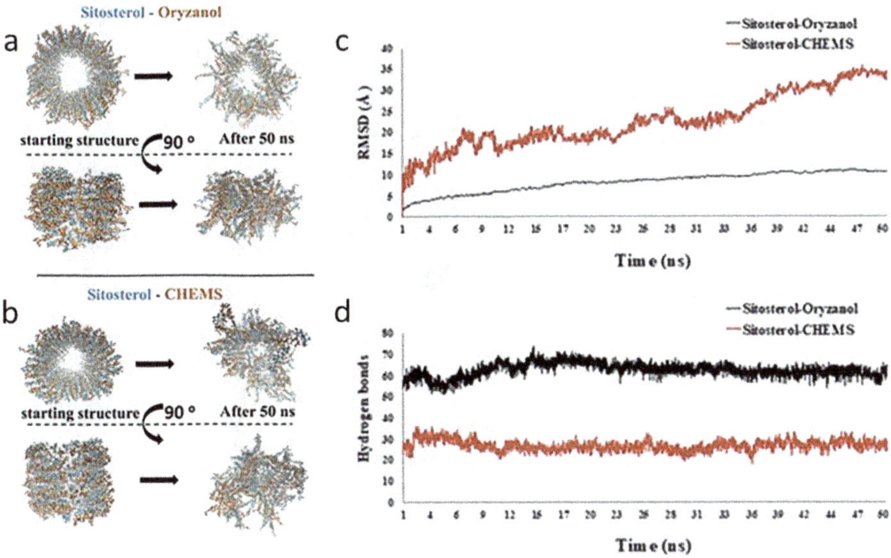

Fig. 22.5 Snapshots of the (**a**) sitosterol–oryzanol and (**b**) CHEMS–oryzanol organogels at the beginning and at the end of the course of the MD simulation. Sitosterol is shown with cyan carbons and oryzanol and CHEMS with orange carbons. (**c**) Time-dependent RMSD from the starting conformations. (**d**) Number of hydrogen bonds between the hydroxyl group of the sitosterol and the carbonyl group of the oryzanol/CHEMS. (Reprinted/Adapted from [8], with permission from American Chemical Society)

Altogether, this leads to a much less stable tubule for sitosterol + CHEMS and explains why we see no evidence for tubules forming in the experimental system [8].

22.2.3 Interactions Between Tubules

To form a gel, sterol tubules interact to form a network structure. In this, the tubules appear to fuse into bundles at junction zones, which bifurcate to allow tubules to interact to form the network that traps oil in the gel.

Figure 22.6 shows an atomic force microscopy (AFM) image of sitosterol–oryzanol gel obtained in phase mode [14]. The bundle structure of tubules can be seen clearly. How these bundles are stabilized is not obvious, but they must be important in the formation of the overall gel network. A consequence of the self-association of the model proposed by Bot et al. [12] for tubule structure based on sitosterol–oryzanol dimers is that the ferulate protrudes out from the tubule surface. In Sect. 22.2.2, we report that our simulations have shown that inter-dimer π–π interactions between the ferulate groups are important in stabilizing the tubule structure. To determine if intra-tubule π–π stacking is also involved in bundle formation, we simulated the surface–surface interaction between two tubules

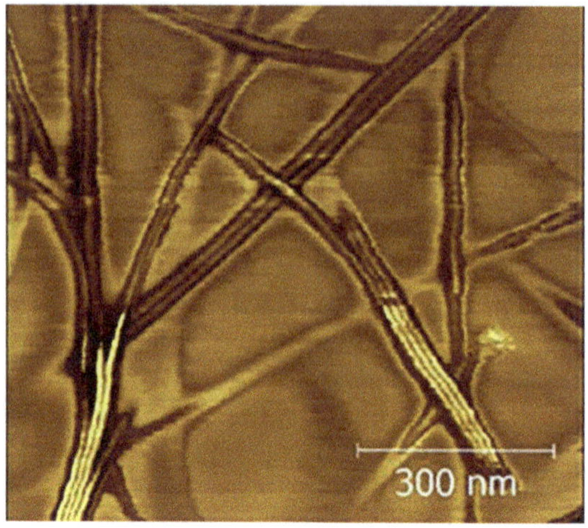

Fig. 22.6 Phase AFM image of a 10% w/w gel sitosterol + oryzanol gel. (Reprinted/Adapted from [14], with permission from American Chemical Society)

(Fig. 22.7). The simulations do indeed suggest a possible mechanism for the interaction between the tubules controlled mainly by van der Waals but also π–π interactions with a lesser role for hydrogen bonding.

22.2.4 *Water Sensitivity of Tubules*

One of the major problems encountered when trying to formulate sterol oleogels in foods is the water sensitivity of the tubules. Small amounts of water in a sitosterol + oryzanol oleogel lead to crystallization of the sitosterol into its hydrate [15]. This limits this type of oleogel to low water activity foods, with aqueous applications such as oil-in-water emulsion-based creams, ice cream among others, currently difficult to formulate [16, 17]. Solving the water sensitivity problem would open up phytosterol-based oleogels to a far wider range of products. To understand why water disrupts sterol oleogel formation, we carried out MD simulations of sitosterol + oryzanol tubules in water solvent rather than in organic solvent [18].

Figure 22.8a shows a simulated tubule after 50 ns in a water solvent. In water, the tubule structure is starting to change after 50 ns. Figure 22.8b shows that the RMSD remains low up to about 5 ns and after that increases rapidly and then more slowly for the rest of the simulation, indicating significant structural changes in the tubule. The radius of gyration (R_g) (Fig. 22.8c) increases after 5 ns and becomes more variable, also indicating structural changes in the tubules. Finally, the number of H-bonds between sitosterol and oryzanol decreases rapidly after 5 ns of simulation (Fig. 22.5d) as the water molecules start to penetrate into the tubule structure. This leads to water forming H-bonds with the hydroxyl group of the sterol and the carbonyl of the sterol ester, disrupting the inter-phytosterol H-bonds that help to stabilize the dimer and tubule.

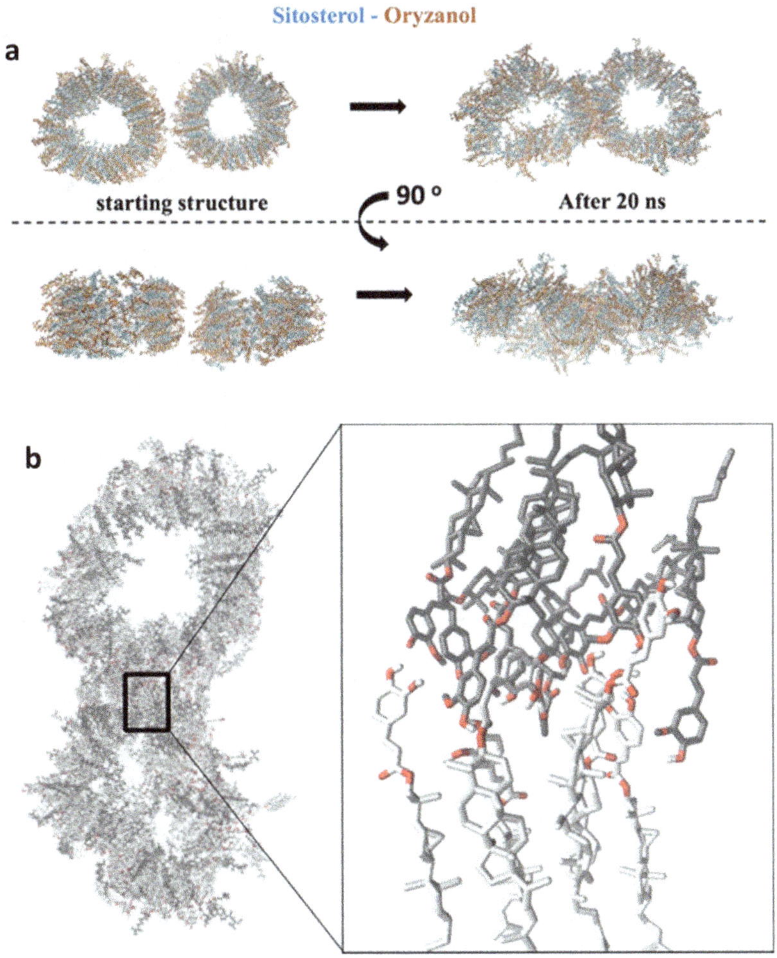

Fig. 22.7 (**a**) Snapshots of two tubules of sitosterol–oryzanol at the beginning and at the end of the course of the MD simulation. Sitosterol is shown with cyan carbons and oryzanol with orange carbons. (**b**) Hydrogen bonds, van der Waals interactions, and π–π contacts between the ferulic acid groups of oryzanol in the interface of two tubules. (Reprinted/Adapted from [8], with permission from American Chemical Society)

To stabilize the tubules in the presence of water, we tried two strategies: addition of a co-solvent glycerol in the polar phase [18] and substitution of the sitosterol with lecithin [19]. We found that with glycerol as the only polar phase component, we could disperse polar phase droplets throughout the oleogel without disruption of the structure. As we added water to the glycerol, the gel strength decreased, but we were still able to form a strong gel with 30% added water. Molecular dynamics simulations of a model tubule in glycerol showed us that it maintained its structure compared to the tubule in water over a 50 ns simulation (Fig. 22.8a). The RMSD

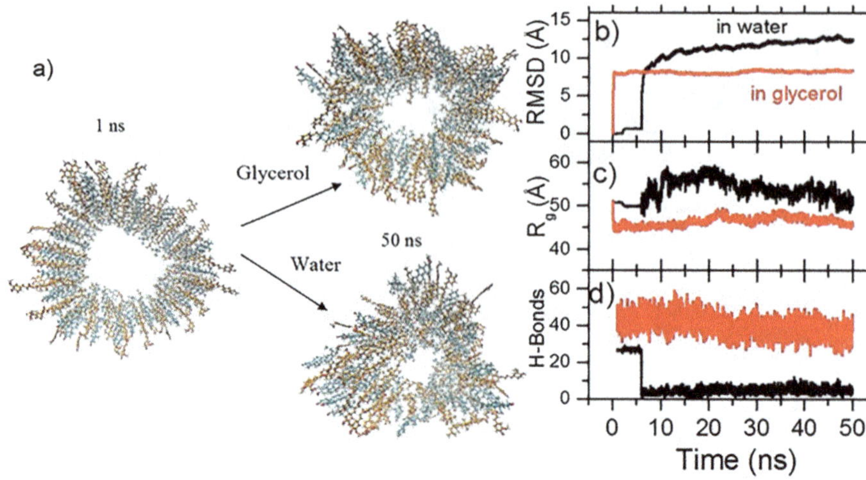

Fig. 22.8 (**a**) Snapshots of the sitosterol–oryzanol organogel with glycerol and water at the beginning and at the end of the course of the MD simulation. Sitosterol is shown with cyan carbons and oryzanol with orange carbons. (**b**) Time-dependent root-mean-square displacement (RMSD) from the starting conformation, (**c**) mass-weighted radius of gyration, and (**d**) number of sitosterol–oryzanol hydrogen bonds measured for the 50 ns MD simulations of the organogel system in water and in glycerol (colored black and red, respectively). Alignments and measurements were performed for the carbon atoms. (Reproduced/Adapted from [18], with permission from The Royal Society of Chemistry)

of the tubule in glycerol did increase rapidly at the start of the simulation (Fig. 22.8b) but then stabilized at a value below that of the tubule in water. The R_g of the tubule changed little from that of the starting conformation (Fig. 22.8c) as did the number of H-bonds in the structure (Fig. 22.8d), which was substantially higher than the tubule in water. These results were surprising as they indicate that even though the glycerol has three hydroxyl groups, it does not interact with the phytosterols in a way that disrupts the tubule. A more detailed analysis of the simulated glycerol tubule shows how this happens.

We find (Fig. 22.9) that glycerol forms H-bonds with and bridges between the methoxy, phenol, and carbonyl groups of the oryzanol and the hydroxyl group of the sitosterol such that the stability of the tubule is enhanced when compared to tubules in water or triglyceride oil. Lecithin, a phospholipid surfactant, is known to be able to form organogels with a range of molecules including steroids with structures like oryzanol [5, 20–25]. Of these, bile salt has been shown to form worm-like micelles with oryzanol in oil that can swell to hold significant quantities of water [20]. Our own work showed that mixed lecithin + oryzanol systems formed gels in triglyceride oil where we were able to dissolve both components at a higher concentration than possible individually [19]. This we surmised was due to the formation of mixed lecithin/oryzanol reversed micelles with a radius of about 4 nm. Fourier transform infra-red spectroscopy suggested that lecithin and oryzanol interact through an

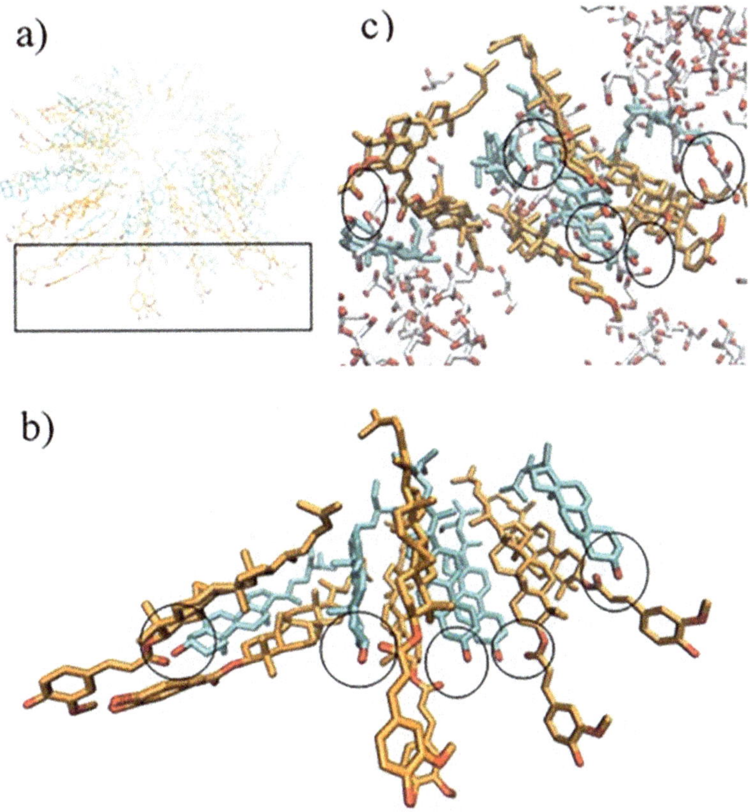

Fig. 22.9 (**a**) Top view of the sitosterol–oryzanol organogel in glycerol after 50 ns of simulation time. (**b**) Hydrogen bond network between the hydroxyl group of sitosterols and the carbonyl group oryzanols. (**c**) Hydrogen bond network between sitosterol, oryzanol, and glycerol. Sitosterol is shown with cyan carbons, oryzanol with orange carbons, glycerol with white carbons, and the hydrogen bonds are in circles. (Reproduced/Adapted from [18], with permission from The Royal Society of Chemistry)

H-bond between the phosphate group of lecithin and the phenol group of the oryzanol.

This was confirmed by molecular docking simulation (Fig. 22.10) revealing a possible underlying mechanism for the self-association into mixed reverse micelles. These reverse micelles were able to take up water into the hydrophilic micellar core. The gelation of the oil proceeded through a network of interacting micelles rather than formation of tubular self-association structures.

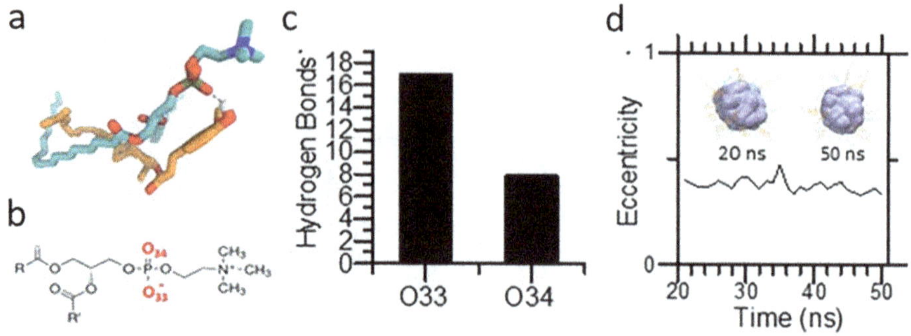

Fig. 22.10 (**a**) Molecular docking simulation of lecithin (carbons are cyan, oxygen is red, nitrogen is blue, and phosphorus is khaki) and oryzanol (orange carbons, oxygen is red, hydrogen is white), showing the presence of an intermolecular hydrogen bond (dashed line). (**b**) The structure of the lecithin's lipid head group to clarify the labeling of oxygen atoms for (**c**) the number of hydrogen bonds of the oxygen atoms of lecithin's phosphate group (25 in total, 1 for each lecithin molecule) with the hydroxyl group of oryzanol. (**d**) Eccentricity of the reverse micelles over the last 30 ns of the simulation. Inset are simulation snapshots corresponding to a configuration of the mixed system at 20 ns of simulation time, and the final configuration of the mixed system after 50 ns. Lecithin's lipid head group and oryzanol's ferulic acid group are rendered as mauve surface, lecithin's tails as cyan lines, and oryzanol's ring system and alkyl chain as orange lines. (Reproduced/Adapted from [19], with permission from The Royal Society of Chemistry)

22.3 Conclusions

Molecular simulation has been applied in various forms to the study of chemical and biological systems since the 1950s, but only over the past decade or so has it received widespread application in food systems. This is in part due to advances in software, methodology, and high-performance compute clusters that can handle the complex composition of foods. The next decade promises further advances in modeling of food systems. For phytosterol oleogels, progress has been made in understanding the structure and stability of the tubules and the effect that water has on these, but there is still scope for further modeling studies. Several technical challenges hinder development of phytosterol oleogel-based foods. First, the complexity of gel formation and interactions between gelator molecules make it difficult to predict gelators that work for different edible oils and especially specific applications [26]. The number of available sterol esters is limited, with γ-oryzanol the only one found in sufficient quantities. To exploit fully the potential of oleogelation by sterol–sterol ester mixtures will require the identification of a wider range of sterols and sterol esters to allow tuning of gel properties for specific food applications. There is a role for molecular docking and molecular dynamics in combination with organic synthesis and experimental studies in developing an answer to which acid moieties and sterols can be utilized in these mixed-molecular materials.

Second, under quiescent conditions, supercooling occurs [11] and nucleation and growth of the sitosterol + oryzanol tubules can take a considerable time after cooling. It is unclear whether supercooling is due to a kinetic barrier to nucleation and growth of tubules, vesicle formation, or whether tubules form but do not interact and gel for some time after. To investigate the kinetics of gelation/nucleation will require a range of techniques such as solid-state nuclear magnetic resonance (NMR), rheology, scattering, and imaging. Interpretation of the results from experimental techniques will require extensive molecular modeling. Formation of clusters in sterol mixtures can be followed by determining the free energy landscape (FEL) [27] of phytosterol dimers, trimers, and larger aggregates, thus identifying any local minima in the FEL that may limit nucleation opportunities. To access longer timescales, which give models that are more realistic for NMR and scattering data interpretation, requires observation of the commencement of nucleation, and the kinetics of the process. For this, coarse-grained MD methodologies, such as those we have developed to study surfactant-induced interfacial crystallization of triglycerides [28], a phenomenon also known to exhibit substantial supercooling effects, would be more suitable. This should allow simulation of tubule formation ab initio, but if this proves not to be the case, the option to use restrained or steered MD to direct self-assembly into tubules can always be explored.

Third, the gelation of phytosterol oleogels is known to be sensitive to shear. Application of oscillatory shear to the system speeds up the gelation process and leads to a greater control over gelation time. Given that shear forces will be applied during processing of oleogels in food applications, a clear understanding of these would need to be developed, although mechanical shear or pulsed waves could be used for robust and reproducible control. Non-equilibrium MD simulations incorporating shear fields, or other simulation methods such as dissipative particle dynamics or lattice-Boltzmann simulations, would be suitable to understand how shear forces influence the kinetics of self-association.

Finally, the high sensitivity to water of β-sitosterol + γ-oryzanol oleogels has already been mentioned [15–19]. This makes it difficult to formulate emulsion-based foods (e.g., creams, ice cream), which are a major industrial target for oleogelating technologies. There is further scope for using simulations to understand the mechanisms of water-induced instability, thus facilitating greater control of oleogel structure, opening up the technology to high water activity foods leading to a greater health impact.

Acknowledgments This work was supported by the BBSRC-DRINC project (BBSRC Grants: BB/M027449/1 [HWU] and BB/M027597/1 [UoE]). This project made use of time on ARCHER granted via the UK High-End Computing Consortium for Biomolecular Simulation (HECBioSim: http://hecbiosim.ac.uk), supported by the Engineering and Physical Sciences Research Council (EPSRC) (grant no. EP/L000253/1).

References

1. Brouwer IA (2020) The public health rationale for reducing saturated fat intakes: is a maximum of 10% energy intake a good recommendation? Nutr Bull 45:271–280
2. Patel AR, Nicholson RA, Marangoni AG (2020) Applications of fat mimetics for the replacement of saturated and hydrogenated fat in food products. Curr Opin Food Sci 33:61–68

3. Parmar V, Sharma R, Sharma S, Singh B (2022) Recent advances in fabrication of food grade oleogels: structuring methods, functional properties and technical feasibility in food products. J Food Meas Charact 16:1–16
4. Dalkas G, Euston SR (2019) Modelling and computer simulation approaches to understand and predict food structure development: structuring by gelation and self-association of biomolecules. In: Lazidis A, Norton I (eds) Handbook of food structure development Spyropoulos F. Royal Society of Chemistry, London, pp 383–401
5. Pernetti M, van Malssen KF, Flöter E, Bot A (2007) Structuring of edible oils by alternatives to crystalline fat. Curr Opin Colloid Interface 12:221–231
6. den Adel R, Heussen PCM, Bot A (2010) Effect of water on self-assembled tubules in β-sitosterol + γ-oryzanol-based organogels. J Phys Conf Ser 247:012025
7. Bot A, den Adel R, Roijers EC, Regkos C (2009) Effect of sterol type on structure of tubules in sterol + γ-oryzanol-based organogels. Food Biophys 4:266–272
8. Dalkas G, Matheson AB, Vass H, Gromov A, Lloyd GO, Koutsos V, Clegg PS, Euston SR (2018) Molecular interactions behind the self-assembly and microstructure of mixed sterol organogels. Langmuir 34:8629–8638
9. Morris GM, Huey R, Lindstrom W, Sanner MF, Belew RK, Goodsell DS, Olson AJ (2009) AutoDock4 and AutoDockTools4: automated docking with selective receptor flexibility. J Comput Chem 30:2785–2791
10. Matheson AB, Dalkas G, Lloyd GO, Hart A, Bot A, den Adel R, Koutsos V, Clegg PS, Euston SR (2022) Exploring how changes to the steroidal core alter oleogelation capability in sterol: γ-oryzanol blends. J Am Oil Chem Soc 99:943–950
11. Bot A, Agterof WGM (2006) Structuring of edible oils by mixtures of γ-oryzanol with β-sitosterol or related phytosterols. J Am Oil Chem Soc 83:513–521
12. Bot A, Gilbert EP, Bouwman WG, Sawalha H, den Adel R, Garamus VM, Venema P, van der Linden E, Floter E (2012) Elucidation of density profile of self-assembled sitosterol + oryzanol tubules with small-angle neutron scattering. Faraday Discuss 158:223–238
13. Gilbert EP (2022) Building blocks of β-sitosterol-γ-oryzanol gels revealed by small-angle neutron scattering and real space modelling. Food Funct 13:7123
14. Matheson AB, Koutsos V, Dalkas G, Euston S, Clegg P (2017) Microstructure of β-sitosterol: γ-oryzanol edible organogels. Langmuir 33:4537–4542
15. Sawalha H, Adel R, Venema P, Bot A, Flöter E, van der Linden E (2012) Organogel-emulsions with mixtures of β-sitosterol and γ-oryzanol: influence of water activity and type of oil phase on gelling capability. J Agric Food Chem 60:3462–3470
16. Matheson A, Dalkas G, Clegg PS, Euston SR (2018) Phytosterol-based edible oleogels: a novel way of replacing saturated fat in food. Nutr Bull 43:189–194
17. Poole J, Bentley J, Barraud L, Samish I, Dalkas G, Matheson A, Clegg P, Euston SR, Kauffman Johnson J, Haacke C, Westphal L (2020) Rising to the challenges: solution-based case studies highlighting innovation and evolution in reformulation. Nutr Bull 45:332–340
18. Matheson AB, Dalkas G, Mears R, Euston SR, Clegg PS (2018) Stable emulsions of droplets in a solid edible organogel matrix. Soft Matter 14:2044–2051
19. Matheson AB, Dalkas G, Gromov A, Euston SR, Clegg PS (2017) The development of phytosterol-lecithin mixed micelles and organogels. Food Funct 8:4547–4554
20. Cautela J, Giustini M, Pavel NV, Palazzo G, Galantini L (2017) Wormlike reverse micelles in lecithin/bile salt/water mixtures in oil. Colloids Surf A Physicochem Eng Asp 532:411–419
21. Tung S-H, Huang Y-E, Raghavan SR (2006) A new reverse wormlike micellar system: mixtures of bile salt and lecithin in organic liquids. J Am Chem Soc 128:5751–5756
22. Lee H-Y, Diehn KK, SW, Tung S-H, Raghavan SR (2010) Can simple salts influence self-assembly in oil? Multivalent cations as efficient gelators of lecithin organosols. Langmuir 26: 13831–13838
23. Han L, Li L, Li B, Zhao L, Liu G-Q, Liu X, Wang X (2014) Structure and physical properties of organogels developed by sitosterol and lecithin with sunflower oil. J Am Oil Chem Soc 91: 1783–1179

24. Kumar R, Ketne AM, Raghavan SR (2010) Nonaqueous photorheological fluids based on light-responsive reverse wormlike micelles. Langmuir 26:5405–5411
25. Shervani Z, Jain TK, Maitra A (1991) Nonconventional lecithin gels in hydrocarbon oils. Colloid Polym Sci 269:720–726
26. Hughes NE, Marangoni AG, Wright AJ, Rogers MA, Rush JWE (2009) Potential food applications of edible oil organogels. Trends Food Sci Technol 20:470–480
27. Zielińska J, Wieczór M, Bączek T, Gruszecki M, Czub J (2016) Thermodynamics and kinetics of amphotericin B self-association in aqueous solution characterized in molecular detail. Sci Rep 6:1–11
28. Green N, Euston SR, Rousseau D (2019) Interfacial ordering of tristearin induced by glycerol monooleate and PGPR: a coarse-grained molecular dynamics study. Colloids Surf B Biointerfaces 179:107–113

Chapter 23
Simulating the Physics of Oleogels: Mathematical Models and Monte Carlo Computer Simulation

David A. Pink and Shajahan G. Razul

Abbreviations

μ	Chemical potential
σ_{jk}	Stress
BC	Boundary condition
CNPs	Crystalline nanoplatelets
DLCA	Diffusion-limited cluster aggregation
ENM	Elastic network model
F	Helmholtz free energy
FT	Fourier transform
H	Enthalpy
MD	Molecular dynamics
MMC	Metropolis Monte Carlo
N	Particle number
P	Pressure
P-B	Poisson-Boltzmann
PDF	Pair distribution function
PMF	Potential of Mean Force
RLCA	Reaction-limited cluster aggregation
S	Entropy
T	Temperature

D. A. Pink (✉)
Department of Physics, St. Francis Xavier University, Antigonish, NS, Canada

Department of Food Science, University of Guelph, Guelph, ON, Canada

S. G. Razul
Department of Chemistry, St. Francis Xavier University, Antigonish, NS, Canada
e-mail: sgrazul@stfx.ca

C. Andrea Palla, F. Valoppi (eds.), *Advances in Oleogel Development, Characterization, and Nutritional Aspects*,
https://doi.org/10.1007/978-3-031-46831-5_23

TAG	Triacylglycerol
U	Internal energy
$V\varepsilon_{kj}$	Volume-weighted strain
V	Volume

23.1 Introduction

This chapter is concerned with: (a) a description of the fundamental concepts and equations needed to create mathematical models of oleogels, and (b) a description of the fundamentals of the Metropolis Monte Carlo (MMC) technique. Commercial programs are available but are untested by us. In carrying out scientific procedures, whether computer simulations or experimental studies, it is essential to be critical of what is being done. The intent of what follows tries to address this by outlining the basic concepts of Statistical Mechanics and Thermodynamics relevant to carrying out computer simulation. Books on a range of levels are by Mandl; Hecht; Pathria; Greiner, Niese, and Stöcker; Baierlein; Reichl; and Parsegian [1–7]. In addition, Pink et al. [8] provide an outline of computer simulation with examples from food science.

23.2 Physical States: Internal Energies, Entropy, Interaction Energies, and Free Energy

A system comprising essentially solid crystals, possibly containing defects, in an oily solvent might be considered an oleogel. The solids can perturb the solvent in their neighborhoods. If no information is missing, then this is a state, a, of the system characterized by its internal energy, E_a, the energy needed to create the solids with their defects, the oil and the properties imposed by the presence of the solids. It is not incorrect to group states with a narrow distribution of energies together for the sake of convenience, and can be considered a single effective state, A, comprising individual states exhibiting an average internal energy, E_A. If N_A is the degeneracy of state A, then the quantity, $S_A = k_B \ell n(N_A)$, is the entropy of A with k_B being Boltzmann's constant and $\ell n(x)$ the natural log of x.

The free energy of a system is of fundamental importance, and there are two forms: (1) The Helmholtz free energy defined as $F = U - TS$, where U is the energy, T is the absolute temperature in units of °Kelvin (°K), and S is the entropy of the system. The volume, V, is required to be constant. (2) The Gibbs free energy, $G = PV + U - TS$, where P is the pressure and V is the volume, and $H = PV + U$ is the enthalpy [9].

23.3 Potential of Mean Force (PMF)

The PMF enables calculating Helmholtz free-energy changes from computer simulations [10]. It is defined as that potential that provides the average force over all configurations of all the particles acting on a particle *i*. The idea can be extended and illustrated with two particles, *i* and *j*, separated by a distance, *r*, in a solvent mixture. The average energy over all the configurations of the solvent along the coordinate, *r*, of particles *i* and *j* is the PMF. In principle, the free-energy changes of any inter- or intra-molecular coordinate may be used. Thus, the free energy is given by the following expression:

$$A(r) = -k_B T \ \ln[g(r)] + \text{constant} \tag{23.1}$$

where the expression in square brackets is the radial distribution function, $g(r)$, which provides a convenient route to calculating the PMF [11]. Each identical molecule of a set possesses a "center." Consider a sphere of radius *r*, with walls of non-zero thickness, and let the center of this sphere coincide with the center of a selected molecule. Count the number of molecule centers (or the centers of atoms, or the centers of any structures at all) inside the wall, and normalize it so that as $r \to \infty$, the number becomes 1. This function of *r* is $g(r)$. The free-energy change here is a function of the separation of two molecules at a distance, *r*. However, for molecular dynamics (MD) simulations that involve processes involving large free-energy barriers or, if the timescales are long or too computationally demanding to be simulated directly, the required PMF may not be calculated by regular MD [12]—adequate sampling of phase space to obtain the appropriate averages required for the calculation of a free-energy profile may not be practicable. To ensure that proper sampling can be achieved, the Umbrella Sampling method, free-energy perturbation, or other enhanced sampling techniques may be used [13, 14, 16]. A review of the Umbrella Sampling technique is by Kastner [15].

The PMF is a remarkable technique, available in Monte Carlo or molecular dynamics simulations, and provides a free-energy profile along a selected path. Thus, the path can be between the centers of two atoms, between the centers of mass of two molecular aggregates, or defined by the torsion angle around a chemical bond. It may calculate free-energy changes as a function of the distance between two residues, or as a protein is pulled through a lipid bilayer.

23.4 Structure Functions

Consider an assembly of *N* points, $\{1, 2, \ldots, j, \ldots\}$, identified by their position vectors $\{r_j\}$. Define the structure function, which is related to the structure factor of crystallography,

$$s(\vec{q}) = \frac{1}{N} \sum_j \sum_k \exp(iq \cdot (r_j - r_k)) \tag{23.2}$$

If we are relating this to a scattering experiment, then $q = k - \bar{k}$, where k is the wave vector of the incident particles (photons, neutrons) and the wave vector, $\bar{k}$, is that of the elastically scattered ($|\bar{k}| = |k|$) particles. Otherwise, it is simply a Fourier analysis of an assembly of points carried out to discover whether there are characteristic structures in the assembly [17, 18]. One can relate spatial lengths, λ, to q, the magnitude of q, $q = |q|$ via $\lambda = 2\pi/q$. We can define a structure function that depends only upon q: $s'(q)$, the Fourier transform (FT) of $g(r)$, the radial distribution function [19].

$$s'(q) = \frac{1}{N}\sum_{j}\sum_{k} \exp(iqr_{jk}) \tag{23.3}$$

where $r_{jk} = |r_j - r_k|$, the distance between points j and k. One can identify the fractal dimension, D, of the assembly of points, for some range of q, via the relation [20–22],

$$s'(q) \propto q^{-D} \tag{23.4}$$

The range of q should be at least one decade in q.

23.5 Phase Space and Ergodicity

Phase space is the set of numbers that define the "generalized coordinates" of a system, analogous to the *x-y-z* coordinates that define the location of an object. As the set of numbers changes, the system moves through phase space. An important concept in statistical mechanics is ergodicity. The Ergodic hypothesis states that for certain systems the time average of their properties is equal to the average over all of phase space.

23.6 Ensembles

All thermodynamic potentials are expressed in terms of conjugate dynamical variables and modeling involves choosing which of those variables to keep constant. Thermodynamics tells us that we can keep three such variables fixed, but none of them may be mutually conjugate. Examples of mutually conjugate variables are absolute temperature (T) and entropy (S), volume (V) and pressure (P), chemical potential (μ) and particle number (N), and stress (σ_{jk}) and volume-weighted strain ($V\varepsilon_{kj}$). The *NPT* and the *NVT* ensembles are popular because they reflect conditions associated with many experiments.

The Microcanonical (*NVE*) ensemble describes a system of fixed volume and fixed number of objects, which is isolated from the external world and is unaware of the surrounding temperature. The Canonical (*NVT*) ensemble describes systems where the number of objects is fixed but which is in thermal contact with an environment (a "heat bath") characterized by temperature, T. The Grand Canonical (μVT) ensemble is useful when there is the exchange of objects (particle exchange) and in thermal contact and participating in particle exchange, must possess the same values of T and chemical potential, μ.

23.7 Boundary Conditions (BCs)

All computer simulations are carried out in finite-size spaces and are accordingly bounded by enclosing mathematical surfaces. Periodic BCs were designed to eliminate their effect. If a simulation volume is bounded by two parallel surfaces separated by a mutually perpendicular vector L, then, if a component of the system possessing velocity $v(p)$ and position $r(p)$ between the surfaces at simulation step (time or MMC step), p, were to move to position $r(p + 1)$ lying outside the surfaces with velocity $v(p + 1)$, then the position is reset to $r(p + 1) - L$ with velocity $v(p + 1)$. In MMC simulations, there is no velocity term and only $r(p)$ is relevant. A BC of major importance at solid–fluid interfaces is the no-slip BC: the requirement that the local speed of the fluid parallel to a stationary surface is zero [23].

23.8 Coarse-Grained Meso-Scale Models: Coarse-Grained Interactions—Nano- to Meso-Scale [23–30, 34]

How we model a system depends upon what spatial, or time, scale interests us. As far as the Coulomb interaction in atomic systems is concerned, we could model a molecule by treating the nuclei as point objects possessing charge and spin, and explicitly describe the electronic states most of which exhibit non-spherical distributions, as well as electronic polarizabilities and spin and orbital angular momentum states. One intent of modeling must be to eliminate those coordinates that undergo motions on timescales much faster than the timescales in which we are interested. We then replace their detailed dynamics by their average values. Such a procedure is called "coarse-graining." Although the references above describe coarse-graining as applied to phospholipid bilayer membranes, they are adequate for illustrating the principle of the procedure. Coarse-graining in computational biology was reviewed by Saunders and Voth [30, 31]. Many useful coarse-grained models, like the elastic network model (ENM) [32, 33] and the MARTINI model (MARTINI CG) [24, 33], exist among others [30].

23.9 Electrostatics [35]

Charged moieties in oleogels generally arise through chemical bond polarization (e.g., C=O bonds) of molecules that are electrically neutral. The interaction between charged moieties in food are generally modeled by treating them as "classical" (non-quantum) charged systems. Oleogels with zero net charge, for example, can manifest short-range electrostatic interactions via the "partial charges" of bond polarization. This section describes classical electrostatics.

23.9.1 Electric Field and Electric Potential: Poisson's Equation

The electric field vector *E*(*r*) at any point, *r*, in the absence of time-dependent magnetic fields, due to a point electric charge, *Q*, located at the origin of coordinates, is

$$E(r) = \frac{1}{4\pi\varepsilon}\frac{Q}{r^2}\widehat{r} \tag{23.5}$$

where $r = |r|$, $\widehat{r}$ is a unit vector pointing radially outward from *Q*, and $\varepsilon = \varepsilon_0\varepsilon_r$ is the permittivity, where ε_0 is the permittivity of free space with ε_r the relative permittivity. Since an infinitesimal electric charge will detect *E*(*r*) as a force, its units are N/C = V/m. It can be more convenient to utilize a scalar quantity, the electric potential $\psi(r)$, from which the field can be calculated, $E(r) = -\nabla\psi(r)$.

The problem of charge–charge interactions reduces to the correct calculation of the electric potential distribution. If the charges are "free" (to move), then one has a self-consistent problem: the movement of the charges is determined by the electrical potential and vice versa. In the latter case, the potential has to be calculated at each instant of time in order to determine how the accelerations—and hence positions—of the charges change with time. The integral and differential forms of Gauss's law can be used to determine the electric potential.

$$\int_S E(r)\cdot ds(r) = \frac{Q}{\varepsilon} \tag{23.6a}$$

$$\nabla^2\psi(r) = -\frac{\rho(r)}{\varepsilon} \tag{23.6b}$$

Equations (23.6a) and (23.6b) are the integral and differential forms of Poisson's equation. Here *Q* is the total charge contained inside the volume bounded by a closed surface *S* over which the integral of the electric field *E*(*r*) is carried out, $\rho(r)$ is the electric charge density at point r, and $\varepsilon = \varepsilon_0 \varepsilon_r$ is the permittivity, where ε_0 is the permittivity of free space and ε_r is the relative permittivity. $ds(r)$ is an infinitesimal area on the closed surface S, at r, pointing outward from S. Equation (23.6a) can be used to calculate the electric field when the system possesses sufficiently high symmetry.

23.9.2 Uniqueness Theorem

An existence theorem states that the electrical potential is unique if it satisfies Poisson's equation and the boundary conditions. This powerful theorem means that we may use *any* technique whatsoever to obtain the electric potential, subject only to those two requirements.

23.9.3 The Superposition Principle

The superposition principle follows from the linearity of Poisson's equation. It means that the total electric potential due to two charges is the sum of the electric potentials due to each of them. This follows from the fact that if ψ_1 and ψ_2 are two solutions of a linear equation, then so is $\psi_1 + \psi_2$.

23.9.4 Boundary Conditions, Polarization, Induced Charges, and Energy

To obtain a correct solution for the potential from Poisson's equation, it is necessary to satisfy the boundary conditions. In the case of a charge Q located at the origin in a uniform dielectric, we obtain as a solution to Eqs. (23.6a) and (23.6b),

$$\psi(r) = \frac{1}{4\pi\varepsilon} \frac{Q}{r} \tag{23.7}$$

Boundary conditions for the case of a non-uniform space are more complex. Examples of this include oleogels, emulsions, bacterial surfaces, and micellar delivery systems. It is unlikely that most food systems do *not* possess interfaces within it. Uncharged interfaces can become electrically polarized, which amounts to inducing electrical dipoles in it. Boundary conditions for non-uniform systems, such as those with interfaces labeled 1 and 2, require that the potential, $\psi_1(r_s) = \psi_2(r_s)$, and the tangential components of the electric field be continuous.

$$[E_2(r_s) - E_1(r_s)] \times \hat{n}(r_s) = 0, \text{or } E_{2T}(r_s) = E_{1T}(r_s) \tag{23.8}$$

And

$$[D_2(r_s) - D_1(r_s)] \cdot \hat{n}(r_s) = \sigma_f(r_s), \text{or } D_{2N}(r_s) - D_{1N}(r_s) = \sigma_f(r_s) \tag{23.9}$$

Where

$$D_m(r) = \varepsilon_m E_m(r) = \varepsilon_0 E_m(r) + P_m(r) \tag{23.10}$$

where $D_m(r)$ is the electric flux density or electric displacement at r in region m. Here, the vector r_s locates a position on the interface, $E_{mT}(r_s)$ is the component of the electric field in region m that is tangential to the interface at position r_s, $D_m(r)$ and $P_m(r)$ are the electric displacement and polarization vectors in region m at r, $\hat{n}(r_s)$ is a unit vector perpendicular to the interface at r_s and pointing outward from the interface, $D_{mN}(r_s)$ is the component of the electric displacement in region m perpendicular to the interface at r_s, $\sigma_f(r_s)$ is the free-charge density at r_s, and ε_m is the permittivity of region m.

For a linear isotropic dielectric, the polarization $P(r_s)$ at point r_s at an interface that is induced by an electric field $E(r)$ is given by

$$P(r_s) = (\varepsilon - \varepsilon_0)E(r_s) \tag{23.11}$$

The induced "bound" surface charge density $\sigma_b(r_s)$ can be obtained from

$$\sigma_b(r_s) = P(r_s) \cdot \hat{n}(r_s) \tag{23.12}$$

This surface charge density is referred to as "bound" because it arises from the polarization of atoms or molecules essentially fixed in space.

For a given charge distribution, the electrostatic energy of the system inside a volume v is

$$U = \frac{1}{2}\int_v \rho_f(r')\psi(r')dv(r') \tag{23.13}$$

where $\rho_f(r')$ is the "free" volume charge density in $dv(r')$. The potential at any point r' is

$$\psi(r) = \frac{1}{4\pi\varepsilon_0}\int_v \frac{\rho_f(\boldsymbol{r}') + \rho_b(\boldsymbol{r}')}{|\boldsymbol{r} - \boldsymbol{r}'|}dv(r') + \frac{1}{4\pi\varepsilon_0}\int_S \frac{\sigma_f(\boldsymbol{r}') + \sigma_b(\boldsymbol{r}')}{|\boldsymbol{r} - \boldsymbol{r}'|}dA(r') \tag{23.14}$$

where $\rho_b(r')$ is the "bound" volume charge density in $dv(r')$, and σ_f and σ_b are the "free" and "bound" surface charge densities in $dA(r')$ and defined above. Although Poisson's equation together with Eqs. (23.8)–(23.14) are sufficient to calculate the energy of a system of charges—*fixed* in position—in the presence of interfaces, one can appreciate the difficulties of applying these to electrostatics in complex systems. Accordingly, we seek an approach that gets around having to calculate the induced bound surface charge density, $\sigma_b(r')$. This is why the method of Image charges was developed.

Image Charges The method of image charges involves finding the electrostatic potential that satisfies Poisson's equation, together with the appropriate boundary conditions, by the artifice of introducing fictitious "images" of the free charges.

23.9.5 (Particle Mesh) Ewald Summation

Ewald summation [36] is a method for calculating the electric potential, $\psi(r) = \psi_{\text{short}}$ $(r < d)$ or $\psi(r) = \psi_{\text{long}}$ $(r \geq d)$, where we have written the potential at r as either a short-range part $(r < d)$ or a long-range part $(r \geq d)$. Let $\varnothing_{\text{long}}(k)$ represent the Fourier transform (FT) of the long-range potential. ψ_{long} $(r \geq d)$ is then evaluated by performing summations to obtain $\varnothing_{\text{long}}(k)$, which converge relatively quickly and so can be truncated, then computing the inverse FT to obtain ψ_{long} $(r \geq d)$. "Mesh" arises because the charge density field in k-space (reciprocal space) is defined on a lattice.

23.9.6 The Poisson-Boltzmann Equation

We can approximate the charge distribution by a continuous function of position, if the spatial scale of the system is such that the distances between the charges is very much less than the characteristic spatial scale. One assumes that the free charges in solution are moving sufficiently rapidly that the stationary charge sees an average "smeared out" distribution of free charges. In many cases, the solution is in thermodynamic equilibrium, characterized by an ambient temperature T, so that the spatial distribution of the free-charge density is determined by the Boltzmann factor $\exp(-\beta U)$, where U is the energy and $\beta = 1/k_BT$. Taking into account the requirement of total electroneutrality, this converts the (linear) Poisson Eq. (23.6b) into the Poisson-Boltzmann (P-B) equation [37], which is non-linear because ψ appears in the exponent:

$$\nabla^2\psi = \frac{e}{\varepsilon_T\varepsilon_0}\sum\nolimits_i z_i n_i^0 \exp\left(\frac{\dot{z}_i e\psi}{k_BT}\right) \tag{23.15}$$

Here e is the elementary proton charge, and n_i^0 and z_i are the concentration and valence of i-th type of ion in the bulk solution.

23.9.7 The Linearized P-B Equation and the Debye Screening Length

In cases where entropic effects (sufficiently high temperature T) dominate the energetics, that is, when the potential ψ is sufficiently small so that $\psi \ll k_B T/e$ ($k_B T \approx 4.0 \times 10^{-21}$ J at room temperature), then the exponential term in Eq. (23.15) can be expanded to first order in $(z_i e\psi/k_B T)$. This results in approximating the non-linear P-B equation by the linearized P-B equation for which analytical solutions can be obtained. In this case, the P-B equation becomes, using the condition of total charge neutrality, $\sum_i z_i n_i = 0$

$$\nabla^2 \psi = \psi \frac{e^2}{\varepsilon k_B T} \sum\nolimits_i z_i^2 n_i^0 = \kappa^2 \psi \tag{23.16}$$

Where

$$\kappa^2 = \frac{e^2}{\varepsilon k_B T} \sum\nolimits_i z_i^2 n_i^0 = \lambda^{-2} \tag{23.17}$$

The units of κ are inverse length (e.g., nm^{-1}). The distance λ can be treated as the effective radius of a perturbed volume in a neutral solution that contains mobile ions and is called the "Debye (screening) length" [37]. In the simplest case of a 1:1 electrolyte, where $n_1^0 = n_2^0 = n_0$, Eq. (23.17) becomes

$$\kappa^2 = \lambda^{-2} = \frac{2e^2 n_0}{\varepsilon k_B T} \tag{23.18}$$

The Debye screening length of a 100 mM solution at room temperature (300 K) is ~1 nm.

23.10 Stochastic Processes: The Metropolis Monte Carlo (MMC) Method [38–43]

The MMC Method computes averages of characteristic variables for systems in thermal equilibrium. Such variables might be the energy, the number of objects in a given physical state, the average end-to-end length of a polymer chain, and the local density of a given type of object. It involves only the interaction, and internal, energies of the model being simulated and is not concerned with kinetic energies and so does not involve forces, accelerations, or velocities [7]. What this implies is that, since we have no forces in this approach, we cannot, in general, use MMC techniques to model the dynamics of a system. At best, the MMC method could be used for cases in which random motions dominate inertial terms in the energies: the surrounding heat bath provides the impetus for changes of state, and the inertial terms involving velocities and accelerations (e.g., $mv^2/2$) are irrelevant. The power of the Monte Carlo method is that it is based upon "importance" sampling so that it is unnecessary to sample all of phase space.

A standard MMC method utilizes the following procedure:

1. The system is set up in any initial state that is convenient. This is defined by the values of a set of variables, a_1, a_2, ..., a_K. One MMC step involves trying to change each of these values to a different value. At any point in the simulation, let the state of the system be labeled A possessing energy E_A and let us try to change the system to state B, possessing energy E_B, by attempting to change one or more of the variables, a_1, ..., a_K. Define $\Delta E = E_B - E_A$. The following example is intended to make the MMC technique clear.

   Assume that we have a system comprising N spheres at temperature, T. Each sphere can be in two states: a ground state (g) with energy E_g, and an excited state (e) with energy $E_e > E_g$. In the unique state g, the sphere possesses a radius, R_g, and possesses degeneracy, $D_g = 1$. The excited state, e, possesses a radius $R_e > R_g$ and a degeneracy, $D_e > D_g$. Two spheres, labeled j and k, interact via an interaction, $V(s_j, s_k, r_{jk})$, where s_j and s_k are the states (g or e) of the two spheres and r_{jk} is the center-to-center distance between them. We visit each sphere and attempt to carry out a change of state and a change of location.

   Let us attempt to move the sphere labeled j and change its state. Let it be, for example, initially in its state e with its center located at r_j. We attempt to change its state to g. In doing so, the internal free energy would change by the amount

$$\Delta E_1 = E_g - E_e - k_B T \ln D_g + k_B T \ln D_e \tag{23.19}$$

   Let the sphere attempt to relocate at r_j'. Compute the interaction energy difference,

$$\Delta E_2 = \sum\nolimits_k' V\left(g, s_k, r_{jk}'\right) - \sum\nolimits_k V\left(e, s_k, r_{jk}\right) \tag{23.20}$$

   where the first sum would be over all spheres that interact with sphere j after it has carried out the attempted move, while the second sum is over all spheres that interact with sphere j before its attempted move. The distance r_{jk}' is the distance between the sphere j in its new location and another sphere k. The sums are not necessarily over the same sets of spheres. The total energy change is then $\Delta E = \Delta E_1 + \Delta E_2$.
2. If $\Delta E \leq 0$, then the change is accepted and the sphere j will now be in its state g with its center located at a new position r_j'.
3. If $\Delta E > 0$, then choose a random number u with $0 \leq u < 1$.
4. If $u \leq \exp(-\beta \Delta E)$, then the change is accepted and the sphere j will now be in its state g with its center located at a new position r_j'.

Otherwise, the system does not change but remains in its initial state.

5. When we have tried to change the states and positions of all spheres in this example, then *one MMC step* has been carried out.
6. There are two aspects to a simulation: (a) initializing the system, which entails permitting the system to come to equilibrium after the start of the simulation, and (b) calculating average values while the system samples its equilibrium states.

   (a) After the simulation is begun, let us keep track of the total energies at each MMC step, 0, 1, 2, ... Let these energies be $E(0)$, $E(1)$, $E(2)$, ... We will notice that these change in a regular way, apart from fluctuations, and that they approach a value, $\langle E \rangle$, which does not change (apart from fluctuations) as the simulation proceeds further. This is the average equilibrium value of the energy and tells us when the process of initialization has been completed.
   (b) The simulation is continued and the average values of other variables are calculated by carrying out a further m MMC steps and computing the average of the set $\{a_n, n = 1, \ldots, K\}$. Let the sequence of values obtained from these m MMC steps be $a_n(1)$, $a_n(2)$, ..., $a_n(m)$. Then the average over the m MMC steps is,

$$\langle a_n \rangle = \frac{1}{m} \sum_{y=1}^{m} a_n(y) \tag{23.21}$$

7. It has been shown that if a simulation is carried out for a sufficient number of MMC steps, then that system will come to thermal equilibrium at the temperature selected for the simulation, and will exhibit average values and fluctuations characteristic of thermal equilibrium.

The use of the MMC technique to compute average values must employ a sufficient number of MMC steps to obtain averages reflecting the characteristics of the system, which generally requires a large number of steps. The number of "repeats" of the simulation in order to compute an average can be small. However, the simulation to compute the average PMF is best utilized very many times for a much smaller number of MMC steps.

23.11 Are MMC Steps Proportional to Time?

What this ask is, if an MMC simulation involves N steps, then, if, instead, the simulation employed is molecular dynamics, applied for some time, t, then is N proportional to t? The answer is "no." Does the trajectory of a system through phase space computed via an MMC simulation mimic a simulation of the same system but utilizing molecular dynamics? Again the answer is "no," and the cartoon of Fig. 23.1 shows how this comes about. It has been proven that an MMC simulation only guarantees that the system will asymptote to occupy states

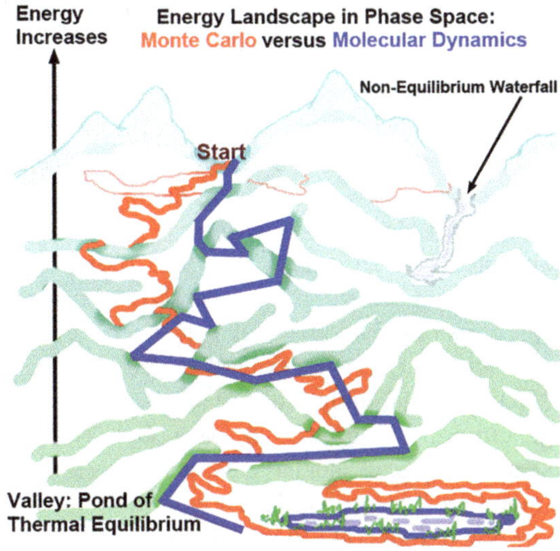

Fig. 23.1 Cartoon comparing MMC (red) and molecular dynamics (blue) trajectories through Phase Space. They both start in a high free-energy state and asymptote to a low free-energy state around the Pond in the Valley. Continuing to run the MMC simulation leads to the system remaining in the neighborhood of the Pond but, in its "motion" through Phase Space, it exhibits fluctuations in its Phase Space "position." The paths of each simulation, and the fluctuations away from a single average MMC path, differ. A fluctuation using MMC has caused the system to follow a trajectory (thin light red line) that initially takes it away from the Valley and brings it to a Waterfall that is not associated with the Pond of Thermal Equilibrium, but is a local free-energy minimum in which the system becomes trapped, and should be eliminated from averaging

characteristic of thermal equilibrium after a sufficiently long simulation. This does not imply that its trajectory through phase space using an MMC simulation is similar to the trajectory utilizing molecular dynamics. The reason is that molecular dynamics involves the kinetic energy term, $\frac{1}{2}mv^2$, which arises from the force, $m\partial v/\partial t$, while an MMC does not use it.

Classical elastic scattering is not treated at each step of an MMC simulation. If one object tries to occupy part of space already occupied by another object, then either the move is forbidden or it might be permitted with an increase in the energy. The effect of this is that, without introducing additional rules for elastic scattering, an object will not cause the displacement of another object in a single MMC step via a seeming collision. This unphysical effect arises because we are attempting to apply the MMC approach beyond its validity. However, over a sufficient number of MMC steps, two objects that overlap, thereby possessing a high energy, will separate.

23.12 Examples of MMC Computer Simulation [44]

Heertje and Neumis [45] identified the smallest stable form of solid fat structures in triacylglycerol (TAG) edible oils at room temperature, to be crystalline nanoplatelets (CNPs). Acevedo and Marangoni [46, 47] also studied CNPs, their rheology, and their abilities for oil binding capacity. To model them, the highly anisotropic CNPs were represented by rigid arrays of close-packed spheres (Fig. 23.2a). The interaction free energy between two spheres is known [7, 48]. The results of MMC simulations together with analyses utilizing structure functions showed that model CNPs formed one-dimensional aggregates (TAGwoods), that TAGwoods aggregated via diffusion-limited cluster aggregation (DLCA) and reaction-limited cluster aggregation (RLCA) to form clusters exhibiting fractal dimensions $D = 1.7 - 2.1$, and that these clusters were, on the average, uniformly distributed in space ($D = 3$) [49].

What might be sharp X-ray lines in a large nanocrystalline sample, become broad lines in a small "amorphous" sample [18]. The question addressed is how do we utilize the pair distribution function (PDF) to deduce nano-scale structure from experimental data? The PDF is the radial distribution function described above. This is clearly described in this chapter and the graphical software used is listed. Thermal motion of materials causes a spectrum of δ-function-like scattering peaks to

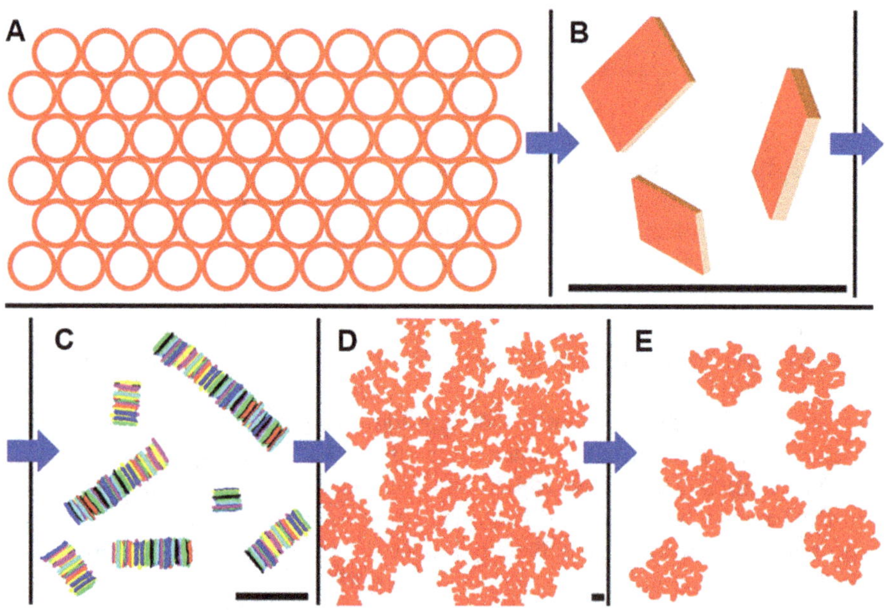

Fig. 23.2 Cartoon illustrating an MMC simulation of an aggregating model of crystalline nanoplatelets (CNPs) in a liquid oil, with black lines showing the same length scale. (**a**) Single model planar CNP, comprising $10 \times 6 \times 1$ spheres. (**b**) Individual model CNPs with spheres omitted. (**c**) Model CNPs aggregated into TAGwoods. (**d**) TAGwoods aggregated into DLCA/RLCA clusters. (**e**) TAGwood clusters uniformly distributed in space. (Adapted from [44], with permission from AIP Publishing)

broaden into Gaussians, and so the PDF can be represented by a sum of independent Gaussian functions. Although only X-ray scattering is referred to, similar comments will hold for neutron scattering.

The work of Yang et al. [50] is a good example of the use of the PMF to compute the free energy as a function of specific reaction coordinates. It shows how the PMF can be computed utilizing Umbrella Sampling to overcome the difficulties of sampling specific regions on the energy landscape where energy barriers are both high and frequent, in order to understand catalytic mechanisms. The work illustrates how many shorter simulation runs yield better results than few longer runs.

An excellent review is that of Flöter et al. [51]. Although it is not involved with mathematical modeling, the range of experimental data and its references are important reading for anyone intending to carry out modeling.

23.13 Conclusions

In this chapter, we have covered essentially all aspects of the implementation of the MMC technique, using a two-state system of interacting identical spheres as an example. We introduced this chapter with brief descriptions of fundamental thermodynamic and mathematical functions, which, however, did not include Phase Transitions, a major field in itself. We have emphasized the significance of the PMF technique, which enables one to calculate a free energy by relating it to the radial correlation function, $g(r)$. We have also discussed, and illustrated, the reasons why Monte Carlo steps do not correspond to the passage of time, although a correspondence can be made in some cases with "elapsed time."

In addition, we have devoted a considerable section to Electrostatics, which is crucial for understanding the fundamental interactions involved in forming oleogels. In addition to solid-forming covalent bonding, Lennard-Jones attractive ($1/r^6$) interactions and electrostatic interactions, even between molecules with zero total charge, can play vital roles in oleogel formation. Such interactions can be strong compared to thermal energies (k_BT) that arise from the surrounding "heat bath" at temperature T.

Finally, we have described four applications related to simulating oleogel-like structures or experimental investigations. These examples provide insight into what needs to be modeled and simulated. The fourth example includes a detailed description of calculating a free energy using the PMF technique.

Acknowledgments It is a pleasure to thank Fernanda Peyronel for her valuable comments on portions of this chapter. This work was supported by St. Francis Xavier University and by NSERC of Canada via grants # R0178050 (DAP) and # R0212010 (SGR).

References

1. Mandl F (1988) Statistical physics, The Manchester physics series, 2nd edn. Wiley, New York
2. Hecht CE (1990) Statistical thermodynamics and kinetic theory. Dover Publications Inc., New York
3. Pathria RK (1996) Statistical mechanics, 2nd edn. Butterworth-Heinemann, London
4. Greiner W, Niese L, Stöcker H (1997) Thermodynamics and statistical mechanics. Springer Verlag, New York
5. Baierlein R (1999) Thermal physics. The University Press, Cambridge
6. Reichl LE (2009) A modern course in statistical physics, 3rd edn. Wiley-VCH Verlag, Weinheim
7. Parsegian VA (2006) Van Der Waals forces: a handbook for biologists, chemists, engineers, and physicists. Cambridge University Press, Cambridge
8. Pink DA, Razul SG, Gordon T, Quinn B, MacDonald AJ (2015) Computer simulation techniques for modelling statics and dynamics of nanoscale structures, Chapter 9. In: Marangoni AG, Pink DA (eds) Edible nanostructures. The Royal Society of Chemistry, Cambridge
9. https://www.linkedin.com/pulse/why-two-free-energies-what-does-helmholtz-energy-nikhilesh-mukherjee/?trk=pulse-article_more-articles_related-content-card
10. Kirkwood JG (1935) Statistical mechanics of fluid mixtures. J Chem Phys 3:300
11. Hansen JP, McDonald IR (1976) Theory of simple liquids. Academic Press, London
12. Leach AR (1996) Molecular modelling: principles and applications. Addison Wesley Longman Ltd., Harlow
13. Roux B (1995) The calculation of the potential of mean force using computer simulations. Comput Phys Commun 91:275–282
14. Torrie GM, Valleau JP (1974) Monte Carlo free energy estimates using non-Boltzmann sampling: application to the sub-critical Lennard-Jones fluid. Chem Phys Lett 28:578–581
15. Kastner J (2011) Umbrella sampling. Wiley Interdiscip Rev Comput Mol Sci 1(6):932–942
16. https://en.wikibooks.org/wiki/Molecular_Simulation/Potential_of_mean_force
17. Sholl DS, Steckel JA (2022) Density functional theory: a practical introduction. Wiley, Hoboken
18. Billinge SJL (2019) The rise of the X-ray atomic pair distribution function method: a series of fortunate events. Phil Trans R Soc A 377:20180413
19. https://en.wikipedia.org/wiki/Radial_distribution_function
20. Jullien R (1992) From Guinier to fractals. J Phys I France 2:759–770
21. Lach-hab M, Gonzalez AE, Blaisten-Barojas E (1998) Structure function and fractal dimension of diffusion-limited colloidal aggregates. Phys Rev E 57:4520–4527
22. Viscek T (1999) Fractal growth phenomena, 2nd edn. World Scientific, Singapore
23. Sochi T (2011) Slip at fluid-solid interface. Polym Rev 51:309–340
24. Müller M, Katsov K, Schick M (2006) Biological and synthetic membranes: what can be learned from a coarse-grained description? Phys Rep 434:113–176
25. Marrink SJ, Risselada HJ, Yefimov S, Tieleman DP, de Vries AH (2007) The MARTINI force field: coarse grained model for biomolecular simulations. J Phys Chem B 111:7812–7824
26. Bennun SV, Hoopes MI, Xing C, Faller R (2009) Coarse-grained modeling of lipids. Chem Phys Lipids 159:59–66
27. Peter C, Kremer K (2009) Multiscale simulation of soft matter systems–from the atomistic to the coarse-grained level and back. Soft Matter 5:4357–4366
28. de Pablo JJ (2011) Coarse-grained simulations of macromolecules: from DNA to composites. Annu Rev Phys Chem 62:555–574
29. Brini E, Algaer EA, Ganguly P, Li C, Rodriguez-Ropero F, van der Vegt NFA (2013) Systematic coarse-graining methods for soft matter simulations – a review. Soft Matter 9: 2108–2119
30. Saunders MG, Voth GA (2013) Coarse-graining methods for computational biology. Annu Rev Biophys 42:73–93

31. https://en.wikipedia.org/wiki/Coarse-grained_modeling
32. Tirion MM (1996) Large amplitude elastic motions in proteins from a single-parameter, atomic analysis. Phys Rev Lett 77:1905–1908
33. Bahar I, Atilgan AR, Erman B (1997) Direct evaluation of thermal fluctuations in proteins using a single-parameter harmonic potential. Fold Des 2:173–181
34. Lopez CA, Rzepiela AJ, de Vries AH, Dijkhuizen L, Hunenberger PH, Marrink SJ (2009) Martini coarse-grained force field: extension to carbohydrates. J Chem Theory Comput 5:3195–3210
35. Kraus JD (1992) Electromagnetics, 4th edn. McGraw-Hill, New York
36. Wells BA, Chaffee AL (2015) Ewald summation for molecular simulations. J Chem Theory Comput 11:3684–3695
37. Israelachvili JN (1992) Intermolecular and surface forces, 2nd edn. Academic Press, London
38. Metropolis N, Rosenbluth AW, Rosenbluth MN, Teller AH, Teller E (1953) Equation of state calculations by fast computing machines. J Chem Phys 21:1087–1092
39. Binder K (1997) Applications of Monte Carlo methods to statistical physics. Rep Prog Phys 60: 487–559
40. Binder K, Hermann D (2010) Monte Carlo simulation in statistical physics: an introduction. Springer, Berlin
41. Baumgärtner A, Binder K, Hansen JP, Kalos MH, Kehr K, Landau DP, Levesque D, Müller-Krumbhaar H, Rebbi C, Saito Y, Schmidt K (2013) Applications of the Monte Carlo method in statistical physics, vol 36. Springer Science & Business Media, Berlin, Heidelberg
42. Feig M (2010) Molecular simulation methods, Chapter 8. In: Computational modelling in lignocellulosic biofuel production, ACS symposium series, vol 1052. American Chemical Society, Washington, DC, pp 155–178
43. Landau DP, Binder K (2005) A guide to Monte Carlo simulations in statistical physics. Cambridge University Press, Cambridge
44. Pink DA, Quinn B, Peyronel F, Marangoni AG (2013) Edible oil structures at low and intermediate concentrations. I. Modeling, computer simulation, and predictions for X ray scattering. J Appl Phys 114:234901
45. Heertje I, Leunis M (1997) Measurement of shape and size of fat crystals by electron microscopy. LWT Food Sci Technol 30(2):141–146
46. Acevedo NC, Marangoni AG (2010) Characterization of the nanoscale in triacylglycerol crystal networks. Cryst Growth Des 10:3327–3333
47. Acevedo NC, Marangoni AG (2010) Toward nanoscale engineering of triacylglycerol crystal networks. Cryst Growth Des 10:3334–3339
48. Hamaker HC (1937) The London—van der Waals attraction between spherical particles. Physica 4:1058–1072
49. Peyronel F, Ilavsky J, Mazzanti G, Marangoni AG, Pink DA (2013) Edible oil structures at low and intermediate concentrations. II. Ultra-small angle X-ray scattering of in situ tristearin solids in triolein. J Appl Phys 114:234902
50. Yang Y, Pan L, Lightstone FC, Merz KM Jr (2016) The role of molecular dynamics potential of mean force calculations in the methods of enzyme analysis, Chapter 1. In: Voth G (ed) Methods in enzymology, vol 577. Elsevier Science, pp 1–30
51. Flöter E, Wettlaufer T, Conty V, Scharfe M (2021) Oleogels—their applicability and methods of characterization. Molecules 26:1673–1692

Chapter 24
In Vitro Digestion of Lipid-Based Gels

Maya Davidovich-Pinhas

Abbreviations

CLSM	Confocal laser scanning microscopy
DAG	Diacylglyceride
EC	Ethyl-cellulose
FFA	Free fatty acid
GI	Gastrointestinal
HIPE	High internal phase Pickering emulsion
HMOG	High-molecular-weight oil gelator
HPLC	High-performance liquid chromatography
LMOG	Low-molecular-weight oil gelator
MAG	Monoacylglyceride
MCT	Medium chain triacylglyceride
PLM	Polarized light microscopy
SDS-PAGE	Sodium dodecyl sulfate–polyacrylamide gel electrophoresis
SGF	Simulated gastric fluid
SIF	Simulated intestine fluid
SSF	Simulated saliva fluid
TAG	Triacylglyceride

M. Davidovich-Pinhas (✉)
Department of Biotechnology and Food Engineering, Technion-Israel Institute of Technology, Haifa, Israel
e-mail: dmaya@technion.ac.il

C. Andrea Palla, F. Valoppi (eds.), *Advances in Oleogel Development, Characterization, and Nutritional Aspects*,
https://doi.org/10.1007/978-3-031-46831-5_24

24.1 Introduction

The food science community is currently working extensively to develop, study, and explore new mimetic systems aiming to replace animal-based products due to their recognized environmental drawbacks. These attempts have led to an increased interest in fat alternatives that will be combined with the protein matrix. Such alternatives are usually manufactured by replacing the animal-based fat with lipid-based gels such as oleogels, emulsion gels, or bigels. Such systems provide structural integrity similar to original animal-fat due to various structuring approaches while maintaining desirable nutritional profile and mouth feel sensation.

Fats and oils are macronutrients that can be classified based on the triacylglycerides (TAGs) content. Each TAG molecule is built of a triple-alcohol glycerol backbone attached to three fatty acids via an ester bond. Fatty acids can be divided into two major categories: saturated versus unsaturated. Saturated fatty acids typically exhibit higher melting temperatures compared to unsaturated fatty acids. Therefore, fats consist mostly of saturated fatty acids, while oils consist of unsaturated ones [1]. The solid texture presented by saturated fats is driven by the ability of the TAG molecules to self-assemble into lamella crystal structures termed nano-crystal platelets [2]. The platelets further associate through one-dimensional stacking to create larger-scale crystal clusters responsible for the final fat structure and properties [3]. This is a temperature-induced process that results in the crystallization of the solid TAG molecules into a crystal network that physically entraps the liquid TAG molecules. Solid fat properties and functionality are directly correlated to its structural building blocks from the nano- to the meso-scale crystal network [4]. Over the past few decades, the deleterious effects of trans and saturated fatty acids, such as an increased risk for coronary heart disease and a metabolic syndrome, have been well established. However, previous studies have shown that health risks are reduced most effectively when trans and saturated fatty acids are replaced with cis unsaturated fatty acids (usually found in vegetable oils) [5].

Targeting minimum saturated fat content while maintaining desirable textural and sensorial attributes led to the development of oleogels, oil-based gels, that consist of high unsaturated fatty acid content (up to 96%) with preferable solid texture. Oleogels are produced by mixing liquid oil with oil structuring agent that self-organize in various architectures to create a stable three-dimensional (3D) network that stabilizes the liquid oil [6]. Several strategies were previously proposed for oil structuring, depending on the type of oil structuring agent used, which can be classified into two major groups: low- and high-molecular-weight oil gelators (LMOGs and HMOGs). Most oil structuring strategies are based on LMOGs that mimic the natural ability of TAGs to self-assemble and crystallize to form an organized fat structure such as monoacylglycerides (MAGs), free fatty acids (FFAs), and waxes [7]. Less common are oleogel systems based on HMOGs, such as biopolymers and proteins. The incorporation of most proteins or polysaccharides directly in oil phase is not possible due to their hydrophilic nature. Therefore, indirect processes, such as emulsion templet and solvent exchange, are used [8, 9].

Several studies explored the incorporation of water phase into the oleogel phase forming a water-in-oleogel or an emulsion gel system [10–12]. Such systems aim to mimic natural butter or margarine systems composed of 20%wt. water [13] or to reduce the amount of total fat used. Other emulsion gel systems exploit the opposite phase behavior where oil-in-water gels are formed [14]. In these systems, the water phase is composed of a biopolymer-based network where oil droplets are embedded inside it. Previous studies have used polysaccharides such as alginate [15] and flaxseed gum [16] and proteins such as whey [17], chickpea [18], and ovalbumin [19].

A relatively new generation of fat mimetic system further advances the concept of water and oil phase combination by structuring both phases, thus producing biphasic gel system termed bigel. In this system, oleogel and hydrogel are combined into a single consistent matrix with unique physiochemical properties. The biphasic system properties can be altered based on each phase composition and the interphase behavior thus producing a versatile system that can be used in a wide range of applications. Previous studies examined the formulation of bigels using various techniques differing by the mixing scheme and conditions [20–25]. The hydrogel formulations included various biopolymers such as Guar gum [25], xanthan gum [26], alginate [20, 27], pectin [22], hydroxypropyl-methylcellulose [21], κ-carrageenan [28], agar–gelatin mixture [24], gelatin [29], and whey protein [30], while the oleogel formulations were based on low-molecular-weight crystalline materials such as waxes [20, 22, 27, 29], stearic acid [24, 30], sorbitan monostearate [25], γ-oryzanol/phytosterol [26], and glyceryl monostearate [22, 28].

Understanding the way these systems hydrolyze and break down under gastrointestinal (GI) conditions is a crucial step in order to utilize these systems in "real foods" and to further develop and improve them in the future.

24.2 The Digestion Routes for Lipids

Lipid digestion occurs along the gastrointestinal (GI) tract, which includes four distinct sub-stages or processes that run in a series comprising unique control gates from one stage to the other [31]. These stages include oral processing, gastric phase, intestinal stage, and fermentation, which occur in the mouth, stomach, small intestine, and large intestine or colon, respectively. Each stage applies different chemical and physical conditions on the food matrix or digesta aiming to maximize the nutrient breakdown and absorption [32].

The digestion process starts in the mouth where the food is broken down through mechanical processing, i.e., chewing, and mixed with saliva that consists of biopolymers such as mucin, enzymes such as amylase that catalyzes the breakdown of starch to produce glucose, and salts such as sodium and calcium. The mixture formed, termed the bolus, is swallowed and transferred into the stomach [31, 33]. In the stomach, the digesta is subjected to a strong acidic environment typically around pH 2 and peristaltic waves that are responsible for structure

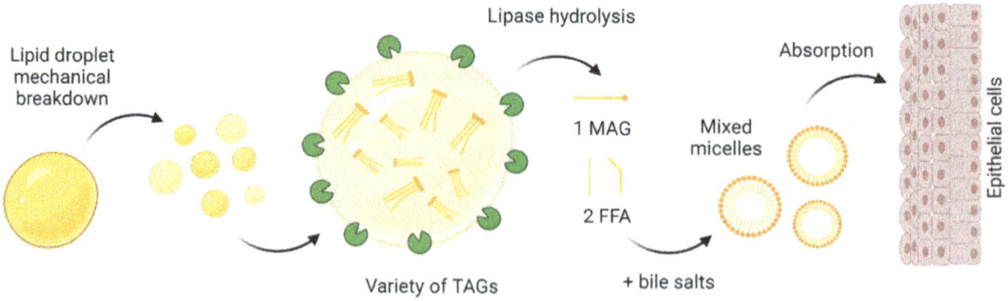

Fig. 24.1 Lipid digestion pathway in the gastrointestinal tract. FFA: Free fatty acid, MAG: monoacylglycerides, TAGs: triacylglycerides (Created with BioRender.com)

breakdown and mixing. Two enzymes are secreted in this stage: pepsin, which is responsible for protein hydrolysis, and gastric lipase, which is responsible for fat lipolysis [34]. The acidic partially broken down digesta is then moved into the intestine where the pH is immediately changed to neutral value, around pH 7, and pancreatin hydrolytic enzyme mixture is secreted in order to break down proteins (protease), fats (pancreatic lipase), and carbohydrates (amylase) [35]. In addition to the hydrolytic enzymes, bile salts are secreted to the intestine to emulsify the fat content, thus contributing to the fat lipolysis. Intestine peristaltic movements are subjected in order to promote content mixing and digesta transport along the intestine. Nutrient absorption mainly occurs in the intestine phase due to the large surface area produced by villi and microvilli structures on the intestine surface [31]. Any hydrolysis products not absorbed in this stage will move to the colon and likely consumed by colonic microbiota. The colon has an anaerobic environment where the microbiota (including probiotic bacteria) can be found. These organisms are responsible for a series of fermentation stages where undigested components such as fibers are broken down mainly to short chain fatty acids, which can be used as a source of energy, and sometimes methane [31, 36].

Lipid digestion is achieved in the stomach and intestine stages, where 10–30% of the fat is digested in the stomach while the rest is digested in the intestine [37]. Overall, lipid digestion involves the breakdown of TAGs into one MAG and two FFAs by three different lipase enzymes (Fig. 24.1). In the stomach, lipolysis is achieved by lingual and gastric lipases that are secreted in the mouth and stomach, respectively. They are only able to catalyze the hydrolysis of one of the fatty acids attached to the glycerol backbone, thus leading to the formation of one FFA and a diacylglyceride (DAG). DAGs are not able to transport through the mucus layer, thus additional breakdown stage is required. Both enzymes are acid stable while their activity rely on their ability to adsorb onto the lipid surface (they do not act with cofactors) [38]. Moreover, the activity of these lipases highly depends on the chain length of attached fatty acids, whereas they are more efficient for short and medium chain length TAGs than long chain length TAGs [37, 39]. In the small intestine pancreatic lipase with the help of a cofactor, co-lipase, catalyzes the hydrolysis of both sn-1 and sn-3 positions of TAGs, thus responsible for the majority of lipid digestion. These byproducts are amphiphilic in nature and thus accumulate at the lipid interface leading to lipase inhibition. Bile salts are important biosurfactants responsible for the solubilization of these byproducts, thus eliminating them from the interface allowing the lipolysis to progress efficiently [38]. These processes lead to the formation of various phases including oil phase, crystalline phase that include calcium soaps, a "viscous isotropic" phase, and a micellar phase [40]. The micellar phase comprises of small and highly dispersed mixed lipid micelles and vesicles (4–8 nm) based on lipid hydrolysates (e.g., FFAs and MAGs), along with bile salts, cholesterol, and phospholipids, that solubilize hydrophobic compounds such as nutrients and drugs [41]. These entities serve as effective shuttles across the viscous unstirred water layer to the intestinal absorptive epithelial cells called enterocytes [42].

24.3 Parameters Affecting Lipid Digestion

The digestion of lipids is a complex process that involves the activity and performance of various entities and phases. Therefore, the process is controlled mainly by parameters such as the lipid phase structure and state, interface behavior, and enzyme activity.

The lipid phase structure and state refer to the TAGs content and type, supramolecular organization that can potentially alter the lipid state, i.e., liquid vs. solid, or macroscopic structures. More specifically, the molecular content of the lipid phase includes saturated, unsaturated fatty acids, or other lipidic components such as phospholipids; the molecular organization in crystal phases, membranes; and the bulk properties such as emulsion, emulsion gels, oleogels, or bigels [43]. Generally, it was shown that solid fats are digested more slowly than liquid oils although both comprise TAGs with different fatty acid content [44].

Lipid hydrolysis is an interfacial process; thus, controlling the physical and chemical properties of the interface as well as the enzyme activity can directly affect the lipid digestion process. Overall, the interface surface area can positively contribute to the lipolysis process while byproducts or bile salts occupying the interface can negatively affect the lipolysis [38]. In addition, the lipase activity along the GI tract plays a major role in the lipid hydrolysis [39]. As discussed above there are three main lipase enzymes responsible for the TAG hydrolysis: lingual, gastric, and pancreatic lipases. The activity of each enzyme can be controlled and manipulated based on its action and environment. Lingual and gastric activities rely on their ability to adsorb directly to the lipid surface and act in acidic environment. Therefore, their activities are mainly governed by the droplet surface area and the interface content and structure. Pancreatic lipase activity, on the other hand, is governed by the action of the cofactor, which allows better anchoring to the lipid interface. Thus, in this case its activity is controlled by the presence of the cofactor and the occupancy of the interface. During the intestine digestion stage FFAs and MAGs occupy the interface, thus hindering the pancreatic lipase activity; thus, bile salts activity as solubilizing agents of the lipolysis byproducts is crucial [38].

24.4 Analysis of Lipid Digestion

Analysis of digestion process raises some challenges relating to the high complexity of the process and the various conditions applied on the food matrix. Investigating food digestion in real life using in vivo models (animals or humans) is complicated, expensive, and sometimes raises ethical issues [45, 46]. Therefore, in order to mimic real digestion conditions, which involves shear and grinding during mastication, dynamic enzyme and bile salts release, physical forces exerted on the digesta, and flow behavior, dynamic in vitro digestion systems were proposed [47]. Several dynamic models were proposed for digestion process, which can be classified into two main groups: mono-compartmental (simulate one compartment of the GI tract) or multicompartmental (simulate a cascade of several compartments) [46]. These models include physicochemical and mechanical processes and temporal changes in luminal conditions as occur in vivo [48]. On the other hand, the static in vitro digestion system allows easy analysis of the digestion process in three successive stages. It suffers from some limitations related to its static nature, which does not reproduce the dynamic processes occurring during human digestion such as gastric emptying, continuous changes in pH, and dynamic ingredients secretion [47, 48]. However, it offers simple, high-throughput, and cost-effective solution that can assist with the initial screening of food hydrolysis during development; thus, most studies up to date use this approach [47].

Over the years various research and review papers dealt with the issue of food in vitro digestion analysis using static model systems [47, 49, 50] and its relation to in vivo results [45]. Various studies implemented various conditions such as sample size, enzyme unit activity, and mixing scheme. Such variation between different publications hindered the ability to compare between studies and provide a reliable progress in the field. Therefore, a standardized procedure was recently published after extensive research and collaboration as part of the COST action INFOGEST [51]. The main protocol developed for static analysis of food digestion involves several important stages required in order to compare and reproduce data between studies [51] (Fig. 24.2). First, all enzymes' activities and bile concentration must be determined prior to analysis in order to maintain comparable conditions. The in vitro analysis should include three stages:

1. *Oral phase*: where food matrix is mixed with simulated saliva fluid (SSF) at 1:1 w/w ratio and incubated while mixing for 2 min at 37 °C and pH 7. SSF solution consists of 1.5 mM $CaCl_2$ and salivary amylase 75 U mL^{-1} (if required).
2. *Gastric phase*: where the oral bolus is mixed with simulated gastric fluid (SGF) at 1:1 v/v ratio and incubated while mixing for 2 h at 37 °C and pH 3. SGF solution consists of 0.15 mM $CaCl_2$, pepsin (2000 U mL^{-1}), and gastric lipase (60 U mL^{-1}).

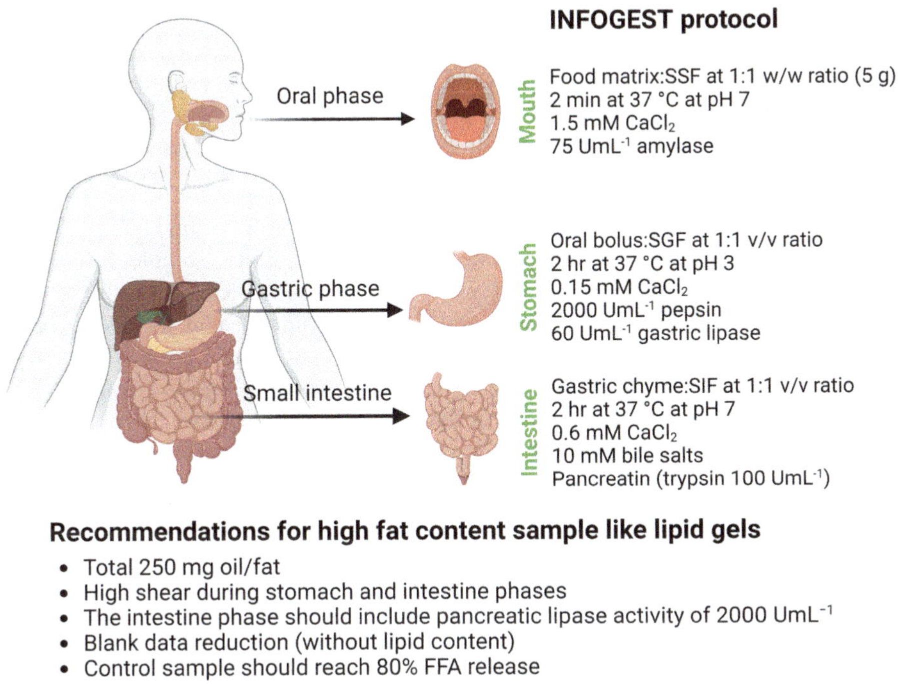

Fig. 24.2 The digestion process stages and conditions used in INFOGEST protocol and recommendation for high-fat lipid gels. (Created with BioRender.com)

3. *Intestinal phase*: where the gastric chyme is mixed with simulated intestine fluid (SIF) at 1:1 v/v ratio and incubated while mixing for 2 h at 37 °C and pH 7. SIF solution consists of 0.6 mM $CaCl_2$, 10 mM bile salts, and pancreatin (trypsin activity 100 U mL^{-1}, or for high-fat-containing foods a pancreatic lipase activity of 2000 U mL^{-1}).

The process analysis is performed based on the research requirements by sampling during the above stages. More specifically, parameters such as protein molecular weight, carbohydrate hydrolysis, particle size, and lipid lipolysis can be analyzed during digestion [52]. While handling lipid systems the main parameters that are usually examined are the FFA release or lipolysis progress, FFA content, droplet size, zeta potential, and micelle size [53].

The standardized protocol suffers from some limitations that can be related to its static nature, which is different from the dynamic natural process involving gradual changes in pH and controlled enzyme and salt secretion. However, the protocol also failed to relate to food systems with high fat content, such as oleogels, emulsion gels, and bigels. Recently, Sabet and coworkers [54] pointed out the limitation of the INFOGEST protocol with respect to oleogel systems and proposed a modification of the protocol for such systems (Fig. 24.2). After thorough examination of various oleogel systems comprising ethyl-cellulose (EC) and various waxes (sunflower seed, rice bran, candelilla, carnauba, and berry waxes) using the in vitro static protocol suggested by INFOGEST and the same protocol with some modifications, important conclusions were obtained. To obtain consistent and reliable data, the authors propose to modify the protocol accordingly:

- The amount of sample should be modified with total oil/fat of maximum 250 mg over 40 mL of total digestion volume (summation of diluted sample, SSF, SGF, and SIF). Therefore, the oleogel/fat sample should be diluted with water to total 5 g sample and then SSF can be added as recommended in the INFOGEST protocol.
- Due to the high stickiness and low dispersibility of the oleogel/fat samples, it is recommended to use high shear during digestion. Moreover, the experiment should be conducted in the static titration chamber with controlled stirring, pH, and temperature to maintain reproducibility of the data and avoid errors.
- Proper blank data (without lipid content) should be subtracted from the data obtained with the lipid content.
- Reliable comparison between results should only be achieved when the control sample (with only oil phase) reaches at least 80% FFA release at the end of the intestinal phase.

24.5 Digestibility of Lipid-Based Gels

Oleogels, emulsion gels, and bigels were proposed for various food applications such as bioactive delivery, food 3D printing, confectionary products, bakery products, spreads, and meats [55]. All these applications involve human consumption and thus physiological breakdown of the lipid components in the GI system. Understanding the lipid lipolysis and the obtained byproducts is therefore crucial for further development of these systems in real food applications.

The first relation for the digestibility behavior was started with the analysis of the lipolysis of EC oleogels using various EC grades (i.e., molecular weight) under in vitro analysis test modulated from lipolysis of emulsions [56]. In this study, EC/canola oil oleogels were prepared with and without β-carotene where the % FFA and β-carotene release were examined. Various EC types and concentration were examined where a linear correlation between the β-carotene transfer and lipolysis extent was found suggesting the release of FFAs and MAGs contributes to the β-carotene micellization. A comparative analysis of the digestibility of various oleogel systems based on different oil structuring agents, which involves different gelation mechanism, was done using EC, MAGs and DAGs mixture (E-471), and β-sitosterol/γ-oryzanol mixture. The results show that in addition to gels' mechanical properties that alter its ability to physically break down, different structuring agents and gelation mechanisms exhibit significantly different susceptibility to digestive lipolysis [57]. Moreover, lipid digestion can be altered using specific structuring agent and concentration in order to tune the susceptibility of a product to digestive degradation [58]. Such approach can be exploited to alter and control the release of hydrophobic nutraceuticals through network stability [59].

The digestibility of various lipid-based gel systems has drawn a lot of attention over the last 5 years, as can be seen in the sharp incline in the number of publications related to this topic (Tables 24.1 and 24.2). It seems that researchers revealed the importance of quantifying the sample functionality with respect to digestibility in addition to other mechanical, thermal, and structural material characterization

Table 24.1 Summary of previous studies using in vitro digestion models used to investigate the digestion and absorption of oleogel systems

Type of oil structuring agent	Oil gelation mechanism	Type of oil	Characterization techniques used for digestibility analysis	References
Ethyl-cellulose	Direct	Canola oil	%FFA release and β-carotene	[56]
E471 (mono- and diglyceride mixture), ethyl-cellulose, β-sitosterol + γ-oryzanol	Direct	Canola oil	%FFA release	[57]
β-sitosterol and lecithin	Direct	Corn oil	%FFA release and curcumin release	[60]
β-sitosterol + γ-oryzanol, saturated MAGs, and rice waxes	Direct	Sunflower oil	%FFA, particle size and zeta potential, and curcumin release	[59]
Candelilla wax	Direct	Nut oils (peanut, pine nut, and walnut oil)	β-carotene release	[61]
Ethyl-cellulose and waxes (sunflower seed, rice bran, candelilla, carnauba, and berry waxes)	Direct	Sunflower oil	FFA content by HPLC and %FFA released	[54]
Gelatin/gellan gum mixture	Indirect	MCT oil	Bioaccessibility of curcumin, %FFA release, and ex vivo everted gut sac-permeability	[62]
Alginate/gelatin	Indirect	Camellia oil	%FFA release	[63]
Hydroxypropyl-methylcellulose	Indirect	MCT oil	Curcumin release and %FFA release; in vivo tests were done by analyzing the curcumin concentration over time in rats	[64]

(continued)

Table 24.1 (continued)

Type of oil structuring agent	Oil gelation mechanism	Type of oil	Characterization techniques used for digestibility analysis	References
Alginate, carboxymethyl cellulose, and pectin	Indirect	MCT oil	Curcumin release and % FFA release	[65]
Soy protein-tannic acid	Indirect	Pine nut oil	%FFA release, lipolysis kinetic analysis, light and confocal micros-copy, particle size and zeta potential, fat bioaccessibility (micellization), fatty acid composition, and oxidative products of the micellar fraction	[66]
Glycerol monostearate	Direct	Sunflower oil	%FFA release and lipol-ysis kinetics, CLSM and PLM, bioaccessibility of astaxanthin, and FFAs composition of micellar fraction	[67]
Soy protein isolate and glycerol monolaurate	Indirect	Soybean oil	%FFA release, particle size, and zeta potential	[68]
Whey protein isolate	Indirect	Sunflower or flaxseed oil	CLSM, protein digest-ibility, %FFA release, and micelle size distribution	[69]
MAGs and waxes (car-nauba wax, beeswax, candelilla wax, rice bran wax)	Direct	High oleic sunflower oil	FFA release of particles, particle size, and zeta potential	[70]
Gelatin	Indirect	Camellia oil	%FFA release, particle size, and zeta potential	[71]
Diosgenin	Direct	Sesame oil, olive oil, and linseed oil	%FFA release	[72]
Bamboo shoot protein and soybean protein isolate	Indirect	Camellia oil	%FFA release, particle size, and zeta potential	[73]
Sorbitan tristearate and nanocellulose	Direct	Coconut oil	5-Aminosalicylic acid release	[74]

CLSM: confocal laser scanning microscopy; FFA: free fatty acid; HPLC: high-performance liquid chromatography; MCT: medium chain triacylglyceride; PLM: polarized light microscopy

Table 24.2 Summary of previous studies using in vitro digestion models used to investigate the digestion and absorption of emulsion gels and bigel systems

Type of oil structuring agent	System type	Type of oil	Characterization techniques used for digestibility analysis	References
Casein hydrogel embedded with milk fat or rapeseed oil	Emulsion gel (oil in hydrogel)	Milk fat or rapeseed oil	Physical breakdown (matrix degradation index) and %FFA release	[77]
Agar hydrogel embedded with soybean oil	Emulsion gel (oil in hydrogel)	Soybean oil	%FFA release, droplet size, and zeta potential during digestion	[78]
Alginate hydrogel embedded with MCT loaded with nobiletin	Emulsion gel (oil in hydrogel)	MCT	%FFA release, nobiletin release and bioaccessibility, and optical and fluorescent microscopy	[15]
κ-carrageenan hydrogel and monoglyceride oleogels	Bigel	Corn oil	β-carotene release	[28]
Chickpea protein cross-linked with transglutaminase embedded with corn oil	Emulsion gel (oil in hydrogel)	Corn oil	Protein breakdown using SDS-PAGE	[18]
Whey protein concentrate hydrogel and stearic acid with/without soy lecithin oleogel	Bigel	Soybean oil	Probiotic protection, % FFA release, and fatty acid content released during digestion	[17]
Whey protein and flaxseed gum hydrogel embedded with corn oil	Emulsion gel (oil in hydrogel)	Corn oil	Astaxanthin release and bioaccessibility	[16]
Ovalbumin hydrogel with sunflower oil	Emulsion gel (oil in hydrogel)	Sunflower oil	Protein digestibility	[19]
Lipophilic protein embedded with soybean oil	Emulsion gel (oil-in-hydrogel) and HIPE	Soybean oil	Particle size and zeta potential, CLSM, FFAs release, and lycopene bioavailability	[79]
Konjac glucomannan/ gelatin hydrogel and stearic acid oleogel	Bigel	Soybean oil	%FFA release, quercetin release, and bioaccessibility	[80]
Ovalbumin hydrogel with different fillers embedded with sunflower oil	Emulsion gel (oil in hydrogel)	Sunflower oil	Protein digestibility	[81]
Myofibrillar protein with carboxymethyl cellulose embedded with soybean oil	Emulsion gel (oil in hydrogel)	Soybean oil	Protein digestibility and % FFAs release	[82]

CLSM: confocal laser scanning microscopy; FFA: free fatty acid; HIPE: high internal phase Pickering emulsion; MCT: medium chain triacylglyceride; SDS-PAGE: sodium dodecyl sulfate–polyacrylamide gel electrophoresis analysis. According to Table 24.1, it is evident that most studies concentrated on the %FFA and drug release; however, more in-depth characterization analyses tools were also suggested such as lipolysis kinetic analysis, light and confocal microscopy, particle size and zeta potential, bioaccessibility (micellization), and fatty acid composition.

Over the last decade the combination of two phases to formulate biphasic lipid-based gels was suggested and explored due to the obvious benefits arising from the presence of two phases, which allows the ability to deliver hydrophilic and hydrophobic drugs, better spreadability, better water and oil binding capacities, protective environment for various drugs, and ability to manipulate the gel mechanical and thermal properties as well as drug release rate [14, 75]. With respect to digestibility process, these systems offer a complex breakdown process arising from the unique composition of each phase. The water phase usually includes water-soluble components that can gel or solidify it by the formation of a polymer-based network such as carbohydrates (polysaccharides, fibers, and starch) and proteins. Oil phase solidification includes low- or high-molecular-weight oil gelators such as waxes, FFAs, MAGs, polysaccharides, and proteins, using direct or indirect gelation routes [76]. Analysis of the in vitro digestibility of these systems usually includes %FFA release, protein hydrolysis, drug release, and microscopy observation analysis (Table 24.2).

24.6 Concluding Remarks

Progress in the field of food design leads to the formulation of various lipid-based gel systems aiming for various edible applications in food, pharmaceutical, and medicine. Such systems can improve food nutritional values by replacing harmful ingredients, deliver hydrophobic micronutrient and drugs through the natural lipid digestion routes thus improving their bioavailability, and protect them from the harsh environment of the stomach. Therefore, understanding the way such systems hydrolyze and break down along the GI tract is crucial. Increased interest in lipid gel systems and their digestibility behavior is evident from the increased amount of publications dealing with this issue in the last 5 years.

Recent standardized protocol for in vitro digestion analysis, the INFOGEST protocol, was published aiming to allow reproducibility and comparability of in vitro studies between different laboratories as well as to compare these data as much as possible to in vivo results. This protocol needs adjustment in order to fit digestion of high-fat food products such as lipid-based gels. Today, most studies dealing with lipid-based gels such as oleogels, emulsion gels, and bigels rely on the INFOGEST protocol with some modifications in order to address the high fat content in lipid gels.

References

1. O'Brien RD, Farr WE, Wan PJ (2000) Introduction to fats and oils technology. AOCS Press, Champaign
2. Acevedo NC, Marangoni AG (2015) Nanostructured fat crystal systems. Annu Rev Food Sci Technol 6:71–96. https://doi.org/10.1146/annurev-food-030713-092400
3. Peyronel F, Pink DA, Marangoni AG (2014) Triglyceride nanocrystal aggregation into polycrystalline colloidal networks: ultra-small angle X-ray scattering, models and computer simulation. Curr Opin Colloid Interface Sci 19:459–470. https://doi.org/10.1016/j.cocis.2014.07.001
4. Marangoni AG (2012) Structure-function analysis of edible fats. AOCS Press, Urbana
5. Mensink RP, Zock PL, Katan MB et al (2003) Effects of dietary fatty acids and carbohydrates on the ratio of serum total to HDL cholesterol and on serum lipids and apolipoproteins: a meta-analysis of 60 controlled trials. Am J Clin Nutr 77:1146–1155. https://doi.org/10.1093/ajcn/77.5.1146
6. Davidovich-Pinhas M (2018) Oleogels. In: Polym gels. Woodhead Publishing, Cambridge, MA, pp 231–249. https://doi.org/10.1016/B978-0-08-102179-8.00008-9
7. Co ED, Marangoni AG (2012) Organogels: an alternative edible oil-structuring method. J Am Chem Soc 89:749–780. https://doi.org/10.1007/s11746-012-2049-3
8. Feichtinger A, Scholten E (2020) Preparation of protein oleogels: effect on structure and functionality. Foods 9(12):1745
9. Davidovich-Pinhas M (2019) Oil structuring using polysaccharides. Curr Opin Food Sci 27:29. https://doi.org/10.1016/j.cofs.2019.04.006
10. Wijarnprecha K, de Vries A, Santiwattana P et al (2019) Microstructure and rheology of oleogel-stabilized water-in-oil emulsions containing crystal-stabilized droplets as active fillers. LWT 115:108058. https://doi.org/10.1016/j.lwt.2019.04.059
11. de Vries A, Jansen D, van der Linden E, Scholten E (2018) Tuning the rheological properties of protein-based oleogels by water addition and heat treatment. Food Hydrocoll 79:100–109. https://doi.org/10.1016/j.foodhyd.2017.11.043
12. da Silva TLT, Arellano DB, Martini S (2019) Effect of water addition on physical properties of emulsion gels. Food Biophys 14:30–40. https://doi.org/10.1007/s11483-018-9554-3
13. Bockisch M (1998) Fats and oils handbook. AOCS Press, Urbana
14. Lin D, Kelly AL, Miao S (2020) Preparation, structure-property relationships and applications of different emulsion gels: bulk emulsion gels, emulsion gel particles, and fluid emulsion gels. Trends Food Sci Technol 102:123–137. https://doi.org/10.1016/j.tifs.2020.05.024
15. Lei L, Zhang Y, He L et al (2017) Fabrication of nanoemulsion-filled alginate hydrogel to control the digestion behavior of hydrophobic nobiletin. LWT Food Sci Technol 82:260–267. https://doi.org/10.1016/j.lwt.2017.04.051
16. Zhang Z, Chen W, Zhou X et al (2021) Astaxanthin-loaded emulsion gels stabilized by Maillard reaction products of whey protein and flaxseed gum: physicochemical characterization and in vitro digestibility. Food Res Int 144:110321. https://doi.org/10.1016/j.foodres.2021.110321
17. Bollom MA, Clark S, Acevedo NC (2021) Edible lecithin, stearic acid, and whey protein bigels enhance survival of probiotics during in vitro digestion. Food Biosci 39:100813. https://doi.org/10.1016/j.fbio.2020.100813
18. Glusac J, Isaschar-Ovdat S, Fishman A (2020) Transglutaminase modifies the physical stability and digestibility of chickpea protein-stabilized oil-in-water emulsions. Food Chem 315:126301. https://doi.org/10.1016/j.foodchem.2020.126301
19. Li R, Xue H, Gao B et al (2022) Physicochemical properties and digestibility of thermally induced ovalbumin–oil emulsion gels: effect of interfacial film type and oil droplets size. Food Hydrocoll 131:107747. https://doi.org/10.1016/j.foodhyd.2022.107747
20. Rehman K, Mohd Amin MCI, Zulfakar MH (2014) Development and physical characterization of polymer-fish oil bigel (hydrogel/oleogel) system as a transdermal drug delivery vehicle. J Oleo Sci 63:961. https://doi.org/10.5650/jos.ess14101

21. Ibrahim MM, Hafez SA, Mahdy MM (2013) Organogels, hydrogels and bigels as transdermal delivery systems for diltiazem hydrochloride. Asian J Pharm Sci 8:48–57. https://doi.org/10.1016/J.AJPS.2013.07.006
22. Lupi FR, Shakeel A, Greco V et al (2016) A rheological and microstructural characterisation of bigels for cosmetic and pharmaceutical uses. Mater Sci Eng C 69:358–365. https://doi.org/10.1016/j.msec.2016.06.098
23. Gonzalez-Gutierrez J, Scanlon MG (2018) Chapter 5 – Rheology and mechanical properties of fats. In: Marangoni A (ed) Structure-function analysis of edible fats, 2nd edn. AOCS Press, Champaign, pp 119–168
24. Wakhet S, Singh VK, Sahoo S et al (2015) Characterization of gelatin–agar based phase separated hydrogel, emulgel and bigel: a comparative study. J Mater Sci Mater Med 26:118. https://doi.org/10.1007/s10856-015-5434-2
25. Singh VK, Banerjee I, Agarwal T et al (2014) Guar gum and sesame oil based novel bigels for controlled drug delivery. Colloids Surf B Biointerfaces 123:582–592. https://doi.org/10.1016/j.colsurfb.2014.09.056
26. Moschakis T, Panagiotopoulou E, Katsanidis E (2016) Sunflower oil organogels and organogel-in-water emulsions (part I): microstructure and mechanical properties. LWT 73:153–161. https://doi.org/10.1016/j.lwt.2016.03.004
27. Martins AJ, Silva P, Maciel F et al (2019) Hybrid gels: influence of oleogel/hydrogel ratio on rheological and textural properties. Food Res Int 116:1298–1305. https://doi.org/10.1016/j.foodres.2018.10.019
28. Zheng H, Mao L, Cui M et al (2020) Development of food-grade bigels based on κ-carrageenan hydrogel and monoglyceride oleogels as carriers for β-carotene: roles of oleogel fraction. Food Hydrocoll 105:105855. https://doi.org/10.1016/j.foodhyd.2020.105855
29. Saffold AC, Acevedo NC (2021) Development of novel rice bran wax/gelatin-based biphasic edible gels and characterization of their microstructural, thermal, and mechanical properties. Food Bioprocess Technol 14:2219–2230. https://doi.org/10.1007/s11947-021-02719-7
30. Bollom MA, Clark S, Acevedo NC (2020) Development and characterization of a novel soy lecithin-stearic acid and whey protein concentrate bigel system for potential edible applications. Food Hydrocoll 101:105570. https://doi.org/10.1016/j.foodhyd.2019.105570
31. Boland M (2016) Human digestion – a processing perspective. J Sci Food Agric 96:2275–2283. https://doi.org/10.1002/jsfa.7601
32. McClements DJ (2016) Food emulsions: principles, practices, and techniques, 3rd edn. CRC Press, Boca Raton
33. Pedersen AM, Bardow A, Jensen SB, Nauntofte B (2002) Saliva and gastrointestinal functions of taste, mastication, swallowing and digestion. Oral Dis 8:117–129. https://doi.org/10.1034/j.1601-0825.2002.02851.x
34. O'Connor A, O'Moráin C (2014) Digestive function of the stomach. Dig Dis 32:186–191. https://doi.org/10.1159/000357848
35. Johnson M, Hillier K (2007) Pancreatin. In: Enna SJ, Bylund DB (eds) xPharm: the comprehensive pharmacology. Elsevier, New York, pp 1–3
36. Carey MC, Small DM, Bliss CM (1983) Lipid digestion and absorption. Annu Rev Physiol 45: 651–677
37. Favé G, Coste TC, Armand M (2004) Physicochemical properties of lipids: new strategies to manage fatty acid bioavailability. Cell Mol Biol (Noisy-le-Grand) 50:815–831. https://doi.org/10.1170/T575
38. Wilde PJ, Chu BS (2011) Interfacial & colloidal aspects of lipid digestion. Adv Colloid Interf Sci 165:14–22. https://doi.org/10.1016/j.cis.2011.02.004
39. Reis P, Holmberg K, Watzke H et al (2009) Lipases at interfaces: a review. Adv Colloid Interf Sci 147–148:237–250. https://doi.org/10.1016/j.cis.2008.06.001
40. Hamosh M (1990) Lingual and gastric lipases: their role in fat digestion. Nutrition 6(6):421–428

41. Chakraborty S, Shukla D, Mishra B, Singh S (2009) Lipid – an emerging platform for oral delivery of drugs with poor bioavailability. Eur J Pharm Biopharm 73:1–15. https://doi.org/10.1016/j.ejpb.2009.06.001
42. Nordskog BK, Phan CT, Nutting DF, Tso P (2001) An examination of the factors affecting intestinal lymphatic transport of dietary lipids. Adv Drug Deliv Rev 50:21–44. https://doi.org/10.1016/S0169-409X(01)00147-8
43. Michalski MC, Genot C, Gayet C et al (2013) Multiscale structures of lipids in foods as parameters affecting fatty acid bioavailability and lipid metabolism. Prog Lipid Res 52:354–373. https://doi.org/10.1016/j.plipres.2013.04.004
44. McClements DJ (2016) Food emulsions principles practices and techniques, 3rd edn. CRC Press Taylor & Francis Group, Boca Raton
45. Bohn T, Carriere F, Day L et al (2018) Correlation between in vitro and in vivo data on food digestion. What can we predict with static in vitro digestion models? Crit Rev Food Sci Nutr 58:2239–2261. https://doi.org/10.1080/10408398.2017.1315362
46. Dupont D, Alric M, Blanquet-Diot S et al (2019) Can dynamic in vitro digestion systems mimic the physiological reality? Crit Rev Food Sci Nutr 59:1546–1562. https://doi.org/10.1080/10408398.2017.1421900
47. Lucas-González R, Viuda-Martos M, Pérez-Alvarez JA, Fernández-López J (2018) In vitro digestion models suitable for foods: opportunities for new fields of application and challenges. Food Res Int 107:423–436. https://doi.org/10.1016/j.foodres.2018.02.055
48. Guerra A, Etienne-Mesmin L, Livrelli V et al (2012) Relevance and challenges in modeling human gastric and small intestinal digestion. Trends Biotechnol 30:591–600. https://doi.org/10.1016/j.tibtech.2012.08.001
49. Minekus M, Alminger M, Alvito P et al (2014) A standardised static in vitro digestion method suitable for food – an international consensus. Food Funct 5:1113–1124. https://doi.org/10.1039/c3fo60702j
50. Egger L, Ménard O, Delgado-Andrade C et al (2016) The harmonized INFOGEST in vitro digestion method: from knowledge to action. Food Res Int 88:217–225. https://doi.org/10.1016/j.foodres.2015.12.006
51. Brodkorb A, Egger L, Alminger M et al (2019) INFOGEST static in vitro simulation of gastrointestinal food digestion. Nat Protoc 14:991–1014. https://doi.org/10.1038/s41596-018-0119-1
52. Bornhorst GM, Gouseti O, Wickham MSJ, Bakalis S (2016) Engineering digestion: multiscale processes of food digestion. J Food Sci 81:R534–R543. https://doi.org/10.1111/1750-3841.13216
53. McClements DJ, Li Y (2010) Review of in vitro digestion models for rapid screening of emulsion-based systems. Food Funct 1:32–59
54. Sabet S, Kirjoranta SJ, Lampi A-M et al (2022) Addressing criticalities in the INFOGEST static in vitro digestion protocol for oleogel analysis. Food Res Int 160:111633. https://doi.org/10.1016/j.foodres.2022.111633
55. Manzoor S, Masoodi FA, Naqash F, Rashid R (2022) Oleogels: promising alternatives to solid fats for food applications. Food Hydrocoll Health 2:100058. https://doi.org/10.1016/j.fhfh.2022.100058
56. O'Sullivan CM, Davidovich-Pinhas M, Wright A et al (2017) Ethylcellulose oleogels for lipophilic bioactive delivery – effect of oleogelation on *in-vitro* bioaccessibility and stability of beta-carotene. Food Funct 8:1438–1451. https://doi.org/10.1039/c6fo01805j
57. Ashkar A, Laufer S, Rosen-Kligvasser J et al (2019) Impact of different oil gelators and oleogelation mechanisms on digestive lipolysis of canola oil oleogels. Food Hydrocoll 97:105218. https://doi.org/10.1016/j.foodhyd.2019.105218
58. Ashkar A, Rosen-Kligvasser J, Lesmes U, Davidovich-Pinhas M (2020) Controlling lipid intestinal digestibility using various oil structuring mechanisms. Food Funct 11:7495–7508. https://doi.org/10.1039/d0fo00223b

59. Calligaris S, Alongi M, Lucci P, Anese M (2020) Effect of different oleogelators on lipolysis and curcuminoid bioaccessibility upon in vitro digestion of sunflower oil oleogels. Food Chem 314:126146. https://doi.org/10.1016/j.foodchem.2019.126146
60. Li L, Wan W, Cheng W et al (2019) Oxidatively stable curcumin-loaded oleogels structured by β-sitosterol and lecithin: physical characteristics and release behaviour in vitro. Int J Food Sci Technol 54:2502–2510. https://doi.org/10.1111/ijfs.14208
61. Li L, Taha A, Geng M et al (2021) Ultrasound-assisted gelation of β-carotene enriched oleogels based on candelilla wax-nut oils: physical properties and *in-vitro* digestion analysis. Ultrason Sonochem 79:105762. https://doi.org/10.1016/j.ultsonch.2021.105762
62. Kavimughil M, Leena MM, Moses JA, Anandharamakrishnan C (2022) 3D printed MCT oleogel as a co-delivery carrier for curcumin and resveratrol. Biomaterials 287:121616. https://doi.org/10.1016/j.biomaterials.2022.121616
63. Li J, Zhang C, Li Y, Zhang H (2022) Fabrication of aerogel-templated oleogels from alginate-gelatin conjugates for in vitro digestion. Carbohydr Polym 291:119603. https://doi.org/10.1016/j.carbpol.2022.119603
64. Chuesiang P, Zhang J, Choi E et al (2022) Observation of curcumin-loaded hydroxypropyl methylcellulose (HPMC) oleogels under in vitro lipid digestion and in situ intestinal absorption in rats. Int J Biol Macromol 208:520–529. https://doi.org/10.1016/j.ijbiomac.2022.03.120
65. Zhang J, Chuesiang P, Kim JT, Shin GH (2022) The role of nanostructured lipid carriers and type of biopolymers on the lipid digestion and release rate of curcumin from curcumin-loaded oleogels. Food Chem 392:133306. https://doi.org/10.1016/j.foodchem.2022.133306
66. Guo Y, Yang X, Bao Y et al (2022) Investigation of the in vitro digestion fate and oxidation of protein-based oleogels prepared by pine nut oil. LWT 164:113660. https://doi.org/10.1016/j.lwt.2022.113660
67. Wang S, Chen K, Liu G (2022) Monoglyceride oleogels for lipophilic bioactive delivery – influence of self-assembled structures on stability and in vitro bioaccessibility of astaxanthin. Food Chem 375:131880. https://doi.org/10.1016/j.foodchem.2021.131880
68. Zou Y, Xi Y, Pan J et al (2022) Soy oil and SPI based-oleogels structuring with glycerol monolaurate by emulsion-templated approach: preparation, characterization and potential application. Food Chem 397:133767. https://doi.org/10.1016/j.foodchem.2022.133767
69. Plazzotta S, Alongi M, De Berardinis L et al (2022) Steering protein and lipid digestibility by oleogelation with protein aerogels. Food Funct 13:10601–10609
70. Li J, Guo R, Wang M et al (2022) Development and characterization of compound oleogels based on monoglycerides and edible waxes. ACS Food Sci Technol 2:302–314. https://doi.org/10.1021/acsfoodscitech.1c00390
71. Li J, Zhang H (2023) Efficient fabrication, characterization, and in vitro digestion of aerogel-templated oleogels from a facile method: electrospun short fibers. Food Hydrocoll 135:108185. https://doi.org/10.1016/j.foodhyd.2022.108185
72. Lu Y, Li J, Ding J et al (2023) Comparison of diosgenin-vegetable oils oleogels with various unsaturated fatty acids: physicochemical properties, *in-vitro* digestion, and potential mechanism. Food Chem 413:135663. https://doi.org/10.1016/j.foodchem.2023.135663
73. Li J, Xi Y, Wu L, Zhang H (2023) Preparation, characterization and in vitro digestion of bamboo shoot protein/soybean protein isolate based-oleogels by emulsion-templated approach. Food Hydrocoll 136:108310. https://doi.org/10.1016/j.foodhyd.2022.108310
74. Qin Z, Kong F (2023) Nanocellulose incorporated oleogel matrix for controlled-release of active ingredients in the lower gastrointestinal tract. Int J Biol Macromol 225:615–624. https://doi.org/10.1016/j.ijbiomac.2022.11.121
75. Shakeel A, Farooq U, Iqbal T et al (2019) Key characteristics and modelling of bigels systems: a review. Mater Sci Eng C 97:932–953. https://doi.org/10.1016/j.msec.2018.12.075
76. Okuro PK, Santos TP, Cunha RL (2021) Compositional and structural aspects of hydro- and oleogels: similarities and specificities from the perspective of digestibility. Trends Food Sci Technol 111:55–67. https://doi.org/10.1016/j.tifs.2021.02.053

77. McIntyre I, Michael O, Dolores O (2017) Altering the level of calcium changes the physical properties and digestibility of casein-based emulsion gels. Food Funct 8:1641–1651
78. Wang Z, Neves MA, Kobayashi I et al (2013) Preparation, characterization, and in vitro gastrointestinal digestibility of oil-in-water emulsion-agar gels. Biosci Biotechnol Biochem 77:467–474. https://doi.org/10.1271/bbb.120659
79. Sun Y, Zhong M, Sun Y et al (2022) Stability and digestibility of encapsulated lycopene in different emulsion systems stabilized by acid-modified soybean lipophilic protein. J Sci Food Agric 102:6146–6155. https://doi.org/10.1002/jsfa.11968
80. Liu L, Abdullah, Tian W et al (2023) Oral sensation and gastrointestinal digestive profiles of bigels tuned by the mass ratio of konjac glucomannan to gelatin in the binary hydrogel matrix. Carbohydr Polym 312:120765. https://doi.org/10.1016/j.carbpol.2023.120765
81. Li R, Wu N, Xue H et al (2023) Effect of filler type on the physicochemical properties, microbial numbers, and digestibility of ovalbumin emulsion gels during storage. Food Funct 14:3779
82. Zhou L, Jiang J, Feng F et al (2023) Effects of carboxymethyl cellulose on the emulsifying, gel and digestive properties of myofibrillar protein-soybean oil emulsion. Carbohydr Polym 309: 120679. https://doi.org/10.1016/j.carbpol.2023.120679

Chapter 25
Preclinical and Clinical Research on Oleogels

Teemu Aitta-aho, Afsane Kazerani García, Saman Sabet, Tiago C. Pinto, and Fabio Valoppi

Abbreviations

^{18}F-FDG	2-Deoxy-2-^{18}F-fluoro-D-glucose
AWS	Automated weighing scales
BAT	Brown adipose tissue
BMI	Body mass index
BW	Body weight
CT	Computer tomography

T. Aitta-aho (✉)
Department of Pharmacology, Faculty of Medicine, University of Helsinki, Helsinki, Finland
e-mail: teemu.aitta-aho@helsinki.fi

A. K. García · T. C. Pinto
Department of Food and Nutrition, University of Helsinki, Helsinki, Finland
e-mail: afsane.a.kazerani@helsinki.fi; tiago.pinto@helsinki.fi

S. Sabet
Department of Food and Nutrition, University of Helsinki, Helsinki, Finland

Helsinki Institute of Sustainability Science, Faculty of Agriculture and Forestry, University of Helsinki, Helsinki, Finland
e-mail: saman.sabet@helsinki.fi

F. Valoppi
Department of Food and Nutrition, University of Helsinki, Helsinki, Finland

Helsinki Institute of Sustainability Science, Faculty of Agriculture and Forestry, University of Helsinki, Helsinki, Finland

Electronics Research Laboratory, Department of Physics, University of Helsinki, Helsinki, Finland

Helsinki Institute of Life Science, University of Helsinki, Helsinki, Finland
e-mail: fabio.valoppi@helsinki.fi

C. Andrea Palla, F. Valoppi (eds.), *Advances in Oleogel Development, Characterization, and Nutritional Aspects*,
https://doi.org/10.1007/978-3-031-46831-5_25

DEXA	Dual energy X-ray absorptiometry
ELISA	Enzyme-linked immunosorbent assay
HDL	High-density lipoprotein
HFD	High-fat diet
HFSD	High fat and sugar diet
HSD	High-sugar diet
LDL	Low-density lipoproteins
MRI	Magnetic resonance imaging
NMR	Nuclear magnetic resonance
PET	Positron emission tomography
TC	Total cholesterol
VLDL	Very-low-density lipoprotein
WAT	White adipose tissue
WHO	World Health Organization

25.1 General Principles in Small Animal Experimentation

25.1.1 Animal Species and Strain Considerations

Laboratory rodent experiments are generally conducted using rats and mice, and more seldom guinea pigs. While historically the rat was the more commonly used rodent species, mice are now used more for several reasons. Mice have advantages over rats in the following factors such as more efficient genetic engineering manipulations, more options on the animal strains enabling a more suitable animal model for a particular study project, and smaller size resulting in lower animal housing and experimental costs.

For typical physiological and nutritional studies, both rats and mice are suitable. When a study plan includes obesity as an endpoint, more consideration is needed to fulfill this specific requirement. Of the most common mouse strains, there is considerable variability in proneness to developing obesity. Parameters to consider include a clear weight gain in a few weeks after initiating the high-fat diet (HFD), an increase in fat mass in comparison to lean mass, growing adipose deposits, and hyperglycemia as monitored by an increase in blood glucose level. Of the typical laboratory mouse strains, C57BL/6 J black mice are a suitable choice in this regard as they fulfill the above criteria. However, based on this common laboratory mouse strain, there are some substrains available, which are not as susceptible to obesity, such as C57BL/6JRj substrain [1]. Other strains prone to obesity are C57BL/6N, DBA/2J, and AKR/J mouse strains [2, 3]. In rats, Zucker rats (wang pmid 35178802) are an obesity-prone model strain with diabetic pathology.

25.1.2 Ethical Permissions

To conduct rodent experiments there is a requirement for ethical permissions, which typically fall into two categories: personal licenses and project licenses. These are both needed prior to the initiation of the laboratory animal study project. Local legislation sets the requirements for both the ethical permission application procedure and taking formal courses before planning and performing the studies. Depending on the local authorities, the time course to fulfilling these requirements can take from a few months to over a year, and therefore it is advised to leave sufficient time for addressing these measures. The reporting of the study results includes a statement of the granted permissions.

25.2 Food Intake, Body Weight, and Metabolism

25.2.1 Food Intake

Overall health assessments often entail doctors and other well-being experts acquainting themselves with the nutritional aspect of the individual under study, as the monitorization of both qualitative and quantitative levels of food intake is vital to understand the effect of alteration in the feeding practices. Due to the high complexity of human studies, rodent models have been historically instrumental in the garnering of numerous insights into fundamental pathophysiological mechanisms that are conserved across species, where mice and rats provide the primary model systems in which the pathophysiology of obesity and its associated pandemic diseases are investigated [4]. To obtain findings that are translatable onto human studies, preparing an experiment based on food intake must take into consideration several factors including the sex, strain, and housing style of the mice. Additionally, animal handlers should be judicious when purchasing, transporting, storing, and handling food to minimize the introduction of diseases, parasites, potential disease vectors such as vermin, and chemical contaminants in animal colonies. The user should know the date of manufacture and other factors that affect the food's shelf life since stale food or food transported and stored inappropriately can become deficient in nutrients [5]. Other considerations must include the type of housing, handling practices, and measurement schedules. For example, natural behavior in multiple-housed cages may bring out the dominant male's tendency to guard over the food source, limiting the ad libitum access to food of the less aggressive ones, whereas individual housing has a strong sex-, strain-, and test-specific effect in mice behavior [6–9].

Quantifying food consumption was explored by Ali and Kravitz [10], who compiled several different methods that mice's food consumption can be measured with. Since these animals' relative daily food intake is small, the cumulative measurement error can be decisive to the representativeness of the data [10]. One way to measure food intake is through manual weighing. This is a simple and cost-effective method, using a scale to measure the food added to the cage before and after feeding periods left in the feeding dish or food hopper. The limiting factor is time and being labor-consuming per measurement. The data obtained will be subjected to errors due to spillage (especially if the food is crumbly) or the presence of urine and feces in the case of placing a feeding dish instead of a food hopper. Another method where continuous food weight data are obtained is through the use of automated weighing scales (AWS), in which highly precise weighing points of the food source are obtained thanks to a load cell or strain gauge [11–13]. Advantages in this method include the obtention of precise data outputs that are representative of feeding patterns [14]. AWS is also suitable for liquid diets and restrictive practices where researchers may regulate the caloric intake based on the weight changes. The downsides of AWS are the need for specialized cages, thus the monetary investment. Additionally, to avoid the common drifting tendencies of the scales, load cells require calibrations and corrections. Spillage may skew results too.

One other method is the use of a pellet dispenser, which is specially indicated to study the learning of operant tasks in rodents, where individuals must perform a certain pre-selected action to obtain a food amount of known mass, although ad lib feeding is also possible with this method [15, 16]. Pellet dispensers can be, for example, activated through a motion sensor that will supply a pellet while recording a timestamp, allowing for high temporal resolution of meal patterns. These devices can be small enough and coded to adjust to requirements, making this method quite cost-effective. The disadvantages of this method include the restriction of the system to deliver only preset-size solid pellets, causing possible jamming of the mechanical device when the pellet size is not uniform or due to dust build-up, thus requiring routine monitoring. This method also allows for hoarding behaviors, where animals retrieve food that is not immediately consumed; this conduct must be monitored and accounted for.

Finally, Ali and Kravitz [10] explored video-based systems. These allow for non-invasive observation and recording of a plethora of behaviors, including but not limited to feeding. The obtained data have a high temporal resolution, and these can be used in home-cages, although adjustments must be done to the camera set and setting. These methods rely on vision algorithms that have not yet been developed to record food consumption in weight terms, and such algorithms must be adapted to each cage's characteristics. The computational and monetary requirements of these methods make them hard to come by, but present very promising potential. Additionally, the video-storing capability of the used systems may limit the length of the study at hand.

25.2.2 *Obesity-Inducing Food Intake*

Obesity is one facet of malnutrition and is now on the rise not only in high-income countries but also worldwide. The World Health Organization (WHO) defines obesity as the abnormal or excessive fat accumulation that may impair health, and its classification is based on the body mass index (BMI), the relation between the weight and the height of an individual. BMI is therefore associated with the amount of fat in the body as well as the increased risks of obesity [17]. It was estimated in 2021 that 39 million infants under 5 are overweight (BMI above 40), closing the gap on their wasting (BMI below 18) counterparts, which accrued to 45 million children globally [18]. As it appears, the everlasting struggle to end world hunger has led society down the equally dangerous road of overconsumption.

Pinpointing the root of this issue is no easy task, since obesity can be the result of a wide array of combined environmental, genetic, and behavioral conditions. One simplistic yet relevant way that food consumption is linked to obesity is through sedentarism, which can be defined as the disproportion between the energy consumed (too much) and the energy expenditure (too little). This is an especially relevant matter when considering animal studies, where researchers have observed that mice and rats housed under traditionally "standard" laboratory conditions, with continuous access to water and food with a limited array of environmental stimulation, became obese and suffered metabolic and cardiovascular changes, such as insulin resistance and higher blood pressure [19, 20]. These standard conditions are applicable to all animals used as controls in basic and translational biomedical research studies. Consequently, all obtained data have been derived from sedentary and obesogenic living conditions, increasing the risk of suffering a premature death. Because of this, alternative housing conditions for rodents have been explored so as to determine how to avoid or mitigate the effect of physical stagnation. Some of the environmental settings under study included limiting food access, increasing the amount and variety of nesting materials, and/or adapting their cages with additional vertical space. Such isolated conditions resulted in lower adiposity, lower glucose intolerance, lower incidence of tumors, and overall enhancing the life quality, without compromising the variability of the data [21, 22]. These alternative housing conditions are of special interest for defining better the health impact of food consumption and should be taken into consideration when defining future animal research plans.

Other studies have solely focused on the genetic predisposition of an individual to obesity, consolidating obesity as a symptom of the expression, through environmental interaction, of an intricate gene collection [23]. Such genetic predisposition was accountable for up to 70% of the obese mice and strains such as C57BL/6J, C57L/J, C3H/HeJ, AKR/J, and A/J; DBA/2J were found to be especially susceptible to obesity when fed a high-fat diet (HFD), whereas other strains like SJL/J, I/STN, and SWR/J were obesity-resistant, indicating that certain combinations of alleles are more obesogenic than others, and these may be particularly expressed when exposed to HFD [24–27]. The convoluted manner that sex, age, metabolic-, embryonic-, and tissue-health interconnect food intake to obesity makes this issue truly Sisyphean. Particularly relevant in this matter is hormonal, as it has been observed in C57BL/6J mice that, unlike their male counterparts, females are generally protected against age-related obesity and insulin resistance, but such defense is eliminated when ovariectomized [28–30]. Equivalently, interfering with other body tissues can result in food intake imbalances and ultimately trigger obesogenic traits. This effect has been studied on rats by King et al. [31], who showed overeating patterns and excessive weight gain after sustaining damage in their posterodorsal aspects of the medial amygdala [31]. Maternal nutrition has been also studied for its effects on their offspring in pigs and rodents, resolving into a wide array of outcomes dependent on the type and timing of early malnutrition [32–34].

Finally, yet outstanding, is the complex role of gut microbiota during digestion and how it is itself affected by all the above. Several pathways link microbial metabolism and dietary regulation, including but not limited to selective fermentation of nutrients, expression of genes associated with health issues, and disruption of intestinal functions [35, 36]. Generally, gut microbes improve energy yield obtained from food by degrading the particles undigested by the host. Microbiota also improves the immune response at the gut lining and the intestinal permeability, deeply impacting the body's health [37, 38]. In humans, intestinal microbes are affected by both long- and short-term dietary regulations. Gut microbiota changes substantially through the aging process along with the host's immune system, forming a delicate balance that becomes more susceptible to infections. When comparing human gut health to mice, there are obvious physical differences as well as quantitative and qualitative microbial disparities, yet the murine digestive system is affected in a similar manner to that of humans by diet, environmental factors, and host genotype [39].

25.2.3 *Diet-Inducing Obesity*

Giving special diets to obesity-prone strains (see previous section) is the go-to method for researchers who want to focus solely on the food's effect on health and the possible treatments for human conditions. This measure is used to expedite the effects of such diets, shortening the length of the study as well as lowering its cost. Some of the most frequent diets contain either a 30% caloric content derived from—mostly refined—sugars also known as high-sugar diets (HSD), or a 35% caloric content from—mostly saturated—fats, high-fat diets (HFD), or a combination of both, that is, high fat and sugar diets (HFSD). However, it should be noted that even normal laboratory animals with ad libitum food intake of normal animal diet low in fat content develop obesity. Therefore, keeping mice or rats in normal laboratory conditions presents a valid method of modeling humans slowly generating obesity syndrome.

25.2.3.1 Genetical Obesity Models

In addition to diet-induced obesity models, there are several genetical models available. These models are based on the genetical deletion of a gene for a factor that has been found important in body weight (BW) regulation. The most used genetical models of obesity include leptin gene-deficient animals (ob/ob), leptin receptor gene-deficient animals (db/db), and melanocortin 4 receptor gene-deficient animals (mc4r) [40]. The validity of these models can be contrasted with that of diet-induced obesity since obesity is unlikely a product of a single gene. Nevertheless, these models offer a possibility to study highly progressive and early obesity. For example, mice with deletions in leptin and leptin receptor gene lead to obesity already from the age of weaning (4-week old, before adolescence), and therefore this correlates well with human leptin gene deficiency syndrome in which obesity begins before adolescence [41].

25.2.4 Body Weight

Adult human's relative body weight (BW) stability is governed by biochemical mechanisms yet to be completely unraveled. An example of inherent BW regulation is the popular "yo-yo effect" (weight cycling), in which people who undergo diet-induced weight reduction regain the lost weight over time. This happens in other species too, as for example, male mice fed ad libitum following caloric restriction-induced weight loss demonstrated gains in both visceral and subcutaneous fat stores. However, female mice were less capable of regaining visceral fat after weight loss [42, 43]. According to Zhang et al. [27], research C57BL/6J mice's initial body fatness was the strongest predictor of the variability in weight gain independent of HFD feeding duration. Additionally, the baseline physical activity and fat-free mass were associated with late-stage body weight gain yet baseline food intake, resting metabolic rate, and body temperature were not significant predictors at any of the timepoints [27].

Encoded genetic and developmental factors play a role in body weight regulation. Level of circulating hormone that inhibits hunger (i.e., leptin), thus regulating energy balance, is a major signal of overall energy availability. The perceived relative leptin insufficiency of weight-reduced individuals is at least partially responsible for a hypometabolic phenotype [44]. Nonetheless, this is an asymmetric regulatory response, as higher leptin levels in the blood do not induce a response from the central nervous system to regulate accordingly the food intake or energy expenditure. This asymmetry is shown in both human and rodent studies, proving how sustaining weight loss encompassed a lower metabolic rate that was sustained over long periods of time if not indefinitely [45, 46]. In contrast, an extended rise in body weight resulted in the preservation of a high level of body fatness compared to never-obese control mice [47].

25.2.5 Energy Expenditure

Rodent metabolism studies include often not only food intake and body composition analyses, but also they are accompanied by studies on energy expenditure, an important part of the metabolism data [48]. With the advent of commercial indirect calorimetric studies, these measurements are straightforward although they require specialized equipment. For the determination of energy expenditure, mice are single-housed, and habituated to the measurement chambers for a few days, minimum one day [48]. Typical systems simultaneously measure oxygen consumption and carbon dioxide production, and automatically calculate energy expenditure based on these values. In addition, commercial systems are often accompanied by built-in scales for food and water intake measurement, and some systems harbor a possibility to measure locomotor activity in the animals. To correctly take the body weight and body composition into account, analysis of energy expenditure should be performed using statistical methods such as a generalized linear model or analysis of covariance to compare groups. With these statistical approaches, several variables such as genotype, gender, body mass, lean and fat mass, or even masses of individual organs can be included as variates or covariates for a more comprehensive and reliable statistical turnover [49].

25.3 In Vivo Body Composition Analysis

Body composition at a macroanatomical level refers to the separation of fat mass from lean mass. At a subanatomical level, fat mass can be detected in specific compartments, such as abdominal fat and subcutaneous fat. More specifically, also the division of white adipose tissue (WAT) and brown adipose tissue (BAT) can be separated depending on the utilized imaging technology. Preclinical laboratory animal studies benefit from longitudinal body composition analysis, which can reveal diet-induced changes in adipose tissue accumulation. Advances in intravital, that is, living animal imaging technology, modern imaging methods have replaced endpoint carcass analysis of fat and lean mass content for body composition analysis [50]. These imaging approaches have benefits such as the possibility to monitor the progress of the experiment and an increase in statistical power due to capabilities of within-subject statistical calculations. These contribute to the first and second *R*s of the "animal experimentation 3Rs ethics of reduction, replacement, and refinement" [51]. The most common imaging methods for fat content analysis are described here.

Magnetic resonance imaging (MRI) dissociates fat and water based on their magnetic resonance characteristics, enabling a macroanatomical level observation of body composition [52]. Spatial resolution suffices for substructural analysis, which permits the separation of subcutaneous and abdominal fat deposits. MRI is capable of distinguishing typical WAT deposits from BAT [53], which localizes to the interscapular adipose area. Intrahepatocellular lipid analyses are also possible with a high correlation to triglyceride content. These features increase the combined versatility of the imaging method.

Magnetic resonance imaging is an imaging application of nuclear magnetic resonance (NMR) [52]. NMR can be utilized on its own for body composition analysis. NMR-based body composition analysis provides a very quick (<2 min), less expensive, and ease-of-use method for non-imaging-based body composition analysis in unanesthetized animals [54]. However, this method does not provide data on anatomical regional fat deposits or separate WAT and BAT.

Computer tomography (CT or micro-CT in small animal terminology) uses X-rays to dissociate fat tissue due to higher carbon/hydrogen content compared to other soft tissues and more dense tissues such as bones [55]. Differential attenuation of X-ray by anatomical structures forms the contrast in the CT images. CT-emitted X-rays are received on a circular detection coil, providing three-dimensional images by computed image processing. Anesthesia is required, and scans are relatively fast (<10 min) depending on the scanned body area. Typically, scan times can be shortened by scanning only the regions of interest. The acquired image displays high-resolution spatial separation, therefore enabling the sorting of the abdominal and subcutaneous fat deposits, and even smaller adipose collections, in a reliable manner [56]. Hepatic steatosis, the collection of fat in the liver, can be measured in CT scanned images by comparison of the liver-to-spleen density ratio [56].

Dual energy X-ray absorptiometry (DEXA) operates on the same principle as CT [57]. Compared to CT scanning, the image provided has two dimensions, as the X-ray is emitted and received in one axis only [55]. Despite these shortcomings, DEXA provides a good measure of the fat and lean tissue mass composition [58]. This method compares to the NMR body composition analysis, although DEXA requires anesthesia and is slower.

Positron emission tomography (PET) operates on radioactive tracers that are typically radiochemical-derived endogenous-like substances such as glucose. Due to this feature, PET imaging has an advantage in imaging highly more specific targets compared to the above discussed methods. On the other hand, there is a requirement for a radiochemical unit to synthesize the radiotracers as well as a need for a skilled animal technician to perform tail-vein tracer injections [59]. This unit should be located logistically close to the PET imager and animal facility, as many of the radiotracers have short radionuclide half-life [60]. For example, a common radiotracer 2-deoxy-2-^{18}F-fluoro-D-glucose (^{18}F-FDG) has a half-life of 110 min [52]. For the metabolic analysis, ^{18}F-FDG is taken up by metabolically active cells and does not externalize before radioactive decay. These properties make it a useful tracer for imaging the metabolic rate and it is also the most important radiotracer for BAT metabolism [61]. For BAT imaging, alternative radiotracer strategies have been adopted by using a combination of tracers: ^{18}F-FDG (glucose uptake), ^{11}C-acetate (oxidative activity), and ^{18}FTHA (non-esterified fatty acids). These could provide a more comprehensive BAT status [62].

25.4 Intestinal Permeability Measurements, Preclinical Chemistry, and Histology

25.4.1 *Intestinal Permeability Measurements*

For a specific analysis of intestinal permeability and integrity, Ussing chamber system provides a feasible method when studying the effect of various nutrients, fat modifications, inflammatory factors, and obesity models [63]. While Ussing chamber method operates on ex vivo biopsy samples, gut integrity can be studied also in vivo using measurements of, for example, sugars or dextran tracers in blood samples or in urine [64]. These in vivo protocols can give an overall picture of gut integrity but do not provide detailed data and lack information on the various regional differences in the intestine. In the Ussing system, the gut biopsy sample is secured so that each side of the biopsy is isolated. This enables the measurement of both chemical tracer movements and electrophysiological readings. Chemical tracer molecule transition from luminal to serosal side offers a readout for gut permeability. Electrophysiological measurements are based on concentrations between isolated sides using Ag/AgCl electrodes. A low transepithelial resistance is considered an indication of increased permeability [63].

25.4.2 *Lipids, Glucose, and Insulin*

Nutritional modifications in experimental research can often affect various biochemical markers such as lipids, glucose, insulin, and hepatic markers. These are measured in a variable manner from blood samples using blood plasma as an analytical matrix. Blood samples can be drawn as the endpoint analysis when sacrificing the mice using cardiac puncture [65], but then only one timepoint can be measured. For longitudinal repeated analyses, saphenous vein blood sampling is advised, while the limiting factor is the low blood volume [66]. In general, the markers are measured in either handheld or tabletop devices with commercial biochemical reagent kits. Usually, blood samples are stored at ≤ -20 °C for long-term storage (several months).

Lipids, lipoproteins, and apolipoproteins such as cholesterol-containing lipoproteins and triglycerides are determined by colorimetric assays, which are based on chemical reactions yielding colored products after chemical reaction between the analyzed lipid and reaction substance. The intensity of the color reaction can be measured spectrophotometrically, with lipid concentration directly correlating with the optical density as determined at a specific wavelength. Triglycerides are measured colorimetrically [67], and typically kits and commercial solutions are available. Cholesterol can be measured as total plasma cholesterol or separated into divisions according to lipoprotein status (high-density lipoprotein [HDL], low-density lipoprotein [LDL], very-low-density lipoprotein [VLDL]). Total cholesterol (TC) in plasma samples may be analyzed by enzymatic methods [68], and commercial systems and kits are offered. For HDL determination, precipitation methods are available [69]. Non-HDL cholesterol can be calculated by subtraction of HDL from the total cholesterol value. Various lipoprotein components (HDL, LDL, VLDL) can be also separated by ultracentrifugation at 40,000 rpm.

Glucose in blood samples can be determined by many different methods; however, a handheld glucometer used normally by diabetic patients is a convenient way for analyses in animal blood samples. For example, a saphenous vein blood sample is sufficient to reach a reliable sample volume for such analysis.

Insulin in mouse blood can be measured using enzyme-linked immunosorbent assay (ELISA). These are based on the immunological determination of insulin in the blood sample, finally producing a color reaction, which can be detected on plate readers.

25.4.3 Histological Analysis of Adipose Tissue, Intestine, Liver, Heart

Histological samples of adipose tissue, intestine, liver, heart, and other tissue provide anatomical dimensions for the data analysis, and therefore have an advantage when compared to biochemical markers. Nonetheless, the biochemical determination of markers offers quantitative readouts of substances, which might be difficult to get from histological samples. Histological samples, on the other hand, can provide quantitative readouts when, for example, the number of cells is counted and statistically analyzed. Histological sampling takes place at the conclusion of the experiment and is not suitable for longitudinal analysis. Samples stored are either fresh frozen, fixed with a suitable fixative, or can be further paraffin-embedded. Tissue samples or smaller specific blocks are cut into slices of variable thickness based on the requirement of the histological method. The slicing procedure can be performed with vibratomes, microtomes for non-frozen tissue, or with cryostat for frozen tissue. The choice of the pre-processing methodology is determined by histological tissue type and histological staining method. Both non-immunological and immunological histological methods are available. Non-immunological methods such as Nissl staining using, for example, thionin are used for contrasting purposes to aid microscopic imaging. Immunohistochemical methods based on specific primary and secondary antibodies followed by color or fluorescent probes allow multiple simultaneous markers to be analyzed in the same tissue section. Histological analysis technology requires microscopic equipment.

25.5 Preclinical and Clinical Studies on Oleogel-Containing Diets

25.5.1 Preclinical Studies on Oleogels

Most in vivo tests on oleogels were preclinical studies on rats. It has been shown that the substitution of a 50% beef tallow containing high-fat diet by beeswax oleogel-containing canola oil significantly reduced body weight in obese diabetic Zucker rats in 2 months [70]. The authors also reported that this change in the diet significantly ($p < 0.05$) reduced the serum triglyceride and increased the serum HDL cholesterol [70]. Their results also suggested that oleogels effectively lower adipogenesis (forming fat cells from stem cells) and improve angiogenesis (forming new blood vessels) in an obese rat model [70], which plays a role in preventing obesity and cardiovascular diseases. Oleogels also improve hepatic enzyme activity and contribute to the prevention of liver steatosis [70].

In another study, it has been shown that if rats are fed with oleogel made of 15 wt % gelators of 2:1 molar ratio of palm stearin with cetyl laurate or palm stearin with cetyl caprylate, the LDL cholesterol and triglyceride serum level were significantly ($p < 0.05$) lower compared to the rats fed with control oils (blend of rice bran oil and flaxseed oil, 4:1 wt:wt) or high trans-fatty-acid-rich vanaspati (vegetable oil that has been hydrogenated and hardened, with about 11.2% trans fatty acid content) [71]. In contrast, the authors showed that the rats fed with oleogels had significantly higher serum levels of HDL. High ratios of total cholesterol (TC) to HDL and LDL to HDL indicate higher cardiovascular risk [72]. In their in vivo study, the authors showed that TC/HDL and LDL/HDL ratios were also significantly ($p < 0.05$) lower in rats fed with oleogels, whereas significantly higher in rats fed with high trans-fatty-acid-rich vanaspati [71]. In addition, the liver lipid profile of rats showed that feed containing oleogels led to a lower level of total cholesterol, LDL, and triglyceride and significantly higher HDL cholesterol than rats fed with control oil and vanaspati [71]. However, the authors concluded that there was no significant ($p > 0.05$) difference between the body weight gain of the group fed with control oil and the group fed with oleogels.

In another study, the health impact of oleogel (5 wt% rice bran wax in rice bran oil) compared to margarine was studied using a high-fat diet feeding for 4 weeks [73]. The results showed the consumption of oleogel caused a decrease in adipose tissue accumulation and liver and serum triglyceride levels, as well as increased excreted triglyceride, total cholesterol, and bile acid contents in feces compared with rats fed with the unstructured oil containing 5 wt% rice bran wax and margarine [73]. The authors showed oleogel decreased serum and liver triglyceride levels by approximately 30% and increased excreted triglyceride levels by nearly 30% compared with the unstructured oil containing 5 wt% rice bran wax. This implied the viscoelastic solid-like network contributes to the decrease in lipid digestibility [73]. They concluded that although the oleogel did not show anti-obesity effects, it showed hypolipidemic-, hypocholesterolemia-, and liver-protective effects [73].

25.5.2 Clinical Studies on Oleogels

Only a few studies have been done to reveal the health impact of the consumption of oleogels on humans. For example, a clinical study on 12 young, healthy individuals (6 male and 6 female) showed that the mean serum triglyceride and free fatty acid levels after the consumption of oleogel (made of 2 wt% 12-hydrostearic acid in canola oil) were significantly (unidentified *p*-value) lower than that after consumption of butter and margarine but not significantly different from liquid canola oil [74]. Another clinical study on 16 healthy young adult males showed that consuming coconut oil in an oleogel form (using 11 wt% ethylcellulose) when co-ingested with a carbohydrate-rich meal reduces the postprandial triglyceride excursion significantly ($p = 0.001$) compared to consumption of liquid coconut oil [75].

25.6 Conclusions

Prior to clinical trials, a wide range of preclinical studies can be executed to help predict the outcome of the clinical studies. Additionally, specific nutritional and metabolic effects of oleogel-containing diets can be investigated in preclinical experimental settings in high detail. Translation and reverse translation from preclinical rodent studies to human trials and vice versa is possible since the methods described in this chapter are available for both preclinical and clinical studies.

Acknowledgments The authors acknowledge Jane and Aatos Erkko Foundation (grant number 200075), the University of Helsinki (decision letter number HY/217/05.01.07/2020), Business Finland (project number 1871/31/2021), Academy of Finland (decision number 350193), and Sigrid Juselius Foundation for their funding. Fabio Valoppi also acknowledges the European Union's Horizon 2020 research and innovation program funding under the Marie Skłodowska-Curie grant agreement No. 836071.

References

1. Kern M, Knigge A, Heiker JT, Kosacka J, Stumvoll M, Kovacs P, Bluher M, Kloting N (2012) C57bl/6jrj mice are protected against diet induced obesity (dio). Biochem Biophys Res Commun 417:717–720. https://doi.org/10.1016/j.bbrc.2011.12.008
2. Podrini C, Cambridge EL, Lelliott CJ, Carragher DM, Estabel J, Gerdin AK, Karp NA, Scudamore CL, Sanger Mouse Genetics, Ramirez-Solis R, White JK (2013) High-fat feeding rapidly induces obesity and lipid derangements in c57bl/6n mice. Mamm Genome 24:240–251. https://doi.org/10.1007/s00335-013-9456-0
3. Wang CY, Liao JK (2012) A mouse model of diet-induced obesity and insulin resistance. 821: 421–433. https://doi.org/10.1007/978-1-61779-430-8_27
4. Even PC, Virtue S, Morton NM, Fromentin G, Semple RK (2017) Editorial: are rodent models fit for investigation of human obesity and related diseases? Front Nutr 4:58. https://doi.org/10.3389/fnut.2017.00058

5. National Research Council (US) Committee for the Update of the Guide for the Care and Use of Laboratory Animals (2011) Guide for the care and use of laboratory animals: Eighth edition. National Academies Press (US), Washington, DC
6. Võikar V, Polus A, Vasar E, Rauvala H (2005) Long-term individual housing in c57bl/6j and dba/2 mice: assessment of behavioral consequences. Genes Brain Behav 4:240–252. https://doi.org/10.1111/j.1601-183X.2004.00106.x
7. Arndt SS, Laarakker MC, van Lith HA, van der Staay FJ, Gieling E, Salomons AR, van't Klooster J, Ohl F (2009) Individual housing of mice – impact on behaviour and stress responses. Physiol Behav 97:385–393. https://doi.org/10.1016/j.physbeh.2009.03.008
8. Hunt C, Hambly C (2006) Faecal corticosterone concentrations indicate that separately housed male mice are not more stressed than group housed males. Physiol Behav 87:519–526. https://doi.org/10.1016/j.physbeh.2005.11.013
9. Späni D, Arras M, König B, Rülicke T (2003) Higher heart rate of laboratory mice housed individually vs in pairs. Lab Anim 37:54–62. https://doi.org/10.1258/002367703762226692
10. Ali MA, Kravitz AV (2018) Challenges in quantifying food intake in rodents. Brain Res 1693: 188–191. https://doi.org/10.1016/j.brainres.2018.02.040
11. Hulsey MG, Martin RJ (1991) A system for automated recording and analysis of feeding behavior. Physiol Behav 50:403–408. https://doi.org/10.1016/0031-9384(91)90086-4
12. Meguid MM, Kawashima Y, Campos ACL, Gelling PD, Hill TW, Chen T-Y, Yang Z-J, Hitch DC, Hammond WG, Mueller WJ (1990) Automated computerized rat eater meter: description and application. 48:759–763. https://doi.org/10.1016/0031-9384(90)90222-P
13. Minematsu S, Hiruta M, Taki M, Fujii Y, Aburada M (1991) Automatic monitoring system for the measurement of body weight, food and water consumption and spontaneous activity of a mouse. J Toxicol Sci 16:61–73. https://doi.org/10.2131/jts.16.61
14. Farley C, Cook JA, Spar BD, Austin TM, Kowalski TJ (2003) Meal pattern analysis of diet-induced obesity in susceptible and resistant rats. Obes Res 11:845–851. https://doi.org/10.1038/oby.2003.116
15. Gill K, Mundl WJ, Cabilio S, Amit Z (1989) A microcomputer controlled data acquisition system for research on feeding and drinking behavior in rats. Physiol Behav 45:741–746. https://doi.org/10.1016/0031-9384(89)90288-6
16. Aponte Y, Atasoy D, Sternson SM (2011) Agrp neurons are sufficient to orchestrate feeding behavior rapidly and without training. Nat Neurosci 14:351–355. https://doi.org/10.1038/nn.2739
17. Revicki DA, Israel RG (1986) Relationship between body mass indices and measures of body adiposity. Am J Public Health 76:992–994. https://doi.org/10.2105/ajph.76.8.992
18. United Nations Children's Fund (UNICEF), World Health Organization & International Bank for Reconstruction and Development/The World Bank (2021) Levels and trends in child malnutrition: key findings of the 2021 edition of the joint child malnutrition estimates. World Health Organization, Geneva
19. Martin B, Ji S, Maudsley S, Mattson MP (2010) "Control" laboratory rodents are metabolically morbid: why it matters. Proc Natl Acad Sci U S A 107:6127–6133. https://doi.org/10.1073/pnas.0912955107
20. Oscai LB, Holloszy JO (1969) Effects of weight changes produced by exercise, food restriction, or overeating on body composition. J Clin Invest 48:2124–2128. https://doi.org/10.1172/jci106179
21. Hatori M, Vollmers C, Zarrinpar A, DiTacchio L, Bushong EA, Gill S, Leblanc M, Chaix A, Joens M, Fitzpatrick JA, Ellisman MH, Panda S (2012) Time-restricted feeding without reducing caloric intake prevents metabolic diseases in mice fed a high-fat diet. Cell Metab 15:848–860. https://doi.org/10.1016/j.cmet.2012.04.019
22. Bailoo JD, Murphy E, Boada-Saña M, Varholick JA, Hintze S, Baussière C, Hahn KC, Göpfert C, Palme R, Voelkl B, Würbel H (2018) Effects of cage enrichment on behavior, welfare and outcome variability in female mice. Front Behav Neurosci 12:232. https://doi.org/10.3389/fnbeh.2018.00232

23. Speakman JR (2004) Obesity: the integrated roles of environment and genetics. J Nutr 134: 2090S–2105S. https://doi.org/10.1093/jn/134.8.2090S
24. Barsh GS, Farooqi IS, O'Rahilly S (2000) Genetics of body-weight regulation. Nature 404:644–651. https://doi.org/10.1038/35007519
25. Hu CC, Qing K, Chen Y (2004) Diet-induced changes in stearoyl-coa desaturase 1 expression in obesity-prone and -resistant mice. Obes Res 12:1264–1270. https://doi.org/10.1038/oby.2004.160
26. West DB, Boozer CN, Moody DL, Atkinson RL (1992) Dietary obesity in nine inbred mouse strains. Am J Physiol Regul Integr Comp Physiol 262:R1025–R1032. https://doi.org/10.1152/ajpregu.1992.262.6.R1025
27. Zhang LN, Morgan DG, Clapham JC, Speakman JR (2012) Factors predicting nongenetic variability in body weight gain induced by a high-fat diet in inbred c57bl/6j mice. Obesity (Silver Spring) 20:1179–1188. https://doi.org/10.1038/oby.2011.151
28. Dalton A, Calzini L, Tuluca A, Ives S, Reynolds T (2018) The effects of age and sex on obesity and insulin action in c57bl/6j mice: 956 board #217 may 30 2. Med Sci Sports Exerc 50:223. https://doi.org/10.1249/01.mss.0000535820.70552.aa
29. Hong J, Stubbins RE, Smith RR, Harvey AE, Núñez NP (2009) Differential susceptibility to obesity between male, female and ovariectomized female mice. Nutr J 8:11. https://doi.org/10.1186/1475-2891-8-11
30. Becerril S, Rodríguez A, Catalán V, Ramírez B, Mentxaka A, Neira G, Gómez-Ambrosi J, Frühbeck G (2022) Sex- and age-dependent changes in the adiponectin/leptin ratio in experimental diet-induced obesity in mice. Nutrients 15:73. https://doi.org/10.3390/nu15010073
31. King BM, Cook JT, Dallman MF (1996) Hyperinsulinemia in rats with obesity-inducing amygdaloid lesions. Am J Phys 271:R1156–R1159. https://doi.org/10.1152/ajpregu.1996.271.5.R1156
32. Barbero A, Astiz S, Lopez-Bote CJ, Perez-Solana ML, Ayuso M, Garcia-Real I, Gonzalez-Bulnes A (2013) Maternal malnutrition and offspring sex determine juvenile obesity and metabolic disorders in a swine model of leptin resistance. PLoS One 8:e78424. https://doi.org/10.1371/journal.pone.0078424
33. Bieswal F, Ahn MT, Reusens B, Holvoet P, Raes M, Rees WD, Remacle C (2006) The importance of catch-up growth after early malnutrition for the programming of obesity in male rat. Obesity (Silver Spring) 14:1330–1343. https://doi.org/10.1038/oby.2006.151
34. Remacle C, Bieswal F, Reusens B (2004) Programming of obesity and cardiovascular disease. Int J Obes Relat Metab Disord 28(Suppl 3):S46–S53. https://doi.org/10.1038/sj.ijo.0802800
35. Kau AL, Ahern PP, Griffin NW, Goodman AL, Gordon JI (2011) Human nutrition, the gut microbiome and the immune system. Nature 474:327–336. https://doi.org/10.1038/nature10213
36. Shulzhenko N, Morgun A, Hsiao W, Battle M, Yao M, Gavrilova O, Orandle M, Mayer L, Macpherson AJ, McCoy KD, Fraser-Liggett C, Matzinger P (2011) Crosstalk between b lymphocytes, microbiota and the intestinal epithelium governs immunity versus metabolism in the gut. Nat Med 17:1585–1593. https://doi.org/10.1038/nm.2505
37. Hooper LV, Midtvedt T, Gordon JI (2002) How host-microbial interactions shape the nutrient environment of the mammalian intestine. Annu Rev Nutr 22:283–307. https://doi.org/10.1146/annurev.nutr.22.011602.092259
38. Ley RE, Bäckhed F, Turnbaugh P, Lozupone CA, Knight RD, Gordon JI (2005) Obesity alters gut microbial ecology. Proc Natl Acad Sci U S A 102:11070–11075. https://doi.org/10.1073/pnas.0504978102
39. Hugenholtz F, de Vos WM (2018) Mouse models for human intestinal microbiota research: a critical evaluation. Cell Mol Life Sci 75:149–160. https://doi.org/10.1007/s00018-017-2693-8
40. Kennedy AJ, Ellacott KL, King VL, Hasty AH (2010) Mouse models of the metabolic syndrome. Dis Model Mech 3:156–166. https://doi.org/10.1242/dmm.003467
41. Montague CT, Farooqi IS, Whitehead JP, Soos MA, Rau H, Wareham NJ, Sewter CP, Digby JE, Mohammed SN, Hurst JA, Cheetham CH, Earley AR, Barnett AH, Prins JB, O'Rahilly S

(1997) Congenital leptin deficiency is associated with severe early-onset obesity in humans. Nature 387:903–908. https://doi.org/10.1038/43185

42. Shi H, Strader AD, Woods SC, Seeley RJ (2007) Sexually dimorphic responses to fat loss after caloric restriction or surgical lipectomy. Am J Physiol Endocrinol Metab 293:E316–E326. https://doi.org/10.1152/ajpendo.00710.2006
43. Chusyd DE, Wang D, Huffman DM, Nagy TR (2016) Relationships between rodent white adipose fat pads and human white adipose fat depots. Front Nutr 3:10. https://doi.org/10.3389/fnut.2016.00010
44. Ravussin Y, LeDuc CA, Watanabe K, Mueller BR, Skowronski A, Rosenbaum M, Leibel RL (2014) Effects of chronic leptin infusion on subsequent body weight and composition in mice: can body weight set point be reset? Mol Metab 3:432–440. https://doi.org/10.1016/j.molmet.2014.02.003
45. Rosenbaum M, Hirsch J, Gallagher DA, Leibel RL (2008) Long-term persistence of adaptive thermogenesis in subjects who have maintained a reduced body weight. Am J Clin Nutr 88: 906–912. https://doi.org/10.1093/ajcn/88.4.906
46. MacLean PS, Higgins JA, Johnson GC, Fleming-Elder BK, Donahoo WT, Melanson EL, Hill JO (2004) Enhanced metabolic efficiency contributes to weight regain after weight loss in obesity-prone rats. Am J Physiol Regul Integr Comp Physiol 287:R1306–R1315. https://doi.org/10.1152/ajpregu.00463.2004
47. Ravussin Y, Gutman R, Diano S, Shanabrough M, Borok E, Sarman B, Lehmann A, LeDuc CA, Rosenbaum M, Horvath TL, Leibel RL (2011) Effects of chronic weight perturbation on energy homeostasis and brain structure in mice. Am J Physiol Regul Integr Comp Physiol 300: R1352–R1362. https://doi.org/10.1152/ajpregu.00429.2010
48. Tschöp MH, Speakman JR, Arch JRS, Auwerx J, Brüning JC, Chan L, Eckel RH, Farese RV, Galgani JE, Hambly C, Herman MA, Horvath TL, Kahn BB, Kozma SC, Maratos-Flier E, Müller TD, Münzberg H, Pfluger PT, Plum L, Reitman ML, Rahmouni K, Shulman GI, Thomas G, Kahn CR, Ravussin E (2012) A guide to analysis of mouse energy metabolism. Nat Methods 9:57–63. https://doi.org/10.1038/nmeth.1806
49. Fernández-Verdejo R, Ravussin E, Speakman JR, Galgani JE (2019) Progress and challenges in analyzing rodent energy expenditure. Nat Methods 16:797–799. https://doi.org/10.1038/s41592-019-0513-9
50. Brommage R (2003) Validation and calibration of dexa body composition in mice. Am J Physiol Endocrinol Metab 285:E454–E459. https://doi.org/10.1152/ajpendo.00470.2002
51. Fenwick N, Griffin G, Gauthier C (2009) The welfare of animals used in science: how the "three rs" ethic guides improvements. Can Vet J 50:523–530
52. Marzola P, Boschi F, Moneta F, Sbarbati A, Zancanaro C (2016) Preclinical in vivo imaging for fat tissue identification, quantification, and functional characterization. Front Pharmacol 7:336. https://doi.org/10.3389/fphar.2016.00336
53. Sbarbati A, Baldassarri AM, Zancanaro C, Boicelli A, Osculati F (1991) In vivo morphometry and functional morphology of brown adipose tissue by magnetic resonance imaging. Anat Rec 231:293–297. https://doi.org/10.1002/ar.1092310302
54. Halldorsdottir S, Carmody J, Boozer CN, Leduc CA, Leibel RL (2009) Reproducibility and accuracy of body composition assessments in mice by dual energy x-ray absorptiometry and time domain nuclear magnetic resonance. Int J Body Compos Res 7:147–154
55. Luu YK, Lublinsky S, Ozcivici E, Capilla E, Pessin JE, Rubin CT, Judex S (2009) In vivo quantification of subcutaneous and visceral adiposity by micro-computed tomography in a small animal model. Med Eng Phys 31:34–41. https://doi.org/10.1016/j.medengphy.2008.03.006
56. Judex S, Luu YK, Ozcivici E, Adler B, Lublinsky S, Rubin CT (2010) Quantification of adiposity in small rodents using micro-ct. Methods 50:14–19. https://doi.org/10.1016/j.ymeth.2009.05.017

57. Toombs RJ, Ducher G, Shepherd JA, De Souza MJ (2012) The impact of recent technological advances on the trueness and precision of dxa to assess body composition. Obesity 20:30–39. https://doi.org/10.1038/oby.2011.211
58. Nagy TR, Clair A-L (2000) Precision and accuracy of dual-energy x-ray absorptiometry for determining in vivo body composition of mice. Obes Res 8:392–398. https://doi.org/10.1038/oby.2000.47
59. Kmiotek EK, Baimel C, Gill KJ (2012) Methods for intravenous self administration in a mouse model. J Vis Exp e3739. https://doi.org/10.3791/3739
60. Pimlott SL, Sutherland A (2011) Molecular tracers for the pet and spect imaging of disease. Chem Soc Rev 40:149–162. https://doi.org/10.1039/B922628C
61. Fueger BJ, Czernin J, Hildebrandt I, Tran C, Halpern BS, Stout D, Phelps ME, Weber WA (2006) Impact of animal handling on the results of 18f-fdg pet studies in mice. J Nucl Med 47: 999–1006
62. Labbé SM, Caron A, Chechi K, Laplante M, Lecomte R, Richard D (2016) Metabolic activity of brown, "beige," and white adipose tissues in response to chronic adrenergic stimulation in male mice. 311:E260–E268. https://doi.org/10.1152/ajpendo.00545.2015
63. Thomson A, Smart K, Somerville MS, Lauder SN, Appanna G, Horwood J, Sunder Raj L, Srivastava B, Durai D, Scurr MJ, Keita ÅV, Gallimore AM, Godkin A (2019) The ussing chamber system for measuring intestinal permeability in health and disease. BMC Gastroenterol 19:98. https://doi.org/10.1186/s12876-019-1002-4
64. Wang L, Llorente C, Hartmann P, Yang A-M, Chen P, Schnabl B (2015) Methods to determine intestinal permeability and bacterial translocation during liver disease. J Immunol Methods 421: 44–53. https://doi.org/10.1016/j.jim.2014.12.015
65. Parasuraman S, Raveendran R, Kesavan R (2010) Blood sample collection in small laboratory animals. J Pharmacol Pharmacother 1:87–93. https://doi.org/10.4103/0976-500x.72350
66. Hem A, Smith AJ, Solberg P (1998) Saphenous vein puncture for blood sampling of the mouse, rat, hamster, gerbil, guineapig, ferret and mink. Lab Anim 32:364–368. https://doi.org/10.1258/002367798780599866
67. Bligh EG, Dyer WJ (1959) A rapid method of total lipid extraction and purification. Can J Biochem Physiol 37:911–917. https://doi.org/10.1139/o59-099
68. Warnick GR (1986) Enzymatic methods for quantification of lipoprotein lipids. In: Albers JJ, Segrest JP (eds) Methods in enzymology. Academic, pp 101–123
69. Puppione DL, Charugundla S (1994) A microprecipitation technique suitable for measuring α-lipoprotein cholesterol. Lipids 29:595–597. https://doi.org/10.1007/BF02536633
70. Issara U, Park S, Lee S, Lee J, Park S (2020) Health functionality of dietary oleogel in rats fed high-fat diet: a possibility for fat replacement in foods. J Funct Foods 70:103979. https://doi.org/10.1016/j.jff.2020.103979
71. Ghosh M, Begg F, Bhattacharyya DK, Bandyopadhya N, Ghosh M (2017) Nutritional evaluation of oleogel made from micronutrient rich edible oils. J Oleo Sci 66:217–226. https://doi.org/10.5650/jos.ess16165
72. Millán J, Pintó X, Muñoz A, Zúñiga M, Rubiés-Prat J, Pallardo LF, Masana L, Mangas A, Hernández-Mijares A, González-Santos P, Ascaso JF, Pedro-Botet J (2009) Lipoprotein ratios: physiological significance and clinical usefulness in cardiovascular prevention. Vasc Health Risk Manag 5:757–765. https://doi.org/10.2147/VHRM.S6269
73. Limpimwong W, Kumrungsee T, Kato N, Yanaka N, Thongngam M (2017) Rice bran wax oleogel: a potential margarine replacement and its digestibility effect in rats fed a high-fat diet. J Funct Foods 39:250–256. https://doi.org/10.1016/j.jff.2017.10.035
74. Stortz TA, Zetzl AK, Barbut S, Cattaruzza A, Marangoni AG (2012) Edible oleogels in food products to help maximize health benefits and improve nutritional profiles. Lipid Technol 24: 151–154. https://doi.org/10.1002/lite.201200205
75. Tan S-Y, Wan-Yi Peh E, Marangoni AG, Henry CJ (2017) Effects of liquid oil vs. Oleogel co-ingested with a carbohydrate-rich meal on human blood triglycerides, glucose, insulin and appetite. Food Funct 8:241–249. https://doi.org/10.1039/C6FO01274D

Chapter 26
Edible Applications

Martina Dominguez and María Elena Carrín

Abbreviations

AA	Adipic acid
BS	β-sitosterol
BW	Beeswax
CB	Cocoa butter
CLW	Candelilla wax
CM	Commercial margarine
CRW	Carnauba wax
DG	Diglycerides
EC	Ethylcellulose
ERCA	Esterified rice flour with citric acid
FAP	Fatty acid profile
GO	γ-Oryzanol
HF	Hard fat
HFM	High-fat margarine
HIU	High intensity ultrasound
HO	Hazelnut oil
HOSO	High oleic sunflower oil
HPMC	Hydroxypropyl methylcellulose
LC	Lecithin
MG	Saturated monoglycerides

M. Dominguez · M. E. Carrín (✉)
Departamento de Ingeniería Química, Universidad Nacional del Sur (UNS), Bahía Blanca, Argentina

Planta Piloto de Ingeniería Química – PLAPIQUI (UNS-CONICET), Bahía Blanca, Argentina
e-mail: mdominguez@plapiqui.edu.ar; mcarrin@plapiqui.edu.ar

C. Andrea Palla, F. Valoppi (eds.), *Advances in Oleogel Development, Characterization, and Nutritional Aspects*,
https://doi.org/10.1007/978-3-031-46831-5_26

MUFA	Monounsaturated fatty acids
O/W	Oil-in-water
OBC	Oil binding capacity
PGE	Polyglycerol esters
PS	Phytosterols
PUFA	Polyunsaturated fatty acids
RBW	Rice bran wax
SFA	Saturated fatty acids
SFC	Solid fat content
SFW	Sunflower wax
SLW	Shellac wax
SMS	Sorbitan monostearate
SSL	Sodium stearyl lactate
STS	Sorbitan tristearate
TFA	*Trans* fatty acids
UFA	Unsaturated fatty acids
UMG	Unsaturated monoglycerides
VOO	Virgin olive oil
W/O	Water-in-oil
XG	Xanthan gum
ΔE	Total color differences with respect to the control

26.1 Introduction

Advances in the knowledge of the harmful effects of some food components on health are driving the demand for healthier food products from different sectors of society. Fats are key ingredients in a wide variety of foods, performing some technological and sensory functions in defining desirable food properties, contributing to their flavor, lubricity, texture, stability, and shelf life and to consumer satiety [1–3]. Traditional, stable, and relatively inexpensive fatty products used to obtain these characteristics, such as butter, margarines, and shortenings, are mainly composed of saturated fatty acids (SFA) or mixtures of SFA and industrially produced *trans* fatty acids (TFA) [4–6]. However, overconsumption of both SFA and TFA has been claimed to cause some serious health concerns [3, 7–10]. Therefore, recommendations by international organizations emerged to eliminate TFA provision through food [11], as well as to decrease SFA consumption to values of at most 10% of total energy needs [12]. In addition, consumers are increasingly expected to demand not only healthy and natural food but also ethical and sustainable food production systems, according to the United Nations Sustainable Development Goals [13].

To meet these demands, the need for lower fat or improved quality products led industries and scientists to search for ingredients that could replace TFA and at least

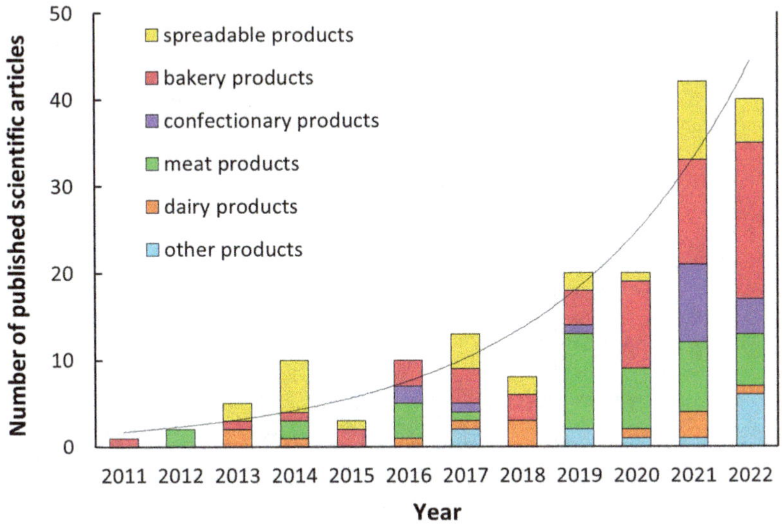

Fig. 26.1 Trend over time of different categories of oleogel-based system food applications. Bars represent the annual data and solid line denotes their tendency. Data were collected from Scopus-Elsevier by searching oleogels or organogels food applications (including each of the specified applications) (update to February 17, 2023)

reduce SFA use. However, this involves a major development and innovation effort in the research and industry sectors to mimic the functionalities that hard fats (HF) provide to each of the different products where they are used. As has been introduced in previous chapters, this requirement has been addressed, among other proposals, through the structuring of vegetable oils with a high content of unsaturated fatty acids (UFA). For this purpose, various gelling agents or structurants have been tested—waxes, HF, lecithin (LC), ethylcellulose (EC), phytosterols (PS), saturated monoglycerides (MG), and diglycerides (DG), or emulsifiers with biopolymers, including proteins and polysaccharides—requiring different preparation methodologies. Therefore, over almost the last decade, some oleogel-based systems have demonstrated their promising applicability as substitutes for fats conventionally used in a variety of foods, which can be grouped into broad categories (Fig. 26.1). Replacement proposals were initially moderate and focused mainly on spreadable products. This trend was changing and currently the most evaluated reformulations are also based on bakery and meat products. Additionally, new technological applications have been emerging lately.

Nevertheless, these systems have still limited commercial-level implementation in the food industry due to some technological challenges arising from process adaptation or consumer sensory demands [4, 14]. In addition, the requirement to use ingredients suitable for human consumption has to be met. For instance, although there are some doubts about the food use of certain wax types, legislation is moving positively in this regard [15, 16].

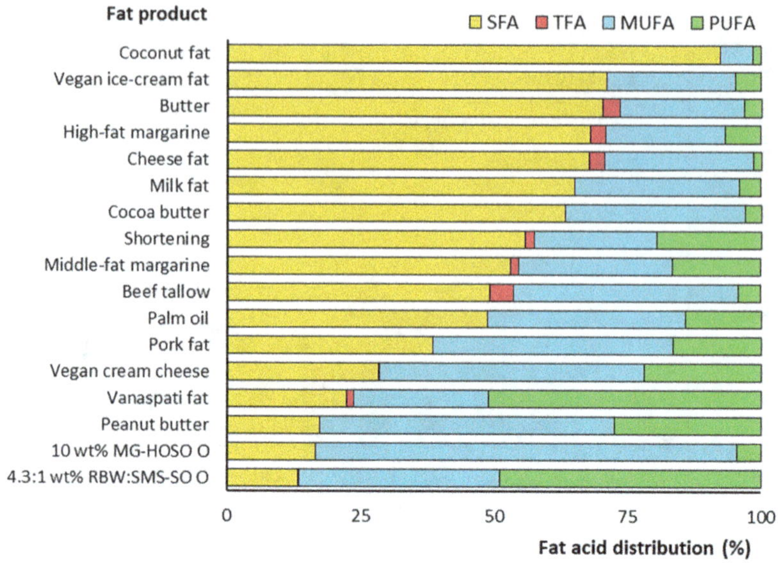

Fig. 26.2 Fatty acid distribution—SFA, TFA, PUFA, and MUFA to saturated, *trans*, polyunsaturated, and monounsaturated fatty acids, respectively—in some commercial fats and oleogel-based systems. *MG* saturated monoglycerides, *HOSO* high oleic sunflower oil, *O* oleogel, *RBW* rice bran wax, *SMS* sorbitan monostearate, *SO* sunflower oil. (Data were collected from different sources [14, 16–24])

In nutritional terms, replacing all or part of the HF with oleogel-based systems significantly impacts the lipid profile of food products, and this will be determined not only by the composition of that system (base oil and other added compounds such as structurants and emulsifiers) but also by the substitution level [14]. Particularly, oleogel lipid profile will be close to that of the base oil used, especially when it is relatively low in structurant. Thus, oleogel-based materials using oils high in monounsaturated fatty acids (MUFA)—MUFA >50%, such as high oleic varieties of sunflower and safflower, canola, hazelnut, peanut, and olive oils—or oils high in polyunsaturated fatty acids (PUFA)—PUFA >50%, such as corn, flaxseed, sunflower, soybean, and walnut oils—have an improved fatty acid profile (FAP) compared to the fats they are intended to replace, as can be observed in Fig. 26.2.

In view of the abovementioned, this chapter explores progress in research carried out to evaluate the suitability of oleogel-based systems for their application in different food products, focusing especially on lipid profile improvement, but also including other functional or technological issues. Also, it attempts to highlight the challenges faced by the food industry in making oleogel applications effective and the prospects for the future.

26.2 Spreadable Products

A variety of fatty products are usually ingested as breakfast spreads or manufactured as spreadable products to be post-processed into other foods. These products contribute undesirable levels of SFA and TFA to the diet, which makes oleogel-based systems an attractive replacement option. A large number of them have been evaluated as alternative spreads, mainly as margarine or shortening substitutes (Table 26.1).

Margarine is commonly used as a breakfast product or as an ingredient in cooking and baking preparations, being a butter substitute in many recipes. It is formed by uniformly dispersing small water droplets in a highly saturated fat phase, thus constituting a stable solid form of a water-in-fat emulsion. Developing stable water-in-oil (W/O) emulsion-based products with reduced SFA and without TFA is the main constraint to overcome. Different oleogel-based systems have been used to emulate margarine, with variations in the components and their proportions and in the preparation method—emulsions from pre-formed oleogels or directly from the molten materials. Although some product differences have been assigned to the preparation method [29], no systematic comparison between both methodologies was found to justify those results.

Waxes from different origins have been the most evaluated structurants for margarine production. Considering that sunflower wax (SFW), rice bran wax (RBW), and candelilla wax (CLW) efficiently structured soybean oil even at low concentrations, the corresponding oleogels were tested in a high-fat (80%) margarine (HFM) [27]. Among them, SFW-based margarines were the best option, having comparable firmness to margarines formulated using a mixture of hydrogenated and conventional soybean oil, a commercial margarine (CM), and some spreads. It is to be noted that the UFA:SFA ratio was improved from 2.5–4.5 to 6 using SFW oleogels. Similar texture results were obtained using other 13 oils [28]. Final product properties were also modified by using waxes or oils from different suppliers, origins, or processing, probably related to minority component presence in dissimilar proportions. Promising results were reported using SFW and RBW to structure hemp seed oil [31]. Using 3 wt% wax in emulsions was enough to achieve commercial spread firmness. However, stick margarine firmness was not obtained even with 7 wt% wax. Similarly, an oleogel with 10 wt% beeswax (BW) was satisfactorily used to replace palm and hydrogenated palm oil in the formulation of a medium-fat margarine (70% fat)—with 28% SFA and 80% TFA reductions—[26]. Although the oleogel-based product had lower solid fat content (SFC) and higher melting point than the control, rheological and textural characteristics were similar. Rapeseed oil was successfully gelled by shellac wax (SLW) and used to formulate emulsifier-free W/O emulsions at different water:oil proportions [46, 48]. Acceptable spreadable properties were obtained when emulating HFM (Fig. 26.3a1), achieving 80% SFA and 40% PUFA reductions and a 158% MUFA increase. However, emulsion stability was difficult to maintain when dealing with low-fat (<60%) products. Additionally, emulsifiers and cold storage were used to favor textural and

Table 26.1 Oleogel applications in spreadable products

Food product	System	Gelator (wt% in system)/ oil	Replaced fat (wt% replacement)	Reference
Margarine	Oleogel	MG (12)/vC-loaded corn	Margarine fat (100)	[25]
	Oleogel	BW (2.5–10)/n.s	Palm oil, partially hydrogenated palm olein oil (100)	[26]
	Oleogel	SFW, RBW, CLW (2–6)/soybean	CM fat (100)	[27]
	Oleogel	SFW (3, 5, 7)/various[a]	Margarine fat (100)	[28]
	Oleogel	CLW:BW (0–7:7–0)/ soybean	Monocomponent wax oleogel (100)	[29]
	W/O emulsion	BW (7)/corn	Margarine (100)	[30]
	W/O emulsion	SFW, RBW (3–7)/hemp seed	Margarine (100)	[31]
	W/O emulsion	MG:LC (22:8)/palm olein	CM (100)	[32]
	W/O emulsion	MG:DG:TG (31.2:42.3: 26.5)/tigernut	CM, butter (100)	[33]
	W/O emulsion	CLW:ICF:MG (3.3:6.1: 2.2)/HOSO	CM (100)	[34]
	W/O emulsion	CLW:FHPO:UMG (0.4–1:0.6–1.3:0.3–0.8)/ soybean	Margarine (100)	[35]
Mayonnaise	O/W emulsion	SFW:Tween20:SMS20 (1.5:2:0.05)/various[b]	Mayonnaise (100)	[36]
Shortening	Oleogel	MG (10.8, 12.4)/HOSO	Low-fat CM (100)	[37]
	Oleogel	MG (6.6, 8.2)/HOSO	Low-fat CM (100)	[38]
	Oleogel	BW, SFW (3, 7, 10)/ olive	Breakfast margarine (100)	[39]
	Oleogel	RBW (5)/rice bran	Margarine, beef tallow (100)	[21]
	Oleogel	RBW:SMS (2–5:1–3)/ sunflower	Vanaspati fat (100)	[23]
	Oleogel	EC:TMS (4:1)/palm stearin:soybean (23:77)	Shortening (100)	[40]
Spread	Oleogel	MG (3, 7, 10)/pomegranate seed	CM fat (100)	[41]
	Oleogel	MG, CRW (3, 7, 10)/ VOO	Breakfast CM fat (100)	[42]
	Oleogel	BW, SFW (5)/VOO, HO	Breakfast CM fat, butter (100)	[43]
	Oleogel	BW, SFW, SLW (1, 3, 5)/*n.a*	Tahini emulsifier (100)	[44]
	Oleogel	SFW (10)/VOO	n.s	[45]
	Oleogel	SLW (6)/rapeseed	n.s	[46]
	Oleogel	WSW (8)/VOO	n.s	[47]

(continued)

Table 26.1 (continued)

Food product	System	Gelator (wt% in system)/ oil	Replaced fat (wt% replacement)	Reference
	W/O emulsion	BW, CLW, RBW, SFW, HCO (0.5–2)/peanut butter	Peanut butter (100)	[16]
	W/O emulsion	SLW (5.5)/rapeseed	n.s	[48]
	W/O emulsion	EC:MG (7:0.5, 7:1)/high oleic safflower	Spread CM (100)	[49]
Chocolate spread	Oleogel	MG, BW, PW (5)/ pomegranate	Palm oil (50)	[50]
	Oleogel	SLW (6.8)/rapeseed: palm	Oil binder (100), palm oil (~27)	[48]
	W/O emulsion	MG (20)/corn	Oleogel (45, 50, 55)	[51]
	Emulsion-templated oleogel	HPMC:XG (1.62:0.97)/ sunflower, olive	Coconut fat (50, 100)	[52]
	Dispersion	Cellulose fibers (5–40)/ rapeseed	Palm oil (100)	[53]

[a]Almond, canola, corn, flaxseed, grapeseed, olive, peanut, pumpkin seed, safflower, sesame, soybean, sunflower, and walnut

[b]Groundnut, sunflower, soybean, sesame, mustard, rice bran, coconut, and palm

BW beeswax, *CLW* candelilla wax, *CRW* carnauba wax, *DG* diglycerides, *EC* ethylcellulose, *FHPO* fully hydrogenated palm oil, *HCO* fully hydrogenated cottonseed oil, *HO* hazelnut oil, *HOSO* high oleic sunflower oil, *HPMC* hydroxypropyl methylcellulose, *ICF* interesterified commercial fat, *LC* lecithin, *MG* saturated monoglycerides, *n.a* not applicable, *n.s* not specified, *PW* propolis wax, *RBW* rice bran wax, *SFW* sunflower wax, *SLW* shellac wax, *SMS* sorbitan monostearate, *TG* triglycerides, *TMS* triglyceryl monostearate, *UMG* unsaturated MG, *vC* vitamin C, *VOO* virgin olive oil, *XG* xanthan gum, *W/O* water-in-oil, *WSW* whale spermaceti wax

rheological properties and stability of margarines based on BW-corn oil oleogels—diminishing 56% SFA and increasing 26% PUFA—[30]. Although some differences in thermal properties were detected between reformulated and commercial products, sensory properties were not examined to confirm or correct the reformulation.

MG have also been used to prepare oleogels for margarine production. After complete fat content replacement of HFM with an MG-corn oil oleogel, lower values of firmness, SFC, and TFA were obtained in comparison with commercial butters [25]. However, oleogel-based margarine had the highest values of adhesiveness, spreadability, and UFA content. Interestingly, although the appearance of the oleogel-based margarine was visually more translucent than that of the butters, consumers rated all samples with high appearance scores declaring their willingness to try or even buy that product.

Some structurants have been tested in combination, either in binary or multicomponent mixtures. However, a few of these systems have been evaluated in margarine formulations, finding that some of their important properties could be tailored by structurant proportions. CLW-BW oleogels were able to form HFM with

Fig. 26.3 Appearance of spreadable products elaborated with different oleogel-based systems: (**a**) margarines from (a1) shellac wax oleogel into a high-fat emulsion, reproduced from [46] with permission from The Royal Society of Chemistry, and tigernut glycerolysis product stored at refrigerated conditions for (a2) 1 week and (a3) 10 months after sample preparation, reproduced

improved or similar textural characteristics compared to that obtained using only one wax type [29]. These binary wax-based margarines had lower melting points and residual SFC at near body temperature, thereby yielding products with features closer to CM. However, other properties required a CLW:BW ratio reduction. An HFM was also formulated using a mixture of CLW, MG, and an interesterified commercial fat in high oleic sunflower oil (HOSO) [34]. Although differences in microstructures and melting curves were detected between the reformulation and CM, the former yielded a softer product of similar color but more resistant to temperature fluctuations than CM. Even with a high unsaturation proportion, the oleogel-based margarine showed acceptable oxidative stability. However, this reformulated margarine could not reach the sensory scores assigned to CM, potentially due to a residual waxy taste. Surprisingly, consumer intention to buy them was similar, suggesting that other attributes should be considered when looking for relationships between product properties and consumer reactions. Low-fat (35%) and middle-fat (60%) margarines were also reformulated with CLW and unsaturated monoglycerides (UMG) in soybean and fully hydrogenated palm oil [35]. Margarines with lower water content showed higher hardness and thermal stability and lower spreadability. Oleogelation made it possible to obtain different margarines with improved PUFA (3–21%), reduced SFA (10–33%), and practically TFA free. MG and LC, besides water, were studied to produce a low-fat margarine with CM-like mechanical properties [32]. An optimized product was obtained with respect to its structural properties, but using a high structurant proportion (30 wt% LC + MG) and 28 wt% water. Thus, although palm oil content was reduced (~24%), SFA increased ~43% because of reformulation. An EC-MG gelled W/O emulsion had higher elasticity and thermal stability than a conventional emulsion (without EC) or EC-MG oleogel [49]. Its ability to maintain MG at the droplet interface could be responsible for increased emulsion stability at higher MG concentrations. In comparison to low-fat CM, the gelled emulsions had lower elasticity. Therefore, higher concentrations of mixed emulsifiers could be a promising strategy to reach CM elasticity, without adding high proportions of SFA or using TFA.

Promising oleogels resulted from the products obtained by unsaturated oil enzymatic glycerolysis [33]. This oleogel type showed satisfactory results to substitute palm and hydrogenated oils in margarine production—plasticity similar to CM and butter and high storage emulsion stability (Fig. 26.3a2, a3)—highly improving UFA

Fig. 26.3 (continued) from [33] with permission from Elsevier; (**b**) mayonnaises from (b1) commercial product and (b2) SFW oleogel-based emulsions and (b3) its application on bread, reprinted with permission from [36] Copyright (2021) American Chemical Society; (**c**) spreads using oleogels from (c1) beeswax (BW)-hazelnut oil (HO), (c2) BW-virgin olive oil (VOO), (c3) sunflower wax (SFW)-HO, and (c4) SFW-VOO, reproduced from [43] with permission from The Royal Society of Chemistry; and (**d**) chocolate pastes using water:oleogel-based emulsions at ratios of (d1) 0:100, (d2) 45:55, (d3) 50:50, and (d4) 55:45, reproduced from [51] with permission from Elsevier.

(110%). It also exhibited a more suitable melting curve than that usually achieved by wax-based oleogel margarines.

Mayonnaise is based on an oil-in-water (O/W) emulsion with a large proportion of oil, which has a characteristic highly viscous and shear-thinning behavior. Looking for conventional mayonnaise alternatives, oleogel-based emulsions were prepared using different vegetable oils structured with SFW and some emulsifiers [36]. Although each oil contributed its own characteristic color to the emulsions formed, all presented a visual mayonnaise-like appearance. Varying oil unsaturation degree significantly affected the product strength, SFC, and oil binding capacity (OBC), while not affecting its microstructure. As the unsaturation degree increases, OBC increases, making emulsions of highly unsaturated oils capable of substituting commercial mayonnaise (Fig. 26.3b).

Shortening is considered any fat that, at room temperature, remains solid and is generally used to make a wide variety of food products. Oleogels are usually presented as a direct healthier replacement for shortenings, although achieving their specific properties is a challenge. In addition to improving FAP, different oleogels have been tested to prevent the formation of water-containing systems. This last point confers oleogels a great advantage in terms of their chemical and physicochemical stability. MG-HOSO oleogels were produced to resemble the properties of a low-fat CM [37, 38, 54]. Both processing conditions and MG concentration (without exceeding control SFA) were optimized to achieve the desired properties, which were maintained unchanged for at least 3 weeks. Oleogels from BW or SFW were formulated to mimic breakfast margarine [39]. Although BW oleogel had thermal properties closer to those of the control, using different concentrations of each wax allowed to reach target textural properties—which remained stable for at least 3 months. RBW-rice bran oil oleogels have also been successfully formulated as substitutes for baking margarine and even beef tallow [21]. These products were used in a digestibility assay, showing that although the oleogel-based diet—highly enhanced in FAP—did not have anti-obesity effects, it did provide some other health benefits, confirming the postulates used in favor of oleogels. RBW was also used in combination with sorbitan monostearate (SMS) to structure sunflower oil as a low-saturated fat substitute for vanaspati, a popular fat in some countries [23]. Oleogel-based vanaspati, presenting nearly 40% SFA reduction, had rheological properties similar to commercial samples, but not some thermal properties.

Spreads are usually pastes that can be applied and adhere properly to food surfaces. Oleogel-based spreads have been widely formulated using virgin olive oil (VOO), based on the benefits that its consumption can bring not only for its FAP but also for its minority components. MG-VOO oleogels showed higher OBC and thermal properties closer to breakfast CM than carnauba wax (CRW)-VOO oleogels, both generating major UFA increments [41, 42]. However, some margarine textural properties were not achieved by any of the oleogels, nor with MG oleogel which retained control SFA level. Sensory properties and consumer acceptance of oleogels from VOO or diacetyl-enriched hazelnut oil (HO) and two types of waxes were evaluated as spreadable fats [43]. Both types of oleogels showed acceptable appearance and sensory attributes, regardless of the wax type used. Although the sample's visual appearance was quite remarkable (Fig. 26.3c), more than half of the consumers expressed their intention to buy these substitutes.

Spices added to oleogels provide antioxidant bioactive compounds. SFW was used to structure VOO enriched with thyme and cumin spices [45]. Spices did not affect gel formation, gel thermal stability, or gelation time, but generated intermediate shear-stable oleogels. Moreover, whale spermaceti wax-VOO oleogels enriched with ground red pepper and turmeric allowed the obtention of spread-type products with satisfactory quality indexes [47]. As expected, VOO and spices contributed to the oleogel color and aromatic profile; however, the spice-enriched oleogels were liked by consumers.

Through gelling, waxes have been highly efficient in improving oil separation in some products. For instance, to reformulate tahini paste, SFW, SLW, and BW were tested as oil stabilizers compared to the commercially used emulsifier, hydrogenated palm stearin [44]. Using SFW and SLW, oil migration was significantly reduced and products similar to the control were obtained, even with improved FAP. Liquid oil in peanut butter spreads is also usually stabilized by a high-SFA material. OBC, firmness, and 6-month oil stabilization were improved by wax oleogelation, mainly with SFW [16].

Chocolate spreads are constituted by a continuous phase of commonly solid fat—such as cocoa butter (CB), palm oil, or coconut oil—into which another phase is dispersed. Reformulation of chocolate spreads lowering their SFC can be especially challenging, as saturated fats significantly support some of the most important properties required in this type of product—creamy texture, glossy appearance, taste, flavor, and mouth-melting behavior—[51]. A chocolate spread was reformulated by replacing oil binder and palm oil with SLW-rapeseed oil oleogel, obtaining a storage stable oleogel-based paste and reducing SFA by ~24% [48]. Oleogels produced from pomegranate seed oil and MG, BW, or propolis wax were also used to reformulate chocolate spreads partially replacing palm oil and, therefore, lowering SFA by at least 39% [50]. Product mechanical parameters were modified depending on the structurant used, with the lowest and highest values found for propolis wax and MG, respectively. However, paste storage caused hardness to vary, tending toward more similar values. Additionally, oleogels from olive or sunflower oil and hydroxypropyl methylcellulose (HPMC) and xanthan gum (XG) were used to reformulate chocolate spreads [52]. Total coconut butter replacement led to inhomogeneous materials, while structures similar to the control were obtained by partial replacement—even strongly reducing SFA. Moreover, when sunflower oil oleogels were used in partial replacement, a sensory evaluation similar to the control was reached. A bamboo fiber-rapeseed oil dispersion—without heating or emulsifying—was used to produce a chocolate spread with good rheological properties, high thermal stability, and healthier nutritional qualities than the palm oil-based spread [53].

The expected mouth-melting behavior of chocolate pastes can be difficult to achieve when high proportions of oleogels are used in reformulations. In this context, a gelled W/O emulsion from an MG-corn oil oleogel was used to formulate low-fat spreads [51]. By using a 45:55 water:oleogel-based emulsion, an acceptable product was obtained, with good sensory scores and rheological and textural properties similar to an oleogel-based chocolate paste (Fig. 26.3d).

26.3 Bakery Products

Fat content reduction is a crucial aspect in bakery product formulation, considering all the characteristics that fats provide to these products during baking, storage, and consumption. Among others, hardness, fracturability, spread ratio, fat migration, and surface color are important product quality attributes because they impact consumer acceptability and willingness to purchase. Table 26.2 shows several examples of different oleogel-based systems proposed for the reformulation of bakery products by substituting traditionally used shortenings and margarines.

Fat and water have an important role in the development of gluten tridimensional structure in bakery products. However, not all of these products require the same gluten net strength. Some bakery goods are characterized as short-dough products—cookies and biscuits—while other products need extended gluten network building—such as bread. Therefore, not only fat/water proportion but also fat type and its distribution in dough must be carefully selected and generated to obtain the desired properties [77, 88]. When baking products based on conventional fats, it is necessary to provide a relatively high SFC to ensure proper dough lubrication, aeration, and hardness and consequently to generate areas strong enough to contain air bubbles during cooking to finally obtain products with appropriate characteristics [77, 78, 85, 87, 88]. Different behaviors have been reported using oleogel-based systems instead of conventional fats. The presence of emulsifiers in the fat phase has been recognized as a key factor to adequately disperse this phase on the dough hydrophilic ingredients, not only disturbing gluten formation but also increasing starch gelatinization temperature and prolonging gas retention time [77].

Biscuits and cookies, usually with about 25% fat, should be formulated from smooth, uniform, and plastic doughs [78]. During ingredient mixing, initial emulsified creams from sugar and commercial fats are typically harder than those from oils and oleogels, but after flour inclusion, the desired dough properties can be reached. However, oil-based creams present difficulties in achieving appropriate air entrapment and adhesion during mechanical working, generating hard cookies [78, 87, 88]. When oil was previously structured, cookie dough hardness approached that obtained with shortenings, while dough extensibility decreased. Compared to the oil-based system, oil structuring by CRW or CLW improved dough air-holding capacity and cookie spread ratio. Nevertheless, shortening-based cookie properties were not fully achieved by replacing shortening with oleogels with up to 5 wt% waxes [88], but they were reached by partial replacement with CLW oleogel while improving FAP [87]. Additionally, a certain waxy aroma was reported after cooking, which could be a negative factor for consumers and, therefore, deserves to be evaluated [88]. Moreover, oil-based biscuits have shown fat migration, which was

Table 26.2 Oleogel applications in bakery products

Food product	System	Gelator (wt% in system)/oil	Replaced fat (wt% replacement)	Reference
Bread	Oleogel	MG, RBW (10)/soybean	Shortening (100)	[55]
	Oleogel	SSL (9)/sunflower	Margarine (100)	[56]
	Oleogel	EC:TMS (4:1)/palm stearin: soybean	Shortening (100)	[40]
Bun	Emulsion-templated oleogel	HPMC:XG (1.6:1)/sunflower, olive	Margarine (100)	[57]
Sweet bread	Oleogel	CLW (10)/rice bran	Butter (25, 50, 75, 100)	[58]
	Oleogel, O/W emulsion	MG (2.7, 5)/sunflower	Palm oil oleogel, emulsion (100)	[59]
Cake	Oleogel	BW, SFW (5), BW:SFWh (2.5:2.5)/canola	Palm fat (100)	[60]
	Oleogel	CLW (5)/canola	Butter (25, 50, 75, 100)	[61]
	Oleogel	BW (10)/sunflower	Shortening (15, 30, 45, 60, 80, 100)	[62]
	Oleogel	CRW (10)/canola	Shortening (25, 50, 75, 100)	[63]
	Oleogel	RBW, BW, CLW (10)/ sunflower	Shortening (100)	[64]
	Oleogel	MG:BW (7:3, 3:7, 5:5, 10:0, 0: 10)/HOSO	Margarine (100)	[65]
	Oleogel	CRW:AA (2:4, 6:0)/soybean	Shortening (50)	[66]
	Oleogel	EC:AA (2:4, 6:0)/soybean	Shortening (50)	[67, 68]
	Oleogel	BS:GO (3.2:4.8), SSAP (12)/ menhaden, SL	Shortening (100)	[69]
	Oleogel, W/O emulsion	CRW (5)/cotton seed, HOSO, blend fat	Shortening (100)	[70]
	W/O emulsion	SLW (5.5)/rapeseed	Margarine (100)	[48]
	O/W emulsion oleogel	Canola proteins (8)/canola	Shortening (100)	[71]
	O/W emulsion oleogel	OSA-KGM (6)/peanut	Margarine (100)	[72]
	O/W emulsion oleogel	XG:GG (1.7:1.7)/sunflower		[73]

(continued)

Table 26.2 (continued)

Food product	System	Gelator (wt% in system)/oil	Replaced fat (wt% replacement)	Reference
			Shortening (25, 50, 75, 100)	
	O/W emulsion oleogel	Gelatin:BLP (2.2:0.4)/soybean	Margarine (100)	[74]
	Foam-templated oleogel	MG (0–10)/soybean	Butter (50)	[75]
	Foam-templated oleogel	Protein:XG (3.1:0.03)/canola	Shortening (100)	[76]
Cookie	Oleogel	MG, SSL (3–15), PGE, SMS (9–18)/corn	Shortening (100)	[77]
	Oleogel	MG (5), CLW (2, 3), RBW (2), BW (5), EC (8)/high oleic rapeseed	Palm oil (100)	[78]
	Oleogel	MG, BS (10), MG:BS (5:5)/ soybean	Shortening (100)	[79]
	Oleogel	MG, CRW (10), BS:BW (2: 8), BS:LC (12.8:3.2)/ sunflower	Margarine (100)	[80]
	Oleogel	BW, SFW, RBW, CLW (8)/ olive, soybean, flaxseed	Margarine (100)	[81]
	Oleogel	BW (6)/sunflower	Oil (100)	[82]
	Oleogel	BW (6)/sunflower	HF (100)	[83]
	Oleogel	BW, CLW, CRW (n.s)/TML	Shortening (100)	[84]
	Oleogel	BW, SFW (5)/hazelnut	Shortening (100)	[85]
	Oleogel	CLW (3, 6)/canola	Shortening (100)	[86]
	Oleogel	CLW (3, 6)/canola	Shortening (30, 60, 100)	[87]
	Oleogel	CRW, CLW (2.5, 5)/sunflower	Shortening (100)	[88]
	Oleogel	RBW (3–9)/soybean	Shortening (100)	[89]
	Oleogel	RBW, BW, CLW, CRW (3, 5, 7, 9)/corn	Shortening (100)	[90]
	Oleogel	EC (1.5)/MCT	Liquid oil (100)	[91]
	Oleogel	RBW:BF (4.5:0.5, 5.5:0.5)/ sunflower	Butter (100)	[92]
	Oleogel, O/W emulsion	ERCA:BW (0–15:2)/soybean	Shortening (50)	[93]
	Oleogel, emulsion-templated or foam-templated oleogel	MG, BW, RBW, HPMC, SSL (6)/corn	Shortening (100)	[94]
	Oleogel, W/O emulsion	RBW (9)/rice bran		[95]

(continued)

Table 26.2 (continued)

Food product	System	Gelator (wt% in system)/oil	Replaced fat (wt% replacement)	Reference
			Margarine (100)	
	Bigel	CRW (5)/canola	Shortening (25–100)	[96]
	Bigel	BW (5)/canola	Shortening (100)	[97]
	Foam-templated oleogel	MG (0–10)/soybean	Butter (50)	[75]
	Emulsion-templated oleogel	TPE (3)/camellia	Butter (25, 50, 75, 100)	[98]
	Emulsion-templated oleogel	LC: Inulin (2:19)/VOO	Butter (20, 40, 50)	[99]
	Emulsion-templated approach + crosslinking	Chitosan: Vanillin (1:3)/canola	Shortening (100)	[100]
Cracker	Oleogel	MG, RBW (10)/soybean	Shortening (100)	[55]
Muffin	Oleogel	MG (6.6, 8.06)/HOSO	CM (100)	[38, 101]
	Oleogel	CLW (10), CLW:MG (7.5: 2.5), CLW:BC (10:0.02), CLW:MG:BC (7.5:2.5:0.02)/ sunflower	Shortening (100)	[102]
	Oleogel	CLW, SFW, BW (5)/rapeseed	Shortening (100)	[103]
	Oleogel	CLW (8)/grape seed	Shortening (25, 50, 75, 100)	[104]
	Foam-templated oleogel	HPMC (4)/sunflower	Shortening (25, 50, 75, 100)	[105]
Puff pastry	O/W emulsion	MG:SFA (9:1.8)/sunflower	CM (100)	[106]
	O/W emulsion	MG:CB (3:10)/VOO	CM (100)	[107]
Tart pastry	Oleogel	EC:MG (12:0, 6:6, 0:12)/avocado:olive (1:1)	Butter, shortening (100)	[108]

AA adipic acid, *BC* β-carotene, *BF* bamboo fiber, *BLP* bayberry leave proanthocyanidins, *BS* β-sitosterol, *BW* beeswax, *CLW* candelilla wax, *CM* commercial margarine, *CRW* carnauba wax, *EC* ethylcellulose, *ERCA* esterified rice flour with citric acid, *GG* guar gum, *GO* γ-oryzanol, *HF* hard fat, *HOSO* high oleic sunflower oil, *HPMC* hydroxypropyl methylcellulose, *LC* lecithin, *MCT* medium chain triglycerides, *MG* saturated monoglycerides, *n.s* not specified, *OSA-KGM* biopolymer from octenyl succinic anhydride and Konjac glucomannan, *O/W* oil-in-water, *PGE* polyglycerol esters, *SFA* saturated fatty acids, *SFW* sunflower wax, *SFWh* sunflower wax hydrolysate, *SL* structured lipids, *SLW* shellac wax, *SMS* sorbitan monostearate, *SSAP* sucrose stearate/ascorbyl palmitate, *SSL* sodium stearyl lactate, *TG* triglycerides, *TML* Tenebrio molitor larvae, *TMS* triglyceryl monostearate, *TPE* tea polyphenol ester, *VOO* virgin olive oil, *W/O* water-in-oil, *XG* xanthan gum significantly diminished by using oleogels [78]. Other studies have reported that cookies made with MG, BW, RBW, CLW, SFW, or SMS oleogels have shown hardness more similar to commercial fat-based products than those with CRW, sodium stearyl lactate (SSL), or polyglycerol esters (PGE) oleogels, even with similar SFA reduction [77, 78, 80, 81, 84–86, 90, 94].

Cookie textural properties depended strongly on the components structuring the oleogel—purity and composition—rather than on oleogel or dough properties [82, 83]. For instance, oleogels from unfractionated BW and different fractionated BW provided some variation in cookie hardness and organoleptic properties compared to those obtained with oil and butter. With little change in cookie appearance and spread ratio, RBW-corn oil oleogels, regardless of the use of crude or refined oil, also were suitable for replacing cookie fat [89]. Nevertheless, the use of unrefined oil in oleogels generated harder cookies [79, 89]. Although important differences have been described between the properties of oleogels based on different types of oils and structurants and the corresponding cookie dough properties, final cookie properties have not been affected in the same way. For instance, different wax oleogel-based systems led to cookies with comparable quality attributes to commercial fat-based cookies [81, 84]. In addition, CLW oleogels were the most similar to palm oil, but MG oleogels produced biscuits more similar to the control product [78]. Moreover, the inclusion of fibers in RBW oleogels reduced dough firmness positively impacting cookie texture parameters and approximating butter-based cookie properties [92]. However, differences in recipes and technological parameters, as well as component quality used, may have led to some of the reported differences in the indicated behaviors.

Looking for more generalized trends to guide cookie reformulation from a theoretical framework, the hardness of cookies reformulated with wax oleogels was more inversely related to SFC and β' crystalline form in the corresponding oleogels, in addition to their rheological parameters, than to crystal size [90]. Furthermore, when other emulsifiers were used (MG, SSL, PGE, and SMS), softer cookies were produced from high-viscosity oleogels using structurants with higher hydrophilic-lipophilic balance values, reflecting their greater ability to withstand processing conditions [77]. However, the hardness of these cookies was driven neither by the amount of crystals or SFC nor by the type of crystalline material in the oleogels, nor by their viscoelasticity, even during heating. This is a very interesting result that should be taken into account when considering oleogel formulation optimization, since increasing structurant concentration usually increases oleogel hardness, but this will not necessarily favor final cookie properties.

Cookie structure of similar hardness to that based on shortening proved to be of homogeneous porosity [77, 94]. Additionally, the increased cookie brightness has been related to a higher oleogel SFC, postulating that the mobility of some browning reaction reagents could be hindered by the higher SFC. Moreover, the generation of more fluid doughs during preparation tended to produce cookies that were more differentiated from conventional fat ones [94]. Some significant total color differences with respect to the control (ΔE) have also been found between reformulated and traditional cookies ($\Delta E > 3$), indicating that consumers would perceive such a change [78, 80, 94]. However, cookies reformulated with MG, RBW, SFW, and BW oleogels scored comparable or even higher than shortening-based cookies in terms of surface color or overall acceptability [85, 94]. Moreover, cookie dimensions and spread ratio were not modified after reformulating the recipe by replacing 50 and 100% of margarine with an emulsion from RBW oleogel [95]. However, less hard doughs and cookies and lighter colored cookies were obtained as fat replacement increased.

Multicomponent oleogels using β-sitosterol (BS)-BW and BS-LC produced doughs with hardness more similar to a margarine-based dough, in comparison to MG or CRW oleogel [80]. However, not only were these differences not maintained after baking, but also these structurant combinations generated cookies with a different hardness than conventional ones. Nevertheless, similar or improved cookie properties were produced using BS-MG oleogels instead of shortening—even diminishing 30% SFA—regardless of using crude or refined soybean oil [79].

Bigels in which oleogels were produced from canola oil and CLW or BW showed great potential to replace conventionally used cookie fat [96, 97]. Although some differences in cookie hardness and color have been detected due to fat replacement (Fig. 26.4a1–a3), improvements in FAP and starch digestibility were remarked [96]. Moreover, emulsion-templated approaches were tested as total or partial substitutes for commercial fats or oils [98, 100]. By replacing oil with a low-vanillin content oleogel (without emulsifier), the spread ratio, hardness, and ΔE of cookies were diminished, resembling shortening-cookie properties [100]. Those oleogel structures probably interact with proteins and starch in a more butter-like than oil-like manner, providing techno-functional properties more similar to a commercial fat—although FAP was notably modified, lowering SFA by 20–80% and increasing UFA by 90–370%. Varying some structurant characteristics, such as the fatty acid chain length of polyphenol ester, modified oleogel properties more significantly than those of cookies [98]. In contrast, fat replacement level was strongly reflected in some product properties, generating harder doughs and cookies with higher break strength and lower overall quality scores as butter replacement was increased [98, 99]. Additionally, when HPMC emulsion-based or foam-templated oleogels were used instead of shortening, harder cookies were obtained, resulting in low consumer acceptance [94]. Some product texture properties were raised by increasing the amount of MG in an HPMC-MG oleogel, which corresponded to the formation of more compact structures (Fig. 26.4a4–a6) generated by reduced dough aeration [75]. However, no differences in spread ratio or overall appearance were observed when esterified rice flour with citric acid (ERCA)-BW oleogels were used, although a 28% UFA increase was achieved [93].

Medium chain triglycerides-EC oleogels obtained through high intensity ultrasound (HIU) gelation were introduced in cookie formulation [91]. HIU-based oleogels allowed to obtain cookies with good texture and not as hard as those produced with non-HIU oleogels. Sensory scores and overall acceptability were improved by using HIU-based oleogels.

Crackers and breads, although typically require low fat content (<10%), still need the properties that fats provide [55, 56]. MG and RBW oleogels were shown to

Fig. 26.4 Appearance of bakery products elaborated with different oleogel-based systems compared with a control product: (**a**) cookies from (a1) commercial shortening, candelilla wax oleogel-based bigel with (a2) 50% and (a3) 100% fat replacement, reproduced from [96] with permission from Elsevier, and from 50% fat and 50% hydroxypropyl methylcellulose (HPMC)-saturated monoglycerides (MG) oleogels with (a4) 0 wt% MG, (a5) 4 wt% MG, and (a6) 10 wt% MG, reproduced from [75] with permission from Elsevier; (**b**) breads from (b1) margarine and (b2, b3) oleogels from two different sodium stearyl lactates, reproduced from [56] with permission from Elsevier; (**c**) puff pastries from (c1) commercial laminating margarine and (c2) MG: saturated fatty acid-dried emulsion, reproduced from [106] with permission from John Wiley and Sons; (**d**) cakes from 50% fat and 50% HPMC-MG oleogels with (d1) 0 wt% MG, (d2) 6 wt% MG, and (d3) 8 wt% MG, reproduced from [75] with permission from Elsevier; and (**e**) muffins from (e1) margarine, (e2) unstructured oil, and (e3) optimized MG oleogel, reproduced from [38] with permission from John Wiley and Sons

be acceptable options to replace shortening in these formulations [55]. MG oleogel was postulated as a retardant of starch gelatinization and retrogradation and as a crumb softener, leading to a better performance than RBW oleogel and commercial shortening during bread but not for cracker production. However, lubrication and gluten network development were similar in the three lipid-based systems and, therefore, in the final product appearance. In low-fat bread doughs, water absorption was not modified by the type of fat used [55], although it was in doughs with higher lipid amounts [55, 109]. An EC-triglyceryl monostearate oleogel-based shortening also displayed superior properties to a commercial fat for producing softer breads with higher specific volume [40]. Breads formulated with SSL oleogels achieved a similar general appearance to those formulated with margarine (Fig. 26.4b) or sunflower oil, even with some textural property differences and a 30% reduction in SFA compared to margarine-based breads [56].

Also, it has been shown that oleogelation can greatly affect the final characteristics of sweet bread, in which 12–19% fat is used. Butter replacement with up to 75% CLW oleogel led to obtaining breads with specific volume and hardness similar to the control, even reducing SFA from 71 to 35% [58]. Moreover, palm oil substitution by sunflower oil or by the corresponding MG oleogel was detrimental to bread specific volume, but the use of an MG hydrogel resulted in a product with quality characteristics comparable to those of the control, although both gels implied a reduction of ~80% SFA [59]. This fact evidenced that MG interacted in different ways depending on the phase or interface in which they are found, going from the generation of an inhomogeneous lipid distribution and the corresponding lower leavening effect by the MG oleogel to a better interaction with the preparation ingredients—water, starch, and gluten—by the MG hydrogel.

Baked or steamed buns were also reformulated by substituting margarine with oleogels based on HPMC and XG [57]. Although implying a great FAP improvement, no differences were detected in crumb structure, bun shape, texture, sensory attributes, or lipid digestibility.

Puff pastries are laminated doughs that require high fat content (~37%) to be properly mechanized to reach rheological fat-dough equilibration. Laminating CM was fully replaced by high internal phase O/W emulsion produced with MG-SFA, achieving satisfactory performance during laminating and baking and a good final appearance (Fig. 26.4c)—with ~7% SFA reduction—[106, 107]. Although an increased oiliness in the mouth was detected, product friability was satisfactorily maintained. Other emulsion formulations based on olive pomace oil, PS, and different gelators—BW, saturated triglycerides, or MG—were postulated as possible alternatives for puff pastry fat [110]. Although acceptable approximations were achieved between the BW emulsion and the commercial product, the final product functionality remains to be tested.

Cakes, with a characteristic spongy structure due to air entrapment, are usually made with 10–25% butter or margarine. When conventional cake fats were replaced by some wax oleogels, batters with lower viscosity and less shear-thinning behavior were obtained, which affected batter aeration [63, 64]. Cake porosity and specific volume decreased using wax oleogels instead of shortening, resulting in harder cakes with higher chewiness [64]. Exceptionally, BW oleogel-based cakes maintained a similar specific volume to the control although doubling in hardness [64], while CLW oleogels increased cake specific volume reducing hardness [61]. Crumb structures with homogeneously distributed fine air cells were obtained using shortening and BW oleogels [64]; however, less amorphous areas and more hydrated and short-range crystallized starch networks were reported using CLW oleogels [61]. Nevertheless, a consumer health problem could be introduced due to the proposed substitution, since the in vitro starch digestibility increased with it. Although total substitution can reduce cake SFA by ~70%, partial replacement might be a better option for maintaining cake properties. Oils structured with CRW—as oleogel or emulsion—were also used to partially replace a high-fat material in cakes [70]. In general, UFA increment did not affect batter's rheological behavior nor cake's textural and sensory properties. Gelled emulsions produced batters with higher consistency and cakes with greater aeration than oleogel-based cakes; however, the latter had the highest overall acceptability by consumers. Full margarine replacement with SLW oleogel-based emulsions also showed that, while the batter exhibited unfavorable properties compared to the control, the cake had acceptable texture and sensory characteristics [48].

Oleofoams from wax oleogels were used to replace palm fat in cakes [60]. Although some waxes such as SFW showed a good ability to generate oleogels, this was not transferred to the behavior of oleofoams nor to those of oleogel-based batters. The different behaviors were related to the varying arrangement of wax crystals in air interphase or bulk oil. However, only minor differences were found in cakes, being BW oleogel the preferred to diminish oil migration. Moreover, although sensory results did not show that cakes were much different, oleogel-based cakes got higher scores.

A gluten-free batter for producing aerated products was reformulated using BW oleogel co-crystallized with a commercial cake shortening [62]. Full replacement produced batters with reduced air-holding ability and cakes with lower porosity and specific volume. However, similar properties to control cakes were obtained with partial replacement.

Oleogels from MG and BW were also used as margarine replacers [65] and as a result of a higher moisture retention capacity, reformulated cakes with higher porosity and dimensional values than the control were obtained. After sensory analyses, an MG-BW-HOSO-based cake was preferred even over control. Moreover, the shelf life of this reformulated cake was extended, diminishing its oil migration. Binary systems from different gelators reinforced with adipic acid (AA) have been used to improve cake properties. Acceptable texture profile, color, and organoleptic properties were obtained by 50% shortening substitution with CRW-AA or EC-AA oleogels, making these results better than those obtained without the reinforcement [66, 67]. Moreover, the partial substitution with EC-AA oleogel resulted in cakes with greater oxidative stability [68]. Some acceptable results were obtained when using oleogels from PS or sucrose stearate/ascorbyl palmitate blends to replace shortening in a low-fat cake recipe, improving cake oxidative stability compared with oil-based cake but not reaching that of shortening one [69].

Polymer-stabilized O/W emulsion-based oleogels have been used to replace commercial fats or oils. Full replacement of margarine, peanut oil, or shortening resulted in softer and spongier cakes with higher adhesiveness [71, 72]. Moreover, the non-uniform pore distribution in oil-based cakes could be partially overcome by using oleogels [72]. Although cake texture was nearly maintained when shortening was partially or even totally substituted with a gum emulsion-based oleogel, sensory analysis showed that the 75% replacement-based cake was preferred over the control [73]. Along with the search for improvements in emulsion stability and shelf life, e.g., by using antioxidant compounds, reformulation effects on final product properties must be considered. By using an optimized emulsion-based oleogel including polyphenols, oil oxidation was retarded and batter and cake properties more similar to margarine-based than oil-based products were obtained [74].

Foam template-based oleogels were used to partially replace commercial butter in a low-gluten cake, achieving improved hardness and chewiness but showing no major effect on the final product appearance (Fig. 26.4d) [75]. Increasing MG content resulted in higher dough density and lower cohesiveness and resilience cakes. Moreover, cakes prepared using protein foam-based oleogels were harder than shortening-based cakes, although similar to oil-based ones, probably because oil and protein oleogel could not avoid the starch-protein adhesion and thus product staling [76]. Adding another structurant may help to overcome this issue. In addition, the cake color was modified by oleogel ingredients, although they were at low concentrations ensuring maintenance of healthy oil composition.

Muffins or cupcakes are a special baked cake type. An optimized MG oleogel proved to be an effective replacement for CM, not only from measured properties but also due to sensory analysis and consumer acceptance (Fig. 26.4e)—with 68 and 100% SFA and TFA reduction, respectively—[38, 101]. By increasing the shortening replacement level with HPMC foaming-templated or CLW oleogels, batter specific gravity tended to increase and muffin air cells were bigger, closer, and not evenly distributed [102, 104, 105]. However, muffin specific volume was not modified with up to 50 or 25% replacement depending on the used system, ensuring an improvement in the UFA:SFA ratio of at least 42%. In addition, when shortening was replaced with MG-enriched CLW oleogels, muffins with similar porosity and improved textural characteristics with respect to CLW oleogels were produced [102]. The addition of β-carotene in CLW-MG oleogels did not adversely affect muffin structural characteristics; however, muffin color changed, and its oxidative stability was improved.

By replacing shortening with different wax oleogels in a gluten-free muffin recipe, water content, physical parameters, and textural properties were not significantly affected, although dough specific gravity was slightly increased and muffin porosity modified [103]. Muffin crust color changed depending on the wax used ($\Delta E < 6$) and lipid fraction was modified, diminishing ~40% SFA and increasing ~45% UFA.

Tart pastries need solid fats to form properly. Oleogels from EC, MG, or EC-MG were used to replace butter and palm shortening [108]. MG oleogels and control fats produced doughs with similar firmness. However, MG oleogel-based tart had lower hardness than the control. On the other hand, EC oleogel-based dough and tart were the firmest and most hard. However, combining EC and MG was a successful strategy to improve textural properties and dough workability.

26.4 Confectioneries

The quality of chocolate, fillings, and related confectioneries is significantly affected by the fat phase. Therefore, its substitution is the main challenge facing the confectionery market due to changes in thermal properties and oil migration that occur during storage. In this regard, different oleogel-based systems have been proposed as fat replacers (Table 26.3), offering promising results.

Cocoa butter constitutes a main ingredient in confectionery due to its fat or triglyceride composition—high in SFA—resulting in unique physical and sensory characteristics of final products. CB equivalents are needed, not only because of excessive fat consumption but also because it is the most expensive confectionery ingredient. Against this background, HPMC structured HOSO to develop an emulsion-templated oleogel with a healthier FAP to be used as a partial or complete CB substitute [111]. Oleogel incorporation resulted in a significant effect on hardness, which decreased as oleogel quantity increased in the system, probably due to the difference in SFC, i.e., SFA contributed more solidity. Although OBC was also affected by the substitution, very high values were still obtained (>93%), showing that strong networks had been formed. Among the reformulated systems, similar rheological and textural characteristics to CB were obtained with 50% replacement.

Chocolate is globally consumed primarily for its sensory stimulation and excitement potential. However, because it is composed mainly of CB or milk fat—both responsible for hardness, temperability, and melting point—it is a high-calorie product containing up to 40% SFA [126]. Consequently, fat replacers for healthier chocolate formulations that maintain the desirable physicochemical and sensorial properties are required. Recently, MG-palm oil oleogels combined with some healthy sweeteners—maltitol, tagatose, and palm sugar—were tested as CB replacers to prepare heat-stable and bloom-resistant chocolate [113]. As a result of the reduced CB content, all reformulated products were significantly less hard. The 50:50 sucrose:palm sugar oleogel-based chocolate presented the polymorphic structure required in this type of product (β crystals) to provide stable crystal networks. In turn, it improved the shape-retention capability, maintained high bloom resistance, and displayed higher melting enthalpy compared to the pure CB chocolate. Moreover, this reformulated chocolate showed the highest overall acceptability, with no waxiness by fast mouth fusion. Three oleogels structured by MG, EC, or BS-LC were used to reformulate dark chocolate with total or partial substitution of CB [114, 115]. Incorporating oleogels resulted in chocolates with soft texture, shear-

Table 26.3 Oleogel applications in confectionery products

Food product	System	Gelator (wt% in system)/oil	Replaced fat (wt % replacement)	Reference
Cocoa butter	Emulsion-templated oleogel	HPMC (1)/HOSO	CB (50, 60, 70, 80, 100)	[111]
	Oleofoam	CB (15, 22, 30)/HOSO	CB (70, 78, 85)	[112]
Chocolate	Oleogel	MG (10)/palm	CB (30)	[113]
	Oleogel	MG, EC (10), BS:LC (8:2)/corn	CB (50)	[114]
	Oleogel	MG, EC (10), BS:LC (8:2)/corn	CB (50, 100)	[115]
	Oleogel	BS: Stearic acid (2.4:9.6), BS:GO (4.8:7.2), BS:LC (9.6:2.4)/corn	CB (50)	[116]
	Foam-templated oleogel	HPMC (2)/sunflower	CB (30, 50, 70, 100)	[117]
	Emulsion-templated oleogel	HPMC (0.5–2)/HOSO	CB (50)	[118]
Filling creams	Oleogel	MG (10)/HOSO	Beef fat (100)	[119]
	Oleogel	BW (1.5–3.5)/rice bran	Palm oil (17, 33, 50)	[120]
	Oleogel	CRW (6)/pumpkin seed	CB (100)	[121]
	Oleogel	MG (3, 6), STS:LC (3:3, 4:4, 5:5)/soybean	Filling fat/coating fat (100)	[122]
	Oleogel	MG:CLW (10:0, 4:6, 0:10)/canola	Shortening (100)	[123]
	Oleogel	MG:CLW:HF (0.35–1.40:0.35–1.05:0.35–1.75)/soybean, HOSO	Shortening (100)	[124]
	Oleogel	BS:GO (5:5, 12.5:12.5)/sunflower	*n.a*	[125]

BS β-sitosterol, *BW* beeswax, *CB* cocoa butter, *CLW* candelilla wax, *CRW* carnauba wax, *EC* ethylcellulose, *GO* γ-oryzanol, *HF* hard fat, *HOSO* high oleic sunflower oil, *HPMC* hydroxypropyl methylcellulose, *LC* lecithin, *MG* saturated monoglycerides, *n.a* not applicable, *STS* sorbitan tristearate

thinning behavior, and a high degree of unsaturation [115]. The polymorphic form and thermal properties were similar to the traditional chocolate. However, the different gelation mechanisms of the three kinds of oleogels resulted in variations in some physicochemical properties of reformulated products. For 50% replacement, EC oleogel-based chocolate had the highest hardness, which was attributed to higher SFC and particle-particle interactions. Nevertheless, MG oleogel was the only one that could form a solid-like chocolate with total CB replacement. Furthermore, these oleogels improved chocolate bloom stability during different storage conditions, even with reduced saturation levels [114]. While an incipient bloom was detected at 1 day of cold storage (Fig. 26.5a1–a4), this was reversed with longer storage time obtaining non-significant differences between samples. Furthermore, under fluctuating temperatures, oleogel-based chocolates exhibited bloom-delaying properties.

Fig. 26.5 Appearance of confectioneries elaborated with different oleogel-based systems compared with a control product: (**a**) chocolates from (a1) cocoa butter, (a2) MG oleogels, (a3) BS-LC oleogels, and (a4) EC oleogels, all stored at 4 °C for 1 day, reproduced from [114] with permission from Elsevier, and from (a5) cocoa butter and hydroxypropyl methylcellulose (HPMC) oleogel with (a6) 0.5 wt% HPMC, (a7) 1.5 wt% HPMC, and (a8) 2 wt% HPMC, reproduced from [118] with permission from Elsevier, under the terms of the Creative Commons CC-BY 4.0 license; and (**b**) filling creams from (b1) commercial fat and variable amounts of saturated monoglyceride oleogel (b2) 26%, (b3) 30%, and (b4) 40%, reproduced from [119] with permission from John Wiley & Sons

Due to fat bloom formation, dark chocolate showed bigger hardness changes during storage than oleogel-based chocolates, which remained thermally and polymorphically stable. Similarly, other binary oleogels from BS combined with γ-oryzanol (GO), stearic acid, or LC were investigated to replace CB in dark chocolate [116]. Although some properties differed between the three oleogels, reformulated chocolates exhibited the required β crystals, as well as similar characteristics as that of dark chocolate.

In addition to oleogels, other replacement systems based on HPMC were addressed to reformulate chocolates. For instance, a foam-templated approach was tested as a partial or total CB substitute [117]. Despite some differences in properties between oleogel-CB blends, these differences diminished when other ingredients were added to oleogel-based chocolate formulations. All reformulated products presented an improved FAP, proportionately to CB replacement. Nonetheless, a sensory analysis indicated that HPMC oleogel could replace CB up to 70%, although a technological perspective would suggest a replacement level of 50%. Other HPMC oleogels obtained by emulsion-templated approach were developed to produce a healthier chocolate with lower fat content and optimum sensory properties [118]. Oleogel-based chocolates had a similar appearance to CB-based ones (Fig. 26.5a5–a8), but softer texture related to the structurant amount in the oleogel. Furthermore, texture and flavor were negatively impacted by the highest HPMC concentration, while more similar sensory properties to those of the control were achieved with the lowest proportions of cellulose. To summarize, the majority of research has concentrated primarily on dark and milk chocolate reformulations, while those for white, compound, and ruby chocolate should also be examined to determine how low-calorie ingredients affect their properties.

Filling creams represent an important ingredient in different food products, such as cookies and filled chocolates. Based on their composition—large amounts of fat and sugar—they are considered unhealthy products and their consumption is concerning. As a result, it is becoming increasingly important to replace or reduce the use of HF in these formulations maintaining the required physical and organoleptic properties. In light of this, an MG oleogel was tested as a fat material in filling creams for sandwich cookies [119]. Creams prepared with variable oleogel amounts resulted in a significant impact on final properties and appearance (Fig. 26.5b). Increasing oleogel concentration, softer creams with higher adhesiveness than the control were obtained, improving cookie adhesion. Wax oleogels have also been proposed to replace filling cream fats. Up to 17% replacement was possible with BW oleogels, maintaining high OBC and significantly lowering SFA content [120]. Incorporating lucuma powder into a CRW oleogel resulted in creams with lower saccharose levels without affecting OBC and water activity, ensuring physical and microbiological stability [121]. In a model praline system, MG oleogels displayed better migration-delaying properties than STC-LC oleogels [122]. Additionally, these oleogels were fully melted at body temperature, not causing waxiness. Moreover, multicomponent oleogels have been used in filling production aiming to obtain a healthier product [123, 124]. For instance, a reduction of at least 53% in SFA and 100% in TFA was achieved with some MG-CLW-HF oleogels, as well as good technological properties [124].

26.5 Meat-Based Products

Meat product final quality is greatly affected by the amount and/or composition of the animal fat employed, as it contributes to textural characteristics—tenderness, palatability, juiciness—as well as flavor and physical appearance. Nonetheless, excessive consumption may be harmful to human health due to its high SFA,

TFA, and cholesterol content [127, 128]. Hence, proper strategies are needed to produce reformulated low-fat meat products with healthier FAP maintaining consumer acceptance. This section assesses the development of a range of meat products whose natural fat profile has been modified with different oleogel-based systems, as shown in Table 26.4.

Burgers are the most popular fast food across genders and cultures, containing between 20 and 25% animal fat. Different oleogels have been proposed as fat replacers for burger formulations, achieving improved FAP. Oxidative stability and hardness of BW oleogel-based patties were lower compared to the control [130, 131], while using EC oleogels enhanced textural properties, although it also reduced oxidative stability [129]. Moreover, sensory analysis revealed high ratings for BW oleogel-based burgers, opposite to EC-SMS ones, which scored below neutral [131]. Patties with some acceptable characteristics were obtained using EC and CRW oleogels reinforced with AA, even significantly better than those formulated with non-reinforced oleogels [66, 67]. Additionally, GO-BS-linseed oil oleogels did not alter burger texture while improving its nutritional profile [132]. Overall, lower fat replacement products were accepted by consumers, despite a marked preference for the control.

Reformulated hamburgers with foam-based oleogels were softer, having an enriched FAP due to a $\sim$ 65% SFA reduction [134]. Recently, bigels using EC oleogels have shown great potential in a meat product [135]. Compared to the control, bigel-based burgers showed improved cooking properties and sensory analysis scores. However, lipid oxidation increased with bigel incorporation, which could be improved by adding antioxidants or other structurants.

Meatballs consist of ground meat rolled into a ball with other ingredients. A healthier meatball was designed based on the partial replacement of beef fat with CRW oleogels, negatively affecting their textural properties [139]. Oxidative stability was affected not only by the amount of fat substitution but also by the type of oil used in the oleogel, improving by raising the MUFA:PUFA ratio. Meatballs with 25% sunflower oil oleogel substitution scored significantly higher than the control in some characteristics and overall acceptability.

Meat batters are made of water, fat, and protein, and thus they are considered emulsions. Consequently, fat and moisture stabilization is essential to the manufacture of these products. Total fat substitution in meat batters has been studied by different replacement systems. Healthier FAP was achieved with structured canola oil systems—kappa-carrageenan emulsions or EC oleogels—by decreasing SFA and the n-6/n-3 ratio and increasing PUFA [140]. Reformulated products showed similar color and texture to the control while reducing lipid oxidation. Additionally, oleogel-based batters showed better matrix stability than emulsion-based ones, with a more uniform microstructure and no fat losses. Multicomponent oleogels from soybean oil also decreased lipid oxidation, but no effects on texture were observed [141]. Although reformulated batters were darker and less red, products were accepted. Moreover, cellulose-based oleogels significantly increased PUFA while keeping total fat content close to control levels.

Table 26.4 Oleogel applications in meat-based products

Food product	System	Gelator (wt% in system)/oil	Replaced fat (wt% replacement)	Reference
Burgers	Oleogel	EC (10)/sesame	Animal fat (25, 50)	[129]
	Oleogel	BW (10)/rapeseed	Beef fat (100)	[130]
	Oleogel	BW (11), EC:SMS (11:3.67)/ olive:linseed:fish (44.39:37.87: 17.4)	Pork backfat (100)	[131]
	Oleogel	EC:AA (2:4)/soybean	Flank and shank fat (50)	[67]
	Oleogel	CRW:AA (2:4)/soybean	Bovine fat (50)	[66]
	Oleogel	GO:BS (4.8:3.2)/linseed	Pork backfat (25, 75)	[132]
	Oleogel	GO:BS (n.s)/linseed	*n.a*	[133]
	Foam-structured oleogel	HPMC (4)/canola	Beef tallow (50, 100)	[134]
	Bigel	EC (10)/sunflower	Animal fat (25, 50, 75)	[135]
	O/W emulsion	Pork skin (20)/olive	Bovine backfat (100)	[136]
	O/W emulsion	Prosella® (6.7)/olive	Pork fat (100)	[137]
	O/W emulsion	Sodium alginate: Carrageenan: Glycerol monostearate (1:1:1)/ olive	Beef backfat (33.3, 66.6, 100)	[138]
Meatballs	Oleogel	CRW (7.5)/sunflower, sunflower: black cumin seed (90:10, 80:20)	Beef fat (25, 50, 75)	[139]
Meat batters	Oleogel	EC (12), EC:MG (12:1.5, 12:3)/ canola	Beef fat (100)	[140]
	Oleogel	EC:Avicel RC-591:α-cellulose (7.37:1.815:1.815)/soybean	Pork backfat (100)	[141]
	O/W emulsion	Kappa-carrageenan (1.5, 3)/ canola	Beef fat (100)	[140]
Meat sauces	Oleogel	MG (0.5, 2.5), LC (2.5)/olive: sunflower (80:20)	Olive: Sunflower (100)	[142]
	Oleogel	MG, fatty alcohols (0.5, 2.5)/ olive:sunflower (80:20)	Olive: Sunflower (100)	[143]
Pâté	Oleogel	BW (8)/linseed	Pork backfat (30, 60)	[144]
	Oleogel	BW (9.12)/linseed	Pork backfat (60, 100)	[145]
	Oleogel	BW (11)/olive:linseed:fish (44.39:37.87:17.4)	Pork backfat (60, 100)	[146]
	Oleogel	EC (14), EC:MG (12–14:1.5–3)/ canola	Pork fat (100)	[147]
	Oleogel	EC:MG (12:3)/canola	Lard (20, 40, 60, 80, 100)	[148]

(continued)

Table 26.4 (continued)

Food product	System	Gelator (wt% in system)/oil	Replaced fat (wt% replacement)	Reference
	Oleogel	EC:SMS (11:3.67)/olive:linseed: fish (44.39:37.87:17.4)	Pork backfat (60, 100)	[146]
Salami	Oleogel	EC:MG, EC:SOSA (6–14: 3), EC:LC (8–12:1)/canola	Pork and beef fat (50)	[149]
Sausages				
Bologna	Oleogel	MG (5)/sunflower, HOSO	Pork backfat (25, 50, 75, 100)	[150]
	Oleogel	RBW (2.5, 10)/soybean, high oleic soybean	Pork backfat (100)	[151]
	O/W emulsion	Pork skin (37.5)/HOSO	Pork backfat (25, 50, 75, 100)	[152]
Breakfast	Oleogel	EC (8–14), EC:SMS (8–14:1.5–3)/canola	Pork backfat (100)	[153]
Fermented	Oleogel	MG (15)/olive	Pork backfat (50)	[154]
	Oleogel	BW (10)/olive:chia (80:20)	Pork fat (80)	[155]
	Oleogel	BW (8), GO:BS (4.8:3.2)/linseed	Pork backfat (20, 40)	[156]
	O/W emulsion	Soy protein isolate: Gelatin (10: 3)/olive:chia (80:20)	Pork fat (80)	[155]
	O/W emulsion	Kappa-carrageenan: Polysorbate 80 (1.5:0.12)/linseed	*n.a*	[157]
Frankfurter	Oleogel	BW (8)/linseed	Pork backfat (25, 50)	[158]
	Oleogel	RBW (2.5, 10)/soybean	Pork backfat (100)	[159]
	Oleogel	EC (10)/canola	Beef fat (100)	[160]
	Oleogel	MG:PS (15:5)/sunflower	Pork backfat (50)	[161]
	Oleogel	EC (8, 10, 12, 14), EC:SMS (8–14:1.5–3)/canola	Beef fat (100)	[162]
	Oleogel	EC:SMS (8:1.5, 8:3, 10:1.5)/ canola	Beef fat (20, 40, 60, 80)	[163]
	Oleogel, O/W emulsion	GO:PS (3:7, 6:4, 6:14, 12:8)/ sunflower	Pork backfat (50)	[164]
Sucuk	Oleogel	SFW, BW (10)/flaxseed	Beef tallow fat (100)	[165]
Thai sweet	Oleogel	RBW (2)/rice bran	Pork backfat (25, 50, 75)	[166]
Venison	W/O emulsion	Soy protein concentrate (0.1)/ olive	High-fat pork meat (15, 25, 35, 45, 55)	[167]

AA adipic acid, *BS* β-sitosterol, *BW* beeswax, *CRW* carnauba wax, *EC* ethylcellulose, *GO* γ-oryzanol, *HOSO* high oleic sunflower oil, *HPMC* hydroxypropyl methylcellulose, *MG* saturated monoglycerides, *n.a* not applicable, *n.s* not specified, *PS* phytosterols, *RBW* rice bran wax, *SFW* sunflower wax, *LC* lecithin, *SMS* sorbitan monostearate, *SOSA* stearyl alcohol/stearic acid

Meat sauce stabilization is a relevant issue due to the difficulty of structuring oil phases. Oleogelation can improve system quality and organoleptic properties as gelators structure the oil phase into a semi-solid gel network linking solid particles, thus preventing or delaying separation [14]. Taking this into account, MG and LC oleogels were tested as replacers for liquid oil in meat suspensions [142]. Both exert a stabilizing action on the sauces, although MG oleogels reduced phase separation more effectively than LC. Even with the lowest MG concentration, a stabilized sauce with similar rheological properties to those of the control was obtained. Likewise, MG or fatty alcohol oleogels were used in meat suspensions to improve their stabilities—being fatty alcohols more efficient than MG, even at low concentrations—[143]. Moreover, both oleogels enhanced meat sauce mechanical stress resistance without altering their consistency.

Pâté is a highly appreciated spreadable meat product known for its flavor and smooth texture, containing 17–50% animal fat [128]. BW oleogels were tested to replace pork backfat in liver pâtés, obtaining products with optimal FAP. Pâté stability, sensory parameters, and texture were not significantly affected by substitution with an olive, linseed, and fish oil oleogel [146]. Conversely, when a linseed oil oleogel was used, a decrease in hardness and adhesiveness was observed [144]. Fat replacement significantly increased lipid oxidation [146], which could be compensated by antioxidant incorporation [145]. Other reformulated pâtés made from EC or EC-MG oleogels exhibited higher oil loss than the control, possibly due to the formation of larger fat globules [147]. Also, they all had similar sensory hardness, oiliness, juiciness, and cohesiveness as the control. Overall, the 12:3 EC: MG oleogel-based pâté demonstrated the best performance in achieving similar characteristics to the control, while also successfully reducing SFA. By substituting up to 60% pork fat with this binary system, the final product demonstrated good oil retention and improved textural properties, without modifying sensory properties and color [148]. An EC-SMS oleogel prepared using a lipid mixture of olive, linseed, and fish oils was also tested, yielding products not significantly different from the control in technological behavior and physicochemical properties, except lipid oxidation [146]. Additionally, although all samples were rated near neutral, sensory parameters showed a negative effect.

Salami, a coarse ground meat product, has large fat particles that play a crucial role in their appearance, texture, and mouthfeel; hence, its replacement without product destabilizing is a challenge. Fat partial replacement with different canola oil oleogels—from EC-MG, EC-LC, or EC-stearyl alcohol/stearic acid—was proposed [149]. Structuring canola oil led to a microstructure with fat globules similar to the control. Specifically, EC-LC oleogels produced larger gelled oil particles that effectively bind to the protein matrix. Texture parameters were not affected in most formulations.

Sausages consist of ground meat and non-meat ingredients which contribute to the quality, taste, and aroma of the final product. Several types of sausage are available, including fresh, smoked, cured, and cooked [128]. *Bologna sausages* have been reformulated with oleogel-based systems from different gelators. MG oleogels were used to replace pork fat, resulting in sausages that were easier to slice than the control, possibly due to a more compact product structure as the amount of pork fat was reduced [150]. Despite some modified textural properties, most replacements produced stable emulsions that were sensorially accepted by consumers with increments in UFA. Oleogels formulated with soybean or high oleic soybean oil and RBW were also tested to totally replace pork backfat [151]. Sausage quality and organoleptic properties were not affected by the oil type, except for FAP, which was improved by the high oleic type resulting in products with lower n-6/n-3 ratios. In addition, no sensory differences were detected in terms of aroma, flavor, texture, and moistness, but the color of the control was more intense. Oleogel-based bologna sausages had fewer and larger lipid globules than unstructured oil-based ones, more closely resembling pork fat sausages. Stable batters were obtained with the use of oleogels instead of control fat, observing water separation for the sausages containing oil without structurant [150, 151].

Breakfast sausages were reformulated using oleogels from canola oil structured with EC or EC-SMS to totally replace pork backfat [153]. SMS incorporation improved hardness values being similar to those of control. However, sensory hardness could not be matched by oleogels, and therefore samples were generally less sensory juiciness and oiliness, while the replacement with ungelled canola oil had comparable values to the control.

Fermented sausages have been produced with different oleogel-based systems to evaluate the effects of pork backfat replacement. An MG-olive oil oleogel showed promising results by replacing 50% fat [154]. Reformulated sausages were sensory acceptable and microbiologically safe, with an enhanced FAP by reducing ~19% cholesterol and ~17% SFA and increasing ~9% MUFA. Similarly, two oleogels from BW or GO-BS and linseed oil were tested as replacers [156]. Gelator type and replacement level led to significant differences in various physicochemical and mechanical properties as well as consumer acceptability—probably due to modifications generated in the drying process from different initial water contents. Sausage nutritional profile was improved with oleogel incorporation achieving the greatest reductions in SFA with the highest level of replacement. However, oleogel-based sausages revealed lower sensory acceptance than the control.

Not only oleogels but also emulsions were tested to replace the fat in this type of sausages. The ability of wax oleogels and emulsion gels formulated with a mixture of olive and chia oils to replace fat was studied [155]. Both replacement strategies enriched FAP with decreased SFA and increased PUFA. Sausage oxidative stability and their microbiological and technological properties were not affected using oleogel or emulsion, except for hardness (oleogel $<$ emulsion $\sim$ control). Reformulated sausage sensory attributes scored similarly, but lower than control. Besides, the lipid source affected the cross-sectional appearance, perfectly distinguishing the animal fat in the meat matrix of the control product, whereas the presence of oleogel or emulsion in the reformulated products was less discernible (Fig. 26.6a). To preserve product quality, a linseed oil emulsion added with a natural antioxidant extract was used [157]. The extract could enhance sausage physicochemical properties and prolong their shelf life, even using a high-PUFA oil.

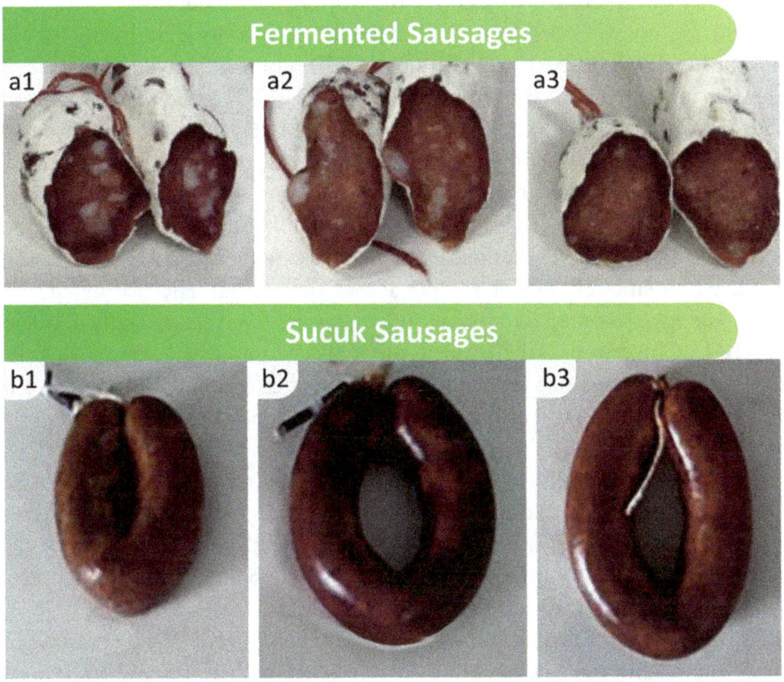

Fig. 26.6 Appearance of meat-based products formulated with different oleogel-based systems compared with a control product: (**a**) low-fat fermented sausages from (a1) animal fat and with 80% fat replacement using wax (a2) oleogel and (a3) emulsion, reproduced with permission from [155] under the terms of an open access Creative Commons CC BY 4.0 license; and (**b**) sucuk sausages from (b1) animal fat, (b2) beeswax oleogel, and (b3) sunflower wax oleogel, reproduced from [165] with permission from Elsevier

Frankfurter sausages have also been successfully manufactured with a variety of oleogel-based systems. For instance, BW-linseed oil oleogels were tested to replace backfat significantly improving sausage FAP by reducing SFA and cholesterol content and balancing the n-6/n-3 ratio [158]. Although some physicochemical, mechanical, and sensory properties were unfavorable, all reformulations scored positively and received acceptance; however, the control received the highest acceptability. Fat total replacement was evaluated using RBW-soybean oil oleogels, significantly reducing frankfurter flavor [159]. Gelator concentration affected product quality, but not enough to distinguish their texture from that of the control. Another monocomponent oleogel—EC-canola oil—was investigated to totally replace fat, resulting in products with improved texture compared to those made with ungelled oil and similar chewiness and hardness to the control [160].

Fat partial replacement was also studied using binary oleogels from PS and a high MG content, resulting in sausages with a PUFA-rich and SFA-low lipid profile [161]. Although some differences in texture were found between oleogel-based frankfurters and the control, sensory analysis indicated that reformulated products were generally accepted. Furthermore, frankfurter reformulation substituting 100% beef fat with EC or EC-SMS oleogels showed the possibility of tailoring textural and sensory sausage properties [162]. Without SMS addition, lower sensory hardness than that for ungelled oil was obtained. When an 8:1.5 or 8:3 EC:SMS ratio was used, hardness was similar to the fat control, while higher EC concentrations resulted in a significant increase in this parameter. Likewise, the same binary system was used to reformulate frankfurters based on the partial replacement of beef fat [163]. Sausage containing 10:1.5 EC:SMS oleogel had the most similar parameters to the control. In addition, juiciness and oiliness increased as replacement levels increased, while smokehouse and water losses decreased. The use of GO-PS-sunflower oil oleogels or their emulsions was also assessed to partially replace pork backfat [164]. Compared to the control, oleogels did not affect any sausage textural parameters, while emulsions reduced chewiness, hardness, and gumminess. Furthermore, all treatments had similar acceptance, except for some emulsion-based sausages.

Sucuk sausages were reformulated with flaxseed oil oleogels from BW or SFW to replace 100% fat [165]. Some color variations were observed, which could be due to differences in fat sources used (Fig. 26.6b). Oleogel-based sausages had lower textural and sensory properties and consumer acceptance than the control, but an $\sim$ 17% increase in PUFA and essential fatty acids.

Thai sweet sausage formulation was redesigned with a RBW-rice bran oil oleogel to replace the traditional fat, resulting in a product with improved FAP and some changed final characteristics [166]. For 75% replacement, softer sausages with the lowest cholesterol level were obtained, while the highest overall acceptance was reached with 50% replacement.

Venison sausages were reformulated by replacing high-fat pork meat with an emulsion from olive oil and soy protein [167]. During ripening, all replacements showed satisfactory physicochemical characteristics and acceptable levels of lipolysis and lipid oxidation. Although all emulsion-based sausages were accepted by consumers, those containing no more than 25% emulsion were preferred.

26.6 Dairy Products

Different reformulations studied to decrease SFA in dairy-based products—and/or to simulate them as vegan products—are presented in Table 26.5.

Cheese, being recognized as a great protein source, is usually formulated with a high content of saturated fats, whether of dairy or vegetable origin [168]. Both fat content and composition affect important cheese properties such as flavor, rheology, texture, and appearance [170]. Therefore, the replacement of high SFA feedstock with oleogel-based systems can drastically impact product quality and consumer acceptance. CRW oleogels were tested as palm oil full replacers in the production of a low SFA cheese mimic, obtaining an increase in product elasticity, hardness, and chewiness and at least 85% SFA reduction [168]. In addition, oleogel color was transferred to the final product, generating slightly darker structures.

Cream cheese is a fresh, fermented, unripened spreadable product with a high moisture and HF (~30%) content and a whitish base color. RBW or EC oleogel-based cream cheesees were successfully structured showing microstructures with similar fat globule size to a full-fat commercial sample and, consequently, were similar in hardness, spreadability, and stickiness [170]. Also, a 25% reduction in total fat and a 120% increase in UFA were achieved. RBW was preferred to EC as the structurant of high oleic soybean oil instead of the regular type. Although no or little differences were detected between texture and mouthfeel of an RBW oleogel-based cream cheese and a control, flavor intensity and bitterness were rated as stronger, warranting some additional reformulation corrections. Additionally, the whole reformulation process had a protective effect against oxidation [169].

A cheese product based on commercial fat partial replacement with RBW or SFW oleogel was proposed [171]. Structural differences in the shapes of fat globules and their connections were associated with possible differences in protein interactions with SFA and UFA. However, increasing wax content favored a better oil

Table 26.5 Oleogel applications in dairy-based products

Food product	System	Gelator (wt% in system)/oil	Replaced fat (wt% replacement)	Reference
Cheese	Oleogel	CRW (3, 6, 9)/canola	Palm oil (100)	[168]
Cream cheese	Oleogel	RBW (10)/high oleic soybean	Commercial cream cheese fat (100)	[169]
	Oleogel	RBW, EC (10)/high oleic soybean, soybean	Commercial cream cheese fat (100)	[170]
Processed cheese	Oleogel	RBW, SFW (5, 10)/soybean	Commercial cheese fat (41.8)	[171]
Ice cream	Oleogel	RBW (10)/HOSO	Milk fat (100)	[172]
	Oleogel	RBW, CLW, CRW (5, 7, 10)/HOSO	Milk fat (100)	[173]
	Oleogel	EC:UMG, EC:MG-DG (10: 3)/HOSO	Coconut fat (100)	[174]
	W/O emulsion	Hydrolyzed collagen:PS (5: 23.75)/n.s	Milk fat cream (50, 100)	[175]
	O/W emulsion	Whey protein (46.03)/LC-added high oleic palm oil	Milk fat cream (100)	[176]
	O/W emulsion	PS:GO (0.128–0.384:0.192–0.576)/sunflower	Milk cream (100)	[177]
Vegan cream	Oleogel	CLW (3)/rapeseed:linseed (1:1)	Milk fat, palm oil (100)	[178]
	Emulsion-templated oleogel	Basil seed gum:Soy protein (1:0, 2:1)/n.s	Fat (100)	[179]

CLW candelilla wax, *CRW* carnauba wax, *DG* diglycerides, *EC* ethylcellulose, *GO* γ-oryzanol, *HOSO* high oleic sunflower oil, *LC* lecithin, *MG* saturated monoglycerides, *n.s* not specified, *PS* phytosterols, *RBW* rice bran wax, *SFW* sunflower wax, *UMG* unsaturated monoglycerides structuration, reaching rheological and texture properties similar to control. Moreover, although reformulated products had similar total fat content, an $\sim$ 30% SFA reduction was obtained.

Ice cream is a frozen colloidal structure composed of air bubbles, fat globules, ice crystals, and an unfrozen whey phase, as well as sugar compounds [175]. A semi-solidified emulsion is required to achieve the fat coalescence necessary in the whipping and freezing processes, which leads to air bubble stabilization and, subsequently, to the appropriate overrun and melt resistance. Ice-cream fat—being 6–16% of product—is usually high in SFA, which is recognized as responsible for contributing to the freezing and melting behavior expected in ice cream [180]. In addition, high SFA content provides a smooth ice-cream texture and serves as an aroma carrier [172, 173, 175]. Therefore, the challenge is to maintain the ice-cream structure by replacing HF with vegetable oils, allowing the formation of a suitable colloidal oil network and inhibiting droplet coalescence. Oleogels from HOSO and EC or waxes, added with UMG or MG-DG, were tested as solid fat replacers. To avoid oil overheating during EC system emulsification, a pre-made oleogel method was successfully applied [174]. Although UMG favored the slow melting of wax oleogel-based ice creams due to their fat globule aggregation capability [172, 173], they destabilized EC oleogel-based emulsified structure [174]. However, small and stable globule sizes were obtained using MG-DG in the EC oleogel emulsification [174]. Although oil drop coalescence decreased with EC inclusion, the overrun could not be improved to reach the expected control value. On the other hand, the use of RBW instead of CLW or CRW improved some properties of oleogel-based ice creams, probably due to differences in wax crystallization in oil droplets [173]. Thus, RBW oleogels could favor ice-cream reformulations with 10 and 15% fat.

A gelled system with PS and hydrolyzed collagen was used to replace ice-cream fat, resulting in whiter creams ($\Delta E < 6$) with physical properties of similar or improved quality compared to the control, probably due to an enhanced protein network formation [175]. Thus, a promising option was obtained to decrease SFA content, even including healthy compounds such as PS and collagen. Additionally, PS-GO oleogel-based emulsions were adequate options to replace milk cream [177]. Increasing PS-GO content improved ice-cream quality characteristics compared to unstructured oil-based ice cream and showed similar or even better parameters than those of the control. Additionally, not only SFA was highly diminished ($\sim$80%), but also the incorporation of PS and GO could be beneficial to offer a healthier product.

Aiming to obtain an adequate ice-cream base from whey protein and LC, composition and process parameters for emulsion formation were optimized [176]. The use of LC-added high oleic palm oil, instead of coconut or soybean oil, and the structurant-oil system microfluidization generated the most suitable structures. Reformulated ice cream presented viscosity and some texture properties similar to those of the control, although SFA content was significantly reduced.

Vegan creams from stable high UFA oleogels were formulated not only to replace milk fat but also to reduce SFA commonly provided by palm oil [178]. Soy-based emulsion physicochemical properties were not substantially modified by processing conditions nor by fat phase. CLW oleogel and milk fat generated emulsions with similar rheological and physical instability parameters, but different from those of dairy cream. However, palm oil-based emulsion was the most structurally unstable. Additionally, oleogels produced by using polysaccharides-protein Pickering emulsions were tested as fat replacers in low-fat (5–15%) vegan cream formulations [179]. Appearance and texture attributes of reformulated creams were rated at acceptable values, being the lowest fat level the most similar to the control.

26.7 Other Food Applications

In addition to mimicking the performance of solid fats as food ingredients, structuring oils with high UFA content has also been explored to provide technological functions, such as shelf-life prolongation of ready-to-eat foods, quality improvements of fried foods, and development of tailor-made foods through 3D printing, among others (Table 26.6).

Coatings are restricted-permeability films usually used to protect perishable foods from different external agents. Among the various coating options, edible films stand out for their reduced waste generation and consumer acceptance. W/O emulsions formulated using vinegar and MG-sorbitan tristearate (STS) oleogels were used to coat roasted eggplants by immersion and subsequent freezing [182]. Coating efficiency was related to film wettability and greasiness, improving with increasing STS:MG ratio, although a high MG amount was needed to enhance emulsion consistency. Film appearance was semitransparent and did not change with the freezing process. Essential oil-enriched EC-MG oleogel-based emulsions were applied by spraying over sausages [183]. Antibacterial protection was demonstrated only for low and medium viscosity EC emulsion-based films, which improved sausage edibility by 100 and 220% or 66 and 166% compared with untreated products or control (unstructured oil-based film), respectively.

Additionally, films can be used to improve packaging technological aspects. In this sense, a BW oleogel-oil bilayer structure was used to promote yogurt sliding out from its container [181]. The oleogel layer on the base material enabled subsequent oil adhesion and, consequently, appropriate oil layer formation, which greatly improved viscous liquid product sliding (Fig. 26.7a). Moreover, this effect was maintained during long-term storage, without affecting product aroma and stability.

Frying mediums are usually preferred with minimum PUFA and middle or high SFA. CRW oleogels from canola or soybean oils were tested as frying mediums for different products such as chicken breast [184], flour-based snacks [185], and instant noodles [186]. In all cases, sample fat content was reduced by at least 16% when frying in structured oil rather than free oil, regardless of the structurant level used. This effect could be attributed to changes induced in the surface texture of the fried product, generating smaller pores by frying in oleogel than in oil and, therefore, achieving lower oil retention capacity in the final product. Oxidative stability was

Table 26.6 Oleogel technological functions

Food application	System	Gelator (wt% in system)/oil	Replaced fat (wt % replacement)	Reference
Coating	Oleogel	BW (10)/sunflower	*n.a*	[181]
	Oleogel	MG:STS (0.6–11:0.6–11)/olive:sunflower (27.0:62.0, 28.7:65.8)	*n.a*	[182]
	O/W emulsion	EC:MG (1.2:2)/cinnamon essential	Cinnamon essential oil (100)	[183]
Deep frying medium	Oleogel	CRW (5, 10)/canola	Canola oil (100)	[184]
	Oleogel	CRW (5, 10, 15)/soybean	Soybean oil (100)	[185]
	Oleogel	CRW (5, 10)/soybean	Soybean or palm oil (100)	[186]
Gummies	Oleogel	Stearic acid (10)/canola	*n.a*	[187]
Foams	Oleogel	MG:PS (3–10:0–10)/canola	*n.a*	[188]
Noodles	Oleogel	CLW (10)/soybean	*n.a*	[189]
3D printing material	Oleogel	BW (2–25)/sunflower	*n.a*	[190]
	Oleogel	MG:PS (10–20:20–50)/HOSO	*n.a*	[191]
	Oleogel	MG:PS (10:20)/HOSO	*n.a*	[192]
	Oleogel	EC:PEG (11:1–5)/MCT	*n.a*	[193]
	Oleogel	PS:LC (6–28:4–20)/sunflower	*n.a*	[194]

BW beeswax, *CLW* candelilla wax, *CRW* carnauba wax, *EC* ethylcellulose, *HOSO* high oleic sunflower oil, *LC* lecithin, *MCT* medium chain triglycerides, *MG* saturated monoglycerides, *n.a* not applicable, *n.s* not specified, *PEG* polyethylene glycol monostearate, *PS* phytosterols, *STS* sorbitan tristearate

maintained or improved with respect to controls. Although there were some differences in color and appearance (Fig. 26.7b), all snacks were generally well accepted by consumers [185].

The use of oleogels in other applications, such as foams [188], noodles [189], gummies (Fig. 26.7c) [187], and inks for 3D printing [190–194] has contributed positively to some product characteristics, demonstrating the wide range of options available for their usage.

26.8 Conclusions

Although a step behind oil structuring, during the last years, the application of oleogel-based systems in some goods has advanced significantly, with particular emphasis on replacing HF to improve FAP, but also applied to maintain or improve some physicochemical, structural, or sensory characteristics of foods and even their containers. Additionally, oleogel-based systems have the advantage of serving as carriers of other nutritional compounds, which is enhanced with greater versatility when such systems are formed by simultaneous hydrophobic and hydrophilic

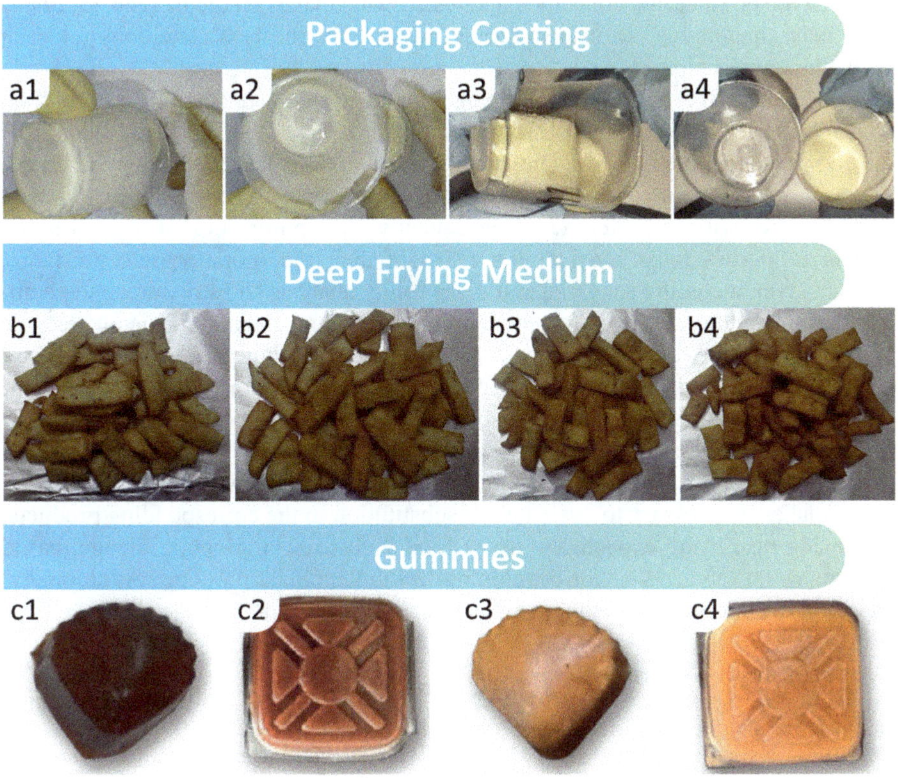

Fig. 26.7 Appearance of products of different oleogel-based system technological applications: (**a**) packaging with (a1, a2) unstructured oil and (a3, a4) coating from beeswax oleogel, reproduced from [181] with permission from Elsevier; (**b**) products fried in (b1) soybean oil and carnauba wax (CRW) oleogels with (b2) 5 wt% CRW, (b3) 10 wt% CRW, and (b4) 15 wt% CRW, reproduced from [185] with permission from Elsevier; and (**c**) gummies from stearic acid oleogel-based systems enriched with (c1, c2) doum juice and (c3, c4) orange juice, reproduced from [187] with permission from Springer-Nature under the terms of the Creative Commons CC BY 4.0 licence

phases—as emulsion, templated-based, or bigel systems. Therefore, current applications will continue to be evaluated as feedback to improve system formulation—with a high expectation on gelling using multicomponents—and processing. Generally, with an appropriate choice of gelling system and fat substitution level, formulations could be tailored to meet specific raw material and/or other end-use requirements and maintain the desired mechanical properties. However, at this point, it is important to emphasize that not all property differences found between structured oils and control fats are transmitted to reformulated foods. The effect of oleogel-based system incorporation on food properties depends not only on lipid replacement level and oleogel structuring system but also on several factors, such as specific formulation and raw materials—type and purity—even their interactions, replaced fat, total lipid content, processing and storage conditions, and temperature at which the products are consumed and tested. This makes predicting general trends extremely challenging, so it is suggested to look for the best conditions for each particular formulation.

While an active evaluation of oleogel-based system applications in the food sector has been demonstrated throughout this chapter, there are some gaps that have yet to be explored in sufficient depth. Instrumentally measured values, although very valuable as a guide for experimentation, do not always have the ability to reflect consumer behavior or preference. Consequently, the optimization of oleogel-based system properties must consider the effect of their incorporation into the target product, even including sensorial and digestibility analysis and structural and chemical stability studies during storage to confirm goal achievement. In general, oleogels and emulsions have been widely studied as fat substitutes in almost all food categories presented here. However, other system applications could offer advantages not yet examined. In addition, some of them deserve specific studies, such as microbiology stability in those where water activity is increased. Also, with a rather limited approach yet, the growing demand for vegan, gluten-free, and/or lactose-free foods makes it necessary to evaluate HF substitution in the corresponding products. Promising results are expected to assist food industries in adopting oleogel-based systems as fat replacers or for developing health-beneficial products. Additionally, based on the current knowledge, new applications may emerge.

Acknowledgments The authors acknowledge the financial support by the Consejo Nacional de Investigaciones Científicas y Técnicas (CONICET, PIP 2021-2023 N°101968), the Agencia Nacional de Promoción Científica y Tecnológica (ANPCyT, PICT 2017-1522), and the Universidad Nacional del Sur (UNS, PGI 24/M172) in Argentina. In addition, the authors are grateful to Desigr. Dutari for her assistance in designing image layouts.

References

1. Martins AJ, Vicente AA, Pastrana LM, Cerqueira MA (2020) Oleogels for development of health-promoting food products. Food Sci Human Wellness 9:31–39. https://doi.org/10.1016/j.fshw.2019.12.001
2. Puscas A, Muresan V, Socaciu C, Muste S (2020) Oleogels in food: a review of current and potential applications. Foods 9:1–27. https://doi.org/10.3390/foods9010070
3. Zambelli A (2021) Current status of high oleic seed oils in food processing. JAOCS 98:129–137. https://doi.org/10.1002/aocs.12450
4. Cui XT, Saleh ASM, Yang S et al (2022) Oleogels as animal fat and shortening replacers: research advances and application challenges. Food Rev Int 00:1–22. https://doi.org/10.1080/87559129.2022.2062769
5. Parmar V, Sharma R, Sharma S, Singh B (2022) Recent advances in fabrication of food grade oleogels: structuring methods, functional properties and technical feasibility in food products. J Food Meas Charact 16:4687–4702. https://doi.org/10.1007/s11694-022-01538-4
6. Silva PM, Cerqueira MA, Martins AJ et al (2022) Oleogels and bigels as alternatives to saturated fats: a review on their application by the food industry. JAOCS 99:911–923. https://doi.org/10.1002/aocs.12637

7. Mozaffarian D, Aro A, Willett WC (2009) Health effects of trans-fatty acids: experimental and observational evidence. Eur J Clin Nutr 63:S5–S21. https://doi.org/10.1038/sj.ejcn.1602973
8. FAO (2010) Fats and fatty acids in human nutrition. Report of an expert consultation. Food Nutr Pap 91:1–166
9. Berg J, Seyedsadjadi N, Grant R (2020) Saturated fatty acid intake is associated with increased inflammation, conversion of kynurenine to tryptophan, and Delta-9 desaturase activity in healthy humans. Int J Tryptophan Res 13:117864692098194. https://doi.org/10.1177/1178646920981946
10. Zhou H, Urso CJ, Jadeja V (2020) Saturated fatty acids in obesity-associated inflammation. J Inflamm Res 13:1–14. https://doi.org/10.2147/JIR.S229691
11. WHO (2020) Health diet. https://www.who.int/news-room/fact-sheets/detail/healthy-diet
12. WHO (2021) REPLACE Trans Fat-Free by 2023. https://www.who.int/teams/nutrition-and-food-safety/replace-trans-fat
13. Economic Commission for Latin America and the Caribbean (ECLAC) (2020) The 2030 agenda for sustainable development in the new global and regional context: scenarios and projections in the current crisis. Santiago
14. Palla CA, Dominguez M, Carrín ME (2022) Recent advances on food-based applications of monoglyceride oleogels. JAOCS 99:985–1006. https://doi.org/10.1002/aocs.12617
15. FDA (U.S. Food and Drug Administration) (2017) GRN No. 720, GRAS notification for Rice bran wax
16. Winkler-Moser JK, Anderson J, Byars JA et al (2019) Evaluation of beeswax, candelilla wax, rice bran wax, and sunflower wax as alternative stabilizers for peanut butter. JAOCS 96:1235–1248. https://doi.org/10.1002/aocs.12276
17. Palmquist DL (2006) Milk fat: origin of fatty acids and influence of nutritional factors thereon. In: Fox PF, McSweeney PL (eds) Advanced dairy chemistry, vol 2 L. Springer, Boston
18. Salas JJ, Bootello MA, Martínez-Force E, Garcés R (2009) Tropical vegetable fats and butters: properties and new alternatives. OCL 16:254–258. https://doi.org/10.1051/ocl.2009.0278
19. Ritvanen T, Putkonen T, Peltonen K (2012) A comparative study of the fatty acid composition of dairy products and margarines with reduced or substituted fat content. Food Nutr Sci 03: 1189–1196. https://doi.org/10.4236/fns.2012.39156
20. Ockerman HW, Basu L (2014) Inedible. Encycl Meat Sci 1:125–136. https://doi.org/10.1016/B978-0-12-384731-7.00032-5
21. Limpimwong W, Kumrungsee T, Kato N et al (2017) Rice bran wax oleogel: a potential margarine replacement and its digestibility effect in rats fed a high-fat diet. J Funct Foods 39: 250–256. https://doi.org/10.1016/j.jff.2017.10.035
22. Paszczyk B, Łuczyńska J (2020) The comparison of fatty acid composition and lipid quality indices in hard cow, sheep, and goat cheeses. Foods 9. https://doi.org/10.3390/foods9111667
23. Emami SZ, Golestan L, Khoshtinat K et al (2021) Optimization of the oleogel-vanaspati formulation based on rice bran wax and sorbitan monostearate. J Food Process Preserv 45:1–10. https://doi.org/10.1111/jfpp.15913
24. Pădure S (2022) The quantification of fatty acids, color, and textural properties of locally produced bakery margarine. Appl Sci 12. https://doi.org/10.3390/app12031731
25. Wang X, Wang S, Nan Y, Liu G (2021) Production of margarines rich in unsaturated fatty acids using oxidative-stable vitamin C-loaded oleogel. J Oleo Sci 70:1059–1068. https://doi.org/10.5650/jos.ess20264
26. Abdolmaleki K, Alizadeh L, Sheikhi Z, Nayebzadeh K (2022) Effects of replacing palm oil with beeswax oleogel on the characteristics of low-saturated fatty acid margarine. Iran J Nutr Sci Food Technol 16
27. Hwang HS, Singh M, Bakota EL et al (2013) Margarine from organogels of plant wax and soybean oil. JAOCS 90:1705–1712. https://doi.org/10.1007/s11746-013-2315-z

28. Hwang HS, Singh M, Winkler-Moser JK et al (2014) Preparation of margarines from organogels of sunflower wax and vegetable oils. J Food Sci 79:C1926–C1932. https://doi.org/10.1111/1750-3841.12596
29. Hwang HS, Winkler-Moser JK (2020) Properties of margarines prepared from soybean oil oleogels with mixtures of candelilla wax and beeswax. J Food Sci 85:3293–3302. https://doi.org/10.1111/1750-3841.15444
30. Wang Y, Deng Z, Mao L (2022) Preparation and characterization of margarine based on beeswax Oleogels. Shipin Kexue/Food Sci 43:43–50. https://doi.org/10.7506/spkx1002-6630-20210705-046
31. Hwang HS, Kim S, Winkler-Moser JK et al (2022) Feasibility of hemp seed oil oleogels structured with natural wax as solid fat replacement in margarine. JAOCS 99:1055–1070. https://doi.org/10.1002/aocs.12619
32. Ghan SY, Siow LF, Tan CP et al (2022) Palm olein organogelation using mixtures of soy lecithin and glyceryl monostearate. Gels 30(8):30. https://doi.org/10.3390/GELS8010030
33. Nicholson RA, Marangoni AG (2021) Lipase-catalyzed glycerolysis extended to the conversion of a variety of edible oils into structural fats. Curr Res Food Sci 4:163–174. https://doi.org/10.1016/j.crfs.2021.03.005
34. da Silva TLT, Chaves KF, Fernandes GD et al (2018) Sensory and technological evaluation of margarines with reduced saturated fatty acid contents using oleogel technology. JAOCS 95: 673–685. https://doi.org/10.1002/aocs.12074
35. Silva TJ, Fernandes GD, Bernardinelli OD et al (2021) Organogels in low-fat and high-fat margarine: a study of physical properties and shelf life. Food Res Int 140:110036. https://doi.org/10.1016/j.foodres.2020.110036
36. Holey SA, Sekhar KPC, Mishra SS et al (2021) Sunflower wax-based Oleogel emulsions: physicochemical characterizations and food application. ACS Food Sci Technol 1:152–164. https://doi.org/10.1021/acsfoodscitech.0c00050
37. Palla C, Giacomozzi A, Genovese DB, Carrín ME (2017) Multi–objective optimization of high oleic sunflower oil and monoglycerides oleogels: searching for rheological and textural properties similar to margarine. Food Struct 12:1–14. https://doi.org/10.1016/j.foostr.2017.02.005
38. Giacomozzi AS, Carrín ME, Palla CA (2018) Muffins elaborated with optimized monoglycerides oleogels: from solid fat replacer obtention to product quality evaluation. J Food Sci 83: 1505–1515. https://doi.org/10.1111/1750-3841.14174
39. Yilmaz E, Öğütcü M (2014) Comparative analysis of olive oil organogels containing beeswax and sunflower wax with breakfast margarine. J Food Sci 79:E1732–E1738. https://doi.org/10.1111/1750-3841.12561
40. Ye X, Li P, Lo YM et al (2019) Development of novel shortenings structured by ethylcellulose oleogels. J Food Sci 84:1456–1464. https://doi.org/10.1111/1750-3841.14615
41. Ölütcü M, Yilmaz E (2015) Comparison of the pomegranate seed oil organogels of carnauba wax and monoglyceride. J Appl Polym Sci 132:10–13. https://doi.org/10.1002/app.41343
42. Öğütcü M, Yilmaz E (2014) Oleogels of virgin olive oil with carnauba wax and monoglyceride as spreadable products. Grasas Aceites 65:e040. https://doi.org/10.3989/gya.0349141
43. Yilmaz E, Ölütcü M (2015) Oleogels as spreadable fat and butter alternatives: sensory description and consumer perception. RSC Adv 5:50259–50267. https://doi.org/10.1039/c5ra06689a
44. Öğütcü M, Arifoğlu N, Yılmaz E (2017) Restriction of oil migration in tahini halva via organogelation. Eur J Lipid Sci Technol 119:1–12. https://doi.org/10.1002/ejlt.201600189
45. Yilmaz E, Demirci Ş (2021) Preparation and evaluation of virgin olive oil oleogels including thyme and cumin spices with sunflower wax. Gels 7:1–18. https://doi.org/10.3390/gels7030095
46. Patel AR, Schatteman D, De Vos WH, Dewettinck K (2013) Shellac as a natural material to structure a liquid oil-based thermo reversible soft matter system. RSC Adv 3:5324–5327. https://doi.org/10.1039/c3ra40934a

47. Yilmaz E, Demirci Ş, Uslu EK (2022) Red pepper and turmeric-flavored virgin olive oil oleogels prepared with whale spermaceti wax. J Oleo Sci 71:187–199. https://doi.org/10.5650/jos.ess21167
48. Patel AR, Rajarethincm PS, Grędowska A et al (2014) Edible applications of shellac oleogels: spreads, chocolate paste and cakes. Food Funct 5:645–652. https://doi.org/10.1039/c4fo00034j
49. García-Ortega ML, Toro-Vazquez JF, Ghosh S (2021) Development and characterization of structured water-in-oil emulsions with ethyl cellulose oleogels. Food Res Int 150:110763. https://doi.org/10.1016/j.foodres.2021.110763
50. Fayaz G, Goli SAH, Kadivar M et al (2017) Potential application of pomegranate seed oil oleogels based on monoglycerides, beeswax and propolis wax as partial substitutes of palm oil in functional chocolate spread. LWT 86:523–529. https://doi.org/10.1016/j.lwt.2017.08.036
51. Tirgarian B, Yadegari H, Bagheri A et al (2023) Reduced-fat chocolate spreads developed by water-in-oleogel emulsions. J Food Eng 337:111233. https://doi.org/10.1016/j.jfoodeng.2022.111233
52. Bascuas S, Espert M, Llorca E et al (2021) Structural and sensory studies on chocolate spreads with hydrocolloid-based oleogels as a fat alternative. LWT 135:110228. https://doi.org/10.1016/j.lwt.2020.110228
53. David A, David M, Lesniarek P et al (2021) Oleogelation of rapeseed oil with cellulose fibers as an innovative strategy for palm oil substitution in chocolate spreads. J Food Eng 292:110315. https://doi.org/10.1016/j.jfoodeng.2020.110315
54. Giacomozzi AS, Carrín ME, Palla CA (2021) Storage stability of oleogels made from monoglycerides and high oleic sunflower oil. Food Biophys 16:306–316. https://doi.org/10.1007/s11483-020-09661-9
55. Zhao M, Rao J, Chen B (2022) Effect of high oleic soybean oil oleogels on the properties of doughs and corresponding bakery products. JAOCS 99:1071–1083. https://doi.org/10.1002/aocs.12594
56. Meng Z, Guo Y, Wang Y, Liu Y (2019) Oleogels from sodium stearoyl lactylate-based lamellar crystals: structural characterization and bread application. Food Chem 292:134–142. https://doi.org/10.1016/j.foodchem.2018.11.042
57. Bascuas S, Morell P, Quiles A et al (2021) Use of oleogels to replace margarine in steamed and baked buns. Foods 10:1781. https://doi.org/10.3390/FOODS10081781
58. Jung D, Oh I, Lee JH, Lee S (2020) Utilization of butter and oleogel blends in sweet pan bread for saturated fat reduction: dough rheology and baking performance. LWT 125:109194. https://doi.org/10.1016/j.lwt.2020.109194
59. Calligaris S, Manzocco L, Valoppi F, Nicoli MC (2013) Effect of palm oil replacement with monoglyceride organogel and hydrogel on sweet bread properties. Food Res Int 51:596–602. https://doi.org/10.1016/j.foodres.2013.01.007
60. Wettlaufer T, Flöter E (2022) Wax based oleogels and their application in sponge cakes. Food Funct 13:9419–9433. https://doi.org/10.1039/d2fo00563h
61. Alvarez-Ramirez J, Vernon-Carter EJ, Carrera-Tarela Y et al (2020) Effects of candelilla wax/canola oil oleogel on the rheology, texture, thermal properties and in vitro starch digestibility of wheat sponge cake bread. LWT 130:109701. https://doi.org/10.1016/j.lwt.2020.109701
62. Demirkesen I, Mert B (2019) Utilization of beeswax oleogel-shortening mixtures in gluten-free bakery products. JAOCS 96:545–554. https://doi.org/10.1002/aocs.12195
63. Kim JY, Lim J, Lee JH et al (2017) Utilization of oleogels as a replacement for solid fat in aerated baked goods: physicochemical, rheological, and tomographic characterization. J Food Sci 82:445–452. https://doi.org/10.1111/1750-3841.13583
64. Oh IK, Amoah C, Lim J et al (2017) Assessing the effectiveness of wax-based sunflower oil oleogels in cakes as a shortening replacer. LWT 86:430–437. https://doi.org/10.1016/j.lwt.2017.08.021

65. Dikhtyar A, Andrieieva S, Fedak N et al (2021) Determining patterns in the formation of functional technological properties of a fatbased semi-finished product in the technology of sponge cake products. Eastern-Eur J Enterp Technol 6:15–31. https://doi.org/10.15587/1729-4061.2021.246006
66. Aliasl Khiabani A, Tabibiazar M, Roufegarinejad L et al (2020) Preparation and characterization of carnauba wax/adipic acid oleogel: a new reinforced oleogel for application in cake and beef burger. Food Chem 333:127446. https://doi.org/10.1016/j.foodchem.2020.127446
67. Adili L, Roufegarinejad L, Tabibiazar M et al (2020) Development and characterization of reinforced ethyl cellulose based oleogel with adipic acid: its application in cake and beef burger. LWT 126:109277. https://doi.org/10.1016/j.lwt.2020.109277
68. Adili L, Roufegarinejad L, Tabibiazar M et al (2021) Effects of shortening replacement with reinforced ethyl cellulose-based oleogels with adipic acid on physicochemical and sensory characteristics of cakes. Iran J Nutr Sci Food Technol 15:71–80
69. Willett SA, Akoh CC (2019) Physicochemical characterization of yellow cake prepared with structured lipid Oleogels. J Food Sci 84:1390–1399. https://doi.org/10.1111/1750-3841.14624
70. Pehlivanoglu H, Ozulku G, Yildirim RM et al (2018) Investigating the usage of unsaturated fatty acid-rich and low-calorie oleogels as a shortening mimetics in cake. J Food Process Preserv 42:1–11. https://doi.org/10.1111/jfpp.13621
71. Tang YR, Ghosh S (2021) Canola protein thermal denaturation improved emulsion-templated oleogelation and its cake-baking application. RSC Adv 11:25141–25157. https://doi.org/10.1039/d1ra02250d
72. Bu N, Huang L, Cao G et al (2022) Stable O/W emulsions and oleogels with amphiphilic konjac glucomannan network: preparation, characterization, and application. J Sci Food Agric 102:6555–6565. https://doi.org/10.1002/jsfa.12021
73. Noshad M, Hojjati M, Hassanzadeh M et al (2022) Edible utilization of xanthan-guar oleogels as a shortening replacement in sponge cake: physicochemical properties. J Chem Heal Risks 12:255–264. https://doi.org/10.22034/jchr.2020.1908257.1169
74. Pan H, Xu X, Qian Z et al (2021) Xanthan gum-assisted fabrication of stable emulsion-based oleogel structured with gelatin and proanthocyanidins. Food Hydrocoll 115:106596. https://doi.org/10.1016/j.foodhyd.2021.106596
75. Jiang Q, Yu Z, Meng Z (2022) Double network oleogels co-stabilized by hydroxypropyl methylcellulose and monoglyceride crystals: baking applications. Int J Biol Macromol 209:180–187. https://doi.org/10.1016/j.ijbiomac.2022.04.011
76. Mohanan A, Tang YR, Nickerson MT, Ghosh S (2020) Oleogelation using pulse protein-stabilized foams and their potential as a baking ingredient. RSC Adv 10:14892–14905. https://doi.org/10.1039/c9ra07614j
77. Li S, Ling Z, Wu G et al (2022) Relationship between the microstructure and physical properties of emulsifier based oleogels and cookies quality. Food Chem 377:131966. https://doi.org/10.1016/j.foodchem.2021.131966
78. Onacik-Gür S, Żbikowska A (2020) Effect of high-oleic rapeseed oil oleogels on the quality of short-dough biscuits and fat migration. J Food Sci Technol 57:1609–1618. https://doi.org/10.1007/s13197-019-04193-8
79. Zhao M, Lan Y, Cui L et al (2020) Physical properties and cookie-making performance of oleogels prepared with crude and refined soybean oil: a comparative study. Food Funct 11:2498–2508. https://doi.org/10.1039/c9fo02180a
80. Tanislav AE, Puşcaş A, Păucean A et al (2022) Evaluation of structural behavior in the process dynamics of oleogel-based tender dough products. Gels 8. https://doi.org/10.3390/gels8050317
81. Hwang HS, Singh M, Lee S (2016) Properties of cookies made with natural wax-vegetable oil Organogels. J Food Sci 81:C1045–C1054. https://doi.org/10.1111/1750-3841.13279
82. Frolova YV, Sobolev RV, Kochetkova AA (2021) Comparative analysis of the properties of cookies containing oleogel based on beeswax and its fractions. IOP Conf Ser Earth Environ Sci 941:012033. https://doi.org/10.1088/1755-1315/941/1/012033

83. Frolova Y, Sarkisyan V, Sobolev R et al (2022) The influence of edible oils' composition on the properties of beeswax-based Oleogels. Gels 8. https://doi.org/10.3390/gels8010048
84. Kim D, Oh I (2022) The characteristic of insect oil for a potential component of oleogel and its application as a solid fat replacer in cookies. Gels 8. https://doi.org/10.3390/gels8060355
85. Yilmaz E, Öǧütcü M (2015) The texture, sensory properties and stability of cookies prepared with wax oleogels. Food Funct 6:1194–1204. https://doi.org/10.1039/c5fo00019j
86. Jang A, Bae W, Hwang HS et al (2015) Evaluation of canola oil oleogels with candelilla wax as an alternative to shortening in baked goods. Food Chem 187:525–529. https://doi.org/10.1016/j.foodchem.2015.04.110
87. Mert B, Demirkesen I (2016) Reducing saturated fat with oleogel/shortening blends in a baked product. Food Chem 199:809–816. https://doi.org/10.1016/j.foodchem.2015.12.087
88. Mert B, Demirkesen I (2016) Evaluation of highly unsaturated oleogels as shortening replacer in a short dough product. LWT 68:477–484. https://doi.org/10.1016/j.lwt.2015.12.063
89. Zhao M, Lan Y, Cui L et al (2020) Formation, characterization, and potential food application of rice bran wax oleogels: expeller-pressed corn germ oil versus refined corn oil. Food Chem 309:125704. https://doi.org/10.1016/j.foodchem.2019.125704
90. Li S, Ling Z, Li X et al (2022) Determination of characteristic evaluation indexes for novel cookies prepared with wax oleogels. J Sci Food Agric 102:5544–5553. https://doi.org/10.1002/jsfa.11909
91. Jadhav HB, Pratap AP, Gogate PR, Annapure US (2022) Ultrasound-assisted synthesis of highly stable MCT based oleogel and evaluation of its baking performance. Appl Food Res 2: 100156. https://doi.org/10.1016/j.afres.2022.100156
92. Principato L, Sala L, Duserm-Garrido G, Spigno G (2021) Effect of dietary fibre and thermal condition on rice bran wax oleogels for biscuits preparation. Chem Eng Trans 87:49–54. https://doi.org/10.3303/CET2187009
93. Kwon UH, Chang YH (2022) Rheological and physicochemical properties of oleogel with esterified Rice flour and its suitability as a fat replacer. Foods 11. https://doi.org/10.3390/foods11020242
94. Li S, Wu G, Li X et al (2021) Roles of gelator type and gelation technology on texture and sensory properties of cookies prepared with oleogels. Food Chem 356:129667. https://doi.org/10.1016/j.foodchem.2021.129667
95. Pandolsook S, Kupongsak S (2017) Influence of bleached rice bran wax on the physicochemical properties of organogels and water-in-oil emulsions. J Food Eng 214:182–192. https://doi.org/10.1016/j.jfoodeng.2017.06.030
96. Barragán-Martínez LP, Román-Guerrero A, Vernon-Carter EJ, Alvarez-Ramirez J (2022) Impact of fat replacement by a hybrid gel (canola oil/candelilla wax oleogel and gelatinized corn starch hydrogel) on dough viscoelasticity, color, texture, structure, and starch digestibility of sugar-snap cookies. Int J Gastron Food Sci 29:100563. https://doi.org/10.1016/j.ijgfs.2022.100563
97. Quilaqueo M, Iturra N, Contardo I et al (2022) Food-grade bigels with potential to replace saturated and trans fats in cookies. Gels 8:1–17. https://doi.org/10.3390/gels8070445
98. Pan LH, Ling WX, Luo SZ et al (2020) Effects of tea polyphenol ester with different fatty acid chain length on camellia oil-based oleogels preparation and its effects on cookies properties. J Food Sci 85:2461–2469. https://doi.org/10.1111/1750-3841.15341
99. Paciulli M, Littardi P, Carini E et al (2020) Inulin-based emulsion filled gel as fat replacer in shortbread cookies: effects during storage. LWT 133:109888. https://doi.org/10.1016/j.lwt.2020.109888
100. Brito GB, Peixoto VODS, Martins MT et al (2022) Development of chitosan-based oleogels via crosslinking with vanillin using an emulsion templated approach: structural characterization and their application as fat-replacer. Food Struct 32:100264. https://doi.org/10.1016/j.foostr.2022.100264

101. Giacomozzi AS, Carrín ME, Palla CA (2022) Muffins made with monoglyceride oleogels: impact of fat replacement on sensory properties and fatty acid profile. JAOCS 1–7:343–349. https://doi.org/10.1002/aocs.12674
102. Jeong S, Lee S, Oh I (2021) Development of antioxidant-fortified oleogel and its application as a solid fat replacer to muffin. Foods 10:1–12. https://doi.org/10.3390/foods10123059
103. Kupiec M, Zbikowska A, Marciniak-Lukasiak K et al (2021) Study on the introduction of solid fat with a high content of unsaturated fatty acids to gluten-free muffins as a basis for designing food with higher health value. Int J Mol Sci 22. https://doi.org/10.3390/ijms22179220
104. Lim J, Jeong S, Lee JH et al (2017) Effect of shortening replacement with oleogels on the rheological and tomographic characteristics of aerated baked goods. J Sci Food Agric 97: 3727–3732. https://doi.org/10.1002/jsfa.8235
105. Oh IK, Lee S (2018) Utilization of foam structured hydroxypropyl methylcellulose for oleogels and their application as a solid fat replacer in muffins. Food Hydrocoll 77:796–802. https://doi.org/10.1016/j.foodhyd.2017.11.022
106. Calligaris S, Plazzotta S, Barba L, Manzocco L (2021) Design of roll-in margarine analogous by partial drying of monoglyceride-structured emulsions. Eur J Lipid Sci Technol 123:1–10. https://doi.org/10.1002/ejlt.202000206
107. Lupi FR, Baldino N, Seta L et al (2011) Sheeting process modelling and rheological analysis of an oliveoil- emulsion-based puff pastry. In: 10th international conference model appl simulation, MAS 2011, held Int Mediterr Lat Am Model Multiconference, I3M 2011 242–248
108. Sim SYJ, Wong KX, Henry CJ (2021) Healthier pineapple tart pastry using oleogel-based solid fat replacement. Malays J Nutr 27:327–333. https://doi.org/10.31246/MJN-2021-0007
109. Huschka B, Challacombe C, Marangoni AG, Seetharaman K (2011) Comparison of oil, shortening, and a structured shortening on wheat dough rheology and starch pasting properties. Cereal Chem 88:253–259. https://doi.org/10.1094/CCHEM-03-10-0041
110. Álvarez MD, Cofrades S, Pérez-Mateos M et al (2022) Development and physico-chemical characterization of healthy puff pastry margarines made from olive-pomace oil. Foods 11:1–18. https://doi.org/10.3390/foods11244054
111. Alvarez MD, Cofrades S, Espert M et al (2021) Thermorheological characterization of healthier reduced-fat cocoa butter formulated by substitution with a hydroxypropyl methylcellulose (Hpmc)-based oleogel. Foods 10. https://doi.org/10.3390/foods10040793
112. Metilli L, Lazidis A, Francis M et al (2021) The effect of crystallization conditions on the structural properties of oleofoams made of cocoa butter crystals and high oleic sunflower oil. Cryst Growth Des 21:1562–1575. https://doi.org/10.1021/acs.cgd.0c01361
113. Chen H, Zhou P, Song C et al (2022) An approach to manufacturing heat-stable and bloom-resistant chocolate by the combination of oleogel and sweeteners. J Food Eng 330:111064. https://doi.org/10.1016/j.jfoodeng.2022.111064
114. Li L, Liu G, Lin Y (2021) Physical and bloom stability of low-saturation chocolates with oleogels based on different gelation mechanisms. LWT 140:110807. https://doi.org/10.1016/j.lwt.2020.110807
115. Li L, Liu G (2019) Corn oil-based oleogels with different gelation mechanisms as novel cocoa butter alternatives in dark chocolate. J Food Eng 263:114–122. https://doi.org/10.1016/j.jfoodeng.2019.06.001
116. Sun P, Xia B, Ni ZJ et al (2021) Characterization of functional chocolate formulated using oleogels derived from β-sitosterol with γ-oryzanol/lecithin/stearic acid. Food Chem 360: 130017. https://doi.org/10.1016/j.foodchem.2021.130017
117. Alvarez MD, Cofrades S, Espert M et al (2021) Development of chocolates with improved lipid profile by replacing cocoa butter with an oleogel. Gels 7:1–20. https://doi.org/10.3390/gels7040220
118. Espert M, Hernández MJ, Sanz T, Salvador A (2021) Reduction of saturated fat in chocolate by using sunflower oil-hydroxypropyl methylcellulose based oleogels. Food Hydrocoll 120: 106917. https://doi.org/10.1016/j.foodhyd.2021.106917

119. Palla CA, Wasinger MF, Carrín ME (2021) Monoglyceride oleogels as fat replacers in filling creams for sandwich cookies. J Sci Food Agric 101:2398–2405. https://doi.org/10.1002/jsfa.10863
120. Doan CD, Patel AR, Tavernier I et al (2016) The feasibility of wax-based oleogel as a potential co-structurant with palm oil in low-saturated fat confectionery fillings. Eur J Lipid Sci Technol 118:1903–1914. https://doi.org/10.1002/ejlt.201500172
121. Borriello A, Miele NA, Masi P, Cavella S (2022) Rheological properties, particle size distribution and physical stability of novel refined pumpkin seed oil creams with oleogel and lucuma powder. Foods 11. https://doi.org/10.3390/foods11131844
122. Si H, Cheong LZ, Huang J et al (2016) Physical properties of soybean oleogels and oil migration evaluation in model praline system. JAOCS 93:1075–1084. https://doi.org/10.1007/s11746-016-2846-1
123. Kim M, Hwang HS, Jeong S, Lee S (2022) Utilization of oleogels with binary oleogelator blends for filling creams low in saturated fat. LWT 155:112972. https://doi.org/10.1016/j.lwt.2021.112972
124. da Silva TLT, Fernandes GD, Arellano DB (2021) Development of reduced saturated fat cookie fillings using multicomponent oleogels. JAOCS 98:1069–1082. https://doi.org/10.1002/aocs.12527
125. Wendt A, Abraham K, Wernecke C et al (2017) Application of β-sitosterol + γ-oryzanol-structured organogel as migration barrier in filled chocolate products. JAOCS 94:1131–1140. https://doi.org/10.1007/s11746-017-3024-9
126. Selvasekaran P, Chidambaram R (2021) Advances in formulation for the production of low-fat, fat-free, low-sugar, and sugar-free chocolates: an overview of the past decade. Trends Food Sci Technol 113:315–334. https://doi.org/10.1016/j.tifs.2021.05.008
127. López-Pedrouso M, Lorenzo JM, Gullón B et al (2021) Novel strategy for developing healthy meat products replacing saturated fat with oleogels. Curr Opin Food Sci 40:40–45. https://doi.org/10.1016/j.cofs.2020.06.003
128. Manzoor S, Masoodi FA, Rashid R et al (2022) Oleogels for the development of healthy meat products: a review. Appl Food Res 2:100212. https://doi.org/10.1016/j.afres.2022.100212
129. Moghtadaei M, Soltanizadeh N, Goli SAH, Sharifimehr S (2021) Physicochemical properties of beef burger after partial incorporation of ethylcellulose oleogel instead of animal fat. J Food Sci Technol 58:4775–4784. https://doi.org/10.1007/s13197-021-04970-4
130. Gao Y, Li M, Zhang L et al (2021) Preparation of rapeseed oil oleogels based on beeswax and its application in beef heart patties to replace animal fat. LWT 149:111986. https://doi.org/10.1016/j.lwt.2021.111986
131. Gómez-Estaca J, Pintado T, Jiménez-Colmenero F, Cofrades S (2019) Assessment of a healthy oil combination structured in ethyl cellulose and beeswax oleogels as animal fat replacers in low-fat, PUFA-enriched pork burgers. Food Bioprocess Technol 12:1068–1081. https://doi.org/10.1007/s11947-019-02281-3
132. Martins AJ, Lorenzo JM, Franco D et al (2019) Omega-3 and polyunsaturated fatty acids-enriched hamburgers using sterol-based oleogels. Eur J Lipid Sci Technol 121:1–8. https://doi.org/10.1002/ejlt.201900111
133. Agregán R, Barba FJ, Gavahian M et al (2019) Fucus vesiculosus extracts as natural antioxidants for improvement of physicochemical properties and shelf life of pork patties formulated with oleogels. J Sci Food Agric 99:4561–4570. https://doi.org/10.1002/jsfa.9694
134. Oh I, Lee JH, Lee HG, Lee S (2019) Feasibility of hydroxypropyl methylcellulose oleogel as an animal fat replacer for meat patties. Food Res Int 122:566–572. https://doi.org/10.1016/j.foodres.2019.01.012
135. Ghiasi F, Golmakani MT (2022) Fabrication and characterization of a novel biphasic system based on starch and ethylcellulose as an alternative fat replacer in a model food system. Innov Food Sci Emerg Technol 78:103028. https://doi.org/10.1016/j.ifset.2022.103028

136. Lopes R, Costa V, Costa M, Paiva-Martins F (2022) Olive oil oleogels as strategy to confer nutritional advantages to burgers. Food Chem 397:133724. https://doi.org/10.1016/j.foodchem.2022.133724
137. Teixeira A, Ferreira I, Pereira E et al (2021) Meat burgers. Effect of Fat Source Foods:1–11
138. Özer CO, Çelegen Ş (2021) Evaluation of quality and emulsion stability of a fat-reduced beef burger prepared with an olive oil oleogel-based emulsion. J Food Process Preserv 45:1–11. https://doi.org/10.1111/jfpp.14547
139. Palamutoglu R (2021) Replacement of beef fat in meatball with oleogels (black cumin seed oil/sunflower oil). J Hell Vet Med Soc 72:3031–3040. https://doi.org/10.12681/JHVMS.28484
140. Alejandre M, Astiasarán I, Ansorena D, Barbut S (2019) Using canola oil hydrogels and organogels to reduce saturated animal fat in meat batters. Food Res Int 122:129–136. https://doi.org/10.1016/j.foodres.2019.03.056
141. Ferrer-González BM, García-Martínez I, Totosaus A (2019) Textural properties, sensory acceptance and fatty acid profile of cooked meat batters employing pumpkin seed paste or soybean oil oleogel as fat replacers. Grasas Aceites 70:1–11. https://doi.org/10.3989/gya.1055182
142. Lupi FR, Gabriele D, Baldino N et al (2012) Stabilization of meat suspensions by organogelation: a rheological approach. Eur J Lipid Sci Technol 114:1381–1389. https://doi.org/10.1002/ejlt.201200212
143. Lupi FR, Gabriele D, Seta L et al (2014) Rheological design of stabilized meat sauces for industrial uses. Eur J Lipid Sci Technol 116:1734–1744. https://doi.org/10.1002/ejlt.201400286
144. Martins AJ, Lorenzo JM, Franco D et al (2020) Characterization of enriched meat-based pâté manufactured with oleogels as fat substitutes. Gels 6:1–14. https://doi.org/10.3390/gels6020017
145. Ramírez-carrasco P, Paredes-toledo J, Romero-hasler P et al (2020) Effect of adding curcumin on the properties of linseed oil organogels used as fat replacers in Pâtés. Antioxidants 9:1–18. https://doi.org/10.3390/antiox9080735
146. Gómez-Estaca J, Herrero AM, Herranz B et al (2019) Characterization of ethyl cellulose and beeswax oleogels and their suitability as fat replacers in healthier lipid pâtés development. Food Hydrocoll 87:960–969. https://doi.org/10.1016/j.foodhyd.2018.09.029
147. Barbut S, Marangoni AG, Thode U, Tiensa BE (2019) Using canola oil organogels as fat replacement in liver Pâté. J Food Sci 84:2646–2651. https://doi.org/10.1111/1750-3841.14753
148. Barbut S, Tiensa BE, Marangoni AG (2021) Partial fat replacement in liver pâté using canola oil organogel. LWT 139:110428. https://doi.org/10.1016/j.lwt.2020.110428
149. Woern C, Marangoni AG, Weiss J, Barbut S (2021) Effects of partially replacing animal fat by ethylcellulose based organogels in ground cooked salami. Food Res Int 147:110431. https://doi.org/10.1016/j.foodres.2021.110431
150. Ferro AC, de Souza PC, Rodrigues Pollonio MA, Lopes Cunha R (2021) Glyceryl monostearate-based oleogels as a new fat substitute in meat emulsion. Meat Sci 174:108424. https://doi.org/10.1016/j.meatsci.2020.108424
151. Tarté R, Paulus JS, Acevedo NC et al (2020) High-oleic and conventional soybean oil oleogels structured with rice bran wax as alternatives to pork fat in mechanically separated chicken-based bologna sausage. LWT 131:109659. https://doi.org/10.1016/j.lwt.2020.109659
152. da Silva SL, Amaral JT, Ribeiro M et al (2019) Fat replacement by oleogel rich in oleic acid and its impact on the technological, nutritional, oxidative, and sensory properties of Bologna-type sausages. Meat Sci 149:141–148. https://doi.org/10.1016/j.meatsci.2018.11.020
153. Barbut S, Wood J, Marangoni A (2016) Quality effects of using organogels in breakfast sausage. Meat Sci 122:84–89. https://doi.org/10.1016/j.meatsci.2016.07.022
154. Zampouni K, Soniadis A, Dimakopoulou-Papazoglou D et al (2022) Modified fermented sausages with olive oil oleogel and NaCl–KCl substitution for improved nutritional quality. LWT 158:113172. https://doi.org/10.1016/j.lwt.2022.113172

155. Pintado T, Cofrades S (2020) Quality characteristics of healthy dry fermented sausages formulated with a mixture of olive and chia oil structured in oleogel or emulsion gel as animal fat replacer. Foods 9. https://doi.org/10.3390/foods9060830
156. Franco D, Martins AJ, López-Pedrouso M et al (2020) Evaluation of linseed oil oleogels to partially replace pork backfat in fermented sausages. J Sci Food Agric 100:218–224. https://doi.org/10.1002/jsfa.10025
157. Noorolahi Z, Sahari MA, Ahmadi Gavlighi H, Barzegar M (2022) Pistachio green hull extract as natural antioxidant incorporated to omega-3 rich kappa-carrageenan oleogel in dry fermented sausage. Food Biosci 50:101986. https://doi.org/10.1016/j.fbio.2022.101986
158. Franco D, Martins AJ, López-Pedrouso M et al (2019) Strategy towards replacing pork backfat with a linseed oleogel in frankfurter sausages and its evaluation on physicochemical, nutritional, and sensory characteristics. Foods 8:366. https://doi.org/10.3390/foods8090366
159. Wolfer TL, Acevedo NC, Prusa KJ et al (2018) Replacement of pork fat in frankfurter-type sausages by soybean oil oleogels structured with rice bran wax. Meat Sci 145:352–362. https://doi.org/10.1016/j.meatsci.2018.07.012
160. Zetzl AK, Marangoni AG, Barbut S (2012) Mechanical properties of ethylcellulose oleogels and their potential for saturated fat reduction in frankfurters. Food Funct 3:327–337. https://doi.org/10.1039/c2fo10202a
161. Kouzounis D, Lazaridou A, Katsanidis E (2017) Partial replacement of animal fat by oleogels structured with monoglycerides and phytosterols in frankfurter sausages. Meat Sci 130:38–46. https://doi.org/10.1016/j.meatsci.2017.04.004
162. Barbut S, Wood J, Marangoni A (2016) Potential use of organogels to replace animal fat in comminuted meat products. Meat Sci 122:155–162. https://doi.org/10.1016/j.meatsci.2016.08.003
163. Barbut S, Wood J, Marangoni AG (2016) Effects of organogel hardness and formulation on acceptance of frankfurters. J Food Sci 81:C2183–C2188. https://doi.org/10.1111/1750-3841.13409
164. Panagiotopoulou E, Moschakis T, Katsanidis E (2016) Sunflower oil organogels and organogel-in-water emulsions (part II): implementation in frankfurter sausages. LWT 73:351–356. https://doi.org/10.1016/j.lwt.2016.06.006
165. Yılmaz E, Toksöz B (2022) Flaxseed oil-wax oleogels replacement for tallowfat in sucuk samples provided higher concentrations of polyunsaturated fatty acids and aromatic volatiles. Meat Sci 192:1–8. https://doi.org/10.1016/j.meatsci.2022.108875
166. Issara U (2022) Improvement of Thai sweet sausage (goon Chiang) properties by oleogel made of rice bran wax and Rice bran oil: a textural, sensorial, and nutritional aspect. IOP Conf Ser Earth Environ Sci 995:012045. https://doi.org/10.1088/1755-1315/995/1/012045
167. Utrilla MC, García Ruiz A, Soriano A (2014) Effect of partial replacement of pork meat with an olive oil organogel on the physicochemical and sensory quality of dry-ripened venison sausages. Meat Sci 97:575–582. https://doi.org/10.1016/j.meatsci.2014.03.001
168. Moon K, Choi KO, Jeong S et al (2021) Solid fat replacement with canola oil-carnauba wax oleogels for dairy-free imitation cheese low in saturated fat. Foods 10. https://doi.org/10.3390/foods10061351
169. Park C, Bemer HL, Maleky F (2018) Oxidative stability of rice bran wax oleogels and an oleogel cream cheese product. JAOCS 95:1267–1275. https://doi.org/10.1002/aocs.12095
170. Bemer HL, Limbaugh M, Cramer ED et al (2016) Vegetable organogels incorporation in cream cheese products. Food Res Int 85:67–75. https://doi.org/10.1016/j.foodres.2016.04.016
171. Huang H, Hallinan R, Maleky F (2018) Comparison of different oleogels in processed cheese products formulation. Int J Food Sci Technol 53:2525–2534. https://doi.org/10.1111/ijfs.13846
172. Zulim Botega DC, Marangoni AG, Smith AK, Goff HD (2013) The potential application of rice bran wax oleogel to replace solid fat and enhance unsaturated fat content in ice cream. J Food Sci 78:1334–1339. https://doi.org/10.1111/1750-3841.12175

173. Zulim Botega DC, Marangoni AG, Smith AK, Goff HD (2013) Development of formulations and processes to incorporate wax oleogels in ice cream. J Food Sci 78:1845–1851. https://doi.org/10.1111/1750-3841.12248
174. Munk MB, Munk DME, Gustavsson F, Risbo J (2018) Using ethylcellulose to structure oil droplets in ice cream made with high oleic sunflower oil. J Food Sci 83:2520–2526. https://doi.org/10.1111/1750-3841.14296
175. da Santos PHS, da Lannes SCS (2022) Application of organogel-like structured system as an alternative for reducing saturated fatty acid and replacing fat in milk ice cream. J Food Process Preserv 46:1–10. https://doi.org/10.1111/jfpp.16932
176. Silva-Avellaneda E, Bauer-Estrada K, Prieto-Correa RE, Quintanilla-Carvajal MX (2021) The effect of composition, microfluidization and process parameters on formation of oleogels for ice cream applications. Sci Rep 11:1–10. https://doi.org/10.1038/s41598-021-86233-y
177. Moriano ME, Alamprese C (2017) Organogels as novel ingredients for low saturated fat ice creams. LWT 86:371–376. https://doi.org/10.1016/j.lwt.2017.07.034
178. Szymańska I, Żbikowska A, Kowalska M, Golec K (2021) Application of oleogel and conventional fats for ultrasound-assisted obtaining of vegan creams. J Oleo Sci 70:1495–1507. https://doi.org/10.5650/jos.ess21126
179. Naji-Tabasi S, Mahdian E, Arianfar A, Naji-Tabasi S (2020) Investigation of oleogel properties prepared by Pickering emulsion-templated stabilized with solid particles of basil seed gum and isolated soy protein as a fat substitute in cream. J Res Innov Food Sci Technol 9:1–14. https://doi.org/10.22101/jrifst.2020.229269.1168
180. Goff HD (1997) Colloidal aspects of ice cream – a review. Int Dairy J 7:363–373. https://doi.org/10.1016/S0958-6946(97)00040-X
181. Hao P, Lou X, Tang L et al (2022) Solvent-free fabrication of slippery coatings from edible raw materials for reducing yogurt adhesion. Prog Org Coatings 162:106590. https://doi.org/10.1016/j.porgcoat.2021.106590
182. Lupi FR, Gabriele D, Baldino N et al (2017) Film edibili a reologia controllata per la ricopertura di vegetali da surgelare. Ind Aliment:15–26
183. Zhang K, Wang W, Wang X et al (2019) Fabrication and physicochemical and antibacterial properties of ethyl cellulose-structured cinnamon oil oleogel: relation between ethyl cellulose viscosity and oleogel performance. J Sci Food Agric 99:4063–4071. https://doi.org/10.1002/jsfa.9636
184. Adrah K, Adegoke SC, Tahergorabi R (2022) Physicochemical and microbial quality of coated raw and oleogel-fried chicken. LWT 154:112589. https://doi.org/10.1016/j.lwt.2021.112589
185. Chauhan DS, Khare A, Lal AB, Bebartta RP (2022) Utilising oleogel as a frying medium for deep fried Indian traditional product (Mathri) to reduce oil uptake. J Indian Chem Soc 99:100378. https://doi.org/10.1016/j.jics.2022.100378
186. Lim J, Jeong S, Oh IK, Lee S (2017) Evaluation of soybean oil-carnauba wax oleogels as an alternative to high saturated fat frying media for instant fried noodles. LWT 84:788–794. https://doi.org/10.1016/j.lwt.2017.06.054
187. Salama HH, Hashim AF (2022) A functional spreadable canola and milk proteins oleogels as a healthy system for candy gummies. Sci Rep 12:1–18. https://doi.org/10.1038/s41598-022-16809-9
188. Truong T, Prakash S, Bhandari B (2019) Effects of crystallisation of native phytosterols and monoacylglycerols on foaming properties of whipped oleogels. Food Chem 285:86–93. https://doi.org/10.1016/j.foodchem.2019.01.134
189. Oh I, Lee S (2020) Rheological, microstructural, and tomographical studies on the rehydration improvement of hot air-dried noodles with oleogel. J Food Eng 268:109750. https://doi.org/10.1016/j.jfoodeng.2019.109750
190. Shi Y, Zhang M, Bhandari B (2021) Effect of addition of beeswax based oleogel on 3D printing of potato starch-protein system. Food Struct 27:100176. https://doi.org/10.1016/j.foostr.2021.100176

191. Cotabarren IM, Cruces S, Palla CA (2019) Extrusion 3D printing of nutraceutical oral dosage forms formulated with monoglycerides oleogels and phytosterols mixtures. Food Res Int 126: 108676. https://doi.org/10.1016/j.foodres.2019.108676
192. De Salvo MI, Palla CA, Cotabarren IM (2023) Effect of printing parameters on the extrusion 3D printing of oleogel-based nutraceuticals. J Food Eng 349:111459. https://doi.org/10.1016/j.jfoodeng.2023.111459
193. Kavimughil M, Leena MM, Moses JA, Anandharamakrishnan C (2022) Effect of material composition and 3D printing temperature on hot-melt extrusion of ethyl cellulose based medium chain triglyceride oleogel. J Food Eng 329:111055. https://doi.org/10.1016/j.jfoodeng.2022.111055
194. Oliveira SM, Martins AJ, Fuciños P et al (2023) Food additive manufacturing with lipid-based inks: evaluation of phytosterol-lecithin oleogels. J Food Eng 341:111317. https://doi.org/10.1016/j.jfoodeng.2022.111317

Chapter 27
Legislation, Industrial Feasibility, and Scalability of Oleogel Production Processes

Maria Scharfe

Abbreviations

ADI	Acceptable daily intake
BWX	Beeswax
CLX	Candelilla wax
CNP	Crystalline nanoplatelets
CRX	Carnauba wax
EC	Ethylcellulose
FDA	Food and Drug Administration
FFDCA	Federal Food, Drug, and Cosmetic Act
MUFA	Monounsaturated fatty acid
NGO	Non-governmental organization
PPP	Tripalmitate
PUFA	Polyunsaturated fatty acid
RBX	Rice bran wax
SAFA	Saturated fatty acid
SCX	Sugarcane wax
SFX	Sunflower wax
TAG	Triacylglycerol
TFA	Trans fatty acid
WHO	World Health Organization

M. Scharfe (✉)
Formo Bio GmbH, Berlin, Germany
e-mail: maria@formo.bio

C. Andrea Palla, F. Valoppi (eds.), *Advances in Oleogel Development, Characterization, and Nutritional Aspects*,
https://doi.org/10.1007/978-3-031-46831-5_27

27.1 Introduction

27.1.1 Current Solutions for Structured Lipids

Structured lipids are an essential component to achieve particular textures in food products. Further, they can stabilize surfaces, generate macroscopic hardness, and influence disintegration characteristics and melting behavior [1, 2]. The fatty acid composition of triacylglycerols (TAGs) dictates their physical state. Oils that are derived from plants are composed of a combination of saturated fatty acids (SAFAs) and mono- and polyunsaturated fatty acids (MUFAs, PUFAs). The quantity and type of these acids decide a fat's state, as saturated fatty acids tend to arrange into crystalline nanoplatelets (CNPs) at room temperature. These CNPs stack and aggregate to create a three-dimensional network, which immobilizes liquid TAGs.

Since the beginning of the 1980s, palm oil has been the primary source of TAGs (triacylglycerol) used to create functional fat phases [3]. In 2020, the annual volume of palm oil used was 75.9 million metric tons, which accounts for about 39% of total edible oil production (data available via FAO STAT). Palm oil is roughly comprised of a 50/50 mix of SAFAs and MUFAs, and PUFAs. In addition, high- and low-melting palm oil fractions, as well as their combinations, are used to deliver desired properties such as hardness or melting temperature and are an exceedingly important ingredient in many food products. In fact, these blends (often including other edible plant oils and fats) are essential trade secrets of edible fat manufacturers and are tailored to specific product needs. All of the aforementioned, combined with a fairly low production price, high oil yield, availability, and unique fatty acid composition, has given palm oil a distinctive position on the global market.

However, palm oil has been met with criticism in some areas due to its environmental impact, as large, diverse rainforest areas have been cut down since the 1980s in order to make room for its production [4]. This has led to a generally negative perception of palm oil by consumers in Western countries [5, 6]. Growing ecological awareness and good campaigning of some non-governmental organization (NGOs) have firmly established this critical view in industrialized countries and resulted in a boom in products labeled as palm oil-free (Fig. 27.1). Therefore, alternatives to palm oil that provide structure to food must be considered.

Hydrogenated plant oils are a cost-effective source of semi-solid fats. However, deliberate partial oil hydrogenation may result in increased levels of *trans* fatty acid (TFA). Studies have indicated that these TFAs increase low-density lipoprotein and simultaneously lower high-density lipoprotein, and this, in turn, boosts the risk of developing coronary heart disease (CHD) by 21% [7, 8]. In light of this, public health authorities worldwide suggested limiting TFA contents in food products and recommended that intake should not exceed 1% of the total nutritional energy [9, 10]. Currently, partial hydrogenation is much less common practice, and the primary intake of TFA is associated with conventional products, fried foods, and baking goods. The European Union (EU) Commission Regulation 2019/649 states that the non-natural TFA content in food products should not exceed 2 g per 100 g

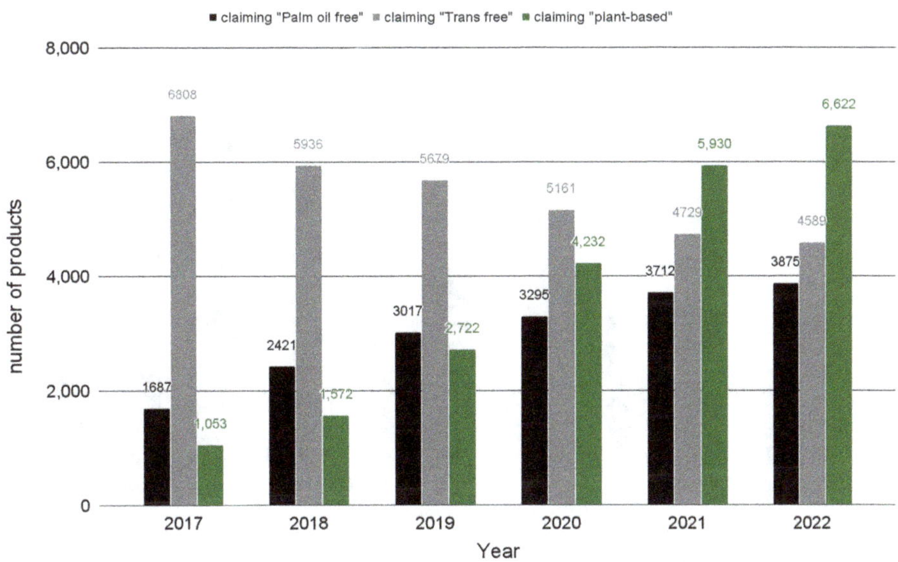

Fig. 27.1 Product launches from 2017 to 2022 claiming either palm oil-free, trans-free, or plant-based. (Data kindly provided by Mintel Group Ltd.)

fat. A similar regulation has been issued in the USA and supported by the World Health Organization (WHO) [10]. Consequently, consumers became more aware of the topic in the past decades, and many food products were labeled "*trans* free" whereas this trend seems to decrease in the last 5 years (Fig. 27.1). In convenience products, baking goods, or fried foods, TFA levels can still exceed the threshold value [11]. However, the TFA intake largely depends on the individual's consumption behavior, and an accurate number is generally difficult to estimate [12–15].

27.1.2 Factors Promoting a Shift to Oleogels

With the growth in demand for organic, sustainable, and regional food products that are also healthy, there is a need to find alternatives to current structuring routes. One potential solution is to use oleogelators, a type of structuring agent, to gel liquid edible oils that are rich in MUFAs and PUFAs.

It has been suggested for many years that a high intake of SAFAs could be correlated to an increased risk for coronary heart disease (CHD). However, recent studies have concluded that a diet that is rich in MUFAs and PUFAs significantly reduces the risk of CHD-related incidents, whereas the negative impact of SAFA is still up for debate [16, 17]. Thus, providing higher amounts of MUFAs and PUFAs through food products can help promote a healthier diet.

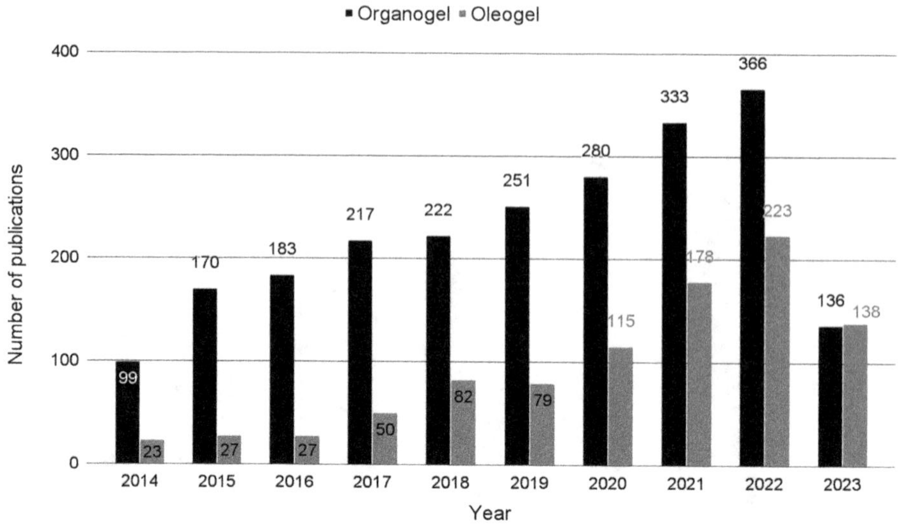

Fig. 27.2 Number of publications between 2014 and March 2023 having the terms oleogel or organogel in their title or abstract

The oleogelator should be able to dissolve in a non-polar liquid like edible oil but have limited solubility, which creates a solid structure known as an oleogel or organogel. This is different from traditional structuring routes as it does not depend on crystalline triacylglycerol (TAGs) but rather on complex structures such as fibers, ribbons, and tubules. The gelling process can be achieved by using four methods: (i) crystalline particles; (ii) self-assembled networks; (iii) polymer networks; and (iv) emulsions.

There are numerous studies addressing all structuring routes mentioned above. More than 4300 publications concerning oleo−/organogel indicate that many research groups consider the topic worthwhile (Fig. 27.2). However, the sum of product launches is still insubstantial, considering the global scientific effort (about 7000 patents, including non-food applications).

If oleogel ought to be a qualified substitute for comparably cheap structured lipids based on crystalline TAGs, they have to satisfy particular needs tailored to meet the individual food product's requirements. For example, they should offer the possibility to design specific product properties such as melting temperature. The list of requirements for oleogels for food production is long. The criteria may include that they are food grade, affordable, provide a (semi-)solid structure, do not negatively interfere with other ingredients in a product while ensuring a similar taste and mouthfeel, and be versatile, according to Marangoni and Garti [18].

Moreover, consumer acceptance is an essential factor for a successful application. It is, therefore, necessary for oleogels to provide similar sensory sensations or to bring some other exceptional benefits to the table. However, they tend to disintegrate when consumed, while lipids based on crystalline TAGs melt, which provides a different experience. As a result, consumer acceptance appears to be one of the most critical and complicated to achieve, yet it is a crucial factor for a switch on an industrial scale, given that oleogels can hardly meet the price tag of crystalline lipids (see Tables 27.4 and 27.5). Speaking of large-scale production, it seems obvious that the processing of oleogels has to be transferable to existing equipment without major changes (investments) in a production line. Additional costs associated with this are difficult to estimate at this point since the application of oleogels in food products has merely exceeded the pilot plant scale.

It should also be mentioned that the year 2022 has shown that growing political tensions and unexpected events—whether they are dramatic climate or political events—can impact our fragile global food supply chain within the blink of an eye. The Russian invasion has caused a global chain reaction in edible oil supply: With Russia and Ukraine being the biggest producers of sunflower oil, fear of shortages has led to a dramatic increase in sunflower oil commodity price from about 1400 $/t to 2400 $/t (ADM, personal communication). As a result, palm oil could be sold for a higher price to compensate, and exports soared. This has led to a shortage of palm oil in one of the main exporting countries of palm oil (Indonesia), whereupon exports of certain palm oil products were banned temporarily, leading to a further price push (palm oil 1300 $/t to 1800 $/t). All of this happened within 2 months. In light of increasing numbers of dramatic climate events and resulting political tensions, it appears reasonable to research alternatives for structured lipids. However, last year's example has shown that no edible oil commodity can be considered independent and that price spikes may happen quickly and affect consumers globally.

Finally, oleogels could play a role in a shifting diet observed in Western countries associated with growing ecological awareness and promote vegetarian or vegan diets. Because of that, many alternatives to conventional meat and dairy products have entered the market in the past years (Fig. 27.1). Besides developing alternatives that mimic animal-based products, new protein sources from plants and fungi or microorganisms offer the opportunity to design new food products. Here, oleogels could become a valuable alternative.

All of the factors mentioned emphasize the challenges that must be met to replace conventional SAFA-based lipid phases effectively. It is a complex task to meet both the functional and economic needs in this situation. The following sections provide a comprehensive overview of the current legislative situation of oleogelators in the USA and EU and discuss their feasibility on an industrial level. To this end, estimated production volumes and prices will be given.

27.2 Legislation

In addition to the functional restraints mentioned in the previous sections, oleogelators are considered to be direct food additives and, therefore, must be approved by the respective state authorities. In Europe, new food additives are approved according to the regulation (EC) No 1333/2008 (December 16, 2008) and subsequently given a numerical E number to inform consumers. Approval is only granted if the additive is not a risk to human health (based on scientific evidence), is technologically necessary, and does not mislead the consumer. The E number system has been adopted and extended by the Codex Alimentarius Commission (an international numbering system for food additives, INS), which provides a set of recognized standards for good food practice, safety, and transparency (created by the FAO and supported by the WHO). The FAO and WHO also regularly release the Joint FAO-WHO Expert Committee (JECFA) Report on food additives.

Prior to approval, the European Food Safety Authority (EFSA) assesses the health and safety of a new food additive or reassesses existing additives in light of new scientific insights. Approval is usually granted for certain food applications and includes an acceptable daily intake (ADI). This assessment is typically based on current research, which uses long-term feeding studies on animals and/or humans. Consequently, the approval of a completely new food additive is a complex process.

Furthermore, the EU has specified the Novel Food legislation (according to EC Nr. 1829/2003) for foods that have a new or modified molecular structure or have undergone a novel process, which results in a change of food structure, composition, or metabolism. This legislation also applies to compounds derived from microorganisms, fungi, algae, or extracts of the aforementioned novel plants, animals, or their extracts. A key aspect of the regulation is that it states that there was no history of "significant" consumption in the European Union prior to May 15, 1997.

Applicants who wish to have their novel food product approved by the European Union (EU) Commission must submit a proposal. This proposal must include details about the food's composition, production processes, and scientific evidence that confirms it does not pose a danger to human health. If the novel food is significantly different from an existing food product or food ingredient, the approval and admission procedure can be lengthy (several years). It is also important to note that any health claims associated with a food product or ingredient must be approved separately according to Regulation EC 1924/2006.

In contrast, novel food products in the USA are regulated by the US Food and Drug Administration (FDA) unless they are listed as "generally recognized as safe" (GRAS). GRAS approval is issued by experts and only for the intended use according to the Federal Food, Drug, and Cosmetic Act (FFDCA). In contrast to the European Union's E number system, food additives in the USA are listed using their Chemical Abstracts Number (CAS) and FDA regulation number (according to the US Code of Federal Regulations).

Authorization of the use of food additives for oleogelation can be conducted by official commissions and committees, countries, or any other interested parties, such as companies. Tables 27.1 and 27.2 outline the current legislation on the use of such additives. Most substances are accepted for specific food products, with the upper limit being dependent upon the application, such as the use of waxes as glazing agents. It is anticipated that rice bran and sunflower wax (limited to cosmetic use) may receive approval for food use based on the data and feeding studies for other approved plant waxes, as their composition is very similar.

Table 27.1 Overview of the legislation of selected oleogelators in the USA

Oleogelator	GRAS[a]	Purpose	ADI[b]
12-Hydroxy stearic acid	No	Not applicable	Not applicable
Rice bran wax	Yes	Chewing gum, confections and frostings, hard candy soft candy, others	No ADI, depends on product
Beeswax	Yes	Lubricant, surface-finishing agent, chewing gum, hard candy	No ADI, depends on product
Candelilla wax	Yes	Anticaking agent, surface-finishing agent, baked goods, chewing gum, processed fruits, soft candy, sauces	None
Carnauba wax	Yes	Grain products, vegetables, fruits, milk products, nuts and seeds, fats and oils, sugar, beverages	None
Ethylcellulose	Yes	Emulsifier or emulsifier salt, flavor enhancer, flavoring agent or adjuvant, lubricant or release agent, masticatory substance, stabilizer or thickener, surface-active agent, texturizer	No ADI, depends on product
Mono-diglycerides	Yes	Meat, poultry, fish, dried beans, pea, nuts and seeds, grain products, fruit and vegetable products, oils, dressings, sugars, sweets, beverages	No ADI, depends on product
Phytosterols/-esters	Yes	Candy, fresh fruits and vegetables, chewing gum, snack bars	2–3 g, recommendation

ADI acceptable daily intake, *GRAS* generally recognized as safe

[a]Data available at https://www.cfsanappsexternal.fda.gov/scripts/fdcc/index.cfm?set=GRASNotices

[b]Data available at https://www.cfsanappsexternal.fda.gov/scripts/fdcc/index.cfm?set=FoodSubstances

However, when used for oil structuring, the oleogelators must go through the legislative processes of the Food and Drug Administration (FDA) and European Food Safety Authority (EFSA) (and Novel Food) again. Depending on the amount of structurant in the final product, additional feeding studies may have to be conducted. Although most feeding studies use exposures much higher than the estimated intended consumption, the NOALs (no-observed-adverse-effect levels) of wax constituents in toxicological studies of plant waxes were 10–50 times higher than the conservative approach suggested by the EFSA (22 mg/kg/day). The limitations imposed by the authorities are based on the metabolism of the respective component, among other factors. The detailed results of these studies for different oleogelators cannot be discussed within this chapter. For further details, the evaluations on food additives published by the EFSA or FDA can be consulted.

Table 27.2 Overview of the legislation of selected oleogelators in the EU

Oleogelator	E number	Purpose	ADI
12-Hydroxy stearic acid	None	Not applicable	Not applicable
Rice bran wax	908 (INS)	Chewing gum, confections and frostings, hard candy, soft candy, others	Not defined
Beeswax	901	Glazing agent, surface treatment fruits, food supplements, color carrier	None [62]
Candelilla wax	902	Glazing agent	None [63]
Carnauba wax	903	Glazing agent	None [64], 7 mg/kg (JECFA)
Ethylcellulose	462	Binder, filler	None [65]
Mono-diglycerides	471	Emulsifier, stabilizer	None, re-evaluation ongoing [66]
Phytosterols/-esters	None, novel food[a]	Candy, fresh fruits and vegetables, chewing gum, snack bars	3 g [67], 40 mg/kg (JECFA)

ADI acceptable daily intake, *INS* international numbering system for food additives, *JECFA* Joint FAO-WHO Expert Committee
[a]Data available at https://eur-lex.europa.eu/legal-content/EN/TXT/?uri=CELEX:32017R2470

27.3 Feasibility and Scalability for Food Applications

In theory, the scalability of oleogels for food applications is a major advantage. They could be used to replace traditional fats and oils in various applications, from baked goods to spreads and dressings. This allows manufacturers to advertise a healthier product or even a health claim.

However, depending on the type of gelators used, very different network structures and interactions and gelation temperatures and disintegration temperatures can be observed. Besides, the gels may vary significantly in their rheology and physical and chemical properties. These properties can be modified to a certain extent by varying the gelator composition, its concentration, the production process, and type and composition of the continuous phase. However, none of the systems discussed can be said to be completely comparable to semi-solid lipid phases based on SAFAs in terms of all aspects.

As previously mentioned, a structuring system's availability and affordability are essential components for its application potential. To better understand this, the prices for a selected structuring system—edible waxes—from different sources will be compared in detail. Waxes are particularly interesting for oil structuring since most of them are derived from waste streams of the food industry, as outlined in the following. Table 27.3 lists the potential production volumes and prices of waxes from different sources (sunflower [SFX], rice bran [RBX], candelilla [CLX], carnauba [CRX], bees [BWX], sugarcane [SCX]). It should be noted that the critical gelation concentration (CGC) can vary greatly, depending on the wax and solvent

Table 27.3 Production volumes, critical gelation concentration (CGC), and prices of commercially available waxes

	SFX	RBX	CLX	CRX	BWX	SCX
Production volume (thousand tons)	1–5[a]	>50[a]	1–2[a]	16–18[a]	67[b] 10–15 for food[a]	3800[c]
CGC (wt%)	0.5–1	1–5	0.75–2	3–5	1–3	3
Structured oil (million metric tons [MM.t])	10	1	0.05	0.3	0.3	126
Wax price (thousand $/t)	–	–	–	5[d]	3–7[e]	–

BWX beeswax, *CGC* critical gelation concentration, *CLX* candelilla wax, *CRX* carnauba wax, *RBX* rice bran wax, *SCX* sugarcane wax, *SFX* sunflower wax
[a]Personal communication Kahlwax GmbH. & Co. K.G
[b]FAO STAT
[c]Estimated, based on sugarcane production worldwide
[d]Bengsten et al. [19]
[e]Range producer price in Mexico, Turkey, Argentina, Spain, Kenya [20]

composition; therefore, potential oleogel volumes were estimated conservatively using the highest CGC found in the literature. Unfortunately, reliable data on global production volumes and prices of waxes are scarce, as the market could be more transparent and can be affected by external factors. Beeswax is an exception, as the FAO offers annual global production volumes (FAO STAT).

In contrast to the potential production of sugarcane wax (SCX), which was calculated based on its core product, the potential production of rice bran wax (RBX) was determined based on the process of rice bran oil production. In 2017, approximately 1900 million tons of sugarcane were harvested worldwide, from which 60 million tons comprised the dry filter cake, containing an average wax content of 9%. This amount of wax could be refined to produce 3.8 million tons of refined sugarcane wax yearly.

At the same time, in 2019, 782 million tons of rice (paddy) were harvested, from which 62.5 million tons were removed during rice polishing, containing an average of 15–30% lipids and 0.4–1.5% wax. This would enable for a potential production of 0.4 million tons of rice bran wax annually. However, the actual production was larger than 50,000 tons since only 1.2 million tons of rice bran oil were produced from rice bran in 2017, while most bran was sold as a low-cost animal feed. This potential and actual production difference highlights that waxes may be more economically attractive than other structuring systems.

The potential amount of oleogel that could be produced using the reported annual wax production volume was calculated for each species of wax based on its respective critical gelation concentration (CGC). The CGC is the minimal concentration at which the mixture of wax and solvent does not flow when turned upside down at a certain temperature (20 °C). This value strongly depends on the wax composition, its minor components, the solvent's composition, and the gel's manufacturing process. For example, the CGC of candelilla wax to structure high

Table 27.4 Production volumes, critical gelation concentration (CGC), and prices of commercially available oleogelators and palm oil and tripalmitate (PPP) derived from palm oil

Structuring material	Production volume (MM.t)	Price ($/t)	Min. wt% for oil structuring	Amount of structured oil (MM.t)	Price per t structured oil ($/t)
Palm oil	71.4[a]	940	Straight	71.4	940
PPP from palm oil	5.7[b]	750[b]	3–4	142.7–190.3	896–903
EC	0.45–0.47[c]	1000[d]	3–4	11.3–15.7	903–913
SFX	$1–5 \cdot 10^{-3}$ [a]	1000[d]	0.5–1.0	10	900.5–901
RBX	0.4[c]	1000[d]	1–5	1–5	901–905
SCX	3.8[c]	1000[d]	3	126	903
β-sitosterol/γ-oryzanol	0.2[c] 0.11–0.34[c]	1000[d] 3330[c]	2–4 (40:60 mass ratio)	4.6–28.3	1361–1919

EC ethylcellulose, *RBX* rice bran wax, *SCX* sugarcane wax, *SFX* sunflower wax
Palm oil and PPP are used as a reference for structured fat
[a]Actual; see Table 27.3
[b]Assuming 8% tripalmitate (PPP) in palm oil; additional costs of double dry fractionation estimated 200 $/t
[c]Potential production volume
[d]Assumption to calculate the price per t structured oil; the actual price is likely much higher

oleic sunflower oil was reported to be as low as 0.75%. In comparison, 2% was needed to solidify refined soybean, olive, and canola oils. Additionally, the CGC of several types of rice bran waxes varied between 0.5% and 5%. These differences in CGC illustrate the complexity of the gelling behavior of waxes and the necessity to perform an in-depth characterization of waxes and solvents to obtain comparable results. Therefore, the volumes stated are likely to be overly optimistic since the CGC does not necessarily represent a wax dosage that warrants food functionality.

Table 27.4 provides a similar overview of other oil structuring systems. Furthermore, as can be seen, prices per ton for structured oil surpass that of palm oil and even dry fractionated tripalmitin. For reasons of comparison, Table 27.4 also includes the data of the potentially most abundant wax species. It is important to note that several estimates were used to arrive at the price calculations in this table, and the actual prices for waxes, for example, are likely much higher, as evidenced by the price for beeswax production (3000–7000 $/t) and carnauba wax market price (5000$/t, 2010). That said, prices will likely decrease once production volumes increase, and manufacturing processes are integrated into existing production lines (e.g., wax extraction from sunflower seed hull). However, as mentioned in the previous sections, price fluctuations due to unforeseeable global events could lead to drastic price fluctuations in edible oil/fat commodities. Moreover, in most cases, edible oil makes up about 95% of the mass of oleogel, therefore impacting the overall price accordingly. Thus, oleogels are subjected to the same price fluctuations as traditionally structured lipids.

However, the health benefits associated with consuming (poly)unsaturated fatty acids could outweigh the additional costs. Therefore, using oleogels in certain food products appears reasonable, although one must consider the functionality, state, and quantity of the lipid phase in the product. For instance, oleogels are not ideal for food products where the lipid phase is responsible for a particular melting sensation, as oleogels tend to disintegrate rather than melt when ingested. On the other hand, partial substitution or the development of new products (e.g., heat-resistant chocolate) may be possible. Table 27.5 provides an overview of application studies of different oleogels divided into food categories.

This shows the variety of their application, but with regard to scalability in industrial processes, the state and functionality of the structured lipid at various stages of the production appear to be crucial. For example, the role of fat in the production of chocolate is multifaceted and not to be underestimated. Not only does it affect the appearance, snap, and melting sensation of the final product, but also it is integral to the process conditions. Precise temperature control is necessary during crystallization to avoid the crystallization of cocoa butter into undesired polymorphs. Additionally, fat needs to be molten during the conching of the cocoa mass (at 60–65 °C for milk chocolate) in order to ensure coating of the particles. This liquid state also has the benefit of decreasing the viscosity of the mass, which allows for the discharge of undesired flavors, the improved mobility of ingredients, and the prevention of particle aggregation.

Given the intricate demands of chocolate production, oleogels have been studied in an attempt to replace or substitute cocoa butter in chocolate-like applications such as filled confectionaries [27, 29, 30], heat-resistant chocolate [21–24, 68], and reduced-fat chocolate [25]. However, these attempts are yet to be successful.

In comminuted meat products, the size of the fat globules and the ratio of fat to protein are both major factors that define the texture of the final product. In order to produce a stable batter, the myosin protein needs to form a network that is able to support the larger animal fat globules (<100 μm) [69]. The beef fat remains semi-solid as the mincing process occurs at lower temperatures (~8–12 °C). This makes it susceptible to mechanical disruption, leading to a wide fat globule size distribution in the final product, which gives it the desired snap and soft texture [33]. Earlier studies have shown that substituting animal fat with edible oils results in a firmer and chewier sausage [69–71]. This is because the smaller droplets (<20 μm) create a much larger fat surface area, resulting in more homogeneous and stiffer protein networks filled with fat globules [35]. However, due to the different sensory properties, products made with vegetable oil are unlikely to meet consumers' expectations. Therefore, oleogels should be added to the batter in the solid state to replace the animal fat while still maintaining the desired fat globule size and texture. This will enable a nutritionally improved fatty acid profile while still delivering a product with the same structural properties.

The solid fat phase in margarine and spreads is essential in order to stabilize the emulsion and trap the liquid oil. Furthermore, the three-dimensional fat crystal network allows the spreadability of the product. Therefore, hardstock fats must be solid at room temperature and yet melt quickly when in contact with the body temperature to avoid a waxy texture. Additionally, when the fat crystal network disintegrates, it leads to the destabilization of the emulsion and the release of flavors from the aqueous phase. For this reason, it is difficult to replace the traditional

Table 27.5 Overview of recent publications on food applications of oleogels sorted by product category

Food category	Fat functionality	Product	Oleogel system	Oleogelator	References
Confectionary products	Snap, appearance, melting sensation	Heat-resistant chocolate	Ethylcellulose (EC)	12-Hydroxy stearic acid (12-HSA)	[21–24]
		Reduced-fat chocolate	EC	Rice bran wax	[25]
		Oil migration/confectionary fillings	12-HSA	Beeswax	[26–28]
			Sterol/sterol-ester	Candelilla wax	[29]
			Wax	Carnauba wax	[30]
Meat products	Snap, texture	Sausages (Frankfurter-, Breakfast-, Bologna-type), hamburger	Sterol/sterol-ester	Ethylcellulose	[31–34]
			EC, Wax	Mono-diglycerides	[34–41]
			Pork skin emulsion	Phytosterols/–esters	[42]
	Texture, spreadability	Pâté	EC, Wax		[43, 44]
			EC + MAG		[45]
Margarine and spreads	Emulsion stabilization, hardness, melting properties, spreadability	Margarine/spreads	Wax		[46–49]
Shortening for baking	Prevent adherence of gluten and starch, air cell stabilization, mouthfeel, flavor release	Cookies	EC, Wax		[50–54]
		(Gluten-free) cake	Wax		[41, 53, 55]
			EC		[40]
Dairy products	Richness, creaminess, scoopability, stabilization	Ice cream	Sterol/sterol-ester		[56–58]
			Wax		[20]
	Richness	Yogurt	Sterol/sterol-ester		[59]
	Richness, spreadability	Cream cheese	Wax		[60, 61]

hardstock in margarine with oleogels. Notably, the emulsion must remain stable throughout the product life cycle. During the production of conventional margarine, the liquid emulsion goes through a series of scraped surface heat exchanger units (SSHE) that subject it to intense shear forces and rapid cooling rates. Oleogels, which do not have a crystalline structure such as EC or sterol/sterol-ester, offer the chance to simplify the margarine production process by eliminating the need for SSHE as the liquid emulsion can be filled directly into containers if it is able to maintain the desired droplet size of the aqueous phase (3–10 μm) during the process of structural development and storage.

Finally, in dairy applications, the application of oleogels has been studied, inter alia, in ice cream. Ice cream is a complex multiphase system that consists of a partially solidified aerated oil in water emulsion. The continuous phase is a highly viscous, concentrated solution of sugars, proteins, and hydrocolloids. In addition to ice crystals, the ice cream contains finely dispersed semi-solid fat droplets that form agglomerates and help to stabilize the air/water interface. This process is further facilitated by the use of emulsifiers such as mono- and diglycerides. It is important to accurately distribute and stabilize the fat droplets by proteins in order to prevent creaming. This is accomplished through homogenization at elevated temperatures (75 °C). After homogenization, the emulsion is typically left to ripen, during which time the fat solidifies within the globules and forms flocs and possibly a network. Before freezing, the emulsion is often whipped to incorporate air bubbles. A higher proportion of air results in softer ice cream but one that is less creamy. The increase in volume is usually expressed as overrun, which is the excess volume of the final product compared to the initial emulsion volume. Oleogels ought to deliver a benefit in such a complex food matrix and would have to be at least able to stabilize the interfaces and provide a similar melting sensation. Due to their crystalline structure, waxes appear as a promising candidate.

Indeed, the studies mentioned in Table 27.5 have revealed that wax-based oleogels can deliver similar structural properties as milk fat in ice cream. This means that the accumulation of partially aggregated fat globules can stabilize the air bubbles. Although the functionality of oleogels was found to be better than liquid oils, the disintegration behavior was still vastly different from the reference. Further research is needed to tune formulations and explore processing options in order to create acceptable oleogel-based ice creams, such as the addition of selected emulsifiers to improve the efficacy of air cell stabilization. However, none of the studies have provided information regarding ice creams' organoleptic properties or storage stability.

In conclusion, it can be said that oleogel applications and, thus, large-scale production are generally promising and feasible. In fat continuous products, stabilization of the interfaces and distribution within the matrix are most important. In water continuous products, the fat is often utilized as a medium. Finally, the functionality and condition of the fat during and following processing must be carefully considered. Fortunately, the wide variety of available structurants at present makes the selection of the most suitable system possible, which potentially will result in manageable costs for food manufacturers. However, interactions with complicated food matrices are difficult to predict, so, in the most unfavorable circumstances, the gel could break apart as it was reported for the sterol/sterol ester system [72, 73].

27.4 Conclusion and Perspective

Implementing a ban on *trans* fatty acids, coupled with the health benefits associated with a diet including MUFAs and PUFAs and the negative public image of palm oil, has led to a considerable push to replace or substitute conventional solid fats with healthier alternatives. Although MUFAs and PUFAs support our cardiovascular system's health, they cannot, in their native state, provide the necessary structure for many food product applications. Oleogels, on the other hand, offer the possibility to improve the nutritional profile of foods, particularly their lipid phases, by immobilizing liquid oil rich in unsaturated fatty acids in a non-TAG three-dimensional network. Various researchers have put much effort into discovering new oleogel systems and understanding their formation and properties over the past decades. However, oleogels are still not part of our daily diet, so the question arises of whether we have lost track of our goal to provide healthier products to the consumer or if we have just set our expectations too high in terms of finding the ideal oleogel system. It could be argued that the price per ton of structured oil is much higher than that of palm oil, which will be reflected in end-user pricing. However, the prices of certain structurants, such as waxes, will likely decrease when production is carried out on a larger scale. Sugarcane and rice bran waxes, as well as ethylcellulose production, for example, have not yet reached their full potential, as noted in this review. Despite this, much of the price difference is due to the markup of seed oils compared to palm oil. The broader implementation of oleogels is currently hindered not only by economic issues but also by the legal situation, which is a bit unclear and difficult (novel food regulation) and will become clearer once products are introduced into the market. Despite this, the ongoing demand for healthier alternatives to SAFAs and the controversial public perception of palm oil should be sufficient to incentivize using oleogels on a larger scale. However, many studies on oleogels lack challenges under real process conditions, sensory evaluations, and storage tests. Furthermore, it is not always assessed if the systems studied offer any benefit over the use of liquid oil. Additionally, studies based on known structurants, in alternative ratios or combinations with similar continuous phases to those used in previous studies, should only be done when they aim to provide specific new insights and not just to replicate previous results. To this end, oleogels need to be functional, available, and affordable to a certain extent. Progress toward successful product applications can only be achieved if the initial approaches provided by scientists are reasonable. Additionally, intricate processes that are unlikely to be available on a larger scale do not offer viable solutions. At present, there is a disconnect between oleogel science and application in the industry. To create a better understanding, scientists investigating food product applications should always contemplate the functionality of the fat and its physical state during processing and in the final product. Consequently, suitable structuring systems satisfying the identified critical characteristics may be chosen. Moreover, the idea of replicating all the unique properties that specific solid TAG networks provide should be abandoned to make room for new developments. This could be achieved by either generating new product characteristics or producing products similar to consumer expectations but with different technological processes.

In the academic community, edible oleogels have been intensively studied for about 25 years. However, for consumers, oleogels are a new class of food ingredient. Although they may provide a healthier and more sustainable alternative to traditional fats and oils, it cannot be assumed that there is a great willingness to buy oleogel-based products. Recently, consumers have been fairly overwhelmed with new processes, ingredients, and products and changing recommendations for dietary habits. For some, this has led to confusion and skepticism toward new products. Therefore, conducting extensive consumer studies before moving toward an industrial scale is crucial. These may include theoretical studies to evaluate the curiosity and readiness toward these products and product tests, ideally among people with varying social backgrounds and dietary preferences.

While the science behind oleogels is important to take steps toward industrial applications, it appears that projects between academic research and food manufacturers are scarce. This is partly because basic research had to be performed to understand oleogels in depth and identify new structuring systems. However, for a transition from lab benches to pilot and industrial plants, science should also focus on providing practical tools for functionality during processing and upscaling. To this end, the physical properties and preparation methods of different oleogels need to be considered in connection with the respective food application, existing technology, and the urge to switch from tropical oils to oleogels.

However, current market trends may lead to an additional push toward the utilization of oleogels due to the development of animal-free meat and dairy alternatives. This also opens the possibility of developing radically new products.

References

1. Hamm W, Hamilton RJ, Calliauw G (2013) Edible oil processing, Second edn. Wiley, Chichester, West Sussex Hoboken
2. Kloek W (1998) Mechanical properties of fats in relation to their crystallization. Wageningen, CIP-Data Koninklijke Bibliotheek, Den Haag, The Netherlands
3. World Health Organisation W (2018) Guidelines: saturated fatty acid and trans-fatty acid intake for adults and children, Geneva
4. Papanikolaou Y, Brooks J, Reider C, Fulgoni VL (2014) U.S. adults are not meeting recommended levels for fish and omega-3 fatty acid intake: results of an analysis using observational data from NHANES 2003–2008. Nutr J 13:31. https://doi.org/10.1186/1475-2891-13-31

5. Mozaffarian D, Clarke R (2009) Quantitative effects on cardiovascular risk factors and coronary heart disease risk of replacing partially hydrogenated vegetable oils with other fats and oils. Eur J Clin Nutr 63:S22–S33. https://doi.org/10.1038/sj.ejcn.1602976
6. Zhang Z, Fulgoni V, Kris-Etherton P, Mitmesser S (2018) Dietary intakes of EPA and DHA omega-3 fatty acids among US childbearing-age and pregnant women: an analysis of NHANES 2001–2014. Nutrients 10:416. https://doi.org/10.3390/nu10040416
7. de Souza RJ, Mente A, Maroleanu A et al (2015) Intake of saturated and trans unsaturated fatty acids and risk of all cause mortality, cardiovascular disease, and type 2 diabetes: systematic review and meta-analysis of observational studies. BMJ:h3978. https://doi.org/10.1136/bmj.h3978
8. Bundesinstitut für Risikobewertung B (2013) Höhe der derzeitigen trans-Fettsäureaufnahme in Deutschland ist gesundheitlich unbedenklich, Berlin
9. Micha R, Khatibzadeh S, Shi P et al (2014) Global, regional, and national consumption levels of dietary fats and oils in 1990 and 2010: a systematic analysis including 266 country-specific nutrition surveys. BMJ 348:g2272–g2272. https://doi.org/10.1136/bmj.g2272
10. World Health Organisation W (2018) REPLACE - Trans fat-free by 2023
11. Gibbs HK, Ruesch AS, Achard F et al (2010) Tropical forests were the primary sources of new agricultural land in the 1980s and 1990s. Proc Natl Acad Sci U S A 107:16732–16737. https://doi.org/10.1073/pnas.0910275107
12. West PC, Gibbs HK, Monfreda C et al (2010) Trading carbon for food: global comparison of carbon stocks vs. crop yields on agricultural land. Proc Natl Acad Sci U S A 107:19645–19648. https://doi.org/10.1073/pnas.1011078107
13. Bonato G (2014) Brazil farmers poised to plant record soy are in 2018–19
14. Guillaume T, Kotowska MM, Hertel D et al (2018) Carbon costs and benefits of Indonesian rainforest conversion to plantations. Nat Commun 9:2388. https://doi.org/10.1038/s41467-018-04755-y
15. Willer H, Lernoud J (2019) The world of organic agriculture. Statistics and emerging trends 2019
16. Mensink RP, Zock PL, Kester AD, Katan MB (2003) Effects of dietary fatty acids and carbohydrates on the ratio of serum total to HDL cholesterol and on serum lipids and apolipoproteins: a meta-analysis of 60 controlled trials. Am J Clin Nutr 77:1146–1155. https://doi.org/10.1093/ajcn/77.5.1146
17. Nettleton JA, Brouwer IA, Geleijnse JM, Hornstra G (2017) Saturated fat consumption and risk of coronary heart disease and ischemic stroke: a science update. Ann Nutr Metab 70:26–33. https://doi.org/10.1159/000455681
18. Marangoni AG, Garti N (2018) Edible oleogels: structure and health implications, Second edn. AOCS Press, Academic, an imprint of Elsevier, London, UK; San Diego
19. Bengsten P (2011) Carnauba wax background paper
20. Zulim Botega DC, Marangoni AG, Smith AK, Goff HD (2013) The potential application of rice bran wax oleogel to replace solid fat and enhance unsaturated fat content in ice cream: rice bran wax oleogel in ice cream.... J Food Sci 78:C1334–C1339. https://doi.org/10.1111/1750-3841.12175
21. Stortz TA, Marangoni AG (2011) Heat resistant chocolate. Trends Food Sci Technol 22:201–214. https://doi.org/10.1016/j.tifs.2011.02.001
22. Stortz TA, Marangoni AG (2013) Ethylcellulose solvent substitution method of preparing heat resistant chocolate. Food Res Int 51:797–803. https://doi.org/10.1016/j.foodres.2013.01.059
23. Stortz TA, De Moura DC, Laredo T, Marangoni AG (2014) Molecular interactions of ethylcellulose with sucrose particles. RSC Adv 4:55048–55061. https://doi.org/10.1039/C4RA12010H
24. Stortz TA, Laredo T, Marangoni AG (2015) The role of lecithin and solvent addition in ethylcellulose-stabilized heat resistant chocolate. Food Biophys 10:253–263. https://doi.org/10.1007/s11483-014-9379-7

25. Do T-AL, Mitchell JR, Wolf B, Vieira J (2010) Use of ethylcellulose polymers as stabilizer in fat-based food suspensions examined on the example of model reduced-fat chocolate. React Funct Polym 70:856–862. https://doi.org/10.1016/j.reactfunctpolym.2010.07.012
26. Marty S, Baker K, Dibildox-Alvarado E et al (2005) Monitoring and quantifying of oil migration in cocoa butter using a flatbed scanner and fluorescence light microscopy. Food Res Int 38:1189–1197. https://doi.org/10.1016/j.foodres.2005.04.008
27. Hughes NE, Marangoni AG, Wright AJ et al (2009) Potential food applications of edible oil organogels. Trends Food Sci Technol 20:470–480. https://doi.org/10.1016/j.tifs.2009.06.002
28. Marty S, Baker KW, Marangoni AG (2009) Optimization of a scanner imaging technique to accurately study oil migration kinetics. Food Res Int 42:368–373. https://doi.org/10.1016/j.foodres.2008.12.017
29. Wendt A, Abraham K, Wernecke C et al (2017) Application of β-Sitosterol + γ-Oryzanol-structured organogel as migration barrier in filled chocolate products. J Am Oil Chem Soc 94:1131–1140. https://doi.org/10.1007/s11746-017-3024-9
30. Doan CD, Patel AR, Tavernier I et al (2016) The feasibility of wax-based oleogel as a potential co-structurant with palm oil in low-saturated fat confectionery fillings: hazelnut fillings from beeswax-oleogels and palm oil. Eur J Lipid Sci Technol 118:1903–1914. https://doi.org/10.1002/ejlt.201500172
31. Moschakis T, Panagiotopoulou E, Katsanidis E (2016) Sunflower oil organogels and organogel-in-water emulsions (part I): microstructure and mechanical properties. LWT 73:153–161. https://doi.org/10.1016/j.lwt.2016.03.004
32. Panagiotopoulou E, Moschakis T, Katsanidis E (2016) Sunflower oil organogels and organogel-in-water emulsions (part II): implementation in frankfurter sausages. LWT 73:351–356. https://doi.org/10.1016/j.lwt.2016.06.006
33. Martins AJ, Lorenzo JM, Franco D et al (2019) Omega-3 and polyunsaturated fatty acids-enriched hamburgers using sterol-based oleogels. Eur J Lipid Sci Technol 121:1900111. https://doi.org/10.1002/ejlt.201900111
34. Franco D, Martins AJ, López-Pedrouso M et al (2020) Evaluation of linseed oil oleogels to partially replace pork backfat in fermented sausages. J Sci Food Agric 100:218–224. https://doi.org/10.1002/jsfa.10025
35. Zetzl AK, Marangoni AG, Barbut S (2012) Mechanical properties of ethylcellulose oleogels and their potential for saturated fat reduction in frankfurters. Food Funct 3:327. https://doi.org/10.1039/c2fo10202a
36. Barbut S, Wood J, Marangoni AG (2016) Effects of organogel hardness and formulation on acceptance of frankfurters: effects of organogel hardness on franks.... J Food Sci 81:C2183–C2188. https://doi.org/10.1111/1750-3841.13409
37. Barbut S, Wood J, Marangoni A (2016) Potential use of organogels to replace animal fat in comminuted meat products. Meat Sci 122:155–162. https://doi.org/10.1016/j.meatsci.2016.08.003
38. Barbut S, Wood J, Marangoni A (2016) Quality effects of using organogels in breakfast sausage. Meat Sci 122:84–89. https://doi.org/10.1016/j.meatsci.2016.07.022
39. Barbut S, Marangoni A (2019) Organogels use in meat processing – effects of fat/oil type and heating rate. Meat Sci 149:9–13. https://doi.org/10.1016/j.meatsci.2018.11.003
40. Adili L, Roufegarinejad L, Tabibiazar M et al (2020) Development and characterization of reinforced ethyl cellulose based oleogel with adipic acid: its application in cake and beef burger. LWT 126:109277. https://doi.org/10.1016/j.lwt.2020.109277
41. Aliasl khiabani A, Tabibiazar M, Roufegarinejad L et al (2020) Preparation and characterization of carnauba wax/adipic acid oleogel: a new reinforced oleogel for application in cake and beef burger. Food Chem 333:127446. https://doi.org/10.1016/j.foodchem.2020.127446
42. da Silva SL, Amaral JT, Ribeiro M et al (2019) Fat replacement by oleogel rich in oleic acid and its impact on the technological, nutritional, oxidative, and sensory properties of Bologna-type sausages. Meat Sci 149:141–148. https://doi.org/10.1016/j.meatsci.2018.11.020

43. Gómez-Estaca J, Herrero AM, Herranz B et al (2019) Characterization of ethyl cellulose and beeswax oleogels and their suitability as fat replacers in healthier lipid pâtés development. Food Hydrocoll 87:960–969. https://doi.org/10.1016/j.foodhyd.2018.09.029
44. Martins AJ, Lorenzo JM, Franco D et al (2020) Characterization of enriched meat-based Pâté manufactured with oleogels as fat substitutes. Gels 6:17. https://doi.org/10.3390/gels6020017
45. Barbut S, Marangoni AG, Thode U, Tiensa BE (2019) Using canola oil organogels as fat replacement in liver Pâté. J Food Sci 84:2646–2651. https://doi.org/10.1111/1750-3841.14753
46. Hwang H-S, Singh M, Bakota EL et al (2013) Margarine from organogels of plant wax and soybean oil. J Am Oil Chem Soc 90:1705–1712. https://doi.org/10.1007/s11746-013-2315-z
47. Hwang H-S, Singh M, Winkler-Moser JK et al (2014) Preparation of margarines from organogels of sunflower wax and vegetable oils: margarines from sunflower wax organogels .... J Food Sci 79:C1926–C1932. https://doi.org/10.1111/1750-3841.12596
48. Fayaz G, Goli SAH, Kadivar M (2017) A novel propolis wax-based organogel: effect of oil type on its formation, crystal structure and thermal properties. J Am Oil Chem Soc 94:47–55. https://doi.org/10.1007/s11746-016-2915-5
49. Winkler-Moser JK, Anderson J, Byars JA et al (2019) Evaluation of beeswax, candelilla wax, rice bran wax, and sunflower wax as alternative stabilizers for peanut butter. J Am Oil Chem Soc 96:1235–1248. https://doi.org/10.1002/aocs.12276
50. Jang A, Bae W, Hwang H-S et al (2015) Evaluation of canola oil oleogels with candelilla wax as an alternative to shortening in baked goods. Food Chem 187:525–529. https://doi.org/10.1016/j.foodchem.2015.04.110
51. Mert B, Demirkesen I (2016) Evaluation of highly unsaturated oleogels as shortening replacer in a short dough product. LWT Food Sci Technol 68:477–484. https://doi.org/10.1016/j.lwt.2015.12.063
52. Mert B, Demirkesen I (2016) Reducing saturated fat with oleogel/shortening blends in a baked product. Food Chem 199:809–816. https://doi.org/10.1016/j.foodchem.2015.12.087
53. Demirkesen I, Mert B (2019) Utilization of beeswax oleogel-shortening mixtures in gluten-free bakery products. J Am Oil Chem Soc 96:545–554. https://doi.org/10.1002/aocs.12195
54. Onacik-Gür S, Żbikowska A (2020) Effect of high-oleic rapeseed oil oleogels on the quality of short-dough biscuits and fat migration. J Food Sci Technol 57:1609–1618. https://doi.org/10.1007/s13197-019-04193-8
55. Alvarez-Ramirez J, Vernon-Carter EJ, Carrera-Tarela Y et al (2020) Effects of candelilla wax/canola oil oleogel on the rheology, texture, thermal properties and in vitro starch digestibility of wheat sponge cake bread. LWT 130:109701. https://doi.org/10.1016/j.lwt.2020.109701
56. Kammegne TD (2015) Beeinflussung der Stabilität organogelbasierter Emulsionen am Beispiel eines Eiscreme-Premixes. Bachelor Thesis, TU Berlin
57. Lubosch K (2015) Weiterführende Untersuchungen Organogelbasierter Emulsionen am Beispiel eines Eiscreme-Premixes. Diploma Thesis, TU Berlin
58. Moriano ME, Alamprese C (2017) Organogels as novel ingredients for low saturated fat ice creams. LWT 86:371–376. https://doi.org/10.1016/j.lwt.2017.07.034
59. Moschakis T, Dergiade I, Lazaridou A et al (2017) Modulating the physical state and functionality of phytosterols by emulsification and organogel formation: application in a model yogurt system. J Funct Foods 33:386–395. https://doi.org/10.1016/j.jff.2017.04.007
60. Bemer HL, Limbaugh M, Cramer ED et al (2016) Vegetable organogels incorporation in cream cheese products. Food Res Int 85:67–75. https://doi.org/10.1016/j.foodres.2016.04.016
61. Park C, Bemer HL, Maleky F (2018) Oxidative stability of rice bran wax oleogels and an oleogel cream cheese product. J Am Oil Chem Soc 95:1267–1275. https://doi.org/10.1002/aocs.12095
62. EFSA (2007) Beeswax (E 901) as a glazing agent and as carrier for flavours - Scientific Opinion of the Panel on Food additives, Flavourings, Processing aids and Materials in Contact with Food (AFC). EFSA J. https://doi.org/10.2903/j.efsa.2007.615

63. EFSA (2012) Scientific Opinion on the re-evaluation of candelilla wax (E 902) as a food additive. EFSA J. https://doi.org/10.2903/j.efsa.2012.2946
64. EFSA (2012) Scientific Opinion on the re-evaluation of carnauba wax (E 903) as a food additive. EFS2 10. https://doi.org/10.2903/j.efsa.2012.2880
65. EFSA (2012) Scientific Opinion on DL-methionine, DL-methionine sodium salt, the hydroxy analogue of methionine and the calcium salt of methionine hydroxy analogue in all animal species; on the isopropyl ester of methionine hydroxy analogue and DL-methionine technically pure protected with copolymer vinylpyridine/styrene in dairy cows; and on DL-methionine technically pure protected with ethylcellulose in ruminants. EFS2 10. https://doi.org/10.2903/j.efsa.2012.2623
66. EFSA, Younes M, Aggett P et al (2017) Re-evaluation of mono- and di-glycerides of fatty acids (E 471) as food additives. EFS2 15. https://doi.org/10.2903/j.efsa.2017.5045
67. EFSA (2012) Scientific Opinion on the safety of stigmasterol-rich plant sterols as food additive. EFSA J. https://doi.org/10.2903/j.efsa.2012.2659
68. Stortz TA, Zetzl AK, Barbut S et al (2012) Edible oleogels in food products to help maximize health benefits and improve nutritional profiles. Lipid Technol 24:151–154. https://doi.org/10.1002/lite.201200205
69. Youssef MK, Barbut S (2009) Effects of protein level and fat/oil on emulsion stability, texture, microstructure and color of meat batters. Meat Sci 82:228–233. https://doi.org/10.1016/j.meatsci.2009.01.015
70. Marquez EJ, Ahmed EM, West RL, Johnson DD (1989) Emulsion stability and sensory quality of beef frankfurters produced at different fat or peanut oil levels. J Food Sci 54:867–870. https://doi.org/10.1111/j.1365-2621.1989.tb07901.x
71. Park J, Rhee KS, Keeton JT, Rhee KC (1989) Properties of low-fat frankfurters containing monounsaturated and omega-3 polyunsaturated oils. J Food Sci 54:500–504. https://doi.org/10.1111/j.1365-2621.1989.tb04637.x
72. den Adel R, Heussen PCM, Bot A (2010) Effect of water on self-assembled tubules in β-sitosterol + γ-oryzanol-based organogels. J Phys Conf Ser 247:012025. https://doi.org/10.1088/1742-6596/247/1/012025
73. Sawalha H, den Adel R, Venema P et al (2012) Organogel-emulsions with mixtures of β-sitosterol and γ-oryzanol: influence of water activity and type of oil phase on gelling capability. J Agric Food Chem 60:3462–3470. https://doi.org/10.1021/jf300313f

Chapter 28
The Future of Oleogels Between Challenges and Opportunities

Fabio Valoppi and Camila Palla

Abbreviations

EFSA	European Food Safety Authority
FDA	Food and Drug Administration
GRAS	Generally recognized as safe
HMWG	High-molecular-weight gelator
LMWG	Low-molecular-weight gelator
TRIZ	Theory of inventive problem solving

Fabio Valoppi and Camila Palla contributed equally

F. Valoppi (✉)
Department of Food and Nutrition, University of Helsinki, Helsinki, Finland

Helsinki Institute of Sustainability Science, Faculty of Agriculture and Forestry, University of Helsinki, Helsinki, Finland

Electronics Research Laboratory, Department of Physics, University of Helsinki, Helsinki, Finland

Helsinki Institute of Life Science, University of Helsinki, Helsinki, Finland
e-mail: fabio.valoppi@helsinki.fi

C. Palla
Departamento de Ingeniería Química, Universidad Nacional del Sur (UNS), Bahía Blanca, Argentina

Planta Piloto de Ingeniería Química – PLAPIQUI (UNS-CONICET), Bahía Blanca, Argentina
e-mail: cpalla@plapiqui.edu.ar

C. Andrea Palla, F. Valoppi (eds.), *Advances in Oleogel Development, Characterization, and Nutritional Aspects*,
https://doi.org/10.1007/978-3-031-46831-5_28

28.1 Introduction

Throughout this book, we have summarized the tremendous progress in the oleogel field over the last 20 years. We explored the evolution of oleogels, from early development to the latest advancements, encompassing formulation and preparation methods, techniques for chemical, physical, and physiological characterization, stability over time, applications, regulations, and industrial feasibility.

Despite great efforts to develop a variety of oleogels that serve as replacers for saturated fats in current food products, the fat-to-oleogel transition has not yet materialized. Among different causes, regulative hurdles, economic barriers, low/poor sustainability of production methods, storage instability, low sensorial acceptability, and limited availability of gelators, to name a few, are currently hindering oleogels from becoming the "fat of the future." In this final chapter, we present a list of the main limitations that affect oleogels. Addressing these limitations is crucial to boost the "fat transition," while simultaneously presenting extraordinary opportunities for future research in the oleogel field. In the following sections, a more exhaustive list with descriptions for each limitation is presented along with some possible opportunities for researchers to focus on in the future.

28.2 Limitations and Challenges to Be Tackled in the Oleogel Field

To organize this complex matter, the reader will find the current limitations we have identified grouped into four categories with relevant subcategories, including a general description followed by a list of specific examples along with their implications. Some examples and suggestions on how to tackle these limitations are also discussed. The reader should keep in mind that additional challenges will emerge as the field evolves.

28.2.1 Oleogel Development, Properties, and Applications

Although the oleogel field is now approaching maturity, there are still limitations in the methods employed for oleogel's development, as well as in their properties, to more effectively mimic the fats they aim to replace. These limitations include the need for a shift in the development approach, a better control over the oleogel's structure, and the possibility to better tailor oleogel's melting profile, mechanical and rheological properties, and physiological effects. In addition to these limitations, further studies are needed to understand the applicability of oleogels in various food products, aiming to develop more universally applicable oleogels.

28.2.1.1 Development

- To develop new oleogels and improve existing ones, it is essential to gain a better understanding of various factors, including the interaction between oil and gelators, the crystallization phenomenon in low-molecular-weight gelator (LMWG) oleogels, the changes in gelator conformation and system composition during drying or oil absorption for high-molecular-weight gelators (HMWG) used in some indirect methods—such as emulsion-, foam-, and aerogel-templated and solvent-exchange methods—, and the role of impurities and other inert ingredients used in oleogel preparation, such as fillers. This effort would require not only extensive systematic experimental work, but also a shift toward computational approaches to develop simulations and models that can deepen our understanding and expand the underlying theoretical knowledge. The development of digital twins and the use of computational tools like artificial intelligence to elucidate complex mechanisms involved in oil gelation would help to move away from the slow and resource-intensive trial-and-error approach, leading to a more rapid progress in the oleogel field.
- An important aspect for further advancement in the oleogel field lies in the control of their microarchitecture, which is important for tailoring their mechanical and rheological properties. Achieving an efficient and responsive microarchitecture requires extensive efforts, not only in determining the proper oleogel formulation but also in identifying and controlling external stimuli/forces and process parameters that would make this control possible, such as cooling rate, application of shear in laminar or turbulent regime, high-intensity ultrasound, and ultrasonic standing waves during cooling. Indeed, these external factors play a crucial role in shaping oleogel architectures, especially when self-assembly of structuring molecules is the driving force in oleogel formation. Their proper application can lead to improved or superior architectures, ultimately enhancing oleogel performance.

28.2.1.2 Properties

- Oleogels often exhibit melting profiles that do not match the fats they are intended to replace. Although some studies have explored the use of specific gelators or mixtures of gelators to solve this problem, the hurdle of accurately mimicking the melting profile of fats remains [1, 2]. On the other hand, oleogels prepared using indirect and semi-direct methods do not possess any melting ability as the constituent building blocks (particles, fibers, biopolymers) do not melt when the temperature of the system is increased. Therefore, although oleogels should be selected based on their application, mimicking the melting profile of fats remains a challenge to be properly addressed, especially when considering restrictions on gelators imposed by regulations.
- The mechanical and rheological properties of oleogels are highly dependent on the preparation method and the type and concentration of gelators. As an

example, matching the hardness of saturated fats can be achieved by increasing the concentration of gelators, but this high concentration might introduce other undesirable side effects like brittleness (oleogels crumble and fracture, and do not show a uniform plastic deformation) as well as negative sensory properties. In general, there is still a need for extensive research to achieve the ability to replicate the deformation profile of saturated fats under compression and shear stress, especially to mimic the plastic deformation exhibited by most fats at large strains, the ability to withstand prolonged mixing, and good thixotropic recovery. In this direction, the mixture of gelators or the addition of plasticizers have been studied [3, 4], as well as the development of composite oleogels containing solid particles along with crystalline gelators [5].
- Although the number of studies on in vitro digestion of oleogels has increased in recent years, in vivo studies to evaluate digestion and physiological effects in animals and human subjects are scarce. The lack of information on the absorption and metabolic pathway of gelators, and more in general the effect of oleogels on human metabolism, microbiome, homeostasis, and health, should be addressed in the future to further elucidate the health benefits and possible limitations of the different oleogels developed over the last two decades. Rigorous clinical trials should be conducted to achieve conclusive findings. This information could also help regulatory authorities to approve gelators for oleogel production.

28.2.1.3 Applications

- Applicability of oleogels should be aimed to be more transversal and less vertical (across several food categories and less within one category). A better understanding of the interaction that oleogels (especially gelators and particles) can establish with other food ingredients such as proteins, carbohydrates, and water, when used in food products, is required. For a more transversal application, efforts should be directed towards the development of an "all-purpose" or universal oleogel, and indeed, some attempts are nowadays emerging in this direction.

28.2.2 Sensory and Consumer Acceptability

A reformulated product must meet consumer expectations in terms of taste, texture, and appearance. Thus, sensory acceptability plays a crucial role in the evaluation of oleogel-based products, reducing the risk of the reformulated product failing when introduced to the market. In addition, consumers are paying more attention to labeling and ingredients, highlighting the need to understand how gelators and other ingredients are perceived when applying oleogels to food products.

- Typically, oleogels have the taste of the underlying oils with a mouthfeel that depends on the type and concentration of gelators, and, in certain cases, the production processes used, while smell depends on all these factors. Therefore, it is important to evaluate the impact of fat replacement in terms of possible sensory changes that could affect the final food product. Specifically, in case oleogels are intended to replace butter, cocoa butter, and coconut oil, which have distinctive smell, taste, and mouthfeel, the changes introduced by the replacement might lead to products with limited sensory characteristics. In such cases, possible addition of flavoring agents in oleogels should be considered. However, the neutral flavor and taste of certain oleogels can be an advantage for some food products. For example, to produce a solid fat substitute for use in bakery products, a neutral-tasting oil such as rapeseed or high oleic sunflower oil might be desirable. Conversely, wax-based oleogels have a waxy mouthfeel when the gelator concentration is sufficiently high (e.g., 3–5%). Candelilla wax has its own smell and taste, which further contributes to the oleogel sensory characteristics. Ethylcellulose-based oleogel, on the other hand, can have a typical smell of rancid oil when prepared under regular atmosphere. This shows that oleogel flavor should be adjusted case-by-case and sensory tests should be carried out on products incorporating oleogels to assess the impact of fat replacement.
- While oleogels have proven to be a promising alternative to traditional fats in some food products, such as bakery goods, there are still limitations to their use as a sole fat source in other products, like chocolates, where the differences in functionality and sensory properties are significant. However, some studies indicate that there is a correlation between consumer acceptance and perception of a product and knowledge of its nutritional characteristics [6]. This means that claiming a product as a free-trans-fat and reduced saturated fat product could result in even greater consumer acceptance.
- Melting of oleogels is still a challenge that researchers are trying to address, as discussed in the previous section. However, physical melting and melting sensation can be interpreted as two different phenomena. Indeed, the melting-in-mouth feeling is difficult to be achieved with oleogels, although some blends of waxes and other gelators have been proposed with this purpose [2]. On the other hand, the reader can also imagine that the melting sensation can be simulated by obtaining a gradual physical release of oil from oleogels during mastication and simultaneous absorption of energy from other components in the oleogel matrix, with the development of an oral lubrication sensation. Therefore, if physical melting is a challenge to be tackled in the future, the melting sensation is another that needs to be overcome through possible other routes.

28.2.2.1 Labeling and Ingredients

- Consumers are increasingly paying attention to the ingredients listed on labels and their origin while trying to reduce the consumption of products containing additives or E-numbered ingredients. Less refined, natural, and organic ingredients are perceived as healthier and certain segments of consumers are willing to pay higher prices for food products that contain such ingredients. Currently, few oleogels are obtained from less refined oils, gelators, or particles. In this area, the development of oleogels using raw materials with minimal or no fractionation and purification, such as unrefined oils extracted from oilseeds by only cold pressing or water/enzymatic extraction, is taking place. Another example is the use of oleosomes extracted from oilseeds for the development of oleogels [7, 8]. Another important aspect to consider is that consumer awareness of sustainability issues is also increasing, posing further challenges on oleogels produced through chemically derived gelators or where extensive energy input for their production is required.

- Emerging nutrition-related labeling such as Nutriscore and Health Star Rating are getting popular among companies in Europe and Australia, respectively, to show the nutritional value and consequently the healthiness of a food product. The score is calculated based on the composition of the product and its calorie content. Oleogels can improve the final nutritional rating of food products when used to replace saturated fats. The magnitude of the improvement is then determined by the composition of the oleogels, which can be further tailored to maximize their impact on these labels.

28.2.3 Stability

To allow the application of oleogels, their stability during processing and storage should be sufficiently long to permit (i) their incorporation into existing food manufacturing processes, (ii) their usage in food products including interactions with other ingredients, and (iii) their consumption over the shelf-life period of the products containing them. More in particular:

- Some concerns related to physical stability during processing and storage, like oil release, low resistance to shearing/mixing, and polymorphic transformation, still need to be addressed to enhance the applicability of oleogels. Storage stability issues of oleogels used as fat ingredients can be addressed by producing oleogels on-site. This means that a food industry can produce oleogels (e.g., by using a direct method) for use in its production lines. In certain cases, oleogels require hours to several days of annealing before use; in other cases, they can be used immediately. This implies that the company should store in warehouses the produced oleogel in case waiting times in the food products are expected, as well as having the equipment for oleogel production. In addition, physical stability of oleogels should also be kept in the food matrix in which they are embedded. This means that the interactions of the oleogels with other food ingredients should not affect, e.g., their ability to retain the oil.
- Oxidative stability of oleogels during both production and storage needs to be carefully considered. Oil oxidation is a major concern when it comes to the

nutritional profile of oleogels, as well as possible off-flavors. Indeed, when heating is involved in the production of oleogels, like in the case of the hot direct method and some indirect methods, it is important to avoid long heating times, and possibly to control the environment (atmosphere) by making it inert (reducing oxygen levels by using, e.g., nitrogen) or by adding antioxidants. Although oxidation levels can be lower in oleogels produced at moderate or low temperatures, such as the cold direct method, some indirect methods, and semi-direct methods, it is worth mentioning that gelators, particles, and equipment used for preparing oleogels could act as or introduce pro-oxidants possibly promoting oil oxidation. Finally, it is also important to remember that the degradation of oils rich in omega-6 and omega-3 fatty acids can undertake the polymerization route, leading to a further decrease of the quality and nutritional properties of oils [9].

28.2.4 Process Scale-Up

One of the first problems that are currently encountered in the oleogel field is related to large-scale production requirements, which include careful consideration of factors such as the availability of gelators, sustainability of methods of production, costs, and regulatory aspects.

- The availability and the production systems of gelators and particles used for structuring purposes are aspects that should be carefully considered when upscaling of oleogel production. Some gelators involve chemical derivatization, which can generate waste. Biotechnological routes such as genetically modified organisms and enzymes can be further developed to provide more sustainable methods of production of gelators.
- As already stated in Chap. 4, oleogels should be produced using sustainable production methods. A substitute for saturated fats that aims to be the solution for future generations should be sustainable, as it is part of a delicate equilibrium between environment and economy. Local production would contribute to this objective. Oleogels are already more sustainable than saturated fats from tropical plants and hydrogenated fats. This is because most production techniques require equipment already available in the food industry and can be produced in situ using by-products and renewable resources through physical processes, thus reducing production and transport costs and waste.
- There are a few types of oleogels currently being produced at pilot and industrial scales, and some commercially available oleogels can be found in the market. The reader can imagine that some companies are already using oleogels in their food products, but it is not always possible to identify the presence of oleogels by just looking at the labels of the products. This is because some ingredients may overlap with those already reported on the label of the original food products (e.g., E471—mono- and diglycerides of fatty acids).

- In general, oleogels are currently more expensive than other fats they aim to replace, like palm oil and its fractions, coconut oil, and margarine. However, the more the demand increases, the more the producers of gelators can reduce prices due to competitors and scaling up of production (large-scale economy). When the production process only involves heating and cooling, without additional steps like homogenization and drying, oleogels have lower production costs. Also, when gelators or co-structurants are commercially available and do not require further modification, the costs are contained.
- Regulation is one of the most important hurdles. Scientists should team up with companies in the effort of getting gelators approved for use by the Food and Drug Administration (FDA) and the European Food Safety Authority (EFSA). For example, among all the gelators studied to date, only a few (e.g., monoglycerides, beta-sitosterol, rice bran wax, and ethylcellulose) can be used under certain conditions (scopes) and in certain countries.
- As with any food product, it is important to ensure that oleogels and oleogel-based products are safe. This includes selecting high-quality ingredients, implementing good manufacturing practices, and conducting quality and safety testing to ensure they are safe for human consumption. This point is particularly important when employing oil structuring methods that involve the use of solvents or non-GRAS (generally recognized as safe) ingredients.

28.3 Opportunities and Potential Future Directions in the Oleogel Field

In the previous section, we discussed the limitations and challenges that need to be overcome to accelerate the transition between saturated fats and oleogels. Solving these limitations would help boosting this transition, where we foresee that some of the most important points to be addressed are as follows: gaining a full understanding of the oil structuring abilities of gelators, developing of computational models for predicting oleogel properties, optimizing the melting profile, creating an "all-purpose" oleogel, better tailoring of the sensory properties of oleogels, improving oxidative stability, and elucidating the physiological effects of oleogels after ingestion. These topics are still open for research and working on them could represent a great opportunity to accelerate the progress in the oleogel field.

To explore new horizons in oleogel development, novel oleogelation methods like capillary suspensions [10, 11] and fiber-based oleogels [12, 13] should be further explored, which could lead to better control over the final product properties, and potentially new opportunities for product development. More in general, it is important to highlight that the advancement of new technologies or the application of existing technologies that have not yet been explored for oleogel production could bring further research opportunities. For example, current oleogelation methods that include a drying step could take advantage of microwave and radio frequency drying technologies, and alternative heating methods like ohmic heating. In addition, other technologies such as ultraviolet (UV) light and pulsed light could be used as initiators of photopolymerization for oleogel production from monomeric gelators, although this solution might lead to the development of oleogels for non-edible applications like lubrication in mechanical engineering and cosmetics.

Future work in the oleogel field should also be focused on scaling up laboratory preparation methods to pilot scale to obtain relevant information on the technical and economic feasibility of oleogel production processes, as well as to establish the optimum operating parameters that will facilitate the subsequent design and construction of industrial-scale production facilities. Collaborating with industrial partners can accelerate this process and facilitate the successful implementation of large-scale oleogel production.

In addition, oleogel-based products with health-enhancing properties (e.g., anti-inflammatory activity), is an exciting avenue to explore. This can be achieved by adding bioactive molecules into the formulation. Oleogels can be used as a matrix for liposoluble compounds, while oleogel-based emulsions and bigels offer the flexibility to incorporate both liposoluble and water-soluble compounds. A simple and already studied example is the development of spreadable systems incorporating phytosterols, curcumin, and resveratrol [14–16]. This approach holds promise for advancing the development of oleogel-based nutraceuticals.

It is clear that although the oleogel field is consolidating, increasing research opportunities are constantly emerging and will possibly lead to new frontiers in the field of oleogels.

28.3.1 Theory of Inventive Problem Solving for Boosting Oleogel Development

As we previously stated, while artificial intelligence is a valuable tool for advancing in the field of oleogels, there are other approaches that can also enhance creativity and introduce systematic engineering methods to the field. Among them, we would like to suggest the reader about the "Theory of Inventive Problem Solving" or TRIZ. In this method, a specific problem is abstracted by performing an in-depth analysis, then by using general and abstract specific tools the abstract problem is solved, eventually the several possible solutions are brought back to reality for the specific case, and a feasibility analysis is performed. TRIZ allows to solve contradictions as they are seen as a fundamental part of the innovation process. With this method, solutions found in other fields can be easily implemented in the oleogel field, possibly equipping researchers with a new mindset that can foster innovation in this field. If the reader finds this approach interesting or if we have ignited a spark of curiosity, we suggest reading books on the topic like *TRIZ for Engineers: Enabling Inventive Problem Solving* by Karen Gadd and *TRIZ: Theory of Inventive Problem Solving* by Vladimir Petrov.

28.4 Conclusions

In this chapter, we have discussed the limitations and challenges in the oleogel field, bringing to the reader some insights on the next possible research topics that need to be explored for its advancement. We have also explored further opportunities beyond the current needs that could potentially open up new frontiers in the oleogel area. We conclude this chapter by emphasizing the importance of using advanced tools such as artificial intelligence and TRIZ in oleogel research. These cutting-edge tools hold great potential to overcome current limitations and find innovative solutions in the field, allowing researchers to drive progress and address challenges more effectively.

Acknowledgments Fabio Valoppi acknowledges the European Union's Horizon 2020 research and innovation program funding under the Marie Skłodowska-Curie grant agreement No. 836071. Camila Palla acknowledges the Consejo Nacional de Investigaciones Científicas y Técnicas (CONICET) of Argentina.

References

1. Choi KO, Hwang HS, Jeong S, Kim S, Lee S (2020) The thermal, rheological, and structural characterization of grapeseed oil oleogels structured with binary blends of oleogelator. J Food Sci 85:3432–3441. https://doi.org/10.1111/1750-3841.15442
2. Tavernier I, Doan CD, Van de Walle D, Danthine S, Rimaux T, Dewettinck K (2017) Sequential crystallization of high and low melting waxes to improve oil structuring in wax-based oleogels. RSC Adv 7:12113–12125. https://doi.org/10.1039/c6ra27650d
3. Rodríguez-Hernández A, Pérez-Martínez J, Gallegos-Infante J, Toro-Vazquez J, Ornelas-Paz J (2021) Rheological properties of ethyl cellulose-monoglyceride-candelilla wax oleogel Vis-a-Vis edible shortenings. Carbohydr Polym 252:117171
4. Gravelle AJ, Davidovich-Pinhas M, Zetzl AK, Barbut S, Marangoni AG (2016) Influence of solvent quality on the mechanical strength of ethylcellulose oleogels. Carbohydr Polym 135: 169–179. https://doi.org/10.1016/j.carbpol.2015.08.050
5. Bhattarai M, Penttilä P, Barba L, Macias-Rodriguez B, Hietala S, Mikkonen KS, Valoppi F (2022) Size-dependent filling effect of crystalline celluloses in structural engineering of composite oleogels. LWT 160:113331. https://doi.org/10.1016/j.lwt.2022.113331
6. Tarancón P, Sanz T, Fiszman S, Tárrega A (2014) Consumers' hedonic expectations and perception of the healthiness of biscuits made with olive oil or sunflower oil. Food Res Int 55:197–206. https://doi.org/10.1016/j.foodres.2013.11.011
7. Mert B, Vilgis TA (2021) Hydrocolloid coated oleosomes for development of oleogels. Food Hydrocoll 119:106832. https://doi.org/10.1016/j.foodhyd.2021.106832
8. Zambrano JC, Vilgis TA (2023) Tunable oleosome-based oleogels: influence of polysaccharide type for polymer bridging-based structuring. Food Hydrocoll 137:108399. https://doi.org/10.1016/j.foodhyd.2022.108399
9. Chen J, Zhang L, Li Y, Zhang N, Gao Y, Yu X (2021) The formation, determination and health implications of polar compounds in edible oils: current status, challenges and perspectives. Food Chem 364:130451. https://doi.org/10.1016/j.foodchem.2021.130451

10. Calabrese V, Gunes DZ, Farrés IF (2021) Rheological control of pea fibre dispersions in oil: the role of particle and water volume fractions. Food Hydrocoll 121:106988. https://doi.org/10.1016/j.foodhyd.2021.106988
11. Jarray A, Feichtinger A, Scholten E (2022) Linking intermolecular interactions and rheological behaviour in capillary suspensions. J Colloid Interface Sci 627:415–426. https://doi.org/10.1016/j.jcis.2022.07.067
12. David A, David M, Lesniarek P, Corfias E, Pululu Y, Delample M, Snabre P (2021) Oleogelation of rapeseed oil with cellulose fibers as an innovative strategy for palm oil substitution in chocolate spreads. J Food Eng 292:110315. https://doi.org/10.1016/j.jfoodeng.2020.110315
13. Valoppi F, Schavikin J, Lassila P, Laidmäe I, Heinämäki J, Hietala S, Haeggström E, Salmi A (2023) Formation and characterization of oleogels obtained via direct dispersion of ultrasound-enhanced electrospun nanofibers and cold milling. Food Struct 37:100338. https://doi.org/10.1016/j.foostr.2023.100338
14. Sawalha H, den Adel R, Venema P, Bot A, Floter E, van der Linden E (2012) Organogel-emulsions with mixtures of beta-sitosterol and gamma-oryzanol: influence of water activity and type of oil phase on gelling capability. J Agric Food Chem 60:3462–3470. https://doi.org/10.1021/jf300313f
15. Tian Y, Acevedo NC (2020) Role of supramolecular policosanol oleogels in the protection of retinyl palmitate against photodegradation. RSC Adv 10:2526–2535. https://doi.org/10.1039/c9ra07820g
16. Li L, Wan W, Cheng W, Liu G, Han L (2019) Oxidatively stable curcumin-loaded oleogels structured by β-sitosterol and lecithin: physical characteristics and release behaviour in vitro. Int J Food Sci Tech 54:2502–2510. https://doi.org/10.1111/ijfs.14208

Index

C. Andrea Palla, F. Valoppi (eds.), *Advances in Oleogel Development, Characterization, and Nutritional Aspects*,
https://doi.org/10.1007/978-3-031-46831-5

Printed in the USA
CPSIA information can be obtained
at www.ICGtesting.com
LVHW050721150224
771922LV00003B/93

9 783031 468308